WORLD *of* EARTH SCIENCE

K. Lee Lerner and Brenda Wilmoth Lerner, *Editors*

Volume 1

A-L

GALE®

THOMSON ™

GALE

Detroit • New York • San Diego • San Francisco • Cleveland • New Haven, Conn. • Waterville, Maine • London • Munich

THOMSON

GALE

World of Earth Science

K. Lee Lerner and Brenda Wilmoth Lerner

Project Editor
Ryan L. Thomason

Editorial
Deirdre S. Blanchfield, Madeline Harris, Kate Kretschmann, Michael D. Lesniak, Kimberley A. McGrath, Brigham Narins, Mark Springer

Permissions
Shalice Shah-Caldwell

Imaging and Multimedia
Robert Duncan, Leitha Etheridge-Sims, Lezlie Light, Kelly A. Quin, Barbara J. Yarrow

Product Design
Michael Logusz, Tracey Rowens

Manufacturing
Wendy Blurton, Evi Seoud

ISBN   0-7876-7739-6 (set)
        0-7876-7740-X (vol. 1)
        0-7876-7741-8 (vol. 2)

LIBRARY OF CONGRESS CATALOGING-IN-PUBLICATION DATA

World of earth science / K. Lee Lerner and Brenda Wilmoth Lerner, editors.
    p. cm.
        Includes bibliographical references and index.
        ISBN 0-7876-7739-6 (set)—ISBN 0-7876-7740-X (v. 1)—ISBN 0-7876-7741-8 (v. 2)
        1. Earth sciences—Encyclopedias. I. Lerner, K. Lee. II. Lerner, Brenda Wilmoth.
    QE5 .W59 2003
    550'.3—dc21   2002012069

Printed in the United States of America
10 9 8 7 6 5 4 3 2 1

CONTENTS

A

Abyssal plains
Acid rain
Adamantine
Adiabatic heating
Advection
Aerodynamics
Africa
Agassiz, Jean Louis
Agricola, Georgius
Air masses and fronts
Alkaline earth metals
Alloy
Alluvial system
Aluminum
Alvarez, Luis
Amorphous
Analemma
Andesite
Anning, Mary
Antarctica
Aphanitic
Aquifer
Archean
Archeological mapping
Area

Aretes
Armstrong, Neil
Arrhenius, Svante August
Artesian
Asia
Asteroids
Asthenosphere
Astrolabe
Astronomy
Atmospheric chemistry
Atmospheric circulation
Atmospheric composition and structure
Atmospheric inversion layers
Atmospheric lapse rate
Atmospheric pollution
Atmospheric pressure
Atom
Atomic mass and weight
Atomic number
Atomic theory
Aurora Borealis and Aurora Australialis
Australia
Avalanche
Aviation physiology

B

Ballard, Robert Duane
Banded iron formation
Barrier islands
Barringer meteor crater
Basalt
Basin and range topography
Batholith

P

Q

R

S

INTRODUCTION

As of June 2002, astronomers had discovered more than 100 other planets orbiting distant suns. With advances in technology, that number will surely increase during the opening decades of the twenty-first century. Although our explorations of the Cosmos hold great promise of future discoveries, among all of the known worlds, Earth remains unique. Thus far it is the only known planet with blue skies, warm seas, and life. Earth is our most tangible and insightful laboratory, and the study of Earth science offers us precious opportunities to discover many of the most fundamental laws of the Universe.

Although Earth is billions of years old, geology—literally meaning the study of Earth—is a relatively new science, having grown from seeds of natural science and natural history planted during the Enlightenment era of the eighteenth and nineteenth centuries. In 1807, the founding of the Geological Society of London, the first learned society devoted to geology, marked an important turning point for the science (some say its nascence). In the beginning, geologic studies were mainly confined to the study of minerals (mineralogy), strata (stratigraphy), and fossils (paleontology), and hotly debated issues of the day included how well new geologic findings fit into religious models of creation. In less than two centuries, geology has matured to embrace the most fundamental theories of physics and chemistry—and broadened in scope to include the diverse array of subdisciplines that comprise modern Earth science.

Modern geology includes studies in seismology (earthquake studies), volcanology, energy resources exploration and development, tectonics (structural and mountain building studies), hydrology and hydrogeology (water-resources studies), geologic mapping, economic geology (e.g., mining), paleontology (ancient life studies), soil science, historical geology and stratigraphy, geological archaeology, glaciology, modern and ancient climate and ocean studies, atmospheric sciences, planetary geology, engineering geology, and many other subfields. Although some scholars have traditionally attempted to compartmentalize geological sciences into subdisciplines, the modern trend is to incorporate a holistic view of broader Earth sci-

ence issues. The incorporation of once-diverse fields adds strength and additional relevance to geoscience studies.

World of Earth Science is a collection of 650 entries on topics covering a diversity of geoscience related interests—from biographies of the pioneers of Earth science to explanations of the latest developments and advances in research. Despite the complexities of terminology and advanced knowledge of mathematics needed to fully explore some of the topics (e.g., seismology data interpretation), every effort has been made to set forth entries in everyday language and to provide accurate and generous explanations of the most important terms. The editors intend *World of Earth Science* for a wide range of readers. Accordingly, *World of Earth Science* articles are designed to instruct, challenge, and excite less experienced students, while providing a solid foundation and reference for more advanced students.

World of Earth Science has attempted to incorporate references and basic explanations of the latest findings and applications. Although certainly not a substitute for in-depth study of important topics, we hope to provide students and readers with the basic information and insights that will enable a greater understanding of the news and stimulate critical thinking regarding current events (e.g., the ongoing controversy over the storage of radioactive waste) that are relevant to the geosciences.

The broader and intellectually diverse concept of Earth science allows scientists to utilize concepts, techniques, and modes of thought developed for one area of the science, in the quest to solve problems in other areas. Further, many geological problems are interrelated and a full exploration of a particular phenomenon or problem demands overlap between subdisciplines. For this reason, many curricula in geological sciences at universities stress a broad geologic education to prepare graduates for the working world, where they may be called upon to solve many different sorts of problems.

World of Earth Science is devoted to capturing that sense of intellectual diversity. True to the modern concept of Earth science, we have deliberately attempted to include some of the most essential concepts to understanding Earth as a dynamic body traveling through space and time.

Although no encyclopedic guide to concepts, theories, discoveries, pioneers, issues and ethics related to Earth science could hope to do justice to any one of those disciplines in two volumes, we have attempted to put together a coherent collection of topics that will serve not only to ground students in the essential concepts, but also to spur interest in the many diverse areas of this increasing critical set of studies.

In addition to topics related to traditional geology and meteorology, we have attempted to include essential concepts in physics, chemistry, and astronomy. We have also attempted to include topical articles on the latest global positioning (GPS), measurement technologies, ethical, legal, and social issues and topics of interest to a wide audience. Lastly, we have attempted to integrate and relate topics to the intercomplexities of economics and geopolitical issues.

Such a multifaceted and "real world" approach to the geosciences is increasingly in demand. In the recent past, geologic employment was dominated by the petroleum industry and related geologic service companies. In the modern world, this is no longer so. Mining and other economic geology occupations (e.g., prospecting and exploration), in former days plentiful, have also fallen away as major employers. Environmental geology, engineering geology, and ground water related jobs are more common employment opportunities today. As these fields are modern growth areas with vast potential, this trend will likely hold true well into the future. Many modern laws and regulations require that licensed, professional geologists supervise all or part of key tasks in certain areas of engineering geologic work and environmental work. It is common for professional geologists and professional engineers to work together on such projects, including construction site preparation, waste disposal, ground-water development, engineering planning, and highway construction. Many federal, state, and local agencies employ geologists, and there are geologists as researchers and teachers in most academic institutions of higher education.

Appropriate to the diversity of Earth science, we attempted to give special attention to the contributions by women and scientists of diverse ethnic and cultural backgrounds. In addition, we have included special articles written by respected experts that are specifically intended to make *World of Earth Science* more relevant to those with a general interest in the historical and/or geopolitical topics aspects of Earth science.

The demands of a dynamic science and the urgency of many questions related to topics such as pollution, global warming, and ozone depletion place heightened demands on both general and professional students of geosciences to increasingly broaden the scope and application of their knowledge.

For example, geological investigations of ancient and modern disasters and potential disasters are important—and often contentious—topics of research and debate among geologists today. Among the focus areas for these studies are earthquake seismicity studies. While much work continues in well-known problem areas like southern California, Mexico City, and Japan, less well-known, but potentially equally dangerous earthquake zones like the one centered near New Madrid, Missouri (not far from Memphis, Tennessee and St. Louis, Missouri) now receive significnt research attention. Geologists cannot prevent earthquakes, but studies can help predict earthquake events and help in planning the design of earthquake-survivable structures. Another focus of study is upon Earth's volcanoes and how people may learn to live and work around them. Some volcanoes are so dangerous that no one should live near them, but others are more predictable. Earthquake prediction and planning for eruptions is going on today by looking at the geologic record of past eruptions and by modern volcano monitoring using thermal imaging and tilt or motion-measuring devices. Other foci of disaster prevention research include river-flood studies, studies of slope stability (prevention of mass movement landslides), seismic sea-wave (tsunami) studies, and studies of possible asteroid or comet impacts.

Aside from geologic studies of disaster, there is a side of geology centered upon providing for human day-to-day needs. Hydrology is an interdisciplinary field within geology that studies the relationship of water, the earth, and living things. A related area, hydrogeology, the study of ground water, has undergone a revolution recently in the use of computer modeling to help understand flow paths and characteristics. These studies of water flow on the surface and in the subsurface connect with other subdisciplines of geology, such as geomorphology (the study of landforms, many of which are formed by water flow), river hydrology, limnology (study of lakes), cave and sinkhole (karst) geology, geothermal energy, etc. Geologic studies related to human and animal health (i.e., medical geology) are becoming very common today. For example, much work is currently devoted to tracing sources of toxic elements like arsenic, radon, and mercury in rock, soil, air, water, and groundwater in many countries, including the United States. There has been a major effort on the part of medical geologists to track down dangerous mineral species of asbestos (not all asbestos is harmful) and determine how best to isolate or remove the material. Atmospheric scientists have been at work for some years on the issue of air-born pathogens, which ride across oceans and continents born on fine soil particles lifted by winds.

Geologists are also focused on study of the past. Today, paleobotany and palynology (study of fossil spores and pollen) complement traditional areas of invertebrate and vertebrate paleontogy. Recent discoveries such as small, feathered dinosaurs and snakes with short legs are helping fill in the ever-shrinking gaps within the fossil record of evolution of life on Earth. Paleontologic studies of extinction, combined with evidence of extraterrestrial bombardment, suggest that mass death and extinction of species on Earth at times in the past has come to us from the sky. In a slightly related area, geoarchaeology, the geologic context of archaeological remains and the geologic nature of archaeological artifacts remains key to interpreting details about the pre-historic human past. Careful study of drilling records of polar ice sheets, deep-sea sediments, and deep lake sediments has recently revealed that many factors, including subtle variations in some of Earth's orbital parameters (tilt, wobble, and shape of orbit around the Sun), has had a profound, cyclical effect upon Earth's climate in the past (and is continuing today). Paleontologic studies, combined with geologic investigations on temperature sensitive ratios of certain isotopes (e.g. O^{16}/O^{18}), have helped unlock mysteries of climate change on Earth (i.e., the greenhouse to icehouse vacillation through time).

Earth science studies are, for the first time, strongly focused on extraterrestrial objects as well. Voyages of modern exploratory spacecraft missions to the inner and outer planets have sent back a wealth of images and data from the eight major planets and many of their satellites. This has allowed a new field, planetary geology, to take root. The planetary geologist is engaged in photo-geologic interpretation of the origin of surface features and their chronology. Planetary geologic studies have revealed some important comparisons and contrasts with Earth. We know, for example, some events that affected our entire solar system, while the effects of other events were unique to certain planets and satellites. In addition, planetary geologists have found that impact-crater density is important for determining relative age on many planets and satellites. As a result, Earth is no longer the only planet with a knowable geologic time scale.

The geosciences have undergone recent revolutions in thought that have profoundly influenced and advanced human understanding of Earth. Akin to the fundamental and seminal concepts of cosmology and nucleosynthesis, beginning during the 1960s and continuing today, the concept of plate tectonics has revolutionized geologic thought and interpretation. Plate tectonics, the concept that the rigid outer part of Earth's crust is subdivided into plates, which move about on the surface (and have moved about on the surface for much of geologic time) has some profound implications for all of geology. This concept helps explain former mysteries about the distribution of volcanoes, earthquakes, and mountain chains. Plate tectonics also helps us understand the distribution of rocks and sediments on the sea floor, and the disparity in ages between continents and ocean floors. Plate motion, which has been documented through geologic time, helps paleontologists explain the distribution of many fossil species and characteristics of their ancient climates. Plate tectonic discoveries have caused a rewriting of historical geology textbooks in recent years.

Although other volumes are chartered to specifically explore ecology related issues, the topics included in *World of Earth Science* were selected to provide a solid geophysical foundation for ecological or biodiversity studies. We have specifically included a few revolutionary and controversial concepts, first written about in a comprehensive way during the 1970s, such as the Gaia hypothesis. Simply put, Gaia is the notion that all Earth systems are interrelated and interconnected so that a change in one system changes others. It also holds the view that Earth functions like a living thing. Gaia, which is really a common-sense philosophic approach to holistic Earth science, is at the heart of the modern environmental movement, of which geology plays a key part.

Because Earth is our only home, geoscience studies relating meteor impacts and mass extinction offer a profound insight into delicate balance and the tenuousness of life. As Carl Sagan wrote in *Pale Blue Dot: A Vision of the Human Future in Space*, "The Earth is a very small stage in a vast cosmic arena." For humans to play wisely upon that stage, to secure a future for the children who shall inherit Earth, we owe it to ourselves to become players of many parts, so that our repertoire of scientific knowledge enables us to use reason and intellect in our civic debates, and to understand the complex harmonies of Earth.

K. Lee Lerner & Brenda Wilmoth Lerner, editors
London
May, 2002

How to Use the Book

The articles in the book are meant to be understandable by anyone with a curiosity and willingness to explore topics in Earth science. Cross-references to related articles, definitions, and biographies in this collection are indicated by **bold-faced type**, and these cross-references will help explain, expand, and enrich the individual entries.

This first edition of *World of Earth Science* has been designed with ready reference in mind:

- **Entries are arranged alphabetically**, rather than by chronology or scientific field.
- **Bold-faced terms** direct reader to related entries.
- **"See also" references** at the end of entries alert the reader to related entries not specifically mentioned in the body of the text.
- A **Sources Consulted** section lists the most worthwhile print material and web sites we encountered in the compilation of this volume. It is there for the inspired reader who wants more information on the people and discoveries covered in this volume.
- The **Historical Chronology** includes many of the significant events in the advancement of the diverse disciplines of Earth science. The most current entries date from just days before *World of Earth Science* went to press.
- A **comprehensive General Index** guides the reader to topics and persons mentioned in the book. Bolded page references refer the reader to the term's full entry.

A detailed understanding of physics and chemistry is neither assumed nor required for *World of Earth Science*. In preparing this text, the editors have attempted to minimize the incorporation of mathematical formulas and to relate physics concepts in non-mathematical language. Accordingly, students and other readers should not be intimidated or deterred by chemical nomenclature. Where necessary, sufficient information regarding atomic or chemical structure is provided. If desired, more information can easily be obtained from any basic physics or chemistry textbook.

For those readers interested in more information regarding physics related topics, the editors recommend Gale's *World of Physics* as an accompanying reference. For those readers interested in a more comprehensive treatment of chemistry, the editors recommend Gale's *World of Chemistry*.

In an attempt to be responsive to advisor's requests and to conform to standard usage within the geoscience community, the editors elected to make an exception to previously used style guidelines regarding geologic time. We specifically adopted the convention to capitalize applicable eons, eras, periods and epochs. For example, Cenozoic Era, Tertiary Period, and Paleocene Epoch are intentionally capitalized.

Advisory Board

In compiling this edition, we have been fortunate in being able to rely upon the expertise and contributions of the following scholars who served as academic and contributing advisors for *World of Earth Science*, and to them we would like to express our sincere appreciation for their efforts to ensure that *World of Earth Science* contains the most accurate and timely information possible:

Cynthia V. Burek, Ph.D.
Environment Research Group, Biology Department
Chester College, England, U.K.

Nicholas Dittert, Ph.D.
Institut Universitaire Européen de la Mer
University of Western Brittany, France

William J. Engle. P.E.
Exxon-Mobil Oil Corporation (Rt.)
New Orleans, Louisiana

G. Thomas Farmer, Ph.D., R.G.
Earth & Environmental Sciences Division,
Los Alamos National Laboratory
Los Alamos, New Mexico

Lyal Harris, Ph.D.
Tectonics Special Research Center, Dept. of Geology & Geophysics
University of Western Australia
Perth, Australia

Alexander I. Ioffe, Ph.D.
Senior Scientist, Geological Institute of the Russian Academy of Sciences
Moscow, Russia

David T. King, Jr., Ph.D.
Professor, Dept. of Geology
Auburn University
Auburn, Alabama

Cherry Lewis, Ph.D.
Research Publicity Officer
University of Bristol
Bristol, England, U.K.

Eric v.d. Luft, Ph.D., M.L.S.
Curator of Historical Collections
S.U.N.Y. Upstate Medical University
Syracuse, New York

Jascha Polet, Ph.D.
Research Seismologist, Caltech Seismological Laboratory,
California Institute of Technology
Pasadena, California

Yavor Shopov, Ph.D.
Professor of Geology & Geophysics
University of Sofia
Sofia, Bulgaria

Acknowledgments

In addition to our advisors and contributing advisors, it has been our privilege and honor to work with the following contributing writers and scientists who represent scholarship in the geosciences that spans five continents:

Molly Bell, Ph.D.; Alicia Cafferty; John Cubit, Ph.D.; Laurie Duncan, Ph.D.; John Engle; Agnes Galambosi; Larry Gilman, Ph.D.; David Goings, Ph.D.; Brooke Hall, Ph.D.; William Haneberg, Ph.D.; Michael Lambert; Adrienne Wilmoth Lerner, (Graduate Student, Department of History, Vanderbilt University), Lee Wilmoth Lerner, Jill Liske, M.Ed.; Robert Mahin, Ph.D.; Kelli Miller; William Phillips, Ph.D.; William Rizer, Ph.D.; Jerry Salvadore, Ph.D.; and David Tulloch, Ph.D.

Many of the academic advisors for *World of Earth Science*, along with others, authored specially commissioned articles within their field of expertise. The editors would like to specifically acknowledge the following special contributions:

Cynthia V. Burek, Ph.D.
History of geoscience: Women in the history of geoscience

Nicholas Dittert, Ph.D.
Scientific data management in Earth sciences

William J. Engle. P.E.
Petroleum, economic uses of
Petroleum extraction

G. Thomas Farmer, Ph.D., R.G.
Hydrogeology

Lyal Harris, Ph.D.
Supercontinents

Alexander I. Ioffe, Ph.D.
Bathymetric mapping

David T. King, Jr., Ph.D.
Geologic time
Stratigraphy
Uniformitarianism

Cherry Lewis, Ph.D.
The biography of Author Holmes

Yavor Shopov, Ph.D.
Paleoclimate

The editors wish to gratefully acknowledge Dr. Eric v.d. Luft for his diligent and extensive research related to his compilation of selected biographies for *World of Earth Science*.

The editors also gratefully acknowledge Dr. Yavor Shopov's generous and significant contribution of photographs for *World of Earth Science* and Dr. David King's guidance, comments, and contributions to the introduction.

The editors thank Ms. Kelly Quin and others representing the Gale Imaging Team for their guidance through the complexities and difficulties related to graphics. Last, but certainly not least, the editors thank Mr. Ryan Thomason, whose dedication, energy, and enthusiasm made a substantial difference in the quality of *World of Earth Science*.

Because Earth is theirs to inherit, the editors lovingly dedicate this book to their children, Adrienne, Lee, Amanda, and Adeline. *Per ardua ad astra.*

Cover

The image on the cover depicts an example of several geologic cross sections of strata, illustrating the fundamental laws of geology.

A

AA FLOW • *see* LAVA

ABLATION OF GLACIERS • *see* GLACIATION

ABSOLUTE AGE • *see* DATING METHODS

ABSOLUTE HUMIDITY • *see* HUMIDITY

ABYSSAL PLAINS

Abyssal plains are the vast, flat, sediment-covered areas of the deep ocean floor. They are the flattest, most featureless areas on Earth, and have a slope of less than one foot of elevation difference for each thousand feet of distance. The lack of features is due to a thick blanket of sediment that covers most of the surface.

These flat abyssal plains occur at depths of over 6,500 ft (1,980 m) below sea level. They are underlain by the oceanic **crust**, which is predominantly basalt—a dark, fine-grained volcanic **rock**. Typically, the **basalt** is covered by layers of sediment, much of which is deposited by deep ocean turbidity currents (caused by the greater density of sediment-laden **water**), or biological materials, such as minute shells of marine plants and animals, that have "rained" down from the ocean's upper levels, or a mixture of both.

Other components of abyssal plain sediment include wind-blown dust, volcanic ash, chemical precipitates, and occasional meteorite fragments. Abyssal plains are often littered with nodules of manganese containing varying amounts of **iron**, nickel, cobalt, and **copper**. These pea to potato-sized nodules form by direct **precipitation** of **minerals** from the seawater onto a bone or rock fragment. Currently, deposits of manganese nodules are not being mined from the sea bed, but it is possible that they could be collected and used in the future.

Of the 15 billion tons of river-carried **clay**, **sand**, and gravel that are washed into the **oceans** each year, only a fraction of this amount reaches the abyssal plains. The amount of biological sediment that reaches the bottom is similarly small. Thus, the rate of sediment accumulation on the abyssal plains is very slow, and in many areas, less than an inch of sediment accumulates per thousand years. Because of the slow rate of accumulation and the monotony of the **topography**, abyssal plains were once believed to be a stable, unchanging environment. However, deep ocean currents have been discovered that scour the ocean floor in places. Some currents have damaged trans-oceanic communication cables laid on these plains.

Although they are more common and widespread in the Atlantic and Indian ocean basins than in the Pacific, abyssal plains are found in all major ocean basins. Approximately 40% of the planet's ocean floor is covered by abyssal plains. The remainder of the ocean floor topography consists of hills, cone-shaped or flat-topped mountains, deep trenches, and **mountain chains** such as the mid-oceanic ridge systems.

The abyssal plains do not support a great abundance of aquatic life, though some species do survive in this relatively barren environment. Deep sea dredges have collected specimens of unusual-looking fish, worms, and clam-like creatures from these depths.

See also Deep sea exploration; Ocean trenches

ACID RAIN

Acid rain is rain with a **pH** (a logarithmic measurement of acidity or alkalinity) of less than 5.7. Acid rain usually results from elevated levels of nitric and sulfuric acids in air pollution. Acidic pollutants that can lead to acid rain are common by-products from burning fossil **fuels** (e.g., oil, **coal**, etc.) and are found in high levels in exhaust from internal combustion engines (e.g., automobile exhaust). Acidic **precipitation** may also occur in other forms such as snow.

Acid rain occurs when polluted gasses become trapped in **clouds**. The clouds may drift for hundreds, even thousands, of miles before finally releasing acidic precipitation. Trees, **lakes**, animals, and even buildings are vulnerable to the slow corrosive effects of acid rain, whose damaging components are emitted by power plants and factories, especially those burning low grades of coal and oil.

Acid rain was first recognized in 1872, approximately 100 years after the start of the Industrial Revolution in England, when an English scientist, Robert Angus Smith (1817–1884), pointed out the problem. Almost another century passed, however, before the public became aware of the damaging effects of acid rain. In 1962, the Swedish scientist Svante Oden brought the acid rain quandary to the attention of the press, instead of the less popular scientific journals. He compiled records from the 1950s indicating that acid rain came from air masses moving out of central and western **Europe** into Scandinavia.

After acid rain was discovered in Europe, scientists began measuring the acidity of rain in **North America**. Initially, they found that the problem was concentrated in the northeastern states of New York and Pennsylvania because the type of coal burned there was more sulfuric. By 1980, most of the states east of the Mississippi, as well as southeastern Canada, were receiving acidic rainfall. Acid rain falls in the West also, although the problem is not as severe. Acid rain in Los Angeles, California is caused primarily by local traffic emissions. Car emissions contain nitrogen oxide, the second highest problematic gas in acid rain after sulfur dioxide.

Acid rain is measured through pH tests that determine the concentration of hydrogen ions. Pure **water** has a neutral pH of approximately 7.0. When the pH is greater than 7, the material is thought to be alkaline. At a pH of 5.7, rain is slightly acidic, but when its pH is further reduced, the rain becomes an increasingly stronger acid rain. In the worst cases, acid rain has shown a pH of 2.4 (about as acidic as vinegar). When pH levels are drastically tipped in **soil** and water, entire lakes and **forests** are jeopardized. Evergreen trees in high elevations are especially vulnerable. Although the acid rain itself does not kill the trees, it makes them more susceptible to other dangers. High acid levels in soil causes **leaching** of other valuable **minerals** such as calcium, magnesium, and potassium. According to the World Watch Institute, in the late 1980s and early 1990s forest damage in Europe ranged from a low of 4% in Portugal to a high of 71% in Czechoslovakia, averaging 35% overall.

Small marine organisms cannot survive in acidic lakes and **rivers**, and their depletion affects larger fish and ultimately the entire marine life food chain. Snow from acid rain is also damaging; snowmelt has been known to cause massive, instant death for many kinds of fish. Some lakes in Scandinavia, for example, are completely devoid of fish. Acid rain also eats away at buildings and metal structures. From the Acropolis in Greece to Renaissance buildings in Italy, ancient structures are showing signs of slow **corrosion** from acid rain.

In some industrialized parts of Poland, trains cannot exceed 40 miles (65 km) per hour because the **iron** railway tracks have been weakened from acidic air pollution.

New power plants in the United States are being built with strict emissions standards, but retrofitting older plants is difficult and expensive. Nevertheless, the United States Environmental Protection Agency requires most of the older and dirtier power plants to install electrostatic precipitators and baghouse filters—devices designed to remove solid particulates. Such devices are required in Canada, in industrialized countries in Western Europe, and in Japan. Scrubbers, or flue-gas desulfurization technology, are also being used because of their effectiveness in removing as much as 95% of a power plant's sulfur dioxide emissions. These devices are expensive, however, and there are clauses in pollution control laws that allow older plants to continue operation at higher pollution levels. Another way to reduce acid rain is for power plants to burn cleaner coal in their plants. This does not require retrofitting but it does increase transportation costs since coal containing less sulfur is mined in the western part of the United States, far away from where it is needed in the midwest and eastern part of the country.

See also Atmospheric pollution; Erosion; Global warming; Groundwater; Petroleum, economic uses of; Rate factors in geologic processes; Weathering and weathering series

ACTUALISM • *see* UNIFORMITARIANISM

ADAMANTINE

Some transparent **minerals** with very high indices of refraction have a non-metallic, brilliant manner of reflecting and transmitting light called an adamantine luster. **Diamond** is the best-known adamantine mineral, and its coveted sparkle is an example of this type of non-metallic luster. A diamond's internal structure of covalently-bound **carbon** atoms in a three-dimensional matrix causes incident light to refract deeply into the crystal, giving the crystal its characteristic clarity. The isometric, or three-dimensionally symmetrical, crystal structure of diamond also causes light to disperse within the mineral giving cut diamonds their spectral "fire." The synthetic diamond substitute, cubic zirconium, or CZ, has an adamantine luster due to its high index of refraction, but its dispersion, though relatively high, leaves this copy without the fire of the real diamond.

The index of refraction, n, for a given material is the ratio between the velocity of light in air, and its velocity in a denser material. Snell's law defines the precise relationship between the angle of incidence (i), and the angle of refraction (r), as sin i/sin r=n, where n is again the index of refraction. Non-metallic minerals with tightly bound, tightly packed atoms in a strong three-dimensional crystal lattice are more likely to have a high index of refraction. They are also more likely to be very hard and to have an adamantine luster. The mineral corundum, whose colored varieties include the **gemstones** ruby and sapphire, has a hardness of nine on the Moh's scale and a vitreous to adamantine luster. The **lead** carbonate mineral, cerussite, and lead sulfate mineral, anglesite, also have adamantine lusters.

See also Crystals and crystallography

ADIABATIC HEATING

Adiabatic processes are those in which there is no net heat transfer between a system and its surrounding environment (e.g., the product of pressure and volume remains constant). Because it is a gas, air undergoes adiabatic heating and cooling as it experiences **atmospheric pressure** changes associated with changing altitudes. Increasing pressure adiabatically heats air masses, falling pressures allow air to expand and cool.

Adiabatic heating and cooling is common in convective atmospheric currents. In adiabatic heating and cooling there is no net transfer of mass or thermal exchange between the system (e.g., volume of air) the external or surrounding environment. Accordingly, the change in **temperature** of the air mass is due to internal changes.

In adiabatic cooling, when a mass of air rises—as it does when it moves upslope against a mountain range—it encounters decreasing atmospheric pressure with increasing elevation. The air mass expands until it reaches pressure equilibrium with the external environment. The expansion results in a cooling of the air mass.

With adiabatic heating, as a mass of air descends in the atmosphere—as it does when it moves downslope from a mountain range—the air encounters increasing atmospheric pressure. Compression of the air mass is accompanied by an increase in temperature.

Because warmer air is less dense than cooler air, warmer air rises. Counter-intuitively, moist air is also lighter than less humid air. The **water**, composed of the elements of **oxygen** and hydrogen is lighter than dominant atmospheric elements of oxygen and nitrogen. For this reason, warm moist air rises and contributes to atmospheric instability.

In the lower regions of the atmosphere (up to altitudes of approximately 40,000 feet [12,192 m]), temperature decreases with altitude at the **atmospheric lapse rate**. Because the atmosphere is warmed by conduction from Earth's surface, this lapse or reduction in temperature normal with increasing distance from the conductive source. The measurable lapse rate is affected by the relative **humidity** of an air mass. Unsaturated or dry air changes temperature at an average rate 5.5°F (3.05°C) per 1,000 feet (304 m). Saturated air—defined as air at 100% relative humidity—changes temperature by an average of 3°F (1.66°C) per 1,000 feet (304 m). These average lapse rates can be used to calculate the temperature changes in air undergoing adiabatic expansion and compression.

For example, as an air mass at 80% relative humidity (dry air) at 65°F (18.3° C) rises up the side of a mountain chain from sea level it will decrease in temperature at rate of 5.5°F (3.05°C) per 1,000 feet (304 m) until the changing temperature changes

the relative humidity (a measure of the moisture capacity of air) to 100%. In addition to cloud formation and **precipitation**, the continued ascension of this now "wet" or saturated air mass proceeds at 3°F (1.66°C) per 1,000 feet (304 m). If the saturation point (the point at which "dry" air becomes "wet" air) is at 4,000 feet (1,219 m), the hypothetical air mass starting at 65°F (18.3°C) would cool 22°F (12.2°C) to 43°F (6.1°C) at an altitude of 4,000 feet (1,219 m). If the air ascended another 6,000 feet (1,829 m) to the top of the mountain chain before starting downslope, the temperature at the highest elevation of 10,000 feet (3,048 m) would measure 25°F (–3.9°C). This accounts for precipitation in the form of snow near mountain peaks even when valley temperatures are well above **freezing**. Because the absolute moisture content of the air mass has been reduced by cloud formation and precipitation, as the air moves downslope and warms it quickly falls below saturation and therefore heats at the dry lapse rate of 5.5°F (3.05°C) per 1,000 feet (304 m). A dry air mass descending 10,000 feet (3,048 m) would increase in temperature by 55°F (30.6°C). In the example given, the hypothetical air mass starting upslope at 65°F (18.3°C), rising 10,000 feet (3,048 m) and then descending 10,000 feet (3,048 m) would measure 80°F (26.7°C) at sea level on the other side of the mountain chain.

Although actual lapse rates do not strictly follow these guidelines, they present a model sufficiently accurate to predict temperate changes. The differential wet/dry lapse rates can result in the formation of hot downslope winds (e.g., Chinook winds, Santa Anna winds, etc).

See also Air masses and fronts; Land and sea breeze; Seasonal winds

ADVECTION

Earth's atmosphere is a dynamic sea of gases in constant motion and Earth's **oceans** contain currents that move **water** across the globe. Advection is a lateral or horizontal transfer of mass, heat, or other property. Accordingly, winds that blow across Earth's surface represent advectional movements of air. Advection also takes place in the ocean in the form of currents. Currently, geologists debate the presence and role of substantial advective processes in Earth's mantle.

Differential pressures and temperatures drive the **mass movement** of air seeking equilibrium (the lowest energy state). Advective winds move from areas of higher **temperature** toward areas of lower temperature. In contrast, convection, the vertical movement of mass or transfer of heat, manifests itself as air currents. Accordingly, winds are a result of advection, while air currents are a result of convection.

Although in a gaseous state, the atmosphere observes fluid-like dynamics. This is an important consideration when considering advection, because advection is usually more pronounced in the movement of fluids. For example, advection also takes place in the oceans where advection is broadened to include the lateral (horizontal) transfer of not only fluid mass and heat, but of other properties such and **oxygen** content and salinity.

In the atmosphere, advection is the sole process of lateral transfer of mass. In contrast, vertical transfer occurs via conduction, convection, and radiation. Just as ocean currents permit heat transfer from areas of warm water to an **area** of water with cooler temperatures, advective winds allow the transfer of both sensible heat and latent heat (a function of **humidity**).

Although advection processes are important heat equilibration mechanisms for both the atmosphere and the oceans, the speed and volume of mass transported differs greatly between the atmosphere and oceans. The magnitude of heat transference depends on heat flux (the rate of heat transport), and flux in turn relates the transfer of heat energy in terms of area and time. Both processes contribute approximately equally because **wind** currents are much faster (higher rate) than ocean currents but ocean currents move substantially denser masses of molecules.

Advection is also responsible for the formation of advection **fog**. Advection fog usually occurs when the atmosphere is very stable so that moist (humid) air near the surface does not mix vertically with an overlying layer of drier air. The advection fog forms as warm and moist air moves horizontally along the cooler surface and the air near the surface is cooled to its **dew point**.

See also Adiabatic heating; Atmospheric circulation; Atmospheric composition and structure; Atmospheric inversion layers; Atmospheric lapse rate; Atmospheric pressure; Convection (updrafts and downdrafts); Insolation and total solar irradiation; Wind chill; Wind shear

ADVECTIONAL FOG · *see* FOG

AERODYNAMICS

Aerodynamics is the science of airflow over airplanes, cars, buildings, and other objects. Aerodynamic principles are used to find the best ways in which airplanes produce lift, reduce drag, and remain stable (by controlling the shape and size of the wing, the angle at which it is positioned with respect to the airstream, and the flight speed). The flight characteristics change at higher altitudes as the surrounding air becomes colder and thinner. The behavior of the airflow also changes dramatically at flight speeds close to, and beyond, the speed of sound. The explosion in computational capability has made it possible to understand and exploit the concepts of aerodynamics and to design improved wings for airplanes. Increasingly sophisticated **wind** tunnels are also available to test new models.

Airflow is governed by the principles of fluid dynamics that deal with the motion of liquids and gases in and around solid surfaces. The viscosity, density, compressibility, and **temperature** of the air determine how the air will flow around a building or a plane. The viscosity of a fluid is its resistance to flow. Even though air is 55 times less viscous than **water**, vis-

cosity is important near a solid surface because air, like all other fluids, tends to stick to the surface and slow down the flow. A fluid is compressible if its density can be increased by squeezing it into a smaller volume. At flow speeds less than 220 mph (354 kph), one third the speed of sound, we can assume that air is incompressible for all practical purposes. At speeds closer to that of sound (660 mph [1,622 kph]) however, the variation in the density of the air must be taken into account. The effects of temperature change also become important at these speeds. A regular commercial airplane, after landing, will feel cool to the touch. The Concorde jet, which flies at twice the speed of sound, can feel hotter than boiling water.

Flow patterns of the air may be laminar or turbulent. In laminar or streamlined flow, air, at any point in the flow, moves with the same speed in the same direction at all times so that smoke in the flow appears to be smooth and regular. The smoke then changes to turbulent flow, which is cloudy and irregular, with the air continually changing speed and direction.

Laminar flow, without viscosity, is governed by **Bernoulli's principle** that states that the sum of the static and dynamic pressures in a fluid remains the same. A fluid at rest in a pipe exerts static pressure on the walls. If the fluid starts moving, some of the static pressure is converted to dynamic pressure, which is proportional to the square of the speed of the fluid. The faster a fluid moves, the greater its dynamic pressure and the smaller the static pressure it exerts on the sides.

Bernoulli's principle works very well far from the surface. Near the surface, however, the effects of viscosity must be considered since the air tends to stick to the surface, slowing down the flow nearby. Thus, a boundary layer of slow-moving air is formed on the surface of an airplane or automobile. This boundary layer is laminar at the beginning of the flow, but it gets thicker as the air moves along the surface and becomes turbulent after a point.

Airflow is determined by many factors, all of which work together in complicated ways to influence flow. Very often, the effects of factors such as viscosity, speed, and turbulence cannot be separated. Engineers have found ingenious ways to get around the difficulty of treating such complex situations. They have defined some characteristic numbers, each of which tells us something useful about the nature of the flow by taking several different factors into account.

One such number is the Reynolds number, which is greater for faster flows and denser fluids and smaller for more viscous fluids. The Reynolds number is also higher for flow around larger objects. Flows at lower Reynolds numbers tend to be slow, viscous, and laminar. As the Reynolds number increases, there is a transition from laminar to turbulent flow. The Reynolds number is a useful similarity parameter. This means that flows in completely different situations will behave in the same way as long as the Reynolds number and the shape of the solid surface are the same. If the Reynolds number is kept the same, water moving around a small stationary airplane model will create exactly the same flow patterns as a full-scale airplane of the same shape, flying through the air. This principle makes it possible to test airplane and automobile designs using small-scale models in wind tunnels.

At speeds greater than 220 mph (354 kph), the compressibility of air cannot be ignored. At these speeds, two different flows may not be equivalent even if they have the same Reynolds number. Another similarity parameter, the Mach number, is needed to make them similar. The Mach number of an airplane is its flight speed divided by the speed of sound at the same altitude and temperature. This means that a plane flying at the speed of sound has a Mach number of one.

The drag coefficient and the lift coefficient are two numbers that are used to compare the forces in different flow situations. Aerodynamic drag is the force that opposes the motion of a car or an airplane. Lift is the upward force that keeps an airplane afloat against **gravity**. The drag or lift coefficient is defined as the drag or lift force divided by the dynamic pressure, and also by the **area** over which the force acts. Two objects with similar drag or lift coefficients experience comparable forces, even when the actual values of the drag or lift force, dynamic pressure, area, and shape are different in the two cases.

There are several sources of drag. The air that sticks to the surface of a car creates a drag force due to skin friction. Pressure drag is created when the shape of the surface changes abruptly, as at the point where the roof of an automobile ends. The drop from the roof increases the **space** through which the air stream flows. This slows down the flow and, by Bernoulli's principle, increases the static pressure. The air stream is unable to flow against this sudden increase in pressure and the boundary layer gets detached from the surface creating an area of low-pressure turbulent wake or flow. Because the pressure in the wake is much lower than the pressure in front of the car, a net backward drag or force is exerted on the car. Pressure drag is the major source of drag on blunt bodies. Car manufacturers experiment with vehicle shapes to minimize the drag. For smooth or "streamlined" shapes, the boundary layer remains attached longer, producing only a small wake. For such bodies, skin friction is the major source of drag, especially if they have large surface areas. Skin friction comprises almost 60% of the drag on a modern airliner.

An airfoil is the two-dimensional cross-section of the wing of an airplane as one looks at it from the side. It is designed to maximize lift and minimize drag. The upper surface of a typical airfoil has a curvature greater than that of the lower surface. This extra curvature is known as camber. The straight line, joining the front tip or the leading edge of the airfoil to the rear tip or the trailing edge, is known as the chord line. The angle of attack is the angle that the chord line forms with the direction of the air stream.

The stagnation point is the point at which the stream of air moving toward the wing divides into two streams, one flowing above and the other flowing below the wing. Air flows faster above a wing with greater camber since the same amount of air has to flow through a narrower space. According to Bernoulli's principle, the faster flowing air exerts less pressure on the top surface, so that the pressure on the lower surface is higher, and there is a net upward force on the wing, creating lift. The camber is varied, using flaps and slats on the wing in order to achieve different degrees of lift during take-off, cruise, and landing.

Because the air flows at different speeds above and below the wing, a large jump in speed will tend to arise when the two flows meet at the trailing edge, leading to a rearward stagnation point on top of the wing. Wilhelm Kutta (1867–1944) discovered that a circulation of air around the wing would ensure smooth flow at the trailing edge. According to the Kutta condition, the strength of the circulation, or the speed of the air around the wing, is exactly as much as is needed to keep the flow smooth at the trailing edge.

Increasing the angle of attack moves the stagnation point down from the leading edge along the lower surface so that the effective area of the upper surface is increased. This results in a higher lift force on the wing. If the angle is increased too much, however, the boundary layer is detached from the surface, causing a sudden loss of lift. This is known as a stall; the angle at which this occurs for an airfoil of a particular shape is known as the stall angle.

The airfoil is a two-dimensional section of the wing. The length of the wing in the third dimension, out to the side, is known as the span of the wing. At the wing tip at the end of the span, the high-pressure flow below the wing meets the low-pressure flow above the wing, causing air to move up and around in wing-tip vortices. These vortices are shed as the plane moves forward, creating a downward force or downwash behind it. The downwash makes the airstream tilt downward and the resulting lift force tilt backward so that a net backward force or drag is created on the wing. This is known as induced drag or drag due to lift. About one third of the drag on a modern airliner is induced drag.

In addition to lift and drag, the stability and control of an aircraft in all three dimensions is important since an aircraft, unlike a car, is completely surrounded by air. Various control devices on the tail and wing are used to achieve this. Ailerons, for instance, control rolling motion by increasing lift on one wing and decreasing lift on the other.

Flight at speeds greater than that of sound are supersonic. Near a Mach number of one, some portions of the flow are at speeds below that of sound, while other portions move faster than sound. The range of speeds from Mach number 0.8 to 1.2 is known as transonic. Flight at Mach numbers greater than five is hypersonic.

The compressibility of air becomes an important aerodynamic factor at these high speeds. The reason for this is that sound waves are transmitted through the successive compression and expansion of air. The compression due to a sound wave from a supersonic aircraft does not have a chance to get away before the next compression begins. This pile up of compression creates a shock wave, which is an abrupt change in pressure, density, and temperature. The shock wave causes a steep increase in the drag and loss of stability of the aircraft. Drag due to the shock wave is known as wave drag. The familiar "sonic boom" is heard when the shock wave touches the surface of the earth.

Temperature effects also become important at transonic speeds. At hypersonic speeds above a Mach number of five, the heat causes nitrogen and **oxygen** molecules in the air to break up into atoms and form new compounds by chemical reactions. This changes the behavior of the air and the simple laws relating pressure, density, and temperature become invalid.

The need to overcome the effects of shock waves has been a formidable problem. Swept-back wings have helped to reduce the effects of shock. The supersonic Concorde that cruises at Mach 2 and several military airplanes have **delta** or triangular wings. The supercritical airfoil designed by Richard Whitcomb of the NASA Langley Laboratory has made air flow around the wing much smoother and has greatly improved both the lift and drag at transonic speeds. It has only a slight curvature at the top and a thin trailing edge. The proposed hypersonic aerospace plane is expected to fly partly in air and partly in space and to travel from Washington to Tokyo within two hours. The challenge for aerodynamicists is to control the flight of the aircraft so that it does not burn up like a meteor as it returns to Earth at several times the speed of sound.

See also Atmosphere; Atmospheric circulation; Atmospheric composition and structure; Atmospheric pressure; Aviation physiology; Bernoulli's equation; Meteorology; Physics; Space physiology; Wind shear

AEROMAGNETICS · *see* MAPPING TECHNIQUES

AFRICA

From the perspective of geologists and paleontologists, Africa takes center stage in the physical history and development of life on Earth. Africa is the world's second largest continent. Africa possesses the world's richest and most concentrated deposits of **minerals** such as gold, diamonds, uranium, chromium, cobalt, and platinum. It is also the cradle of human **evolution** and the birthplace of many animal and plant species, and has the earliest evidence of reptiles, dinosaurs, and mammals.

Present-day Africa, occupying one-fifth of Earth's land surface, is the central remnant of the ancient southern supercontinent called Gondwanaland, a landmass once made up of **South America**, **Australia**, **Antarctica**, India, and Africa. This massive supercontinent broke apart between 195 million and 135 million years ago, cleaved by the same geological forces that continue to transform Earth's **crust** today.

Plate tectonics are responsible for the rise of mountain ranges, the gradual drift of continents, earthquakes, and **volcanic eruptions**. The fracturing of Gondwanaland took place during the **Jurassic Period**, the middle segment of the **Mesozoic Era** when dinosaurs flourished on Earth. It was during the Jurassic that flowers made their first appearance, and dinosaurs like the carnivorous Allosaurus and plant eating Stegasaurus lived.

Geologically, Africa is 3.8 billion years old, which means that in its present form or joined with other continents as it was in the past, Africa has existed for four-fifths of Earth's 4.6 billion years. Africa's age and geological continuity are unique among continents. Structurally, Africa is composed of five cratons (structurally stable, undeformed regions of Earth's crust). These cratons, in south, central, and west Africa are mostly igneous **granite**, **gneiss**, and **basalt**, and formed separately between 3.6 and 2 billion years ago, during the **Precambrian** Era.

The Precambrian, an era which comprises more than 85% of the planet's history, was when life first evolved and the earth's atmosphere and continents developed. Geochemical analysis of undisturbed African rocks dating back 2 billion years has enabled paleoclimatologists to determine that Earth's atmosphere contained much higher levels of **oxygen** than today.

Africa, like other continents, "floats" on a plastic layer of Earth's upper mantle called the **asthenosphere**. The overlying rigid crust or **lithosphere** can be as thick as 150 mi (240 km) or under 10 mi (16 km), depending on location. The continent of Africa sits on the African plate, a section of the earth's crust bounded by mid-oceanic ridges in the Atlantic and Indian **Oceans**. The entire plate is creeping slowly toward the northwest at a rate of about 0.75 in (2 cm) per year.

The African plate is also spreading or moving outward in all directions, and therefore Africa is growing in size. Geologists state that sometime in the next 50 million years, East Africa will split off from the rest of the continent along the East African rift which stretches 4,000 mi (6,400 km) from the Red Sea in the north to Mozambique in the south.

Considering its vast size, Africa has few extensive mountain ranges and fewer high peaks than any other continent. The major ranges are the Atlas Mountains along the northwest coast and the Cape ranges in South Africa. Lowland plains are also less common than on other continents.

Geologists characterize Africa's **topography** as an assemblage of swells and basins. Swells are **rock** strata warped upward by heat and pressure, while basins are masses of lower lying crustal surfaces between swells. The swells are highest in East and central West Africa where they are capped by volcanic flows originating from the seismically active East African rift system. The continent can be visualized as an uneven tilted plateau, one that slants down toward the north and east from higher elevations in the west and south.

During much of the **Cretaceous Period**, from 130 million to 65 million years ago, when dinosaurs like tyrannosaurus, brontosaurus, and triceratops walked the earth, Africa's coastal areas and most of the Sahara **Desert** were submerged underwater. **Global warming** during the Cretaceous Period melted **polar ice** and caused ocean levels to rise. Oceanic organic sediments from this period were transformed into the **petroleum** and **natural gas** deposits now exploited by Libya, Algeria, Nigeria, and Gabon. Today, oil and natural gas drilling is conducted both on land and offshore on the **continental shelf**.

The continent's considerable geological age has allowed more than enough time for widespread and repeated **erosion**, yielding soils leached of organic nutrients but rich in **iron** and **aluminum** oxides. Such soils are high in mineral deposits such as bauxite (aluminum ore), manganese, iron, and gold, but they are very poor for agriculture. Nutrient-poor **soil**, along with deforestation and desertification (expansion of deserts) are just some of the daunting challenges facing African agriculture in modern times.

The most distinctive and dramatic geological feature in Africa is undoubtedly the East African rift system. The rift opened up in the **Tertiary Period**, approximately 65 million years ago, shortly after the dinosaurs became extinct. The same tectonic forces that formed the rift valley and which threaten to eventually split East Africa from the rest of the continent have caused the northeast drifting of the Arabian plate, the opening of the Red Sea to the Indian Ocean, and the volcanic uplifting of Africa's highest peaks including its highest, Kilimanjaro in Tanzania. Mount Kibo, the higher of Kilimanjaro's two peaks, soars 19,320 ft (5,796 m) and is permanently snowcapped despite its location near the equator.

Both Kilimanjaro and Africa's second highest peak, Mount Kenya (17,058 ft; 5,117 m) sitting astride the equator, are actually composite volcanos, part of the vast volcanic field associated with the East African rift valley. The rift valley is also punctuated by a string of **lakes**, the deepest being Lake Tanganyika with a maximum depth of 4,708 ft (1,412 m). Only Lake Baikal in Eastern Russia is deeper at 5,712 ft (1,714 m).

Seismically, the rift valley is very much alive. **Lava** flows and volcanic eruptions occur about once a decade in the Virunga Mountains north of Lake Kivu along the western stretch of the rift valley. One **volcano** in the Virunga **area** in eastern Zaire which borders Rwanda and Uganda actually dammed a portion of the valley formerly drained by a tributary of the Nile River, forming Lake Kivu as a result.

On its northern reach, the 4,000-mi (6,400-km) long rift valley separates Africa from **Asia**. The rift's eastern arm can be traced from the Gulf of Aqaba separating Arabia from the Sinai Peninsula, down along the Red Sea, which divides Africa from Arabia. The East African rift's grabens (basins of crust bounded by fault lines) stretch through the extensive highlands of central Ethiopia which range up to 15,000 ft (4,500 m) and then along the Awash River. Proceeding south, the rift valley is dotted by a series of small lakes from Lake Azai to Lake Abaya and then into Kenya by way of Lake Turkana.

Slicing through Kenya, the rift's grabens are studded by another series of small lakes from Lake Baringo to Lake Magadi. The valley's trough or basin is disguised by layers of volcanic ash and other sediments as it threads through Tanzania via Lake Natron. However, the rift can be clearly discerned again in the elongated shape of Lake Malawi and the Shire River Valley, where it finally terminates along the lower Zambezi River and the Indian Ocean near Beira in Mozambique.

The rift valley also has a western arm which begins north of Lake Mobutu along the Zaire-Uganda border and continues to Lake Edward. It then curves south along Zaire's eastern borders forming that country's boundaries with Burundi as it passes through Lake Kivu and Tanzania by way of Lake Tanganyika.

The rift's western arm then extends toward Lake Nysasa (Lake Malawi). Shallow but vast, Lake Victoria sits in a trough between the rift's two arms. Although the surface altitude of the rift valley lakes like Nyasa and Tanganyika are

hundreds of feet above sea level, their floors are hundreds of feet below due to their great depths.

The eastern arm of the rift valley is much more active than the western branch, volcanically and seismically. There are more volcanic eruptions in the crust of the eastern arm with intrusions of **magma** (subterranean molten rock) in the middle and lower crustal depths. Geologists consider the geological forces driving the eastern arm to be those associated with the origin of the entire rift valley and deem the eastern arm to be the older of the two.

It was in the great African rift valley that hominids, or human ancestors, arose. Hominid **fossils** of the genus *Australopithicus* dating 3–4 million years ago have been unearthed in Ethiopia and Tanzania. And the remains of a more direct ancestor of man, *Homo erectus*, who was using fire 500,000 years ago, have been found in Olduvai Gorge in Tanzania as well as in Morocco, Algeria, and Chad.

Paleontologists, who study fossil remains, employ radioisotope dating techniques to determine the age of hominid and other species' fossil remains. This technique measures the decay of short-lived radioactive isotopes like **carbon** and argon to determine a fossil's age. This is based on the radioscope's atomic **half-life**, or the time required for half of a sample of a radioisotope to undergo radioactive decay. Dating is typically done on volcanic ash layers and charred wood associated with hominid fossils rather than the fossils themselves, which usually do not contain significant amounts of radioactive isotopes.

Present-day volcanic activity in Africa is centered in and around the East African rift valley. Volcanoes are found in Tanzania at Oldoinyo Lengai and in the Virunga range on the Zaire-Uganda border at Nyamlagira and Nyiragongo. There is also volcanism in West Africa. Mount Cameroon (13,350 ft; 4,005 m) along with smaller volcanos in its vicinity, stand on the bend of Africa's West Coast in the Gulf of Guinea, and are the exception. They are the only active volcanoes on the African mainland not in the rift valley.

However, extinct volcanoes and evidence of their activity are widespread on the continent. The Ahaggar Mountains in the central Sahara contain more than 300 volcanic necks that rise above their surroundings in vertical columns of 1,000 ft or more. Also, in the central Sahara, several hundred miles to the east in the Tibesti Mountains, there exist huge volcanic craters or calderas. The Trou au Natron is 5 mi (8 km) wide and over 3,000 ft (900 m) deep. In the rift valley, the Ngorongoro Crater in Tanzania, surrounded by teeming wildlife and spectacular scenery, is a popular tourist attraction. Volcanism formed the diamonds found in South Africa and Zaire. The Kimberly **diamond** mine in South Africa is actually an ancient volcanic neck.

The only folded mountains in Africa are found at the northern and southern reaches of the continent. Folded mountains result from the deformation and uplift of the earth's crust, followed by deep erosion. Over millions of years this process built ranges like the Atlas Mountains, which stretch from Morocco to Algeria and Tunisia.

Geologically, the Atlas Mountains are the southern tangent of the European Alps, geographically separated by the

Strait of Gibraltar in the west and the Strait of Sicily in the east. The Atlas are strung across northwest Africa in three parallel arrays; the coastal, central, and Saharan ranges. By trapping moisture, the Atlas Mountains carve out an oasis along a strip of northwest Africa compared with the dry and inhospitable Sahara Desert just to the south.

The Atlas Mountains are relatively complex folded mountains featuring horizontal thrust faults and ancient crystalline cores. On the other hand, the Cape ranges are older, simpler structures, analogous in age and erosion to the Appalachian Mountains of the eastern United States. The Cape ranges rise in a series of steps from the ocean to the interior, flattening out in plateaus and rising again to the next ripple of mountains.

For a continent of its size, Africa has very few islands lying off its coast. The major Mediterranean islands of Corsica, Sardinia, Sicily, Crete, and Cyprus owe their origins to the events that formed Europe's Alps, and are a part of the Eurasian plate, not Africa. Islands lying off Africa's Atlantic Coast like the Canaries, Azores, and even the Cape Verde Islands near North Africa are considered Atlantic structures. Two islands in the middle of the South Atlantic, Ascension and St. Helena, also belong to the Atlantic. Islands belonging to Equatorial Guinea as well as the island country of Sao Tome and Principe at the sharp bend of Africa off of Cameroon and Gabon are related to volcanic peaks of the Cameroon Mountains, the principal one being Mount Cameroon.

Madagascar, the world's fourth largest island after Greenland, New Guinea, and Borneo, is a geological part of ancient Gondwanaland. The island's eastern two-thirds are composed of crystalline **igneous rocks**, while the western third is largely sedimentary. Although volcanism is now quiescent on the island, vast lava flows indicate widespread past volcanic activity. Madagascar's unique plant and animal species testify to the island's long separation from the mainland.

Marine fossils, notably tribolites dating from the **Cambrian Period** (505–570 million years ago; the first period of the **Paleozoic Era**) have been found in southern Morocco and Mauritania. Rocks from the succeeding period, the Ordovician (500–425 million years ago) consist of sandstones with a variety of fossilized marine organisms; these rocks occur throughout northern and western Africa, including the Sahara.

The Ordovician Period was characterized by the development of brachiopods (shellfish similar to clams), corals, starfish, and some organisms that have no modern counterparts, called sea scorpions, conodonts, and graptolites. At the same time, the African crust was extensively deformed. The continental table of the central and western Sahara was lifted up almost a mile (1.6 km). The uplifting alternated with crustal subsidings, forming valleys that were periodically flooded.

During the **Ordovician Period**, Africa, then part of Gondwanaland, was situated in the southern hemisphere on or near the South Pole. It was toward the end of this period that huge **glaciers** formed across the present-day Sahara and the valleys were filled by **sandstone** and glacial deposits. Although Africa today sits astride the tropics, it was once the theater of the Earth's most spectacular glacial activity. In the next period, the Silurian (425–395 million years ago), further marine sediments were deposited.

The Silurian was followed by the Devonian, Mississippian, and Pennsylvanian Periods (408–286 million years ago), the time interval when insects, reptiles, amphibians, and **forests** first appeared. A continental collision between Africa (Gondwanaland) and the North American plate formed a super-supercontinent (Pangaea) and raised the ancient Mauritanide mountain chain that once stretched from Morocco to Senegal. During the late **Pennsylvanian Period**, layer upon layer of fossilized plants were deposited, forming seams of **coal** in Morocco and Algeria.

When Pangaea and later Gondwanaland split apart in the Cretaceous Period (144–66 million years ago), a shallow sea covered much of the northern Sahara and Egypt as far south as the Sudan. Arabia, subjected to many of the same geological and climatic influences as northern Africa, was thrust northward by tectonic movements at the end of the Oligocene and beginning of the Miocene Epochs (around 30 million years ago). During the Oligocene and Miocene (5–35 million years ago; segments of the modern **Cenozoic Era**) bears, monkeys, deer, pigs, dolphins, and early apes first appeared.

Arabia at this time nearly broke away from Africa. The Mediterranean swept into the resulting rift, forming a gulf that was plugged by an isthmus at present-day Aden on the Arabian Peninsula and Djibouti near Ethiopia. This gulf had the exact opposite configuration of today's Red Sea, which is filled by waters of the Indian Ocean.

As the **Miocene Epoch** ended about five million years ago, the isthmus of Suez was formed and the gulf (today's Red Sea) became a saline (salty) lake. During the Pliocene (1.6–5 million years ago) the Djibouti-Aden isthmus subsided, permitting the Indian Ocean to flow into the rift that is now the Red Sea.

In the **Pleistocene Epoch** (11,000–1.6 million years ago), the Sahara was subjected to humid and then to dry and arid phases, spreading the Sahara desert into adjacent forests and green areas. About 5,000–6,000 years ago in the post glacial period of the modern epoch, the Holocene, a further succession of dry and humid stages, further promoted desertification in the Sahara as well as the Kalahari in southern Africa.

Earth scientists state the expansion of the Sahara is still very much in evidence today, causing the desertification of farm and grazing land and presenting the omnipresent specter of famine in the Sahel (Saharan) region.

Africa has the world's richest concentration of minerals and gems. In South Africa, the Bushveld Complex, one of the largest masses of igneous rock on Earth, contains major deposits of strategic **metals** such as platinum, chromium, and vanadium—metals that are indispensable in tool making and high tech industrial processes. The Bushveld complex is about 2 billion years old.

Another spectacular intrusion of magmatic rocks composed of **olivine**, augite, and hypersthene occurred in the **Archean** Eon over 2.5 billion years ago in Zimbabwe. Called the Great Dyke, it contains substantial deposits of chromium, asbestos, and nickel. Almost all of the world's chromium reserves are found in Africa. Chromium is used to harden alloys, to produce stainless steels, as an industrial catalyst, and to provide **corrosion** resistance.

Unique eruptions that occurred during the Cretaceous in southern and central Africa formed kimberlite pipes—vertical, near-cylindrical rock bodies caused by deep **melting** in the upper mantle. Kimberlite pipes are the main source of gem and industrial diamonds in Africa. Africa contains 40% of the world's diamond reserves, which occur in South Africa, Botswana, Namibia, Angola, and Zaire.

In South Africa, uranium is found side-by-side with gold, thus decreasing costs of production. Uranium deposits are also found in Niger, Gabon, Zaire, and Namibia. South Africa alone contains half the world's gold reserves. Mineral deposits of gold also occur in Zimbabwe, Zaire, and Ghana. Alluvial gold (eroded from soils and rock strata by **rivers**) can be found in Burundi, Côte d'Ivoire, and Gabon.

As for other minerals, half of the world's cobalt is in Zaire and a continuation into Zimbabwe of Zairian cobalt-bearing geological formations gives the former country sizable reserves of cobalt as well. One quarter of the world's aluminum ore is found in a coastal belt of West Africa stretching 1,200 mi (1,920 km) from Guinea to Togo, with the largest reserves in Guinea.

Major coal deposits exist in southern Africa, North Africa, Zaire, and Nigeria. North Africa is awash in petroleum reserves, particularly in Libya, Algeria, Egypt, and Tunisia. Nigeria is the biggest petroleum producer in West Africa, but Cameroon, Gabon, and the Congo also contain oil reserves. There are also petroleum reserves in southern Africa, chiefly in Angola.

Most of Africa's iron reserves are in western Africa, with the most significant deposits in and around Liberia, Guinea, Gabon, Nigeria, and Mauritania. In West Africa as well as in South Africa where iron deposits are also found, the ore is bound up in Precambrian rock strata.

Africa, like other continents, has been subjected to gyrating swings in **climate** during the Quarternary Period of the last 2 million years. These climatic changes have had dramatic affects on **landforms** and vegetation. Some of these cyclical changes may have been driven by cosmic or astronomical phenomena including asteroid and comet collisions.

But the impact of humankind upon the African environment has been radical and undeniable. Beginning 2,000 years ago and accelerating to the present day, African woodland belts have been deforested. Such environmental degradation has been exacerbated by overgrazing, agricultural abuse, and man-made changes, including possible global warming partially caused by the buildup of man-made **carbon dioxide**, chlorofluorocarbons (CFCs), and other **greenhouse gases**.

Deforestation, desertification, and soil erosion pose threats to Africa's man-made lakes and thereby Africa's hydro-electric capacity. Africa's multiplying and undernourished populations exert ever greater demands on irrigated agriculture, but the continent's **water** resources are increasingly taxed beyond their limits. To stabilize Africa's ecology and safeguard its resources and mineral wealth, many earth scientists argue that greater use must be made of sustainable agricultural and pastoral practices. Progress in environmental and resource management, as well as population control is also vital.

See also Earth (planet)

AGASSIZ, LOUIS (1807-1873)
Swiss-born American naturalist

Jean Louis Rodolphe Agassiz was born in Motieren-Vuly, Switzerland, and grew up appreciating the beauty of the Swiss Alps. Agassiz's childhood was supervised by his minister father, who believed that supernatural powers created all natural wonders. Agassiz followed his family's wishes and pursued a degree in medicine. After attending the universities in Munich and Heidelberg, Germany, and Zurich, Switzerland, he eventually earned his Ph.D. in 1829.

Upon his graduation from the University of Munich, Agassiz published a monograph on the fish of Brazil that sparked the attention of the noted French anatomist **Georges Cuvier**. Although he possessed a strong interest in zoology, Agassiz went on to earn a medical degree. In 1832, he went to Paris to serve as an apprentice to Cuvier during that renowned scientist's last years.

Agassiz then accepted his first professional position as a professor of natural history at Neuchatel in Switzerland. For his first project, he published a five-volume work on fossil fish. This work helped establish his reputation as a naturalist and earned him the Wollaston Prize.

Agassiz then shifted his attention to the study of **glaciers**. Among many others, Agassiz was fascinated with the extreme heights of the Alps and the occasional sight of huge boulders that were thought to have been created by glacial movement. He spent his vacations in 1836 and 1837 exploring the glacial formations of Switzerland and compared them with the **geology** of England and central **Europe**.

The question of whether or not glaciers moved intrigued Agassiz, who discovered the answer in 1839 at a cabin that had been built on a glacier approximately 10 years earlier. In one decade it had moved nearly 1 mi (1.6 km) down the glacier from its original site. In a unique experiment, Agassiz drove a straight line of stakes deeply into the **ice** across the glacierhill and then observed their movement. After moving, the stakes formed a U shape as middle stakes had moved more quickly than the side ones. Agassiz concluded that the center stakes moved faster since the glacier was held back at the edges by friction with the mountain wall.

This experiment demonstrated not only that glacier moved, but that many thousands of years before massive ice blocks had probably moved across a great deal of the European land masses that now lacked the massive ice formations. The resulting conclusions led to the term Ice Age, which purported that glacial movement is responsible for modern geological configurations. One of the most significant developments that came out of his observations resulted when his discovery helped provide answers to studies pursued by such naturalists as **Charles Darwin** and **Charles Lyell**.

These two men concluded that **glaciation** was a primary mechanism in causing the geographical distribution and apparent similarities of flora and fauna that were otherwise inexplicably separated by land and **water** masses. Despite the evidence with which he was presented, Agassiz's background prevented him from agreeing with such conclusions, and he continued to believe that supernatural forces were responsible for the similarities.

See also Glacial landforms; Ice ages

AGRICOLA, GEORGIUS (1494-1555)

German physician and geologist

Georgius Agricola was born Georg Bauer, but later Latinized his name to Georgius Agricola, as was the custom of the time. (The German word bauer and the Latin word agricola both mean farmer.) His research and publications on a wide range of geologic topics, including **mineralogy**, paleontology, **stratigraphy**, mountains, earthquakes, volcanoes, and **fossils** have led some biographers to describe him as the forefather of **geology** or one or more of its many branches.

After earning a medical degree in Italy, Agricola was appointed physician in the town of Joachimstahl in Bohemia (now Jachymov, Czechoslovakia). It was in this important mining center that he began some of his earliest research into mining and a lifelong love of geologic studies. When he later relocated to the mining center of Chemnitz (now in modern day Germany) he continued a remarkably systematic and meticulous research into the many facets of mining.

One of his earliest works, published in 1546, was *De Natura Fossileum* (On the nature of fossils). In this book, he summarized much of what was known by the ancient Greeks and Romans about fossils, **minerals**, and **gemstones**. In Agricola's time the meaning of the word fossil encompassed all three of these terms. It wasn't until early in the nineteenth century that the word was given its modern meaning. However, unlike other sixteenth century researchers, he did not accept ancient wisdom as fact. He ridiculed the mystical properties that many ancient scientists, physicians, and philosophers had assigned to fossils, minerals, and gemstones and derided Greek and Roman methods of classification. Instead of using alphabetical listings or groupings by supposed magical traits, Agricola developed a system that that relied on such properties as odor, taste, color, combustibility, shape, origin, brittleness, and cleavage. This classification system has endured for more than 450 years.

In another book, *De Ortu et Causis Subterraneum* (Of subterranean origins and sources), published the same year as *De Naturea Fossileum*, Agricola attempted to explain the existence of mountains, volcanoes, and earthquakes. Although the scientific equipment and knowledge of his time made some explanations impossible, Agricola recognized the power of **wind** and **water** as an erosive force, and associated the hot interior of Earth with volcanoes and earthquakes.

Despite the remarkable observations made by Agricola in *De Natura Fossileum, De Ortu et Causis Subteraneum,* and

at least four other books, it is his seventh and final book, *De Re Metallica* (On the subject of **metals**), published in 1556 (one year after his death) that many geologists consider to be his finest work.

De Re Metallica was the culmination of years of careful and patient research in the two mining towns where Agricola served as a physician. Unlike many scientists before and during his time, he did not rely on hearsay or the work of others. According to Agricola, everything he wrote about he observed first hand or learned from reliable sources.

De Re Metallica was handsomely printed and illustrated with over 250 woodcuts on assaying (analyzing ores for their metallic content), pumps for removing water from mines, machinery for digging, and processes for smelting. The text included such topics as stratigraphy, minerals, finding and identifying ores, administrating mining operations, surveying, and even diseases related to mining.

Given the remarkable depth and scope of Georgius Agricola's research and publications it is not surprising that many modern geologists and historians consider his contributions essential to the early development of the science of geology.

AIR MASSES AND FRONTS

An air mass is an extensive body of air that has a relatively homogeneous **temperature** and moisture content over a significant altitude. Air masses typically cover areas of a few hundred, thousand, or million square kilometers. A front is the boundary at which two air masses of different temperature and moisture content meet. The role of air masses and fronts in the development of **weather** systems was first appreciated by the Norwegian father and son team of Vilhelm and Jacob Bjerknes in the 1920s. Today, these two phenomena are still studied intensively as predictors of future weather patterns.

Air masses form when a body of air comes to rest over an **area** large enough for it to take on the temperature and **humidity** of the land or **water** below it. Certain locations on the earth's surface possess the topographical characteristics that favor the development of air masses. The two most important of these characteristics are topographic regularity and atmospheric stability. Deserts, plains, and **oceans** typically cover very wide areas with relatively few topographical irregularities. In such regions, large masses of air can accumulate without being broken apart by mountains, land/water interfaces, and other features that would break up the air mass.

The absence of consistent **wind** movements also favors the development of an air mass. In regions where cyclonic or anticyclonic storms are common, air masses obviously cannot develop easily.

The system by which air masses are classified reflects the fact that certain locations on the planet possess the topographic and atmospheric conditions that favor air mass development. That system uses two letters to designate an air mass. One letter, written in upper case, indicates the approximate **lat-**

itude (and, therefore, temperature) of the region: A for arctic; P for polar; E for equatorial; T for tropical. The distinctions between arctic and polar on the one hand, and equatorial and tropical on the other are relatively modest. The first two terms (arctic and polar) refer to cold air masses, and the second two (equatorial and tropical) to warm air masses.

A second letter, written in lower case, indicates whether the air mass forms over land or sea and, hence, the relative amount of moisture in the mass. The two designations are c for continental (land) air mass and m for maritime (water) air mass.

The two letters are then combined to designate both temperature and humidity of an air mass. One source region of arctic air masses, for example, is the northern-most latitudes of Alaska, upper Canada, and Greenland. Thus, air masses developing in this source region are designated as cA (cold, land) air masses. Similarly, air masses developing over the **Gulf of Mexico**, a source region for maritime tropical air masses, are designated as mT (warm, water) air masses.

The movement of air masses across the earth's surface is an important component of the weather that develops in an area. For example, weather patterns in **North America** are largely dominated by the movement of about a half dozen air masses that travel across the continent on a regular basis. Two of these air masses are the cP and cA systems that originate in Alaska and central Canada and sweep down over the northern United States during the winter months. These air masses bring with them cold temperatures, strong winds, and heavy **precipitation**, such as the snowstorms commonly experienced in the **Great Lakes** states and New England. The name "Siberian Express" is sometimes used to describe some of the most severe storms originating from these cP and cA air masses.

From the south, mT air masses based in the Gulf of Mexico, the Caribbean, and western Atlantic Ocean move northward across the southern states, bringing hot, humid weather that is often accompanied by thunderstorms in the summer.

Weather along the western coast of North America is strongly influenced by mP air masses that flow across the region from the north Pacific Ocean. These masses actually originate as cP air over Siberia, but are modified to mP masses as they move over the broad expanse of the Pacific, where they often pick up moisture. When an mP mass strikes the west coast of North America, it releases its moisture in the form of showers and, in northern regions, snow.

The term front was suggested by the Bjerkneses because the collisions of two air masses reminded them of a battlefront during a military operation. That collision often results in warlike weather phenomena between the two air masses.

Fronts develop when two air masses with different temperatures and, usually, different moisture content come into contact with each other. When that happens, the two bodies of air act almost as if they are made of two different materials, such as oil and water. Imagine what happens, for example, when oil is dribbled into a **glass** of water. The oil seems to push the water out of its way and, in return, the water pushes back on the oil. A similar shoving match takes place between warm and cold air masses along a front. The exact nature of that shoving match depends on the relative temperature and moisture content of the two air masses and the relative movement of the two masses.

One possible situation is that in which a mass of cold air moving across the earth's surface comes into contact with a warm air mass. When that happens, the cold air mass may force its way under the warm air mass like a snow shovel wedging its way under a pile of snow. The cold air moves under the warm air because the former is denser. The boundary formed between these two air masses is a cold front.

Cold fronts are usually accompanied by a falling barometer and the development of large cumulonimbus **clouds** that bring rain showers and thunderstorms. During the warmer **seasons**, the clouds form as moisture-rich air inside the warm air mass, which is cooled as it rises; water subsequently condenses out as precipitation. Cold fronts are represented on weather maps by means of solid lines that contain solid triangles at regular distances along them. The direction in which the triangles point shows the direction in which the cold front is moving.

A situation opposite to the preceding is one in which a warm air mass approaches and then slides up and over a cold air mass. The boundary formed in this case is a warm front. As the warm air mass meets the cold air mass, it is cooled and some of the moisture held within it condenses to form clouds. In most cases, the first clouds to appear are high cirrus clouds, followed sometime later by stratus and nimbostratus clouds.

Warm fronts are designated on weather maps by means of solid lines to which are attached solid half circles. The direction in which the half circles point shows the direction in which the warm front is moving.

A more complex type of front is one in which a cold front overtakes a slower-moving warm front. When that happens, the cold air mass behind the cold front eventually catches up and comes into contact with the cold air mass underneath the warm front. The boundary between these two cold air masses is an occluded front. A distinction can be made depending on whether the approaching cold air mass is colder or warmer than the second air mass beneath the warm front. The former is called a cold-type occluded front, while the latter is a warm-type occluded front. Once again, the development of an occluded front is accompanied by the formation of clouds and, in most cases, by steady and moderate precipitation. An occluded front is represented on a weather map by means of a solid line that contains, alternatively, both triangles and half circles on the same side of the line.

In some instances, the collision of two air masses results in a stand-off. Neither mass is strong enough to displace the other, and essentially no movement occurs. The boundary between the air masses in this case is known as a stationary air mass and is designated on a weather map by a solid line with triangles and half circles on opposite sides of the line. Stationary fronts are often accompanied by fair, clear weather, although some light precipitation may occur.

See also Atmospheric circulation; Atmospheric composition and structure; Atmospheric pressure; Clouds and cloud types; Weather forecasting methods; Weather forecasting

AIR POLLUTION • *see* ATMOSPHERIC POLLUTION

ALKALINE EARTH METALS

On the **Periodic table**, Group 2 (IIA) consists of beryllium, magnesium, calcium, strontium, barium, and radium. This family of elements is known as the alkaline earth **metals**, or just the alkaline earths. Although early chemists gave the name "earths" to a group of naturally occurring substances that were unaffected by heat and insoluble in **water**, the alkaline earth metals are also usually found in the continental **crust**. In contrast, Group 1 compounds and ions tend to concentrate in the ocean.

Calcium carbonate is geologically evident as **limestone**, **marble**, coral, pearls, and chalk—all derived mainly from the shells of small marine animals. The **weather**[1]**ng** of calcium silicate rocks over millions of years converted the insoluble calcium silicate into soluble calcium salts, which were carried to the **oceans**. The dissolved calcium was used by marine organisms to form their shells. When the organisms died, the shells were deposited on the ocean floor where they were eventually compressed into sedimentary **rock**. Collisions of tectonic plates eventually allow this rock to rise above the ocean floor to become "land-based" limestone deposits.

Caverns throughout the world are formed by the action of atmospheric carbonic acid (water plus **carbon dioxide**) on limestone to form the more soluble calcium bicarbonate. When the solution of calcium bicarbonate reaches the open cavern and the water evaporates, **carbon** dioxide is released and calcium carbonate remains. The calcium carbonate is deposited as stalagmites if the drops hit the ground before evaporating, or as stalactites if the water evaporates while the drop hangs from above.

Other **minerals** of alkaline earth metals are beryllium **aluminum** silicate (beryl), calcium magnesium silicate (asbestos), potassium magnesium chloride (carnallite), calcium magnesium carbonate (**dolomite**), magnesium sulfate (epsomite), magnesium carbonate (magnesite), hydrogen magnesium silicate (talc), calcium fluoride (fluorspar), calcium fluorophosphate (fluorapatite), calcium sulfate (**gypsum**), strontium sulfate (celestite), strontium carbonate (strontianite), barium sulfate (barite), and barium carbonate (witherite). Radium compounds occur in pitchblende, which is primarily uranium oxide, because radium is a product of the radioactive disintegration of U-238. Most pitchblende in the United States is found in Colorado.

The alkaline earth metals, like the alkali metals, are too reactive to be found in nature except as their compounds; the two valence electrons completing an s-subshell are readily lost, and ions with +2 charges are formed. The alkaline earth metals all have a silver luster when their surfaces are freshly cut, but, except for beryllium, they tarnish rapidly. Like most metals, they are good conductors of **electricity**.

Only magnesium and calcium are abundant in Earth's crust. Magnesium is found in seawater and as the mineral carnallite, a combination of potassium chloride and magnesium chloride. Calcium carbonate exists as whole mountain ranges of chalk, limestone, and marble. Its most abundant mineral is **feldspar**, which accounts for two-thirds of the earth's crust. Beryllium is found as the mineral beryl, a beryllium aluminum silicate. With a chromium-ion impurity, beryl is known as emerald. If **iron** ions are present, the gemstone is blue-green and known as aquamarine.

Beryllium is lightweight and as strong as steel. It is hard enough to scratch **glass**. Beryllium is used for windows in x-ray apparatus and in other nuclear applications, allowing the rays to pass through with minimum absorption.

Because beryllium is rather brittle, it is often combined with other metals in alloys. Beryllium-copper alloys have unusually high tensile strength and resilience, which makes them ideal for use in springs and in the delicate parts of many instruments. The **alloy** does not spark, and so finds use in tools employed in fire-hazard areas. Because beryllium-nickel alloys resist **corrosion** by salt water, they are used in marine engine parts.

Magnesium, alone or in alloys, replaces aluminum in many construction applications because the supply of this metal from seawater is virtually unlimited. Magnesium is soft and can be machined, cast, and rolled. Magnesium-aluminum alloys (trade name Dowmetal) are often used in airplane construction.

Magnesium hydroxide is used as milk of magnesia for upset stomachs. Epsom salts are magnesium sulfate. Soapstone, a form of talc, is used for laboratory table tops and laundry tubs. Magnesium oxide is used for lining furnaces.

Slaked lime, or calcium hydroxide, is the principal ingredient in plaster and mortar, in which the calcium hydroxide is gradually converted to calcium carbonate by reaction with the carbon dioxide in the air. Slaked lime is an important flux in the reduction of iron in blast furnaces. It is also used as a mild germ-killing agent in buildings that house poultry and farm animals, in the manufacture of cement and sodium carbonate, for neutralizing acid **soil**, and in the manufacture of glass.

Calcium carbide, made by reacting calcium oxide with carbon in the form of coke, is the starting material for the production of acetylene. Calcium propionate is added to foods to inhibit mold growth. Calcium carbonate and calcium pyrophosphate are ingredients in toothpaste.

Plaster of Paris is $2CaSO_4 \cdot H_2O$, which forms $CaSO_4 \cdot 2H_2O$ (gypsum), as it sets. Gypsum is used to make wallboard, or sheet rock. Asbestos—no longer used as a building material in the United States because of concerns that exposure to asbestos fibers can cause cancer—is a naturally occurring mineral, a calcium magnesium silicate. Calcium and magnesium chlorides, byproducts of sodium chloride purification, are used in the de-icing of roads. Calcium chloride absorbs water from the air, so is used in the prevention of dust on roads, **coal**, and tennis courts and as a drying agent in the laboratory.

Florapatite, a calcium fluorophosphate, is an important starting material in the production of phosphoric acid, which, in turn, is used to manufacture fertilizers and detergents. The mines in Florida account for about one-third of the world's supply of this phosphate rock. Fluorspar, or calcium fluoride,

Steel is an alloy of iron and carbon. *© Wolfgang Kaeler/Corbis. Reproduced by permission.*

is used as a flux in the manufacture of steel. It is also used to make hydrofluoric acid, which is then used to make fluorocarbons such as Teflon.

Calcium is involved in the function of nerves and in blood coagulation. Muscle contraction is regulated by the entry or release of calcium ions by the cell. Calcium phosphate is a component of bones and teeth. Hydroxyapatite, calcium hydroxyphosphate, is the main component of tooth enamel. Cavities are formed when acids decompose this apatite coating. Adding fluoride to the diet converts the hydroxyapatite to a more acid-resistant coating, fluorapatite or calcium fluorophosphate. Magnesium is the metal ion in chlorophyll, the substance in plants that initiates the photosynthesis process in which water and carbon dioxide are converted to sugars. Calcium ions are needed in plants for cell division and cell walls. Calcium pectinate is essential in holding plant cells together. Calcium and magnesium ions are required by living systems, but the other Group 2 elements are generally toxic.

The word barium comes from the Greek *barys*, meaning heavy. Barium salts are opaque to x rays, and so a slurry of barium sulfate is ingested in order to outline the stomach and intestines in x-ray diagnosis of those organs. Although barium ions are poisonous, the very low solubility of barium sulfate keeps the concentration low enough to avoid damage.

Both barium and strontium oxides are used to coat the filaments of vacuum tubes, which are still used in some applications. Because these elements act to remove traces of **oxygen** and nitrogen, a single layer of barium or strontium atoms on a filament may increase the efficiency more than a hundred million times.

Radium is a source of radioactive rays traditionally used in cancer treatment, though other radioactive isotopes are now more commonly used. A radioactive isotope of strontium, strontium-90, is a component of nuclear fallout.

The alkaline earths and their compounds burn with distinctive colors. The green of barium, the red of strontium, and the bright white of magnesium are familiar in fireworks. Strontium is also used in arc lamps to produce a bright red light for highway flares.

See also Chemical bonds and physical properties; Chemical elements; Geochemistry; Stalactites and stalagmites

ALLOY

An alloy is a mixture of two or more elements, at least one of which is metallic, that itself has metallic properties (ductility, conductivity, etc.). Compounds that involve **metals** but do not have metallic properties are not alloys. Alloying occurs naturally; most raw gold, for example, is alloyed with silver, and natural nickel-iron alloys occur both in terrestrial rocks and as a common ingredient of meteorites. However, all alloys used for modern technological purposes are created industrially. This is necessary both because most raw metals exist as chemical compounds in rocks and because the balance of ingredients in a useful alloy must be precise.

In a given alloy, one metal is usually present in higher concentration than any other element; this is termed the parent metal or solvent of the alloy. Most alloys are solid at room **temperature**, and are assumed to be in the solid state when their properties are specified. Three common alloys are steel (parent metal **iron**, main additive **carbon**), bronze (parent metal copper, main additive tin), and brass (parent metal copper, main additive zinc).

The nature of the mixing in an alloy depends on the chemical properties of its ingredients. The atoms of the different elements in an alloy can be roughly classed as indifferent to each other, as attracting each other, or as repelling each other. If all atoms in an alloy are indifferent to each other, they mix randomly and produce an alloy that is uniform at all levels above the atomic. Such an alloy is termed a random solid solution. If the atoms of unlike elements in an alloy attract each other, some orderly pattern develops when the alloy cools from its molten to its solid state. Such a solid is termed a superlattice or ordered solid solution. For example, a half-copper, half-aluminum alloy is an ordered solid solution in which planes of **aluminum** atoms alternate with planes of copper atoms. However, if the unlike atoms in a substance are attracted by strong electrical forces, the result is not an ordered solid solution with metallic properties but a true chemical compound. Salt, for example (sodium chloride, NaCl), is considered an ionic compound, not an alloy of sodium.

If the unlike atoms in an alloy attract each other less than the like atoms, the elements tend to segregate into distinct crystal domains upon solidification. The alloy is then a mass of pure, microscopic **crystals** of its component elements and is termed a phase mixture.

See also Crystals and crystallography; Industrial minerals; Metals; Precious metals; Phase state changes

ALLUVIAL SYSTEM

An alluvial system is a landform produced when a stream or river, that is, some channelized flow (geologists call them all streams no matter what their scale) slows down and deposits sediment that was transported either as bedload or in suspension. The basic principle underlying alluvial deposits is that the more rapidly **water** is moving, the larger the particles it can hold in suspension and the farther it can transport those particles.

For example, suppose that a river is flowing across a mountainous region, eroding **rock**, **sand**, gravel, silt, and other materials from the stream bed. As long as the stream is flowing rapidly, a considerable quantity of materials such as these

can be transported, either along the bottom or as particles suspended in the water column. But then imagine that the stream rushes out of the mountainous region and onto a valley floor. As the river slows down, suspended materials begin to be deposited. The larger bedload materials (for example, rocks and stones) accumulate first, and the lighter suspended materials (sand, silt, and **clay**) later. Any collection of materials deposited by a process such as this is known as *alluvium*. The conditions under which an alluvial system forms are found in both arid and humid climates, and in areas of both low slope (river deltas or swamps) and high slope (mountain streams).

Although the system mentioned above was in a mountainous setting, any river or stream is part of an alluvial system. Many stream systems consist of several common features including channels, heads, mouths, meanders, point bars and cut banks, **floodplains**, levees, oxbow **lakes**, and stream terraces.

The channel is the sloping trough-like depression down which water flows from the stream's origin, or head, to its destination, or mouth. All channels naturally curve, or meander. At the outside of a bend in a channel meander, the flow is concentrated and so **erosion** causes undercutting, and a cutbank forms. On the inside of the meander, flow decreases, so deposition occurs; a sand bar, or point bar, forms.

When a stream **floods**, several processes naturally follow. As the water flows out of its channel, it immediately begins to slow down because it spreads out over a large **area**, increasing the resistance to flow. Coarser sediments are therefore deposited very close to the channel. This forms a very gently sloping lump of alluvium that parallels the channel, known as a natural levee. As the natural levee builds up over thousands of years, it helps prevent flooding. That is why humans build man-made levees—to emulate natural levees. Finer sediments flow with the stream water out onto the flat area behind the levee, known as the floodplain.

During the same flood, if the water is especially high, or the channel is highly meandering, the flood may cut a new channel, connecting two closely positioned meanders, a neck, in what is called a neck cut-off. Once the neck is cut, the channel is much straighter, and the meander is abandoned to become a part of the floodplain. This abandoned meander then forms a lake known as an oxbow.

Another common feature of alluvial systems is the stream terrace. A stream terrace is simply an old floodplain that is now abandoned. Abandonment occurred when the erosive power of the stream increased and it began to rapidly downcut to a lower elevation. The stream did not have time to erode its old floodplain by meandering over it, so it was preserved. The abandoned floodplain, or stream terrace, can be seen well above the new stream channel elevation. Multiple terraces can sometimes be seen, resembling steps in a giant staircase.

When an alluvial system operates over a long period of time, perhaps millions of years, it works to flatten the surrounding landscape, and significantly decrease its average elevation. Areas that were originally mountainous can be worn down to rolling hills, and eventually produce extensive plains composed of alluvial sediment. The sediment is eroded from highlands that may be tens, hundreds, or perhaps thousands of miles from the coast, and the alluvium serves to bury existing coastal features beneath a blanket of sediment. During periods of lower sea level in the geologic past, coastal plains extended far out on the margins of the continents. Today, these alluvial sediments are hundreds of feet below sea level.

As a stream emerges from a mountain valley, its waters are dispersed over a relatively wide region of valley floor. Such is the case, for example, along the base of the Panamint Mountains that flank California's Death Valley. A stream flowing down a mountain side tends to deposit heavier materials near the foot of the mountain, somewhat lighter materials at a greater distance from the mountain, and the lightest materials at a still greater distance from the mountain.

Often, the flow of water ends within the deposited material itself. This material tends to be very porous, so water is more likely to soak into the ground than to flow across its surface. Thus, there is no preferred direction of deposition from side to side at the mouth of the stream, and as the alluvium accumulates it forms a cone-shaped pattern on the valley floor known as an alluvial fan.

The idealized model described above would suggest that an alluvial fan should have a gradually changing composition, with heavier materials such as rocks and small stones at the base of the mountain and lighter materials such as sand and silt at the base (toe) of the fan. In actual fact, alluvial fans seldom have this idealized structure. One reason for the more varied structure found in a fan is that stream flows change over time. During flows of low volume, lighter materials are deposited close to the mountain base on top of heavier materials deposited during earlier flows of high volume. During flows of high volume, heavier materials are once more deposited near the base of the mountain, now on top of lighter materials. A vertical cross-section of an alluvial fan is likely to be more heterogeneous, therefore, than would be suggested by an idealized depositional model.

Alluvial fans tend to have small slopes that may be no more than a foot every half a mile (a few tenths of a meter per kilometer). The exact slope of the fan depends on a number of factors. For example, streams that drain an extensive area, that have a large volume of water, or that carry suspended particles of smaller size are more likely to form fans with modest slopes than are streams with the opposite characteristics.

Under some circumstances, a river or stream may continue to flow across the top of an alluvial fan as well as soak into it. For example, the volume of water carried during floods may cause water to cut across an alluvial fan and empty onto the valley floor itself. Also, over time, sediments may become compacted within the fan, and it may become less and less porous. Then, the stream or river that feeds the fan may begin to cut a channel through the fan itself and to lay down a new fan at the base of the older fan. As the fans in a valley become more extensive, their lateral edges may begin to overlap each other. This feature is known as a bajada or piedmont alluvial plain.

In some regions, piedmont alluvial plains have become quite extensive. The city of Los Angeles, for example, is largely constructed on such a plain. Other extensive alluvial systems can be found in the Central Valley of California and along the base of the Andes Mountains in Paraguay, western Argentina, and eastern Bolivia.

Alluvial fans have certain characteristics that make them attractive for farming. In the first place, they generally have a somewhat reliable source of water (except in a **desert**): the stream or river by which they were formed. Also, they tend to be relatively smooth and level, making it easy for planting, cultivating, and harvesting.

Deltas are common alluvial features, and can be found at the mouths of most streams that flow into a lake or ocean. When **rivers** and streams flow into standing water, their velocity decreases rapidly. They then deposit their sediment load, forming a fan-shaped, sloping deposit very similar to an alluvial fan, but located in the water rather than on dry land. This is known as a **delta**. Deltas show a predictable pattern of decreasing sediment size as you proceed farther and farther from shore.

The Mississippi River Delta is the United States' best known delta. Other well-known deltas are the Nile Delta of northern **Africa** and the Amazon Delta of **South America**. When Aristotle observed the Nile Delta, he recognized it was shaped like the Greek letter, delta, hence the name. Most deltas clog their channels with sediment and so must eventually abandon them. If the river then flows to the sea along a significantly different path, the delta will be abandoned and a new delta lobe will form. This process, known as delta switching, helps build the coastline outward, forming new land for agriculture, as well as other uses.

ALPINE GLACIER · *see* GLACIERS

ALTOCUMULOUS CLOUD · *see* CLOUDS AND CLOUD TYPES

ALTOSTRATUS CLOUD · *see* CLOUDS AND CLOUD TYPES

ALUMINUM

Aluminum is the third most abundant element in the earth's **crust**, ranking only behind **oxygen** and **silicon**. It makes up about 9% of the earth's crust, making it the most abundant of all **metals**. The chemical symbol for aluminum, Al, is taken from the first two letters of the element's name.

Aluminum has an **atomic number** of 13 and an **atomic mass** of 26.98. Aluminum is a silver-like metal with a slightly bluish tint. It has a **melting** point of 1,220°F (660°C), a boiling point of about 4,440°F (2,450°C), and a density of 2.708 grams per cubic centimeter. Aluminum is both ductile and malleable.

Aluminum is a very good conductor of **electricity**, surpassed only by silver and copper in this regard. However, aluminum is much less expensive than either silver and copper. For that reason, engineers are currently trying to discover new ways in which aluminum can be used to replace silver and copper in electrical wires and equipment.

Aluminum occurs in nature as a compound, never as a pure metal. The primary commercial source for aluminum is the mineral bauxite, a complex compound consisting of aluminum, oxygen, and other elements. Bauxite is found in many parts of the world, including **Australia**, Brazil, Guinea, Jamaica, Russia, and the United States. In the United States, aluminum is produced in Montana, Oregon, Washington, Kentucky, North Carolina, South Carolina, and Tennessee.

Aluminum is extracted from bauxite in a two-step process. In the first step, aluminum oxide is separated from bauxite. Aluminum metal is produced from aluminum oxide.

At one time, The extraction of pure aluminum metal from aluminum oxide was very difficult. The initial process requires that aluminum oxide first be melted, then electrolyzed. This is difficult and expensive because aluminum oxide melts at only very high temperatures. An inexpensive method for carrying out this operation was discovered in 1886 by Charles Martin Hall, at the time, a student at Oberlin College in Ohio. Hall found that aluminum oxide melts at a much lower **temperature** if it is first mixed with a mineral known as cryolite. Passing electric current through a molten mixture of aluminum oxide and cryolite, produces aluminum metal.

At the time of Hall's discovery, aluminum was a very expensive metal. It sold for about $10 per pound—so rare and was displayed at the 1855 Paris Exposition next the French crown jewels. As a result of Hall's research, the price of aluminum dropped to less than $.40 per pound).

Aluminum was named for one of its most important compounds, alum, a compound of potassium, aluminum, sulfur, and oxygen. The chemical name for alum is potassium aluminum sulfate, $KAl(SO_4)_2$.

Alum has been widely used by humans for thousands of years. It was mined in ancient Greece and then sold to the Turks who used it to make a beautiful red dye known as Turkey red. Alum has also been long used as a mordant in dyeing. In addition, alum was used as an astringent to treat injuries.

Eventually, chemists began to realize that alum might contain a new element. The first person to actually produce aluminum from a mineral was the Danish chemist and physicist Hans Christian Oersted (1777-1851). Oersted was not very successful, however, in producing a very pure form of aluminum.

The first pure sample of aluminum metal was not made until 1827 when the German chemit Friedrich Wöhler heated a combination of aluminum chloride and potassium metal. Being more active, the potassium replaces the aluminum, leaving a combination of potassium chloride and aluminum metal.

Aluminum readily reacts with oxygen to form aluminum oxide: $4Al + 3O_2 \rightarrow 2Al_2O_3$. Aluminum oxide forms a thin, whitish coating on the aluminum metal that prevents the metal from reacting further with oxygen (i.e., **corrosion**).

The largest single use of aluminum alloys is in the transportation industry. Car and truck manufacturers use aluminum alloys because they are strong, but lightweight. Another important use of aluminum alloys is in the packaging industry. Aluminum foil, drink cans, paint tubes, and contain-

ers for home products are all made of aluminum alloys. Other uses of aluminum alloys include window and door frames, screens, roofing, siding, electrical wires and appliances, automobile engines, heating and cooling systems, kitchen utensils, garden furniture, and heavy machinery.

Aluminum is also made into a large variety of compounds with many industrial and practical uses. Aluminum ammonium sulfate, $Al(NH_4)(SO_4)_2$, is used as a mordant, in **water** purification and sewage treatment systems, in paper production and the tanning of leather, and as a food additive. Aluminum borate is used in the production of **glass** and ceramics.

One of the most widely used compounds is aluminum chloride ($AlCl_3$), employed in the manufacture of paints, antiperspirants, and synthetic rubber. It is also important in the process of converting crude **petroleum** into useful products, such as gasoline, diesel and heating oil, and kerosene.

See also Chemical elements; Minerals

ALVAREZ, LUIS (1911-1988)
American physicist

Luis Alvarez proposed a controversial theory involving the possibility of a massive collision of a meteorite with the earth 65 million years ago, an event that Alvarez believed may account for the disappearance of the dinosaurs. After a varied and illustrious career as a Nobel Prize-winning physicist, Alvarez shared his last major scientific achievement with his son Walter, who was then a professor of **geology** at The University of California at Berkeley. In 1980, the Alvarezes accidentally discovered a band of sedimentary **rock** in Italy that contained an unusually high level of the rare metal iridium. Dating techniques set the age of the layer at about 65 million years. The Alvarezes hypothesized that the iridium came from an asteroid that struck the earth, thereby sending huge volumes of smoke and dust (including the iridium) into the earth's atmosphere. They suggested that the cloud produced by the asteroid's impact covered the planet for an extended period of time, blocked out sunlight, and caused the widespread death of plant life on Earth's surface. The loss of plant life in turn, they theorized, brought about the extinction of dinosaurs, who fed on the plants. While the theory has found favor among many scientists and has been enhanced by additional findings, it is still the subject of scientific debate.

Luis Walter Alvarez was born in San Francisco, California. His father, Dr. Walter Clement Alvarez, was a medical researcher at the University of California at San Francisco and also maintained a private practice. Luis' mother was the former Harriet Skidmore Smythe. Alvarez's parents met while studying at the University of California at Berkeley.

Alvarez attended grammar school in San Francisco and enrolled in the city's Polytechnic High School, where he avidly studied science. When his father accepted a position at the prestigious Mayo Clinic, the family moved to Rochester, Minnesota. Alvarez reported in his autobiography *Alvarez: Adventures of a Physicist*, that his science classes at Rochester High School were "adequately taught [but] not very interesting." Dr. Alvarez noticed his son's growing interest in **physics** and hired one of the Mayo Clinic's machinists to give Luis private lessons on weekends. Alvarez enrolled at the University of Chicago in 1928 and planned to major in **chemistry**. He was especially interested in organic chemistry, but soon came to despise the mandatory chemistry laboratories. Alvarez "discovered" physics in his junior year and enrolled in a laboratory course, "Advanced Experimental Physics: Light" about which he later wrote in his autobiography: "It was love at first sight." He changed his major to physics and received his B.S. in 1932. Alvarez stayed at Chicago for his graduate work and his assigned advisor was Nobel Laureate Arthur Compton, whom Alvarez considered "the ideal graduate advisor for me" because he visited Alvarez's laboratory only once during his graduate career and "usually had no idea how I was spending my time."

Alvarez earned his bachelor's, master's, and doctoral degrees at the University of Chicago before joining the faculty at the University of California at Berkeley, where he remained until retiring in 1978. His doctoral dissertation concerned the diffraction of light, a topic considered relatively trivial, but his other graduate work proved to be more useful. In one series of experiments, for example, he and some colleagues discovered the "east-west effect" of cosmic rays, which explained that the number of cosmic rays reaching the earth's atmosphere differed depending on the direction from which they came. The east-west effect was evidence that cosmic rays consist of some kind of positively charged particles. A few days after passing his oral examinations for the Ph.D. degree, Alvarez married Geraldine Smithwick, a senior at the University of Chicago, with whom he later had two children. Less than a month after their wedding, the Alvarezes moved to Berkeley, California, where Luis became a research scientist with Nobel Prize-winning physicist Ernest Orlando Lawrence, and initiated an association with the University of California that was to continue for forty-two years.

Alvarez soon earned the title "prize wild idea man" from his colleagues because of his involvement in such a wide variety of research activities. Within his first year at Berkeley, he discovered the process of K-electron capture, in which some atomic nuclei decay by absorbing one of the electrons in its first orbital (part of the nuclear shell). Alvarez and a student, Jake Wiens, also developed a mercury vapor lamp consisting of the artificial isotope mercury–198. The U.S. Bureau of Standards adopted the wavelength of the light emitted by the lamp as an official standard of length. In his research with Nobel Prize-winning physicist Felix Bloch, Alvarez developed a method for producing a beam of slow moving neutrons, a method that was used to determine the magnetic moment of neutrons (the extent to which they affect a **magnetic field**). Just after the outbreak of World War II in **Europe**, Alvarez discovered tritium, a radioactive isotope (a variant **atom** containing a different number of protons) of hydrogen.

World War II interrupted Alvarez's work at Berkeley. In 1940, he began research for the military at Massachusetts Institute of Technology's (MIT's) radiation laboratory on radar (radio detecting and ranging) systems. Over the next three years, he was involved in the development of three new types of radar systems. The first made use of a very narrow radar

beam to allow a ground-based controller to direct the "blind" landing of an airplane. The second system, code-named "Eagle," was a method for locating and bombing objects on the ground when a pilot could not see them. The third invention became known as the microwave early-warning system, a mechanism for collecting images of aircraft movement in overcast skies.

In 1943, Alvarez left MIT to join the Manhattan Project research team working in Los Alamos, New Mexico. His primary accomplishment with the team was developing the detonating device used for the first plutonium bomb. Alvarez flew in the B–29 bomber that observed the first test of an atomic device at Alamogordo, south of Los Alamos. Three weeks later, Alvarez was aboard another B–29 following the bomber "Enola Gay" as it dropped the first atomic bomb on Hiroshima, Japan. Like most scientists associated with the Manhattan Project, Alvarez was stunned and horrified by the destructiveness of the weapon he had helped to create. Nonetheless, he never expressed any doubts or hesitation about the decision to use the bombs, since they brought a swift end to the war. Alvarez felt strongly that the United States should continue its nuclear weapons development after the war and develop a fusion (hydrogen) bomb as soon as possible.

After the war, Alvarez returned to Berkeley where he had been promoted to full professor. Determining that the future of nuclear physics lay in high-energy research, he focused his research on powerful particle accelerators—devices that accelerate electrons and protons to high velocity. His first project was to design and construct a linear accelerator for use with protons. Although his machine was similar in some ways to the electron accelerators that had been available for many years, the proton machine posed a number of new problems. By 1947, however, Alvarez had solved those problems and his forty-foot-long proton accelerator began operation.

Over the next decade, the science of particle physics (the study of atomic components) developed rapidly at Berkeley. An important factor in that progress was the construction of the 184-inch synchrocyclotron at the university's radiation laboratory. The synchrocyclotron was a modified circular particle accelerator capable of achieving much greater velocities than any other type of accelerator. The science of particle physics involves two fundamental problems: creation of particles to be studied in some type of accelerator and detection and identification of those particles. After 1950, Alvarez's interests shifted from the first to the second of these problems, particle detection, because of a chance meeting in 1953 with University of Michigan physicist Donald Glaser. Glaser had recently invented the bubble chamber, a device that detects particles as they pass through a container of superheated fluid. As the particles move through the liquid, they form ions that act as nuclei on which the superheated material can begin to boil, thereby forming a track of tiny bubbles that shows the path taken by the particles. In talking with Glaser, Alvarez realized that the bubble chamber could be refined and improved to track the dozens of new particles then being produced in Berkeley's giant synchrocyclotron. Among these particles were some with very short lifetimes known as resonance states.

Improving Glaser's original bubble chamber involved a number of changes. First, Alvarez decided that liquid hydrogen would be a more sensitive material to use than the diethyl ether employed by Glaser. In addition, he realized that sophisticated equipment would be needed to respond to and record the resonance states that often lasted no more than a billionth of a second. The equipment he developed included relay systems that transmitted messages at high speeds and computer programs that could sort out significant from insignificant events and then analyze the former. Finally, Alvarez aimed at constructing larger and larger bubble chambers to record a greater number of events. Over a period of about five years, Alvarez's chambers grew from a simple one-inch **glass** tube to his most ambitious instrument, a 72-in (183 cm) chamber that was first put into use in 1959. With these devices, Alvarez eventually discovered dozens of new elementary particles, including the unusual resonance states.

The significance of Alvarez's work with bubble chambers was recognized in 1968 when he was awarded the Nobel Prize for physics. At the awards ceremony in Stockholm, the Swedish Academy of Science's Sten von Friesen stated that, because of his work with the bubble chamber, "entirely new possibilities for research into high-energy physics present themselves....Practically all the discoveries that have been made in this important field [of particle physics] have been possible only through the use of methods developed by Professor Alvarez." Alvarez attended the Nobel ceremonies with his second wife, Janet Landis, whom he married in 1958. The couple had two children.

Advancing years failed to reduce Alvarez's curiosity on a wide range of topics. In 1965 he was in charge of a **joint** Egyptian-American expedition whose goal was to search for hidden chambers in the pyramid of King Kefren at Giza. The team aimed high-energy muons (subatomic particles produced by cosmic rays) at the pyramid to look for regions of low density, which would indicate possible chambers. However, none were found. Alvarez's hobbies included flying, golf, music, and inventing. He made his last flight in his Cessna 310 in 1984, almost exactly 50 years after he first learned to fly. In 1963, he assisted the Warren Commission in the investigation of President John F. Kennedy's assassination. Among his inventions were a system for color television and an electronic indoor golf-training device developed for President Eisenhower. In all, he held 22 patents for his inventions. Alvarez died of cancer in Berkeley, at the age of 77.

AMORPHOUS

Solid substances fall into two general classes, crystalline and amorphous. Those whose atoms show long-range order, like the squares on a chessboard or the loops in a chain-link fence, are crystalline; those whose atoms are arranged in no particular, repeating pattern are amorphous. Naturally occurring amorphous solids are also termed mineraloids.

Amorphous solids are made of the same elements that produce crystalline solids, often mixed in the same ratios. For example, pure **silicon** dioxide (silica; SiO_2) occurs both in a crystalline form (e.g., **quartz**); and in an amorphous form (e.g., **glass**). The difference between the two forms is one of atomic-level organization. Given sufficient time, as when precipitating **atom** by atom from a hydrothermal solution or solidifying slowly from a pure melt, silicon and **oxygen** atoms assume an orderly, crystalline arrangement because it is a lower-energy state and therefore more stable, as a pencil lying on its side has less energy and is more stable than a pencil balanced on its eraser. However, if cooled suddenly, the silicon and oxygen atoms in, for example, molten silica have no time to line up in orderly crystalline ranks but are trapped in a random solid arrangement. Natural glasses (lechatelierites) are in fact produced in large quantities when silica-rich **lava** is quenched suddenly in air or, as during undersea eruptions, in **water**.

Although few amorphous solids beside glasses occur naturally, an amorphous form of virtually any substance can be manufactured by sufficiently rapid quenching of the liquid phase or by depositing atoms from the vapor phase directly onto a cool substrate. Vapor deposition is used to build up the amorphous silicon films found in all integrated electronic circuit chips.

Most natural amorphous solids are formed by fast quenching, but not all. The precious stone opal ($SiO_2 \cdot nH_2O$) is a mineraloid formed by the solidification of a colloidal solution (fine-particle mixture) of silica and water—in essence, opal is very firm silica jello. **Minerals** formed by solidification of colloids, like opal, are termed gel minerals. Limonite ($Fe_2O_3 \cdot nH_2O$) is another gel mineral.

A crystalline solid may be transformed into an amorphous solid by alpha-particle radiation emitted by uranium or thorium atoms contained in the crystal itself. Each alpha particle that passes through the crystal strikes a tiny but violent blow against its atomic structure, slightly scrambling the orderly ranks of atoms. A once-crystalline mineral whose crystal structure has been obliterated by alpha radiation is termed a metamict mineral.

See also Chemical bonds and physical properties; Crystals and crystallography

ANALEMMA

The earth's orbit around the **Sun** is not a perfect circle. It is an ellipse, albeit not a very flattened one, and this leads to a number of interesting observational effects. One of these is the analemma, the apparent path traced by the Sun in the sky when observed at the same time of day over the course of a year. The path resembles a lopsided figure eight (which is printed on some globes).

When measuring the Sun's position in the sky every day precisely at noon, over the course of a year the Sun appears to move higher in the sky as summer approaches and then move lower as winter approaches. This occurs because the tilt of Earth's axis causes the Sun's apparent celestial **latitude**, or *declination*, to change over the course of the year.

Mount St. Helens erupted explosively because its magma is andesitic, with increased amounts of trapped gas. *AP/Wide World. Reproduced by permission.*

However, at some times of the year, the Sun will appear slightly farther west in the sky than it does at other times, as if it were somehow gaining time on a watch. This results from the ellipticity of Earth's orbit. According to Kepler's second law of motion, planets moving in an elliptical orbit will move faster when they are closer to the Sun than when they are farther away. Therefore, Earth's speed in its orbit is constantly changing, decelerating as it moves from perihelion (its closest point to the Sun) to aphelion (its farthest point from the Sun), and accelerating as it then "falls" inward toward perihelion again.

It would be nearly impossible for watchmakers to try to make a clock that kept actual solar time. The clock would have to tick at different rates each day to account for Earth's changing velocity about the Sun. Instead, watches keep what is called mean solar time, which is the average value of the advance of solar time over the course of the year. As a result, the Sun gets ahead of, and behind, mean solar time by up to 16 minutes at different times of the year. In other words, if one measured the position of the Sun at noon mean solar time at one time of year, the Sun might not reach that position until 12:16 P.M. at another time of year.

Now all the elements are in place to explain the analemma's figure eight configurations. The tilt of Earth's orbital axis causes the Sun to appear higher and lower in the sky at different times of year; this forms the vertical axis of the eight. The ellipticity of Earth's orbit causes the actual solar time to first get ahead of, and then fall behind, mean solar time. This makes the Sun appear to slide back and forth across the vertical axis of the eight, forming the rest of the figure.

The shape of the analemma depends upon a particular planet's orbital inclination and ellipticity. The Sun would appear to trace a unique analemma for any of the planets in the **solar system**; the analemmas thus formed are fat, thin, or even teardrop-shaped variants on the basic figure eight.

See also Celestial sphere: The apparent movements of the Sun, Moon, planets, and stars

ANDESITE

Andesite is the most common volcanic **rock** after **basalt**. It is porphyritic, that is, consists of coarse **crystals** (phenocrysts)

embedded in a granular or glassy matrix (groundmass). Having a silica content of 57%, it is in the intermediate category (52–66% silica) of the silicic–mafic scale. The large volcanic **mountain chains** of North and **South America**, including the Andes (for which andesite is named), are composed largely of andesite. Indeed, andesite is common in all the mountain-building zones that rim the Pacific Ocean. The transition from the oceanic **crust** of the main basin of the Pacific to the andesitic rocks around its perimeter is termed the andesite line. The crust on the deep-sea side of the andesite line is a product of **seafloor spreading**, and the andesitic mountains on the other side are a product of orogenic volcanism. The andesite line thus marks the geological border of the true Pacific basin.

The primary ingredient of most andesites is andesine, a **feldspar** of the **plagioclase** series. Smaller amounts of **quartz** or **minerals** rich in **iron** and magnesium such as **olivine**, pyroxene, biotite, or hornblende are also present. Andesites are ordered in three classes according to the identity of their non-feldspar components: from most **silicic** to most **mafic**, these are (1) quartz-bearing andesites, (2) pyroxene andesites, and (3) biotite and hornblende andesites. All are intermediate in composition between diorite (an intrusive igneous rock consisting mostly of plagioclase feldspar) and **rhyolite**, a volcanic rock having the same composition as **granite** (i.e., feldspar plus quartz). In other words, andesites are higher in feldspar than rhyolite but lower in feldspar than diorites.

Andesite's character usually results from the **melting** and assimilation of rock fragments by **magma** rising to the surface. Rocks nearer the surface tend to be higher in **silicon**, because silicon is less dense than iron and magnesium, components that increase at greater depths. An andesite can thus be viewed very roughly as a basalt contaminated with excess silicon (and perhaps other ingredients). Indeed, many olivine-bearing andesites are so close to basalt in appearance that they can only be distinguished on the strength of chemical analysis.

Andesites high in quartz—dacites—are sometimes classed as a separate group.

See also Bowen's reaction series; Crust; Earth, interior structure; Minerals; Volcanic eruptions

ANGULAR UNCONFORMITIES • *see* UNCONFORMITIES

ANION • *see* CHEMICAL BONDS AND PHYSICAL PROPERTIES

ANNING, MARY (1799-1847)
English paleontologist

Mary Anning, a self-educated fossil hunter and collector, was eventually credited with the first discovery of the plesiosaur.

Anning was born in Lyme Regis in Dorset, England and remained single all her life. Lyme Regis is famous for its Jurassic ammonites and dinosaur remains. Her claim to fame was firmly established when she, along with her brother,

found and extracted a complete ichthyosaur skeleton, subsequently sent to a London Museum. At the time she was only 12 years old. Anning also found a nearly complete skeleton of a plesiosaur in 1823, and made her third great discovery in 1828 of the anterior sheath and ink bag of Belemnospia. In 1929, she discovered the fossil fish Squaloraja, thought to be an ancestor of the shark and the ray. Her last major discovery in 1830 was the Plesiosaurus macrocephalus, named by Professor William Buckland.

Aning was born to a poor family and taught her initial trade by her father who was killed in an accident when she was 11 years old. However, over the next 35 years Anning knew and became known to most of the famous geologists of the time by collecting and running her fossil shop first with her mother, and then later on her own. Her knowledge of ammonites, dinosaur bones and other marine **fossils** found on the beach at Lyme Regis gave her fame, but not fortune. Towards the end of her life she was however, granted a government research grant in 1838 to help with her work. Because she was a woman, Anning was never allowed to present her work to the Geological Society of London.

Anning also never published any of her findings. Many of her discoveries are now displayed in museums although her name is rarely mentioned, as most the fossils carry the name of the donator, not the discoverer.

During her lifetime, in 1841 and 1844, Anning had two fossils named after her by **Louis Agassiz**, the Swiss exponent of the **Ice** Age theory. After her death, Anning was recognized by the very society that had failed to admit her during her lifetime, the Geological Society of London. She was an accomplished paleontologist, largely self-educated, and a highly intelligent woman even teaching herself French so that she could read Georges Cuvier's work in the original French. Today, scientists recognize Anning as an authority on British dinosaur anatomy.

See also Fossil record; Fossils and fossilization; Jurassic

ANNULAR ISLANDS • *see* GOYOTS AND ATOLLS

ANTARCTIC CIRCLE • *see* SOLAR ILLUMINATION: SEASONAL AND DIURNAL PATTERNS

ANTARCTIC OCEAN • *see* OCEANS AND SEAS

ANTARCTICA

Among the seven continents on planet Earth, Antarctica lies at the southernmost tip of the world. It is the coldest, driest, and windiest continent. **Ice** covers 98% of the land, and its 5,100,000 sq mi (13,209,000 sq km) occupy nearly one-tenth of Earth's land surface, or the same **area** as **Europe** and the United States combined. Despite its barren appearance, Antarctica and its surrounding waters and islands teem with life all their own, and the continent plays a significant role in the **climate** and health of the entire planet.

Seventy percent of the world's fresh **water** is frozen atop continental Antarctica. These icecaps reflect warmth from the **Sun** back into the atmosphere, preventing planet Earth from overheating. Huge **icebergs** break away from the stationary ice and flow north to mix with warm water from the equator, producing currents, **clouds**, and complex **weather** patterns. Creatures as small as microscopic phytoplankton and as large as whales live on and around the continent, including more than 40 species of birds. Thus, the continent provides habitats for vital links in the world's food chain.

Geologists believe that, millions of years ago, Antarctica was part of a larger continent called Gondwanaland, based on findings of similar **fossils**, rocks, and other geological features on all of the other southern continents. About 200 million years ago, Gondwanaland broke apart into the separate continents of Antarctica, **Africa**, **Australia**, **South America**, and India (which later collided with **Asia** to merge with that continent). Antarctica and these other continents drifted away from each other as a result of shifting of the plates of the earth's **crust**, a process called continental drift that continues today. The continent is currently centered roughly on the geographic South Pole, the point where all south latitudinal lines meet. It is the most isolated continent on Earth, 600 mi (1,000 km) from the southernmost tip of South America and more than 1,550 mi (2,494 km) away from Australia.

Antarctica is considered both an island and a continent. The land itself is divided into east and west parts by the Transantarctic Mountains. The larger side, to the east, is located mainly in the eastern longitudes. West Antarctica is actually a group of islands held together by permanent ice.

Almost all of Antarctica is under ice, in some areas by as much as 2 mi (3 km). The ice has an average thickness of about 6,600 ft (2,000 m), which is higher than many mountains in warmer countries. This grand accumulation of ice makes Antarctica the highest continent on Earth, with an average elevation of 7,500 ft (2,286 m).

While the ice is extremely high in elevation, the actual landmass of the continent is, in most places, well below sea level due to the weight of the ice. If all of this ice were to melt, global sea levels would rise by about 200 ft (65 m), flooding the world's major coastal ports and vast areas of low-lying land. Even if only one-tenth of Antarctica's ice were to slide into the sea, sea levels would rise by 20 ft (6 m), severely damaging the world's coastlines.

Under all the ice, the Antarctic continent is made up of mountains. The Transantarctic Mountains are the longest range on the continent, stretching 3,000 mi (4,828 km) from Ross Sea to Weddell Sea. Vinson Massif, at 16,859 ft (5,140 m), is the highest mountain peak. The few areas where mountains peek through the ice are called nunataks.

Among Antarctica's many mountain ranges lie three large, moon-like valleys—the Wright, Taylor, and Victoria Valleys—which are the largest continuous areas of ice-free land on the continent. Known as the "dry valleys," geologists estimate that it has not rained or snowed there for at least one million years. Any falling snow evaporates before it reaches

the ground, because the air is so dry from the ceaseless winds and brutally cold temperatures. The dryness also means that decomposition is slow, and seal carcasses there have been found to be more than 1,000 years old. Each valley is 25 mi (40 km) long and 3 mi (5 km) wide and provides rare glimpses of the rocks that form the continent and the Transantarctic Mountains.

Around several parts of the continent, ice forms vast floating shelves. The largest, known as the Ross Ice Shelf, is about the same size as Texas. The shelves are fed by **glaciers** on the continent, so the resulting shelves and icebergs are made up of fresh frozen water. Antarctica hosts the largest glacier on Earth; the Lambert Glacier on the eastern half of the continent is 25 mi (40 km) wide and more than 248 mi (400 km) long.

Gigantic icebergs are a unique feature of Antarctic waters. They are created when huge chunks of ice separate from an ice shelf, a cliff, or glacier in a process known as calving. Icebergs can be amazingly huge; an iceberg measured in 1956 was 208 mi (335 km) long by 60 mi (97 km) wide (larger than some small countries) and was estimated to contain enough fresh water to supply London, England, for 700 years. Only 10–15% of an iceberg normally appears above the water's surface, which can create great dangers to ships traveling in Antarctic waters. As these icebergs break away from the continent, new ice is added to the continent by snowfall.

Icebergs generally flow northward and, if they do not become trapped in a bay or inlet, will reach the Antarctic Convergence, the point in the ocean where cold Antarctic waters meet warmer waters. At this point, ocean currents usually sweep the icebergs from west to east until they melt. An average iceberg will last several years before **melting**.

Three **oceans** surround Antarctica—the Atlantic, Pacific, and Indian Oceans. Some oceanographers refer to the parts of these oceans around Antarctica as the Southern Ocean. While the **saltwater** that makes up these oceans does not usually freeze, the air is so cold adjacent to the continent that even the salt and currents cannot keep the water from **freezing**. In the winter months, in fact, the ice covering the ocean waters may extend over an area almost as large as the continent. This ice forms a solid ring close to the continent and loose chunks at the northern stretches. In October (early spring) as temperatures and strong winds rise, the ice over the oceans breaks up, creating huge icebergs.

Because of the way the Earth tilts on its axis as it rotates around the Sun, both polar regions experience long winter nights and long summer days. At the South Pole itself, the sun shines around the clock during the six months of summer and virtually disappears during the cold winter months. The tilt also affects the angle at which the Sun's radiation hits the Earth. When it is directly overhead at the equator, it strikes the polar regions at more indirect angles. As a result, the Sun's radiation generates much less heat, even though the polar regions receive as much annual daylight as the rest of the world.

Even without the **wind chill**, the continent's temperatures can be almost incomprehensible to anyone who has not visited there. In winter, temperatures may fall to –100°F (–73°C). The world's record for lowest **temperature** was recorded on Antarctica in 1960, when it fell to –126.9°F (–88.3°C).

The coastal regions are generally warmer than the interior of the continent. The Antarctic Peninsula may get as warm as 50°F (10°C), although average coastal temperatures are generally around 32°F (0°C). During the dark winter months, temperatures drop drastically, however, and the warmest temperatures range from –4 to –22°F (–20 to –30°C). In the colder interior, winter temperatures range from –40 to –94°F (–40 to –70°C).

The strong winds that constantly travel over the continent as cold air races over the high ice caps and then flows down to the coastal regions, are called katabatic winds. Winds associated with Antarctica **blizzards** commonly gust to more than 120 mi (193 km) per hour and are among the strongest winds on Earth. Even at its calmest, the continent's winds can average 50–90 mi (80–145 km) per hour. Cyclones occur continually from west to east around the continent. Warm, moist ocean air strikes the cold, dry polar air and swirls its way toward the coast, usually losing its force well before it reaches land. These cyclones play a vital role in the exchange of heat and moisture between the tropical and the cold polar air.

Surprisingly, with all its ice and snow, Antarctica is the driest continent on Earth based on annual **precipitation** amounts. The constantly cold temperatures have allowed each year's annual snowfall to build up over the centuries without melting. Along the **polar ice** cap, annual snowfall is only 1–2 in (2.5–5 cm). More precipitation falls along the coast and in the coastal mountains, where it may snow 10–20 in (25–51 cm) per year.

Few creatures can survive Antarctica's brutal climate. Except for a few mites and midges, native animals do not exist on Antarctica's land. Life in the sea and along the coast of Antarctica and its islands, however, is often abundant. A wide variety of animals make the surrounding waters their home, from zooplankton to large birds and mammals. A few fish have developed their own form of antifreeze over the centuries to prevent ice **crystals** from forming in their bodies, while others have evolved into cold-blooded species to survive the cold.

Because the emperor penguin is one of the few species that lives on Antarctica year-round, researchers believe it could serve as an indicator to measure the health of the Antarctic ecosystem. The penguins travel long distances and hunt at various levels in the ocean, covering wide portions of the continent. At the same time, they are easily tracked because the emperor penguins return to their chicks and mates in predictable ways. Such indicators of the continent's health become more important as more humans travel to and explore Antarctica and as other global conditions are found to affect the southernmost part of the world.

A wide variety of research is continuing on Antarctica, primarily during the relatively warmer summer months from October to February when temperatures may reach a balmy 30–50°F (–1–10°C). The cold temperatures and high altitude of Antarctica allow astronomers to put their telescopes above the lower atmosphere, which lessens blurring. During the summer months, they can study the Sun around the clock, because it shines 24 hours a day. Antarctica is also the best

place to study interactions between solar **wind** and Earth's **magnetic field**, temperature circulation in the oceans, unique animal life, **ozone** depletion, ice-zone ecosystems, and glacial history. Buried deep in Antarctica's ice lie clues to ancient climates, which may provide answers to whether the earth is due for **global warming** or the next ice age.

Scientists consider Antarctica to be a planetary bellwether, an early indicator of negative changes in the entire planet's health. For example, they have discovered that a hole is developing in the ozone layer over the continent, a protective layer of gas in the upper atmosphere that screens out the ultraviolet light that is harmful to all life on Earth. The ozone hole was first observed in 1980 during the spring and summer months, from September through November. Each year, greater destruction of the layer has been observed during these months, and the first four years of the 1990s have produced the greatest rates of depletion thus far. The hole was measured to be about the size of the continental United States in 1994, and it lasts for longer intervals each year. Scientists have identified various chemicals created and used by humans, such as chlorofluorocarbons (CFCs), as the cause of this destruction, and bans on uses of these chemicals have begun in some countries.

Researchers have also determined that a major climate change may have occurred in Antarctica in the 1980s and 1990s, based on recorded changes in ozone levels and an increase in cloudiness over the South Pole. This, coupled with a recorded weakening of the ozone shield over **North America** in 1991, has led scientists to conclude that the ozone layer is weakening around the entire planet.

Others are studying the ice cap on Antarctica to determine if, in fact, the earth's climate is warming due to the burning of fossil **fuels**. The global warming hypothesis is based on the atmospheric process known as the **greenhouse effect**, in which pollution prevents the heat energy of the earth from escaping into the outer atmosphere. Global warming could cause some of the ice cap to melt, flooding many cities and lowland areas. Because the polar regions are the engines that drive the world's weather system, this research is essential to identify the effect of human activity on these regions.

Most recently, a growing body of evidence is showing that the continent's ice has fluctuated dramatically in the past few million years, vanishing completely from the continent once and from its western third at least several times. These collapses in the ice structure might be triggered by climatic change, such as global warming, or by far less predictable factors, such as **volcanic eruptions** under the ice. While the east Antarctic ice sheet has remained relatively stable because it lies on a single tectonic plate, the western ice sheet is a jumble of small plates whose erratic behavior has been charted through **satellite** data.

See also Atmospheric pollution; Freshwater; Glacial landforms; Glaciation; Greenhouse gases and greenhouse effect; Ice ages; Ice heaving and wedging; Ozone layer and hole dynamics; Polar axis and tilt

ANTICLINE • *see* SYNCLINE AND ANTICLINE

APHANITIC

Crystalline rocks with mineral grains that cannot be distinguished from one another without magnification have an aphanitic igneous texture. **Igneous rocks** form by crystallization of **minerals** from liquid **magma** rising into the upper portion of Earth's **crust** from the lower crust and underlying mantle. Igneous **rock** texture indicates the rate of magmatic cooling. Crystallization takes place either slowly in deeply buried intrusions called plutons, or rapidly at the earth's surface where magma has been extruded as **lava** by volcanic activity. Igneous rocks are therefore classified as either intrusive (plutonic) or extrusive (volcanic). Slow, undisturbed cooling in a well-insulated pluton is conducive to orderly arrangement of atoms and molecules into large, well-formed **crystals**. Rapid cooling from a lava flow is not. Intrusive igneous rocks thus have coarse-grained, or phaneritic, textures with visible crystals, and extrusive igneous rocks have fine-grained, or aphanitic, texture. Volcanic **glass**, called **obsidian**, forms when lava is quenched and solidified so quickly that the silicate ions in the melt form no orderly atomic structure.

Texture indicates the rate at which an igneous rock cooled, but it has no relationship to the chemical or mineralogical composition of the rock. Aphanitic, extrusive, igneous rocks therefore have coarse-grained, intrusive counterparts with the same chemical and mineral composition. For example, the silica-rich extrusive rock, **rhyolite**, common in continental volcanic regions, is the fine-grained equivalent of intrusive **granite**. Both rock types are composed mainly of the silicate minerals **quartz** and orthoclase **feldspar**, but the crystals in the rhyolite are too small to see without a microscope. **Basalt**, the **iron** and magnesium-rich extrusive igneous rock that comprises the majority of the sea floor, has the same composition as the intrusive rock gabbro. The intermediate-composition extrusive igneous rock, **andesite**, is common in volcanic arcs above subduction zones, and its coarse-grained equivalent, diorite, is found in plutons along these same convergent plate tectonic margins.

See also Pluton and plutonic bodies

AQUIFER

An aquifer is a body of **sand** or porous **rock** capable of storing and producing significant quantities of **water**. An aquifer may be a layer of loose gravel or sand, a layer of porous **sandstone**, a **limestone** layer, or even an igneous or metamorphic body of rock. An aquifer may be only a few feet to hundreds of feet thick. Aquifers occur near the surface or buried thousands of feet below the surface. It may have an aerial extent of thousands of square miles or a few acres. The key requirements are that the layer or body has sufficient **porosity** to store the water,

sufficient **permeability** to transmit the water, and be at least partly below the **water table**. The water table is the elevation of the top of the completely saturated (phreatic) zone. Above the water table is the vadose or **unsaturated zone** where the pore spaces are only partially saturated and contain a combination of air and water.

Porosity and permeability are important measures of producibility in aquifers. Porosity is the ratio of the volume of voids in a rock or **soil** to the total volume. Porosity determines the storage capacity of aquifers. In sand or **sedimentary rocks**, porosity is the space between grains and the volume of open space (per volume) in fractures. In dense rocks such as **granite**, porosity is contained largely within the crack and/or fracture system. Permeability is the capacity of a rock for transmitting a fluid, and is a measure of the relative ease with which a fluid can be produced from an aquifer.

A rock that yields large volumes of water at high rates must have many interconnected pore spaces or cracks. A dense, low porosity rock such as granite can be an adequate aquifer only if it contains an extensive enough system of connected fractures and cracks to be permeable. In the shallow subsurface, this is common because nearly all (indurate) rocks are fractured, often heavily. For that reason, caution should be exercised before assuming a low porosity rock will be an aquitard (impermeable body) and not an aquifer.

Fluid pressure, measured in pounds per square inch (psi), in an aquifer depends on whether it is unconfined or confined. An unconfined aquifer is one that is hydraulically open or connected to the surface. Examples would include sand bodies on or near the surface and more deeply buried layers of rock or sand connected to the surface by fractures and/or faults. The fluid pressure in unconfined aquifers is equivalent to what one would measure at a point in a standing body of water and would increase linearly (at a constant rate) with depth. The elevation of the top surface of an unconfined aquifer is free to fluctuate with rainfall.

A confined aquifer is one that is surrounded on all sides by an aquitard, a formation that does not transmit fluid. The pressure in a confined aquifer can be different from that of an unconfined aquifer at the same elevation. A body of sand surrounded on all sides by a soft, impermeable **clay** or shale serves as a typical example.

See also Hydrogeology; Saturated zone; Water table

ARCHEAN

The Archean is the period in the earth's history from about 3.8 to 2.5 billion years ago (Ga). The term was derived from the Latin word for first because the beginning of the Archean is defined as the age of the oldest rocks identified on Earth. As the study of these rocks continues and older rocks are discovered, some scientists now expand the Archean back to 4 billion years to include recently dated rocks. The Archean is part of the **Precambrian** Era, the entire time span between the formation of the earth 4.56 billion years ago and the beginning of the

Cambrian Era, 544 million years ago. The Archean is preceded by the Hadean Eon, a little used term for the period from which no rocks are preserved (4.56–3.8 Ga), and is followed by the Proterozoic (2.5–0.54 Ga).

Archean aged rocks are found mostly in the interior of continents. They provide evidence that Earth during the Archean was a very active geologic environment. Most of the rocks from the early Archean are highly regional metamorphic rocks. These granitic-gneiss versions of either sedimentary or igneous parent rocks suggest a high degree of lithospheric recycling. Later in the Archean, vast **lava** flows were erupted from undersea rift zones as pillow basalts. Subsequent **metamorphism** altered the basalts into greenstones. Some **sedimentary rocks** are preserved from the Archean and are largely coarse and poorly sorted sandstones and conglomerates. These observations suggest that the Archean Earth was very active tectonically with volcanic activity and movement along plate boundaries occurring at a much higher rate than today. The Archean mantle was much hotter than the modern Earth's interior, resulting in heavy mantle convection and crustal turbulence.

The active tectonics of the Archean produced numerous, relatively small continental landmasses that were very mobile as they floated on the turbulent mantle. Toward the end of the Archean, however, these minicontinents had begun to coalesce. By about 2.5 billion years ago when the Archean eon came to an end a more tectonically stable supercontinent had formed from the accreted landmasses. About 70% of modern continents are Archean in age and were derived from this single large landmass. This supercontinent had a much thicker **crust** than the earlier, smaller crusts and heat flow from the mantle had begun to subside. As a result volcanic and tectonic activity within and along the margins of the supercontinent, were reduced significantly by the start of the Proterozoic.

The first fossilized signs of life appeared in the Archean. Although life probably developed 3.8–3.6 billion years ago as non-photosynthetic bacteria, the oldest evidence of life on Earth are 3.5 Ga old stromatolite **fossils** from **Australia**. Stromatolites are finely layered, mound-shaped accumulations of mud trapped by growing mats of blue-green algae. Other early Archean fossils include 3.5 Ga microscopic filamentous structures resembling modern blue-green algae from Australia and cells apparently in various stages of division from South **Africa** in rocks that are 3.0 Ga old. The Archean atmosphere in which the primordial organisms developed was likely a reducing atmosphere of methane and ammonia. As the Archean progressed and photosynthetic organisms spread, the atmosphere became more **oxygen** rich.

See also Craton; Greenstone belt

ARCHEOLOGICAL MAPPING

An archeological map is used to relate the findings of an examination of an **area** (usually a dig or **remote sensing** analysis) to a particular set of geophysical coordinates (e.g., **latitude and longitude**). Archeological maps usually relate findings to three dimensions, including depth or altitude. Archeological

maps help archeologists maintain accurate records that allows them to relate the results of their examination, to **GIS** maps and other specialized data processing and data depiction programs.

Before any excavation is begun at a site, the archeologist must prepare a survey map of the site. Site mapping may be as simple as a sketch of the site boundaries, or as complex as a topographic map complete with details about vegetation, artifacts, structures, and features on the site. By recording the presence of artifacts on the site, the site map may reveal information about the way the site was used, including patterns of occupational use. Contour maps may shed light on ways in which more recent environmental activity may have changed the original patterns of use. In cases where structural remains are visible at a site, the site map can provide a basis for planning excavations.

When staking out a site to be excavated, the archeologist typically lays out a square grid that will serve as a reference for recording data about the site. The tools required to construct the grid may be as simple as a compass, a measuring tape, stakes, and a ball of twine. After the grid has been laid out, the archeologist draws a representation of it on graph paper, being careful to note the presence of any physical landmarks such as trees, **rivers**, and large rocks. Once the excavation is underway, each artifact recovered is mapped into the square in the grid and layer in which it was found.

As artifacts are removed from each layer in the site, their exact positions are plotted on a map. At the end of the excavation, a record of the site will exist in the form of maps for each excavated layer at the site. Photographs are also taken of each layer for comparison with the maps.

To facilitate artifact recovery, deposited material at the site may be screened or sifted to make sure materials such as animal bones, snails, seeds, or chipping debris are not overlooked. When a screen is used, clumps of **soil** are thrown on it so that the dirt sifts through the screen, leaving any artifacts behind. In some cases, the deposit may be submerged in a container filled with plain or chemically treated **water**. When the water is agitated, light objects such as seeds, small bones, and charred plant material rise to the top of the container.

Prior to shipment to the laboratory for processing, artifacts are placed in a bag that is labeled with a code indicating the location and stratigraphic layer in which the artifacts were found. Relevant information about each artifact is recorded in the field notes for the site.

Many **mapping techniques** developed for use on land have also been adapted for underwater archeology. Grids can be laid out to assist in mapping and drawing the site, and to assist the divers who perform the excavation. In this case, however, the grids must be weighted to keep them from floating away, and all mapping, recording, and photographing must be done with special equipment designed for underwater use.

Most modern archeologists will attempt to place data taken from a site into archeological context by mapping the spatial and stratigraphic dimensions of the site.

Spatial dimensions include the distribution of artifacts, and other features in three dimensions. The level of detail given in the spatial description typically depends on the goals of the research project. One hundred years ago, finds were

Archeologists mapping a dig site. © Layne Kennedy/Corbis.
Reproduced by permission.

recorded much less precisely than they are today; it might have been sufficient to map an object's location to within 25 sq yd (7 sq m). Today, the location of the same artifact might be recorded to the nearest centimeter. Modern archeologists still use maps to record spatial information about a site. Such information includes the spatial distribution of artifacts, features, and deposits, all of which are recorded on the map. Measuring tools range from simple tapes and plumb bobs to highly accurate and precise **surveying instruments** called laser theodolites.

The accuracy of a map is the degree to which a recorded measurement reflects the true value; the precision of the map reflects the consistency with which a measurement can be repeated. Although the archeologist strives for accuracy in representing the site by the map, the fact that much of what is recorded represents a subjective interpretation of what is present makes any map a simplification of reality. The levels of accuracy and precision that will be deemed acceptable for the project must be determined by the archeologists directing the investigation.

The second technique involved in recording the archeological context of a site is stratigraphic mapping. Any process that contributed to the formation of a site (e.g., dumping, flooding, digging, **erosion**, etc.) can be expected to have left some evidence of its activity in the stratification at the site. The sequential order these processes contribute to the formation of a site must be carefully evaluated in the course of an excavation. The archeologist records evidence of ordering in any deposits and interfaces found at the site for the purposes of establishing a relative chronology of the site and interpreting the site's history. In order to document the stratification at the site, the archeologist may draw or photograph vertical sections in the course of an excavation. Specific graphing techniques have been developed to aid archeologists in recording this information. Finally, the archeologist typically notes such details as soil color and texture, and the presence and size of any stones, often with the aid of reference charts to standardize the descriptions.

See also Bathometric mapping; Drainage calculations and engineering; Field methods in geology; Geologic map; GIS; Scientific data management in Earth Sciences; Topography and topographic maps

ARCMINUTES • *see* LATITUDE AND LONGITUDE

ARCSECONDS • *see* LATITUDE AND LONGITUDE

ARCTIC CIRCLE • *see* SOLAR ILLUMINATION: SEASONAL AND DIURNAL PATTERNS

ARCTIC OCEAN • *see* OCEANS AND SEAS

AREA

An accurate map requires precise geographic characterization of the land surface it represents. The two-dimensional extent of a region, or its area, is essential information for scientists whose studies include a geographic component. Land area measurement, however, is particularly critical for governments, industries, and individuals concerned with land management.

Human societies, beginning from the first agronomic civilizations of northern **Africa**, the Middle East, and China, and continuing with the modern geopolitical array of countries and cultures, have parceled their land between individuals, industries, cities, and nations. Geographers, from the ancient Egyptians and Greeks, to present-day remote **satellite remote sensing** and geographic information systems (**GIS**) specialists, have worked to devise methods of measuring land area, and of surveying land parcel boundaries.

Because the solid Earth has **topography**, and the two-dimensional plane of the Earth's curved surface is defined by three variables—longitude, **latitude**, and elevation—accurate calculation of land area is often quite complex. An area measurement of a topographic surface requires summation of the areas of measured rectangles small enough to capture areal variations introduced by variations in elevation. This summation can be accomplished by measuring and adding a sufficient number of small land areas, or by using integral calculus to compute the area of a three-dimensional surface. Both methods require precise measurement of geographic coordinates; the second also requires measurement of an elevation value at each survey point. In the Roman Empire, surveying (*limitatio*) and erection of measured survey markers (*terminatio*), preceded construction of geographic systems called *cadastres* that were composed of linear structures like roads and canals, and measured in *actus* (an *actus*) equaled 120 Roman feet, or 35.5 meters.

Cartographers during the fifteenth and sixteenth centuries, and European colonial surveyors in the seventeenth and eighteenth centuries, added surveyed elevations to their geographic systems by beginning surveys at sea level and calculating relative gain and loss of elevation at benchmarks. Today, satellite-aided global positioning (**GPS**), aeronautical and **space** remote sensing, and computer-assisted mapping of geographical information (GIS), have greatly enhanced the accuracy of land area measurements. However, most present-day land area surveys, including real estate appraisals and assessments of agricultural and forestry lands, measure land area by projecting the earth's three dimensional surface on to a flat surface. So, as it was in Rome, it is today; a hilly modern acre covers more area than a flat acre.

See also Archeological mapping; History of exploration II (Age of exploration); Physical geography; Surveying instruments

ARETES

Aretes are mountain (alpine) structures carved by glacial **ice**. To be more exact, they are carved by the continual action of cirques wearing away the tops of high elevation mountains. Consequently, it is necessary to understand the formation and erosional properties of cirques before the identification of an arete can occur.

Nivation, or the process of snow becoming compacted ice, begins at the mouth of small, high elevation valleys. As the ice continues to grow, its increased weight bears down on the surrounding **rock** and grinds the solid base into smaller debris. Cycles of daily melt and nighttime freeze produce frost-wedging, which in turn, loosens more surrounding rock. Meltwater carries the loosened debris down steep tributaries and away from growing deposits of ice. This ceaseless carving of the mountain forms a small bowl-like depression of ice that, over time, advances up the valley slope. At this point, the ice has identifiable properties and structure. Geologists call these structures cirques.

Cirques do not generally appear in isolation on higher elevation peaks. They are usually found in groups. In mountain ranges where ice sheets have not covered the surface and sheared the peaks, the action of encroaching cirques is free to carve distinctive patterns. As cirques grow and increase in number, they join or coalesce to form a continuous ridge of ice. When a ridge of cirques exists on the opposite side of the peak, the dual carving action meets to form a steep and ser-

rated-looking crest of rock. This knifelike edge is called an arete. These imposing structures can be very treacherous and difficult to access. They are characteristically found in the Alps, Himalayas, and Andes mountains and give the ranges their distinct and rugged appearance. When three aretes meet on a particular peak, a distinctive type of tip emerges and is named a matterhorn peak. The tip is formed by the meeting of three aretes, which form a trapezoidal figure. The most famous matterhorn is found in the Alps where it is still being carved by cirques and defined by aretes.

See also Glacial landforms; Mountain chains

ARMSTRONG, NEIL (1930-)
American astronaut

Neil Armstrong was the first human to stand on the **Moon**. The former test pilot's lunar stroll on July 20, 1969 marked the pinnacle of the most ambitious engineering project ever undertaken. Afterwards, Armstrong pursued a career in aerospace teaching, research, and business.

Neil Alden Armstrong was fascinated by flying from the time of his first airplane ride when he was a six-year-old boy in Ohio. He was the son of Stephen Armstrong, an auditor who moved his family several times during Armstrong's childhood. When Neil was 13, Stephen and his wife, the former Viola Louise Engel, along with Neil and his younger brother and sister, settled in the town of Wapakoneta. Armstrong earned his pilot's license before his driver's license, and at sixteen was not only flying airplanes, but also experimenting with a **wind** tunnel he had built in his basement. He worked a variety of jobs to pay for his flying lessons and also played in a jazz band, pursuing the musical interest that remained a hobby throughout his life. Armstrong earned a Navy scholarship to Purdue University, which he entered in 1947. His schooling was interrupted when the Navy called him to active duty. Armstrong soon qualified as a Navy pilot, and he was flying combat missions in Korea at the age of 20. He flew 78 missions, earning three air medals.

After the Korean conflict, Armstrong left the navy and returned to Purdue. In 1955, he earned his bachelor's in aerospace engineering. In 1956, he married fellow Purdue student Janet Shearon. By then, Armstrong was a test pilot for the National Advisory Committee for Aeronautics, the forerunner of the National Aeronautics and Space Administration (NASA). At NACA's facility at Edwards Air Force Base in California, Armstrong flew a variety of aircraft under development. In 1960, Armstrong made his first of seven trips to the fringes of space in the X–15 rocket plane. The X–15, a sleek craft air-launched from a B–52 bomber and landed on Edwards's famous dry lake bed, gathered data about high-speed flight and atmospheric reentry that influenced many future designs, including the **space shuttle**.

When the astronaut program was first announced, Armstrong discounted it, believing that the winged X–15 design and not the *Mercury* capsule was the better approach to space. After John Glenn made the first U.S. orbital flight in

1962, Armstrong changed his mind and applied for NASA's astronaut corps. He was accepted into the second group of astronauts, becoming the first civilian to be chosen. In March, 1966, after serving as a backup for the *Gemini-Titan 5* mission, Armstrong made his first space flight as commander of *Gemini-Titan 8.* On this mission, Armstrong's capsule achieved the first docking between spacecraft in orbit. After docking the Gemini spacecraft to the Agena target vehicle, however, the combined vehicles began to tumble uncontrollably. Armstrong and co-astronaut David Scott disengaged the *Agena* and found the problem was a thruster on their capsule that was firing continuously. They had to shut down the flight control system to stop it, an action that forced the two astronauts to abort their flight.

Armstrong moved on to the moon-bound *Apollo* program. He was instrumental in adding a system that, in the event of a failure of the *Saturn 5* booster's guidance system, would allow the astronauts to fly the enormous vehicle manually. Armstrong was on the backup crew for *Apollo 8,* and in January, 1969, was selected to command *Apollo 11.* The crew included lunar module pilot Edwin "Buzz" Aldrin Jr. and command module pilot Michael Collins. Armstrong carried with him a piece of fabric and a fragment of a propeller from American aviators Wilbur and Orville Wrights' first airplane.

On July 20, 1969, the spider-shaped lunar module *Eagle* carried Armstrong and Aldrin toward the Sea of Tranquility. The pre-selected landing **area** turned out to be much rougher than thought, and Armstrong was forced to guide the *Eagle* over the terrain until he found a vacant site. The two men finally brought their craft to a soft landing with approximately thirty seconds' worth of fuel remaining. "The *Eagle* has landed," Armstrong reported. Almost seven hours later, he climbed down the ladder and took the epochal first step on the moon. Television viewers around the world watched as the astronaut in his bulky white suit uttered the words, "That's one small step for a man, one giant leap for mankind." (Viewers did not hear the word "a"; Armstrong later explained that his voice-operated microphone, which "can lose you a syllable," failed to transmit the word.)

Joined by Aldrin, Armstrong spent nearly three hours walking on the moon. The astronauts deployed experiments, gathered samples, and planted an American flag. They also left a mission patch and medals commemorating American and Russian space explorers who had died in the line of duty, along with a plaque reading, "Here men from the planet Earth first set foot upon the Moon. We came in peace for all mankind." Then the three men took their command module *Columbia* safely back to Earth. Armstrong and the other *Apollo 11* astronauts then traveled around the world for parades and speeches. The mission brought honors including the Presidential Medal of Freedom, the Harmon International Aviation Trophy, the Royal Geographic Society's Hubbard Gold Medal, and other accolades from a total of seventeen nations. Armstrong became a Fellow of the Society of Experimental Test Pilots, the American Astronautical Society, and the American Institute of Aeronautics and Astronautics.

Apollo 11 was Armstrong's final space mission. He moved to NASA's Office of Advanced Research and

Neil Armstrong on the Moon.

Technology, where he served as deputy associate administrator for aeronautics. One of his major priorities in this position was to further research into controlling high-performance aircraft by computer. In 1970, he earned his master's degree in aerospace engineering from the University of Southern California.

A quiet man who values his privacy, Armstrong rejected most opportunities to profit from his fame. He left NASA in 1971, and moved his family back to Ohio to accept a position at the University of Cincinnati. There he spent seven years engaged in teaching and research as a professor of aerospace engineering. He took special interest in the application of space technology to challenges on Earth such as improving medical devices and providing data on the environment. In 1978, Armstrong was one of the first six recipients of the Congressional Space Medal of Honor, created to recognize astronauts whose "exceptionally meritorious efforts" had contributed to "the welfare of the Nation and mankind."

ARRHENIUS, SVANTE AUGUST (1859-1927)

Swedish chemist

Svante August Arrhenius was awarded the 1903 Nobel Prize in chemistry for his research on the theory of electrolytic dissociation, a theory that had won the lowest possible passing grade for his Ph.D. two decades earlier. Arrhenius's work with chemistry was often closely tied to the science of **physics**, so much so that the Nobel committee was not sure in which of the two fields to make the 1903 award. In fact, Arrhenius is regarded as one of the founders of physical chemistry—the field of science in which physical laws are used to explain chemical phenomena. In the last decades of his life Arrhenius became interested in theories of the **origin of life** on Earth, arguing that life had arrived on our planet by means of spores blown through **space** from other inhabited worlds. He was also one of the first scientists to study the heat-trapping ability of **carbon dioxide** in the atmosphere in a phenomenon now known as the **greenhouse effect**.

Arrhenius was born on February 19, 1859, in Vik (also known as Wik or Wijk), in the district of Kalmar, Sweden. His mother was the former Carolina Thunberg, and his father was Svante Gustaf Arrhenius, a land surveyor and overseer at the castle of Vik on Lake Mälaren, near Uppsala. Young Svante gave evidence of his intellectual brilliance at an early age. He taught himself to read by the age of three and learned to do arithmetic by watching his father keep books for the estate of which he was in charge. Arrhenius began school at the age of eight, when he entered the fifth-grade class at the Cathedral School in Uppsala. After graduating in 1876, Arrhenius enrolled at the University of Uppsala.

At Uppsala Arrhenius concentrated on mathematics, chemistry, and physics, and he passed the candidate's examination for the bachelor's degree in 1878. He then began a graduate program in physics at Uppsala, but left after three years of study. He was said to be dissatisfied with his physics advisor, Tobias Thalén, and felt no more enthusiasm for the only advisor available in chemistry, Per Theodor Cleve. As a result he obtained permission to do his doctoral research in absentia with the physicist Eric Edlund at the Physical Institute of the Swedish Academy of Sciences in Stockholm.

The topic Arrhenius selected for his dissertation was the electrical conductivity of solutions. In 1884 Arrhenius submitted his thesis on this topic. He hypothesized that when salts are added to **water** they break apart into charged particles now known as ions. What was then thought of as a molecule of sodium chloride, for example, would dissociate into a charged sodium **atom** (a sodium ion) and a charged chlorine atom (a chloride ion). The doctoral committee that heard Arrhenius's presentation in Uppsala was unimpressed by his ideas. Among the objections raised was the question of how electrically charged particles could exist in water. In the end the committee granted Arrhenius his Ph.D., but with a score so low that he did not qualify for a university teaching position.

Convinced that he was correct, Arrhenius had his thesis printed and sent it to a number of physical chemists on the continent, including Rudolf Clausius, Jacobus van't Hoff, and Wilhelm Ostwald. These men formed the nucleus of a group of researchers working on problems that overlapped chemistry and physics, developing a new discipline that would ultimately be known as physical chemistry. From this group Arrhenius received a much more encouraging response than he had received from his doctoral committee. In fact Ostwald

came to Uppsala in August 1884 to meet Arrhenius and to offer him a job at Ostwald's Polytechnikum in Riga. Arrhenius was flattered by the offer and made plans to leave for Riga, but eventually declined for two reasons. First, his father was gravely ill (he died in 1885), and second, the University of Uppsala decided at the last moment to offer him a lectureship in physical chemistry.

Arrhenius remained at Uppsala only briefly, however, as he was offered a travel grant from the Swedish Academy of Sciences in 1886. The grant allowed him to spend the next two years visiting major scientific laboratories in **Europe**, working with Ostwald in Riga, Friedrich Kohlrausch in Würzburg, Ludwig Boltzmann in Graz, and van't Hoff in Amsterdam. After his return to Sweden, Arrhenius rejected an offer from the University of Giessen, Germany, in 1891 in order to take a teaching job at the Technical University in Stockholm. Four years later he was promoted to professor of physics there. In 1903, during his tenure at the Technical University, Arrhenius was awarded the Nobel Prize in chemistry for his work on the dissociation of electrolytes.

Arrhenius remained at the Technical University until 1905 when, declining an offer from the University of Berlin, he became director of the physical chemistry division of the Nobel Institute of the Swedish Academy of Sciences in Stockholm. He continued his association with the Nobel Institute until his death in Stockholm on October 2, 1927.

Although he will be remembered best for his work on dissociation, Arrhenius was a man of diverse interests. In the first decade of the twentieth century, for example, he became especially interested in the application of physical and chemical laws to biological phenomena. In 1908 Arrhenius published a book entitled *Worlds in the Making* in which he theorized about the transmission of life forms from planet to planet in the universe by means of spores.

Arrhenius's name has also surfaced in recent years because of the work he did in the late 1890s on the greenhouse effect. He theorized that **carbon** dioxide in the atmosphere has the ability to trap heat radiated from the Earth's surface, causing a warming of the atmosphere. Changes over time in the concentration of carbon dioxide in the atmosphere would then, he suggested, explain major climatic variations such as the glacial periods. In its broadest outlines, the Arrhenius theory sounds similar to current speculations about **climate** changes resulting from **global warming**.

See also Atmospheric chemistry; Greenhouse gases and greenhouse effect

ARTESIAN

Artesian refers to a condition in which **groundwater** flows from a well without the aid of a pump or other artificial means. One can speak of artesian wells, artesian aquifers, or artesian **water**. Artesian conditions arise when the energy per unit weight possessed by groundwater is great enough to force the water from a deeply buried **aquifer** to the ground surface in the event that the aquifer is tapped by a well. Artesian wells were used by ancient Egyptians, and the word artesian comes from the French province of Artois, where the first European artesian well was constructed in 1126.

The energy per unit weight of groundwater is known as hydraulic head and consists of two main components, elevation head and pressure head. Elevation head is the potential energy per unit weight due to the elevation of the groundwater, whereas pressure head is the energy per unit weight arising as water flows downward and is compressed by the weight of the overlying water. Flowing groundwater also possesses kinetic energy proportional to the square of its velocity, but groundwater generally moves so slowly that its velocity head is virtually nonexistent. Hydraulic head has units of length and is measured relative to some reference elevation, typically sea level; in practical terms, it is defined as the elevation to which groundwater will rise in a specially constructed well known as a piezometer. Thus, the hydraulic head of artesian groundwater must be equal to or greater than the elevation of the ground surface to which it is flowing.

Artesian aquifers are confined, meaning that they are sandwiched between lower **permeability** aquitards. Artesian water enters confined aquifers at high elevations and flows downward towards areas of lower hydraulic head. Although elevation head decreases as groundwater flows downward within an aquifer, pressure head increases because the aquifer is confined and energy must be conserved. Artesian groundwater, therefore, has nearly the same hydraulic head deep underground as it did when it entered the confined aquifer at a higher elevation. When a well is drilled into the artesian aquifer, the hydraulic head of the groundwater will be great enough that the water will rise to nearly the elevation at which it entered the aquifer.

See also Hydrogeology; Hydrologic cycle; Hydrostatic pressure

ASIA

Asia is the world's largest continent, encompassing an **area** of 17,177,000 sq mi (44,500,000 sq km), 29.8% of the world's land area. The Himalaya Mountains, which are the highest and youngest mountain range in the world, stretch across the continent from Afghanistan to Burma. The highest of the Himalayan peaks, Mount Everest, reaches an altitude of 29,028 ft (8,848 m). There are many famous deserts in Asia, including the Gobi **Desert**, the Thar Desert, and Ar-Rub'al-Khali ("the Empty Quarter"). The continent has a wide range of climatic zones, from the tropical jungles of the south to the Arctic wastelands of the north in Siberia.

The continent of Asia encompasses such an enormous area and contains so many countries and islands that its exact borders remain unclear. In the broadest sense, it includes central and eastern Russia, the countries of the Arabian Peninsula, the Far Eastern countries, the Indian subcontinent, and numerous island chains. It is convenient to divide this huge region into five categories: the Middle East, South Asia, Central Asia, the Far East, and Southeast Asia.

Rice is a staple food for much of Asia. © Roger Ressmeyer/Corbis. Reproduced by permission.

The Middle Eastern countries lie on the Arabian Peninsula, southwest of Russia and northeast of **Africa**, separated from the African continent by the Red Sea and from **Europe** in the northwest by the Mediterranean Sea. This area stretches from Turkey in the northwest to Yemen in the south, which is bordered by the Arabian Sea. In general, the **climate** is extremely dry, and much of the area is still a desert wilderness. **Precipitation** is low, so the fertile regions of the Middle East lie around the **rivers** or in valleys that drain the mountains. Much of the coastal areas are arid, and the vegetation is mostly desert scrub.

Saudi Arabia is the largest of the Middle Eastern countries. In the west it is bordered by the Red Sea, which lies between Saudi Arabia and the African continent. The Hijaz Mountains run parallel to this coast in the northwest, rising sharply from the sea to elevations ranging from 3,000 to 9,000 ft (910 to 2,740 m). In the south is another mountainous region called the Asir, stretching along the coast for about 230 mi (370 km) and inland about 180–200 mi (290–320 km). Between the two ranges lies a narrow coastal plain called the Tihamat ash-Sham. East of the Hijaz Mountains are two great plateaus called the Najd, which slopes gradually downward over a range of about 3,000 ft (910 m) from west to east, and the Hasa, which is only about 800 ft (240 m) above sea level. Between these two plateaus is a desert region called the Dahna.

About one third of Saudi Arabia is estimated to be desert. The largest of these is the Ar-Rub'al-Khali, which lies in the south and covers an area of about 250,000 sq mi (647,500 sq km). In the north is another desert, called the An-Nafud. The climate in Saudi Arabia is generally very dry; there are no **lakes** and only seasonally flowing rivers. Saudi Arabia, like most of the Middle Eastern countries, has large oil reserves; also found here are rich gold and silver mines which are thought to date from the time of King Solomon.

Israel contains three main regions. Along the Mediterranean Sea lies a coastal plain. Inland is a hilly area that includes the hills of Galilee in the north and Samaria and Judea in the center. In the south of Israel lies the Negev Desert, which covers about half of Israel's land area. The two bodies of **water** in Israel are the Sea of Galilee and the Dead Sea. The latter, which takes its name from its heavy salinity, lies 1,290 ft (393 m) below sea level, and is the lowest point on the earth's landmasses. It is also a great resource for potassium chloride, magnesium bromide, and many other salts. Jordan borders on Israel in the east near the Dead Sea. To the east of the Jordan River, which feeds the Dead Sea, is a plateau region. The low hills gradually slope downward to a large desert, which occupies most of the eastern part of the country.

Lebanon borders Israel in the north and is divided up by its steep mountain ranges. These have been carved by **erosion** into intricate clefts and valleys, lending the landscape an unusual rugged beauty. On the western border, which lies along the Mediterranean Sea, is the Mount Lebanon area. These mountains rise from sea level to a height of 6,600–9,800 ft (2,000–3,000 m) in less than 25 mi (40 km). On the eastern border is the Anti-Lebanon mountain range, which separates Lebanon from Syria. Between the mountains lies Bekaa Valley, Lebanon's main fertile region.

Syria has three major mountain ranges. In the southwest, the Anti-Lebanon mountain range separates the country geographically from Lebanon. In the southeast is the Jabal Ad-Duruz range, and in the northwest, running parallel to the Mediterranean coast, are the Ansariyah Mountains. Between these and the sea is a thin stretch of coastal plains. The most fertile area is in the central part of the country east of the Anti-Lebanon and Ansariyah mountains; the east and northeastern part of Syria is made up of steppe and desert region.

Turkey, at the extreme north of the Arabian Peninsula, borders on the Aegean, the Mediterranean, and the Black **Seas**. Much of the country is cut up by mountain ranges, and the highest peak, called Mount Ararat, reaches an altitude of 16,854 ft (5,137 m). In the northwest is the Sea of Marmara, which connects the Black Sea with the Aegean Sea. Most of this area, called Turkish Thrace, is fertile and has a temperate climate. In the south, along the Mediterranean, there are two fertile plains called the Adana and the Antalya, which are separated by the Taurus Mountains.

The two largest lakes in Turkey are called Lake Van, which is close to the border with Iraq, and Lake Tuz, which lies in the center of the country. Lake Tuz has such a high level of salinity that it is actually used as a source of salt. Turkey is a country of seismic activity, and earthquakes are frequent.

Most of the Far Eastern countries are rugged and mountainous, but rainfall is more plentiful than in the Middle East, so there are many forested regions. Volcanic activity and **plate tectonics** have formed many island chains in this region of the world, and nearly all the countries on the coast include some of these among their territories.

China, with a land area of 3,646,448 sq mi (9,444,292 sq km), is an enormous territory. The northeastern part of the country is an area of mountains and rich forestland, and its mineral resources include **iron**, **coal**, gold, oil, **lead**, copper, and magnesium. In the north, most of the land is made up of fertile plains. It is here that the Yellow (Huang) River is found, which has been called "China's sorrow" because of its great flooding. The northwest of China is a region of mountains and highlands, including the cold and arid steppes of Inner Mongolia. It is here that the Gobi Desert, the fifth largest desert in the world, is found. The Gobi was named by the Mongolians, and its name means "waterless place." It encompasses an area of 500,000 sq mi (1,295,000 sq km), and averages 2–4 in (5–10 cm) of rainfall a year. In contrast, central China is a region of fertile land and temperate climate. Many rivers, including the great Chang (Yangtze) River, flow through this region, and there are several **freshwater** lakes. The largest of these, and the largest in China, is called the Poyang Hu. In the south of China the climate becomes tropical, and the land is very fertile; the Pearl (Zhu or Chu) River **delta**, which lies in this region, has some of the richest agricultural land in China. In the southwestern region, the land becomes mountainous in parts, and coal, iron, phosphorous, manganese, **aluminum**, tin, **natural gas**, copper, and gold are all found here. In the west, before the line of the Himalayas which divides China from India, lies Tibet, which is about twice as large as Texas and makes up about a quarter of China's land area. This is a high plateau region, and the climate is cold and arid. A little to the north and east of Tibet lies a region of mountains and grasslands where the Yangtze and Yellow Rivers arise.

Japan consists of a group of four large islands, called Honshu, Hokkaido, Kyushu, and Shikoku, and more than 3,000 smaller islands. It is a country of intense volcanic activity, with more than 60 active volcanoes, and frequent earthquakes. The terrain is rugged and mountainous, with lowlands making up only about 29% of the country. The highest of the mountain peaks is an extinct **volcano** found on Honshu called Mount Fuji. It reaches an altitude of 12,388 ft (3,776 m). Although the climate is generally mild, tropical cyclones usually strike in the fall, and can cause severe damage.

Central Asia includes Mongolia and central and eastern Russia. This part of Asia is mostly cold and inhospitable. While only 5% of the country is mountainous, Mongolia has an average elevation of 5,184 ft (1,580 m). Most of the country consists of plateaus. The **temperature** variation is extreme, ranging from –40 to 104°F (–40 to 40°C). The Gobi Desert takes up about 17% of Mongolia's land mass, and an additional 28% is desert steppe. The remainder of the country is forest steppe and rolling plains.

North of China and Mongolia lies Russian Siberia. This region is almost half as large as the African continent, and is usually divided into the eastern and western regions. About the top third of Siberia lies within the Arctic Circle, and the climate is very harsh. The most extreme temperatures occur in eastern Siberia, where it falls as low as –94°F (–70°C), and there are only 100 days a year when it climbs above 50°F (10°C). Most of the region along the east coast is mountainous, but in the west lies the vast West Siberian Plain.

The most important lake in this area, and one of the most important lakes in the world, is called Lake Baikal. Its surface area is about the size of Belgium, but it is a mile deep and contains about a fifth of the world's fresh water supply. The diversity of aquatic life found here is unparalleled; it is the only habitat of 600 kinds of plants and 1,200 kinds of animals, making it the home of two-thirds of the freshwater species on Earth.

Southeast Asia includes a number of island chains as well as the countries east of India and south of China on the mainland. The area is quite tropical, and tends to be very humid. Much of the mountainous regions are extremely rugged and inaccessible; they are taken up by forest and jungle and have been left largely untouched; as a result, they provide habitat for much unusual wildlife.

Thailand, which is a country almost twice the size of Colorado, has a hot and humid tropical climate. In the north, northeast, west, and southeast are highlands that surround a central lowland plain. This plain is drained by the river Chao Phraya, and is rich and fertile land. The highlands are mostly covered with **forests**, which include tropical rainforests, deciduous forests, and coniferous pine forests. Thailand also has two coastal regions; the largest borders on the Gulf of Thailand in the east and southeast, and on the west is the shore of the Andaman Sea.

South of the mainland countries lie the island chains of Malaysia, Indonesia, and the Philippines. The latter two are both sites of much volcanic activity; Indonesia is estimated to have 100 active volcanoes. These islands, in particular Malaysia, are extremely fertile and have large regions of tropical rain forests with an enormous diversity in the native plant and wildlife.

South Asia includes three main regions: the Himalayan mountains, the Ganges Plains, and the Indian Peninsula.

The Himalayas stretch about 1,860 mi (3,000 km) across Asia, from Afghanistan to Burma, and range from 150 to 210 mi (250 to 350 km) wide. They are the highest mountains in the world, and are still being pushed upward at a rate of about 2.3 in (6 cm) a year. This great mountain range originated when the Indian subcontinent collided with Asia, which occurred due to the subduction of the Indian plate beneath the Asian continent. The Himalayas are the youngest mountains in the world, which accounts in part for their great height. At present they are still growing as India continues to push into the Asian continent at the rate of about 2.3 in (6 cm) annually. The Indian subcontinent is believed to have penetrated at least 1,240 mi (2,000 km) into Asia thus far. The range begins in Afghanistan, which is a land of harsh climate and rugged environment.

Bordered by China, several former Russian breakaway republics, Pakistan, and Iran, Afghanistan is completely landlocked. High, barren mountains separate the northern plains of Turan from the southwestern desert region, which covers most of Afghanistan's land area. This desert is subject to violent sandstorms during the winter months. The mountains of Afghanistan, which include a spur of the Himalayas called the Hindu Kush, reach an elevation of more than 20,000 ft (6,100 m), and some are snow-covered year-round and contain **glaciers**. The rivers of the country flow outward from the mountain range in the center of the country; the largest of these are the Kabul, the Helmand, the Hari Rud, and the Kunduz. Except for the Kabul, all of these dry up soon after flowing onto the dry plains.

To the east of Afghanistan and separated from it by the Hindu Kush, lies Pakistan. In the north of the country are the mountain ranges of the Himalayas and the Karakoram, the highest mountains in the world. Most of the peaks are over 15,000 ft (4,580 m) and almost 70 are higher than 22,000 ft (6,700 m). By comparison, the highest mountain in the United States, Mount McKinley in Alaska, is only 20,321 ft (6,194 m). Not surprisingly, many of the mountains in this range are covered with glaciers.

In the west of the country, bordering on Afghanistan, is the Baluchistan Plateau, which reaches an altitude of about 3,000–4,000 ft (900–1,200 m). Further south, the mountains disappear, replaced by a stony and sandy desert. The major rivers of Pakistan are the Kabul, the Jhelum, the Chenab, the Ravi, and the Sutlej; all of these drain into the Indus River, which flows into the Arabian Sea in the south of Pakistan.

Also found in the Himalaya Mountains are Nepal and the kingdom of Bhutan. Both of these countries border on the fertile Ganges Plains, so that in the south they are densely forested with tropical jungles; but most of both territories consist of high mountains. It is in Nepal that the highest peak in the world, called Mount Everest, is found; it is 29,028 ft (8,848 m) high.

South of the Himalaya Mountains, India is divided into two major regions. In the north are the Ganges Plains, which stretch from the Indus to the Ganges River delta. This part of India is almost completely flat and immensely fertile; it is thought to have alluvium reaching a depth of 9,842 ft (3,000 m). It is fed by the snow and **ice** from the high peaks, and streams and rivers from the mountains have carved up the northern edge of the plains into rough gullies and crevices. Bangladesh, a country to the north and east of India, lies within the Ganges Plains. The Ganges and the Brahmaputra flow into Bangladesh from India, and they are fed by many tributaries, so the country is one of the most well-watered and fertile regions of Asia. However, it is also close to sea level, and plagued by frequent flooding.

ASTEROID BELT · *see* SOLAR SYSTEM

ASTEROIDS

Asteroids are rocky material left over from the formation of the **solar system** that orbit the **Sun**, but are too small to be

Asteroid 243 Ida and its moon, photograph. *Corbis Corporation. Reproduced by permission.*

viewed as planets. Most asteroids are composed of stone, **iron**, nickel, or a combination of the three ingredients, and resemble terrestrial rocks in appearance. Asteroids can range in size from pebble-sized rocks up to almost 1,000 km in diameter. Asteroids whose orbits will eventually cause them to collide with Earth are known as **meteoroids**. When the heat and friction of entering Earth's atmosphere at high velocity causes the meteoroid to burn brightly in its path across the sky, it is known as a meteor. Particles or chunks of the meteor that survive the atmospheric entry and fall to Earth are meteorites. Asteroids are classified according to their composition, size, or location. Although Near-Earth Asteroids (NEAs) have been observed in Earth's orbit, the vast majority of asteroids, including the largest asteroid Ceres, are located in the Main Asteroid Belt between Mars and Jupiter.

The astronomer **Johannes Kepler** (1571–1630) was the first to postulate the existence of a hidden planet between Mars and Jupiter, a theory long considered by future astronomers, and in the region now known to contain the solar system's Main Asteroid Belt. In 1766, Johannes Titius (1729–1796), a professor of mathematics and **physics** in

Germany, developed a formula for calculating planetary distances that also suggested a planet belonged between Mars and Jupiter. When the planet Uranus was discovered in 1781, it fit into the formula, causing many scientists to be even more certain that the hidden planet existed. One astronomer, Franz Xaver, proposed the formation of a society of astronomers that would be responsible for looking in assigned areas of the sky for the mystery planet.

Father Giuseppe Piazzi (1746–1826), was involved in such a search at this time. During the night of New Year's Eve, 1800, he saw a small star in Taurus. Because he couldn't find it listed in star catalogues, he observed it over several nights. Piazzi discovered that the body moved relative to the fixed stars, so it had to be an object that belonged to the solar system. Discovering the largest asteroid in the solar system, Piazzi gave this object the name of Ceres, the patron goddess of Sicily. Piazzi was unable, however, to calculate Ceres's orbit from so few observations. A German mathematician, Carl Friedrich Gauss, became intrigued with the problem and invented a new method for orbit calculations. Using his technique, the small object was rediscovered in the winter of

1801–02. That same winter, another German, Heinrich Olbers (1758–1840), found a second planetoid: Pallas.

This second discovery sparked a debate: were these two objects remnants of some planet's catastrophe, or did they always exist in their present form? It is now known that all the asteroids together would produce an object much smaller than our **moon**, so it is unlikely they were ever in one piece. Scientists generally agree that asteroids are leftovers from the formation of the solar system out of the solar nebula.

In 1804 and 1807, two more asteroids were found. The third was called Juno, and the fourth was dubbed Vesta. These were the only planetoids found until the mid-1800s, when telescopic equipment and techniques improved. From 1854 until 1870, five new asteroids were discovered every year. The all-time champion asteroid hunter in the days before photography was Johann Palisa (1848–1932) who found 53 by 1900, and added many more before his death.

In 1891, the German astronomer Maximilian Wolf (1863–1932) began using photographic techniques to search for asteroids. He had his **telescope** set up to follow the apparent motion of the stars, so that any other object like an asteroid would produce a short line in a photographic image rather than a dot like the stars. There had been about 300 asteroids found up until his time, but the use of photography opened the floodgates. Wolf alone discovered 228 asteroids. Astronomers now estimate that roughly 100,000 asteroids exist that are bright enough to appear on photographs taken from Earth.

Asteroids are not uniformly distributed in **space**. The huge planet Jupiter has captured some planetoids, called Trojan asteroids, which are found in two clusters ahead and behind the giant planet. They gather at these two points because of the gravitational forces of the Sun and Jupiter. In addition to these, there are other asteroids that have odd orbits that bring them into the inner regions of the solar system. A few have come close to the earth: in 1937, Hermes swept within 600,000 miles of the earth (only twice the distance from the Earth to the Moon); in 1989, another asteroid came within 500,000 miles of our planet. There is evidence that occasionally an asteroid, or a piece of one, has collided with the earth; one of the best-preserved impact craters can be seen in Arizona.

Because they are remnants of the beginnings of our solar system, asteroids can provide astronomers with valuable information about the conditions under which the solar system was formed.

See also Barringer meteor crater; Celestial sphere: The apparent movements of the Sun, Moon, planets, and stars; Comets; Hubble Space Telescope; Meteoroids and meteorites; Solar system

ASTHENOSPHERE

The asthenosphere is the layer of Earth that lies at a depth 60–150 mi (100–250 km) beneath Earth's surface. It was first named in 1914 by the British geologist J. Barrell, who divided Earth's overall structure into three major sections: the **lithosphere**, or outer layer of rock-like material; the asthenosphere; and the centrosphere, or central part of the planet. The asthenosphere gets its name from the Greek word for weak, asthenis, because of the relatively fragile nature of the materials of which it is made. It lies in the upper portion of Earth's structure traditionally known as the mantle.

Geologists are somewhat limited as to the methods by which they can collect information about Earth's interior. For example, they may be able to study rocky material ejected from volcanoes and **lava** flows for hints about properties of the interior regions. But generally speaking, the single most dependable source of such information is the way in which seismic waves are transmitted through Earth's interior. These waves can be produced naturally as the result of earth movements, or they can be generated synthetically by means of explosions, air guns, or other techniques.

Seismic studies have shown that a type of wave known as S-waves slow down significantly as they reach a depth of about 62 mi (100 km) beneath Earth's surface. Then, at a depth of about 155 mi (250 km), their velocity increases once more. Geologists have taken these changes in wave velocity as indications of the boundaries for the region now known as the asthenosphere.

The material of which the asthenosphere is composed can be described as plastic-like, with much less rigidity than the lithosphere above it. This property is caused by the interaction of **temperature** and pressure on asthenospheric materials. Any **rock** will melt if its temperature is raised to a high enough temperature. However, the **melting** point of any rock is also a function of the pressure exerted on the rock. In general, as the pressure is increased on a material, its melting point increases.

The temperature of the materials that make up the asthenosphere tend to be just below their melting point. This gives them a plastic-like quality that can be compared to **glass**. As the temperature of the material increases or as the pressure exerted on the material increases, the material tends to deform and flow. If the pressure on the material is sharply reduced, so will be its melting point, and the material may begin to melt quickly. The fragile melting point pressure balance in the asthenosphere is reflected in the estimate made by some geologists that up to 10% of the asthenospheric material may actually be molten. The rest is so close to being molten that relatively modest changes in pressure or temperature may cause further melting.

In addition to loss of pressure on the asthenosphere, another factor that can bring about melting is an increase in temperature. The asthenosphere is heated by contact with hot materials that make up the **mesosphere** beneath it. Obviously, the temperature of the mesosphere is not constant. It is hotter in some places than in others. In those regions where the mesosphere is warmer than average, the extra heat may actually increase the extent to which asthenospheric materials are heated and a more extensive melting may occur.

The asthenosphere is now thought to play a critical role in the movement of plates across the face of Earth's surface. According to plate tectonic theory, the lithosphere consists of a relatively small number of very large slabs of rocky material. These plates tend to be about 60 mi (100 km) thick and many thousands of miles wide. They are thought to be very rigid themselves but capable of flowing back and forth on top of the asthenosphere. The collision of plates with each other, their lateral sliding past each other, and their separation from each other are thought to be responsible for major geologic features and events such as volcanoes, lava flows, mountain building, and deep-sea rifts.

In order for plate tectonic theory to seem sensible, some mechanism must be available for permitting the flow of plates. That mechanism is the semi-fluid character of the asthenosphere itself. Some observers have described the asthenosphere as the lubricating oil that permits the movement of plates in the lithosphere.

Geologists have now developed sophisticated theories to explain the changes that take place in the asthenosphere when plates begin to thin or to diverge from or converge toward each other. For example, suppose that a region of weakness has developed in the lithosphere. In that case, the pressure exerted on the asthenosphere beneath it is reduced, melting begins to occur, and asthenospheric materials begin to flow upward. If the lithosphere has not actually broken, those asthenospheric materials cool as they approach Earth's surface and eventually become part of the lithosphere itself.

On the other hand, suppose that a break in the lithosphere has actually occurred. In that case, the asthenospheric materials may escape through that break and flow outward before they have cooled. Depending on the temperature and pressure in the region, that outflow of material (**magma**) may occur rather violently, as in a **volcano**, or more moderately, as in a lava flow.

Pressure on the asthenosphere may also be reduced in zones of divergence, where two plates are separating from each other. Again, this reduction in pressure may allow asthenospheric materials in the asthenosphere to begin melting and to flow upward. If the two overlying plates have actually separated, asthenospheric material may flow through the separation and form a new section of lithosphere.

In zones of convergence, where two plates are flowing toward each other, asthenospheric materials may also be exposed to reduced pressure and begin to flow upward. In this case, the lighter of the colliding plates slides upward and over the heavier of the plates, which dives down into the asthenosphere. This process is called subduction. Since the lithospheric material is more rigid than the material in the asthenosphere, the latter is pushed outward and upward. During this movement of plates, pressure on the asthenosphere is reduced, melting occurs, and molten materials flow upward to Earth's surface. In any one of the examples cited here, the asthenosphere supplies new material to replace lithospheric materials that have been displaced by some other tectonic or geologic mechanism.

See also Continental drift theory; Continental shelf; Crust; Earth, interior structure; Plate tectonics

ASTROLABE

The astrolabe is an ancient astronomical instrument, dating back more than 2,000 years, used to observe the positions of the stars. With modifications it has also been used for time-keeping, navigation, and surveying.

Astrolabes depict the visual reference points of stars on the night sky as a function of time. As such, an observer can also set the time to predict the visible star pattern expected. The most common type of astrolabe, the planispheric astrolabe, consists of a star map (the rete) engraved on a round sheet of metal. With regard to the rete, only the angular relationship of the stars needs to be accurate to ensure proper functioning of the astrolabe. A metal ring is moved across the map to represent the position of the local horizon. An outer ring is adjusted to allow for the apparent **rotation** of the stars around the North Star, using prominent stars as reference points.

Astrolabes were forerunners of mechanical clocks, and looked somewhat like watches. With a set of tables, the observer could determine the day and hour for a fixed location by the position of the stars. With the addition of a sighting-rule, called an alidade, an astrolabe could be used as a surveying instrument. The rule could be moved across a scale to measure elevation. Navigational astrolabes marked celestial altitudes (the altitude in degrees above the horizon).

Although there is evidence to support the assertion that ancient Greek culture had astrolabes, it is certain that the Arabs perfected and made regular use of the astrolabe. With the clear **desert** sky at their constant disposal, the Arab people excelled in **astronomy** and used the stars to navigate across the **seas** of sand. Regular use of astrolabes continued into the 1800s. The newer prismatic astrolabe continues to be used for precision surveying.

Modern versions of stellar charts and bowls with adjustable time and date markings on sliding rings are based upon earlier astrolabe construction and design principles.

See also Celestial sphere: The apparent movements of the Sun, Moon, planets, and stars; History of exploration I (Ancient and classical); History of exploration II (Age of exploration)

ASTRONOMY

Astronomy, the oldest of all the sciences, seeks to describe the structure, movements, and processes of celestial bodies.

Some ancient ruins provide evidence that the most remote ancestors observed and attempted to understand the workings of the Cosmos. Although not always fully understood, these ancient ruins demonstrate that early man attempted to mark the progression of the **seasons** as related to the changing positions of the **Sun**, stars, planets, and **Moon** on the celestial sphere. Archaeologists speculate that such observation made more reliable the determination of times for planting and harvest in developing agrarian communities and cultures.

The regularity of the heavens also profoundly affected the development of indigenous religious beliefs and cultural practices. For example, according to Aristotle (384-322 B.C.), Earth occupied the center of the Cosmos, and the Sun and planets orbited Earth in perfectly circular orbits at an unvarying rate of speed. The word astronomy is a Greek term for star arrangement. Although heliocentric (Sun centered) theories were also advanced among ancient Greek and Roman scientists, the embodiment of the geocentric theory conformed to prevailing religious beliefs and, in the form of the Ptolemaic model subsequently embraced by the growing Christian church, dominated Western thought until the rise of empirical science and the use of the **telescope** during the Scientific Revolution of the sixteenth and seventeenth centuries.

In the East, Chinese astronomers carefully charted the night sky, noting the appearance of "guest stars" (**comets**, novae, etc.). As early as 240 B.C., the records of Chinese astronomers record the passage of a "guest star" known now as Comet Halley, and in A.D. 1054, the records indicate that one star became bright enough to be seen in daylight. Archaeoastronmers argue that this transient brightness was a supernova explosion, the remnants of which now constitute the Crab Nebula. The appearance of the supernova was also recorded by the Anasazi Indians of the American Southwest.

Observations were not limited to spectacular celestial events. After decades of patient observation, the Mayan peoples of Central America were able to accurately predict the movements of the Sun, Moon, and stars. This civilization also devised a calendar that accurately predicted the length of a year, to what would now be measured to be within six seconds.

Early in the sixteenth century, Polish astronomer **Nicolas Copernicus** (1473–1543) reasserted the heliocentric theory abandoned by the Greeks and Romans. Although sparking a revolution in astronomy, Copernicus' system was deeply flawed by an insistence on circular orbits. Danish astronomer Tycho Brahe's (1546–1601) precise observations of the celestial movements allowed German astronomer and mathematician **Johannes Kepler** (1571–1630) to formulate his laws of planetary motion that correctly described the elliptical orbits of the planets.

Italian astronomer and physicist **Galileo Galilei** (1564–1642) was the first scientist to utilize a newly invented telescope to make recorded observations of celestial objects. In a prolific career, Galileo's discoveries, including phases of Venus and moons orbiting Jupiter, dealt a death blow to geocentric theory.

In the seventeenth century, English physicist and mathematician Sir Isaac Newton's (1642–1727) development of the laws of motion and gravitation marked the beginning of Newtonian **physics** and modern astrophysics. In addition to developing calculus, Newton made tremendous advances in the understanding of light and optics critical to the development of astronomy. Newton's seminal 1687 work, *Philosophiae Naturalis Principia Mathematica* (Mathematical principles of natural philosophy) dominated the Western intellectual landscape for more than two centuries and proved the impetus for the advancement of celestial dynamics.

Hubble image of the Eagle Nebula, "the Pillars of Creation." *U.S. National Aeronautics and Space Administration (NASA).*

Theories surrounding celestial mechanics during the eighteenth century were profoundly shaped by important contributions by French mathematician Joseph-Louis Lagrange (1736–1813), French mathematician Pierre Simon de Laplace (1749–1827), and Swiss mathematician Leonhard Euler (1707–1783) that explained small discrepancies between Newton's predicted and the observed orbits of the planets. These explanations contributed to the concept of a clockwork-like mechanistic universe that operated according to knowable physical laws.

Just as primitive astronomy influenced early religious concepts, during the eighteenth century, advancements in astronomy caused significant changes in Western scientific and theological concepts based upon an unchanging, immutable God who ruled a static universe. During the course of the eighteenth century, there developed a growing scientific disregard for understanding based upon divine revelation and a growing acceptance of an understanding of Nature based upon the development and application of scientific laws. Whether God intervened to operate the mechanisms of the universe through miracles or signs (such as comets) became a topic of lively philosophical and theological debate. Concepts of the divine became increasingly identified with the assumed eternity or infinity of the Cosmos. Theologians argued that the assumed immutability of a static universe, a concept shaken by the discoveries of Copernicus, Kepler, Galileo, and Newton, offered proof of the existence of God. The clockwork universe viewed as confirmation of the existence of God of infinite power who was the "prime mover" or creator of the universe. For many scientists and astronomers, however, the revelations of a mechanistic universe left no place for the influence of the Divine, and they discarded their religious views. These philosophical shifts sent sweeping changes across the political and social landscape.

In contrast to the theological viewpoint, astronomers increasingly sought to explain "miracles" in terms of natural phenomena. Accordingly, by the nineteenth century, the appearances of comets were no longer viewed as direct signs from God but rather a natural, explainable and predictable result of a deterministic universe. Explanations for catastrophic events (e.g., comet impacts, extinctions, etc.) increasingly came to be viewed as the inevitable results of time and statistical probability.

The need for greater accuracy and precision in astronomical measurements, particularly those used in navigation, spurred development of improved telescopes and pendulum driven clocks that greatly increased the pace of astronomical discovery. In 1781, improved mathematical techniques, combined with technological improvements, along with the proper application of Newtonian laws, allowed English astronomer William Herschel to discover the planet Uranus.

Until the twentieth century, astronomy essentially remained concerned with the accurate description of the movements of planets and stars. Developments in electromagnetic theories of light and the formulation of quantum and relativity theories, however, allowed astronomers to probe the inner workings of the celestial objects. Influenced by German-American physicist Albert Einstein's (1879–1955) theories of relativity and the emergence of **quantum theory**, Indian-born American astrophysicist Subrahmanyan Chandrasekhar (1910-1995) first articulated the **evolution** of stars into supernova, white dwarfs, neutron stars and accurately predicted the conditions required for the formation of black holes subsequently found in the later half of the twentieth century. The articulation of the stellar evolutionary cycle allowed rapid advancements in cosmological theories regarding the creation of the universe. In particular, American astronomer Edwin Hubble's (1889–1953) discovery of red shifted spectra from stars provided evidence of an expanding universe that, along with increased understanding of **stellar evolution**, ultimately led to the abandonment of static models of the universe and the formulation of big bang based cosmological models.

In 1932, American engineer **Karl Jansky** (1905–1945) discovered existence of radio waves of emanating from beyond the earth. Janskey's discovery led to the birth of radio astronomy that ultimately became one of the most productive means of astronomical observation and spurred continuing studies of the Cosmos across all regions of the **electromagnetic spectrum**.

Profound questions regarding the birth and death of stars led to the stunning realization that, in a real sense, because the heavier atoms of which he was comprised were derived from nucleosynthesis in dying stars, man too was a product of stellar evolution. After millennia of observing the Cosmos, by the dawn of the twenty-first century, advances in astronomy allowed humans to gaze into the night sky and realize that they were looking at the light from stars distant in **space** and time, and that they, also, were made from the very dust of stars.

ATLANTIC OCEAN • *see* OCEANS AND SEAS

ATMOPHILE • *see* CHALCOPHILES, LITHOPHILES, SIDEROPHILES, AND ATMOPHILES

ATMOSPHERIC CHEMISTRY

Man lives at the bottom of an ocean of air. We may ordinarily take the atmosphere for granted and focus much more concern on the **weather**. This ocean of air, however, has profound consequences for life on Earth.

The surface density of air is about 0.074 lb/ft^3 (1.184 g/l) and surface pressure is about 14 lb/ft^2 (1 atm). This mass of air presses downward at all times. At a higher altitude, however, both the pressure and the density of air decrease. This explains why passenger jets, which often fly near 40,000 ft (12,192 m) to take advantage of the thin or low-density air, require pressurized cabins. Without them, passengers would not be able to take in enough **oxygen** with each breath.

The atmosphere is generally divided into four zones or layers. Starting at sea level and increasing in altitude, they are the **troposphere** (0–10 mi [0–16.1 km]), the **stratosphere** (10–30 mi [16.1–48.3 km]), the **mesosphere** (30–60 mi [48.3–96.6 km]), and the **thermosphere** (beyond 60 mi [96.6 km]). These altitudes are approximate and depend upon a variety of conditions, and are clearly distinct in both their physical properties (e.g., **temperature**) and their **chemistry**.

The troposphere is the region of air closest to the ground. It is where the **clouds** and storm systems are to be found, and where our weather occurs. The troposphere is in direct contact with effluent chemicals generated by living things. These can range from the **carbon dioxide** and **water** vapor we exhale to industrial or automotive pollutants. In the absence of such compounds, atmospheric chemistry is very simple. Since the splitting of both the nitrogen and oxygen molecules requires a great deal of energy, the atmospheric composition is fairly constant at sea level, and without interfering compounds.

Smog is the term applied to the mixture of nitrous oxides, spent **hydrocarbons**, **carbon** monoxide, and **ozone** that is generated by automobiles and industrial combustion. Smog is the thick brown haze that hovers over large populated areas. This combination of gases is reactive. The addition of water vapor or raindrops, for example, can result in the scrubbing of these compounds from the air but also the generation of nitrous, nitric, and carbonic acid. Ozone is a powerful oxidizing agent and results in the degradation of plastics and other materials. However, it is also capable of reacting with spent hydrocarbons to generate noxious chemicals.

Industrial pollutants, such as sulfur dioxide generated by coal-burning power plants, can generate **acid rain** as the sulfur dioxide is converted to sulfurous and sulfuric acid. Even forest fires contribute a large variety of chemical compounds into the atmosphere and induce chemical reactions. And, the largest of all natural disasters, a volcanic eruption, spews tons of chemical compounds into the troposphere where they react to produce acids and other compounds.

The stratosphere is the home of the ozone layer, which is misleading as it implies a distinct region in the atmosphere that has ozone as the major constituent. Ozone is never more than a minor constituent of the atmosphere, although it is a significant minor constituent. The concentration of ozone achieves its maximum in the stratosphere. It is here that the chemistry occurs that blocks incoming ultraviolet radiation.

The complete spectrum of radiation from the **sun** contains a significant amount of high energy ultraviolet light and the energy of these photons is sufficient to ionize atoms or molecules. If this light penetrated to Earth's surface, life as we know it could not exist as the ionizing radiation would continually break down complex molecules.

Within the ozone layer, this ultraviolet energy is absorbed by a delicate balance of two chemical reactions. The first is the photolytic reaction of molecular oxygen to give atomic oxygen, which subsequently combines with another oxygen molecule to give ozone. The second reaction is the absorption of another photon of ultraviolet light by an ozone molecule to give molecular oxygen and a free oxygen **atom**.

$$h\nu$$
$$O_2 \rightarrow O + O$$
$$O + O_2 \rightarrow O_3$$
$$h\nu$$
$$O_3 \rightarrow O_2 + O$$

It is the combination of these two reactions that allows the ozone layer to protect the planet. These two reactions actually form an equilibrium with the forward reaction being the formation of ozone and the backwards reaction being the depletion.

$$h\nu$$
$$3O_2 \leftrightarrows 2O_3$$

The ozone concentration is thus at a constant and relatively low level. It occurs in the stratosphere because this is where the concentration of gases is not so high that the excited molecules are deactivated by collision, but not so low that the atomic oxygen generated can not find a molecular oxygen with which to react.

In the last half of the twentieth century, the manufacture of chlorofluorocarbons (CFCs) for use as propellants in aerosol sprays and refrigerants has resulted in a slow mixing of these compounds with the stratosphere. Upon exposure to high-energy ultraviolet light, the CFCs break down to atomic chlorine, which interferes with the natural balance between molecular oxygen and ozone. The result is a shift in the equilibrium and a depletion of the ozone level. The occurrence of ozone depletion was first noted over **Antarctica**. Subsequent investigations have demonstrated that the depletion of ozone also occurs over the Arctic, resulting in higher than normal levels of ultraviolet radiation reaching many heavily populated regions of **North America**. This is, perhaps, one of the most important discoveries in atmospheric chemistry and has lead to major changes in legislation in all countries in an attempt to stop ozone depletion.

Beyond the stratosphere, the energy levels increase dramatically and the available radiation is capable of initiating a wide variety of poorly characterized chemical reactions. Understanding all of the complexities of atmospheric chemistry is subject for much ongoing research.

See also Atmospheric composition and structure; Atmospheric pollution; Global warming

ATMOSPHERIC CIRCULATION

The **troposphere**, the lowest 9 mi (15 km) of Earth's atmosphere, is the layer in which nearly all **weather** activity takes place. Weather is the result of complex air circulation patterns that can best be described by going from the general to more localized phenomena.

The prime mover of air above Earth's surface is the unequal heating and cooling of Earth by the **Sun**. Air rises as it is heated and descends as it is cooled. The differences in air pressure cause air to circulate, which results in the creation of **wind**, **precipitation**, and other weather related features.

Earth's **rotation** also plays a role in air circulation. Centrifugal force, friction and the apparent Coriolis force are responsible for the circular nature of its flow, as well as for erratic eddies and surges.

On a global scale, there are three circulation belts between the equator and each pole. From 0° to 30° **latitude**, the trade winds, or tropical easterlies, flow toward the equator and are deflected to the west by the earth's rotation as they move across the earth's surface. The winds then rise at the equator, then flow poleward at the tropopause, the boundary between the troposphere and the **stratosphere**. The trade winds descend back to the surface at 30° latitude. At the equator, where air from both trade wind belts rises, the lack of cross-surface winds results in the doldrums, an **area** of calm, which historically has been a bane to sailing vessels.

Between 30° and 60° are the mid-latitude, or prevailing westerlies. The circulation pattern of these wind belts is opposite that of the trades. They flow poleward at the earth's surface, deflecting eastward. They rise at 60°, flow back to the equator, then descend at 30°.

As with the equatorial calm, the earth's surface at 30° North and South has little lateral wind movement since the circulation of the tropical and mid-latitude belts is downward, then outward at this latitude. These calm regions are referred to as the horse latitudes because sailors who were stranded for lack of wind either had to eat their horses or throw them overboard to lighten the load.

The third set of circulation belts, the polar easterlies, range from 60° to 90° latitude at both ends of the earth and flow in the same pattern as the tropical easterlies.

This global circulation scheme is only the typical model. Other forces complicate the actual flow. Differences in the type and elevation of surface features have widespread effects.

The jet streams, high-speed winds blowing from the west near the tropopause, play a significant role in determining the weather. The northern and southern hemispheres each have two **jet stream** wind belts. The polar front jet stream is the stronger of the two. It flows eastward to speeds of 250 mph

(400 kph) at the center and receives its energy from an accumulation of solar radiation. The subtropical jet stream is weaker and receives its force from an accumulation of westerly momentum.

The monsoons of **Asia** are a result of a combination of influences from the large Asian land mass and the movements of the inter-tropical front, which straddles the equator. From June to September, when the front runs north of the equator, warm moist winds are drawn northward, bringing heavy rains to India and Southeast Asia. From December to February, the front runs slightly south of the equator, drawing dry cooler air off of the Himalayas and out to sea.

On a more local level, air movements occur in the form of interacting air masses and frontal systems. Low-pressure cyclones and high-pressure anticyclones travel from the west to east. Low-pressure cells are responsible for instability in the weather, with cold and warm fronts radiating from the center of the cell. These fronts represent the interface of cold and warm air masses, which develop into storms.

Cold fronts are more active than warm fronts. The upward angle of the cold front line opposes the direction in which it moves, creating friction between the surface and the air, and causing a steeper pressure gradient. The rain band is narrower, but the cumulonimbus **clouds** that form hold a greater amount of energy and a greater potential for violent weather than the altostratus clouds associated with warm front activity.

Within each cyclonic system are even smaller cyclones. Each storm cell along a front is a cyclone in its own right. In addition to producing heavy rain, hail, high winds and electrical activity, these cells occasionally can produce tornadoes—destructive, whirling funnel-shaped clouds that stretch from the base of a storm cell to the ground. Tornadoes are the most powerful cyclones known on Earth.

Independent of air mass and frontal systems are hurricanes, also known as typhoons or cyclones. These tropical cyclones generate over warm moist ocean surfaces. The rising heat and moisture builds into a massive storm that can extend 1000 mi (1,600 km).

A hurricane tracks westward and will decay when the creative factors are eliminated. This occurs rapidly as the storm travels over land or more gradually as it encounters lower ocean surface temperatures. A lower tropopause in higher latitudes can also reduce the storm's mass.

An accurate understanding of atmospheric circulation began to emerge during the 1830s when Gustave de Coriolis put forth the theory that as Earth rotates, an object will appear to move in a deflected path. About twenty years later, American William Ferrel mathematically proved the Coriolis theory, establishing what became known as **Ferrel's law**.

The ability to make regular unmanned balloon soundings of the atmosphere in the late 1890s and early 1900s made it possible for new details to emerge. A group of Scandinavian meteorologists under the guidance of Vilhelm Bjerknes took full advantage of this new knowledge to develop mathematical and laboratory models of air mass properties.

Bjerknes first proposed the existence of air masses. His son Jacob went on to demonstrate the frontal systems that separate the air masses. Carl-Gustaf Rossby discovered the jet streams and hypothesized detailed movements and counter-movements in the circulation complex.

Atmospheric circulation is a simple process with complex results. It is a system of cells within cells. When we observe leaves swirling in the shadow of a building or a bird soaring on an updraft of warm air, the same principles are at work as with larger global units of the same circulation system. It is a system that is worldwide, that reacts to everything it encounters, and that is even interactive with itself.

See also Air masses and fronts; Atmospheric composition and structure; Atmospheric inversion layers; Atmospheric lapse rate; Atmospheric pollution; Atmospheric pressure

ATMOSPHERIC COMPOSITION AND STRUCTURE

During most of history, Earth's atmosphere was regarded as little more than a mass of air and **clouds**. Simple observations from the ground yielded little more than a basic understanding of the atmosphere's characteristics.

Manned balloon ascents in the late 1700s and early 1800s were restricted to about 5 miles (8 km), the limit of life-supporting **oxygen**. There were also risks to human life. From a practical standpoint, it was difficult to make widespread observations over **space** and time.

Breakthroughs in atmospheric research came as new inventions made it possible to obtain information from unmanned balloon flights. The first of these was the theodolite, a viewing instrument used to survey distances and angles, invented by Gustave Hermite in 1896. This device increased the range to which a ground observer could follow a balloon's ascent pattern. The information-gathering packages delivered to the atmosphere by balloons became known as radiosondes, with each balloon journey referred to as a sounding. The maximum height at which the atmosphere will sustain a balloon is about 28 miles (35 km).

In 1893, George Besancon developed a recording thermometer and barometer capable of making in-flight observations during unmanned ascents. Beginning in 1897, Léon Phillipe Teisserenc de Bort made improvements to these instruments at an observatory he established near Paris. Of equal importance, this was the first organized effort to obtain repeated readings of high-altitude phenomena.

It was generally known that temperatures decreased with elevation at a rate of about 33.8°F (18.6°C) per 590 feet (180 m). Until de Bort, it was assumed that this rate continued out into space. His observations revealed, however, that temperatures first level off, then increase, beginning at about 8 miles (14 km). This warm region is located only a few kilometers beyond the highest mountain peaks and the upper limit of regular cloud formation.

In 1908, de Bort's observations led him to divide the atmosphere into two layers, the lower being the **troposphere** and the upper being the **stratosphere**. The **area** where the two meet, where the **temperature** begins to modify, he named the *tropopause*.

Eventually, scientists discovered that the atmosphere consists of several layers, not just two. The warm region, it has been found, continues to a level for about 28 miles (45 km). The temperature starts at about –76°F (–60°C) at the tropopause and exceeds 32°F (0°C) at about 28 miles (45 km). Beyond this, the temperature drops again to about –130°F (–90°C) at about 50 miles (80 km). This area has been called the stratopause, but it is also known as the **mesosphere**. This area is where most meteors disintegrate as they approach the earth.

Above the stratopause, at about 62 miles (100 km), the temperature rises sharply again to 212°F (100°C) and continues to rise to levels that can only be theoretically estimated, perhaps 1200° K. This area is called the **thermosphere**.

In the thermosphere, attention shifts from temperature variations to other phenomena. This layer is characterized by highly energized particles (cosmic rays), which enter from outer space and became electronically charged. The result is the **aurora borealis** in the Northern Hemisphere and the aurora australis in the Southern Hemisphere. Both occur at about 20°–25° **latitude** and at heights of 50–190 miles (80–300 km).

The **ionosphere** coincides with, but is not confined to, the thermosphere. Its only distinguishing property is its layers of ionized gases that reflect radio waves, as opposed to other atmospheric regions, which do not. It exists from 50 miles (80 km) upward to 620 miles (1,000 km) and beyond. The ionosphere is influenced greatly by solar activity, which can rearrange or eliminate the reflective layers.

The exosphere begins at about 310 miles (500 km). Here the atmospheric components lose their molecular structure and become atomic in nature. These components cannot be considered gaseous beyond this point.

Beyond 620 miles (1,000 km), particle structures further degrade into electrons and protons. The earth's **gravity** gives way to magnetic fields as the dominant distributor of particles. The Van Allen belts at 2,500 miles (4,000 km) and at 12,400 miles (20,000 km) mark the outer limit of the magnetosphere, the most remote known sphere of Earth's influence.

Far beyond the limits of balloon flight, knowledge of the upper atmosphere has been made possible by increasingly high airplane flights and orbiting satellites.

Overall, the atmosphere's gaseous composition consists of 78% nitrogen, 21% oxygen, and a 1% mixture of minor gases dominated by argon. This composition not only sustains life, but is also determined by it. Also, there is a general distribution of dust particles carried from Earth's surface or entering from space.

Recently, scientists have been giving attention to two areas of atmospheric research. Concern has been raised over the destruction of the ozone layer (near the stratospheric warm region), which absorbs ultraviolet radiation, by the introduction of chlorofluorocarbons (CFCs) by man.

A new area of research involves exploration of atmospheres of neighboring planets. With an expanding **solar system**, the dense atmosphere of Venus may hold clues to Earth's atmospheric past, while the thin atmosphere of Mars may be a clue to its future.

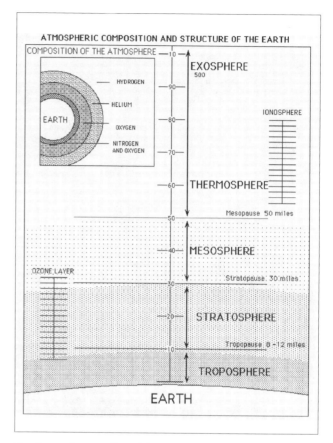

Structure and composition of the atmosphere.

See also Atmospheric circulation; Atmospheric inversion layers; Atmospheric lapse rate; Atmospheric pollution; Atmospheric pressure; Ozone layer and hole dynamics

ATMOSPHERIC INVERSION LAYERS

Whenever an anomaly exists in the atmosphere in which an increase in **temperature, humidity,** or **precipitation** occurs where a decrease would be expected, there is an inversion, or reversal. An atmospheric inversion most commonly refers to temperature inversion where the temperature increases rather than decreases with increasing altitude.

Normally air temperature decreases with altitude at a rate of about 33.8° F (1° C) per 59 feet (180 m) because since the Sun's heating effect is greatest at the Earth's surface. There are three factors that alter this rate, causing the temperature to rise within the first few hundred meters of the ground. Inversions can occur as a result of radiative, or direct, cooling from the earth's surface. This occurs at night when the ground cools more rapidly than the air above it. The effects of an inversion are thus greatest during early morning, usually the coolest part of the day. Inversions also occur as a result of subsidence (sinking) of air in an anticyclone, or high pressure system, where the descending air warms adiabatically, that is,

•

Denver's "brown cloud," the haze of air pollution that hangs over the city, is kept in place by atmospheric inversion layers. © Ted Spiegel/Corbis. Reproduced by permission.

within itself, while the ground remains cool. High pressure systems have the stability that inversion layers require. Finally, movement of air can create an advective inversion. For instance, if a warm air mass moves over a body of **water** or over snow cover, an inversion will occur.

Inversion layers block the upward movement of air, trapping moisture and natural and man-made pollutants near the ground. The result is **fog** and **smog**. The lower the inversion ceiling, the more concentrated the accumulation of moisture and particulates. Some of the most serious episodes of smog or fog occur in mountainous areas, especially where a city (e.g., Denver, Colorado) or industrial site is located. In places in the San Fernando Valley in California, the polluted air is trapped both vertically and horizontally.

The mere presence of a city or factory often creates a microclimate of its own, creating a pocket of warm air within the cool ground layer. Smoke from a stack, instead of escaping upward or laterally, will descend to the ground, delivering a direct dose of pollution to residents of the **area**.

See also Atmospheric composition and structure; Atmospheric lapse rate; Atmospheric pollution; Environmental pollution; Greenhouse gases and greenhouse effect; Meteorology; Troposphere and tropopause

ATMOSPHERIC LAPSE RATE

The atmospheric lapse rate describes the reduction, or lapse of air **temperature** that takes place with increasing altitude. Lapse rates related to changes in altitude can also be developed for other properties of the atmosphere.

In the lower regions of the atmosphere (up to altitudes of approximately 40,000 feet [12,000 m]), temperature decreases with altitude at a fairly uniform rate. Because the atmosphere is warmed by conduction from Earth's surface, this lapse or reduction in temperature normal with increasing distance from the conductive source.

Although the actual atmospheric lapse rate varies, under normal atmospheric conditions the average atmospheric lapse rate results in a temperature decrease of 3.5°F (1.94°C) per 1,000 feet (304 m) of altitude.

The measurable lapse rate is affected by the moisture content of the air (**humidity**). A dry lapse rate of 5.5°F (3.05°C) per 1,000 feet (304 m) is often used to calculate temperature changes in air not at 100% relative humidity. A wet lapse rate of 3°F (1.66°C) per 1,000 feet (304 m) is used to calculate the temperature changes in air that is saturated (i.e., air at 100% relative humidity). Although actual lapse rates do not strictly follow these guidelines, they present a model sufficiently accurate to predict temperate changes associated with updrafts and **downdrafts**. This differential lapse rate (dependent upon both difference in conductive heating and adiabatic expansion and compression) results in the formation of warm downslope winds (e.g., Chinook winds, Santa Anna winds, etc.).

The atmospheric lapse rate, combined with adiabatic cooling and heating of air related to the expansion and compression of atmospheric gases, present a unified model explaining the cooling of air as it moves aloft and the heating of air as it descends downslope.

Atmospheric stability can be measured in terms of lapse rate (i.e., the temperature differences associated with vertical movement of air). A high lapse rate indicates a greater than normal change of temperature associated with a change in altitude and is characteristic of an unstable atmosphere.

Although the atmospheric lapse rate (also known as the environmental lapse rate) is most often used to characterize temperature changes, many properties (e.g., **atmospheric pressure**) can also be profiled by lapse rates.

See also Air masses and fronts; Land and sea breeze; Seasonal winds

ATMOSPHERIC POLLUTION

Atmospheric pollution (also commonly called air pollution) is derived chiefly from the spewing of gasses and solid particulates into the atmosphere. Many pollutants—dust, pollen, and **soil** particles—occur naturally, but most air pollution, as the term is most commonly used and understood, is caused by human activity. Although there are countless sources of air pollution, the most common are emissions from the burning of **hydrocarbons** or fossil **fuels** (e.g., **coal** and oil products). Most of the world's industrialized countries rely on the burning of fossil fuels; power plants heat homes and provide **electricity**, automobiles burn gas, and factories burn materials to create products.

Air pollution is a serious global problem, and is especially problematic in large urban areas such as Mexico City, Mexico, and Athens, Greece. Many people suffer from serious illnesses caused by **smog** and air pollution in these areas. Plants, buildings, and animals are also victims of a particular type of air pollution called **acid rain**. Acid rain is caused by airborne sulfur from burning coal in power plants and can be transported in rain droplets for thousands of miles. Poisons are then deposited in streams, **lakes**, and soils, causing damage to wildlife. In addition, acid rain eats into concrete and other solid structures, causing buildings to slowly deteriorate.

Scientists study air pollution by breaking the particulates into two different categories of gasses: permanent and variable. The most common of the stable gasses are nitrogen at 78%, and **oxygen** at 21% of the total atmosphere. Other highly variable gasses are **water** vapor, **carbon dioxide**, methane, **carbon** monoxide, sulfur dioxide, nitrogen dioxide, **ozone**, ammonia, and hydrogen sulfide.

Output of variable gasses increases with the growth of industrialization and population. The benefits of progress cost people billions of dollars each year in repairing and preventing air pollution damage. This includes health care and the increased maintenance of structures such as the Great Pyramids of Egypt that are crumbling, in part due to air pollution.

The effects of air pollution have to be carefully measured because the build-up of particulates depends on atmospheric conditions and a specific area's emission level. Once pollutants are released into the atmosphere, **wind** patterns make it impossible to contain them to any particular region. This is why the effects of pollution from major oil fires in the Middle East are measurable in **Europe** and elsewhere. On the other hand, terrestrial formations such as mountain ridges can act as natural barriers. The terrain and **climate** of a particular **area** (e.g., Denver, Houston, and Los Angeles) can also help promote or deflect air pollution. Specifically, **weather** conditions called thermal inversions can trap the impurities and cause them to build up until they have reached dangerous levels. A thermal inversion is created when a layer of warm air settles over a layer of cool area closer to the ground. It can stay until rain or wind dissipates the layer of stationary warm air.

The United States government plays an active role in establishing safe and acceptable levels of clean air. In 1967, Congress passed the Air Quality Act that set forth outlines for air quality standards. The Environmental Protection Agency released the first nationwide survey on air pollution in 1989 after Congress passed a law requiring the report. In most cases, it is up to individual states, however, to enforce air pollution controls and meet federally mandated goals. In addition, states may set their own clean air standards that are more strict than those established at the federal level. For example, in 1989, California adopted a radical air pollution reduction plan that essentially requires each region to drastically reduce current levels of air pollution. Even as early as 1970, California adopted more stringent standards for motor-vehicle emissions.

Government regulations have shown moderate success. Since 1970, emissions of sulfur oxide, carbon monoxide, **lead**, and hydrocarbons have decreased by approximately 30% while nitrogen oxide output has been reduced by approximately 10%. Cars are now required to have pollution-control devices called catalytic converters, and most power plants are equipped with filters called scrubbers to remove sulfur oxides.

In addition to atmospheric pollution, indoor air pollution also poses special hazards. Some man-made sources of indoor air pollutants include asbestos particulates and formaldehyde vapors—once common building materials now thought to cause cancer. Lead paint is also a problem in older buildings, but its use has been phased out. Other sources of man-made indoor air pollution include improperly vented stoves and heaters, tobacco smoke, and emissions or spillage from pesticides, aerosol sprays, solvents, and disinfectants.

See also Atmosphere; Geochemistry; Global warming; Greenhouse gases and greenhouse effect; Petroleum, economic uses of; Rate factors in geologic processes; Weathering and weathering series

ATMOSPHERIC PRESSURE

Aristotle, whose teachings sometimes otherwise inhibited the advancement of science, was right on target in his belief that the atmosphere surrounding the Earth had weight. Moreover, Aristotle stated that as air density decreased, it would be possible for an object to move faster. However, he did not believe in the concept of a vacuum because the absence of an atmosphere meant an object could move infinitely fast, and since infinite speed was not possible, a vacuum that allowed infinite speed was not considered possible either. Galileo disputed some of Aristotle's contentions. In 1638, Galileo published a book in which he asserted a vacuum was possible. But Galileo did not hold that air had a weight that could exert a pressure, even though his own experiments showed clearly that air exerted a force on objects. This was perhaps because he discounted everything Aristotle said, even when he happened to be right. Consequently, the thermometer Galileo invented was inaccurate because it did not take the effect of air pressure into account. **Otto von Guericke** became interested in air pressure because of Galileo's comments on the subject. In a public demonstration in 1657, Guericke became the first to use an air pump and create a vacuum, thus ending the debate on whether one could exist.

In 1643, in Florence, Italy, **Evangelista Torricelli** furthered Guericke's work. Filling a narrow tube with mercury and upending it in a bowl of mercury, Torricelli found that only a portion of the tube emptied. He correctly surmised that the atmospheric pressure upon the mercury in the bowl kept the tube from draining completely, and the vacant **area** at the top of the tube was a vacuum. He noticed the height of the column of mercury fluctuated from day to day, indicating that the atmospheric pressure changed. The barometer, a device to measure the pressure of the atmosphere, was born, yet that name wouldn't exist for another 20 years.

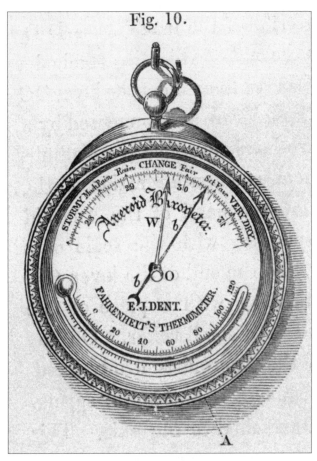

Fig. 10.

Aneroid barometer. *National Oceanic and Atmospheric Administration (NOAA).*

ATOM

Atoms and the subatomic particles that comprise them, are the elementary building blocks of material substances. Although the term atom, derived from the Greek word *atomos*, meaning indivisible, would seem inappropriate for an entity that, as science has established, is divisible, the word atom still makes sense, because, depending on the context, atoms can still be regarded as indivisible. Namely, once the nucleus is split, the atom loses its identity and subatomic particles. Protons, electrons, neutrons, are all the same—regardless of the type of atom or element—it is only their numbers and unique combinations that make for different atoms. Accordingly, an atom is the smallest particle of an element.

Atoms share many characteristics with other material objects: they can be measured, and they also have mass and weight. Because traditional methods of measuring are difficult to use for atoms and subatomic particles, scientists have created a new unit, the atomic mass unit (amu), which is defined as the one-twelfth of the mass of the average **carbon** atom.

The principal subatomic particles are the protons, neutrons, and the electrons. The nucleus, the atom's core, consists of protons, which are positively charged particles, and neutrons, particles without any charge. Electrons are negatively charged particles with negligible mass that orbit around the nucleus. An electron's mass is so small that it is usually given a 0 amu value in atomic mass units, compared to the value of 1 amu assigned to neutrons and protons (neutrons do carry slightly more mass than protons and neither exactly equals 1 amu—but for purposes of this article the approximate values will suffice). In fact, as the nucleus represents more than 99% of an atom's mass, it is interesting to note that an atom is mostly **space**. For example, if a hydrogen atom's nucleus were enlarged to the size of a **marble**, the atom's diameter (to the electron orbit) would be around 0.5 mi (800 m).

At one time, scientists asserted that electrons circled around the nucleus in planet-like orbits. However, because all subatomic particles, including electrons, exhibit wave-like properties, it is makes no sense to conceptualize the movement of electrons as like planetary **rotation**. Scientists therefore prefer terms like "electron cloud patterns," or "shells," indicating an electron's position and/or pattern of movement in relation to the nucleus. Thus, for example, hydrogen has one electron in its innermost, lowest energy shell (a shell is also an energy level); lithium—with three electrons—has two shells, with inner most, lowest energy shell contains two electrons that one electron exists in a more distant shell or higher potential energy level. The elements exhibit four distinctive shapes of shell—designated *s, p, d,* and *f* orbitals.

While subatomic particles are generic and interchangeable, in combination they determine an atom's identity. For example, we know that an atom with a nucleus consisting of one proton must be hydrogen (H). An atom with two protons

Mathematician Blaise Pascal duplicated the experiments of Torricelli, and he expanded on them. In 1648, Pascal, who suffered from ill health, had his brother-in-law make measurements of air pressure at various altitudes on a mountain. As expected, the higher the altitude, the less pressure registered on the barometer. Obviously, the weight of the air at the surface of the Earth was greater because it has to support the atmosphere above it. Robert Boyle duplicated Torricelli's experiment as well. In 1660, Boyce placed his mercury-filled tube in a container and removed the surrounding air, creating a vacuum. As the air was removed, the column of mercury dropped. When completely evacuated, the mercury showed zero air pressure in the container. It was Boyle who coined the word "barometer" in 1665.

Today, it is known that the weight of Earth's atmosphere is more than five quadrillion tons. The weight of air pressing down on one's shoulders is about one ton, but we aren't flattened because we are supported on all sides by an equal amount of pressure. The normal barometric pressure at sea level, equal to one atmosphere (1 atm) of pressure equals 14.7 psi (pounds per square inch), or 760 mm (29.92 inches) of mercury.

See also Air masses and fronts; Weather forecasting

is always a helium (He) atom. Thus, we see that the key to an atom's identity is to be found in the atom's inner structure. In addition, a electrically balanced chemical element is an instance of atomic electronic equilibrium: for example, in an electrically balanced chemical element, the number of positively charged particles (protons) always equals the number of negatively charged particles (electrons). A loss or gain of electrons results in a net charge and the atom becomes an ion.

Although the number of protons determines the name (type) of atom, each atom may be heavier or lighter depending on the number of neutrons present. Atoms of the same element with different mass (reflecting differing numbers of neutrons) are isotopes.

Research into the atom's nucleus has uncovered a variety of subatomic particles, including quarks and gluons. Considered by some researchers the true building blocks of matter, quarks are the particles that form protons and neutrons. Gluons hold smaller clusters of quarks together.

The atom is best characterized by the laws and terminology or quantum **physics**. On a larger scale, chemists study reactions, the behavior of elements in interaction, and reactions, such as those leading to the formation of chemical compounds. Such reactions involve the transfer of electrons and/or the sharing of electrons in atomic bonds.

For example, the formation of sodium chloride, also known as table salt, would be impossible without specific changes at a subatomic level. The genesis of sodium chloride (NaCl) starts when a sodium (Na) atom, which has 11 electrons, loses an electron. With 10 electrons, the atom now has one more proton than electrons and thus becomes a net positively charged sodium ion Na^+ (a positively charged ion is also known as a cation. Chlorine becomes a negatively charged anion by accepting a free electron to take on a net negative charge. The newly acquired electron goes into the outer shell, also known as the valence shell that already contains seven electrons. The addition of the eighth electron to the chlorine atom's outmost shell fulfills the octet rule and allows the atom—although now a negatively charged chlorine ion (Cl^-)—to be more stable. The electrical attraction of the sodium cations for the chlorine anion results in an ionic bond to form salt. **Crystals** of table salt consist of equal numbers of sodium cations and chlorine anions, cation-anion pairs being held together by a force of electrical attraction.

The octet rule is used to describe the attraction of elements toward having, whenever possible, eight valence-shell electrons (four electron pairs) in their outer shell. Because a full outer shell with eight electrons is relatively stable, many atoms lose or gain electrons to obtain an electron configuration like that of the nearest noble gas. Except for helium (with a filled 1s shell), noble gases have eight electrons in their valence shells.

Interestingly, not long after scientists realized that at the level of the nucleus an atom is divisible, transmutation, or the old alchemic dream of turning one substance into another, became a reality. Fission and fusion are tranformative processes that, by altering the nucleus, alter the element. For example, scientists even succeeded in creating gold by bombarding platinum-198 with neutrons to create platinum-199

that then decays to gold-199. Although clearly demonstrating the reality of transmutation, this particular transmutation (a change in the nuclear structure that changes one element into another) is by no means an easy or cheap method of producing gold. Quite the contrary, because platinum, particularly the platinum-199 isotope, is more expensive than gold produced. Regardless, the symbolic value of the experiment is immense, as it shows that the idea, developed by ancient alchemists and philosophers, of material transmutation—accomplished at the nuclear level—does not essentially contradict our understanding of the atom.

Natural transformations also exist—as with the decay of Carbon-14 to nitrogen—accomplished by the nuclear transformation of one Carbon-14 neutron into a proton.

See also Atomic mass and weight; Atomic number; Atomic theory; Chemical bonds and physical properties; Chemical elements

ATOMIC MASS AND WEIGHT

The atomic mass of an **atom** (i.e., a specific isotope of an element) is measured in comparison with the mass of one atom of carbon-12 (^{12}C) that is assigned a mass of 12 atomic mass units (amu). Atomic mass is sometimes erroneously confused with atomic weight—the obsolete term for relative atomic mass. Atomic weights, however, are still listed on many Periodic tables.

A mole of any element or compound (i.e., 6.022×10^{23}—Avogadro's number—atoms or molecules) weighs its total unit atomic mass (formerly termed atomic weight) in grams. For example, **water** (H_2O) has a molar mass (the mass of 6.022×10^{23} water molecules) of approximately 18 grams (the sum of 2 hydrogen atoms, each with an atomic mass of 1.0079 amu, bonded with one **oxygen** atom with an atomic mass of 15.9994 amu).

In general usage if a specific isotope or isotope distribution is specified when using atomic mass, the natural percentage distribution of isotopes of that element is assumed. Periodic tables, for example usually list the atomic weights of individual elements based upon the natural distribution of isotopes of that element.

Mass is an intrinsic property of matter. Weight is a measurement of gravitational force exerted on matter.

In a series of papers published between 1803 and 1805 English physicist and chemist John Dalton (1766–1844) emphasized the importance of knowing the weights of atoms and outlined an experimental method for determining those weights.

The one problem with Dalton's suggestion was that chemists had to know the formulas of chemical compounds before they could determine the weights of atoms. But they had no way of knowing chemical formulas without a dependable table of atomic weights.

Dalton himself had assumed that compounds always had the simplest possible formula: HO for water (actually H_2O), NH for ammonia (actually NH_3), and so on. Although incorrect, this assumption allowed him to develop the concept of atomic weights, but, because his formulas were often wrong, his work inevitably resulted in incorrect values for most of the atomic weights. For example, he reported 5.5 for the atomic weight of monatomic oxygen and 4.2 for monatomic nitrogen. The correct values for those weights are closer to 16 and 14.

The first reasonably accurate table of atomic weights was produced by Swedish chemist Jöns Jacob Berzelius (1779–1848) in 1814. This table had been preceded by nearly a decade of work on the chemical composition of compounds. Once those compositions had been determined, Berzelius could use this information to calculate correct atomic weights.

In this process, Berzelius was faced with a decision that confronted anyone who tried to construct a table of atomic weights: What element should form the basis of that table and what would be the atomic weight of that standard element?

The actual weights of atoms are, of course, far too small to use in any table. The numbers that we refer to as atomic weights are all *ratios*. To say that the atomic weight of oxygen is 16, for example, is only to say that a single oxygen atom is 16 times as heavy as some other atom whose weight is somehow chosen as 1, or eight times as heavy as another atom whose weight has been chosen as 2, or one-half as heavy as another atom whose weight was selected to be 32, and so on.

Dalton had made the logical conclusion to use hydrogen as the standard for his first atomic table and had assigned a value of 1 for its atomic weight. Because hydrogen is the lightest element, this decision assures that all atomic weights will be greater than one.

The problem with Dalton's choice was that atomic weights are determined by measuring the way elements combine with each other, and hydrogen combines with relatively few elements. So, using Dalton's system, determining the atomic weight of another element might require a two-or three-step process.

Berzelius thought it made more sense to choose oxygen as the standard for an atomic weight table. Oxygen forms compounds with most other elements whose atomic weights can, therefore, be determined in a single step. He arbitrarily assigned a value of 100 as the atomic weight of oxygen. Other chemists agreed that oxygen should be the atomic weight standard, but used other values for its weight.

Berzelius continued working on atomic weights until, in 1828, he produced a table with values very close to those accepted today.

With the introduction of the concept of molecules (e.g., that the correct formula for water was H_2O) by Stanislao Cannizarro in 1858, it also became possible to calculate molecular weights. The molecular weight of any compound is equal to the sum of the weights of all the atoms in a molecule of that compound.

The most precise work on atomic weights during the nineteenth century was that of the Belgian chemist Jean Servais Stas (1813–1891). For over a decade, Stas recalculated Berzelius' weights, producing results that were unchallenged for nearly half a century.

An even higher level of precision was reached in the work of the American chemist Theodore William Richards

(1868–1918). Richards spent more than 30 years improving methods for the calculation of atomic weights and redetermining those weights. Richards was awarded the Nobel Prize in chemistry in 1914 for these efforts.

The debate as to which element was to be used as the standard for atomic weights extended into the twentieth century, with the most popular positions being hydrogen with a weight of 1 or oxygen with a weight of 16. Between 1893 and 1903, various chemical societies finally agreed on the latter standard.

The controversy over standards was complicated by the fact that, over time, physicists and chemists began to use different standards for the atomic weight table and, thus, recognized slightly different values for the atomic weights of the elements. This dilemma was finally resolved in 1961 when chemists and physicists agreed to set the atomic weight of the carbon-12 isotope as 12.0000 as the standard for all atomic weights.

See also Atomic number; Atomic theory; Chemical bonds and physical properties; Chemical elements; Chemistry; Minerals

ATOMIC NUMBER

Atomic number is defined as the number of protons in the nucleus of an **atom**. This concept was historically important because it provided a theoretical basis for the periodic law. Dmitri Mendeleev's discovery of the periodic law in the late 1860s was a remarkable accomplishment. It provided a key-organizing concept for the chemical sciences. One problem that remained in Mendeleev's final analysis was the inversion of certain elements in his **periodic table**. In three places, elements arranged according to their chemical properties, as dictated by Mendeleev's law, are out of sequence according to their atomic weights.

The solution to this problem did not appear for nearly half a century. Then, it evolved out of research with x rays, discovered in 1895 by Wilhelm Röntgen. Roentgen's discovery of this new form of electromagnetic radiation had inspired a spate of new research projects aimed at learning more about x rays themselves and about their effects on matter. Charles Grover Barkla, a physicist at the Universities of London and Cambridge, initiated one line of x-ray research. Beginning in 1903, he analyzed the way in which x rays were scattered by gasses, in general, and by elements, in particular. He found that the higher an element was located in the periodic table, the more penetrating the rays it produced. He concluded that the x-ray pattern he observed for an element was associated with the number of electrons in the atoms of that element.

Barkla's work was brought to fruition only a few years later by the English physicist H.G.J. Moseley. In 1913, Moseley found that the x-ray spectra for the elements changed in a simple and regular way as one moved up the periodic table. Moseley, like Barkla, attributed this change to the number of electrons in the atoms of each element and, thus, to the total positive charge on the nucleus of each atom. (Because atoms are electrically neutral, the total number of positive charges on the nucleus must be equal to the total number of negatively charged electrons.)

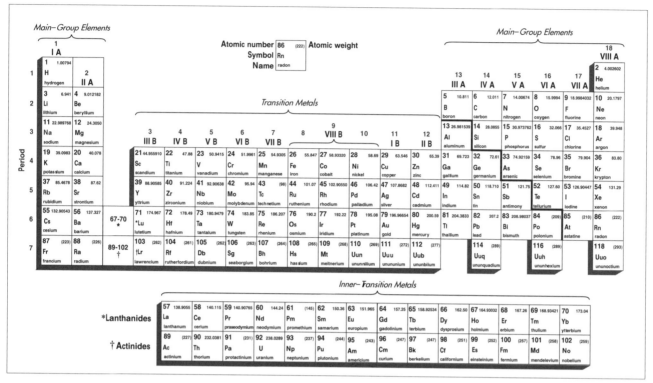

The periodic table.

Moseley devised the concept of atomic number and assigned atomic numbers to the elements in such a way as to reflect the regular, integral, linear relationship of their x-ray spectra. It soon came to be understood that the atomic number of an atom is equal to the number of protons in the atom's nucleus.

Moseley's discovery was an important contribution to the understanding of Mendeleev's periodic law. Mendeleev's law was a purely empirical discovery. It was based on properties that could be observed in a laboratory. Moseley's discovery provided a theoretical basis for the law. It showed that chemical properties were related to atomic structure (number of electrons and nuclear charge) in a regular and predictable way.

Arranging the periodic table by means of atomic number also resolved some of the problems remaining from Mendeleev's original work. For example, elements that appeared to be out of place when arranged according to their atomic weights appeared in their correct order when arranged according to their atomic numbers.

See also Atomic mass and weight; Atomic theory; Bohr model; Chemical bonds and physical properties; Geochemistry

ATOMIC THEORY

One of the points of dispute among early Greek philosophers was the ultimate nature of matter. The question was whether the characteristics of matter that can be observed with the five senses are a true representation of matter at its most basic level. Some philosophers thought that they were. Anaxagoras of Klazomenai (c. 498–428 B.C.), for example, taught that matter can be sub-divided without limit and that it retains its characteristics no matter how it is divided.

An alternative view was that of Leucippus of Miletus (about 490 B.C.) and his pupil, Democritus of Abdera (c. 460–370 B.C.). The views of these scholars are preserved in a few fragments of their writings and of commentaries on their teachings. Some writers doubt that Leucippus even existed. In any case, the ideas attributed to them are widely known. They thought that all matter consists of tiny, indivisible particles moving randomly about in a void (a vacuum). The particles were described as hard, with form and size, but no color, taste, or smell. They became known by the Greek word *atomos*, meaning "indivisible." Democritus suggested that, from time to time, atoms collide and combine with each other by means of hook-and-eye attachments on their surfaces.

Perhaps the most effective popularizer of the atomic theory was the Roman poet and naturalist, Lucretius. In his poem *De Rerum Natura* (On the nature of things) Lucretius states that only two realities exist, solid, everlasting particles and the void. This atomistic philosophy was in competition with other ideas about the fundamental nature of matter. Aristotle, for example, rejected Democritus' ideas because he could not accept the concept of a vacuum nor the idea that particles could move about on their own.

In addition, debates between atomists and anti-atomists quickly developed religious overtones. As the natural philosophy of Aristotle was adopted by and incorporated into early Christian theology, anti-atomism became acceptable and "correct," atomism, heretical. In fact, one objective of Lucretius' poem was to provide a materialistic explanation of the world designed to counteract religious superstition rampant at the time.

In spite of official disapproval, the idea of fundamental particles held a strong appeal for at least some philosophers through the ages. The French philosopher Pierre Gassendi (1592–1655) was especially influential in reviving and promoting the idea of atomism. Robert Boyle and Isaac Newton were both enthusiastic supporters of the theory.

Credit for the first modern atomic theory goes to the English chemist, John Dalton. In his 1808 book, *A New System of Chemical Philosophy*, Dalton outlined five fundamental postulates about atoms: 1. All matter consists of tiny, indivisible particles, which Dalton called atoms. 2. All atoms of a particular element are exactly alike, but atoms of different elements are different. 3. All atoms are unchangeable. 4. Atoms of elements combine to form "compound atoms" (i.e., molecules) of compounds. 5. In chemical reactions, atoms are neither created nor destroyed, but are only rearranged.

A key distinguishing feature of Dalton's theory was his emphasis on the weights of atoms. He argued that every **atom** had a specific weight that could be determined by experimental analysis. Although the specific details of Dalton's proposed mechanism for determining atomic weights were flawed, his proposal stimulated other chemists to begin research on atomic weights.

Dalton's theory was widely accepted because it explained so many existing experimental observations and because it was so fruitful in suggesting new lines of research. But the theory proved to be wrong in many of its particulars. For example, in 1897, the English physicist Joseph J. Thomson showed that particles even smaller than the atom, electrons, could be extracted from atoms. Atoms could not, therefore, be indivisible. The discovery of **radioactivity** at about the same time showed that at least some atoms are not unchangeable, but instead, spontaneously decay into other kinds of atoms.

By 1913, the main features of the modern atomic theory had been worked out. The work of Ernest Rutherford, **Niels Bohr** and others, suggested that an atom consists of a central core, the nucleus, surrounded by one or more electrons, arranged in energy levels each of which can hold some specific number of electrons.

Bohr's atomic model marked the beginning of a new approach in constructing atomic theory. His work, along with that of **Erwin Schrödinger**, Louis Victor de Broglie, Werner Karl Heisenberg, Paul Adrien Maurice Dirac, and others showed that atoms could be understood and represented better through mathematics than through physical models. Instead of drawing pictures that show the location and movement of particles within the atoms, modern scientists tend to write mathematical equations that describe the behavior of observed atomic phenomena.

See also Atomic mass and weight; Atomic number; Chemical bonds and physical properties

AURORA BOREALIS AND AURORA AUSTRALIALIS

The Aurora Borealis and Aurora Australialis are electromagnetic phenomena that occur near Earth's polar regions. The Aurora Borealis—also known as the "northern lights" (boreal derives from Latin for "north")—occurs near the northern polar regions. The Aurora Australialis is a similar phenomenon that occurs in southern polar regions.

Auroras are colored and twisting ribbons of light that appear to twist and gyrate in the atmosphere.

Auroras result from the interaction of Earth's **magnetic field** with ionic gas particles, protons, and electrons streaming outward from the **Sun**. Solar storms result in magnetic disturbances that lead to coronal mass ejections (CMEs) of ionic charged particles in solar "winds." As the magnetic particles pass Earth, the plasma streams (streams of charged particles) interact with Earth's magnetosphere (magnetic field). The magnetic interactions excite electron transitions that result in the emission of visible light.

Charged particles may also travel down Earth's magnetic field lines into Earth's **ionosphere**. As the charged particles interact with charged atmospheric gases in Earth's ionosphere, the electrons in the gases move to higher energy states. As the excited electrons return to their ground state, light photons are emitted. The colors of light correspond to particular frequencies and wavelengths generated by the energy of particular electron orbital transmissions, and are unique to different gaseous compounds. **Oxygen** atoms tend to give off red and greenish light. Nitrogen tend to produce wavelengths light in the bluish region of spectrum.

Although they may form anywhere, auroras are usually found in ring like regions (auroral rings or auroral ovals) that surround Earth's poles. Auroral rings or ovals are readily visible from **space**. Auroras are normally associated with the polar regions because it there where the magnetic field lines of Earth converge and are of the highest density.

The auroras also generate high levels of **electricity** that can exceed 100,000 megawatts within a few hours and sometimes interfere with communications equipment and/or signal transmission or reception.

See also Atmospheric chemistry; Atmospheric composition and structure; Atomic theory; Atoms; Bohr model; Chemical elements; Coronal ejections and magnetic storms; Electricity and magnetism; Electromagnetic spectrum; Quantum electrodynamics (QED); Solar sunspot cycles

AUSTRALIA

Of the seven continents, Australia is the flattest, smallest, and except for **Antarctica**, the most arid. Including the southeastern island of Tasmania, the island continent is roughly equal in **area** to the United States, excluding Alaska and Hawaii. Millions of years of geographic isolation from other landmasses accounts for Australia's unique animal species, notably

Aurora Borealis, one of the great wonders of the natural world. *FMA. Reproduced by permission.*

marsupial mammals like the kangaroo, egg laying mammals like the platypus, and the flightless emu bird. Excluding folded structures (areas warped by geologic forces) along Australia's east coast, patches of the northern coastline and the relatively lush island of Tasmania, the continent is mostly dry and inhospitable.

Australia has been less affected by seismic and orogenic (mountain building) forces than other continents during the past 400 million years. Although seismic (**earthquake**) activity persists in the eastern and western highlands, Australia is the most stable of all continents. In the recent geological past, it has experienced none of the massive upheavals responsible for uplifting the Andes in **South America**, the Himalayas in south **Asia**, or the European Alps. Instead, Australia's **topography** is the result of gradual changes over millions of years.

Australia is not the oldest continent, a common misconception arising from the continent's flat, seemingly unchanged expanse. Geologically, it is the same age as the Americas, Asia, **Africa, Europe**, and Antarctica. Australia's **crust**, however, has escaped strong Earth forces in recent geo-

logical history, accounting for its relatively uniform appearance. As a result, the continent serves as a window to early geological ages.

About 95 million years ago, tectonic forces (movements and pressures of Earth's crust) split Australia from Antarctica and the southern supercontinent of Gondwanaland. Geologists estimate that the continent is drifting northward at a rate of approximately 18 in (28 cm) per year. They theorize that south Australia was joined to Antarctica at the Antarctic regions of Wilkes Land, including Commonwealth Bay. Over a period of 65 million years, beginning 160 million years ago, Australia's crust was stretched hundreds of miles by tectonics before it finally cleaved from Antarctica.

Testimony to the continental stretching and splitting includes Kangaroo Island off South Australia, made up of volcanic basalts, as well as thick layers of sediment along the coast of Victoria. Other signs are the similar **geology** of the Antarctic Commonwealth Bay and the Eyre Peninsula of South Australia, especially equivalent rocks, particularly gneisses (metamorphic rocks changed by heat and pressure) of identical age. The thin crust along Australia's southern flank in the Great Australian Bight also points to continental stretch.

As it drifts north, the Australian plate is colliding with the Pacific and Eurasian plates, forming a **subduction zone** (an area where one continental plate descends beneath another). This zone, the convergence of the Australian continental plate with Papua New Guinea and the southern Indonesian islands, is studded with volcanos and prone to earthquakes. Yet, Australia is unique in that it is not riven by subduction zones like other continents. There are no upwelling sections of the earth's mantle below Australia (the layer below the crust), nor are there intracontinental rift zones like the East African Rift System which threatens to eventually split Africa apart.

Furthermore, Australia and Antarctica are dissimilar to other landmasses; their shapes are not rough triangles with apexes pointing southward like South America, Africa, and the Indian subcontinent, Gondwanaland's other constituent parts. However, like its sister continents, Australia is composed of three structural units. These include in Western Australia a stable and ancient block of basement **rock** or **craton** as geologists call it, an ancient fold mountain belt (the Great Dividing Range along the east coast), and a flat platform-like area in-between composed of crystalline or folded rocks overlaid by flat-lying or only gently deformed sediments.

Millions of years of **erosion** have scoured Australia's surface features. One notable exception to Australia's flat topography is the Great Dividing Range stretching 1,200 mi (1,931 km) along Australia's east coast. The Great Dividing Range was thrust up by geological folding like the Appalachian Mountains in the eastern United States. The mountains are superimposed on larger geological structures including the Tasman and Newcastle geosynclines, troughs of older rocks upon which thick layers of sediment have been deposited. Those sediments in turn have been transformed by folding as well as magmatic and volcanic forces.

Twice, during a 125 million year period beginning 400 million years ago, the geosynclines were compressed, forming mountains and initiating volcanoes. Volcanic activity recurred along the Great Dividing Range 20–25 million years ago during the **Miocene Epoch** when early apes evolved as well as seals, dolphins, sunflowers, and bears. However, over millions of years the volcanic cones from this epoch have been stripped down by erosion. Still, volcanic activity persisted in South Australia until less than a million years ago. In Queensland, near Brisbane in the south and Cairns in the north of the state, the Great Dividing Range hugs the coast, creating beautiful Riviera-like vistas.

East of the Great Dividing Range, along Australia's narrow eastern coastal basin are its two largest cities, Sydney and Melbourne, as well as the capital, Canberra. The Dividing Range tends to trap moisture from easterly **weather** fronts originating in the Pacific Ocean. **Rivers** and streams also course the Range. West of the Range the landscape becomes increasingly forbidding and the weather hot and dry.

Although unrelated to geological forces, the world's largest coral formation, the **Great Barrier Reef** stretches for 1,245 mi (2,003 km) along Australia's northeast coast. Most of

Australia is referred to as outback—desert and semi-desert flatness, broken only by scrub, salt **lakes** which are actually dry lakebeds most of the year, and a few spectacular **sandstone** proturburances like Uluru (also known as Ayers Rock) and the Olgas (Kata Tjuta).

In 1991, geologists discovered a subterranean electrical current in Australia, the longest in the world, which passes through more than 3,700 mi (6,000 km) across the Australian outback. The current is conducted by **sedimentary rocks** in a long horseshoe arc that skirts a huge mass of older igneous and **metamorphic rock** comprising most of the Northern Territory. It begins at Broome in Western Australia near the Timor Sea and then dips south across South Australia before curling northward through western Queensland where it terminates in the Gulf of Carpenteria.

A side branch runs from Birdsville in South Australia near the Flinders Ranges into Spencer Gulf near Adelaide. Geologists say the current is induced by the Earth's ever-changing **magnetic field** and that it runs along fracture zones in sedimentary basins that were formed as the Earth's ancient plates collided. Although the fracture zones contain alkaline fluids that are good conductors of **electricity**, the current is weak and cannot even light a lamp. Geologists say the current might provide clues to deposits of oil and gas and help explain the geological origins of the Australian continent.

Australian topography is also punctuated by starkly beautiful mountain ranges in the middle of the continent like the McDonnell and Musgrave Ranges, north and south respectively of Uluru (Ayers Rock). Uluru, the most sacred site in the country for Australia's aborigines, is a sandstone monolith of which two-thirds is believed to be below the surface. Ayers Rock is about 2.2 mi (3.5 km) long and 1,131 ft (339 m) high. Also in the center of the country, near Alice Springs, are the Henbury Meteorite craters, one of the largest clusters of meteorite craters in the world. The largest of these depressions, formed by the impact of an extraterrestrial rock, is about 591 ft (177 m) long and 49 ft (15 m) deep.

The continent's oldest rocks are in the Western Australian shield in southwest Australia. The basement (underlying) rocks in this area have not been folded since the **Archean** eon over three billion years ago, when the planet was still very young. The nucleus of this shield (called the Yilgarn craton) comprising 230,000 sq mi (59,570,000 ha), consists mostly of **granite** with belts of metamorphic rock like greenstones, rich in economic mineral deposits as well as intrusions of formerly molten rock.

The Yilgarn craton does not quite extend to the coast of Western Australia. It is bounded on the west by the Darling Fault near Perth. To the south and east the Frazer Fault sets off the craton from somewhat younger rocks that were metamorphosed between 2.5 billion and two billion years ago. Both fault lines are 600 mi (960 km) long and are considered major structures on the continent.

Along the north coast of Western Australia near Port Hedland is another nucleus of ancient rocks, the Pilbara Craton. The Pilbara craton is composed of granites over three billion years old as well as volcanic, greenstone, and sedimentary rocks. The Hammersley Range just south of the

Ayers Rock, Australia. *Popperfoto/Archive Photos, Inc. Reproduced by permission.*

Pilbara Craton is estimated to contain billions of tons of **iron** ore reserves.

Other ancient rock masses in Australia are the Arunta Complex north of Alice Springs in the center of Australia which dates to 2.25 billion years ago. The MacArthur Basin, southwest of the Gulf of Carpenteria in the Northern Territory is a belt of sedimentary rocks that are between 1.8 billion and 1.5 billion years old.

The Musgrave block near the continent's center, a component of the Adelaidian geosyncline, was formed by the repeated intrusion of molten rocks between 1.4 billion to one billion years ago during the **Proterozoic Era** when algae, jellyfish, and worms first arose. At the same time, the rocks that underlay the Adelaidian geosyncline were downwarped by geological pressures, with sediments building up through mid-Cambrian times (about 535 million years ago) when the area was inundated 400 mi (640 km) by the sea inland of the present coastline.

The rocks of the Adelaidian geosyncline are as thick as 10 mi (16 km) with sediments that have been extensively folded and subjected to faulting during late **Precambrian** and early Paleozoic times (about 600 million to 500 million years

ago). Some of the rocks of the Adelaidian geosyncline, however, are unaltered. These strata show evidence of a major glacial period around 740 million years ago and contain some of the continent's richest, most diverse fossil records of soft-bodied animals.

This **glaciation** was one manifestation of global cooling that caused glacial episodes on other continents. Geologists say this Precambrian glacial episode was probably one of coldest, most extensive cooling periods in Earth history. They also consider the Adelaide geosyncline to be the precursor of another downwarp related to the most extensive folded belts on the continent, namely the Tasman geosyncline along Australia's east flank.

Victoria is also characterized by a belt of old rocks upon which sediments have been deposited called the Lachlan geosyncline. Marine rocks were deposited in quiet **water** to great thicknesses in Victoria, forming black shales. Some of the sediment was built up by mud-laden currents from higher areas on the sea floor. These current-borne sediments have produced muddy sandstones called graywackes.

At the end of the Ordovician and early Silurian Periods (about 425 million years ago) there was widespread folding of the Lachlan geosyncline called the Benambran **orogeny**. The folding was accompanied by granite intrusions and is thought to be responsible for the composition and texture of the rocks of the Snowy Mountains in Victoria, including Mt. Kosciusko, Australia's tallest peak at 7,310 ft (2,193 m).

In eastern Australia, **Paleozoic Era** volcanic activity built up much of the rock strata. Mountain glaciation during the late Carboniferous period when insects, amphibians, and early reptiles first evolved also transformed the landscape. Mountain building in eastern Australia culminated during the middle and later **Permian Period** (about 250 million years ago) when a huge mass of **magma** (underground molten rock) was emplaced in older rocks in the New England area of northeastern New South Wales. This huge mass, or **batholith**, caused extensive folding to the west and ended the **sedimentation** phase of the Tasman geosyncline. It was also the last major episode of orogeny (mountain building) on the continent.

In parts of Western Australia, particularly the Carnarvon Basin at the mouth of the Gascoyne River, glacial sediments are as thick as 3 mi (5 km). Western Australia, particularly along the coast, has been inundated repeatedly by the sea and has been described by geologists as a mobile shelf area. This is reflected in the alternating strata of deposited marine and non-marine layers.

In the center of Australia is a large sedimentary basin or depression spanning 450 mi (720 km) from east to west and 160 mi (256 km) north to south at its widest point. Sedimentary rocks of all varieties can be found in the basin rocks which erosion shaped into spectacular scenery including Ayres Rock and Mt. Olga. These deposits are mostly of Precambrian age (over 570 million years old), while sediment along the present-day coastline including those in the Eucla Basin off the Great Australian Bight are less than 70 million years old. North of the Eucla Basin is the Nullarbor (meaning treeless) Plain which contains many unexplored **limestone** caves.

Dominating interior southern Queensland is the Great **Artesian** Basin, which features non-marine sands built up during the **Jurassic Period** (190 million to 130 million years ago), sands which contain much of the basin's artesian water. Thousands of holes have been bored in the Great Artesian Basin to extract the water resources underneath but the salt content of water from the basin is relatively high and the water supplies have been used for livestock only.

The Sydney basin formed over the folded rocks of the Tasman geosyncline and is also considered to be an extension of the Great Artesian Basins. Composed of sediments from the Permian and Triassic Periods (290 million to 190 million years old) it extends south and eastward along the **continental shelf**. The sandstone cliffs around Sydney Harbor, often exploited for building stones, date from Triassic sediments.

Minerals in Australia have had a tremendous impact on the country's human history and patterns of settlement. Alluvial gold (gold sediments deposited by rivers and streams) spurred several gold fevers and set the stage for Australia's present demographic patterns. During the post-World War II period there has been almost a continuous run of mineral discoveries, including gold, bauxite, iron, and manganese reserves as well as opals, sapphires, and other precious stones.

It is estimated that Australia has 24 billion tons (22 billion tones) of **coal** reserves, over one-quarter of which (7 billion tons/6 billion tones) is anthracite or black coal deposited in Permian sediments in the Sydney Basin of New South Wales and in Queensland. Brown coal suitable for electricity production in found in Victoria. Australia meets its domestic coal consumption needs with its own reserves and exports the surplus.

Australia supplies much of its oil consumption needs domestically. The first Australian oil discoveries were in southern Queensland near Moonie. Australian oil production now amounts to about 25 million barrels per year and includes pumping from oil fields off northwestern Australia near Barrow Island, Mereenie in the southern Northern Territory, and fields in the Bass Strait. The Barrow Islands, Mereenie, and Bass Strait fields are also sites of **natural gas** production.

Australia has rich deposits of uranium ore, which is refined for use for fuel for the nuclear power industry. Western Queensland, near Mount Isa and Cloncurry contains three billion tons (2.7 billion tones) of uranium ore reserves. There are also uranium deposits in Arnhem Land in far northern Australia, as well as in Queensland and Victoria.

Australia is also extremely rich in zinc reserves, the principal sources for which are Mt. Isa and Mt. Morgan in Queensland. The Northern Territory also has **lead** and zinc mines as well as vast reserves of bauxite (**aluminum** ore), namely at Weipa on the Gulf of Carpenteria and at Gove in Arnhem Land.

Gold production in Australia has declined from a peak production of four million fine ounces in 1904 to several hundred thousand fine ounces. Most gold is extracted from the Kalgoorlie-Norseman area of Western Australia. The continent is also well known for its precious stones, particularly white and black opals from South Australia and western New South Wales. There are sapphires and topaz in Queensland and in the New England District of northeastern New South Wales.

Because of its aridity, Australia suffers from leached, sandy, and salty soils. The continent's largely arid land and marginal water resources represent challenges for conservation and prudent environmental management. One challenge is to maximize the use of these resources for human beings while preserving ecosystems for animal and plant life.

See also Beach and shoreline dynamics; Continental drift theory; Desert and desertification; Earth (planet); Industrial minerals; Plate tectonics; Weathering and weathering series

AVALANCHE

An avalanche is a rapid downslope movement of some combination of **rock**, **regolith**, snow, slush, and **ice**. The movement can occur by any combination of sliding, falling, and rolling of pieces within the avalanche mass, but is generally very rapid. Avalanche velocities can reach tens to hundreds of kilometers per hour.

This avalanche in the Swiss ski resort of Evolene left at least two dead and more than a dozen missing. © *AFP/Corbis Bettman. Reproduced by permission.*

The term avalanche is generally associated with snow and ice. In its most general form, however, it can refer to the cascading of **sand** grains down the leeward face of a dune or the rapid downslope movement of largely disaggregated rock without snow or ice. Rock avalanches, for example, are very rapid and **catastrophic mass movements** of **bedrock** that has been broken into innumerable pieces either before or during movement.

Snow avalanches, hereafter referred to simply as avalanches, are classified according to whether they move across existing snow layers (surface avalanche) or the ground (ground avalanche), whether they are dry or wet, whether they move through the air or over ground and snow, and whether they consist of loose snow or intact slabs. Like landslides, avalanches begin when the weight of snow above some potential sliding surface exceeds the shear strength along that surface. In many cases, sliding occurs along a former snow surface that is quickly buried by new snow during a storm. The **physics** of slip surface formation, however, are more complicated for avalanches than most landslides because the snow and ice in an avalanche prone slope are near their **melting** points. Thus,

phase transitions and metamorphosis of snow and ice **crystals** can alter the strength of snow and ice slopes in a way that does not occur in **soil** or rock slopes. Melting can also trigger avalanches. Although it is not proven that loud noises such as shouting can trigger avalanches, the vibrations caused by explosives can do so—and explosives are often used to deliberately trigger avalanches under controlled conditions as a safety measure.

The aftermath of an avalanche is an avalanche track or chute, which is commonly marked by bent or broken trees and significant amounts of **erosion**. The track can be either a channel-like or open feature. The rock and debris carried by an avalanche can be deposited as an avalanche cone when the avalanche comes to rest, and the rock debris deposited at the base of a cliff or other steep slope by an avalanche is known as avalanche talus.

See also Catastrophic mass movements; Freezing and melting; Ice; Landslide; Mass movement; Mass wasting; Phase state changes

AVIATION PHYSIOLOGY

Aviation physiology deals with the physiological challenges encountered by pilots and passengers when subjected to the environment and stresses of flight.

Human physiology is evolutionarily adapted to be efficient up to about 12,000 feet above sea level (the limit of the physiological efficiency zone). Outside of this zone, physiological compensatory mechanisms may not be able to cope with the stresses of altitude.

Military pilots undergo a series of exercises in high altitude simulating hypobaric (low pressure) chambers to simulate the early stages of hypoxia (**oxygen** depletion in the body). The tests provide evidence of the rapid deterioration of motor skills and critical thinking ability when pilots undertake flight above 10,000 feet above sea level without the use of supplemental oxygen. Hypoxia can also **lead** to hyperventilation as the body attempts to increase breathing rates.

Altitude-induced decompression sickness is another common side effect of high altitude exposure in unpressurized or inadequately pressurized aircraft. Although the percentage of oxygen in the atmosphere remains about 21% (the other 79% of the atmosphere is composed of nitrogen and a small amount of trace gases), there is a rapid decline in **atmospheric pressure** with increasing altitude. Essentially, the decline in pressure reflects the decrease in the absolute number of molecules present in any given volume of air.

Pressure changes can adversely affect the middle ear, sinuses, teeth, and gastrointestinal tract. Any sinus block (barosinusitis) or occlusions that inhibit equalization of external pressure with pressure within the ear usually results in severe pain. In severe cases, rupture of the tympanic membrane may occur. Maxillary sinusitis may produce pain that is improperly perceived as a toothache. This is an example of referred pain. Pain related to trapped gas in the tooth itself (barondontalgia) may also occur.

Ear block (barotitis media) also causes loss of hearing acuity (the ability to hear sounds across a broad range of pitch and volume). Pilots and passengers may use the Valsalva maneuver (closing the mouth and pinching the nose while attempting to exhale) to counteract the effects of **water** pressure on the Eustachian tubes and to eliminate pressure problems associated with the middle ear. When subjected to pressure, the tubes may collapse or fail to open unless pressurized. Eustachian tubes connect the corresponding left and right middle ears to the back of the nose and throat, and function to allow the equalization of pressure in the middle ear air cavity with the outside (ambient) air pressure. The degree of Eustachian tube pressurization can be roughly regulated by the intensity of abdominal, thoracic, neck, and mouth muscular contractions used to increase pressure in the closed airway.

Rapid changes in altitude allow trapped gases to cause pain in joints in much the same way—although to a far lesser extent—that the bends causes pain in scuba divers. Lowered outside atmospheric pressure creates a strong pressure gradient that permits dissolved nitrogen and other dissolved or "trapped" gases within the body to attempt to "bubble off" or leave the blood and tissues in an attempt to move down the concentration gradient toward a region of lower pressure.

Spatial disorientation trainers demonstrate the disorientation and loss of balance (vestibular disorientation) that can be associated with flight at night—or in clouds—where the pilot losses the horizon as a visual reference frame. Balance and the sense of turning depend upon the ability to discriminate changes in the motion of fluids within the semicircular canals of the ear. When turns are gradual, the changes become imperceptible because the fluids are moving at a constant velocity. Accordingly, without visual reference, pilots can often enter into steep turns or dives without noticing any changes. Spatial disorientation chambers allows pilots to learn to "trust their instruments" as opposed to their error-prone sense of balance when flying in IFR (Instrument Flight Rules) conditions.

In addition to vestibular disorientation, spatial disorientation can also lead to motion sickness.

Because of the highly repetitive nature of the active pilot scan of instruments, fatigue is a chronic problem for pilots. Fatigue combined with low oxygen pressures may induce strong and disorienting visual illusions.

Although not often experienced in general aviation, military pilots operate at high speeds and undertake maneuvers that subject them to high "g" (gravitational) forces. In a vertical climb, the increased g forces (called positive "g" forces because they push down on the body) tend to force blood out of the circle of Willis supplying arterial blood to the brain. The loss of oxygenated blood to the brain eventually causes pilots to lose their field of peripheral vision. Higher forces cause "blackouts" or temporary periods of unconsciousness. Pilots can use special abdominal exercises and "g" suits (essentially adjustable air bladders that can constrict the legs and abdomen) to help maintain blood in the upper half of the body when subjected to positive "g" forces.

In a dive, a pilot experiences increased upward "g" forces (termed negative "g" forces) that force blood into the arterial circle of Willis and cerebral tissue. The pilot tends to experience a red out. Increased arterial pressures in the brain can lead to stroke. Although pilots have the equipment and physical stamina to sustain many positive "g" forces (routinely as high as five to nine times the normal force of **gravity**) pilots experience red out at about 2–3 negative "g's." For this reason, maneuvers such as loop, rolls, and turns are designed to minimize pilot exposure to negative "g" forces.

See also Aerodynamics; Atmospheric composition and structure

AXIS • *see* POLAR AXIS AND TILT

B

BALLARD, ROBERT DUANE (1942-   )
American oceanographer and archaeologist

Robert Ballard has participated in over 100 deep-sea expeditions during his career. Ballard is perhaps most well known for leading the 1985 French-American expedition that discovered the wreckage of the RMS *Titanic*. However, Ballard has made many great contributions to the fields of **oceanography**, marine **geology**, and underwater archaeology. He is a pioneer in the use of underwater submersibles in the location and survey of deep-water subjects.

Ballard was born in Wichita, Kansas, but his family soon moved to San Diego, California. He developed a life-long love of the ocean as a child. When he was a teenager, he traded studying creatures in tidal pools for SCUBA lessons. Ballard decided to pursue ocean research as a career when he entered college. He attended the University of California, earning dual undergraduate degrees in geology and **chemistry** in 1965. He trained dolphins for a local marine theme park while pursuing postgraduate studies at the University of Hawaii. Ballard was a member of the Army Reserve Officer Training Corps, but petitioned for transfer to the Navy in 1967. The U.S. Navy granted his request and assigned him to the Deep Submergence Laboratory at Woods Hole, Massachusetts. He completed his graduate studies at the University of Rhode Island, receiving a Ph.D. in geophysics and marine geology in 1974.

The first major research expedition of Ballard's career was the first manned exploration of the Mid-Atlantic Ridge, a large underwater mountain range in the Atlantic Ocean, in 1973–1974. The survey mapped some of the most varied terrain of the ocean floor. In 1977, Ballard was a member of research team that used small submersibles to explore the waters near the Galapagos Islands. Dr. Ballard and his crew observed ecosystems that developed around underwater hot **springs**. Two years later, off the coast of Baja California, Ballard found underwater volcanoes that ejected hot, mineral rich fluids. Ballard and his team studied the effects of these vents on deep marine life and ocean **water** chemistry.

In 1985, Ballard and his team turned their attention to finding one of the most famous shipwrecks, the RMS *Titanic*, a British luxury steamship that sank in the North Atlantic in 1912. Experienced using small manned submersibles, such as ALVIN, Ballard designed other survey apparatus, such as the ARGO-JASON, a remote controlled deep-sea imaging system. In order to gain access to the most sophisticated equipment for his search for the *Titanic*, Ballard was first assigned to conduct deep-sea reconnaissance work for the United States Navy, finding and evaluating the site of a sunken U.S. nuclear submarine. After completing his work for the Navy, Ballard used the remaining expedition time for his join French and American research team to locate the *Titanic*. The team located the wreckage of the British steamer just days before their voyage was to end. Ballard used the submersible JASON to photograph the site. In addition, Ballard and his team designed a small accessory robotic device, named JASON Jr., which could explore the inside of the ship by remote control. Using this array of sophisticated diving equipment, and small submersibles, Ballard also found the wrecks of the USS *Yorktown*, the German battleship *Bismarck*, and part of the lost fleet of Guadalcanal. He also led an expedition to photograph and explore the British luxury liner, *Lusitania*.

Remaining on the cutting edge of under-sea research and exploration, Ballard left Woods Hole in 1997 to pursue career interests in underwater archaeology. Combining his knowledge of deep-water oceanography, and a passion for historic preservation, Ballard accepted a post to head the Institute for Exploration in Mystic, Connecticut. That year, Ballard, using the Navy's nuclear research submarine, NR-1, explored a complex of 2,000 year-old shipwrecks in the Mediterranean Sea. Because of the depth at which the wrecks settled, the site remained perfectly preserved. In 2000, Ballard was named National Geographic's Explorer-in-Residence.

See also Deep sea exploration

BANDED IRON FORMATIONS

Banded **iron** formations (BIFs) are chemically precipitated **sedimentary rocks**. They are composed of alternating thin (millimeter to centimeter scale) red, yellow, or cream colored layers of **chert** or jasper and black to dark gray iron oxides (predominantly magnetite and hematite), and/or iron carbonate (siderite) layers. Banded iron formations have greater than 15% sedimentary iron content. Banded iron formations are of economic interest as they host the world's largest iron ore deposits and many gold deposits.

Algoma-type banded iron formations were deposited as chemical sediments along with other sedimentary rocks (such as greywacke and shale) and volcanics in and adjacent to volcanic arcs and spreading centers. Iron and silica were derived from hydrothermal sources associated with volcanic centres. Algoma-type iron formations are common in **Archean** greenstone belts, but may also occur in younger rocks.

Lake Superior-type banded iron formations were chemically precipitated on marine continental shelves and in shallow basins. They are commonly interlayered with other sedimentary or volcanic rocks such as shale and **tuff**. Most Lake Superior-type banded iron formations formed during the Paleoproterozoic, between 2.5 and 1.8 billion years ago. Prior to this, Earth's primitive atmosphere and **oceans** had little or no free **oxygen** to react with iron, resulting in high iron concentrations in seawater. Iron may have been derived from the **weathering** of iron-rich rocks, transported to the sea as water-soluble Fe^{+2}. Alternatively, or in addition, both iron and silica may have been derived from submarine magmatic and hydrothermal activity. Under calm, shallow marine conditions, the iron in seawater combined with oxygen released during photosynthesis by Cyanobacteria (primitive blue-green algae, which began to proliferate in near-surface waters in the Paleoproterozoic) to precipitate magnetite (Fe_3O_4), which sank to the sea floor, forming an iron-rich layer.

It has been proposed that during periods when there was too great a concentration of oxygen (in excess of that required to bond with the iron in the seawater) due to an abundance of blue-green algae, the blue-green algae would have been reduced in numbers or destroyed. A temporary decrease in the oxygen content of the seawater then eventuated. When magnetite formation was impeded due to a reduction in the amount of oxygen in seawater, a layer of silica and/or carbonate was deposited. With subsequent reestablishment of Cyanobacteria (and thus renewed production of oxygen), **precipitation** of iron recommenced. Repetitions of this cycle resulted in deposition of alternating iron-rich and silica- or carbonate-rich layers. Variations in the amount of iron in seawater, such as due to changes in volcanic activity, may have also led to rhythmic layering. The large lateral extent of individual thin layers implies changes in oxygen or iron content of seawater to be regional, and necessitates calm depositional conditions. Iron and silica-rich layers, originally deposited as **amorphous** gels, subsequently lithified to form banded iron formations. The distribution of Lake Superior-type banded iron formations of the same age range in **Precambrian** cratons worldwide suggests that they record a period of global change in the oxygen content of the earth's atmosphere and oceans. Also, the worldwide abundance of large, calm, shallow platforms where cyanobacterial mats flourished and banded iron formations were deposited may imply a global rise in sea level.

Primary carbonate in banded iron formations may be replaced by silica during diagenesis or deformation. The pronounced layering in banded iron formations may be further accentuated during deformation by pressure solution; silica and/or carbonate are dissolved and iron oxides such as hematite may crystallize along pressure solution (stylolite) surfaces. Banded iron formations are highly anisotropic rocks. When shortened parallel to their layering, they deform to form angular to rounded **folds**, kink bands, and box folds. Folds in banded iron formations are typically doubly plunging and conical. Banded iron formations may interact with hot fluids channeled along faults and more permeable, interbedded horizons such as **dolomite** during deformation. This may remove large volumes of silica, resulting in concentration of iron. Iron, in the form of microplaty hematite can also crystallize in structurally controlled sites such as fold hinges and along detachment faults. If there is sufficient enrichment, an iron ore body is formed. Iron may also be leached, redeposited and concentrated during weathering to form supergene iron ore deposits. Fibrous growth of **quartz** and **minerals** such as crocidolite (an amphibole, also known as asbestiform riebeckite) commonly occurs in banded iron formations during deformation due to dilation between layers, especially in fold hinges. Replacement of crocidolite by silica produces shimmering brown, yellow and orange "tiger-eye," which is utilized in jewelry and for ornamental use.

See also Industrial minerals

BARCHAN DUNES • *see* DUNES

BAROMETER • *see* ATMOSPHERIC PRESSURE

BARRIER ISLANDS

A barrier island is a long, thin, sandy stretch of land, oriented parallel to the mainland coast that protects the coast from the full force of powerful storm waves. Between the barrier island and the mainland is a calm, protected **water** body such as a lagoon or bay. Barrier islands are dynamic systems, constantly on the move, migrating under the influence of changing sea levels, storms, waves, **tides**, and longshore currents. In the United States, barrier islands occur offshore where gently sloping sandy coastlines, as opposed to rocky coastlines, exist. Consequently, most barrier islands are found along the Gulf Coast and the Atlantic Coast as far north as Long Island, New York. Some of the better known barrier islands include Padre Island of Texas, the world's longest; Florida's Santa Rosa Island, composed of sugar-white **sand**; Cape Hatteras of North Carolina, where the first airplane was flown; and Assateague Island near Maryland, home of wild ponies.

Barrier islands are young in geologic terms. They originated in the **Holocene Epoch**, about 4,000–6,000 years ago. During this time, the rapid rise in sea level, associated with **melting glaciers** from the last **ice** age, slowed significantly. Although the exact mechanisms of barrier island formation aren't fully understood, this slowdown of sea level rise allowed the islands to form.

In order for barrier islands to form, several conditions must be met. First, there must be a source of sand to build the island. This sand may come from coastal deposits or offshore deposits (called shoals); in either case, the sand originated from the **weathering** and **erosion** of **rock** and was transported to the coast by **rivers**. In the United States, much of the sand composing barrier islands along Florida and the East Coast came from the Appalachian Mountains. Next, the **topography** of the coastline must have a broad, gentle slope. From the coastal plains of the mainland to the edge of the **continental shelf**, this condition is met along the Atlantic and Gulf Coasts. Finally, the forces of waves, tides and currents must be strong enough to move the sand, and of these three water movement mechanisms, waves must be the dominant force.

Several explanations for barrier island development have been proposed. According to one theory, coastal sand was transported shoreward as sea level rose, and once sea level stabilized, wave and tidal actions worked the sand into a barrier island. Another possibility is that sand was transported to its present location from shoals. Barrier islands may have formed when low-lying areas of spits, extensions of beaches that protrude into a bay as a result of deposition of sediment carried by longshore currents, were breached by the sea. Finally, barrier islands may have formed from sandy coastal ridges that became isolated from low-lying land and formed islands as sea level rose.

Once formed, barrier islands are not static **landforms**; they are dynamic, with winds and waves constantly reworking and moving the barrier island sand. Changes in sea level also affect these islands. Most scientists agree that sea level has been gradually rising over the last thousand years, and this rise could be accelerating today due to **global warming**. Rising sea level causes existing islands to migrate shoreward.

Barrier islands do not stand alone in a geologic sense. A whole system of islands develops along favorable coastlines. The formation of these islands allows other landforms to develop, each characterized by their dominant sediment type and by the water that helps form them. For example, each barrier island has a shoreline that faces the sea and receives the full force of waves, tides, and currents. This shoreline is often called the beach. The beach zone extends from slightly offshore (subtidal, or underwater) to the high water line. Coarser sands and gravels are deposited here, with finer sands and silts carried farther offshore.

Behind the beach are sand **dunes. Wind** and plants (such as sea oats) help form dunes, but occasionally dunes are inundated by high water and may be reworked by storm surges and waves. On wide barrier islands, the landscape behind the foredunes gently rolls as dunes alternate with low-lying swales (marshy wet areas). If the dunes and swales are well developed, distinct parallel lines of dune ridges and swales can be seen from overhead. These differences in topography allow some **soil** to develop and nutrients to accumulate despite the porous, sandy base. Consequently, some barrier islands are host to trees (which are often stunted), bushes, and herbaceous plants. Other narrower or younger barrier islands may be little more than loose sand with few plants.

On the shoreward side of the main body of the island is the back-barrier. Unlike the beach, this zone does not bear the full force of ocean waves. Instead, the back-barrier region consists of a protected shoreline and lagoon, which is more influenced by tides than waves. Occasionally, during storms, water may rush over the island carrying beach and dune sand and deposit the sand in the lagoon. This process, called rolling over, is vital to the existence of barrier islands and is the method by which a barrier island migrates landward. Characteristic sand washover fans in the lagoons are evidence of rolling over. Because the back-barrier region is sheltered, salt marsh, sea grass, and mudflat communities develop. These communities teem with plant and animal life and their muddy or sandy sediments are rich with organic matter.

Finally, barrier islands are characterized by tidal inlets and tidal deltas. Tidal inlets allow water to move into and out of bays and lagoons with rising and falling tides. Tidal inlets also provide a path for high water during storms and hurricanes. As water moves through an inlet, sand is deposited at both ends of the inlet's mouth, forming tidal deltas. Longshore currents may also deposit sand at the **delta**. Eventually the deltas fill in with sand and the inlet closes, only to appear elsewhere on the barrier island, usually at a low-lying spot. The size and shape of the inlet are determined by various factors, including the size of the associated lagoon and the tidal range, or the vertical height between high and low tide for the **area**. A large tidal range promotes the formation of numerous inlets, thereby creating shorter and wider barrier islands referred to as drumsticks. In addition, the larger the lagoon and the greater the tidal range, the deeper and wider the inlet due to the large quantity of water moving from ocean to lagoon and back. Deep, wide inlets occur where the main source of energy shaping the coastal area is tides or tides in conjunction with waves. In contrast, wave-dominated areas form long barrier islands with narrow bays and narrow, shallow inlets.

See also Beach and shoreline dynamics; Gulf of Mexico; Offshore bars; Tropical cyclone; Wave motions

BARRINGER METEOR CRATER

The Barringer Meteor Crater in Arizona was the first recognized terrestrial **impact crater**. The confirmation of a meteor impact (subsequently identified as the **Canyon** Diablo meteorite) at the site proved an important stepping-stone for advances in **geology** and **astronomy**. In solving the mystery surrounding the origin of the Barringer crater, geologists and astronomers made substantial progress in understanding the dynamic interplay of gradual and cataclysmic geologic processes both on Earth and on extra-terrestrial bodies.

Barringer Meteor Crater, Arizona. *U.S. National Aeronautics and Space Administration (NASA).*

The Barringer Meteorite Crater (originally named Coon Butte or Coon Mountain) rises 150 feet above the floor of the surrounding Arizona **desert**. The impact crater itself is almost a mile wide and 570 feet deep. Among geologists, two competing theories were most often asserted to explain the geologic phenomena. Before the nature of hot spots or plate tectonic theory would have convinced them otherwise, many geologists hypothesized that the crater resulted from volcanic activity. A minority of geologists asserted that the crater must have resulted from a meteor impact.

In the last decade of the nineteenth century, American geologist Grove Karl Gilbert, then the head of the U.S. Geological Survey, set out to determine the origin of the crater. Gilbert assumed that for a meteor to have created such a large crater, it was necessary for it to remain intact through its fiery plunge through the earth's protective atmosphere. Moreover, Gilbert assumed that most of the meteor survived its impact with Earth. Gilbert, therefore, assumed that if a meteor collision was responsible for the crater, substantial pieces of the meteor should still exist and there should be ample and direct physical evidence of the size of the meteor. When upon observation it became apparent that there was no substantial mass inside the crater, Gilbert assumed that the meteor might have been covered with the passage of time. Assuming the meteor to be like other known meteorites and similar in percentage of **iron** composition to the smaller meteors found around the crater, Gilbert looked for magnetic evidence in a effort to find the elusive meteor. Gilbert's repeated tests found no evidence for such a buried mass. After carefully examining the crater, Gilbert concluded that, in the absence of the evidence he assumed would be associated with a meteor impact, the crater had resulted from subterranean activity.

In 1902, Daniel Moreau Barringer, an American entrepreneur and mining engineer, began a study of the Arizona crater and took up the opposing view. After discovering that small meteors made of iron had been found at or near the rim of the crater, Barringer was convinced that only a large iron meteor could be the cause of such a geologic phenomenon.

Acting more like a businessman or miner trying to stake a claim, and before doing any studies on the potential masses and energies that would have to be involved in such an impact, Barringer seized the opportunity to form company with the intent of mining the iron from the presumed meteor for commercial profit. Without actually visiting the crater, Barringer formed the Standard Iron Company and sought mining permits.

For nearly the next thirty years, Barringer became the sword and shield of often-rancorous scientific warfare regarding the origin of the crater. In bitter irony, Barringer won the scientific battle, the proof eventually accumulated that the crater resulted from a meteor impact, but lost his financial gamble. In the end, the meteor that caused the impact proved much smaller than hypothesized by either Gilbert or Barringer, and the nature of the impact obliterative. On the heels of these finding in 1929, Barringer died of a heart attack. His lasting legacy was in the attachment of his name to the impact crater.

The debate over the origin of the Great Barringer Meteor Crater came at a time when geology itself was reassessing its methodologies. Within the geologic community there was often vigorous debate over how to interpret geologic data. In particular, debates ranged regarding the scope and extent of **uniformitarianism**. In its simplest form, uniformitarianism asserted only that the laws of **physics** and **chemistry** remained unchanged during the geologic history of the Earth. Debate centered on whether the predominantly dominant gradualism (similar to evolutionary gradualism) of geologic processes was significantly affected by catastrophic events.

Barringer confidently asserted that the Coon Butte crater supported evidence of catastrophic process. Although he argued selective evidence, Barringer turned out to be correct when he asserted that the finely pulverized silica surrounding the crater could have only been created in a process that created instantaneously great pressures. Beyond the absence of volcanic rocks, Barringer argued that there were too many of the iron fragments around the crater to have come from gradually accumulated separate meteor impacts. Moreover, Barringer noticed that instead of defined strata (layers) there was a randomized mixture of the fragments and ejecta (native **rock** presumable thrown out of the crater at the time of impact). Such a random mixture could only have resulted from a cataclysmic impact.

Barringer's cause gained support of mainstream geologists when American geologist George P. Merrill tested rocks taken from the rim and floor of the crater. Merrill concluded that the quartz-like **glass** found in abundance in the presumed eject could only have been created by subjecting the native sands to intense heat. More importantly, Merrill concluded that the absence of sub-surface fusions proved the heat could not have come from below the surface.

The evidence collected by Barringer also influenced astronomers seeking, at that time to explain large, round craters on the **moon**. Once again, the debate moved between those championing extra-terrestrial volcanic activity (gradualism) versus those who favored an impact hypothesis (cataclysm). This outcome of these debates had enormous impact in both geology and astronomy.

One fact that perplexed astronomers was that it appeared that all of the lunar impact craters were generally round. If meteors struck the Moon at varying angles, it was argued, then the craters should have assumed a variety of oblique shapes. Barringer and his 12-year-old son set out to experiment with the formation of such craters by firing bullets into clumps of rock and mud. Regardless of the firing angle, the Barringers demonstrated (and published their results in both popular and scientific magazines) that the resulting craters were substantially round. More definitive proof was subsequently provided in 1924 by calculations of astronomers who determined that forces of impact at astronomical speeds likely resulted in the explosive destruction of the impacting body. Importantly, regardless of the angle of impact, the result of such explosions would leave rounded craters.

The confirmation that a meteor weighing about 300,000 tons (less than a tenth of what Barringer had estimated) and traveling in excess of 35,000 mph at impact could have produced the energy and catastrophic phenomena observed proved a double edge sword for Barringer. In one stroke, his hypothesis that the crater was caused by a meteor impact gained widespread support while, at the same time Barringer's hopes of profitably mining the meteor vaporized like much of the exploded meteor itself.

The scientific debate on the origin of Barringer crater was essentially closed when it was dramatically demonstrated that meteor impacts could impart such large energies far above even the tremendous power of nuclear weapons. In the 1960s, American astronomer and geologist Eugene Shoemaker found distinct similarities between the fused rocks found at Barringer crater and those found at atomic test sites. In addition, unique geologic features termed "shattercones" created by the instantaneous application of tremendous pressure pointed to a tremendous explosion at or above the impact crater. Determinations made by later, more sophisticated dating techniques placed the age of the crater at roughly 50,000 years.

Scientists subsequently understood that massive cataclysmic collisions result in what is now termed **shock metamorphism**. These shock metamorphic effects have been shown to be exclusively associated with meteorite impact craters. No other natural process on earth can account for the observed results.

A great deal of such evidence methodologies derived from the Barringer crater controversy now points to a catastrophic astronomical collision at the end of the **Cretaceous Period** 66 million years ago. The effects of this collision are thought to have precipitated the widespread extinction of large species, including the dinosaurs. The enigmatic Tunguska explosion of 1908, which devastated an vast **area** of Siberian forest, may have been Earth's most recent significant encounter with an impacting object vaporized so as to leave little physical remains beyond the manifest effects of a tremendous explosion.

Methods used to confirm Barringer crater as a meteor impact crater have been used to identify many other impact sites around the world. Once scientists became aware of the tremendous energies involved in astronomical impacts, large terrestrial impacts, often hidden by erosive effects, became a focus of study. With more than 150 such impact sites identified, impacts have taken on an important role in understanding the Earth's geologic history. The accumulated evidence led to a synthesis of gradualism and catastrophic theory. In accord with uniformitarianism, the gradual and inexorable shaping processes taking place over **geologic time** were understood to be punctuated with catastrophic events.

BASALT

Basalt is a **mafic** volcanic **rock** consisting primarily of **plagioclase feldspar** and pyroxene **minerals**. Common accessory minerals can include other pyroxenes, **olivine**, **quartz**, and nepheline. Basalt is the volcanic equivalent of the plutonic rock gabbro, and as such has a low silica content (48%–52%). Like other volcanic rocks, basalt cools quickly after it erupts and therefore generally contains less than 50% visible **crystals** floating in a matrix of **glass** or microscopic crystals. Pillow basalt, consisting of lobes of **lava** emplaced and solidified on top of each other, is the result of undersea eruptions such as those along divergent oceanic plate boundaries. Basalt is also known to occur on the **moon**.

Because of its low silica content, which translates into a high **melting** point and low viscosity, basaltic lava erupts at a higher **temperature** (2,012–2,192°F; 1,100–1,250°C) and flows more easily across low slopes that do more **silicic** lava types. Under some conditions, basaltic lava can flow more than 12.5 miles (20 km) from the point of eruption. The low viscosity of molten basalt also means that dissolved volcanic gasses can escape relatively easily as the **magma** travels to the surface and confining pressure is reduced. Thus, basalt eruptions tend to be quiet and effusive (as typified by Hawaiian volcanoes) as compared to the explosive eruptions often associated with more viscous and silica-rich lava (as typified by Mount St. Helens). Lava fountains can, however, reach heights of several hundred meters during basaltic eruptions.

Lava flows that solidify with a smooth or ropy surface are often described using the Hawaiian term *pahoehoe*, whereas those which solidify with a jagged or blocky surface are described by the Hawaiian term *aa*. The former is pronounced "pa-hoy-hoy" and the latter is pronounced "ah-ah."

Another characteristic of many basalt flows is the presence of polygonal columnar joints, which are understood to form by contraction of the lava as it cools. The result is a system of nearly vertical joints that form a polygonal pattern when viewed from above and break the rock into slender prismatic columns.

See also Divergent plate boundary; Extrusive cooling; Joint and jointing; Rate factors in geologic processes; Rifting and rift valleys; Sea-floor spreading

Devil's Tower National Monument in Wyoming is a column of basalt that has resisted weathering, unlike the less-resistent rock that once surrounded it. © Dave G. Houser/Corbis. Reproduced by permission.

BASIN AND RANGE TOPOGRAPHY

Basin and range **topography** is characterized by tilted fault blocks forming sub-parallel mountain ranges and intervening sediment-filled basins. These elements are typical of the basin and range physiographic topography in the western United States. This province is bounded on the east by the Colorado Plateau, the Columbia and Snake River Plateaus to the north, the Sierra Nevada to the west, and extends southward through eastern California and southern Arizona into northern Mexico. Nearly the entire state of Nevada and western Utah exhibit features distinctive to basin and range topography.

Within the basin and range province, steep mountain ranges are bounded by normal faults, with ground motion along the faults resulting in the relative uplift of the mountains and dropping of the valleys. The longitudinal mountain ranges lie generally parallel to each other and trend northward, leading one early geologist to compare a map view of the ranges within the province to a group of caterpillars crawling slowly north. The bare mountains are cut by numerous drainages that carry the products of **weathering** into the basins below. Sediment in the resultant alluvial fans eventually fills the intermontane basin. In some cases, as much as 10,000 feet of

sand, gravel, and **clay** has accumulated. Ultimately, these relatively featureless alluvial slopes conceal the majority of the fault-block mountains and the faults from which they were formed.

The fundamental structural element of the region is the ever-present north-south trending normal fault. The presence of the normal faults is indicative of tensional stress over the region oriented in an east-west direction. This stress has produced dramatic crustal extension in this same direction, thus allowing the valley blocks to drop between the bounding normal faults as the ranges were stretched apart. Directly related to the tensional stress is the general tectonic uplift of the entire region than began approximately 15 million years ago.

Although geologists have an understanding of the mechanisms by which the basin and range formed, the thinner **crust** and higher heat flow of the **area** present strong evidence that conditions within the upper mantle are responsible. The most common explanation for these conditions is related to the subduction of a crustal plate producing chemical and physical changes within the mantle. Upwelling of heated, lower-density material caused the crust to bow upward. This, in turn, produced the high heat flow, uplift, crustal thin-

Titcomb Basin, Wind River Range. © Richard Hamilton Smith/Corbis. Reproduced by permission.

ning, and regional extension observed in basin and range topography.

See also Alluvial system; Faults and fractures; Plate tectonics

BATHOLITH

Batholiths are large bodies of intrusive igneous **rock**. Formed when **magma** cools and crystallizes beneath Earth's surface, batholiths are the largest type of **pluton**. By definition, a batholith must cover at least 39 mi² (100 km²), although most are even larger. Many batholiths cover hundreds to thousands of square miles. The Idaho batholith, for example, has a surface **area** of over 15,500 mi² (40,000 km²).

Batholiths are generally not comprised of one continuous magmatic intrusion; rather, they are produced by repeated intrusions, and most batholiths are made up of multiple individual plutons. Intruded rock cools and solidifies, later to be exposed at the surface through **erosion**. Because they cool beneath Earth's surface, batholiths have a coarse grained texture, and most are granitic in composition.

Usually associated with mountain building, batholiths are often emplaced near continental margins during periods of subduction. As the subducting slab descends, it begins to melt, and multiple plutons are intruded beneath the continent to form the core of the volcanic arc. The Sierra Nevada Mountains, for example, are comprised of a granitic batholith, which is made up of hundreds of individual plutons intruded over a period of several million years. Emplacement of the Sierra Nevada batholith occurred during a mountain building episode known as the Nevadan **orogeny**, initiated during the Jurassic. Uplift and erosion of the area later exposed the batholith, which now forms the spine of the famous mountains.

The Sierra Nevada batholith not only forms a major mountain chain, but also was responsible for driving the California gold rush. Precious **minerals** including gold are commonly associated with granitic batholiths. As mineral-rich solutions move along cracks in the rock body, gold, copper, and other minerals, especially **quartz**, precipitate out. Gold may be mined from deposits known as quartz veins that form along the fractures. The Mother Lode in the Sierra Nevada is possibly the most famous of such deposits.

Determining the mechanism for batholith emplacement has been a topic of much debate. When gigantic batholiths are intruded, the surrounding rock, known as the **country rock**, must somehow make room for the intrusion. Several models have been suggested, but most geologists now agree that a mechanism known as forceful injection is probably responsible for emplacement. Apparently, as the body of magma rises, it deforms the country rock, pushing it out of the way.

See also Intrusive cooling; Mineral deposits; Pluton and plutonic bodies

BATHYMETRIC MAPPING

Bathymetric mapping refers to construction of ocean and sea maps—bathymetric maps (BM). Bathymetric maps represent the ocean (sea) depth depending on geographical coordinates, just as topographic maps represent the altitude of Earth's surface in different geographic points.

The most popular kind of bathymetric maps is one on which lines of equal depths (isobaths) are represented. Like geographical maps of the surface of Earth, bathymetric maps are constructed in definite **cartography projection**. Mercator projection is used perhaps more often in constructing bathymetric maps, and has been used for a long time in constructing sea charts that are used for sailing in all latitudes except Polar ones.

The creation of a bathymetric map of a given region depends above all on the amount of depth measurement data for that region. Before the invention of the echosounder in 1920's, ocean (sea) depth could be measured only by **lead**. Such measurements were quite rare; these measurements were made only in isolated points, and creation of bathymetric map-

William Beebe and Otis Barton posing with their invention, the bathysphere. © Ralph White/Corbis. Reproduced by permission.

ping was practically impossible. Thus, the structure of the ocean floor was virtually unknown. It should be noted, for example, that the most important structure in Atlantic, the Middle-Atlantic ridge, was discovered and began to be investigated only after World War II. Another important factor for creating bathymetric mapping is determining geographical coordinates of the point where the depth measurement is made. It is evident that when these determinations are more precise, then the maps are better. Today, the **GPS** (Global Positioning System) is used for determining the coordinates of the measurement points.

When constructing topographic maps of land, one can always measure the altitude of any point of the surface precisely. However, when constructing a bathymetric map, it is practically impossible to determine the exact depth of any point of the bottom of the sea. Obviously, bathymetric maps are more precise when more data of depth measurement per surface **area** unit in the given region are available. Nowadays, the most precise and detailed bathymetric maps result from using data from multibeam echosounding. The multibeam echosounder is a special kind of echosounder, which is located on board the vessel and measures the depth simultaneously in several points of the bottom. These points are located on the straight line perpendicular to the vessel track. These points themselves are determined by the reflection of several acoustical pulses (beams) directed from one point at different angles to the vertical. The determination of depth in this method is performed regularly within periods of several seconds during the vessel motion. The measurement data are stored in a computer and using them the map of isobath of narrow bottom stripe can be represented periodically, or these data can be represented on a monitor.

It should be noted that in addition to the multibeam echosounder, other devices that measure depths simultaneously in several points of the ocean bottom have been devel-

oped, but all of them are based on the reflection of sound signals from the bottom.

If there is a lot of measurement data (more precisely this means that the average amount of measurement data per surface area unit is relatively big, and the measurement points themselves are located uniformly on the surface investigated), then computer methods of isobath construction are used. In this case, two stages of the work are executed. First, using the measurement data obtained in arbitrary points of the surface, the values of the depth in knots of regular grid are calculated (sometimes this stage is known as digital surface model construction). Then using these grid values, coordinates of different isobaths are determined (grid values are used also for other forms of bathometric mapping representations, 3-D views, for example). There are many algorithms of digital model creation, such as the least mean square method, and so-called Kriging method, as well as algorithms of constructing an isobath of its own using depth grid values. To construct a precise map of the region it is necessary to perform echosounding surveying on it in such a manner that map stripes, obtained in different vessel tracks, would be as close to each other as possible, or even overlap. After performing such surveying, all data are joined together, and the map of the entire region is constructed.

It should be noted that presently, only a small part of Earth's ocean bottom (several percent) is covered by such precise measurements. In some places, little data is available in a study area, obtained by one beam echosounder, or there is no data at all. In these cases, scientists try to use results of other geophysical measurements, first of all gravimetric measurements, to determine ocean depth. For example, methods of determination of ocean bottom **topography** using **satellite** altimetry or marine gravimetry data are useful. Even with using otherwise accurate satellite technology, indirect geophysical methods for determining the ocean bottom depth can always contain a mistake. Earth and its surface are very complex formations, so the precise value of the ocean depth in a given point should be determined if necessary only by direct measurement.

In the case when depth measurement data are small in numbers for a given region, indirect methods are used in constructing bathymetric mapping, such as geomorphology analysis, for example. Scientists also take into account geological considerations and even human intuition, which can at times be useful.

Several international organizations are currently working on bathymetric mapping of the world's **oceans**. The *General Bathymetric Chart of the Oceans* (GEBCO, in the scale 1:5000000), which may be considered a reference map, is one example. In this map, data of many regional bathymetric maps are collected, taking into account the different methods of their construction. There is also a digital version of this map (on CD), where files are represented in different formats, and in ASCII codes in particular, and where isobaths are represented in the so-called vector format.

Bathymetric mapping is finding increasing scientific and commercial use. For example, bathymetric maps are important in forging different underwater communications.

Also, bathymetric maps are important tools for formulating theories about how Earth developed, along with theories of sea and ocean formation.

See also Abyssal plains; Deep sea exploration; Ocean trenches; Oceans and seas

BEACH AND SHORELINE DYNAMICS

The coast and beach, where the continents meet the sea, are dynamic environments where agents of **erosion** vie with processes of deposition to produce a set of features reflecting their complex interplay and the influences of changes in sea level, **climate**, or sediment supply. "Coast" usually refers to the larger region of a continent or island which is significantly affected by its proximity to the sea, whereas "beach" refers to a much smaller region, usually just the areas directly affected by wave action.

The earth is constantly changing. Mountains are built up by tectonic forces, weathered, and eroded away. The erosional debris is deposited in the sea. In most places these changes occur so slowly that they are barely noticeable, but at the beach we can often watch them progress.

Most features of the beach environment are temporary, steady-state features. To illustrate this, consider an excavation in **soil**, where **groundwater** is flowing in, and being pumped out by mechanical pumps. The level of the **water** in the hole is maintained because it is being pumped out just as fast as it is coming in. It is in a steady state, but changing either rate will promptly change the level of the water. A casual observer may fail to notice the pumps, and erroneously conclude that the water in the hole is stationary. Similarly, a casual observer may think that the **sand** on the beach is stationary, instead of in a steady state. The size and shape of a spit, which is a body of sand stretching out from a point, parallel to the shore, is similar to the level of the water in this example. To stay the same, the rate at which sand is being added to the spit must be exactly balanced by the rate at which it is being removed. Failure to recognize this has often led to serious degradation of the coastal environment.

Sea level is the point from which we measure elevation, and for good reason. A minor change in elevation high on a mountain is undetectable without sophisticated surveying equipment. The environment at 4,320 feet above sea level is not much different from that at 4,310 feet. The same 10-foot change in the elevation of a beach would expose formerly submerged land, or inundate formerly exposed land, making it easy to notice. Not only is the environment different, the dominant geologic processes are different: Erosion occurs above sea level, deposition occurs below sea level. As a result, coasts where the land is rising relative to sea level (emergent coasts) are usually very different from those where the land is sinking relative to sea level (submergent coasts).

If the coast rises, or sea level goes down, areas that were once covered by the sea will emerge and form part of the landscape. The erosive action of the waves will attack surfaces that previously lay safely below them. This wave attack occurs right at sea level, but its effects extend from there. Waves may undercut a cliff, and eventually the cliff will fall into the sea, removing material from higher elevations. In this way the cliff retreats, while the beach profile is extended at its base. The rate at which this process continues depends on the material of the cliff and the profile of the beach. As the process continues, the gradual slope of the bottom extends farther and farther until most waves break far from shore and the rate of cliff retreat slows, resulting in a stable profile that may persist for long periods of time. Eventually another episode of uplift is likely to occur, and the process repeats.

Emergent coasts, such as the coast along much of California, often exhibit a series of terraces, each consisting of a former beach and wave cut cliff. This provides evidence of both the total uplift of the coast, and its incremental nature.

Softer rocks erode more easily, leaving resistant **rock** that forms points of land called headlands jutting out into the sea. Subsurface depth contours mimic that of the shoreline, resulting in wave refraction when the change in depth causes the waves to change the direction of their approach. This refraction concentrates wave energy on the headlands, and spreads it out across the areas in between. The "pocket beaches" separated by jagged headlands, which characterize much of the scenic coastline of Oregon and northern California, were formed in this way. Wave refraction explains the fact that waves on both sides of a headland may approach it from nearly opposite directions, producing some spectacular displays when they break.

If sea level rises, or the elevation of the coast falls, formerly exposed **topography** will be inundated. Valleys carved out by **rivers** will become estuaries like the Chesapeake Bay. Hilly terrains will become collections of islands, such as those off the coast of Maine.

The ability of rivers to transport sediment depends on their velocities. When they flow into a deep body of water they slow down and deposit their sediment in what will eventually become a **delta**. Thus, the flooding of estuaries causes deposition further inland. As the **estuary** fills in with sediment the depth of the water will decrease, and the velocity of the water flowing across the top of the delta will increase. This permits it to transport sediment further, and the delta builds out toward, and eventually into, the sea. The additional load of all the sediment may cause the **crust** of the earth to deform, submerging the coast further.

Wave action moves incredible amounts of sand. As waves approach shallow water, however, they slow down because of friction with the bottom, get steeper, and finally break. It is during this slowing and breaking that sand gets transported. When waves reach the shore they approach it almost straight on, so that the wave front is nearly parallel to the shore as it breaks. The wave front is not exactly parallel to the shore, however, and it is this difference which moves sand along the beach.

When a breaking wave washes up onto the beach at a slight angle it moves sand on the beach with it. This movement is mostly towards shore, but also slightly down the beach. When the water sloshes back, it goes directly down the slope, without any oblique component. As a result, sand moves in a

The beach is an area of constant change, with processes of erosion and deposition constantly warring with each other. *M. Woodbridge Williams. National Park Service.*

zigzag path with a net motion parallel to the beach. This is called "longshore drift." Although most easily observed and understood in the swash zone, the **area** of the beach which gets alternately wet and dry with each passing wave, **longshore drift** is active in any water shallow enough to slow waves down.

Many features of sandy coasts are the result of longshore drift. Spits build out from projecting land masses, sometimes becoming hooked at their end, as sand moves parallel to the shore. At Cape Cod, Massachusetts, glacial debris, deposited thousands of years ago, is still being eroded and redistributed by wave action.

An artificial jetty or "groin" can trap sand on one side of it, broadening the beach there. On the other side, however, wave action will transport sand away. Because of the jetty it will not be replenished, and erosion of the beach will result.

The magnitude and direction of transport of longshore drift depends on the strength and direction of approach of waves, and these may vary with the season. A beach with a very gentle slope, covered with fine sand every July may be a steep pebble beach in February.

Long, linear islands parallel to the shore are common along the Atlantic coast. Attractive sites for resorts and real estate developments, these **barrier islands** are in flux. A hurricane can drive storm waves over low spots, cutting islands in two. Conversely, migration of sand can extend a spit across the channel between two islands, merging them into one.

Interruptions in sand supply can result in erosion. This has happened off the coast of Maryland, where Assateague Island has gotten thinner and moved shoreward since jetties were installed at Ocean City, just to the north.

Often, the steady-state nature of the beach environment has not been properly respected. At higher elevations, where rates of erosion and deposition are so much slower, man can construct huge hills to support interstate highways, level other hills to make parking lots, etc., expecting the results of the work to persist for centuries, or at least decades. But in a beach environment, modifications are ephemeral. Maintaining a parking lot where winds would produce a dune requires

removal of tons of sand-every year. Even more significantly, because the flow of sediment is so great, modifications intended to have only a local, beneficial effect may influence erosion and deposition far down the beach, in ways that are not beneficial. Tossing a drain plug into a bucket of water raises the level of the water by just a tiny amount. Putting the same drain plug into the drain of a bathtub, into which water is flowing steadily, will change the level in the tub in very substantial ways. Similarly, it may be possible to protect the beach in front of a beach house by installing a concrete barrier, but this might result in eroding the supports to the highway that provide access to the beach house.

See also Continental shelf; Drainage basins and drainage patterns; Drainage calculations and engineering; Dunes; Ocean circulation and currents; Offshore bars; Wave motions

BEAUFORT, SIR FRANCIS (1774-1857)
Irish admiral

Sir Francis Beaufort, British admiral and hydrographer to the Royal Navy, was the first in 1805 to introduce and describe a scale of **wind** for estimating wind strengths without the use of instruments, a system based on subjective observations of the sea. Because expansions to land conditions were later added to the **Beaufort wind scale**, and quantitative wind speed values were also supplemented to each category in 1926, the scale is still widely used to describe the wind's speed and strength. As well as the Beaufort wind scale, the Beaufort Sea in the Arctic Ocean is named after Sir Francis Beaufort.

Sir Francis Beaufort was born in County Meath, Ireland. His father was well known in the areas of geography and **topography**: he published one of the earliest detailed maps of Ireland. Sir Francis Beaufort's nautical career began at age 13 as a cabin boy in the British Navy. Three years later, he became interested in the **weather**, and started to write down short comments about the general weather. He was only 22 years old when he was promoted to a lieutenant. In 1805, he was given his first command on the naval ship H.M.S. *Woolwich*, and he was assigned a hydrographic survey in **South America**.

During these years, Beaufort developed the first version of his wind force scale and weather notation coding, which he used for his meteorological journals. Because Beaufort's weather journal entries were written daily, eventually even as frequently as every two hours, he needed a simple yet effective system of abbreviations for the weather conditions. He created a notation consisting of the wind force number from his wind force scale, and a one, two, or three-character alphabetical code describing the state of the sky and weather, even describing cloud conditions and **precipitation** types. He continued writing these meteorological journals until the end of his life.

Beaufort's next assignments were for a hydrographic study of the Eastern Mediterranean, and a patrol mission. He did a major surveying and charting around the Turkish coast, but in 1812, he was wounded by sniper fire during a conflict with local pashas, and later that year the Admiralty ordered him home due to his injury. In 1817, he wrote his experiences about this expedition in a book titled *Karamania*. Although he remained in the British Navy until he was 81, he did not returned to active sea duty. In 1829, Beaufort became hydrographer to the Admiralty, where he promoted hydrographic studies for several British expeditions.

Between 1831 and 1836, on the voyage of the *Beagle,* Beaufort's scale of wind force was used officially for the first time. In 1833, after some slight modifications, the Admiralty prescribed Beaufort's weather notation for all log entries in the British Navy. In 1838, the Admiralty also officially adopted the Beaufort wind scale for all ships. Beaufort became a Rear Admiral in 1846, and he was bestowed the title Knight Commander of the Bath in 1848. After 68 years of service, Sir Francis Beaufort retired from the Admiralty in 1855, and he died two years later in 1857.

Originally, the Beaufort wind scale was meant for ships, specifying the amount of sail that a full-rigged ship should carry under the various wind conditions. It consisted of 13 different degrees of wind strength, ranging from calm to hurricane. In 1838, the use of the Beaufort wind scale became mandatory for all log entries on the ships of the British Admiralty. When steamboats replaced sail ships, certain modifications were necessary to make for the international use in meteorological descriptions. In 1874, the International Meteorological Committee revised the original scale, mainly for usage in international weather telegraphs. The original Beaufort scale numbers needed to be changed such that instead of the sails on a frigate, they referred to states of the sea or degrees of motion of trees. This change was still not satisfactory, since some ambiguities soon arose. The last modification came in 1946, when the International Meteorological Committee extended the scale to 17 values by adding five values to refine the hurricane-force winds, and defined the scale values by ranges of the wind speed as measured at a height of 10 meters above the surface for each category. This concluded the transformation of the Beaufort wind force scale into the Beaufort wind speed scale.

Sir Francis Beaufort was an accomplished hydrographer, making thorough surveys of uncharted coasts. Some of his charts are still used, even almost 200 years after he produced them. However, Beaufort's even more important achievements are the invention of the Beaufort wind scale, which today is still used worldwide, and the usage of the Beaufort weather notation code, which after several modifications, later became the basis for modern-day **meteorology** codes.

See also Cartography; Hydrogeology; Wave motions

BEAUFORT WIND SCALE

In 1805, to standardize nautical observations, **Sir Francis Beaufort**, an Irish hydrographer and member of the British Admiralty, created a scale for judging the strength of **wind** at sea. His scale is still a useful standard for the determination of wind force.

Each of the Beaufort Scale's 12 wind-force levels, ranging from calm to hurricane force, includes a description of the effect of the wind on readily observable, common objects. Beaufort's original purpose in devising the scale was to create a common reference for sailors to estimate and easily convey the effect of wind and sea upon their ships. Thus, the scale gives observers a means of estimating wind force.

The Beaufort Wind Scale numbers correspond to the following states (with estimated wind speeds in knots): Beaufort No.0 calm (less than 1 knot of wind); Beaufort No.1 light air (1-3 knot wind); Beaufort No.2 light breeze (4-6 knot wind); Beaufort No.3 gentle breeze (7-10 knot wind); Beaufort No.4 moderate breeze (11-16 knot wind); Beaufort No.5 fresh breeze (17-21 knot wind); Beaufort No.6 strong breeze (22-27 knot wind); Beaufort No.7 near gale (28-33 knot wind); Beaufort No.8 gale (34-40 knot wind); Beaufort No.9 strong gale (41-47 knot wind); Beaufort No.10 storm breeze (48-55 knot wind); Beaufort No.11 violent storm breeze (56-63 knot wind); Beaufort No.12 hurricane (64 or greater knot wind).

Originally limited to a description of the effects of wind on a sailing vessel's canvas (force 12, for instance, was "that which no canvas could withstand"), the scale was revised in 1939 by the International Meteorological Committee to include the effect of wind on land features. The numbers from the Beaufort scale were used on **weather** maps until 1955, when a system of wind feathers, which show wind direction and intensity, was adopted.

See also Air masses and fronts; Weather forcasting; Weather forecasting methods

BECQUEREL, ANTOINE-HENRI (1852-1908)

French physicist

Antoine-Henri Becquerel's landmark research on x rays and his discovery of radiation laid the foundation for many scientific advances of the early twentieth century. X rays were discovered in 1895 by the German physicist Wilhelm Conrad Röntgen, and in one of the most serendipitous events in science history, Becquerel discovered that the uranium he was studying gave off radiation similar to x rays. Becquerel's student, **Marie Curie**, later named this phenomenon **radioactivity**. His later research on radioactive materials found that at least some of the radiation produced by unstable materials consisted of electrons. For these discoveries Becquerel shared the 1903 Nobel Prize in physics with Marie and **Pierre Curie**. Becquerel's other notable research included the effects of **magnetism** on light and the properties of luminescence.

Becquerel was born in Paris on December 15, 1852. His grandfather, Antoine-César Becquerel, had fought at the Battle of Waterloo in 1815 and later earned a considerable reputation as a physicist. He made important contributions to the study of electrochemistry, **meteorology**, and agriculture. Antoine-Henri's father, Alexandre-Edmond Becquerel, was also scientist, and his research included studies on photography, heat, the conductivity of hot gases, and luminescence.

During his years at the Ecole des Ponts et Chaussées, Becquerel became particularly interested in English physicist Michael Faraday's research on the effects of **magnetism** on light. Faraday had discovered in 1845 that a plane-polarized beam of light (one that contains light waves that vibrate to a specific pattern) experiences a **rotation** of planes when it passes through a **magnetic field**; this phenomenon was called the Faraday effect. Becquerel developed a formula to explain the relationship between this rotation and the refraction the beam of light undergoes when it passes through a substance. He published this result in his first scientific paper in 1875, though he later discovered that his initial results were incorrect in some respects.

Although the Faraday effect had been observed in solids and liquids, Becquerel attempted to replicate the Faraday effect in gases. He found that gases (except for **oxygen**) also have the ability to rotate a beam of polarized light. Becquerel remained interested in problems of magneto-optics for years, and he returned to the field with renewed enthusiasm in 1897 after Dutch physicist Pieter Zeeman's discovery of the Zeeman effect—whereby spectral lines exposed to strong magnetic fields split—provided new impetus for research.

In 1874 Becquerel married Lucie-Zoé-Marie Jamin, daughter of J.-C. Jamin, a professor of physics at the University of Paris. She died four years later in March 1878, shortly after the birth of their only child, Jean. Jean later became a physicist himself, inheriting the chair of physics held by his father, grandfather, and great-grandfather before him. Two months prior to Lucie's death, Becquerel's grandfather died. At that point, his son and grandson each moved up one step, Alexandre-Edmond to professor of physics at the Musée d'Histoire Naturelle, and Antoine-Henri to his assistant. From that point on, Becquerel's professional life was associated with the Musée, the Polytechnique, and the Ponts et Chaussées.

In the period between receiving his engineering degree and discovering radioactivity, Becquerel pursued a variety of research interests. In following up his work on Faraday's magneto-optics, for example, he became interested in the effect of the earth's magnetic field on the atmosphere. His research determined how the earth's magnetic field affected **carbon** disulfide. He proposed to the International Congress on Electric Units that his results be used as the standard of electrical current strength. Becquerel also studied the magnetic properties of a number of materials and published detailed information on nickel, cobalt, and **ozone** in 1879.

In the early 1880s Becquerel began research on a topic his father had been working on for many years—luminescence, or the emission of light from unheated substances. In particular he made a detailed study of the spectra produced by luminescent materials and examined the way in which light is absorbed by various **crystals**. Becquerel was especially interested in the effect that polarization had on luminescence. For this work Becquerel was awarded his doctoral degree by the University of Paris in 1888, and he was once again seen as an

active researcher after years of increasing administrative responsibility.

When his father died in 1891 Becquerel was appointed to succeed him as professor of physics at the museum and at the conservatory. The same year he was asked to replace the ailing Alfred Potier at the Ecole Polytechnique. Finally, in 1894, he was appointed chief engineer at the Ecole des Ponts et Chaussées. Becquerel married his second wife, Louise-Désirée Lorieux, the daughter of a mine inspector, in 1890; the couple had no children.

The period of quiescence in Becquerel's research career ended in 1895 with the announcement of Röntgen's discovery of x rays. The aspect of the discovery that caught Becquerel's attention was that x rays appeared to be associated with a luminescent spot on the side of the cathode-ray tube used in Röntgen's experiment. Given his own background and interest in luminescence, Becquerel wondered whether the production of x rays might always be associated with luminescence.

To test this hypothesis Becquerel wrapped photographic plates in thick layers of black paper and placed a known luminescent material, potassium uranyl sulfate, on top of them. When this assemblage was then placed in sunlight, Becquerel found that the photographic plates were exposed. He concluded that sunlight had caused the uranium salt to luminesce, thereby giving off x rays. The x rays then penetrated the black paper and exposed the photographic plate. He announced these results at a meeting of the Academy of Sciences on February 24, 1896.

Through an unusual set of circumstances the following week, Becquerel discovered radioactivity. As usual, he began work on February 26 by wrapping his photographic plates in black paper and taping a piece of potassium uranyl sulfate to the packet. However, because it wasn't sunny enough to conduct his experiment, Becquerel set his materials aside in a dark drawer. He repeated the procedure the next day as well, and again a lack of sunshine prompted him to store his materials in the same drawer. On March 1 Becquerel decided to develop the photographic plates he had prepared and set aside. It isn't clear why he did this—for, according to his hypothesis, little or no exposure would be expected. Lack of sunlight had meant that no luminescence could have occurred; hence, no x rays could have been emitted.

Surprisingly, Becquerel found that the plates had been exposed as completely as if they had been set in the **sun**. Some form of radiation—but clearly not x rays—had been emitted from the uranium salt and exposed the plates. A day later, according to Oliver Lodge in the *Journal of the Chemical Society,* Becquerel reported his findings to the academy, pointing out: "It thus appears that the phenomenon cannot be attributed to luminous radiation emitted by reason of phosphorescence, since, at the end of one-hundredth of a second, phosphorescence becomes so feeble as to become imperceptible."

With the discovery of this new radiation Becquerel's research gained a new focus. His advances prompted his graduate student, Marie Curie, to undertake an intensive study of radiation for her own doctoral thesis. Curie later suggested the name radioactivity for Becquerel's discovery, a phenomenon that had until that time been referred to as Becquerel's rays.

Becquerel's own research continued to produce useful results. In May 1896, for example, he found uranium metal to be many times more radioactive than the compounds of uranium he had been using, and he began to use it as a source of radioactivity. In 1900 he also found that at least part of the radiation emitted by uranium consists of electrons, particles that were discovered only three years earlier by Joseph John Thomson. For his part in the discovery of radioactivity Becquerel shared the 1903 Nobel Prize in physics with Curie and her husband Pierre.

Honors continued to come to Becquerel in the last decade of his life. On December 31, 1906, he was elected vice president of the French Academy of Sciences, and two years later he became president of the organization. On June 19, 1908, he was elected one of the two permanent secretaries of the academy, a post he held for less than two months before his death on August 25, 1908, at Le Croisic, in Brittany. Among his other honors and awards were the Rumford Medal of the Royal Society in 1900, the Helmholtz Medal of the Royal Academy of Sciences of Berlin in 1901, and the Barnard Medal of the U.S. National Academy of Sciences in 1905.

See also Geochemistry

BED OR TRACTION LOAD

Bed load, sometimes referred to as traction load, is the material that is transported by sliding, rolling, and saltating (skipping) along the bed of a stream. Particles comprising bed load can range in size from **sand** to boulders. The movement of bed load is responsible for **bedforms** that change in time and space along a stream bed.

Particles along a stream bed begin to move when the shear stress exerted by the flowing **water** exceeds a critical value. The critical shear stress depends on a combination of the particle diameter, the slope of the stream channel, the difference between the density of individual particles and that of water (particle buoyancy), and the degree to which the particles are packed together. As a result, particles of different mineralogical composition and size will have different critical shear stresses. Heavy **minerals** such as gold can be concentrated in stream beds because gold nuggets or flakes are left behind while lighter particles move around them. Likewise, small particles may move while large particles of the same mineral or **rock** type are left in place. Water density is proportional to the **suspended load** being carried. Muddy water high in suspended sediment will therefore increase the particle buoyancy and thereby reduce the critical shear stress required to move particles of a given size and composition.

The shear stress exerted by the flowing water, which is proportional to both water depth and stream channel slope, also controls the movement of bed load. Large or heavy particles that have high critical shear stress values may move as bed load when the water is unusually deep during infrequent **floods** and remain stationary between those times.

Once a particle begins to move, the current above the bed may be strong enough to lift it off the bed and into the flowing water. When the entire weight of a particle is borne by water instead of other particles beneath it, that particle ceases to be part of the bed load and becomes part of the suspended load. Conversely, if the current slows a particle may fall out of suspension and become part of the bed load. The distinction between bed load and suspended load in a stream can therefore change continuously through time.

See also Erosion; Rivers; Saltation; Sedimentation; Stream valleys, channels, and floodplains

BEDDING

The term bedding (also called stratification) ordinarily describes the layering that occurs in **sedimentary rocks** and sometimes the layering found in **metamorphic rock**. Bedding may occur when one distinctly different layer of sediment is deposited on an older layer, such as **sand** and pebbles deposited on silt or when a layer of exposed sedimentary **rock** has a new layer of sediments deposited on it. Such depositions of sediments produce a clear division between beds called the bedding plane.

The variation among different sedimentary rock layers (usually referred to as beds or strata) may range from subtle to very distinct depending upon color, composition, cementation, texture, or other factors. One of the best examples may be seen in Arizona's Grand **Canyon** where red, green, white, gray, and other colors heighten the contrast between beds.

The bedding found in metamorphic rock that formed from sedimentary rock is evidence of extreme heat and pressure and is often quite distorted. Distortions may change the sedimentary bedding by compressing, inclining, folding, or other changes.

One of the most common types of bedding is called graded bedding. These beds display a gradual grading from the bottom to the top of the bed with the coarsest sediments at the bottom and the finest at the top. Graded bedding often occurs when a swiftly moving river gradually slows, dropping its heaviest and largest sediments first and lightest last. Changes in a river's speed may be caused by a number of factors, including storm **runoff** or the entry of a river into a lake or an ocean.

Bedding is usually found in horizontal layers called parallel bedding. But bedding may be inclined or have a swirly appearance. Inclined bedding may occur when sediments are deposited on a slope, such as a sand dune, or when beds are tilted from their original horizontality by forces within the earth. Bedding with a swirly appearance, called cross bedding, may indicate that the sediments making up the rock were deposited by strong **desert** winds or turbulence in a river.

The origin, composition, and interpretation of variations in bedding are one of the geologist's most important tools in studying Earth's history. It is for this reason that **stratigraphy**, the study which includes the interpretation of sedimentary and metamorphic beds, was an essential part of even the earliest days of geologic research.

See also Superposition

BEDFORMS (RIPPLES AND DUNES)

Centimeter to meter-scale layering, or **bedding**, is a defining characteristic of **sedimentary rocks**. Non-horizontal depositional beds are called bedforms, and geologists refer to the lithified remains of bedforms in sedimentary rocks as cross bedding. Patterns of stratification, including cross bedding, allow sedimentary geologists called stratigraphers to deduce the dynamic processes that occurred in ancient sedimentary environments.

Atmospheric and aqueous currents create bedforms. When **wind** or **water** carries loose grains across a horizontal bed of unconsolidated sediment, regular geometric patterns develop on the surface of the bed. These structures vary in size from very small ripples, to medium-sized waves and megaripples, to large and very large **dunes**. The velocity, direction, constancy, and homogeneity of the current, or flow field, determines the size and shape of the resulting bedform field. For example, a one-directional flow field like a river current tends to create asymmetrical bedforms, whereas bidirectional currents like **tides** or waves deposit symmetrical ripples and dunes.

Bedforms migrate over time. Sometimes the flow field is strong enough, or the bedforms are small enough, that migration occurs by erasure of the entire bedform field, and formation of a new pattern of ripples or dunes. Ripples typically migrate by this type of wholesale reshaping. Ripples preserved in sedimentary rocks suggest that the ripples froze in their final configuration, either because the flow field waned to a point where it could no longer transport sediment, or perhaps that rapid deposition of an overlying bed buried them. Larger bedforms, including aqueous and wind-formed, or Eolian, dunes, migrate by a process in which the water or wind picks up grains from the upstream face of the bedform and deposits them on deposits them on the downstream face. This process creates a composite bedform with inclined internal beds. When dunes are lithified, this internal stratification is preserved as cross bedding.

See also Eolian processes; Sedimentation; Stratigraphy

BEDROCK

Bedrock (also termed Bed **rock**) is a layer of undisturbed rock usually located beneath a surface layer of **soil** or other material. In areas of high **erosion**, bedrock may become exposed to the surface. Bedrock can be of igneous, sedimentary, or metamorphic origin and forms the upper surface of the rocky foundation that composes the earth's **crust**.

A surface exposure of bedrock is called an outcrop. Bedrock is only rarely exposed, or crops out, where sediment accumulates rapidly, for example, in the bottom of **stream**

valleys and at the base of hills or mountains. Outcrops are common where erosion is rapid, for example, along the sides of steep stream channels and on steep hill or mountain slopes. Deserts and mountain tops above the treeline also host good bedrock exposures due to the scarcity of vegetation, and resulting rapid erosion. Man-made outcrops are common where roadways cut through mountains or hilltops, in quarries, and in mines

Generally, the more rock resists erosion, the more likely it is to crop out. **Granite** and **sandstone** commonly form well-exposed outcrops. Natural exposures of shale and claystone, both soft, fine-grained rocks, are rare—especially in humid climates.

In addition to the occasional mineral crystal or **fossils**, all outcrops contain through-going fractures called joints. These form during the application of stresses to bedrock on a regional scale, for example, during mountain building. Even greater stresses may cause faulting movement of the rock on the sides of a fracture. An example is the large-scale bedrock movement that occurs along the San Andreas Fault in California. When stresses cause plastic rather than brittle deformation of bedrock, it **folds** rather than faulting.

Bedrock is distributed in a predictable pattern. Generally in the central **area** of a continent, geologists find very ancient (one billion years or more) **mountain chains**, consisting of igneous and **metamorphic rock**, eroded to an almost flat surface. This area, called a continental shield, typically contains the oldest continental bedrock. Shields have experienced multiple episodes of deformation so they are intensely folded and faulted. These ancient igneous and metamorphic rocks, called basement rocks, compose much of the continental crust. However, on the shield margins, thick sequences of relatively undeformed, **sedimentary rocks** cover the basement rocks. These deposits, called the continental platform, commonly exceed 1 mi (1.6 km) in thickness and 100 million years in age.

Together, the shield and platform make up the bedrock area known as the continental **craton**. The craton is considered more or less stable, that is, it is not currently experiencing significant deformation. On the margins of the craton, there may be areas of geologically active bedrock, called orogens, from the Greek word for mountain. Orogens are relatively young mountain belts where uplift, folding, faulting, or volcanism occurs. The bedrock here varies in age from **lava** flows that may be only days old to igneous, sedimentary, and metamorphic rock that are hundreds of millions of years old. All bedrock belongs to the continental shield, platform, or the orogens.

See also Earth, interior structure; Faults and fractures; Pluton and plutonic bodies; Soil and soil horizons; Weathering and weathering series

BEEBE, CHARLES WILLIAM (1887-1962)
American explorer

Charles William Beebe (1877–1962), explorer, writer, ornithologist, and deep-sea pioneer, was born in Brooklyn, New York and grew up in East Orange, New Jersey. He is remembered today primarily for his record-breaking 1934 descent off the coast of Bermuda with American engineer Otis Barton. Barton and Beebe dove in a diving machine of their own invention, the bathysphere, to a depth of 3,028 feet (923 m).

Beebe's parents were fascinated by natural history, and so in childhood he was a frequent visitor to the American Museum of Natural History in New York City. As a teenager, Beebe taught himself taxidermy and became friends with the president of the museum, Henry Osborn. Osborn helped him gain admittance to Columbia University in 1896. In 1899, Beebe left college (without receiving a degree) to work as an assistant curator of ornithology (the study of birds) at the zoo then being opened by the New York Zoological Society. He was soon promoted to full curator.

In 1902, Beebe married Mary Rice, whom he was to divorce in 1913. The Beebes made ornithological expeditions to Mexico, Trinidad, and Venezuela and published popular accounts of their experiences. In 1909–1911 they traveled to the Far East on a 17-month expedition sponsored by the New York Zoological Society having the sole purpose of studying pheasants. After years of further labor Beebe published the results of this expedition in a magisterial four-volume work entitled *A Monograph of the Pheasants,* (1918), still in print. While preparing his monograph Beebe also made expeditions to **Asia**, Central and **South America**, the Galapagos Islands, and other regions. In 1916 he established a research station on the coast of British Guiana (today Guyana) on behalf of the New York Zoological Society, and in 1919 was made director of the Society's Department of Tropical Research.

In the mid 1920s Beebe's main interest turned from birds to deep-sea life, which he studied by trawling for specimens and by diving in pressure suits. However, the suits were limited in depth range and the creatures brought to the surface by Beebe's nets were invariably dead. Wishing to observe undamaged specimens alive in their natural habitat, Beebe publicized his need for a practical deep-sea vessel design.

In 1928, Beebe was approached by Otis Barton with his design for the bathysphere (derived from the Greek word for deep, *báthys*), a steel ball filled with breathable air that would be lowered on a cable from a barge. The bathysphere was equipped with two **quartz** portholes 8 inches (0.2 m) wide and with an umbilical hose providing telephone and power. **Oxygen** was supplied from on-board tanks and **carbon dioxide** was removed from the air by trays of soda lime. The bathysphere was a tiny craft—only four feet, nine inches (1.5 m) across (outside diameter), with walls several inches thick. Its interior would have been a tight squeeze for a single person, but Barton and Beebe occupied it along with the oxygen tanks, soda lime trays, and other gear.

Barton and Beebe made a number of bathysphere descents starting in 1930. The pre-bathysphere dive record was 525 feet (160 m); on August 15, 1934, Barton and Beebe dove to 3,028 feet (923 m)—over half a mile. Beebe described the descent in a book published later that year, *Half Mile Down*. Barton and Beebe's bathyspheric dives were the first diving expeditions to penetrate to depths beyond the effective reach of sunlight; below 2,000 feet (610 m), they observed, the

ocean was lightless even with a brilliant tropical **sun** shining on calm **seas** above. The dives were widely popularized by the *National Geographic* magazine, in Beebe's own colorful writings, and for one dive in 1932 by live radio broadcast in the United States and United Kingdom. Even before their record dive in 1934, Barton and Beebe were international celebrities.

Despite its successes, the bathysphere was inherently dangerous. Surface waves could easily subject the suspension cable to breaking strain. Later generations of deep-sea vessels have therefore been built as self-propelled submarines.

Barton and Beebe's 1934 diving record remained unbroken until 1949, when Barton descended to 4,500 feet (1,370 m) in another vessel of his own design, the Benthoscope. Beebe retired from the directorship of the New York Zoological Society's Department of Tropical Research in 1952 and died of natural causes in Bermuda in 1962.

The original bathysphere resides in the New York Aquarium in New York City.

See also Deep sea exploration; History of exploration III (Modern era); Oceanography

BENETT, ETHELDRED (1776-1845)
English geologist

Etheldred Benett, arguably the first female geologist, was born in England in 1776, the same year the American Revolutionary War began. Benett lived in Wiltshire county, southern England, and contributed to the founding of biostratigraphy.

Benett's understanding of the context of **fossils** put her in touch with many of the famous geologists of the day. She corresponded with and met many, from Professor William Buckland at Oxford and the famous Sussex paleontologist, Gideon Mantell to **Charles Lyell**, founder of the principle of **uniformitarianism**, and **William Smith**, the father of **stratigraphy** and producer of the first map of Britain in 1815.

Benett's contributions to **geology** lie in four areas. First, she commissioned the first recorded measured section at the Upper Chicksgrove Quarry, Tisbury in Wiltshire. This was donated to the Geological Society of London Library and signed by her in 1815. Second, she was a recognized expert regarding fossil mollusks and sponges of Wiltshire, as attested to by her contributions to Sowerby's publication. Third, the Czar of Russia gave her a medal for her contribution to his fossil collection because he thought she was a man. She also received a Diploma of appointment as a member of the Imperial Natural History Society of Moscow to which she makes the comment in a letter, "In this diploma I am called Dominum Etheldredum Benett and Mr Lyell told me that he had been written to by foreigners to know if Miss Benett was not a gentleman." The Latinized suffix "um" in her name implies that the sender thought she was male. Finally, she pushed forward the boundaries of biostratigraphy. Etheldred Benett published a classic volume in 1831, *Organic Remains of the County of Wiltshire* with extensive drawings, which she herself produced. She also contributed generously and extensively to Sowerby's *Mineral Conchology* (published in 1816). She gave the second highest number of specimens to this volume of any contributor.

Etheldred Benett never married, instead devoting her life to her fossil collection until she died at the age of 69. Her extensive collection of thousands of labeled Jurassic and Cretaceous fossils was thought to be so valuable a resource that when she died, most of her collection was bought by former Englishman and physician Thomas Wilson of Newark, Delaware; it then was taken to America. The collection was subsequently donated to the Philadelphia Academy of Natural Science between 1848 and 1852. The collection contains some of the first fossil bivalves to have their soft parts preserved.

Etheldred Benett was, therefore, at the forefront of paleontology and biostratigraphy at a time when many people still assumed that fossils were deposited from catastrophic acts of religious significance (such as Noah's flood), and that scientific investigation should be left solely to men.

See also Fossil record; Fossils and fossilization; Historical geology

BENIOFF ZONE

Benioff zones are dipping, roughly planar zones of increased **earthquake** activity produced by the interaction of a downgoing oceanic crustal plate with an overriding continental or oceanic plate. They occur at boundaries of crustal plates called subduction zones. The earthquakes can be produced by slip along the subduction thrust fault or by slip on faults within the downgoing plate, as a result of bending and extension as the plate is pulled into the mantle. The zones have dips typically ranging from 40 to 60 degrees. The zones are also known as the Wadati-Benioff zone.

During the past century, improvements in seismic acquisition and processing led to the observation that the world's earthquakes are not randomly distributed over the earth's surface. Rather, they tend to be concentrated in narrow zones along the boundaries of continental and oceanic crustal plates. According to the plate tectonic theory, the **crust** of the earth is broken into a mosaic of seven major rigid plates floating over a much less rigid mantle. The plates are not static but are in constant motion. Most of the tectonic activity, such as the formation of mountain belts, earthquakes, and volcanoes, occurs at the plate boundaries. There are four different types of these seismic zones corresponding to the four main types of plate boundary interactions: subduction zones as along the western coast of **South America**; strike-slip (transform) zones like the San Andreas system along the west coast of **North America**; zones of seismic activity along midocean ridges like the mid-Atlantic Ridge system; and continental-continental collision zones such as the Himalyan where the Indian subcontinent is ploughing into **Asia**.

Benioff zones are found in subduction zones that form by the collision of two crustal plates of dissimilar density and thickness, for example an oceanic and continental plate. The heavier (thinner) crust of the oceanic plate is thrust or sub-ducted under the lighter and much thicker crust of the continental plate. A deep ocean trench is produced where these two plates meet. Along the Peru-Chile trench, the Pacific plate is being subducted under the South American plate, which responds by crumpling to form the Andes. The earthquake zones that parallel the great oceanic trenches are typically inclined from 40 to 60 degrees from the horizontal and extend several hundred kilometers into the mantle along trends that reach thousands of kilometers in length. These zones are sometimes called Wadati-Benioff zones after two of the seismologists who first recognized them, Kiyoo Wadati of Japan and Hugo Benioff of the United States.

Benioff zones are the seismic expression of the deformation produced by the subduction of one plate under another. The subduction or "destruction" of the oceanic crust compensates for the creation of new ocean crust at the ocean ridges. The compensating result of both processes explains why the earth may not have significantly increased in size since its formation 4.6 billion years ago.

See also Continental drift theory; Plate tectonics; Sea-floor spreading

BENTHIC FORAMINIFERA · *see* CALCAREOUS OOZE

BERNER, ROBERT A. (1935-)
American geochemist

Robert A. Berner's research in sedimentary **geochemistry** led to the application of mathematical models to describe the physical, chemical, and biological changes that occur in ocean sediment. Berner, a professor of **geology** and geophysics at Yale University, also developed a theoretical approach to explain larger geochemical cycles, which led to the creation of a model for assessing atmospheric **carbon dioxide** levels and the **greenhouse effect** over geological time. A prolific researcher, Berner has written many scientific journal articles and is one of the most frequently quoted earth scientists in the *Science Citation Index*.

Robert Arbuckle Berner was born in Erie, Pennsylvania, on November 25, 1935, to Paul Nau Berner and Priscilla (Arbuckle) Berner. As a young man, Berner decided to become a scientist because of his propensity for logical thinking. "Science forces you to seek the truth and see both sides of an argument," he told Patricia McAdams. Berner began his academic studies at the University of Michigan where he earned his B.S. in 1957 and his M.S. a year later. He then went to Harvard University and earned his Ph.D. in geology in 1962. He married fellow geology graduate student Elizabeth Marshall Kay in 1959; the couple have three children.

Berner began his professional career at the Scripps Institute of **Oceanography** in San Diego, where he won a fellowship in oceanography after graduating from Harvard. In 1963, he was appointed assistant professor at the University of

Chicago, and two years later he became an associate professor of geology and geophysics at Yale University. Since 1968, Berner has also served as associate editor or editor of the *American Journal of Science*. He was promoted to full professor at Yale in 1971, and in 1987 he became the Alan M. Bateman Professor of geology and geophysics.

Principles of Chemical Sedimentology, which Berner published in 1971, reflects the interest that has fueled much of his research. Berner sees the application of chemical thermodynamics and kinetics as a valuable tool in unveiling the secrets of sediments and **sedimentary rocks**. Thus, Berner's is an unconventional approach to sedimentology (the chemical study of sediments rather than the study of chemical sediments). Berner identifies his goal in *Principles of Chemical Sedimentology* as illustrating "how the basic principles of physical **chemistry** can be applied to the solution of sedimentological problems." Berner's *Early Diagenesis,* published in 1980, is a study of the processes over geological time whereby sedimentary materials are converted into **rock** through chemical reactions or compaction. Because of the frequency with which *Early Diagenesis* has been quoted, it was declared a Science Citation Classic by the Institute for Scientific Information.

Berner observes in *Scientific American* that "the familiar biological **carbon** cycle—in which atmospheric carbon is taken up by plants, transformed through photosynthesis into organic material and then recovered form this material by respiration and bacterial decomposition—is only one component of a much larger cycle: the geochemical carbon cycle." Berner has studied an aspect of this geochemical carbon cycle that is analogous to the transfer of carbon between plants, animals, and their habitats—the "transfer of carbon between sedimentary rocks at or near the earth's surface and the atmosphere, **biosphere** and oceans." Carbon dioxide is vital to both these aspects of the geochemical carbon cycle, as carbon is primarily stored as carbon dioxide in the atmosphere. Berner's research has contributed to the "BLAG" model (named after Berner and his associates Antonio L. Lasaga and Robert M. Garrels) for assessing the changes in atmospheric levels of carbon dioxide throughout the earth's geological eras. First published in 1983 and subsequently refined, the BLAG model quantifies factors such as degassing (whereby carbon dioxide is released from beneath the earth), carbonate and silicate rock **weathering**, carbonate formation in **oceans**, and the rate at which organic matter is deposited on and buried in the earth that enable scientists to assess the climactic conditions of the planet's previous geological eras.

Berner's research on atmospheric carbon dioxide levels includes the study of the greenhouse effect, whereby carbon dioxide and other gases trap excessive levels of radiated heat within Earth's atmosphere, leading to a gradual increase in global temperatures. Since the nineteenth-century industrial revolution, this phenomenon has increased primarily because of the burning of fossil **fuels** such as **coal**, oil, and **natural gas**; also because of deforestation. Berner reports in *Scientific American* that "slow natural fluctuations of atmospheric carbon dioxide over time scales of millions of years may rival or even exceed the much faster changes that are predicted to arise

from human activities." Thus, the study of the carbon cycle is essential to an objective evaluation of the greenhouse effect within larger geological processes. In 1986, Berner published the textbook *The Global Water Cycle: Geochemistry and Environment* which he co-authored with his wife Elizabeth, who is also a geochemist. *The Global Water Cycle* reviews the properties of **water**, marine environments, and water/energy cycles, and includes a discussion of the greenhouse effect. Berner's research has since focused on Iceland where he is investigating how volcanic rock is broken down by weathering and by the plant-life that gradually takes root on it.

Berner enjoys traveling that is associated with his research and likes to help students learn to think creatively for themselves. "I'm very proud of the...graduate students that have received Ph.D.s working with me. I've learned as much from them as they have from me," he told McAdams. Berner served as president of the Geochemical Society in 1983, and he is a member of the National Academy of Sciences, the American Academy of Arts and Sciences, the Geological Society of America, and the Mineral Society. He has chaired the Geochemical Cycles Panel for the National Research Council and served on the National Committee on Geochemistry, the National Science Foundation Advisory Committees on Earth Sciences and Ocean Sciences, and the National Research Council Committee on Oceanic Carbon. He has received numerous awards, including an honorary doctorate from the Université Aix-Marseille III in France in 1991 and Canada's Huntsman Medal in Oceanography in 1993. His hobbies include Latin American music, tennis, and swimming.

See also Greenhouse gases and greenhouse effect; Weathering and weathering series

BERNOULLI, DANIEL (1700-1782)
Dutch-born Swiss physicist

Daniel Bernoulli's work on fluids pioneered the sciences of hydrodynamics and **aerodynamics**. Born in the Netherlands and spending most of his life in Switzerland, Bernoulli was one of a large family of scientists and mathematicians that included his father, Jean Bernoulli, and uncle, Jacques Bernoulli.

Ignoring his family's pleas to enter the world of business, Bernoulli pursued a degree in medicine and then, after graduation, a career as a professor of mathematics. He began teaching in 1725 at a college in St. Petersburg, Russia, eventually returning to Switzerland in 1732. While a professor at the University of Basel, he became the first scientist outside of Great Britain to fully accept Newtonian **physics**. It was also here that Bernoulli performed the research on fluid behavior that would make him famous.

The 1738 publication *Hydrodynamica* developed the prominent theories of hydrodynamics, or the movement of **water**. Paramount among these was the fact that, as the velocity of a fluid increases, the pressure surrounding it will decrease. Called **Bernoulli's principle**, this pressure drop was also shown to occur in moving air, and it is the reason boats and planes experience lift as water or air passes around them. This effect is easily shown by blowing between two pieces of paper; the drop in pressure will cause the papers to bend toward each other. Bernoulli's research marked the first attempt to explain the connection pressure and **temperature** have with the behavior of gas and fluids.

Bernoulli's experiments with fluids caused him to devise a series of hypotheses about the nature of gases. He was certainly one of the first to formulate principles dealing with gases as groups of particles, which later became the basis for atomic theories. As groundbreaking as this work was, it was paid little attention by his peers, and subsequently it was nearly a century before the **atomic theory** rose again.

See also Atmospheric pressure; Atomic theory; Hydrostatic pressure

BERNOULLI'S PRINCIPLE

Bernoulli's principle describes the relationship between the pressure and the velocity of a moving fluid (i.e., air or **water**). Bernoulli's principle states that as the velocity of fluid flow increases, the pressure exerted by that fluid decreases.

During the late eighteenth century, **Daniel Bernoulli** pioneered the basic tenet of kinetic theory, that molecules are in motion. He also knew that flowing fluids exerted less pressure, but he did not connect these ideas logically. In *Hydrodynamica*, Bernoulli's logic that flow reduced pressure was obscure, and his formula was awkward. Bernoulli's father Johann, amid controversy, improved his son's insight and presentation in *Hydraulica*. This research was centered in St. Petersburg where Leonard Euler, a colleague of Bernoulli and a student of Johann, generalized a rate-of-change dependence of pressure and density on speed of flow. Bernoulli's principle for liquids was then formulated in modern form for the first time.

In this same group of scientists was D'Alembert, who found paradoxically that fluids stopped ahead of obstacles, so frictionless flow did not push.

Progress then seems to have halted for about a century and a half until Ludwig Prandtl or one of his students solved Euler's equation for smooth streams of air in order to have a mathematical model of flowing air for designing wings. Here, speed lowers pressure more than it lowers density because expanding air cools, and the ratio of density times degrees-kelvin divided by pressure is constant for an ideal-gas.

More turbulent flow, as in atmospheric winds, requires an alternative solution of Euler's equation because mixing keeps air-temperature fixed.

Bernoulli's principle is regarded by many as a paradox because currents and winds upset things, but standing a stick in a stream of water helps to clarify the enigma. One can observe calm, smooth, level water ahead of the stick and a cavity of reduced pressure behind it. Calm water pushes the stick, as lower pressure downstream fails to balance the upsetting force.

Bernoulli's principle never acts alone; it also comes with molecular entrainment. Molecules in the lower pressure of faster flow aspirate and whisk away molecules from the higher pressure of slower flow. Solid obstacles such as airfoils carry a thin stagnant layer of air with them. A swift low-pressure airstream takes some molecules from this boundary layer and reduces molecular impacts on that surface of the wing across which the airstream moves faster.

See also Atmospheric chemistry; Hydrostatic pressure

BESSEL, FRIEDRICH (1784-1846)
German astronomer

Friedrich Bessel was a self-taught astronomer. Born in Minden, Germany in 1784, he became an accountant in Bremen, but his true interests were **astronomy** and mathematics. In fact, in 1806, at the age of 20, he recalculated the orbit of Halley's comet, which was due to reappear in 1835. This so impressed astronomer Heinrich Olbers (1758–1840) that Olbers helped Bessel obtain a post at the observatory.

Bessel worked laboriously. He produced a new star catalogue of over 50,000 stars and introduced improvements to astronomical calculations, developing a method of mathematical analysis along the way that can be applied to many problems not related to astronomy. He oversaw construction of the first large German observatory and served as its director from 1813 until his death in 1846.

Bessel's greatest achievement was in determining the parallax of a star. As the earth orbits the **sun**, our position relative to any star shifts by a maximum of 186 million miles (299,274,000 km, the diameter of the earth's orbit). Thus, the apparent position of any star in the sky will change slightly through the year. The amount of observed shift is the parallax. Knowing an object's parallax, it is possible to calculate the distance to it.

In 1838, Bessel announced he had obtained the parallax for a star called 61 Cygni. He had chosen this star because it had shown the largest proper motion of any known star. The large degree of movement, he assumed, was because the star was relatively close, and the closer an object, the greater its parallax would be.

Bessel's calculations showed that 61 Cygni was about 10 light-years from Earth. (This was the introduction of the term light-year.) Although that is actually very close for a star, the distance was mind boggling in 1838. The earlier astronomer **Johannes Kepler** had believed the stars were 0.1 light-year away, and Isaac Newton had risked enlarging that to two light-years.

Discovering the parallax of a star also put another nail in the coffin of the "Earth-centered" concept of the universe. Since parallax could be obtained from a *moving* Earth, Nicholas Copernicus's assertion that the Earth orbited around the sun was further strengthened.

In 1841, Bessel drew a remarkable conclusion about the stars Sirius and Procyon. He had noticed displacements in their motion that could not be attributed to parallax. A parallax shift shows smooth motion; these two stars seemed to be wobbling. He concluded that there had to be invisible companions in orbit around each star. The gravitational tug of the companion would account for the observed wobble. This theory later turned out to be correct.

Bessel was responsible for encouraging astronomers to direct their attention to the stars beyond the **solar system**.

BIG BANG THEORY

Big bang theory describes the origin of the knowable universe and the development of the laws of **physics** and **chemistry** some 15 billion years ago.

During the 1940s Russian-born American cosmologist and nuclear physicist George Gamow (1904–1968) developed the modern version of the big bang model based upon earlier concepts advanced by Russian physicist Alexander (Aleksandr Aleksandrovich) Friedmann (also spelled as Fridman, 1888–1925) and Belgian astrophysicist and cosmologist Abbé Georges Lemaître (1894–1966). Big bang based models replaced static models of the universe that described a homogeneous universe that was the same in all directions (when averaged over a large span of **space**) and at all times. Big bang and static cosmological models competed with each other for scientific and philosophical favor. Although many astrophysicists rejected the steady state model because it would violate the law of mass-energy conservation, the model had many eloquent and capable defenders. Moreover, the steady state model was interpreted by many to be more compatible with many philosophical, social, and religious concepts centered on the concept of an unchanging universe. The discovery of **quasars** and a permeating cosmic background radiation eventually tilted the cosmological argument in favor of big bang theory models.

Before the twentieth century, astronomers could only assume that the universe had existed forever without change, or that it was created in its present condition by divine action at some arbitrary time. Evidence that the universe was evolving did not begin to accumulate until the 1920s. The theory that all matter in the universe was created from a gigantic explosion called the "big bang" is widely accepted by students of **cosmology**.

It was German-American physicist Albert Einstein's (1879–1955) theory of relativity, published in 1915, that set the stage for the conceptual development of an expanding universe. Einstein had designed his theory to fit a static universe of constant dimensions. In 1919, a Dutch astronomer, Willem de Sitter, showed Einstein's theory could also describe an expanding universe. Mathematically, de Sitter's solution for Einstein's equation was sound, but observational evidence of expansion was lacking, and Einstein was skeptical.

In 1929, American astronomer Edwin Powell Hubble made what has been called the most significant astronomical discovery of the century. He observed large red shifts in the spectra of the galaxies he was studying; these red-shifts indicated that the galaxies are continually moving apart at tremendous velocities. Vesto Melvin Slipher, who took photographs

of the red-shift of many of the same galaxies, also drew similar conclusions.

Like de Sitter, Lemaître, who worked with Hubble in 1924, developed out a simple solution to Einstein's equations that described a universe in expansion. Hubble's stunning observation provided the evidence Lemaître was seeking for his theory. In 1933, Lemaître clearly described the expansion of the universe. Projecting *back* in time, he suggested that the universe had originated as a great "cosmic egg," expanding outward from a central point. He did not, however, consider whether an explosion actually took place to initiate this expansion. George Gamow further investigated the origin of the universe in 1948. Because the universe is expanding outward, he reasoned, it should be possible to calculate backward in time to its beginning. If all the mass of the universe was compressed into a small volume 10–15 billion years ago, its density and **temperature** must have been phenomenal. A tremendous explosion would have caused the start of the expansion, left a "halo" of background radiation, and formed the atomic elements that are heavier than the abundant hydrogen and helium. Physicists Ralph A. Alpher and Robert C. Herman established a model to show how such heavier particles could form under these conditions.

Gamow's theory implied there was a specific beginning and end to the universe. However, a number of other scientists, including Fred Hoyle, Thomas Gold, and Hermann Bondi felt that the theory of expansion required no beginning or end. Their model, called the steady state theory, suggested that matter was being continuously created throughout the universe. As galaxies drifted apart, matter would "condense" to form new ones in the void left behind. For nearly two decades, supporters of the competing theories seemed to be on equal footing.

In 1965 Robert H. Dicke made calculations relative to the cooling-off period after the initial big bang explosion. His results indicated that Gamow's residual radiation should be detectable. During the intervening eons it would have cooled to about 5 K (five kelvins above absolute zero). Unknown to him, radio engineers **Arno Penzias** and **Robert W. Wilson** already detected such radiation at 3 K in 1964 while looking for sources of **satellite** communication interference. This was the most convincing evidence yet gathered in support of the big bang theory, and it sent the steady-state theory into decline.

No theory exists today that can account for the extreme conditions that existed at the moment of the big bang. The theory of relativity does not apply to objects as dense and small as the universe must have been prior to the big bang. Cosmologists can project only as far back as 0.01 seconds after the explosion, when the cosmos was a seething mass of protons and neutrons. (It is possible there were many exotic particles that later became important as dark matter.) Based on their theories, cosmologists suggest that during this time neutrinos were produced.

It is argued that the laws of physics and chemistry—manifested in the properties of the fundamental forces of **gravity**, the strong force, electromagnetism, and the weak force (electromagnetism and the weak force are now known to be different manifestations of a more fundamental electroweak force)—formed in the first few fractions of a second of the big bang. Protons and neutrons began to form atomic nuclei about three minutes and 46 seconds after the explosion, when the temperature was a mere 900,000,000 K. After 700,000 years hydrogen and helium formed. About one billion years after the big bang, stars and galaxies began to appear from the expanding mass. Countless stars would condense from swirling nebulae, evolve and die, before our **Sun** and its planets could form in the Milky Way galaxy.

Although the big bang theory accounts for most of the important characteristics of the universe, it still has weaknesses. One of the biggest of these involves the "homogeneity" of the universe. Until 1992, measurements of the background radiation produced by the big bang have shown that matter in the early universe was very evenly distributed. This seems to indicate that the universe evolved at a constant rate following the big bang. But if this is the case, the clumps of matter that we see (such as stars, galaxies, and clusters of galaxies) should not exist.

To remedy this inconsistency, Alan Guth proposed the inflationary theory, which suggests that the expansion of the universe initially occurred much faster. This concept of accelerated expansion allows for the formation of the structures we see in the universe today.

In April 1992, NASA made an electrifying announcement: its Cosmic Background Explorer (COBE), looking 15 billion light-years into space (hence, 15 billion years into the past), detected minute temperature fluctuations in the cosmic background radiation. It is believed these ripples are evidence of gravitational disturbances in the early universe that could have resulted in matter to clumping together to form larger entities. This finding lends support to Guth's theory of inflation.

See also Astronomy; Atom; Atomic theory; Catastrophism; Cosmic microwave background radiation; Cosmology; Earth (planet)

BIOGEOCHEMICAL CYCLES

The term biogeochemical cycle refers to any set of changes that occur as a particular element passes back and forth between the living and non-living worlds. For example, **carbon** occurs sometimes in the form of an atmospheric gas (**carbon dioxide**), sometimes in rocks and **minerals** (**limestone** and **marble**), and sometimes as the key element of which all living organisms are made. Over time, chemical changes occur that convert one form of carbon to another form. At various points in the carbon cycle, the element occurs in living organisms and at other points it occurs in the earth's atmosphere, **lithosphere**, or hydrosphere.

The universe contains about ninety different naturally occurring elements. Six elements—carbon, hydrogen, **oxygen**, nitrogen, sulfur, and phosphorus—make up over 95% of the mass of all living organisms on Earth. Because the total amount of each element is essentially constant, some cycling process must take place. When an organism dies, for example, the elements of which it is composed continue to move through a cycle, returning to the earth, to the air, to the ocean, or to another organism.

All biogeochemical cycles are complex. A variety of pathways are available by which an element can move among hydrosphere, lithosphere, atmosphere, and **biosphere**. For instance, nitrogen can move from the lithosphere to the atmosphere by the direct decomposition of dead organisms or by the reduction of nitrates and nitrites in the **soil**. Most changes in the nitrogen cycle occur as the result of bacterial action on one compound or another. Other cycles do not require the intervention of bacteria. In the sulfur cycle, for example, sulfur dioxide in the atmosphere can react directly with compounds in the earth to make new sulfur compounds that become part of the lithosphere. Those compounds can then be transferred directly to the biosphere by plants growing in the earth.

Most cycles involve the transport of an element through all four parts of the planet—hydrosphere, atmosphere, lithosphere, and biosphere. The phosphorous cycle is an exception since phosphorus is essentially absent from the atmosphere. It does move from biosphere to the lithosphere (when organisms die and decay) to the hydrosphere (when phosphorous-containing compounds dissolve in **water**) and back to the biosphere (when plants incorporate phosphorus from water).

Hydrogen and oxygen tend to move together through the planet in the **hydrologic cycle**. **Precipitation** carries water from the atmosphere to the hydrosphere and lithosphere. It then becomes part of living organisms (the biosphere) before being returned to the atmosphere through respiration, transpiration, and **evaporation**.

All biogeochemical cycles are affected by human activities. As fossil **fuels** are burned, for example, the transfer of carbon from a very old reserve (decayed plants and animals buried in the earth) to a new one (the atmosphere, as carbon dioxide) is accelerated. The long-term impact of this form of human activity on the global environment, as well as that of other forms, is not yet known. Some scientists assert, however, that those affects can be profound, resulting in significant **climate** changes far into the future.

See also Atmospheric composition and structure; Chemical elements; Dating methods; Evolution, evidence of; Evolutionary mechanisms; Fossil record; Fossils and fossilization; Geochemistry; Geologic time; Global warming; Greenhouse gases and greenhouse effect; Hydrologic cycle; Origin of life; Petroleum; Stellar life cycle

BIOLOGICAL PURIFICATION • *See* WATER POLLUTION AND BIOLOGICAL PURIFICATION

BIOSPHERE

The biosphere is the **space** on and near Earth's surface that contains and supports living organisms and ecosystems. It is typically subdivided into the **lithosphere**, atmosphere, and hydrosphere. The lithosphere is the earth's surrounding layer composed of solid **soil** and **rock**, the atmosphere is the surrounding gaseous envelope, and the hydrosphere refers to liquid environments such as **lakes** and **oceans**, occurring between the lithosphere and atmosphere. The biosphere's creation and continuous **evolution** result from physical, chemical, and biological processes. To study these processes a multi-disciplinary effort has been employed by scientists from such fields as **chemistry**, biology, **geology**, and ecology.

The Austrian geologist Eduard Suess (1831–1914) first used the term biosphere in 1875 to describe the space on Earth that contains life. The concept introduced by Suess had little impact on the scientific community until it was resurrected by the Russian scientist Vladimir Vernadsky (1863–1945) in 1926 in his book, *La biosphere*. In that work, Vernadsky extensively developed the modern concepts that recognize the interplay between geology, chemistry, and biology in biospheric processes.

For organisms to live, appropriate environmental conditions must exist in terms of **temperature**, moisture, energy supply, and nutrient availability.

Energy is needed to drive the functions that organisms perform, such as growth, movement, waste removal, and reproduction. Ultimately, this energy is supplied from a source outside the biosphere, in the form of visible radiation received from the **Sun**. This electromagnetic radiation is captured and stored by plants through the process of photosynthesis. Photosynthesis involves a light-induced, enzymatic reaction between **carbon dioxide** and **water**, which produces **oxygen** and glucose, an organic compound. The glucose is used, through an immense diversity of biochemical reactions, to manufacture the huge range of other organic compounds found in organisms. Potential energy is stored in the **chemical bonds** of organic molecules and can be released through the process of respiration; this involves enzymatic reactions between organic molecules and oxygen to form **carbon** dioxide, water, and energy. The growth of organisms is achieved by the accumulation of organic matter, also known as biomass. Plants and some microorganisms are the only organisms that can form organic molecules by photosynthesis. Heterotrophic organisms, including humans, ultimately rely on photosynthetic organisms to supply their energy needs.

The major elements that comprise the chemical building blocks of organisms are carbon, oxygen, nitrogen, phosphorus, sulfur, calcium, and magnesium. Organisms can only acquire these elements if they occur in chemical forms that can be assimilated from the environment; these are termed available nutrients. Nutrients contained in dead organisms and biological wastes are transformed by decomposition into compounds that organisms can reutilize. In addition, organisms can utilize some mineral sources of nutrients. All of the uptake, excretion, and transformation reactions are aspects of nutrient cycling.

The various chemical forms in which carbon occurs can be used to illustrate nutrient cycling. Carbon occurs as the gaseous molecule carbon dioxide, and in the immense diversity of organic compounds that make up living organisms and dead biomass. Gaseous carbon dioxide is transformed to solid organic compounds (simple sugars) by the process of photosynthesis, as mentioned previously. As organisms grow they deplete the

atmosphere of carbon dioxide. If this were to continue without carbon dioxide being replenished at the same rate as the consumption, the atmosphere would eventually be depleted of this crucial nutrient. However, carbon dioxide is returned to the atmosphere at about the same rate that it is consumed, as organisms respire their organic molecules, and microorganisms decompose dead biomass, or when wildfire occurs.

During the long history of life on Earth (about 3.8 billion years), organisms have drastically altered the chemical composition of the biosphere. At the same time, the biosphere's chemical composition has influenced which life forms could inhabit its environments. Rates of nutrient transformation have not always been in balance, resulting in changes in the chemical composition of the biosphere. For example, when life first evolved, the atmospheric concentration of carbon dioxide was much greater than today, and there was almost no free oxygen. After the evolution of photosynthesis there was a large decrease in atmospheric carbon dioxide and an increase in oxygen. Much of carbon once present in the atmosphere as carbon dioxide now occurs in fossil fuel deposits and **limestone** rock.

The increase in atmospheric oxygen concentration had an enormous influence on the evolution of life. It was not until oxygen reached similar concentrations to what occurs today (about 21% by volume) that multicellular organisms were able to evolve. Such organisms require high oxygen concentrations to accommodate their high rate of respiration.

Most research investigating the biosphere is aimed at determining the effects that human activities are having on its environments and ecosystems. Pollution, fertilizer application, changes in land use, fuel consumption, and other human activities affect nutrient cycles and damage functional components of the biosphere, such as the ozone layer that protects organisms from intense exposure to solar ultraviolet radiation, and the **greenhouse effect** that moderates the surface temperature of the planet.

For example, fertilizer application increases the amounts of nitrogen, phosphorus, and other nutrients that organisms can use for growth. An excess nutrient availability can damage lakes through algal blooms and fish kills. Fuel consumption and land clearing increases the concentration of carbon dioxide in the atmosphere, and may cause **global warming** by intensifying the planet's greenhouse effect.

Recent interest in long-term, manned space operations has spawned research into the development of artificial biospheres. Extended missions in space require that nutrients be cycled in a volume no larger than a building. The Biosphere-2 project, which received a great deal of popular attention in the early 1990s, has provided insight into the difficulty of managing such small, artificial biospheres. Human civilization is also finding that it is challenging to sustainably manage the much larger biosphere of planet Earth.

See also Atmospheric pollution; Earth (planet); Environmental pollution; Evolution, evidence of; Evolution, mechanisms of; Foliation and exfoliation; Forests and deforestation; Fossil record; Fossils and fossilization; Freshwater; Gaia hypothesis; Solar energy

BLACKETT, PATRICK MAYNARD STUART
(1897-1974)
English physicist

Patrick Maynard Stuart Blackett was a physicist with wide-ranging scientific and personal interests. He is best known for his improvements to the Wilson cloud chamber leading to important discoveries about fundamental particles and cosmic rays. His contributions to the study of **magnetism** helped confirm **continental drift theory**. Throughout his career he was admired as an ingenious experimenter. Blackett was involved in British military defense strategies during World War II, but remained an outspoken critic of Western nuclear policies to the end of his life. For his work with the Wilson cloud chamber, Blackett was awarded the 1948 Nobel Prize in physics.

Blackett was born in London, England to Arthur Stuart and Caroline Frances Maynard Blackett. His grandfather had been Anglican vicar of Croydon, Surrey, and his father was a stockbroker. As a child he developed a strong interest in nature, especially birds. Intending a naval career, Blackett attended Osborne Royal Naval College and Dartmouth Royal College. He began active naval duty when World War I broke out in 1914.

After the war, while still in the navy, he studied for six months at Magdalene College, Cambridge. This experience, coupled with his sense that the navy was unlikely to pursue technological innovations, convinced him to pursue a scientific career. He left the service, graduating from Cambridge with a B.A. in physics in 1921. In 1924, Blackett married Costanza Bayon, with whom he had a daughter and a son.

Cambridge's Cavendish Laboratory under Ernest Rutherford's direction was one of the world's foremost centers of theoretical physics after World War I. When Blackett received a fellowship to continue studying there, Rutherford put him to work with the Wilson cloud chamber. A cloud chamber is a device that makes it possible to track the movements of fundamental particles. It consists of a transparent cylinder filled with supersaturated **water** vapor. The cylinder is set between the poles of an electromagnet. When charged particles are fired into it, the water vapor condenses on the resulting ions and creates trails which can be photographed.

Blackett made improvements to a cloud chamber he inherited from a previous student, and by 1924, was able to confirm Rutherford's prediction that one element could be transmuted into another artificially. By filling the cloud chamber with nitrogen gas and water vapor and bombarding the mixture with alpha particles (helium atoms), Blackett produced a hydrogen **atom** and an **oxygen** isotope.

In 1932, Blackett began a productive collaboration studying cosmic rays with the Italian physicist Giuseppe P. S. Occhialini. Cosmic rays were known to reach earth from extraterrestrial sources, but their exact composition was unclear. At the time, very few fundamental particles were postulated, and cosmic rays were thought to be high-energy photons, or light quanta.

Blackett and Occhialini further modified the cloud chamber by combining it with two Geiger counters so that they could obtain more continuous photographs of the particle tracks. After three years and many thousands of photographs, they were able to confirm the existence of the first antimatter particle, the positron, which had been predicted by Carl Anderson. The American physicist Robert A. Millikan had thought this positively charged particle was a proton, but Blackett and Occhialini showed that the particle had the same mass as an electron, and that positrons occurred in "showers" paired with equal numbers of electrons. Blackett also noted a curious high energy component of cosmic rays later found to be the meson.

In 1937, Blackett replaced W. L. Bragg at the University of Manchester and began to build a strong research facility there. With the onset of World War II, he was tapped by the British government to assist in defense measures. He served on the Tizard Committee from 1935 to 1936, and became Director of Naval Operational Research, where he made statistical analyses of the predicted results of differing military strategies.

Blackett, however, opposed Britain's efforts to develop its own nuclear weapons, and though he supported the American bomb project, he was highly critical of Allied nuclear policy during and after the war. He decried the bombing of German civilians and the use of atomic bombs at Hiroshima and Nagasaki. In 1948, his book *Military and Political Consequences of Atomic Energy* appeared (published in America as *Fear, War und the Bomb*). That same year he was also awarded the Nobel Prize, but public hostility to his political views in the climate of the early Cold War overshadowed the acclaim accompanying the prize. Blackett did not return to public service until the election of a Labor government in 1964.

In the late 1940s, Blackett became interested in **magnetism** and the **rotation** of massive bodies. The idea that all rotating bodies generate magnetism had been discussed for many years, and if confirmed would have been a major new physical theory. Based on his study of existing observations of the magnetism of the **Sun**, the earth, and some stars, Blackett thought the hypothesis was plausible. In order to test it, he devised a magnetometer that was ten thousand times more sensitive than any previous such instrument. Ultimately Blackett decided the theory was incorrect, but his interest in geological magnetism continued. He investigated the history of changes in the earth's **magnetic field** and came to support the theory of continental drift, which postulates that the earth's continents are made of crustal plates that slowly move atop a layer of molten **rock (magma)**. His magnetometer proved to be very useful in the study of the magnetic fields of small rocks, which eventually helped to confirm continental drift.

In 1965, Blackett became president of the Royal Society, which under his leadership became international in focus. Though he was happy to be welcomed back into the public mainstream, Blackett continued making his political views known, describing himself as a Fabian Socialist and advocating a closer solidarity between scientists and the working class. He also devoted several years to studying scientific, political, and economic conditions in India. Blackett's last academic post was at London's Imperial College of Science and Technology from 1953 to 1965. During his career he received numerous awards in addition to the Nobel Prize, including twenty honorary degrees. In 1969, he was made a life peer, Baron Blackett of Chelsea. Blackett died in London at the age of 76.

See also Continental shelf; Quantum theory and mechanics

BLIZZARDS AND LAKE EFFECT SNOWS

A blizzard is a severe storm, potentially life threatening, caused by wind-driven snow. Although many blizzards involve heavy snow falls, smaller snow amounts may still be driven to blizzard conditions of low visibility and extreme **wind chill**.

The United States National **Weather** Service (NWS) takes a broader approach to the designation of a blizzard. NWS classifies a storm a blizzard if it manifests large amounts of snowfall, or has blowing snow in near gale force winds (generally about 35 mph, or 30 knots) or a combination of **wind** and snow that reduces visibility for more than a few hours. Severe conditions that are not quite blizzard-like are classified as sever winter storms.

Many areas near the **Great Lakes** in the United States and Canada are subject to frequent and severe blizzards due to lake effect snow. Lake effect snow is a meteorological phenomena created by the collision of Artic cold fronts sweeping generally west to east through Canada and the northern portion of the United States, with the relatively warmer air overlying the Great **Lakes**. Although lake effect snow can occur over any large body of **water** in the world, in **North America** lake effect snows are most frequently associated with the Great Lakes.

The combination of moist, unstable air and artic cold can produce locally heavy snows—especially on areas immediately east of the advancing cold front. Lake effect snowstorms are unique because they can manifest from otherwise dry cold fronts that produce clear cold weather in other parts of the country. Lake effect snow storms are not associated with advancing cells of low pressure, but rather dense high pressure Artic cold fronts.

Lake effect snow may contribute to more than half the annual snowfall for some areas on the east or southeast side of the Great Lakes. If winds are light enough to move the falling snow onshore, but not strong enough to blow the developing system over an **area** too quickly, snowfalls measuring 4–6 feet are possible.

Generally, the greater the **temperature** differential between the relatively warmer air over the lake and the advancing cold front, the more pronounced the lake effect snowfall. If the differential is great enough, and the moisture of the rising unstable air high enough, thundersnow may develop (a thunderstorm with snow instead of rain).

See also Land and sea breezes; Precipitation; Seasons; Weather forecasting methods; Weather forecasting

BLOUNT DUNES • *see* DUNES

BLOWING DUST • *see* DUNE FIELDS

BLUESCHIST • *see* SCHIST

BOHR MODEL

The Bohr model of atomic structure was developed by Danish physicist and Nobel laureate **Niels Bohr** (1885–1962). Published in 1913, Bohr's model improved the classical atomic models of physicists J. J. Thomson and Ernest Rutherford by incorporating **quantum theory**. While working on his doctoral dissertation at Copenhagen University, Bohr studied physicist Max Planck's quantum theory of radiation. After graduation, Bohr worked in England with Thomson and subsequently with Rutherford. During this time Bohr developed his model of atomic structure.

Before Bohr, the classical model of the **atom** was similar to the Copernican model of the **solar system** where, just as planets orbit the **Sun**, electrically negative electrons moved in orbits about a relatively massive, positively charged nucleus. The classical model of the atom allowed electrons to orbit at any distance from the nucleus. This predicted that when, for example, a hydrogen atom was heated, it should produce a continuous spectrum of colors as it cooled because its electron, moved away from the nucleus by the heat energy, would gradually give up that energy as it spiraled back closer to the nucleus. Spectroscopic experiments, however, showed that hydrogen atoms produced only certain colors when heated. In addition, physicist James Clark Maxwell's influential studies on electromagnetic radiation (light) predicted that an electron orbiting around the nucleus according to Newton's laws would continuously lose energy and eventually fall into the nucleus. To account for the observed properties of hydrogen, Bohr proposed that electrons existed only in certain orbits and that, instead of traveling between orbits, electrons made instantaneous quantum leaps or jumps between allowed orbits.

In the Bohr model, the most stable, lowest energy level is found in the innermost orbit. This first orbital forms a shell around the nucleus and is assigned a principal quantum number (n) of n = 1. Additional orbital shells are assigned values n = 2, n = 3, n = 4, etc. The orbital shells are not spaced at equal distances from the nucleus, and the radius of each shell increases rapidly as the square of n. Increasing numbers of electrons can fit into these orbital shells according to the formula $2n^2$. The first shell can hold up to two electrons, the second shell (n = 2) up to eight electrons, and the third shell (n = 3) up to 18 electrons. Subshells or suborbitals (designated s, p, d, and f) with differing shapes and orientations allow each element a unique electron configuration.

As electrons move farther away from the nucleus, they gain potential energy and become less stable. Atoms with electrons in their lowest energy orbits are in a "ground" state, and those with electrons jumped to higher energy orbits are in an "excited" state. Atoms may acquire energy that excites electrons by random thermal collisions, collisions with subatomic particles, or by absorbing a photon. Of all the photons (quantum packets of light energy) that an atom can absorb, only those that have energy equal to the energy difference between allowed electron orbits are absorbed. Atoms give up excess internal energy by giving off photons as electrons return to lower energy (inner) orbits.

The electron quantum leaps between orbits proposed by the Bohr model accounted for Plank's observations that atoms emit or absorb electromagnetic radiation only in certain units called quanta. Bohr's model also explained many important properties of the photoelectric effect described by **Albert Einstein**.

According to the Bohr model, when an electron is excited by energy it jumps from its ground state to an excited state (i.e., a higher energy orbital). The excited atom can then emit energy only in certain (quantized) amounts as its electrons jump back to lower energy orbits located closer to the nucleus. This excess energy is emitted in quanta of electromagnetic radiation (photons of light) that have exactly the same energy as the difference in energy between the orbits jumped by the electron. For hydrogen, when an electron returns to the second orbital (n = 2) it emits a photon with energy that corresponds to a particular color or spectral line found in the Balmer series of lines located in the visible portion of the electromagnetic (light) spectrum. The particular color in the series depends on the higher orbital from which the electron jumped. When the electron returns all the way to the innermost orbital (n = 1), the photon emitted has more energy and forms a line in the Lyman series found in the higher energy, ultraviolet portion of the spectrum. When the electron returns to the third quantum shell (n = 3), it retains more energy and, therefore, the photon emitted is correspondingly lower in energy and forms a line in the Paschen series found in the lower energy, infrared portion of the spectrum.

Because electrons are moving charged particles, they also generate a **magnetic field**. Just as an ampere is a unit of electric current, a magneton is a unit of magnetic dipole moment. The orbital magnetic moment for hydrogen atom is called the Bohr magneton.

Bohr's work earned a Nobel Prize in 1922. Subsequently, more mathematically complex models based on the work of French physicist Louis Victor de Broglie (1892–1987) and Austrian physicist **Erwin Schrödinger** (1887–1961) that depicted the particle and wave nature of electrons proved more useful to describe atoms with more than one electron. The standard model incorporating quark particles further refines the Bohr model. Regardless, Bohr's model remains fundamental to the study of **chemistry**, especially the valence shell concept used to predict an element's reactive properties.

The Bohr model remains a landmark in scientific thought that poses profound questions for scientists and philosophers. The concept that electrons make quantum leaps from one orbit to another, as opposed to simply moving between orbits, seems counter-intuitive, that is, outside the human experience with nature. Bohr said, "Anyone who is not shocked by quantum theory has not understood it." Like much

of quantum theory, the proofs of how nature works at the atomic level are mathematical.

See also Atomic mass and weight; Atomic number; Quantum electrodynamics (QED); Quantum theory and mechanics

BOHR, NIELS (1885-1962)
Danish physicist

Niels Bohr received the Nobel Prize in physics in 1922 for the quantum mechanical model of the **atom** that he had developed a decade earlier, the most significant step forward in scientific understanding of atomic structure since English physicist John Dalton first proposed the modern **atomic theory** in 1803. Bohr founded the Institute for Theoretical Physics at the University of Copenhagen in 1920, an Institute later renamed for him. For well over half a century, the Institute was a powerful force in the shaping of atomic theory. It was an essential stopover for all young physicists who made the tour of Europe's center of theoretical physics in the mid-twentieth century. Also during the 1920s, Bohr thought and wrote about some of the fundamental issues raised by modern **quantum theory**. He developed two basic concepts, the principles of complementarity and correspondence, both of which he held must direct all future work in physics. In the 1930s, Bohr became interested in problems of the atomic nucleus and contributed to the development of the liquid-drop model of the nucleus, a model used in the explanation of **nuclear fission**.

Niels Henrik David Bohr was born on in Copenhagen, Denmark, the second of three children of Christian and Ellen Adler Bohr. Bohr's early upbringing was enriched by a nurturing and supportive home atmosphere. His mother had come from a wealthy Jewish family involved in banking, government, and public service. Bohr's father was a professor of physiology at the University of Copenhagen. His closest friends met every Friday night to discuss events, and often, young Niels listened to the conversations during these gatherings.

Bohr became interested in science at an early age. His biographer, Ruth Moore, has written in her book *Niels Bohr: The Man, His Science, and the World They Changed* that as a child he "was already fixing the family clocks and anything else that needed repair." Bohr received his primary and secondary education at the Gammelholm School in Copenhagen. He did well in his studies, although he was apparently overshadowed by the work of his younger brother Harald, who later became a mathematician. Both brothers were also excellent soccer players.

On his graduation from high school in 1903, Bohr entered the University of Copenhagen, where he majored in physics. He soon distinguished himself with a noteworthy research project on the surface tension of **water** as evidenced in a vibrating **jet stream**. For this work, he was awarded a gold medal by the Royal Danish Academy of Science in 1907. In the same year, he was awarded his bachelor of science degree, to be followed two years later by a master of science degree. Bohr then stayed on at Copenhagen to work on his doctorate, which he gained in 1911. His doctoral thesis dealt with the electron theory of **metals** and confirmed the fact that classical physical principles were sufficiently accurate to describe the qualitative properties of metals but failed when applied to quantitative properties. Probably the main result of this research was to convince Bohr that classical electromagnetism could not satisfactorily describe atomic phenomena. The stage had been set for Bohr's attack on the most fundamental questions of atomic theory.

Bohr decided that the logical place to continue his research was at the Cavendish Laboratory at Cambridge University. The director of the laboratory at the time was English physicist J. J. Thomson, discoverer of the electron. Only a few months after arriving in England in 1911, however, Bohr discovered that Thomson had moved on to other topics and was not especially interested in Bohr's thesis or ideas. Fortunately, however, Bohr met English physicist Ernest Rutherford, then at the University of Manchester, and received a much more enthusiastic response. As a result, he moved to Manchester in 1912 and spent the remaining three months of his time in England working on Rutherford's nuclear model of the atom.

On July 24, 1912, Bohr boarded ship for his return to Copenhagen and a job as assistant professor of physics at the University of Copenhagen. Also waiting for him was his bride-to-be Margrethe Nørlund, whom he married on August 1. The couple later had six sons. One son, Aage, earned a share of the 1975 Nobel Prize in physics for his work on the structure of the atomic nucleus.

The field of atomic physics was going through a difficult phase in 1912. Rutherford had only recently discovered the atomic nucleus, which had created a profound problem for theorists. The existence of the nucleus meant that electrons must have been circling it in orbits somewhat similar to those traveled by planets in their motion around the **Sun**. According to classical laws of electrodynamics, however, an electrically charged particle would continuously radiate energy as it traveled in such an orbit around the nucleus. Over time, the electron would spiral ever closer to the nucleus and eventually collide with it. Although electrons clearly must be orbiting the nucleus, they could not be doing so according to classical laws.

Bohr arrived at a solution to this dilemma in a somewhat roundabout fashion. He began by considering the question of atomic spectra. For more than a century, scientists had known that the heating of an element produces a characteristic line spectrum; that is, the specific pattern of lines produced is unique for each specific element. Although a great deal of research had been done on spectral lines, no one had thought very deeply about what their relationship might be with atoms, the building blocks of elements.

When Bohr began to attack this question, he decided to pursue a line of research begun by the German physicist Johann Balmer in the 1880s. Balmer had found that the lines in the hydrogen spectrum could be represented by a relatively simple mathematical formula relating the frequency of a particular line to two integers whose significance Balmer could not explain. It was clear that the formula gave very precise values for line frequencies that corresponded well with those observed in experiments.

Niels Bohr. *Library of Congress.*

When Bohr's attention was first attracted to this formula, he realized at once that he had the solution to the problem of electron orbits. The solution that Bohr worked out was both simple and elegant. In a brash display of hypothesizing, Bohr declared that certain orbits existed within an atom in which an electron could travel without radiating energy; that is, classical laws of physics were suspended within these orbits. The two integers in the Balmer formula, Bohr said, referred to orbit numbers of the "permitted" orbits, and the frequency of spectral lines corresponded to the energy released when an electron moved from one orbit to another.

Bohr's hypothesis was brash because he had essentially no theoretical basis for predicting the existence of "allowed" orbits. To be sure, German physicist Max Planck's quantum hypothesis of a decade earlier had provided some hint that Bohr's "quantification of space" might make sense, but the fundamental argument for accepting the hypothesis was simply that it worked. When his model was used to calculate a variety of atomic characteristics, it did so correctly. Although the hypothesis failed when applied to detailed features of atomic spectra, it worked well enough to earn the praise of many colleagues.

Bohr published his theory of the "planetary atom" in 1913. That paper included a section that provided an interesting and decisive addendum to his basic hypothesis. One of the apparent failures of the Bohr hypothesis was its seeming inability to predict a set of spectral lines known as the Pickering series, lines for which the two integers in the Balmer formula required half-integral values. According to Bohr, of course, no "half-orbits" could exist that would explain these values. Bohr's solution to this problem was to suggest that the Pickering series did not apply to hydrogen at all, but to helium atoms that had lost an electron. He rewrote the Balmer formula to reflect this condition.

Within a short period of time spectroscopists in England had studied samples of helium carefully purged of hydrogen and found Bohr's hypothesis to be correct. Although a number of physicists were still debating Bohr's theory, at least one—Rutherford—was convinced that the young Danish physicist was a highly promising researcher. He offered Bohr a post as lecturer in physics at Manchester, a job that Bohr eagerly accepted and held from 1914 to 1916. He then returned to the University of Copenhagen, where a chair of theoretical physics had been created specifically for him. Within a few years he was to become involved in the planning for and construction of the University of Copenhagen's new Institute for Theoretical Physics, of which he was to serve as director for the next four decades.

In many ways, Bohr's atomic theory marked a sharp break between classical physics and a revolutionary new approach to natural phenomena made necessary by quantum theory and relativity. He was very much concerned about how scientists could and should now view the physical world, particularly in view of the conflicts that arose between classical and modern laws and principles. During the 1920s and 1930s, Bohr wrote extensively about this issue, proposing along the way two concepts that he considered to be fundamental to the "new physics." The first was the principle of complementarity that says, in effect, that there may be more than one true and accurate way to view natural phenomena. The best example of this situation is the wave-particle duality discovered in the 1930s, when particles were found to have wavelike characteristics and waves to have particle-like properties. Bohr argued that the two parts of a duality may appear to be inconsistent or even in conflict and that one can use only one viewpoint at a time, but he pointed out that both are necessary to obtain a complete view of particles and waves.

The second principle, the correspondence principle, was intended to show how the laws of classical physics could be preserved in light of the new quantum physics. We may know that quantum mechanics and relativity are essential to an understanding of phenomena on the atomic scale, Bohr said, but any conclusion drawn from these principles must not conflict with observations of the real world that can be made on a macroscopic scale. That is, the conclusions drawn from theoretical studies must correspond to the world described by the laws of classical physics.

In the decade following the publication of his atomic theory, Bohr continued to work on the application of that theory to atoms with more than one electron. The original theory had dealt only with the simplest of all atoms, hydrogen, but it was clearly of some interest to see how that theory could be extended to higher elements. In March, 1922, Bohr published a summary of his conclusions in a paper entitled "The Structure of the Atoms and the Physical and Chemical Properties of the Elements." Eight months later, Bohr learned that he had been awarded the Nobel Prize in physics for his theory of atomic structure, by that time universally accepted among physicists.

During the 1930s, Bohr turned to a new, but related, topic: the composition of the atomic nucleus. By 1934, scientists had found that the nucleus consists of two kinds of particles, protons and neutrons, but they had relatively little idea how those particles are arranged within the nucleus and what its general shape was. Bohr theorized that the nucleus could be compared to a liquid drop. The forces that operate between protons and neutrons could be compared in some ways, he said, to the forces that operate between the molecules that make up a drop of liquid. In this respect, the nucleus is no more static than a droplet of water. Instead, Bohr suggested, the nucleus should be considered to be constantly oscillating and changing shape in response to its internal forces. The greatest success of the Bohr liquid-drop model was its later ability to explain the process of nuclear fission discovered by German chemist Otto Hahn, German chemist Fritz Strassmann, and Austrian physicist Lise Meitner in 1938.

Bohr continued to work at his Institute during the early years of World War II, devoting considerable effort to helping his colleagues escape from the dangers of Nazi Germany. When he received word in September 1943 that his own life was in danger, Bohr decided that he and his family would have to leave Denmark. The Bohrs were smuggled out of the country to Sweden aboard a fishing boat and then, a month later, flown to England in the empty bomb bay of a Mosquito bomber. The Bohrs then made their way to the United States, where both Bohr and his son became engaged in work on the Manhattan Project to build the world's first atomic bombs.

After the War, Bohr, like many other Manhattan Project researchers, became active in efforts to keep control of atomic weapons out of the hands of the military and under close civilian supervision. For his long-term efforts on behalf of the peaceful uses of atomic energy, Bohr received the first Atoms for Peace Award given by the Ford Foundation in 1957. Meanwhile, Bohr had returned to his Institute for Theoretical Physics and become involved in the creation of the European Center for Nuclear Research (CERN). He also took part in the founding of the Nordic Institute for Theoretical Atomic Physics (Nordita) in Copenhagen. Nordita was formed to further cooperation among and provide support for physicists from Norway, Sweden, Finland, Denmark, and Iceland.

Bohr reached the mandatory retirement age of seventy in 1955 and was required to leave his position as professor of physics at the University of Copenhagen. He continued to serve as director of the Institute for Theoretical Physics until his death in Copenhagen at the age of 77.

Bohr was held in enormous respect and esteem by his colleagues in the scientific community. American physicist **Albert Einstein**, for example, credited him with having a "rare blend of boldness and caution; seldom has anyone possessed such an intuitive grasp of hidden things combined with such a strong critical sense." Among the many awards Bohr received were the Max Planck Medal of the German Physical Society in

1930, the Hughes (1921) and Copley (1938) medals of the Royal Society, the Franklin Medal of the Franklin Institute in 1926, and the Faraday Medal of the Chemical Society of London in 1930. He was elected to more than twenty scientific academies around the world and was awarded honorary doctorates by a dozen universities, including Cambridge, Oxford, Manchester, Edinburgh, the Sorbonne, Harvard, and Princeton.

See also Bohr Model

BORA · *see* SEASONAL WINDS

BOULDER · *see* ROCK

BOWEN'S REACTION SERIES

Bowen's reaction series describes the formation of **minerals** as **magma** cools. Rocks formed from magma are **igneous rocks**, and minerals crystallize as magma cools. The **temperature** of the magma and the rate of cooling determine which minerals are stable (i.e., which minerals can form) and the size of the mineral **crystals** formed (i.e., texture). The slower a magma cools, the larger crystals can grow.

Named after geologist Norman L. Bowen (1887–1956), Bowen's reaction series allows geologists to predict chemical composition and texture based upon the temperature of a cooling magma.

Bowen's reaction series is usually diagramed as a "Y" with horizontal lines drawn across the "Y." The first horizontal line—usually placed just above the top of the "Y"—represents a temperature of 3,272°F (1,800°C). The next horizontal line, represents a temperature of 2,012°F (1,100°C) and is located one-third of the way between the top of the "Y" and the point where the two arms join the base. A third line representing a temperature of 1,652°F (900°C) is located two-thirds of the way from the top of the "Y" to juncture of the upper arms. A fourth horizontal line—representing a temperature of 1,112deg;F (600°C)—intersects the triple point junction where the upper arms of the "Y" meet the base portion.

The horizontal temperature lines divide the "Y" into four compositional sections. Mineral formation is not possible above 3,272°F (1,800°C). Between 2,012°F (1,100°C) and 3,272°F (1,800°C), rocks are ultramafic in composition. Between 1,652°F (900°C) and 2,012°F (1,100°C), rocks are **mafic** in composition. Between 1,112°F (600°C) and 1,652°F (900°C), rocks are intermediate in composition. Below 1,112°F (600°C), **felsic** rocks form.

The upper arms of the "Y" represent two different formation pathways. By convention, the left upper arm represents the discontinuous arm or pathway. The upper right arm represents the continuous arm or continuous path of formation. The discontinuous arm represents mineral formations rich in **iron** and magnesium. The first mineral to form is olivine—it is the only mineral stable at or just below 3,272°F (1,800°C). As the temperature decreases, pyroxene becomes stable. The general

chemical compositional formula—used throughout this article and not to be confused with a balanced molecular or empirical chemical formula—at the highest temperatures includes iron, magnesium, **silicon**, and **oxygen** (FeMgSiO, but no **quartz**). At approximately 2,012°F (1,100°C), calcium containing minerals (CaFeMgSiO) become stable. As the temperature lowers to 1,652°F (900°C), amphibole (CaFeMgSiOOH) forms. As the magmas cools to 1,112°F (600°C), biotite (KFeMgSiOOH) formation is stable.

The continuous arm of Bowen's reaction series represents the formation of **feldspar (plagioclase)** in a continuous and gradual series that starts with calcium rich feldspar (Ca-feldspar, CaAlSiO) and continues with a gradual increase in the formation of sodium containing feldspar (Ca-Na-feldspar, CaNaAlSiO) until an equilibrium is established at approximately 1,652°F (900°C). As the magmas cool and the calcium ions are depleted, the feldspar formation becomes predominantly sodium feldspar (Na-feldspar, NaAlSiO). At 1,112°F (600°C), the feldspar formation is nearly 100% sodium feldspar (Na-feldspar, NaAlSiO).

At or just below 1,112°F (600°C), the upper arms of the "Y" join the base. At this point in the magma cooling, K-feldspar or orthoclase (KAlSiO) forms and as the temperature begins to cool further, muscovite (KAlSiOOH) becomes stable. Just above the base of the "Y," the temperature is just above the point where the magma completely solidifies. At these coolest depicted temperatures (just above 392°F [200°C]), quartz (SiO) forms.

The time that the magma is allowed to cool will then determine whether the **rock** will be **pegmatite** (produced by extremely slow cooling producing very large crystals), phaneritic (produced by slow cooling that produces visible crystals), **aphanitic** (intermediate cooling times that produce microscopic crystals), or glassy in texture (a product of rapid cooling without crystal formation). When magmas experience differential cooling conditions, they produce porphyritic rock, a mixture of crystal sizes and exhibit either a phaneritic or aphanitic groundmass.

Although the above temperature and percentage composition data are approximate, simplified (e.g., the formation of hornblende has been omitted), and idealized, Bowen's reaction series allows the prediction of mineral content in rock and, by examination of rock, allows the reverse determination of the conditions under which the magma cooled and igneous rock formed.

See also Chemical bonds and physical properties; Crystals and crystallography; Magma chamber; Mineralogy; Rate factors in geologic processes; Temperature and temperature scales

BRAHE, TYCHO (1546-1601)
Danish astronomer

Tycho Brahe was one of the most colorful astronomers in history. Born in Denmark, Brahe was "adopted" (some say kidnapped) by his childless uncle at the age of one. Either way, his father, a Swedish nobleman, did not pursue the matter

Tycho Brahe. *New York Public Library Picture Collection.*

(Brahe's given name was "Tyge," but the Latinized version is more common.)

Brahe received an excellent education. At the age of 13 he entered the University of Copenhagen, where he studied rhetoric and philosophy. He was well on his way toward a career in politics when he witnessed an **eclipse** of the **Sun** on August 21, 1560. Brahe spent the next two years studying mathematics and **astronomy**. He moved on to the University of Leipzig in 1562 where a tutor tried to influence him to study law, but Brahe refused to be diverted.

In August 1563, he made his first recorded observation, a close grouping between Jupiter and Saturn. (It was not until many years that Galileo first used a **telescope** to make astronomical observations; Brahe's precise work was done with the naked eye.) This was the turning point of his career. He was perturbed to note that this event occurred a month before its predicted date, and he began to buy astronomical instruments that would allow him to make very precise measurements so he could produce more accurate tables of data. He also developed interests in alchemy and astrology (which he considered a science) and began to cast horoscopes that, if nothing else, generated some income.

In November 1572, a supernova burst into view in the constellation of Cassiopeia, and Brahe was enthralled. The new star became brighter than Venus and was visible for eighteen months. He described it (along with its astrological "sig-

nificance") with such detail in a book, the new star became known as "Tycho's star."

The book did three things: the title *De Nova Stella* (Concerning the new star) linked the name nova to all exploding stars. In addition, Brahe had been unable to make a parallax measurement for the nova. That indicated that it was much more distant than the **Moon**, which was a crushing blow to Aristotle's teachings that the heavens were perfect and unchanging. The third accomplishment was in establishing Brahe's reputation as an astronomer. The book was almost not produced. Initially, Brahe felt it was beneath his dignity as a nobleman to publish, but he was soon convinced otherwise.

Brahe's arrogance was legendary. At the age of nineteen, he was involved in a duel over a mathematical point, during which he lost his nose. He spent the rest of his life wearing a prosthesis. Fortunately, one of the few individuals not alienated by Brahe was Frederick II, the king of Denmark. In 1576, this patron of science gave Brahe a small island called Hveen, subsidized the building of an observatory there, and endowed Brahe with an annual payment. This became the first real astronomical observatory in history, and Brahe, always mindful of his noble background, saw to it that no expense was spared. The principal building, *Uraniborg* (Castle of the heavens), was the main residence; next to it was built the main observatory, *Stjerneborg* (Castle of the stars).

In 1577, a bright comet was visible, and Brahe observed it with great care. Measurements showed that it, too, was further than the moon and could not be atmospheric phenomena as Aristotle taught. Worse, Brahe reluctantly came to the conclusion that the path of the comet was not circular but elongated. This meant it would have to pass through the "spheres" that carried the planets around the sky, which would be impossible unless the spheres did not exist.

This concept troubled Brahe, who rejected the Sun-centered theory of Nicholas Copernicus because it not only violated scripture, it contradicted the teachings of **Ptolemy**. Brahe also reasoned that if Copernican theory was correct, he should have been able to detect stellar parallax as the year passed, but he could not.

Brahe tried to reconcile his beliefs with his observations by proposing a **solar system** in which all the planets orbited around the Sun, but the Sun orbited around the earth (to account for a year), and the celestial sphere made a single **rotation** each day. This would follow Copernicus's theory, do away with the Greeks' planetary spheres, and still keep the earth at its preeminent position. The Tychonic Theory was almost entirely ignored.

Brahe spent 20 years at Hveen, making exceptionally accurate observations. He used devices such as a huge quadrant with a radius of 6 feet (1.83 m), sextants, a bipartite arc, astrolabes, and various armillae. His measurements were the most precise that could be made without the aid of a telescope. He made corrections for nearly every known astronomical measurement and made Pope Gregory's calendar reform in 1582 possible. (Brahe himself did not adopt the new calendar until 1599.)

Frederick II died in 1588. His son, Christian IV, was only 11, so the country was ruled by regents, who left Brahe to his own devices. When Christian came of age in 1596, he quickly lost patience with the expensive, haughty astronomer, and Brahe was relieved of his royal duties the following year.

Brahe moved to Prague, where he resumed observing. As an assistant he employed a young German named **Johannes Kepler**, to whom he gave all his observations on Mars and the task of preparing tables of planetary motion. This would turn out to be the most significant decision of his life, as Kepler used the data to determine the elliptical nature of planetary motion.

BRAUN, WERNHER VON (1912-1977)

German-born American aerospace engineer

Wernher von Braun was the most famous rocket engineer of his time, noted promoter of **space** flight. Teams under his direction designed the V–2, Redstone, Jupiter, and Pershing missiles, as well as the Jupiter C, Juno, and Saturn launch vehicles that carried most of the early U.S. satellites and spacecraft beyond the earth's atmosphere and ultimately to the **moon**. He became both a celebrity and a national hero in the United States, winning numerous awards, including the first **Robert H. Goddard** Memorial Trophy in 1958, the Distinguished Federal Civilian Service Award (presented by President Dwight D. Eisenhower) in 1959, and the National Medal of Science in 1977. As President Jimmy Carter stated at the time of his death: "To millions of Americans, [his] name was inextricably linked to our exploration of space and to the creative application of technology. He was not only a skillful engineer but also a man of bold vision; his inspirational leadership helped mobilize and maintain the effort we needed to reach the Moon and beyond."

The second of three children (all male), Wernher Magnus Maximilian von Braun was born in the east German town of Wirsitz (later, Wyrzysk, Poland). He was the son of Baron Magnus Alexander Maximilian von Braun—then the principal magistrate (*Landrat*) of the governmental district and later (1932–early 1933) the minister of nutrition and agriculture in the last two governments of the Weimar Republic before Hitler rose to power in Germany—and of Emmy (von Quistorp) von Braun, a well-educated woman from the Swedish-German aristocracy with a strong interest in biology and **astronomy**. She inspired her son's interest in space flight by supplying him with the science fiction works of Jules Verne and H. G. Wells and by giving him a **telescope** as a gift upon his confirmation into the Lutheran church in his early teens, instead of the customary watch or camera. Despite these influences, the young von Braun was initially a weak student and was held back one year in secondary school because of his inability in math and **physics**. Due to his interest in astronomy and rockets, he obtained a copy of space pioneer Hermann Oberth's book *Die Rakete zu den Planeträumen* ("Rockets to planetary space") in 1925. Appalled that he could not understand its complicated mathematical formulas, he determined to master his two weakest subjects. Upon completion of secondary school, von Braun

Wernher von Braun. *Library of Congress.*

entered the Berlin-Charlottenburg Institute of Technology, where he earned a bachelor of science in mechanical engineering and aircraft construction in 1932.

In the spring of 1930, von Braun found time to work as part of the German Society for Space Travel, a group founded in part by Hermann Oberth which experimented with small, liquid-fueled rockets. Although Oberth returned to a teaching position in his native Romania, von Braun continued working with the society. When the group ran short of funds during the Depression, von Braun, then twenty, reluctantly accepted the sponsorship of the German military. In 1932 he went to work for the German army's ordnance department at Kummersdorf near Berlin, continuing to develop liquid-fueled rockets. Entering the University of Berlin about this same time, he used his work at Kummersdorf as the basis for his doctoral dissertation and received his Ph.D. in physics in 1934.

Von Braun's staff at Kummersdorf eventually grew to some eighty people, and in early 1937, the group moved to Peenemünde, a town on the Baltic coast where the German army together with the air force had constructed new facilities. Before the move, engineers at Kummersdorf had begun developing ever-larger rockets, and in 1936 they completed the preliminary design for the A–4, better known as the V–2. This was an exceptionally ambitious undertaking, since the missile was to be 45 feet long, deliver a 1-ton warhead to a target some 160 miles distant, and employ a rocket motor that could deliver a 25-ton thrust for 60 seconds, compared to the 1.5 tons of thrust supplied by the largest liquid-fueled rocket motors then available. Von Braun's team encountered numerous difficulties—perfecting the injection system for the propellants, mastering the aerodynamic properties of the missile, and especially in developing its guidance and control system. Thus, even with the assistance of private industry and universities, the first successful launch of the A–4 did not occur at Peenemünde until October 3, 1942. Despite this success, failed launches continued to plague the project, and as a result the first fully operational V–2s were not fired until September 1944. Between then and the end of the war, approximately 6,000 rockets were manufactured at an underground production site named *Mittelwerk,* using the slave labor of concentration camp inmates and prisoners of war. Although several thousand V–2s struck London, Antwerp, and other allied targets, they were not strategically significant in the German war effort. Their importance lies in the technological advances they brought to the development of rocketry.

As the war drew to a close in **Europe** in the early months of 1945, von Braun organized the move of hundreds of people from Peenemünde to Bavaria so they could surrender to the Americans rather than the Soviets. Subsequently, about 120 of them went to Fort Bliss near El Paso, Texas, as part of a military operation called Project Paperclip. They worked on rocket development and employed captured V–2s for high altitude research at the nearby White Sands Proving Ground in New Mexico. In the midst of these efforts, von Braun returned to Germany to marry, returning with his wife to Texas after the wedding. In 1950, the von Braun team transferred to the Redstone Arsenal near Huntsville, Alabama, where between April 1950 and February 1956, it developed the Redstone medium-range ballistic missile under his technical direction. Deployed in 1958, the Redstone was basically an offshoot of the V–2 but featured several modifications including an improved inertial guidance system. The Redstone also served as a launch vehicle, placing Alan B. Shephard and Virgil I. "Gus" Grissom in suborbital flight in May and July 1961, respectively. Meanwhile, in February 1956, von Braun became the director of the development operations division of the newly established Army Ballistic Missile Agency (ABMA) in Huntsville. While located there, he and his wife raised three children. Von Braun himself became a U.S. citizen on April 14, 1955.

Undoubtedly the greatest claim to fame of von Braun and his team was the powerful Saturn family of rockets, which propelled Americans into lunar orbit and landed 12 of them on the moon between July 1969 and January 1971. Development of these launch vehicles began under ABMA and was completed during the decade after July 1, 1960, when von Braun and over 4,000 ABMA personnel transferred to the National Aeronautics and Space Administration (NASA), forming the George C. Marshall Space Flight Center, which von Braun directed until February 1970. The Saturn I and Ib were developmental rockets leading to the massive Saturn V that actually launched the astronauts of the Apollo program. Propelled by liquid **oxygen** and kerosene in its first stage, liquid oxygen and liquid hydrogen for the two upper stages, the Saturn V stood

363 feet high, six stories above the level of the Statue of Liberty. Its first stage constituted the largest **aluminum** cylinder ever produced; its valves were as large as barrels, its fuel pumps larger than refrigerators.

As von Braun repeatedly insisted, he and his team were not alone responsible for the success of the Saturn and Apollo programs. In fact, the engineers at Marshall often urged more conservative solutions to problems occurring in both programs than NASA ultimately adopted. To von Braun's credit, he invariably accepted and supported the more radical approaches once he was convinced they were right. One example involved the debate over all-up versus step-by-step testing of Saturn V. Having experienced numerous rocket system failures going back to the V–2 and beyond, the German engineers favored testing each stage of the complicated rocket. At NASA headquarters, however, administrator George Mueller preferred the Air Force approach, which relied much more heavily on ground testing. He therefore insisted upon testing Saturn V all at once in order to meet President John F. Kennedy's ambitious goal of landing an American on the moon before the end of the decade. Ever cautious, von Braun hesitated but finally concurred in the ultimately successful procedure.

Beyond his role as an engineer, scientist, and project manager, von Braun was also an important advocate for space flight, publishing numerous books and magazine articles, serving as a consultant for television programs and films as well as testifying before Congress. Perhaps most important in this regard were his contributions, with others, to a series of *Collier's* articles from 1952 to 1953 and to a Walt Disney television series produced by Ward Kimball from 1955 to 1957. Both series were enormously influential and, along with the fears aroused by the Soviet space program, galvanized American efforts to conquer space.

See also History of manned space exploration; Spacecraft, manned

BRECCIA

Breccias are rocks composed of angular clasts (fragments). In monomictic breccias, clasts have the same composition, whereas polymictic breccias contain clasts of different compositions. Sedimentary breccias comprise more than 30% gravel-size (>2mm) angular clasts produced by mechanical **weathering** or brittle deformation of nearby rocks. Their angular shape implies minimal transport. Sedimentary breccias develop at the base of talus slopes or in proximity to active faults. Karst breccia forms during **erosion**, dissolution and collapse of **limestone**. Pressure solution due to high local stresses at contacts between angular fragments of limestone, **marble**, or **chert** can result in interpenetration of clasts. Breccias can form during the emplacement of igneous bodies by explosive exsolution of volatile phases and/or explosive interaction of **magma** with **groundwater**. Intrusive breccias (such as associated with kimberlite pipes) often contain fragments of both intrusive and host rocks. Igneous breccia dykes may contain a wide range of **rock** fragments sampled during magma ascent and thus, provide information about the composition of rocks at deeper levels. Volcanic breccias containing lithic (rock) and vitric (**glass** and **pumice**) fragments form near subaerial volcanic vents.

Fault brecciation (or tectonic comminution) can occur due to the development and linkage of a network of fractures during faulting in the upper **crust**. The size of breccia fragments is highly variable. Milling or wear abrasion during displacement on faults may result in further brecciation and size reduction. Fracturing occurs when the applied stress exceeds the brittle resistance of the material or by transient elevation of fluid pressure (hydraulic or fluid-assisted brecciation). The interaction of hydrothermal fluids with tectonically brecciated rock produces hydrothermal breccias common in ore deposits. Brecciation may also occur due to implosion of a vein resulting from a sudden decrease in pressure (critical fracturing) in response to a sudden opening of **space** generated by rapid slip or intersection between different veins. When fault slip is extremely rapid, melt generated by frictional heating is injected along fractures to produce veins of black glass (pseudotachylite) surrounding angular fragments of the surrounding rock.

Impact melt-breccias form by the fracturing and fusion of rocks under extreme pressures and temperatures rapidly induced during meteorite impacts. Impact melt-breccias contain partially or completely melted clasts of basement rocks within a cryptocrystalline glass, **feldspar** and calcium-pyroxene-rich matrix. Impact-melt breccias containing clastic debris and glass fragments produced by meteorite bombardment have been collected from the surface of the **Moon** during Apollo missions.

BUFFON, GEORGES-LOUIS LECLERC, COMTE DE (1707-1788)

French naturalist

Georges Louis Leclerc, Comte de Buffon was an eighteenth century naturalist who advocated the idea that natural forces worked to shape Earth in a gradual and ongoing process. By rejecting the widely-held notion of his time that Earth was shaped by catastrophic divine acts, Buffon inspired later geologists and naturalists to investigate and define the process of natural **evolution**.

Buffon was born to an aristocratic family in Montbard, France. His affluent background allowed him to travel extensively and pursue a number of fields before he developed a passionate interest in natural history. After studying at the Jesuit College in Dijon, France, Buffon obtained a law degree in 1726. The intellectual life of Dijon was active but not oriented toward science, so Buffon went off to Angers, a city in northwestern France, to study medicine, mathematics, botany, and **astronomy**. The threat of a duel forced him to leave

Angers in 1730, but he seized the opportunity to travel through France, England, and Italy. While he was traveling, Buffon's mother died and left him a sizable fortune.

Buffon had been so impressed with the upsurge of science in England that he dedicated the next couple of years to scientific endeavors. His first project, at the request of the French navy, was to write about the tensile strength of timber so that the government could improve the construction of war vessels. Next, he undertook a study of probability theory, *Mémoire sur le jeu du franc-carreau*, a project that contributed to his election to the Royal Society in 1730 and his admission to the Académie Royale des Sciences in 1734.

Buffon began to take an interest in botany and forestry. He wrote numerous dissertations and translated several works into French, including Stephen Hales' works on plants, *Vegetable Statiks*, and Isaac Newton's work on calculus. By this time, his work in the sciences began to elevate his standing, and he was advanced and transferred from the mechanical to the botanical section of the Académie Royale.

Nevertheless, Buffon's interest in natural history remained casual until he was appointed to the prestigious position of keeper of the Jardin du Roi, the French botanical gardens. This opportunity enabled him, for the next 50 years, to spend summers at the estate and return to Paris for the winters. During this time, he published 44 volumes of his *Historie Naturelle* (Natural history), famous as the first modern work that attempted to treat nature as a whole. It was essentially the first encyclopedia on natural history to encompass both plant and animal kingdoms. Assisted by several eminent naturalists of the time, Buffon organized the often-confusing wealth of material into a coherent form. Moreover, in the work, he included suggestions on how the earth might have originated, and he challenged the then-popular belief that the earth was only 6,000 years old. Besides proposing that the earth might be much older, he also suggested that the fact that animals retain parts that serve no known purpose to them is evidence that animals have evolved.

Buffon's popularity increased dramatically due to this work, and he remained a well-known scientific figure until his death in 1788. His prestige earned him an invitation to become a member of many academic societies, including those in Berlin, Germany, and St. Petersburg, Russia. Members of the aristocracy bestowed gifts upon Buffon and King Louis XV made him a count, commissioning a famous sculptor to create a bust of him.

See also Evolution, evidence of; Evolutionary mechanisms

BUTTE • *see* LANDFORMS

BUTTERFLY EFFECT • *see* CHAOS THEORY (METEOROLOGICAL ASPECTS)

C

CALCAREOUS OOZE

Calcareous ooze is the general term for layers of muddy, calcium carbonate ($CaCO_3$) bearing soft **rock** sediment on the seafloor. Of all the distinct types of veneers covering the Earth's crust—be it **soil**, sediment, snow, or ice—none are more widespread than red-clay and calcareous ooze. Only a small proportion of calcareous ooze is precipitated inorganically. For the most part, calcareous ooze comprises the fossil hard parts of planktic (Greek *planktos* = floating around) and benthic (Greek *benthos* = the deep) single-celled marine organisms whose calcium carbonate skeletons are discarded upon death or reproduction. Calcareous ooze is distinguished by its main biogenic component into foraminiferal ooze, coccolithophore ooze, or pteropod ooze, respectively. However, coccolithophorids and planktic foraminifera form the largest part of the pelagic calcareous ooze with less contribution due to pteropods, calcareous dinoflagellates, and lithothamnium.

Foraminiferal ooze contains foraminifera (in Latin, *foramen* = hole; *ferre* = bearing), large, mainly marine protozoans that bear a shell perforated with small holes through which temporary cytoplasmic protrusions (pseudopodia) project. Foraminifera are divided into planktic and benthic foraminifera that inhabit the upper few hundred meters or the bottom of the world **oceans**, respectively. Their global distribution through passive transport by ocean currents, coupled with their prolific productivity and sensitivity to environmental variations, has led to their utilization for interpreting marine sediments. Despite their low number of modern species, their vast quantities produce a sediment cover that occupies roughly one third of the entire earth's surface.

Coccolithophorid ooze contains coccolithophorids (in Greek, *kokkos* = grain; *lithos* = stone; *phoros* = bearing), belong to marine nanno-phytoplankton (algae) whose cells (the so-called coccospheres) are covered by calcite platelets (the so-called coccoliths). However, the ratio of coccoliths per cell varies for different species. Coccolithophorids live in all oceans and depending on geographical zonation, species dom-

inance is changing. Ecological strategies are likely to enable certain species to adapt to different **temperature**, nutrient, light, or energy regimes. Once dead, coccolithophorids disintegrate into single coccoliths that lastly are preserved as coccolith (ophorid) ooze. Coccolithophorids (and even more their coccoliths) may be small in size, but they occur in huge numbers in the sediment.

Pteropod ooze contains pteropods (in Greek *pteron* = wing; *pod* = foot), marine gastropod mollusks adapted to pelagic life that have a foot with wing-shaped lobes used as swimming organs. They are abundant in all oceans, although most species seem to prefer the circum-global tropical and subtropical regions. Distribution of pteropods is limited by **water** depth, temperature, salinity, **oxygen** content, and nutrient supply. They form very thin and fragile shells that hardly preserve under biochemical (e.g., dissolution) or physical (e.g., ingestion) attack. For this reason, preservation of pteropod ooze is mostly restricted to shallow parts of the oceans, i.e., **continental shelf**, slopes, ridges and rises.

Calcium carbonate consists of calcium (Ca^{2+}), inorganic **carbon** (C^{4+}), and oxygen (O^{2-}) ions. Calcium ions are derived from **weathering** of continental calcareous hard rocks and are available in excess. In marine **geochemistry**, carbonate is expressed as total dissolved inorganic carbon and carbonate alkalinity. However, bicarbonate (HCO_3^-) and carbonate (CO_3^{2-}) ions are the predominant forms of dissolved CO_2 in sea water. The simplified calcification reaction:

$$2\ HCO_3^- + Ca^{2+} \leftrightarrows CaCO_3 + CO_2 + H_2O$$

shows that dissolved inorganic carbon (and carbonate alkalinity) lower while, in turn, CO_2 is released to the atmosphere. Consequently, $CaCO_3$ **precipitation** by marine organisms acts as one source for CO_2. Depending on the mineral structure, $CaCO_3$ is called calcite (trigonal structure) or aragonite (rhombic structure).

Due to a complex carbonate **chemistry**, calcareous ooze begins to dissolve below the calcium carbonate lysocline in the water column. Below the calcium carbonate compensation depth (**CCD**) calcareous ooze is completely dissolved.

$CaCO_3$-bearing hard parts carry unique geochemical signals, namely the naturally fractionated isotopes of the elements oxygen (i.e., ^{16}O, ^{18}O) and carbon (i.e., ^{12}C, ^{13}C, ^{14}C). In carbonated water, oxygen and carbon in the dissolved CO_2 and in the surrounding water exchange until there are set amounts of each isotope (^{16}O:^{18}O and ^{12}C:^{13}C:^{14}C, respectively) in CO_2 and H_2O. These amounts are determined by the bonding properties of each molecule type for each isotope, and are a function of temperature. The ^{16}O:^{18}O ratio gets higher the colder the water is from which it precipitates. Since marine organisms use ambient HCO_3^- and CO_3^{2-} ions to build their hard parts, we have knowledge of the isotopic composition of total CO_2 in sea water by measuring CO_2 in calcium carbonate precipitates.

The discovery of calcareous ooze in the deep-sea during the *H.M.S. Challenger* expedition (1872–76) stimulated crucial modern **climate** research. Calcareous ooze became a reliable recorder of past environmental conditions on Earth containing information on ancient **biosphere**, hydrosphere, and atmosphere properties. Among them are abundance and distribution patterns of organism assemblages, oxygen and carbon isotope signatures of calcareous hard parts, bulk sediment properties, etc. (the so-called proxy parameters). Past ocean characteristics (temperature, productivity, etc.) can therefore be deduced by determining the different proxies for any chosen sample. A set of consecutively dated samples (e.g., **chronostratigraphy** by means of ^{14}C, Rb/Sr, K/Ar) consequently yields time series of fossil records that can be transformed into successions of environmental conditions.

Investigating calcareous ooze of modern (i.e., Holocene) and ancient (e.g., Pleistocene) oceans means to elucidate the role that oceanic processes play in global climate change during various **geologic time** intervals and at different levels of precision. Among them, the global carbon cycle is one of the topics that many scientific disciplines have turned their attention to, since short and long term variations of CO_2 in the atmosphere can be attributed—at least partially—to fossil fuel emissions and other human activities.

These calcareous sediment records also contain information relating to the history of adjacent land masses, providing insight into the history of climate and vegetation cover on the continents.

See also Dating methods; Limestone

CALDERA

A caldera is a large, usually circular depression at the summit of a **volcano**. Most calderas are formed by subsidence or sinking of the central part of the volcano; a rare few are excavated by violent explosions.

Craters and calderas are distinct structures. Both are circular depressions at the tops of volcanoes, but a crater is much smaller than a caldera and is formed by the building up of material around a vent rather than by the subsidence of material below a cone.

A volcano's summit may subside in two ways. First, eruptions of large volumes of **pumice** or **magma**, or subterranean drainage of the latter to other areas, may empty a chamber beneath the volcano into which a portion of the cone collapses. Second, the summit of the volcano may act as a thin roof over a large **magma chamber** that breaks under its own weight and sinks, partly or wholly, into the magma beneath. The term cauldron is sometimes reserved for calderas formed by the foundering of a cone summit in underlying magma.

The largest volcanic structures in the world are resurgent calderas. Resurgent calderas form following intense **volcanic eruptions** comparable in violence to asteroid impacts. (None has occurred during historical times.) During such an eruption, vast ejections of volcanic material—in some cases, thousands of cubic miles of pumice and ash—excavate very wide underground chambers, much wider than the volcano itself. Large calderas, up to hundreds of square miles in extent, collapse into these chambers. After settling, the caldera floor resurges or bulges up again, lifted by the refilling magma chamber below. Is in the case of the 22 mile (35 km) wide Cerro Galan caldera in Argentina, which is visible as a whole only from orbit, resurgence has raised the center of the caldera to almost a mile (1500 m) above the point of lowest subsidence.

Caldera complexes—overlapping calderas, some swallowing parts of others—are sometimes formed by repeated episodes of partial subsidence. Calderas and caldera complexes are common not only on Earth but on other bodies in the **Solar System** where volcanoes have erupted in the past or are presently erupting, including Mars, Venus, and Io.

See also Silicic

CALICHE

Caliche and calcrete are obsolete terms for well-developed calcic horizons that are common to soils in arid and semi-arid areas, and which are now known to **soil** scientists and geomorphologists as Bk or K horizons. Caliche is also a colloquial term that has many different uses among miners in Spanish speaking countries.

Calcic horizons form by the gradual **precipitation** of calcium carbonate ($CaCO_3$) and, to a lesser extent, magnesium carbonate ($MgCO_3$) within the B horizon of a soil and follow several well-documented stages of development ranging from I to VI. Stage I calcic soil horizons consist of partial carbonate coatings over the bottoms of gravel particles in the B horizons of coarse grained soils and thin carbonate filaments in the B horizons of fine grained soils. By stage III, carbonate is continuous throughout the zone of accumulation, and the zone of carbonate accumulation is known as the K horizon. Stages IV through VI are characterized be complete carbonate cementation of the former soil and, ultimately, brecciation. These most highly developed calcic horizons are sometimes referred to as petrocalcic because of their rock-like nature, and often form cap rocks atop bluffs and escarpments in arid to semi-arid regions such as the southwestern United States.

The primary source of carbonate in calcic soils is atmospheric, both as carbonate rich dust and rainwater that percolates through the soils carrying dissolved bicarbonate ions. In rare cases, calcic horizons can be formed by other processes such as the upward wicking of carbonate-rich **water** from shallow water tables. Gypsic or halic soils are formed in arid environments when **gypsum** ($CaSO_4 \cdot H_2O$) or halite (NaCL) are precipitated instead of carbonates.

Rates of soil development are controlled by many factors, so universal conclusions about the time required to form calcic soil horizons cannot be drawn. Studies in southern New Mexico have shown, however, that Stage I calcic soils can be hundreds to thousands of years old, State II calcic soils can be thousands to tens of thousands of years old, and Stage III and higher calcic soils can be tens to hundreds of thousands of years old.

See also Breccia; Desert and desertification; Limestone; Soil and soil horizons

CALIFORNIA CURRENT • *see* OCEAN CIRCULATION AND CURRENTS

CALVING OF GLACIERS • *see* GLACIATION

CAMBRIAN PERIOD

Cambrian is the name given to a period of time in Earth's history (i.e., Cambrian Period), which spanned 570–510 million years ago. The proper name Cambrian is also given to all the rocks that formed during that time (i.e., Cambrian System). In other words, the Cambrian System is the **rock** record of events that occurred—and organisms that lived—during an interval of geological time called Cambrian Period. Cambrian is the initial period of the **Paleozoic Era**.

Cambrian is a name derived from the Roman name for Wales, which was *Cambria*. Wales was the original study location for sedimentary rock formed during this interval of Earth history. The term Cambrian was first used in 1835 by Professor Adam Sedgwick (1785–1873) of Cambridge University, who was studying the lower part of what was then called Transition strata (the oldest known **sedimentary rocks**) in Wales. Sedgwick was working in the same general **area** as another prominent stratigrapher of the day, Roderick Merchison (1792–1871), whose focus was upon the overlying Silurian System. Merchison eventually showed that there was some overlap in the original concept of Sedgwick's Cambrian System and his own Silurian System, and ultimately advocated (c. 1852) that the Cambrian System was in fact part of the Silurian. It was not until a comprehensive study of the Cambrian-Silurian overlap problem produced the intervening Ordovician System (1879), that the Cambrian System was fully accepted by all geologists. Since their recognition and definition during the nineteenth century, Cambrian strata have been mapped on all the world's continents.

During Cambrian, the breakup of the supercontinent of Gondwana began with the separation of some landmasses including part of **Asia** and the ancient continents called Baltica and Laurentia (i.e., proto-North America). During its separation from the main Gondwana land mass, Laurentia had a collision with the southern end of what is now **South America** (specifically western Argentina), which resulted in some crustal deformation and mountain building. At this time, there was essentially a single world ocean, which is referred to as Panthalassa.

During most of Cambrian, global sea levels were at relatively high elevations as compared with most of the balance of Earth's history. The world's continents were mainly low-lying deserts and alluvial plains, and the rising Cambrian sea—in what is known as the Sauk transgression—encroached upon these areas, thus forming vast epicontinental **seas**. For example, during most of Cambrian, sea level was so high that an epicontinental sea covered **North America** except for a series of low islands running southwest-northeast along the elevated middle part of the continent (i.e., the Transcontinental Arch) and some parts of the low-lying Canadian cratonic shield region.

Cambrian was a time of rising global temperatures and Cambrian global **climate** ultimately became warmer than today. During Cambrian, there were essentially no polar or high-altitude **glaciers**. Further, there were no continents located at polar positions. The Cambrian Earth likely had more equitable climates than present because of the large amount of surficial seawater (approximately 85% or more, compared to approximately 70 % at present) and lack of significant topographic **relief**. Winds were likely confined to rather well-defined belts, and there is good evidence of persistent trade winds preserved in vast cross-bedded Cambrian sandstones.

Cambrian life in the **oceans** was very plentiful, but rather primitive by modern standards. The transition of pre-Cambrian life forms (mainly soft-body impressions in rock) to Cambrian life (shell-bearing **fossils** and other fossils with hard parts) has been referred to as the "Cambrian explosion." This explosion is more apparent than real, as the main change was the advent of preservable hard parts and shells, which seem to suddenly appear at a level near the onset of Cambrian **sedimentation**. Cambrian faunas include some very unusual creatures that may represent extinct phyla of organisms or organisms so primitive that they are not easily assigned to extant phyla. The most famous of fossil localities with such Cambrian fossils is at Mount Wapta, British Columbia, Canada (i.e., Burgess Shale outcrops). In these strata, the earliest known chordate (spinal cord-bearing animal), *Pikaia*, was first found. Other marine creatures of Cambrian seas included the archaeocyathids and stromatoporoids (two extinct, sponge-like organisms that formed reefs), primitive sponges and corals, simple pelecypods and brachiopods (two kinds of bivalves), simple molluscs, primitive echinoderms and jawless fishes, nautiloids, and a diverse group of early arthropods (including many species of trilobites). Trilobites were particularly abundant and diverse, and over 600 genera of Cambrian trilobites are known. Some species of trilobites were the first organisms to develop complex eye structures. Numerous Cambrian reefs, patch reefs, and shallow-water mounds were formed by stromatolites, a layered mass of sediment formed by the daily trapping and binding action of a symbiotic growth of blue-green algae and bacteria.

Cambrian life on land was probably quite limited. There is evidence that stromatolitic growth of blue-green algae and bacteria covered rocks and formed sediment layers at or near oceanic shorelines and lake margins. However, complex life forms are not found in Cambrian terrestrial sediments. It is possible that some arthropods may have lived partially or entirely upon land at this time, but this is speculative in the absence of fossil evidence. There were no land plants at this time, and thus Cambrian landscapes were at the mercy of **wind** and **water erosion** without any protection from vegetation. The minimal level of photosynthetic activity before and during Cambrian raised **oxygen** levels in Earth's atmosphere to approximately 10% of that found in the modern atmosphere.

The end of Cambrian came gradually with falling sea levels and the onset of slightly cooler global temperatures. During Late Cambrian, trilobite species became the first organisms known to experience widespread mass extinction. In several events during Late Cambrian, trilobite faunas were wiped out over vast areas for causes that are not completely understood. Reasons proposed for the mass extinctions include competition with other organisms and rapid shifts in global **temperature** and/or sea-level changes. Trilobites persisted into Late Paleozoic, but not as prominently as they did in Cambrian seas. Ordovician succeeded Cambrian life and conditions. During Ordovician, plant and animal life continued to diversify, tectonic activity began to be more extensive, and global climate change became more intensive.

See also Stratigraphy; Supercontinents

CANNON, ANNIE JUMP (1863-1941)

American astronomer

Annie Jump Cannon developed a stellar classification system that is considered by many astronomers to be the foundation of stellar **spectroscopy**. The science of stellar spectroscopy analyzes the photographic spectrum of a star. The spectrum, a series of colors that can range from violet to red depending on the star's **temperature**, is produced by using a **telescope** to collect a star's light and to pass it through a spectroscope. The same effect occurs when sunlight passes through raindrops to produce a **rainbow**. The spectroscope also produces a series of narrow dark lines within the spectrum known as spectral lines. These lines give further clues to a star's temperature as well as composition, motion, and other information. Because the position of the spectral lines on the spectrum may vary greatly from one star to another, scientists sometimes refer to this spectrographic data as the fingerprint of a star and consider this information crucial to all stellar theories.

In devising her classification system, Cannon recognized the atmospheric temperature of a star as the most important of the various factors that determine the intensity of a

star's spectral lines. Her method classified stars from hottest to coolest using capital letters to designate each major type of star. The letters O, B, A, F, G, K, and M represent the seven major categories. Cannon identified further distinctions within a major category by placing a number from zero to nine after each letter. For example, the **Sun** is a G–2 star.

Born in Dover, Delaware, Annie Jump Cannon was encouraged by her mother in the study of **astronomy** from an early age. Later, under the tutelage of two American astronomers, Wellesley professor Sarah Frances Whiting (1846–1927) and the Harvard College Observatory director Edward C. Pickering (1846–1919), Cannon became an expert in the relatively new field of astronomical spectroscopy.

It was Pickering who hired Cannon, along with a number of other women astronomers, to collect and catalog spectrographic data about the stars. Before attempting this enormous project, the astronomers needed a system by which hundreds of thousands of stars could be easily classified. Prior to Cannon's arrival at Harvard two other American astronomers, Wilamina Fleming (1857–1911) and Antonia Maury (1866–1952), had joined Pickering in devising two different classification systems. However, Pickering deemed both systems to be too complex or theoretical. Cannon borrowed from these earlier attempts in developing her own unique system of classification. Using her system, Cannon and her colleagues were able to classify the spectrum of over 300,000 stars. This information was published in the *Henry Draper Catalog* (1918–1924) and its extension (1925–1936). The catalog is considered to be a standard for stellar spectroscopy.

Cannon's 43-year career did not go unrewarded. In addition to receiving credit for the Draper catalogs, she also received recognition for discovering over 300 variable stars (which she incorporated into a catalog) and five novae. Cannon was the first woman to receive a doctor of astronomy degree from Groningen University (1921) and an honorary doctorate from Oxford University (1925). She was also the first woman to hold an office in the American Astronomical Society and to receive the Draper Award from the National Academy of Sciences (1931).

Scientists have used the work of Cannon and her successors to derive such information about stars as their motion, composition, brightness, and temperature. In turn, this information has led to theories about stellar life cycles. Modern astronomers, equipped with superior spectroscopes and computers have improved upon Cannon's work and are now able to sort stars into several hundred spectrographic categories. Evolving theories based upon such data are the legacy of Annie Jump Cannon's pioneering work in astronomy.

See also Stellar life cycle

CANYON

A canyon is a narrow, steep-walled, and deep valley with or without a perennial stream at the bottom. It is larger than, but otherwise similar to, a gorge. Canyon walls are commonly

Grand Canyon, Arizona. *Robert J. Huffman. Field Mark Publications. Reproduced by permission.*

composed of **bedrock** with little or no **regolith** and those with nearly vertical walls, particularly in the southwestern United States, are often referred to as slot canyons.

Canyons are characteristic of high plateaus and mountainous regions that have experienced rapid tectonic uplift, for example the Colorado Plateau physiographic province of the southwestern United States. As a region is raised due to tectonic activity, streams will adjust themselves to the change by cutting deeper valleys. Rising or falling sea levels, particularly during glacial periods, can also affect stream incision rates and valley shapes although generally not to the same degree as tectonic activity.

Canyon formation is common in arid and semi-arid climates because bedrock weathers slowly in the absence of **water**. Therefore, canyon-forming streams are able to cut vertically much more rapidly than their valleys can be widened by **mass wasting** or other erosional processes. A canyon eroded into relatively uniform bedrock, for example the Grand Canyon of the Yellowstone River or the Black Canyon of the Gunnison River, will have a generally uniform valley wall profile with no abrupt changes in slope. Canyons eroded into layered rocks with differing degrees of resistance to **erosion**, however, will have irregular or stair-stepped valley wall profiles. This is the case in the Grand Canyon in Arizona, where hard **sandstone** and **limestone** layers, as well as the metamorphic rocks of the inner gorge, form steep cliffs whereas softer shale layers form gentle slopes or benches.

Rivers running through canyons are unable to develop broad **floodplains** because they are not free to migrate laterally and deposit alluvium. Stream terraces, where they do occur, are likely to be highly localized and discontinuous. Most of the sediment delivered to canyon bottoms arrives by mass wasting processes such as **rockfall** or by debris flows when rockfall debris along side channels is mobilized during rainstorms. Rockfall accumulations and alluvial fans formed when debris flows enter the main canyon can in turn restrict stream flow and create the alternating pools and **rapids** characteristic of many canyons.

See also Alluvial system; Bedrock; Channel patterns; Drainage basins and drainage patterns; Landscape evolution; Rapids and waterfalls; Rivers; Stream valleys, channels, and floodplains

CARBON

Carbon is the non-metallic chemical element of **atomic number** 6 in Group 14 of the **periodic table**, symbol C, atomic weight 12.01, specific **gravity** as **graphite** 2.25, as **diamond** 3.51. Its stable isotopes are ^{12}C (98.90%) and ^{13}C (1.10%). The weight of the ^{12}C **atom** is the international standard on which atomic weights are based. It is defined as weighing exactly 12.00000 **atomic mass** units.

Carbon has been known since prehistoric times. It gets its name from *carbo*, the Latin word for charcoal, which is almost pure carbon. In various forms, carbon is found not only on Earth, but in the atmospheres of other planets, in the **Sun** and stars, in **comets** and in some meteorites.

On Earth, carbon can be considered the most important of all the **chemical elements**, because it is the essential element in practically all of the chemical compounds in living things. Carbon compounds are what make the processes of life work. Beyond Earth, carbon-atom nuclei are an essential part of the **nuclear fusion** reactions that produce the energy of the Sun and of many other stars. Without carbon, the Sun would be cold and dark.

In the form of chemical compounds, carbon is distributed throughout the world as **carbon dioxide** gas, CO_2, in the atmosphere and dissolved in all the **rivers**, **lakes** and **oceans**. In the form of carbonates, mostly calcium carbonate ($CaCO_3$), it occurs as huge rocky masses of **limestone**, **marble** and chalk. In the form of **hydrocarbons**, it occurs as great deposits of **natural gas**, **petroleum** and **coal**. Coal is important not only as a fuel, but because it is the source of the carbon that is dissolved in molten **iron** to make steel.

All plants and animals on Earth contain a substantial proportion of carbon. After hydrogen and **oxygen**, carbon is the most abundant element in the human body, making up 10.7% of all the body's atoms.

Carbon is found as the free (uncombined) element in three different allotropic forms-different geometrical arrangements of the atoms in the solid. The two crystalline forms (forms containing very definite atomic arrangements) are graphite and diamond. Graphite is one of the softest known materials, while diamond is one of the hardest.

There is also a shapeless, or **amorphous**, form of carbon in which the atoms have no particular geometric arrangement. Carbon black, a form of amorphous carbon obtained from smoky flames, is used to make rubber tires and inks black. Charcoal—wood or other plant material that has been heated in the absence of enough air to actually burn—is mostly amorphous carbon, but it retains some of the microscopic structure of the plant cells in the wood from which it was made. Activated charcoal is charcoal that has been steam-purified of all the gummy wood-decomposition products, leaving porous grains of pure carbon that have an enormous microscopic surface **area**. It is estimated that one cubic inch of activated charcoal contains 200,000 ft² (18,580 m²) of microscopic surface. This huge surface has a stickiness, called adsorption, for molecules of gases and solids; activated charcoal is therefore used to remove impurities from **water** and air, such as in home water purifiers and in gas masks.

Graphite is a soft, shiny, dark gray or black, greasy-feeling mineral that is found in large masses throughout the world, including the United States, Brazil, England, western **Europe**, Siberia, and Sri Lanka. It is a good conductor of **electricity** and resists temperatures up to about 6,332°F (3,500°C), which makes it useful as brushes (conductors that slide along rotating parts) in electric motors and generators, and as electrodes in high-temperature electrolysiscells. Because of its slipperiness, it is used as a lubricant. For example, powdered graphite is used to lubricate locks, where oil might be too viscous. The "lead" in pencils is actually a mixture of graphite, **clay,** and wax. It is called "lead" because the metallic element **lead** (Pb) leaves gray marks on paper and was used for writing in ancient times. When graphite-based pencils came into use, they were called "lead pencils."

The reason for graphite's slipperiness is its unusual crystalline structure. It consists of a stack of one-atom-thick sheets of carbon atoms, bonded tightly together into a hexagonal pattern in each sheet, but with only very weak attractions—much weaker than actual chemical bonds—holding the sheets together. The sheets of carbon atoms can therefore slide easily over one another; graphite is slippery in the same way as layers of wet leaves on a sidewalk.

Diamond, the other crystalline form of pure carbon, is the world's hardest natural material, and is used in industry as an abrasive and in drill tips for drilling through **rock** in oil fields and human teeth in dentists' offices. On a hardness scale of one to ten, which mineralogists refer to as the Mohs scale of hardness, diamond is awarded a perfect ten. But that's not why diamonds are so expensive. They are the most expensive of all gems, and are kept that way by supply and demand. The supply is largely controlled by the De Beers Consolidated Mines, Inc. in South **Africa**, where most of the world's diamonds are mined, and the demand is kept high by the importance that is widely attributed to diamonds.

A diamond can be considered to be a single huge molecule consisting of nothing but carbon atoms that are strongly bonded to each other by covalent bonds, just as in other molecules. A one-carat diamond "molecule" contains 10^{22} carbon atoms.

The beauty of gem-quality diamonds comes from their crystal clarity, their high refractivity (ability to bend light rays) and their high dispersion—their ability to spread light of different colors apart, which makes the diamond's **rainbow** "fire." Skillful chipping of the gems into facets (flat faces) at carefully calculated angles makes the most of their sparkle. Even though diamonds are hard, meaning that they can't be scratched by other materials, they are brittle—they can be cracked.

Carbon is unique among the elements because its atoms can form an endless variety of molecules with an endless variety of sizes, shapes and chemical properties. No other element

can do that to anywhere near the degree that carbon can. In the **evolution** of life on Earth, nature has always been able to "find" just the right carbon compound out of the millions available, to serve just about any required function in the complicated **chemistry** of living things.

Carbon-containing compounds are called organic compounds, and the study of their properties and reactions is called organic chemistry. The name organic was originally given to those substances that are found in living organisms—plants and animals. Almost all of the chemical substances in living things are carbon compounds (water and **minerals** are the obvious exceptions), and the name organic was eventually applied to the chemistry of all carbon compounds, regardless of where they come from.

Having the atomic number six, every carbon atom has a total of six protons. Therefore, all carbon atoms with a neutral charge also have a total of six electrons. Two are in a completed inner orbit, while the other four are valence electrons-outer electrons that are available for forming bonds with other atoms. An ion is an atom with either a negative or positive charge has either less or more electrons than the number of protons (respectively), and is referred to as either an anion (negatively charged) or a cation (positively charged).

It is impossible to summarize the properties of carbon's millions of compounds. Organic compounds can be classified into families that have similar properties, because they have certain groupings of atoms in common.

See also Carbon dating; Chemical bonds and physical properties; Chemical elements; Gemstones; Geochemistry; Historical geology

CARBON DATING

Carbon dating is a technique used to determine the approximate age of once-living materials. It is based on the decay rate of the radioactive carbon isotope ^{14}C, a form of carbon taken in by all living organisms while they are alive.

Before the twentieth century, determining the age of ancient **fossils** or artifacts was considered the job of paleontologists or paleontologists, not nuclear physicists. By comparing the placement of objects with the age of the **rock** and silt layers in which they were found, scientists could usually make a general estimate of their age. However, many objects were found in caves, frozen in **ice**, or in other areas whose ages were not known; in these cases, it was clear that a method for dating the actual object was necessary.

In 1907, the American chemist Bertram Boltwood (1870–1927) proposed that rocks containing radioactive uranium could be dated by measuring the amount of **lead** in the sample. This was because uranium, as it underwent radioactive decay, would transmute into lead over a long span of time. Thus, the greater the amount of lead, the older the rock. Boltwood used this method, called radioactive dating, to obtain a very accurate measurement of the age of Earth. While the uranium-lead dating method was limited (being only applicable to samples containing uranium), it was proved to scientists that radioactive dating was both possible and reliable.

The first method for dating organic objects (such as the remains of plants and animals) was developed by another American chemist, Willard Libby (1908–1980). He became intrigued by carbon–14, a radioactive isotope of carbon. Carbon has isotopes with atomic weights between 9 and 15. The most abundant isotope in nature is carbon–12, followed in abundance by carbon–13. Together carbon–12 and carbon–13 make up 99% of all naturally occurring carbon. Among the less abundant isotopes is carbon–14, which is produced in small quantities in the earth's atmosphere through interactions involving cosmic rays. In any living organism, the relative concentration of carbon–14 is the same as it is in the atmosphere because of the interchange of this isotope between the organism and the air. This carbon–14 cycles through an organism while it is alive, but once it dies, the organism accumulates no additional carbon–14. Whatever carbon–14 was present at the time of the organism's death begins to decay to nitrogen–14 by emitting radiation in a process known as beta decay. The difference between the concentration of carbon–14 in the material to be dated and the concentration in the atmosphere provides a basis for estimating the age of a specimen, given that the rate of decay of carbon–14 is well known. The length of time required for one-half of the unstable carbon–14 nuclei to decay (i.e., the **half-life**) is 5,730 years.

Libby began testing his carbon–14 dating procedure by dating objects whose ages were already known, such as samples from Egyptian tombs. He found that his methods, while not as accurate as he had hoped, were fairly reliable. He continued his research and, through improvements in his equipment and procedures, was eventually able to determine the age of an object up to 50,000 years old with a precision of plus-or-minus 10%. Libby's method, called radiocarbon or carbon–14 dating, gave new impetus to the science of radioactive dating. Using the carbon–14 method, scientists determined the ages of artifacts from many ancient civilizations. Still, even with the help of laboratories worldwide, radiocarbon dating was only accurate up to 70,000 years old, since objects older than this contained far too little carbon–14 for the equipment to detect.

Starting where Boltwood and Libby left off, scientists began to search for other long-lived isotopes. They developed the uranium-thorium method, the potassium-argon method, and the rubidium-strontium method, all of which are based on the transformation of one element into another. They also improved the equipment used to detect these elements, and in 1939, scientists first used a cyclotron particle accelerator as a mass spectrometer. Using the cyclotron, carbon–14 dating could be used for objects as old as 100,000 years, while samples containing radioactive beryllium could be dated as far back as 10–30 million years. A newer method of radioactive tracing involves the use of a new clock, based on the radioactive decay of ^{235}uranium to ^{231}protactinium.

See also Fossils and fossilization; Geochemistry

CARBON DIOXIDE

Carbon dioxide was the first gas to be distinguished from ordinary air, perhaps because it is so intimately connected with the cycles of plant and animal life. Carbon dioxide is released during respiration and combustion. When plants store energy in the form of food, they use up carbon dioxide. Early scientists were able to observe the effects of carbon dioxide long before they knew its function.

About 1630, Flemish scientist Jan van Helmont discovered that certain vapors differed from air that was then thought to be a single substance or element. Van Helmont coined the term gas to describe these vapors and collected the gas given off by burning wood, calling it gas sylvestre. Today, it is known that this gas is carbon dioxide, and van Helmont is credited with its discovery. In 1756, Joseph Black proved that carbon dioxide, which he called fixed air, is present in the atmosphere and that it combines with other chemicals to form new compounds. Black also identified carbon dioxide in exhaled breath, determined that the gas is heavier than air, and characterized its chemical behavior as that of a weak acid. The pioneering work of van Helmont and Black soon led to the discovery of other gases. As a result, scientists began to realize that gases must be weighed and accounted for in the analysis of chemical compounds, just like solids and liquids.

In 1783, French physicist Pierre Laplace (1749–1827) used a guinea pig to demonstrate quantitatively that **oxygen** from the air is used to burn carbon stored in the body and produce carbon dioxide in exhaled breath. Around the same time, chemists began drawing the connection between carbon dioxide and plant life. Like animals, plants breathe using up oxygen and releasing carbon dioxide. Plants, however, also have the unique ability to store energy in the form of carbohydrates, our primary source of food. This energy-storing process, called photosynthesis, is essentially the reverse of respiration. It uses up carbon dioxide and releases oxygen in a complex series of reactions that also require sunlight and chlorophyll (the green substance that gives plants their color). In the 1770s, Dutch physiologist Jan Ingen Housz established the principles of photosynthesis.

English chemist John Dalton guessed in 1803 that the molecule contains one carbon **atom** and two oxygen atoms (CO_2); this was later proved correct. The decay of all organic materials produces carbon dioxide slowly, and Earth's atmosphere contains a small amount of the gas (about 0.033%). Spectroscopic analysis has shown that in our **solar system**, the planets of Venus and Mars have atmospheres rich in carbon dioxide. The gas also exists in ocean **water**, where it plays a vital role in marine plant photosynthesis.

In modern life, carbon dioxide has many practical applications. For example, fire extinguishers use CO_2 to control electrical and oil fires that cannot be put out with water. Because carbon dioxide is heavier than air, it spreads into a blanket and smothers the flames. Carbon dioxide is also an effective refrigerant. In its solid form, known as dry **ice**, it is used to chill perishable food during transport. Many industrial processes are also cooled by carbon dioxide, which allows faster production rates. For these commercial purposes, carbon dioxide can be obtained from either **natural gas** wells, fermentation of organic material, or combustion of fossil **fuels**.

Recently, carbon dioxide has received negative attention as a greenhouse gas. When it accumulates in the upper atmosphere, it traps the Earth's heat, eventually causing **global warming**. Since the beginning of the industrial revolution in the mid 1800s, factories and power plants have significantly increased the amount of carbon dioxide in the atmosphere by burning **coal** and other fossil fuels. This effect was first predicted by **Svante August Arrhenius**, a Swedish physicist, in the 1880s. Then in 1938, British physicist G. S. Callendar suggested that higher CO_2 levels had caused the warmer temperatures observed in America and **Europe** since Arrhenius's day. Modern scientists have confirmed these views and identified other causes of increasing carbon dioxide levels, such as the clearing of the world's **forests**. Because trees extract CO_2 from the air, their depletion has contributed to upsetting the delicate balance of gases in the atmosphere.

In rare circumstances, carbon dioxide can endanger life. In 1986, a huge cloud of the gas exploded from Lake Nyos, a volcanic lake in northwestern Cameroon, and quickly suffocated more than 1,700 people and 8,000 animals. Scientists have attempted to control this phenomenon by slowly pumping the gas up from the bottom of the lake.

See also Atmospheric chemistry; Atmospheric composition and structure; Atmospheric pollution; Forests and deforestation; Global warming; Greenhouse gases and greenhouse effect

CARBON MONOXIDE • *see* ATMOSPHERIC POLLUTION

CARBONIFEROUS PERIOD • *see* MISSISSIPPIAN PERIOD

CARMICHAEL STOPES, MARIE CHARLOTTE (1880-1958)

Scottish geologist, paleobotanist, and social reformer

Although best known for her later work on birth control issues, Marie Stopes began her career as a geologist and paleobotantist. Stopes advanced the classification of coal-associated macerals (microscopic organic portions of **coal**) through an identification system. Stopes' work on petrography (the classification of coal and other **petroleum** related deposits) contributed to the modern system of identification based upon color, reflecting ability, and general morphology. Stopes' was an accomplished expert on the subject of coal balls (roundish nodules composed of mineral and plant deposits).

Stopes was born in Edinburgh, Scotland to the English architect, archeologist, and geologist Henry Stopes (1852–1902) and his feminist wife, Charlotte Carmichael (1841–1929), one of the first women to attend a Scottish university. Stopes and her younger sister, Winnie, were raised in London in a curious mixture of socially progressive scientific

thought and stern Scottish Protestantism. Her authoritarian mother trusted the Bible, but supported woman suffrage, clothing reform, and free thought. Stopes' father cared mainly for science. As a young girl Stopes met many of her father's friends in the British Association for the Advancement of Science, including Francis Galton (1822–1911), Thomas Henry Huxley (1825–1895), Norman McColl (1843–1904), and Charles Edward Sayle (1864–1924). Through them came Stopes' interest in Charles Darwin, **evolution**, and eventually, eugenics.

Stopes enrolled at University College, London, in 1900 on a science scholarship, graduating with a B.Sc. in 1902 with honors in botany and **geology**. She did graduate work there until 1903, then at the University of Munich, where she received her Ph.D. in paleobotany in June 1904. In October of the same year, Stopes became the first woman scientist on the faculty of the University of Manchester. In 1905, University College made her the youngest Briton of either gender to earn the D.Sc. She studied at the Imperial University of Tokyo from 1907 to 1908, then returned to Manchester in 1909. Stopes married botanist and geneticist Reginald Ruggles Gates (1882–1962) in 1911, but obtained an annulment five years later.

Inspired by meeting Margaret Sanger (1879–1966) in 1915, Marie Stopes began crusading for sexual freedom and birth control. With her second husband, Humphrey Verdon Roe (1878–1949), she opened the first birth control clinic in Great Britain, "The Mothers' Clinic" in Holloway, North London, on 17 March 1921.

Devoted to eugenics, Stopes founded the Society for Constructive Birth Control and Racial Progress in 1921, and after 1937 was a Life Fellow of the British Eugenics Society. Stopes become controversial, in part because she advocated the involuntary sterilization of anyone she deemed unfit for parenthood, including the mentally impaired, addicts, subversives, criminals, and those of mixed racial origin. At one time, Stopes persecuted her son, Harry Stopes Roe (b. 1924), for marrying a woman with bad eyesight. While Sanger's main motivation in promoting birth control was to relieve the misery of the poor, Stopes campaigned vigorously and often flamboyantly for birth control to prevent "inferior" women from reproducing, and to allow all women to lead sexually fulfilling lives without fear of pregnancy. Stopes made enemies on all sides of the issue. Havelock Ellis Sanger (1859–1939), and other left-leaning rivals within the birth control movement accused her of anti-Semitism, political conservatism, and egomania. Stopes' strongest opposition came from the Roman Catholic Church, especially because, unlike most other early advocates of birth control, she did not oppose abortion.

By her own account Stopes had three distinct careers, a scientist until about 1914, a social reformer until the late 1930s, and a poet thereafter. Among her books are *Married Love: A New Contribution to the Solution of Sex Difficulties* (1918), *Wise Parenthood: A Sequel to "Married Love": A Book for Married People* (1919), *Radiant Motherhood: A Book for Those who are Creating the Future* (1920), *Contraception (Birth Control): Its Theory, History and Practice* (1923), *The Human Body* (1926), *Sex and the Young* (1926), *Enduring Passion: Further New Contributions to the*

Solution of Sex Difficulties (1928), *Mother England: A Contemporary History* (1929), *Roman Catholic Methods of Birth Control* (1933), *Birth Control To-Day* (1934), *Marriage in my Time* (1935), *Change of Life in Men and Women* (1936), and *Your Baby's First Year* (1939). Stopes died quietly at her home near Dorking, Surrey, England.

See also Petroleum detection; Petroleum, economic uses of; Petroleum extraction; Petroleum, history of exploration

CARSON, RACHEL (1907-1964)
American marine biologist

Rachel Carson is best known for her 1962 book, *Silent Spring,* which is often credited with beginning the modern environmental movement in the United States. The book focused on the uncontrolled and often indiscriminate use of pesticides, especially dichlorodiphenyltrichloroethane (commonly known as DDT), and the irreparable environmental damage caused by these chemicals. The public outcry Carson generated by the book motivated the United States Senate to form a committee to investigate pesticide use. Her eloquent testimony before the committee altered the views of many government officials and helped lead to the creation of the Environmental Protection Agency (EPA).

Rachel Louise Carson, the youngest of three children, was born in Springdale, Pennsylvania, a small town twenty miles north of Pittsburgh. Her parents, Robert Warden and Maria McLean Carson, lived on 65 acres and kept cows, chickens, and horses. Although the land was not a true working farm, it had plenty of woods, animals, and streams, and here, near the shores of the Allegheny River, Carson learned about the relationship between the land and animals.

Carson's mother instilled in Rachel a love of nature, and taught her the intricacies of music, art, and literature. Carson's early life was one of isolation; she had few friends besides her cats, and she spent most of her time reading and pursuing the study of nature. She began writing poetry at age eight and published her first story, "A Battle in the Clouds," in *St. Nicholas* magazine at the age of 10. She later claimed that her professional writing career began at age 11, when *St. Nicholas* paid her a little over three dollars for one of her essays.

Carson planned to pursue a career as a writer when she received a scholarship in 1925 from the Pennsylvania College for Women (now Chatham College) in Pittsburgh. At the college she fell under the influence of Mary Scott Skinker, whose freshman biology course altered her career plans. In the middle of her junior year, Carson switched her major from English to zoology, and in 1928, she graduated magnum cum laude. "Biology has given me something to write about," she wrote to a friend, as quoted in *Carnegie* magazine. "I will try in my writing to make animals in the woods or waters, where they live, as alive to others as they are to me."

With Skinker's help, Carson obtained first a summer fellowship at the Marine Biology Laboratory at Woods Hole in Massachusetts and then a one-year scholarship from the Johns Hopkins University in Baltimore. While at Woods Hole over the summer, she saw the ocean for the first time and encountered her first exotic sea creatures, including sea anemones and sea urchins. At Johns Hopkins, she studied zoology and genetics. Graduate school did not proceed smoothly; she encountered financial problems and experimental difficulties but eventually managed to finish her highly detailed master's dissertation, "The Development of the Pronephoros during the Embryonic and Early Larval Life of the Catfish (*Inctalurus punctaltus*)." In June 1932, she received her master's degree.

Before beginning her graduate studies at Johns Hopkins, Carson had arranged an interview with Elmer Higgins, who was head of the Division of Scientific Inquiry at the U.S. Bureau of Fisheries. Carson wanted to discuss her job prospects in marine biology, and Higgins had been encouraging, though he then had little to offer. Carson contacted Higgins again at this time, and she discovered that he had an opening at the Bureau of Fisheries for a part-time science writer to work on radio scripts. The only obstacle was the civil service exam, which women were then discouraged from taking. Carson not only did well on the test, she outscored all other applicants. She went on to become only the second woman ever hired by the bureau for a permanent professional post.

At the Bureau of Fisheries, Carson wrote and edited a variety of government publications—everything from pamphlets on how to cook fish to professional scientific journals. She earned a reputation as a ruthless editor who would not tolerate inconsistencies, weak prose, or ambiguity. One of her early radio scripts was rejected by Higgins because he found it too "literary." He suggested that she submit the script in essay form to the *Atlantic Monthly*, then one of the nation's premier literary magazines. To Carson's amazement, the article was accepted and published as "Undersea" in 1937. Her jubilation over the article was tempered by personal family tragedy. Her older sister, Marian, died at age forty that same year, and Carson had to assume responsibility for Marian's children, Marjorie and Virginia Williams.

The *Atlantic Monthly* article attracted the notice of an editor at the publishing house of Simon & Schuster, who urged Carson to expand the four-page essay into book form. Working diligently in the evenings, she was able to complete the book in a few years; it was published as *Under the Sea-Wind*. Unfortunately, the book appeared in print in 1941, just one month before the Japanese attacked Pearl Harbor. Despite favorable, even laudatory reviews, it sold fewer than 1,600 copies after six years in print. It did, however, bring Carson to the attention of a number of key people, including the influential science writer William Beebe. Beebe published an excerpt from *Under the Sea-Wind* in his 1944 compilation *The Book of Naturalists*, including Carson's work alongside the writings of Aristotle, Audubon, and Thoreau.

The poor sales of *Under the Sea-Wind* compelled Carson to concentrate on her government job. The Bureau of Fisheries merged with the Biological Survey in 1940, and was reborn as the Fish and Wildlife Service. Carson quickly moved up the professional ranks, eventually reaching the position of biologist and chief editor after World War II. One of her postwar assignments, a booklet about National Wildlife Refuges called *Conservation in Action*, took her back into the field. As part of her research, she visited the Florida Everglades, Parker River in Massachusetts, and Chincoteague Island in the Chesapeake Bay.

After the war, Carson began work on a new book that focused on **oceanography**. She was now at liberty to use previously classified government research data on oceanography, which included a number of technical and scientific breakthroughs. As part of her research, she did some undersea diving off the Florida coast during the summer of 1949. She battled skeptical administrators to arrange a deep-sea cruise to Georges Bank near Nova Scotia aboard the Fish and Wildlife Service's research vessel, the *Albatross III*.

Entitled *The Sea around Us*, her book on oceanography was published on July 2, 1951. It was an unexpected success, and remained on the *New York Times* bestseller list for 86 weeks. The book brought Carson numerous awards, including the National Book Award and the John Burroughs Medal, as well as honorary doctorates from her alma mater and Oberlin College. Despite her inherent shyness, Carson became a regular on the lecture circuit. With financial security no longer the overriding concern it had been, she retired from government service and devoted her time to writing.

Carson began work on another book, focusing this time on the intricacies of life along the shoreline. She took excursions to the mangrove coasts of Florida and returned to one of her favorite locations, the rocky shores of Maine. She fell in love with the Maine coast and in 1953 bought a summer home in West Southport on the shore of Sheepscot Bay. *The Edge of the Sea* was published in 1955 and earned Carson two more prestigious awards, the Achievement Award of the American Association of University Women and a citation from the National Council of Women of the United States. The book remained on the bestseller list for 20 weeks, and RKO Studios bought the rights to it. In Hollywood, the studio sensationalized the material and ignored scientific fact. Carson corrected some of the more egregious errors but still found the film embarrassing, even after it won an Oscar as the best full-length documentary of 1953.

From 1955 to 1957, Carson concentrated on smaller projects, including a telescript, "Something about the Sky," for the *Omnibus* series. She also contributed a number of articles to popular magazines. In July 1956, Carson published "Help Your Child to Wonder" in the *Woman's Home Companion*. The article was based on her own real-life experiences, something rare for Carson. She intended to expand the article into a book and retell the story of her early life on her parent's Pennsylvania farm. After her death, the essay reappeared in 1965 as the book *The Sense of Wonder*.

In 1956, one of the nieces Carson had raised died at age 36. Marjorie left her son Roger; Carson now cared for him in addition to her mother. She legally adopted Roger that same year and began looking for a suitable place to rear the child. She built a new winter home in Silver Spring, Maryland, on an uncultivated tract of land, and she began another project shortly after the home was finished. The luxuriant setting inspired her to turn her thoughts to nature once again. Carson's next book grew out of a long-held concern about the overuse of pesticides. She had received a letter from Olga Owens

Huckins, who related how the aerial spraying of DDT had destroyed her Massachusetts bird sanctuary. Huckins asked her to petition federal authorities to investigate the widespread use of such pesticides, but Carson thought the most effective tactic would be to write an article for a popular magazine. When her initial queries were rejected, Carson attempted to interest the well-known essayist E. B. White in the subject. White suggested she write the article herself, in her own style, and he told her to contact William Shawn, an editor at the *New Yorker*. Eventually, after numerous discussions with Shawn and others, she decided to write a book instead.

The international reputation Carson now enjoyed enabled her to enlist the aid of an array of experts. She consulted with biologists, chemists, entomologists, and pathologists, spending four years gathering data for her book. When *Silent Spring* first appeared in serial form in the *New Yorker* in June 1962, it drew an aggressive response from the chemical industry. Carson argued that the environmental consequences of pesticide use underscored the futility of humanity's attempts to control nature, and she maintained that these efforts to assume control had upset nature's delicate balance. Although the message is now largely accepted, the book caused controversy in some circles, challenging the long-held belief that humans could master nature. The chemical companies, in particular, attacked both the book and its author; they questioned the data, the interpretation of the data, and the author's scientific credentials. One early reviewer referred to Carson as a "hysterical woman," and others continued this sexist line of attack. Some chemical companies attempted to pressure Houghton Mifflin, the book's publisher, into suppressing the book, but these attempts failed.

The general reviews were much kinder and *Silent Spring* soon attracted a large, concerned audience, both in America and abroad. A special CBS television broadcast, "The Silent Spring of Rachel Carson," which aired on April 3, 1963, pitted Carson against a chemical company spokesman. Her cool-headed, commonsense approach won her many followers and brought national attention to the problem of pesticide use. The book became a cultural icon and part of everyday household conversation. Carson received hundreds of invitations to speak, most of which she declined due to her deteriorating health. She did find the strength to appear before the Women's National Press Club, the National Parks Association, and the Ribicoff Committee—the U.S. Senate committee on environmental hazards.

In 1963, Carson received numerous honors and awards, including an award from the Izaak Walton League of America, the Audubon Medal, and the Cullen Medal of the American Geographical Society. That same year, she was elected to the prestigious American Academy of Arts and Sciences. Carson died of heart failure at the age of 56. In 1980, President Jimmy Carter posthumously awarded her the President's Medal of Freedom. A Rachel Carson stamp was issued by the U.S. Postal Service in 1981.

See also Environmental pollution

CARTOGRAPHY

Cartography is the creation, production, and study of maps. It is considered a subdiscipline of geography, the study of spatial distribution of various phenomena. Cartographers are often geographers who particularly enjoy the combination of art, science, and technology employed in the making and studying of maps.

A map is a generalized two-dimensional representation of the spatial distribution of one or more phenomena. For example, a map may show the location of cities, mountain ranges, and **rivers**, or may show the types of **rock** in a given region. Maps are flat, making their production, storage, and handling relatively easy. Maps present their information to the viewer at a reduced scale. They are smaller than the **area** they represent, using mathematical relationships to maintain proportionally accurate geographic relationships between various phenomena. Maps show the location of selected phenomena by using symbols that are identified in a legend.

There are many different types of maps. A common classification system divides maps into two categories, general and thematic. General maps are maps that show spatial relationships between a variety of geographic features and phenomena, emphasizing their location relative to one another. Thematic maps illustrate the spatial variations of a single phenomenon, or the spatial relationship between two particular phenomena, emphasizing the pattern of the distribution.

Many maps can be either general or thematic, depending on the intent of the cartographer. For example, a cartographer may produce a vegetation map, one that shows the distribution of various plant communities. If the cartographer shows the location of various plant communities in relation to a number of other geographic features, the map is properly considered a general map. The map is more likely to be considered thematic if the cartographer uses it to focus on something about the relationship of the plant communities to each other, or to another particular phenomenon or feature, such as the differences in plant communities associated with changes in elevation or changes in **soil** type.

Some examples of general maps include large-scale and medium-scale topographic maps, planometric maps, and charts. Topographic maps show all-important physical and cultural features, including **relief**. Relief is the difference in elevation of various parts of the earth's surface. Planometric maps are similar to topographic maps, but omit changes in elevation. Charts are used by the navigators of aircraft and seagoing vessels to establish bearings and plot positions and courses. World maps on a small- or medium-scale showing physical and cultural features, such as those in atlases, are also considered general maps.

Although the subject matter of thematic maps is nearly infinite, cartographers use common techniques involving points, lines, and aerial photos to illustrate the structure of spatial distribution. Isarithmic maps use lines to connect points of equal value; these lines are called isopleths, or isolines. Isopleths used for a particular phenomenon may have a particular name; for example, isotherms connect points of equal **temperature**, **isobars** connect points of equal air pressure, and

Cartographer at work. © *Christopher Cormack/Corbis. Reproduced by permission.*

isohyets connect points of equal **precipitation**. Isopleths indicating differences in elevation are called contour lines. Isopleths are used to show how certain quantities change with location.

A topographic map is a good example of how isopleths are used to present information. Topographic maps use isopleths called contour lines to indicate variations in relief. Each contour line connects points of the same elevation. Adjacent lines indicate variations in relief; these variations are called contour intervals. The contour interval is indicated in the map legend. A contour interval of 20 ft (6.1 m) means that there is a 20 ft (6.1 m) difference in elevation between the points connected by one contour line and the points connected by the adjacent contour line. The closer the lines are to each other, the more dramatic the change in elevation.

Chloropleth maps are another type of thematic map. They use areas of graduated gray tones or a series of gradually intensifying colors to show spatial variations in the magnitude of a phenomenon. Greater magnitudes are symbolized by either darker gray tones or more intense colors; lesser magnitudes are indicated by lighter gray tones or less intense colors.

Cartographers traditionally obtained their information from navigators and surveyors. Explorations that expanded the geographical awareness of a map-making culture also resulted

●

in increasingly sophisticated and accurate maps. Today, cartographers incorporate information from aerial photography and **satellite** imagery in the maps they create.

Modern cartographers face three major design challenges when creating a map. First, they must decide how to accurately portray that portion of Earth's surface that the map will represent; that is, they must figure out how to represent three-dimensional objects in two dimensions. Second, cartographers must represent geographic relationships at a reduced size while maintaining their proportional relationships. Third, they must select which pieces of information will be included in the map, and develop a system of generalization, which will make the information presented by the map useful and accessible to its readers.

When creating a flat map of a portion of the earth's surface, cartographers first locate their specific area of interest using **latitude and longitude**. They then use map **projection** techniques to represent the three-dimensional characteristics of that area in two dimensions. Finally, a grid, called a rectangular coordinate system, may be superimposed on the map, making it easier to use.

Distance and direction are used to describe the position of something in **space**, its location. In conversation, terms like right and left, up and down, or here and there are used to indicate direction and distance. These terms are useful only if the location of the speaker is known; in other words, they are relative. Cartographers, however, need objective terms for describing location The system of latitude and longitude, a geographical coordinate system developed by the Greeks, is used by cartographers for describing location.

Earth is a sphere, rotating around an axis tilted approximately 23.5 degrees off the perpendicular. The two points where the axis intersects the earth's surface are called the poles. The equator is an imaginary circle drawn around the center of the earth, equidistant from both poles. A plane that sliced through the earth at the equator would intersect the axis of the earth at a right angle. Lines drawn around the earth to the north and south of the equator and at right angles to the earth's axis are called parallels. Any point on the earth's surface is located on a parallel.

An arc is established when an angle is drawn from the equator to the axis and then north or south to a parallel. Latitude is the measurement of this arc in degrees. There are 90 degrees from the equator to each pole, and sixty minutes in each degree. Latitude is used to determine distance and direction north and south of the equator.

Meridians are lines running from the north pole to the south pole, dividing the earth's surface into sections, like those of an orange. Meridians intersect parallels at right angles, creating a grid. Just as the equator acts as the line from which to measure north or south, a particular meridian, called the prime meridian, acts as the line from which to measure east or west. There is no meridian that has a natural basis for being considered the prime meridian. The prime meridian is established by international agreement; currently, it runs through the Royal Observatory in Greenwich, England. Longitude is the measurement in degrees of the arc created by an angle drawn from the prime meridian to the earth's axis and

then east or west to a meridian. There are 180° west of the prime meridian and 180° east of it. The international date line lies approximately where the 180th meridian passes through the Pacific Ocean.

Using the geographical coordinate system of latitude and longitude, any point on the earth's surface can be located with precision. For example, Buenos Aires, the capital of Argentina, is located 34° 35 minutes south of the equator and 58° 22 minutes west of the prime meridian; Anchorage, the largest city of the state of Alaska, is located 61° 10 minutes north of the equator and 149°s 45 minutes west of the prime meridian.

After locating their area of interest using latitude and longitude, cartographers must determine how best to represent that particular portion of the earth's surface in two dimensions. They must do this in such a way that minimal amounts of distortion affect the geographic information the map is designed to convey.

Cartographers have developed map projections as a means for translating geographic information from a spherical surface onto a planar surface. A map projection is a method for representing a curved surface, such as the surface of the earth, on a flat surface, such as a piece of paper, so that each point on the curved surface corresponds to only one point on the flat surface.

There are many types of map projections. Some of them are based on geometry, others are based on mathematical formulas. None of them, however, can accurately represent all aspects of the earth's surface; inevitably there will be some distortion in shape, distance, direction or area. Each type of map projection is intended to reduce the distortion of a particular spatial element. Some projections reduce directional distortion, others try to present shapes or areas in as distortion-free a manner as possible. The cartographer must decide which of the many projections available will provide the most distortion-free presentation of the information to be mapped.

Maps present various pieces of geographical information at a reduced scale. In order for the information to be useful to the map reader, the relative proportions of geographic features and spatial relationships must be kept as accurate as possible. Cartographers use various types of scales to keep those features and relationships in the correct proportions.

No single map can accurately show every feature on the earth's surface. There is simply too much spatial information at any particular point on the earth's surface for all of the information to be presented in a comprehensible, usable format. In addition, the process of reduction has certain visual effects on geographic features and spatial relationships. Because every feature is reduced by the ratio of the reduction, the distance between features is reduced, crowding them closer together and lessening the clarity of the image. The width and length of individual features are also reduced.

See also Earth (planet); Topography and topographic maps; Weather forecasting methods

CATASTROPHIC MASS MOVEMENTS

Catastrophic mass movements are large and rapid **mass wasting** events such as landslides, rockslides, and **rock** avalanches. Although they are often believed to occur with no warning, catastrophic mass movements are often preceded by subtle changes such as rock **creep** that foreshadow their occurrence. Because of their speed and size, catastrophic mass movements are often fatal events.

One of the most notable catastrophic mass movements to have occurred during recorded times was associated with the May 18, 1980 eruption of Mount St. Helens. **Magma** movement produced a bulge on the north side of the **volcano** that failed as a series of three large landslides during a magnitude 5.2 **earthquake** on May 18, and was immediately followed by the well-known eruption. The volume of the material removed by the landslides is estimated to have been about 2.3 km^3. The landslides broke apart as they began to move and traveled downhill as a rock **avalanche**, which is a common form of catastrophic **mass movement**.

Other rock avalanches buried the towns of Frank, Alberta in 1903 and Elm, Switzerland in 1881; both of these events were triggered by miners undercutting steep slopes above the towns. In 1963 a large rockslide that traveled into the Vaiont reservoir in Italy produced a wave that overtopped the dam and killed many downstream residents. The Vaiont **landslide** was triggered by changes in the reservoir level as it was filled and emptied each year after its completion in 1960, which affected the **groundwater** pressure within the adjacent slopes. Earthquakes can also trigger catastrophic landslides without human intervention, but catastrophic landslides rarely appear to be triggered by rainfall. A notable example of an earthquake-triggered catastrophic landslide during recent times occurred in 1959, when a magnitude 7.5 earthquake triggered a landslide that dammed the Madison River in Montana, killing 26 people and creating Earthquake Lake. The geologic record contains evidence of even more catastrophic events, including prehistoric rock avalanches that involved as much as 20 km^3 of rock and traveled tens of kilometers from their points of origin.

One of the most perplexing aspects of catastrophic rock avalanches is that in many cases they begin as normal landslides but travel much longer distances than would be predicted by solving the simple **physics** problem of rock sliding along rock. The coefficients of friction necessary for rock avalanches larger than 10^6 m^3 to have traveled their observed distances decrease significantly as a function of the avalanche volume. A typical coefficient of friction for one piece of rock sliding past another might be about 0.55, and this is a value calculated for many small rockslides; in large rock avalanches, however, the coefficient of friction necessary to explain the travel distance can be as low as 0.05–0.10. In essence, rock avalanches move as though they are fluids rather than solid masses of rock and often run up the opposing sides of valleys before coming to rest. One early explanation of this phenomenon, based upon studies of the prehistoric Blackhawk landslide in California, was that

Mudslides can hit quickly with devastating results, as seen here. *Jim Sugar. Jim Sugar Photgraphy/Corbis-Bettmann. Reproduced by permission.*

rock avalanches glide atop pockets of air trapped beneath the rock mass. Other proposed friction reducing mechanisms have included the frictional heating of **water** to generate steam that would fluidize the avalanche and the **melting** of rock to produce a layer of liquid **glass** along the base of the avalanche. The discovery of rock avalanche deposits on Mars and the **Moon**, however, cast doubt on mechanisms such as air entrapment and steam generation because neither air nor water would have been available. A process known as acoustic fluidization has been proposed to explain the behavior of rock avalanches without requiring air pockets or steam generation. Acoustic fluidization occurs when elastic waves travel through a rock avalanche as it moves downhill, breaks into pieces, and is jostled by the underlying **topography**.

Because catastrophic rock avalanches are so rare and short lived, and because their remnants contain little evidence of the dynamic processes that occurred during movement, it is likely that explanations of their unusual mechanical behavior will always be inferences based largely on theoretical possibilities rather than empirical observations.

See also Catastrophism; Debris flow; Lahar

In 1908 over Tunguska, Russia, an object that is believed to have been either a comet or a stony meteorite exploded with the force of a nuclear bomb. If it had happened over an urban area instead of over Siberian wilderness, the loss of life would have been immense. *AP/Wide World. Reproduced by permission.*

CATASTROPHISM

Catastrophism is the argument that Earth's features—including mountains, valleys, and lakes—primarily formed and shaped as a result of the periodic but sudden forces as opposed to gradual change that takes place over a long period of time.

Although geologists may argue about the extent of catastrophism in shaping the earth, modern geologists interpret many formations and events as resulting from an interplay catastrophic and uniform forces that result in more slowly evolving change.

For example, according to strict catastrophe theory, one might interpret the origins of the Rocky Mountains or the Alps, as resulting from a huge **earthquake** that uplifted them quickly. When viewing the Yosemite Valley in California a catastrophist might not assert they were carved by **glaciers**,

but rather the floor of the valley collapsed over 1,000 ft (305 m) to its present position in one giant plunge. Strict catastrophic theory also argues for long periods of inactivity following catastrophic events.

In terms of modern geoscience, strict catastrophic theory (e.g., a world shaped by large single **floods**, or massive earthquakes) finds little evidence or support. Catastrophism developed in the seventeenth and eighteenth centuries when, by tradition and even by law, scientists used the Bible and other religious documents as a scientific documents.

For example, when a prominent theologian, Irish biblical scholar Bishop James Ussher in the mid-1600's work, *Annals of the World*, counted the ages of people in the Bible and proclaimed that Earth was created in 4004 B.C. (In fact, Ussher even pronounced an actual date of creation as the evening of October 22), geologists tried to work within a time frame that encompassed only around six thousand years. (Current research estimates Earth at 4.5 billion years old.) In its original form, catastrophism eventually fell from grace with the scientific community as they reasoned more logical explanations for natural history. A new concept, known as **uniformitarianism**, eventually replaced catastrophism. Uniformitarianism is the argument that mountains are uplifted, valleys carved, and sediments deposited over immense time periods by the same physical forces and chemical reactions in evidence today.

Modern catastrophism—increasingly popular since the late 1970s—argues evidence that catastrophic forces can have a profound influence on shaping Earth. For example, modern catastrophic theory argues that large objects from **space** (**Asteroids**, **Comets**, etc.) periodically collide with Earth and that these collisions can have profound effects on both the **geology** and biology of Earth. Based on the extrapolation of experimental data and the observation of large-scale events (e.g., major **volcanic eruptions**), scientists speculate that when these objects strike, they clog the atmosphere with sunlight-blocking dust and gases, ignite forest fires, and trigger volcanism. One hypothesis advances that a large asteroid impact lead to the extinction of dinosaurs roughly 65 million years ago.

See also Cambrian Period; Fossil record; Fossils and fossilization; Geologic time; Historical geology; Impact crater; K-T event; Origin of life; Orogeny; Plate tectonics; Precambrian; Torino scale

CATION · *see* CHEMICAL BONDS AND PHYSICAL PROPERTIES

CAVE

A cave is a naturally occurring hollow **area** inside the earth. Most caves are formed by some type of erosional process. The most notable exception is hollow **lava** tubes such as those found in the Hawaiian Islands. The formation of caves depends upon geologic, topographic, and hydrologic factors. These factors determine where and how caves develop, as well as their structure and shape. The study of caves is called speleology. Some caves may be small hillside openings, while others consist of large chambers and interconnecting tunnels and mazes. Openings to the surface may be large gaping holes or small crevices.

Caves hosted in rocks other than **limestone** are usually formed by **water** erosional processes. For example, **rivers** running through canyons with steep walls erode the **rock** at points where the current is strong. Such caves usually have large openings and are not too deep. Caves of this type can be found in the southwestern United States and were at one time inhabited by prehistoric American Indians known as Cliff Dwellers. Sea caves are formed by waves continually crashing against cliffs or steep walls. Often these caves can only be entered at low tide. **Ice** caves are also formed in **glaciers** and **icebergs** by meltwater that drains down crevices in the ice.

Lava caves, which are often several miles long, form when the exterior of a lava flow hardens and cools to form a roof, but lava below the surface flows out, leaving a hollow tube. **Wind** or aeolian caves usually form in **sandstone** cliffs as wind-blown **sand** abrades the cliff face. They are found in **desert** areas, and occur in a bottleneck shape with the entrance much smaller than the chamber. Talus caves are formed by boulders that have piled up on mountain slopes. The most common, largest, and spectacular caves are solution caves.

Solution caves form by chemical **weathering** of the surrounding **bedrock** as **groundwater** moves along fractures in the rock. These caves produce a particular type of terrain called karst. Karst terrain primarily forms in bedrock of calcium carbonate, or limestone, but can develop in any soluble sedimentary rock such as **dolomite**, rock **gypsum**, or rock salt. The host rock extends from near the earth's surface to below the **water table**. Several distinctive karst features make this terrain easy to identify. The most common are **sinkholes**, circular depressions where the underlying rock has been dissolved away. Disappearing streams and natural bridges are also common clues. Entrances to solution caves are not always obvious, and their discovery is sometimes quite by accident.

Formation of karst involves the chemical interaction of air, **soil**, water, and rock. As water flows over and drains into the earth's surface, it mixes with **carbon dioxide** from the air and soil to form carbonic acid (H_2CO_3). The groundwater becomes acidic and dissolves the calcium carbonate in the bedrock, and seeps or percolates through naturally occurring fractures in the rock. With continual water drainage, the fractures become established passageways. The passageways eventually enlarge and often connect, creating an underground drainage system. Over thousands, perhaps millions of years, these passages evolve into the caves seen today.

During heavy rain or flooding in a well-established karst terrain, very little water flows over the surface in stream channels. Most water drains into the ground through enlarged fractures and sinkholes. This underground drainage system sometimes carries large amounts of water, sand, and mud through the passageways and further erodes the bedrock. Sometimes ceilings fall and passageways collapse, creating new spaces and drainage routes.

Not all solution caves form due to dissolution by carbonic acid. Some caves form in areas where hydrogen sulfide gas is released from the earth's **crust** or from decaying organic material. Sulfuric acid forms when the hydrogen sulfide comes in contact with water. It chemically weathers the limestone, similar to **acid rain**.

The deep cave environment is often completely dark, has a stable atmosphere, and the **temperature** is rather constant, varying only a few degrees throughout the year. The **humidity** in limestone caves is usually near 100%. Many caves contain unique life forms, underground streams and **lakes**, and have unusual mineral formations called speleothems.

When groundwater seeps through the bedrock and reaches a chamber or tunnel, it meets a different atmosphere. Whatever mineral is in solution reacts with the surrounding atmosphere, precipitates out, and is deposited in the form of a crystal on the cave ceiling or walls. Calcite, and to a lesser

Postojna Cave, Slovakia. *Embassy of the Republic of Slovenia. Reproduced by permission.*

degree, aragonite, are the most common **minerals** of speleothems. The amount of mineral that precipitates out depends upon how much gas was dissolved in the water. For example, water that must pass through a thick layer of soil becomes more saturated with **carbon** dioxide than water that passes through a thin layer. This charges the water with more carbonic acid and causes it to dissolve more limestone from the bedrock. Later, it will form a thicker mineral deposit in the cave interior as a result.

Water that makes its way to a cave ceiling hangs as a drop. When the drop of water gives off carbon dioxide and reaches chemical equilibrium with the cave atmosphere, calcite starts to precipitate out. Calcite deposited on the walls or floors in layers is called flowstone.

Sometimes water runs down the slope of a wall, and as the calcite is deposited, a low ridge is formed. Subsequent drops of water follow the ridge, adding more calcite. Constant buildup of calcite in this fashion results in the formation of a large sheet-like formation, called a curtain, hanging from the ceiling. Curtain formations often have waves and **folds** in them and have streaks of various shades of off-white and browns. The streakiness results from variations in the mineral and **iron** content of the precipitating solution.

Often, a hanging drop falls directly to the ground. Some calcite is deposited on the ceiling before the drop falls. When the drop falls, another takes its place. As with a curtain formation, subsequent drops will follow a raised surface and a buildup of calcite in the form of a hanging drop develops. This process results in icicle-shaped speleothems called stalactites. The water that falls to the floor builds up in the same fashion, resembling an upside down icicle called a stalagmite.

Of course, there are variations in the shape of speleothems depending on how much water drips from the ceiling, the temperature of the cave interior, rates and directions of air flow in the cave, and how much dissolved limestone the water contains. Speleothems occur as tiered formations, cylinders, cones, some join together, and occasionally **stalactites and stalagmites** meet and form a tower. Sometimes, when a stalactite is forming, the calcite is initially deposited in a round ring. As calcite builds up on the rim and water drips through the center, a hollow tube called a straw develops. Straws are often transparent or opaque and their diameter may be only that of a drop of water.

Stalagmites and stalactites occur in most solution caves and usually, wherever a stalactite forms, there is also a stalagmite. In caves where there is a great deal of seepage, water may drip continuously. Speleothems formed under a steady drip of water are typically smooth. Those formed in caves where the water supply is seasonal may reveal growth rings similar to those of a tree trunk. Stalactites and stalagmites grow by only a fraction of an inch or centimeter in a year, and since some are many yards or meters long, one can appreciate the time it takes for these speleothems to develop.

The most bizarre of speleothems are called helictites. Helictites are hollow, cylindrical formations that grow and twist in a number of directions and are not simply oriented according to the gravitational pull of a water drop. Other influences such as crystal growth patterns and air currents influence the direction in which these speleothems grow. Helictites grow out from the side of other speleothems and rarely grow larger than 4 in (8.5 cm) in length.

Speleothems called anthodites are usually made of aragonite. Calcite and aragonite are both forms of calcium carbonate, but crystallize differently. Anthodites grow as radiating, delicate, needle-like **crystals**. Pools of seepage water that drain leave behind round formations called cave popcorn. Cave pearls are formed in seepage pools by grains of sand encrusted with calcite; flowing water moves the grains about and they gather concentric layers of calcite.

See also Erosion; Stalactites and stalagmites

CAVE MINERALS AND SPELEOTHEMS

Cave minerals are secondary minerals formed inside a cave resulting from one or several of the following processes: reprecipitation of the **bedrock**, supersaturation of solutions, dehydration, biogenic processes (or human activity), **hydrothermal processes**, hypergenic processes (**weathering** or metasomatoze), reaction of karst solutions with minerals of non-karstic bedrock, or by eruptive processes (fumarolic or magmatic) due to crystallization of volcanic gases or their reactions with minerals or solutions.

Most speleothems (dripstones, including stalagmites and **stalactites**) are formed by hydrocarbonate reprecipitation of carbonate bedrock. **Groundwater** saturated with carbonate dioxide dissolves calcium carbonate ($CaCO_3$ from the bedrock and reprecipitates it inside the cave when carbonate dioxide volatilizes. Most caves are developed in **limestone** or **marble**, so $CaCO_3$ forms most speleothems. Ninety-five percent of all speleothems are formed by calcite, 2–3% by aragonite (the second polymorphic form of $CaCO_3$) and less than 2% are formed by the rest of about 250 cave minerals, most frequently **gypsum** (as some caves are developed in gypsum bedrock).

More than 250 different cave minerals have been identified so far. Many are found only in caves, and some are found only in one cave in the world.

Characteristic red luminescence of hydrothermal calcite. © *Y. Shopov. Reproduced by permission.*

Air **temperature** and **humidity** are exceptionally stable in deep parts of caves. This allows preservation of highly unstable minerals and speleothems, which tend to transform and disintegrate within months if taken out of the cave. Such minerals are crystallohydrates with high numbers of **water** molecules. These minerals dehydrate at room air temperature and humidity, and their mineral aggregates disintegrate to powder.

Speleothems are the form of appearance (aggregates) of secondary cave minerals. Aggregates of primary minerals or detrital materials in caves are not speleothems. Most commonly, speleothems grow from the dripping of flowing water. Stalagmites and stalactites are formed by dripping water. Flowstones are smooth and sheet-like speleothems, which form from films of flowing water. Crystal orientation inside flowstones is perpendicular to the surface, and usually has distinctive layers. Shields are oval speleothems consisting of two parallel spherical plates separated by a medial planar crack and are formed by water seeping thought the medial crack. Cave pearls are small spheroidal solid polycrystalline speleothems. Sometimes they are polished by **rotation** due to dripping of water producing turbulent flows in small pools. Thus, they rotate and polish pearls in the pool. Cave pearls resemble real pearls from the sea, but are usually composed of calcite. They become fractured quickly if removed from the cave.

A variety of speleothems grow from cave pools and **lakes**. Rimstones are deposits that form around the bottom and walls of a cave pool. Rimstone dams (gours) are secondarily deposited barriers obstructing flowing of a cave stream or pool. Shelfstones are flat-ledge or shelf-like deposits around the edges of a cave pool or a speleothem submerged in it. Cave rafts are thin layers of crystalline material that float on the surface of a cave pool. Cave bubbles are crystalline material deposited on the surface of a gas bubble in a cave pool.

Large cave ice skeleton crystal in Crow's Nest Pass. © Y. Shopov. Reproduced by permission.

Phosphorescence of calcite cave pearls. © Y. Shopov. Reproduced by permission.

Bubbles are hollow, with diameters less than 1 cm and have wall thickness of less than 0.2 mm.

Spar is formed by high amounts of water at stable temperature, humidity, and **evaporation** rate. Spar is a speleothem consisting of translucent **crystals** with a vitreous luster, and may be formed by hydrothermal, epithermal or infiltration waters.

Thin films of water form typically form coralloides and anthodites. Coralloides (corallites) are the most common type of speleothem after stalactites, stalagmites, and flowstone. They are nodular, globular or coral-like in shape. Anthodites are speleothems composed of radiating, spiky, quill-like crystal clusters. Frostwork anthodites consist of needle-like crystals. They grow from capillary water moving over the surface of the crystals. Sometimes anthodites are tubular, but never frostwork. Christmas trees are deposits of frostwork over stalagmites.

Helictites are twisted and worm-like speleothems growing via a tiny (0.008–0.5 mm) capillary canal in their center. Helictites twist in any direction, independently of **gravity**. Many helictites resemble a living species. Usually, helictites composed of calcite are transparent, while those of aragonite are composed of bunches of fine crystals, and are opaque.

Cave hair (angel hair) is fibrous speleothem composed of single crystal fibers resembling thin strands of hair. It is formed by gypsum or by highly soluble sulfate or nitrate minerals. Cave hair are very fast growing, but highly unstable (seasonal speleothems), and can be dissolved by increasing of the air humidity or by a heavy rain. Cave flowers are speleothems with crystal petals that curve radially outwards from a common center. They consist of an aggregate of branching and curving bundles of parallel crystals loosely packed together. Cave flowers are usually formed by gypsum.

All such fibrous speleothems grow from their base extruding crystals through the pores of the bedrock or **clay** substrate.

Moonmilk is an aggregate of microcrystalline cave minerals precipitated quickly under highly metastable conditions. Moonmilk resembles cream cheese when wet, but is crumbly and powdery like chalk when dry. Cave balloons are pearly, thin-walled, free-hanging speleothems associated with moonmilk substrate resembling a small, inflated balloon. Usually, they are composed wholly or partly of hydromagnesite.

One-hundred and two cave minerals are known to form coatings and crusts, 57 form **stalactites and stalagmites**, 23 form moonmilk, 15 form anthodites, 14 form helictites, 12 form Angel hair, 7 form coralloides and pearls, and 6 form cave balloons.

Cave minerals and speleothems are so unique in appearance that they are considered natural heritage objects, and laws in most countries prohibit their collection, mining and selling.

See also Stalactites and stalagmites

CCD (CARBONATE COMPENSATION DEPTH)

In **oceanography**, the depth where carbonate ions under saturation in the **water** column or in the sediment pore and the water interface is large enough so that the rate of calcium carbonate ($CaCO_3$) **sedimentation** is totally compensated for by the rate of calcium carbonate dissolution, reaches the carbonate compensation depth (CCD). Alternately stated, the CCD is the depth at which calcareous skeletons of marine animals accumulate at the same rate at which they dissolve. Depending on the mineral structure of $CaCO_3$, the CCD is called calcite

Stalactites in Cueva Nerja Cave, Spain. © Y. Shopov. Reproduced by
permission.

compensation depth (trigonal structure) or aragonite compensation depth (rhombic structure), respectively.

Foraminifera, coccolithophorids, pteropods, and a few other benthic and planktic organisms build calcium carbonate shells or skeletons. Upon death or reproduction, the shells are discarded and sink to the sea-floor. Within the water column, calcium (Ca^{2+}) content varies little, hence the calcium carbonate saturation state (CSS) is controlled by concentration of carbonate (CO_3^{2-}) ions, **pH**, water pressure, **temperature**, and salinity:

$$CSS = (Ca^{2+}) \times (CO_3^{2-}) \div K'_{sp}$$

whereas K'_{sp} is the equilibrium solubility product for the mineral phase of calcite or aragonite, respectively. It is CSS: supersaturated > 1 = saturated = 1 > undersaturated with carbonate ions. Position and thickness of the saturation horizon in the water column can be defined as the difference ΔCO_3^{2-} between the concentration of carbonate ions *in situ* and the concentration of saturated carbonate ion for the respective mineral phase. Since the concentration of carbonate ions cannot be measured directly, it is calculated using the dissociation constants of carbonic acid (H_2CO_3), and measurable parameters such as total inorganic **carbon dioxide** ($\Sigma\,CO_2$) dissolved in sea water, alkalinity, pH, and partial pressure of **carbon** dioxide exerted by sea water (pCO_2).

The water depth at which the sea water carbonate ion content and the concentration of carbonate ions in equilibrium with sea water for calcite or aragonite mineral phase intercept is called hydrographic calcite or aragonite lysocline, respectively. Below the lysocline, calcium carbonate dissolution begins and becomes progressively more intense in proportion to the fourth power of carbonate ion undersaturation. An undersaturation of about 10 µmol/kg is enough to dissolve almost all the calcite descending to the sea floor. At the CCD, the rate of calcium carbonate sedimentation is totally compen-

sated for by the rate of calcium carbonate dissolution. These deepest parts of the global ocean, the **abyssal plains**, where depths exceed about 17,500 feet (4,500 meters), the bottom is no more covered with **calcareous ooze** but with a layer of red-clay that contains no **fossils** at all.

From the surface water as the place of life-history down to the sedimentological archive, diversity of calcareous organism assemblages changes qualitatively and quantitatively due to carbonate dissolution. Each type of calcium carbonate shell architecture yields different crash behavior regarding foraminifera, coccolithophorids, pteropods, and other planktic and benthic calcium carbonate skeleton bearing organisms. For example, planktic foraminifer *Globigerinoides ruber* is rapidly dismantling into single chambers along sutures whereas *Neogloboquadrina pachyderma* is undergoing long lasting ultrastructural breakdown before the final smash. That is, calcium carbonate dissolution is a progress affecting individuals (of any level in Linné's system) in different extent.

Disintegration of calcium carbonate skeleton bearing organisms is a valuable tool to reconstruct modern (i.e., Holocene) and ancient (e.g., Pleistocene) deep and bottom water currents that are traceable through their different CO_2 accumulation. CCD is found deepest in the North Atlantic Ocean (50°N) at about 5,000 m moving upwards continuously in the water column to 3,000 m in the Atlantic sector of the Southern Ocean (60°S) and, in turn, CCD is found deepest in the Pacific sector of the Southern Ocean (60°S) at about 4,500 m moving upwards to 3,000 m in the North Pacific Ocean (50°N). In more general terms, the CCD appears to coincide with the calcium carbonate saturation state of 0.75 in the Atlantic and 0.65 in the Pacific.

Reconstructing CCD of modern and ancient **oceans** means to elucidate the role that oceanic deep-water processes play in global **climate** change during various **geologic time** intervals and at different levels of precision. The location of CCD, lysocline and saturation horizon determine deep water CO_3^{2-} concentrations and thus the pCO_2 of surface waters. Hence, the ocean's ability to take up atmospheric pCO_2 is influenced by the balance of production and dissolution of calcium carbonate, and lifting or lowering the CCD has important consequences on the short and long term variations of CO_2 in the atmosphere.

See also Ocean circulation and currents

CELESTIAL SPHERE: THE APPARENT MOTIONS OF THE SUN, MOON, PLANETS, AND STARS

The celestial sphere is an imaginary **projection** of the **Sun**, **Moon**, planets, stars, and all astronomical bodies upon an imaginary sphere surrounding Earth. The celestial sphere is a useful mapping and tracking remnant of the geocentric theory of the ancient Greek astronomers.

Although originally developed as part of the ancient Greek concept of an Earth-centered universe (i.e., a geocentric model of the Universe), the hypothetical celestial sphere provides an important tool to astronomers for fixing the location and plotting movements of celestial objects. The celestial sphere describes an extension of the lines of **latitude and longitude**, and the plotting of all visible celestial objects on a hypothetical sphere surrounding the earth.

The ancient Greek astronomers actually envisioned concentric crystalline spheres, centered around Earth, upon which the Sun, Moon, planets, and stars moved. Although heliocentric (Sun-centered) models of the universe were also proposed by the Greeks, they were disregarded as "counter-intuitive" to the apparent motions of celestial bodies across the sky.

Early in the sixteenth century, Polish astronomer Nicolaus Copernicus (1473–1543) reasserted the heliocentric theory abandoned by the Ancient Greeks. Although sparking a revolution in **astronomy**, Copernicus' system was deeply flawed by the fact that the Sun is certainly not the center of the universe, and Copernicus insisted that planetary orbits were circular. Even so, the heliocentric model developed by Copernicus fit the observed data better than the ancient Greek concept. For example, the periodic "backward" motion (retrograde motion) in the sky of the planets Mars, Jupiter, and Saturn, and the lack of such motion for Mercury and Venus was more readily explained by the fact that the former planets' orbits were outside of Earth's. Thus, the earth "overtook" them as it circled the Sun. Planetary positions could also be predicted much more accurately using the Copernican model.

Danish astronomer Tycho Brahe's (1546–1601) precise observations of movements across the "celestial sphere" allowed German astronomer and mathematician **Johannes Kepler** (1571–1630) to formulate his laws of planetary motion that correctly described the elliptical orbits of the planets.

The modern celestial sphere is an extension of the **latitude** and **longitude** coordinate system used to fix terrestrial location. The concepts of latitude and longitude create a grid system for the unique expression of any location on Earth's surface. Latitudes—also known as parallels—mark and measure distance north or south from the equator. Earth's equator is designated 0° latitude. The north and south geographic poles respectively measure 90° north (N) and 90° south (S) from the equator. The angle of latitude is determined as the angle between a transverse plane cutting through Earth's equator and the right angle (90°) of the **polar axis**. Longitudes—also known as meridians—are great circles that run north and south, and converge at the north and south geographic poles.

On the celestial sphere, projections of lines of latitude and longitude are transformed into declination and right ascension. A direct extension of Earth's equator at 0° latitude is the celestial equator at 0° declination. Instead of longitude, right ascension is measured in hours. Corresponding to Earth's **rotation**, right ascension is measured from zero hours to 24 hours around the celestial sphere. Accordingly, one hour represents 15 angular degrees of travel around the 360° celestial sphere.

Declination is further divided into arcminutes and arcseconds. In 1° of declination, there are 60 arcminutes (60') and in one arcminute there are 60 arcseconds (60"). Right ascension hours are further subdivided into minutes and seconds of time.

On Earth's surface, the designation of 0° longitude is arbitrary, an international convention long held since the days of British sea superiority. It establishes the 0° line of longitude—also known as the Prime Meridian—as the great circle that passes through the Royal National Observatory in Greenwich, England (United Kingdom). On the celestial sphere, zero hrs (0 h) right ascension is also arbitrarily defined by international convention as the line of right ascension where the ecliptic—the apparent movement of the Sun across the celestial sphere established by the plane of the earth's orbit around the Sun—intersects the celestial equator at the vernal equinox.

For any latitude on Earth's surface, the extended declination line crosses the observer's zenith. The zenith is the highest point on the celestial sphere directly above the observer. By international agreement and customary usage, declinations north of the celestial equator are designated as positive declinations (+) and declinations south of the celestial equator are designated as negative declinations (–) south.

Just as every point on Earth can be expressed with a unique set of latitude and longitude coordinates, every object on the celestial sphere can be specified by declination and right ascension coordinates.

The polar axis is an imaginary line that extends through the north and south geographic poles. The earth rotates on its axis as it revolves around the Sun. Earth's axis is tilted approximately 23.5 degrees to the plane of the ecliptic (the plane of planetary orbits about the Sun or the apparent path of the Sun across the imaginary celestial sphere). The tilt of the polar axis is principally responsible for variations in solar illumination that result in the cyclic progressions of the **seasons**. The polar axis also establishes the principal axis about which the celestial sphere rotates. The projection of Earth's geographic poles upon the celestial sphere creates a north celestial pole and a south celestial pole. In the Northern Hemisphere, the star Polaris is currently within approximately one degree (1°) of the north celestial pole and thus, from the Northern Hemisphere, all stars and other celestial objects appear to rotate about Polaris and, depending on the latitude of observation, stars located near Polaris (circumpolar stars) may never "set."

For any observer, the angle between the north celestial pole and the terrestrial horizon equals and varies directly with latitude north of the equator. For example, at 30° N latitude an observer views Polaris at +30° declination, at the terrestrial North Pole (90° N), Polaris would be directly overhead (at the zenith) at +90° declination.

The celestial meridian is an imaginary arc from the north point on the terrestrial horizon through the north celestial pole and zenith that terminates on the south point of the terrestrial horizon.

Regardless of location on Earth, an observer's celestial equator passes through the east and west points of the terrestrial horizon. In the Northern Hemisphere, the celestial equator is displaced southward from the zenith (the point directly over the observer's head) by the number of degrees equal to the observer's latitude.

Rotation about the polar axis results in a diurnal cycle of night and day, and causes the apparent motion of the Sun across the imaginary celestial sphere. The earth rotates about the polar axis at approximately 15 angular degrees per hour and makes a complete rotation in 23.9 hours. This corresponds to the apparent rotation of the celestial sphere. Because the earth rotates eastward (from west to east), objects on the celestial sphere usually move along paths from east to west (i.e., the Sun "rises" in the east and "sets" in the west). One complete rotation of the celestial sphere comprises a diurnal cycle.

As the earth rotates on its polar axis, it makes a slightly elliptical orbital revolution about the Sun in 365.26 days. Earth's revolution about the Sun also corresponds to the cyclic and seasonal changes of observable stars and constellations on the celestial sphere. Although stars grouped in traditional constellations have no proximate spatial relationship to one another (i.e., they may be billions of light years apart) that do have an apparent relationship as a two-dimensional pattern of stars on the celestial sphere. Accordingly, in the modern sense, constellations establish regional location of stars on the celestial sphere.

A tropical year (i.e., a year of cyclic seasonal change), equals approximately 365.24 mean solar days. During this time, the Sun appears to travel completely around the celestial sphere on the ecliptic and return to the vernal equinox. In contrast, one orbital revolution of Earth about the Sun returns the Sun to the same backdrop of stars—and is measured as a sidereal year. On the celestial sphere, a sidereal day is defined as the time it takes for the vernal equinox—starting from an observer's celestial median—to rotate around with the celestial sphere and recross that same celestial median. The sidereal day is due to Earth's rotational period. Because of precession, a sidereal year is approximately 20 minutes and 24 seconds longer than a tropical year. Although the sidereal year more accurately measures the time it takes Earth to completely orbit the Sun, the use of the sidereal year would eventually cause large errors in calendars with regard to seasonal changes. For this reason the tropical year is the basis for modern Western calendar systems.

Seasons are tied to the apparent movements of the Sun and stars across the celestial sphere. In the Northern Hemisphere, summer begins at the summer solstice (approximately June 21) when the Sun is reaches its apparent maximum declination. Winter begins at the winter solstice (approximately December 21) when the Sun's highest point during the day is its minimum maximum daily declination. The changes result from a changing orientation of Earth's polar axis to the Sun that result in a change in the Sun's apparent declination. The vernal and autumnal equinox are denoted as the points where the celestial equator intersects the ecliptic.

The location of sunrise on the eastern horizon, and sunset on the western horizon also varies between a northern most maximum at the summer solstice to a southernmost maximum at the winter solstice. Only at the vernal and autumnal equinox does the Sun rise at a point due east or set at a point due west on the terrestrial horizon.

During the year, the moon and planets appear to move in a restricted region of the celestial sphere termed the zodiac. The zodiac is a region extending outward approximately 8° from each side of the ecliptic (the apparent path of the Sun on the celestial sphere). The modern celestial sphere is divided into twelve traditional zodiacal constellation patterns (corresponding to the pseudoscientific astrological zodiacal signs) through which the Sun appears to travel by successive eastwards displacements throughout the year.

During revolution about the Sun, the earth's polar axis exhibits parallelism to Polaris (also known as the North Star). Although observing parallelism, the orientation of Earth's polar axis exhibits precession—a circular wobbling exhibited by gyroscopes—that results in a 28,000-year-long precessional cycle. Currently, Earth's polar axis points roughly in the direction of Polaris (the North Star). As a result of precession, over the next 11,000 years, Earth's axis will precess or wobble so that it assumes an orientation toward the star Vega.

Precession causes an objects celestial coordinates to change. As a result, celestial coordinates are usually accompanied by a date for which the coordinates are valid.

Corresponding to Earth's rotation, the celestial sphere rotates through 1° in about four minutes. Because of this, sunrise, sunset, moonrise, and moonset all take approximately two minutes because both the Sun and Moon have the same apparent size on the celestial sphere (about 0.5°). The Sun is, of course, much larger, but the Moon is much closer. If measured at the same time of day, the Sun appears to be displaced eastward on the star field of the celestial sphere by approximately 1° per day. Because of this apparent displacement, the stars appear to "rise" approximately four minutes earlier each evening and set four minutes later each morning. Alternatively, the Sun appears to "rise" four minutes earlier each day and "set" four minutes earlier each day. A change of approximately four minutes a day corresponds to a 24-hour cycle of "rising" and "setting" times that comprise an annual cycle.

In contrast, if measured at the same time each day, the Moon appears to be displaced approximately 13° eastward on the celestial sphere per day and therefore "rises" and "sets" almost one hour earlier each day.

Because the earth is revolving about the Sun, the displacement of the earth along it's orbital path causes the time it takes to complete a cycle of lunar phases—a synodic month—and return the Sun, Earth, and Moon to the same starting alignment to be slightly longer than the sidereal month. The synodic month is approximately 29.5 days.

Earth rotates about its axis at approximately 15 angular degrees per hour. Rotation dictates the length of the diurnal cycle (i.e., the day/night cycle), and creates "time zones" with differing local noons. Local noon occurs when the Sun is at the highest point during its daily skyward arch from east to west (i.e., when the Sun is at its zenith on the celestial meridian). With regard to the solar meridian, the Sun's location (and reference to local noon) is described in terms of being ante meridian (am)—east of the celestial meridian—or post meridian (pm) located west of the celestial meridian.

See also Astrolabe; Geographic and magnetic poles; Latitude and longitude; Revolution and rotation; Year, length of

CEMENTATION • *see* LITHIFICATION

CENOZOIC ERA

In **geologic time**, the Cenozoic Era, the third era in the **Phanerozoic Eon**, follows the **Mesozoic Era** and spans the time between roughly 65 million years ago (mya) and present day. On the geologic time scale, Earth is currently in the Cenozoic Era of the Phanerozoic Eon.

The Cenozoic Era contains two geologic time periods, including the **Tertiary Period** (65 mya to approximately 1.8 mya) and the current **Quaternary Period** (1.8 mya to present day). The Tertiary Period is also sometimes referred to in terms of a Paleogene Period and a Neogene Period. When referred to in terms of a Paleogene Period and a Neogene Period, the Paleogene Period extends from approximately 65 mya to 23 mya and the Neogene Period from 23 mya to 2.6 mya. The Quaternary Period is also termed the Anthropogene Period. These periods are further subdivided into six different major epochs, including the **Paleocene Epoch**, **Oligocene Epoch**, **Miocene Epoch**, and **Pliocene Epoch**, **Pleistocene Epoch** and current **Holocene Epoch**.

The onset of the Cenozoic Era is marked by the K-T boundary or K-T event—the mass extinction of non-avian dinosaurs marking the boundary between the **Cretaceous Period** of the Mesozoic Era and the Tertiary Period of the Cenozoic Era.

At the start of the Cenozoic Era, **North America** and **Europe** were separated by a widening ocean basin spreading along a prominent mid-oceanic ridge. North America and **South America** were separated by a confluence of the future Pacific Ocean and Atlantic Ocean, and extensive flooding submerged much of what are now the eastern and middle portions of the United States. By start of the Cenozoic Era, **water** separated South America from **Africa**, and seafloor spreading continued to push the continents apart. The Australian and Antarctic continents were clearly articulated and the Antarctic continent had begun a southward migration toward its present position in the south polar region. At the outset of the Cenozoic Era, the Indian plate and subcontinent remained far south of the Eurasian plate and continent.

By 30 mya, the modern continental arrangement was easily recognizable. Although still separated by water, the land bridge between North and South America began to reemerge. **Antarctica** assumed a polar position and extensive **ice** accumulation began on the continent. The Indian plate drove rapidly northward of the equator to close with the Eurasian plate. Although still separated by a shallow straight of water, the impending collision of the plates that would eventually form the Himalayan mountain chain had begun. The gap between North America and Europe continued to widen at a site of **seafloor spreading** along a prominent mid-Atlantic ridge. By mid-Tertiary Period, the mid-Atlantic ridge was apparent in a large suture-like extension into the rapidly widening South Atlantic Ocean that separated South America from Africa.

Well into the Cenozoic Era, by the start of the Quaternary Period some 2.6 million years ago, Earth's continents assumed their modern configuration. The Pacific Ocean separated **Asia** and **Australia** from North America and South America, the Atlantic Ocean separated North and South America from Europe (Euro-Asia) and Africa. Separated by the straits of Indonesia, the Indian Ocean filled the basin between Africa, India, Asia, and Australia. The Indian plate driving against and under the Eurasian plate uplifted rapid mountain building. As a result of the collision, ancient oceanic **crust** bearing marine **fossils** was uplifted into the Himalayan chain. The collision between the Indian and Eurasian plate continues with a resulting slow—but measurable—increase in the altitude of the highest Himalayan Mountains (e.g., Mt. Everest) each year. Although glacial sheets advance and recede in cyclic patterns (i.e., reestablish new terrain altering ice ages, the basic patterns of **glaciation** evident today were established during the Quaternary Period.

Many geologists and paleontologists assert that the K-T extinction resulted from a cataclysmic asteroid impact in an **area** now located underwater near the Yucatan Peninsula of Mexico. The impact caused widespread primary damage due to the blast impact and firestorms. The major damage to Earth's ecosystem occurred when the debris and smoke from the collision and subsequent fires moved into the atmosphere to block a sufficient amount of light from the **Sun** that photosynthesis was greatly slowed. The resulting climatic changes and food shortages led to extinction of the largest life forms (those with the greatest energy needs), including the dinosaurs.

Although mammals evolved before the Cenozoic Era, the reduction in predator species allowed land mammals to dominate and thrive—eventually setting the stage from the **evolution** of homo sapiens (humans).

See also Archean; Cambrian Period; Dating methods; Devonian Period; Eocene Epoch; Evolution, evidence of; Fossil record; Fossils and fossilization; Geologic time; Historical geology; Jurassic Period; Mississippian Period; Ordovician Period; Paleozoic Era; Pennsylvanian Period; Precambrian; Proterozoic Era; Quaternary Period; Silurian Period; Supercontinents; Tertiary Period; Triassic Period

CENTIGRADE SCALE • *see* TEMPERATURE AND TEMPERATURE SCALES

CHALCOPHILES, LITHOPHILES, SIDEROPHILES, AND ATMOPHILES

Chalcophiles, lithophiles, siderophiles, and atmophiles are classes of elements based upon similar geochemical properties and reactive affinities. The classes were originally advanced by Swiss-born **Victor Goldschmidt** (1888–1947) and are terms still widely used by geologists and geochemists. The key factor in determining an element's class is the type of **chemical bonds** that the element forms.

Chalcophile elements have a high bonding affinity—usually in the form of covalent bonds—with sulfur, and are, accordingly, usually abundant in sulfides. Chalcophiles also exhibit a bonding affinity with selenium, tellurium, arsenic, and antimony and therefore also exhibit high levels of derivatives of these elements. When sulfur is abundant, chalcophile elements readily form sulfide **minerals** as they precipitate from the **magma**. This process partially explains the formation of extensive deposits of iron-nickel-copper sulfides.

Lithophiles have a high bonding affinity with **oxygen**. Lithophiles have an affinity to form ionic bonds and are represented by silicates (**silicon** and oxygen) in the **crust** and mantle. Other lithophile elements include magnesium, **aluminum**, sodium, potassium, **iron**, and calcium.

Siderophiles exhibit a weak affinity to both oxygen and sulphur. Siderophiles have an affinity for iron and a distinguishing characteristic of siderophiles is that they exhibit high solubility in molten iron. Siderophile elements generally have a low reactivity and exhibit an affinity to form metallic bonds. As a result, siderophiles are most often found in their native state. Not abundant in the core or mantle, most siderophiles are thought to be richest at Earth's core. Platinum (Pt) group **metals**, including Ruthium (Ru), Rhodium (Rd), Palladium (Pd), Osmium (Os), and Iridium (Ir), show exhibit a strong siderophile tendency.

Atmophiles are a related fourth class of elements characterized by their ability to form van der Waals bonds. Atmophiles are also highly volatile.

Chalcophiles, lithophiles, siderophiles, and atmophiles have differing densities. Accordingly, after formation from the molten state, these differential densities tend to separate the classes. For example, siderophiles have a greater average density than lithophiles and thus lithophiles would tend to "rise" in the molten state relative to siderophiles. Although the element classes were derived, in part, from an attempt to explain the distribution of elements, the density differences do not always result in the expected distribution of classes in the earth's core, mantle and crust.

Because geochemical reactivity is a function of electron structure—especially the number of electrons available for bonding—element classes tend to follow groupings or trend as related to the **periodic table**. The difference in the classification of elements can also be linked to differing valence states.

It is possible for some elements to be assigned to more than one group. The reactivity of an element can also be driven by the relative amounts of elements surrounding it. For example, iron, when in an oxygen deprived environment (e.g., at the earth's core) acts as a siderophile. In the more oxygen rich environment of the crust and mantle, iron acts as a lithophile or chalcophile, and in this form is commonly found in **igneous rocks**. When sulfur is present is found in sulfide deposits. Siderophiles when surrounded by sulfur, arsenic, and antimony may act as chalcophiles.

See also Chemical bonds and physical properties; Chemical elements

CHALLENGER EXPEDITION

The British Navy vessel H.M.S. *Challenger* circumnavigated the world between December 1872 and May 1876, conducting history's first systematic, scientific investigation of the world's **oceans**. The *Challenger* expedition gathered a body of data that has been matched by few voyages of discovery. The science of modern **oceanography** essentially began with the *Challenger* expedition.

The *Challenger* was a 200 foot (67 m), three-masted, square-rigged wooden sailing ship equipped with an auxiliary steam engine. Fifteen of its 17 gun bays were rebuilt as laboratories, workrooms, and storage spaces for scientific equipment. It carried a crew of five scientists, an official artist, 20 officers, and about 200 sailors.

The *Challenger* began its voyage by crossing the Atlantic four times, discovering the Mid-Atlantic Ridge in the process. It then visited **Africa**, **Antarctica**, New Zealand, New Guinea, China, Japan, Hawaii, the South **Seas**, and the tip of **South America**, studying not only the sea itself but the fauna, flora, and geography of numerous islands. The *Challenger* team made 362 regularly-spaced midocean measurements of depth, **temperature**, and currents and used special dredges to collect samples of life, ooze, and rocks from the ocean floor. This expedition produced the first global cross-section of the ocean's depth profile and identified over 4,700 ocean-dwelling animal species never before known.

The *Challenger* returned triumphantly to **Europe** freighted with tens of thousands of photographs, drawings, measurements, and biological and geological samples. Publication of the results took 20 years and required 50 thick volumes totaling almost 30,000 pages. Data from the *Challenger* expedition are still cited occasionally in modern scientific literature.

A century after the first *Challenger* expedition, the research drillship *Glomar Challenger* (1968–1983) cruised the world's oceans gathering data that were also to prove revolutionary for Earth sciences. Its deep-sea core samples confirmed the theory of continental drift and revealed for the first time that the oceanic **crust** is extremely young compared to the continental crust.

See also Deep sea exploration

CHANNEL PATTERNS

Channel patterns are types of sedimentary deposits formed by streams and **rivers**. Collectively, they are called fluvial deposits. Their shape and sediment characteristics are easily identified and enormously complex. Understanding of fluvial deposits is essential to economic **geology** because many of these ancient deposits are a good source of **petroleum**. Extremely old fluvial deposits are found extensively on land and often indicate much different environments. For example, a large river deposit is located outside of Flagstaff, Arizona where there now exists nothing but high **desert**. There is no indication of the source of vast channel cuts, **sand** bars, point bars, and cut banks seen captured in the sediments.

In order to understand and identify channel deposits, the types and natures of streams and rivers must first be learned. Each river has its own unique settings in which it flowed. No

Channel pattern of the Niger River, Timbuktu, Mali. © Wolfgang Kaehler. UPI/Corbis-Bettmann. Reproduced by permission.

two rivers have ever been the same. There are some general characters of rivers, however, that are easily observed. Rivers that originate and flow down steep slopes are usually straight and deep compared to a meandering river that flows across relatively flat ground. Meanders are wide curves in rivers that follow the path of least resistance along plains and valleys. In aerial view, the sinuous curves seem to wander over the **topography**. This wavy appearance gives rise to their name meandering rivers. Braided channels occur where steep rivers meet flat lying valleys. Their river's sediment load is rapidly dumped, building a flat surface along which the **water**, previously contained in a single channel, spreads out forming a series of interlocking shallow channels. The channels cut back and forth across the flat plain that gains them the name "braided" channels. The final large type of river group is the anastomosing river where different channels of the river are separated by permanent alluvial islands. From the air, the rivers have many channels that eventually coalesce to form one large possibly meandering river.

The effect of these variable types of waterways is a wide variety of sedimentary deposits. All the depositional types are a result of the increase and decrease of the force of

flowing water. As velocities change, so does the sediment-carrying capacity of the water. The stronger the velocity of water, as in a steep channel, the greater the size and amount of sediment the river can transport. Large boulders are not often moved unless storm conditions exist where velocities can reach dangerous levels. Even cobble-sized stones need immense water velocities to be moved. However, sand and silt are much more easily moved along river bottoms. Many studies on the **physics** of water and its carrying capacity have produced information from which geologists can infer the amount of water in a channel at a particular time in history.

Channel bars are longitudinal deposits of mostly sand that accumulates in the centers of rivers. Their development is constant as new sands accumulate on top of old. The moving water is constantly rolling sand grains along the bottoms of rivers and on top of bars. Because the bars have elevation in the rivers, they act as a sort of brake for the water. When the water slows down as it moves around the bars, it deposits its load of sand, increasing the size of the bar and further slowing the water. To further complicate the picture, the channel bottom acts as a drag on the water column. The rough surface of the sand creates friction and slows the water immediately above it while the surface of the water is not slowed. The result is that lighter sediments not lying near the bottom are carried farther and longer in the water column and down the river.

Bars are rarely composed of any sediment larger than sand-sized particles. No type of sand bars is stable. They can migrate over the bottom surface of the channel with changing **seasons** and water flows. The bars can migrate from side to side in the channel and down the river as the water carries more sediment over their tops. When covered by additional sediment, these bars are buried in the channel and remain as a geological facies (specific identifiable geological pattern).

Meanders in rivers produce a variety of fluvial structures. As the water flows along in a meander, the point at which it first hits a turn in the river course receives the most force from the water. As the water continually strikes this bank, it erodes the sediments in front of it. A sharp and well-excavated corner is formed and appropriately called the cutbank. As the water turns the corner, most of its force is absorbed by the cutbank and it loses much of its velocity around the bend. At this point in the semi-circular water pattern, the water has the least velocity and unloads its sediment. The result is a build-up of sand called a point bar. Point bars will often have finer sediments than channel bars because the water loses so much velocity on the outer edges of the point bar that it cannot hold anything but the finest grained particles or clasts.

Because the water now has so little velocity, its course is easily turned sideways where it regains velocity and begins another meander. From the air, a meandering river is an extraordinary view since all the previous channels and turns can be seen as a result of this depositional pattern of cutbanks, channels bars, and point banks. Older and more massive rivers, such as the Mississippi, have an extensive history that is easily seen from an airplane.

As a meandering river ages, it develops interesting features called oxbow **lakes**. The cutting of the banks around the corners continues until one bank nearly touches the other. It looks something like an omega sign (Ω) from the Greek alphabet. Eventually the corners cut enough land away to meet. When this happens, the flow of the river is stronger where the channel is straighter. As a result, the river takes a new course. The abandoned meander does not receive any new sediment and the remaining water is sealed by new point bank bars along the new course. The unique lake that is formed is named an oxbow lake. The lake will eventually fill and die. Trees may overgrow it and a whole new cycle of river meandering leaves it stranded until the river cuts its way back.

These sedimentary structures are just a few of the interesting patterns formed by rivers and streams. They are found in a large range of size scales and are important to geologists worldwide.

See also Petroleum, detection; Stream valleys, channels, and floodplains

Chaos theory (meteorological aspects)

Chaos theory attempts to identify, describe, and quantify order in apparently unpredictable and/or highly complex systems (i.e., atmospheric dynamics, **weather** systems, etc.) in which, out of seemingly random, disordered (e.g., aperiodic) processes there arise processes that are deterministic and predictable.

Complex phenomena are those generally regarded as having too many variables (or too many possible conditions or states) to yield to conventional quantitative analysis. The motion of molecules in swirling smoke or the turbulent hydraulics of a river current, for example, are systems that exhibit such chaotic complexity.

According to the laws of thermodynamics, all natural processes—when considering both system and surroundings—exhibit a tendency toward net movement from the ordered state to a more the chaotic (disordered) state. Conversely, according to some chaos theory models, chaotic, unpredictable, and irreversible processes may, evolve into or produce ordered states. Entropy is a measure of thermodynamic equilibrium used to explain irreversibility in physical and chemical processes. The second law of thermodynamics specifies that in an isolated system, increasing entropy corresponds to changes in the system over time and that entropy tends toward (a statistical mechanical concept) maximization. The second law of thermodynamics dictates that in natural processes, without work being done on a system, there is a movement from order to disorder.

Because entropy in natural processes increases over time, even very straightforward linear-type relationships must eventually take on a degree of irregularity (i.e., of seemingly disordered complexity). In accord with the second law of thermodynamics, apparently chaotic phenomena arise from initially ordered (i.e., lower entropy) systems. This dual tendency toward increasing entropy and chaos from an initially stable state can take place spontaneously. Small perturbations in initial conditions intensify these tendencies. Chaos theorists describe such departures as the butterfly effect.

The study of such mathematical irregularities involving chaos and order remained a relatively unnoticed corner of advanced mathematics until the advent of the digital computer. In 1956, Edward Lorenz, a professor of **meteorology** at the Massachusetts Institute of Technology was studying the numerical solution to a set of three differential equations in three unknowns, a highly simplified version of the types of equations meteorologists then in use to describe atmospheric phenomena. Lorenz came to the conclusion that his set of differential equations displayed a sensitive dependence on initial conditions, a sensitivity of the same type that French mathematician Jules-Henri Poincaré (1854–1912) had discovered for the Newtonian equations when those equations were applied to celestial dynamics. Lorenz, however, gave this phenomenon a new and highly appealing name, the butterfly effect, suggesting that, in the extreme, the flapping of a butterfly's wings in Kansas might be responsible for a monsoon in India a month later.

Along with quantum and relativity theories, chaos theory—with its inclusive concepts of chaos and order—is widely regarded as one of the great intellectual leaps of the twentieth century. The modern physical concepts of chaos and order, however, actually trace their roots to classical mechanical concepts introduced in English physicist Sir Isaac Newton's (1642–1727) 1686 work, *Philosophy Naturalis Principia Mathematica* (Mathematical principles of natural philosophy). It was Newton, one of the inventors of the calculus, who revolutionized **astronomy** and **physics** by showing that the behavior of all bodies, celestial and terrestrial, was governed by the same laws of motion, which could be expressed as differential equations. These differential equations relate the rates of change of physical quantities to the values of those quantities themselves. Such calculated predictability of physical phenomena led to the concept of a mechanistic, clockwork universe that operated according to deterministic laws. The idea that the universe operated in strict accord with physical laws was profoundly influential on science, philosophy and theology.

Most physical models are devoted to the understanding of simple systems (e.g., kinetic molecular theories often rely on concepts related to a ball bouncing in a box). From fundamental laws, using easily quantifiable behavior of such simple systems, theorists often attempt to project the behavior of more complex systems (e.g., the collision and dynamics of hundreds of balls bouncing in a box). It was long thought by physicists that, with regard to these types of models, the complexity of a system simply veiled an underlying fundamental simplicity.

For example, according to classical deterministic concepts, the accurate analysis and prediction of complex systems (e.g., the determination of the momentum of a particular ball among hundreds of other balls bouncing and colliding in a box) could be calculated only if the initial or starting conditions were accurately known. The fact that it is usually impossible to predict the exact condition or behavior of a system (especially considering that such interactions or measurements of a systems must also alter the system itself) is usually explained away as the result of a lack of knowledge regarding starting conditions or a lack of calculating vigor (e.g., inadequate computing power).

See also Atmospheric circulation; Weather forecasting methods; Weather forecasting

CHATOYENCY • *see* GEMSTONES

CHEMICAL BONDS AND PHYSICAL PROPERTIES

Chemical bonds are the electrical forces of attraction that hold atoms or ions together to form molecules. Different types of chemical bonds and their varying intensity are directly respon-

sible for some of the physical properties of **minerals** such as hardness, **melting** and boiling points, solubility, and conductivity. Chemical bonds also influence such other properties as crystal symmetry and cleavage. Stronger bonds between atoms make them more difficult to separate and, in general, stronger chemical bonds result in greater hardness, higher melting and boiling points, and smaller coefficients of expansion. There are four principal types of chemical bonds found in minerals: ionic, covalent, metallic, and van der Waals.

An ionic bond is the result of the electrostatic attraction between two oppositely charged ions. Ionic bonds exist because some elements tend to capture or lose one or more electrons resulting in a net positive or negative charge. These are called ions. An ion that bears a positive charge is a cation. One with a negative charge is an anion. Ions may carry a single charge, such as Na^+ and Cl^-, or may have multiple charges, such as Ca^{2+} or Fe^{3+}. Oppositely charged ions tend to attract one another because the cation can transfer electrons to the anion, allowing each ion to achieve better stability. For example, Na^+ and Cl^- readily combine to form NaCl, halite (salt). Most minerals are held together by some form of ionic bond.

In order for an ionically bonded solid to melt, some of the bonds, but not all of them, must be broken. For boiling to occur, all of the bonds must be broken. As a result, ionic bonds produce moderate to high melting and boiling points. Ionic bonds are moderate in strength and so result in moderately hard minerals. The electrical conductivity is generally low and minerals with ionic bonds tend to dissolve better in **water**. In addition, because the charge on ions is evenly distributed around the surface of the **atom**, or nondirectional, a cation tends to evenly distribute as many anions as possible over its entire surface **area**. This often yields a high degree of crystal symmetry in minerals. Halite (salt) and fluorite are two common ionically bonded minerals.

Covalent bonds are different from ionic bonds in that electrons are shared between atoms of similar charge as opposed to electrons being donated by a cation to an anion. Covalent bonds form when the electron **clouds** of separate atoms draw near and overlap, enabling electrons to be shared. In covalent bonds, each participating atom usually contributes electrons, resulting in a strong bond. Covalent bonds are common between atoms and ions of the same element such the noble gasses.

Minerals with covalent bonds tend to be hard and insoluble. **Diamond** is one example. Covalent bonds produce high melting and boiling points and low conductivities. The forces that bind the atoms tend to be localized in the vicinity of the shared electrons and so are highly directional. This often yields a lower degree of crystal symmetry.

As the name suggests, metallic bonds are found in pure metallic minerals. Metallic bonds form when an atom of a metallic element, which usually contains loosely held electrons in the outer shell, shares these electrons with closely packed atoms of the same element. The shared electrons pass freely among all the metal atoms. The result may be described as a weak covalent bond. It is different from the true covalent bond, however, in that there are too few electrons to be shared continuously by all atoms simultaneously. The electrons are extremely mobile as they roam within the lattice of positive metal ions. The mobility of the electrons results in the high thermal and electrical conductivity of **metals**. The weakness of the bonds results in the lower hardness, low melting and boiling points, and high ductility so often observed in metallic minerals such as gold and copper.

Van der Waals bonds arise from very minor charge polarities that can develop on molecules that are already bonded together. For example, the directional characteristic of covalent bonds can produce a weak negative charge where the electron clouds overlap with a corresponding weak positive charge opposite the area of overlap. These dipoles may attract each other to form a very weak chemical bond known as Van der Waals. Van der Waals bonds are not common in minerals, but when present result in low hardness and easily cleaved zones. **Graphite** owes its greasy feel to the Van der Waals forces that link sheets of covalently bonded **carbon** atoms, allowing them to easily slip apart.

Most minerals are held together by a combination of chemical bonds. Often, distinct molecular units, consisting of strongly bonded atoms, are linked by weaker bonds, as in the graphite example above. Micas, which cleave perfectly into sheets, are another example. They are composed of covalent-bonded silica tetrahedral sheets joined together by ionic bonds. The ionic bonds tend to break first, separating into the more robust sheets. A single chemical bond in a mineral may also display the properties of more than one bond type. A common example is the silica tetrahedron, which consists of one **silicon** atom, Si^{+4}, surrounded by four **oxygen** atoms, O^{-2}. The bond that binds silicon and oxygen together arises out of an ionic attraction, but it also involves overlapping electron clouds and subsequent sharing of electrons, so it is part covalent as well.

See also Atomic structure; Chemical elements; Crystals and crystallography

CHEMICAL ELEMENTS

By the end of the nineteenth century, the elements and matter comprising all things could no long be viewed as immutable. The dramatic rise of scientific methodology and experimentation during the later half of the eighteenth century set the stage for the fundamental advances in **chemistry** and **physics** made during the nineteenth century. In less than a century, European society moved from an understanding of the chemical elements grounded in mysticism to an understanding of the relationships between elements found in a modern **periodic table**. During the eighteenth century, there was a steady march of discovery with regard to the chemical elements. Isolations of hydrogen and **oxygen** allowed for the formation of **water** from its elemental components. Nineteenth century scientists built experiments on new-found familiarity with elements such as nitrogen, beryllium, chromium and titanium.

By the mid-nineteenth century, chemistry was in need of organization. New elements were being discovered at an increasing pace. Accordingly, the challenge for chemists and physicists was to find a key to understand the increasing volume of experimental evidence regarding the properties of the elements. In 1869, the independent development of the periodic law and tables by the Russian chemist **Dmitry Mendeleev** (1834–1907) and German chemist Julius Meyer (1830–95) brought long sought order and understanding to the elements.

Mendeleev and Meyer did not work in a vacuum. English chemist J.A.R. Newlands (1837–1898) had already published several works that ventured relationships among families of elements, including his "law of octaves" hypothesis. Mendeleev's periodic chart of elements, however, spurred important discoveries and isolation of chemical elements. Most importantly, Mendeleev's table provided for the successful prediction of the existence of new elements and these predictions proved true with the discovery of gallium (1875), scandium (1879) and germanium (1885).

By the end of the nineteenth century, the organization of the elements was so complete that British physicists Lord Rayleigh (born John William Strutt, 1842–1919) and William Ramsay (1852–1916) were able to expand the periodic table and to predict the existence and properties of the noble gases argon and neon.

Nineteenth century advances were, however, not limited to mere identification and isolation of the elements. By 1845, German chemist Adolph Kolbe (1818–84) synthesized an organic compound and, in 1861, another German chemist Friedrich Kekule(1829–1896) related the properties of molecules to their geometric shape. These advances led to the development of wholly new materials (e.g., plastics, celluloids) that had a dramatic impact on a society in midst of industrial revolution.

The most revolutionary development with regard to the elucidation of the elements during the nineteenth century came in the waning years of the century. In 1895, Wilhelm Röntgen (1845–1923) published a paper titled: "On a New Kind of Rays." Röntgen's work offered the first description of x rays and offered compelling photographs of photographs of a human hand. The scientific world quickly grasped the importance of Röntgen's discovery. At a meeting of the French Academy of Science, Henri Becquerel (1852–1908) observed the pictures taken by Röntgen of bones in the hand. Within months Becquerel presented two important reports concerning "uranium rays" back to the Academy. Becquerel, who was initially working with phosphorescence, described the phenomena that later came to be understood as **radioactivity**. Less than two years later, two other French scientists, Pierre (1859–1906) and **Marie Curie** (born in Poland, 1867–1934) announced the discovery of the radioactive elements polonium and radium. Marie Curie then set out on a systematic search for radioactive elements and was able, eventually, to document the discovery of radioactivity in uranium and thorium **minerals**.

As the nineteenth century drew to a close, Ernest Rutherford (1871–1937), using an electrometer, identified two types of radioactivity, which he labeled alpha radiation and beta radiation. Rutherford actually thought he had discovered a new type of x ray. Subsequently alpha and beta radiation were understood to be particles. Alpha radiation is composed of alpha particles (the nucleus of helium). Because alpha radiation is easily stopped, alpha radiation-emitting elements are usually not dangerous to biological organisms (e.g., humans) unless the emitting element actually enters the organism. Beta radiation is composed of a stream of electrons (electrons were discovered by J. J. Thomson in 1897) or positively charged particles called positrons.

The impact of the discovery of radioactive elements produced immediate and dramatic impacts upon society. Within a few years, high-energy electromagnetic radiation in the form of x rays, made possible by the discovery of radioactive elements, was used by physicians to diagnose injury. More importantly, the rapid incorporation of x rays into technology established a precedent increasingly followed throughout the twentieth century. Although the composition and nature of radioactive elements was not fully understood, the practical benefits to be derived by society outweighed scientific prudence.

Italian scientist Alessandro Volta's (1745–1827) discovery, in 1800, of a battery using discs of silver and zinc gave rise to the voltaic pile or the first true batteries. Building on Volta's concepts, English chemist Humphry Davy (1778–1829) first produced sodium from the electrolysis of molten sodium hydroxide in 1807. Subsequently, Davy isolated potassium, another alkali metal, from potassium hydroxide in the same year. Lithium was discovered in 1817.

Studies of the spectra of elements and compounds spawned further discoveries. German chemist Robert Bunsen's (1811–1999) invention of the famous laboratory burner that bears his name allowed for the development of new methods for the analysis of the elemental structure of compounds. Working with Russian-born scientist Gustav Kirchhoff (1824–1887) Bunsen's advances made possible flame analysis (a technique now commonly known as atomic emission **spectroscopy** [AES]) and established the fundamental principles and techniques of spectroscopy. Bunsen examined the spectra (i.e., component colors), emitted when a substance was subjected to intense flame. Bunsen's keen observation that flamed elements that emit light only at specific wavelengths—and that each element produces a characteristic spectra—along with Kirchhoff's work on black body radiation set the stage for subsequent development of **quantum theory**. Using his own spectroscopic techniques, Bunsen discovered the elements cesium and rubidium.

Using the spectroscopic techniques pioneered by Bunsen, other nineteenth century scientists began to deduce the chemical composition of stars. These discoveries were of profound philosophical importance to society because they proved that Earth did not lie in a privileged or unique portion of the universe. Indeed, the elements found on Earth, particularly those associated with life, were found to be commonplace in the cosmos. In 1868, French astronomer P.J.C. Janssen (1824–1907) and English astronomer, Norman Lockyer (1836–1920), used spectroscopic analysis to identify helium on the **Sun**. For the first time an element was first discovered outside the confines of Earth.

See also Atomic mass and weight; Atomic number; Big Bang theory; Stellar life cycle

CHEMISTRY

Chemistry deals with the study of the properties and reactions of atoms and molecules. In particular, chemistry deals with reaction processes and the energy transition. Major divisions of chemistry include inorganic chemistry, organic chemistry (chemistry of **carbon** based compounds), physical chemistry, analytical chemistry, and biochemistry. **Geochemistry** deals with the reaction unique to geological processes.

The origin of the modern science of chemistry is often attributed to the work of French physicist and chemist Antoine Lavoisier (1743–1794). In 1774, Lavoisier demonstrated that **oxygen** is a critical component of air needed for combustion. This observation led into a better understanding of the changes in composition and structure of matter. Lavoisier's publication of the first list of elemental substances eventually evolved into the **Periodic table** of the elements. Other important contributions to early chemistry include British chemist and physicist John Dalton's (1766–1844) **atomic theory**; Italian physicist and chemist Amedeo Avogadro's (1776–1856) theory that molecules are made up of atoms; and Sir Edward Frankland's (1825–1899) descriptions of chemical reactions. These observations and theories all led to the portrayal of chemistry as the architecture of molecules.

Each discipline of chemistry (e.g., inorganic, analytical, physical chemistry, etc.) studies a different facet of the structure and composition of materials and their changes in composition and energy. As molecules and scientific problems become more complex, the traditional areas of chemical investigation begin to overlap with other physical sciences.

Organic chemistry is the study of compounds that contain carbon atoms. The term organic was first introduced by the Swedish scientist, Jöns Jacob Berzelius (1779–1848) to refer to substances isolated from living systems. Inorganic compounds, a call predominant in geological processes, are those isolated from nonliving sources. At the time, it was believed that a "vital force" only present in living systems was necessary for the preparation of organic compounds. In 1828, German chemist Friedrich Wöhler (1800–1882) first synthesized urea, an organic compound isolated from urine, by evaporating a **water** solution of the inorganic compound ammonium cyanate. Eventually, the vital force theories (e.g., those based on the idea that life and the chemistry of life depended upon an undefined vital force peculiar to living organisms) were discarded and organic chemistry became the investigation of the over seven million carbon-containing compounds. Today, organic chemists work primarily to synthesize new molecules to be used in pharmaceuticals, surfactants, paints, and coatings. They are also involved in scaling reactions from grams to tons in industrial research laboratories.

Inorganic compounds, at the time of the vital force theories, were those materials isolated from nonliving sources. Now, inorganic chemistry is the chemistry of all the elements except for carbon. This includes the chemistry of transition **metals** which coordinate with organic ligands and make up hemoglobin; the very reactive alkali metals used to make organometallic compounds in the manufacture of pharmaceutical materials; and also, the semi-metallic elements that have unusual electronic properties used in solar cells for the conversion of light into **electricity**. Inorganic chemists find employment in the production of **glass**, ceramics, semi-conductors, and advanced synthetic catalysts.

In 1909, German scientist Wilhelm Ostwald (1853–1932) was awarded the Nobel Prize in Chemistry for his work with catalysis, a very useful technique in industrial manufacturing. Ostwald is often referred to as the father of physical chemistry, a branch of chemistry devoted to the investigation of the underlying physical processes responsible for chemical properties and phenomena. Physical chemistry describes the influence of **temperature**, pressure, concentration, and catalyst used in organic and inorganic reactions. These data give important insight into the mechanisms of the chemical change and predict the best experimental methodology for a specific manufacturing process. Physical chemists are employed in industrial, academic, and governmental laboratories to study and calculate the fundamental properties of elements and molecular compounds. The application of physical chemistry is critical to the development of efficient devices, new applications of chemicals and better methods for measuring chemical phenomena.

Analytical chemistry is the branch of chemistry involved with the measurement and characterization of materials. Chemical analysis is divided into classical and instrumental methods. Wet or classical chemical analysis is the oldest form of analytical chemistry and involves the use of chemical reactions utilizing gravimetric and volumetric methodology to analyze material compositions. The use of instrumental methods for analytical analysis provides comprehensive information about chemical structure. Instrumental techniques include methods for measuring molecular **spectroscopy** such as infrared spectroscopy (IR), nuclear magnetic resonance spectroscopy (NMR), mass spectroscopy (MS), and x-ray crystallography. Gas chromatography, liquid chromatography, and electrophoresis are examples of separation methods used by analytical chemists. There is a need for analytical chemists in governmental, industrial, and academic research organizations to characterize new materials and determine the chemical composition of materials.

Chemists often work with geologists and geophysicists, in an effort to identify specific geologic reactions or to help characterize a specific geologic formation.

See also Atmospheric chemistry; Bowen's reaction series; Dating methods; Petroleum detection; Petroleum, economic uses of; Petroleum extraction; Weathering and weathering series

CHEMOAUTOTROPHIC AND CHEMOLITHOTROPHIC BACTERIA AS WEATHERING AGENTS

The inorganic processes associated with chemoautotrophic and chemolithotrophic bacteria may make these bacteria one of the most important sources of **weathering** and **erosion** of rocks on Earth.

Autotrophic bacteria obtain the **carbon** that they need to sustain survival and growth from **carbon dioxide** (CO_2). To process this carbon source, the bacteria require energy. Chemoautotrophic bacteria and chemolithotrophic bacteria obtain their energy from the oxidation of inorganic (non-carbon) compounds. That is, they derive their energy from the energy already stored in chemical compounds. By oxidizing the compounds, the energy stored in **chemical bonds** can be utilized in cellular processes. Examples of inorganic compounds that are used by these types of bacteria are sulfur, ammonium ion (NH^{4+}), and ferrous **iron** (Fe^{2+}).

The designation autotroph means "self nourishing." Indeed, both chemoautotrophs and chemolithotrophs are able to grow on medium that is free of carbon. The designation lithotrophic means "rock eating," further attesting to the ability of these bacteria to grow in seemingly inhospitable environments.

Most bacteria are chemotrophic. If the energy source consists of large chemicals that are complex in structure, as is the case when the chemicals are derived from once-living organisms, then it is the chemoautotrophic bacteria that utilize the source. If the molecules are small, as with the elements listed above, they can be utilized by chemolithotrophs.

Only bacteria are chemolithotrophs. Chemoautotrophs include bacteria, fungi, animals, and protozoa.

There are several common groups of chemoautotrophic bacteria. The first group is the colorless sulfur bacteria. These bacteria are distinct from the sulfur bacteria that utilize sunlight. The latter contain the compound chlorophyll, and so appear colored. Colorless sulfur bacteria oxidize hydrogen sulfide (H_2S) by accepting an electron from the compound. The acceptance of an electron by an **oxygen atom** creates **water** and sulfur. The energy from this reaction is then used to reduce carbon dioxide to create carbohydrates. An example of a colorless sulfur bacteria is the genus *Thiothrix*.

Another type of chemoautotroph is the "iron" bacteria. These bacteria are most commonly encountered as the rusty colored and slimy layer that builds up on the inside of toilet tanks. In a series of chemical reactions that is similar to those of the sulfur bacteria, iron bacteria oxidize iron compounds and use the energy gained from this reaction to drive the formation of carbohydrates. Examples of iron bacteria are *Thiobacillus ferrooxidans* and *Thiobacillus thiooxidans*. These bacteria are common in the **runoff** from **coal** mines. The water is very acidic and contains ferrous iron. Chemoautotrophs thrive in such an environment.

A third type of chemoautotrophic bacteria includes the nitrifying bacteria. These chemoautotrophs oxidize ammonia (NH_3) to nitrate (NO_3^-). Plants can use the nitrate as a nutrient source. These nitrifying bacteria are important in the operation of the global nitrogen cycle. Examples of chemoautotrophic nitrifying bacteria include Nitrosomonas and Nitrobacter.

The **evolution** of bacteria to exist as chemoautotrophs or chemolithotrophs has allowed them to occupy niches that would otherwise be devoid of bacterial life. For example, in recent years scientists have studied a **cave** near Lovell,

Wyoming. The **groundwater** running through the cave contains a strong sulfuric acid. Moreover, there is no sunlight. The only source of life for the thriving bacterial populations that adhere to the rocks are the rocks and the **chemistry** of the groundwater.

The energy yield from the use of inorganic compounds is not nearly as great as the energy that can be obtained by other types of bacteria. Regardless, chemoautotrophs and chemolithotrophs do not usually face competition from other microorganisms, so the energy they are able to obtain is sufficient to sustain their existence.

The ability of chemoautotrophic and chemolithotrophic bacteria to thrive through the energy gained by inorganic processes is the basis for the metabolic activities of the so-called extremophiles. These are bacteria that live in extremes of **pH**, **temperature**, or pressure, as three examples. Moreover, it has been suggested that the metabolic capabilities of extremophiles could be duplicated on extraterrestrial planetary bodies.

See also Erosion; Geochemistry; Weathering and weathering series

CHERT

Chert, or cryptocrystalline **quartz**, is a microcrystalline variety of the mineral quartz (SiO_2) that is chemically or biochemically precipitated from seawater. Chert is just one of the many types, or polymorphs, of quartz, a mineral composed of three-dimensionally bonded silicate tetrahedra. Chert is very fine-grained, so it does not occur as the 6-sided, prismatic **crystals** typical of such coarsely crystalline varieties of quartz as **rock** crystal, amethyst, smoky quartz and citrine. It does, however, have the mineral properties of quartz. It has a glassy, or vitreous, luster, it is number seven on the Moh's scale of hardness, and it breaks along uneven, shell-shaped planes, a property called conchoidal fracture. Other microcrystalline varieties of quartz include chalcedony, agate, onyx and jasper. Dark grey, unbanded chert is better known as flint. Early hunters and warriors exploited chert's characteristic hardness and conchoidal fracture to create sharp-edged, durable tools and weapons; they also discovered that the hard, smooth surface of a flint nodule or shard can be used to strike a spark.

Silica is mainly extracted from seawater by the biochemical actions of marine organisms. Microscopic aquatic plants, called **diatoms**, take in silicate (SiO_4^-) ions from seawater and use them to create siliceous hard parts, or frustules. When the plants die, the frustules fall to the seafloor, creating layers of uncrystallized siliceous "ooze." Compaction and cementation of these layers of silica creates chert. Opal is solid marine silica that has not yet bonded into a rigid crystal framework. Chert usually occurs as bands or nodules in **limestone**, a marine sedimentary rock that forms by the same mechanism of biological mineral **precipitation**. Organisms like corals and foraminifera, which contribute their hard parts to limestone use calcium (Ca^{2+}) and carbonate (CO_3^{2-}) ions to create their skeletons and shells. Limestone, accordingly, is composed of

the mineral calcite ($CaCO_3$) instead of quartz, but like chert, it is a geological remnant of a biologically productive marine environment.

See also Calcareous Ooze; Sedimentary Rocks

CHICXULUB CRATER · *see* K-T EVENT

CHINOOK WINDS · *see* SEASONAL WINDS

CHLOROFLUOROCARBON (CFC)

A chlorofluorocarbon (CFC) is an organic compound typically consisting of chlorine, fluorine, **carbon**, and hydrogen. Freon, a trade name, is often used to refer to CFCs, which were invented in the 1930s and have been used widely as aerosol propellants, refrigerants, and solvents. Odorless, colorless, nontoxic, and nonflammable, CFCs are considered valuable industrial products and have proven an especially safe and reliable aid in food preservation. However, the accumulation of CFCs in the **stratosphere** that may be linked to **ozone** depletion has generated considerable public debate and has led to legislation and international agreements (such as the Montreal Protocol and its amendments, signed by 148 countries) that banned the production of most CFCs by the year 2000. One of the substitutes developed by industry, the hydrochlorofluorocarbons (HCFCs), still contain enough chlorine to interfere with atmospheric ozone **chemistry**, although in much lesser amounts than CFCs. The Copenhagen amendment to the Montreal Protocol calls for the cessation of HCFC production by 2030. Hydrofluorocarbons (HFCs) are currently considered a safer substitute due to prevent ozone loss due to their lack of chlorine and shorter reactive time. As of 2002, new automobiles in the United States contain HFC refrigerant products in the air conditioners.

In the late 1920s, researchers had been trying to develop a coolant that was both nontoxic and nonflammable. At that time, methyl chloride was used, but if it leaked from the refrigerator, it could explode. This danger was demonstrated in one case when methyl chloride gas escaped, causing a disastrous explosion in a Cleveland hospital. Sulfur dioxide was sometimes used as an alternative coolant because its unpleasant odor could be easily noticed in the event of a leak. The problem was brought to the attention of Thomas Midgley Jr., a mechanical engineer at the research laboratory of General Motors. He was asked by his superiors to try to manufacture a safe, workable coolant. (At that time, General Motors was the parent company to Frigidaire.) Midgley and his associate chemists thought that fluorine might work because they had read that carbon tetrafluoride had a boiling point of 5°F (−15°C). The compound, as it turns out, had accidentally been referenced. Its actual boiling point is 198°F (92.2°C), not nearly the level necessary to produce refrigeration. Nevertheless, the incident proved useful because it prompted Midgley to look at other carbon compounds containing both fluorine and chlorine. Within three days, Midgley's team discovered the right mix: dichlorodifluoromethane, a compound whose molecules contain one carbon, two chlorine, and two fluorine atoms. It is now referred to as CFC-12 or F-12 and was marketed as Freon—as were a number of other compounds, including trichlorofluoromethane, dichlorotetrafluoroethane, and chlorodifluoromethane.

Midgley and his colleagues had been correct in guessing that CFCs would have the desired thermal properties and boiling points to serve as refrigerant gases. Because they remained unreactive, and therefore safe, CFCs were seen as ideal for many applications. Through the 1960s, the widespread manufacture of CFCs allowed for accelerated production of refrigerators and air conditioners. Other applications for CFCs were discovered as well, including their use as blowing agents in polystyrene foam. Despite their popularity, CFCs became the target of growing environmental concern by certain groups of researchers. In 1972, two scientists from the University of California, **F. Sherwood Rowland** and Mario Jose Molino, conducted tests to determine if the persistent characteristics of CFCs could pose a problem by remaining indefinitely in the atmosphere. Soon after, their tests confirmed that CFCs do indeed persist until they gradually ascend into the stratosphere, break down due to ultraviolet radiation, and release chlorine, which in turn affects ozone production. Their discovery set the stage for vehement public debate about the continued use of CFCs. By the mid-1970s, the United States government banned the use of CFCs as aerosol propellants but it resisted a total ban for all industries. Instead, countries and industries began negotiating the process of phasing out CFCs. As CFC use is allowed in fewer and fewer applications, a black market has been growing for the chemical. In 1997, the United States Environmental Protection Agency and Customs Service, along with other governmental agencies, initiated enforcement actions to prevent (CFC) smuggling in the United States.

See also Atmospheric pollution; Global warming; Greenhouse gases and greenhouse effect; Ozone layer and hole dynamics; Ozone layer depletion

CHRONOSTRATIGRAPHY

The term chronostratigraphy refers to that aspect of the field of **stratigraphy** dealing with temporal (time) relations and ages of **rock** bodies. Chronostratigraphic classification in the field of stratigraphy organizes rocks on the basis of their age or the time of their genesis.

Chronostratigraphic units are defined as bodies of rock—stratified and non-stratified—that formed during a specific interval of **geologic time**. Chronostratigraphic units are thus special rock bodies that are conceptual, as well as being material. They can be thought of as the subset of rocks formed during a specified geologic time interval.

For example, the Devonian System is the set of all rocks (sedimentary as well as igneous and metamorphic), wherever they occur on Earth, formed during the **Devonian Period**. The boundaries of this conceptual set of rocks are synchronous (i.e., are the same age everywhere) and the Devonian System is isochronous (i.e., the same age and age span everywhere). When written with a proper noun, e.g., Devonian System, both parts of the name of any chronostratigraphic unit are capitalized.

Chronostratigraphic units, like the system, are the basis for the Phanerozoic time scale. Chronostratigraphic units have a hierarchy, wherein there are corresponding geochronologic units. The chronostratigraphic hierarchy (with corresponding geochronologic term and example proper name) is as follows:

- Erathem—Corresponding geochronologic term: Eon; Example: Phanerozoic
- Eonothem—Corresponding geochronologic term: Era; Example: Paleozoic
- System—Corresponding geochronologic term: Period; Example: Devonian
- Series—Corresponding geochronologic term: Epoch; Example: Late Devonian
- Stage—Corresponding geochronologic term: Age; Example: Frasnian
- Substage—Corresponding geochronologic term: Subage

Because chronostratigraphic units are potentially vast vertical section of rock, geologists observe the following conventions with regard to reference markers placed at agreed sites, which represent the best reference examples of the lower boundaries of chronostratigraphic units. These sites, called Global Stratotype Sections and Points (GSSPs), help define chronostratigraphic units. Not all the necessary GSSPs have been assigned yet, and the work continues under the auspices of the International Union of Geological Sciences (IUGS).

The system is the fundamental chronostratigraphic unit, meaning that it is the most commonly used and referenced chronostratigraphic unit. Further, the system was the original unit conceived of in early chronostratigraphic classification. The system is a major subdivision within the hierarchy of chronostratigraphic units, and the largest system spans approximately 152 million years of Earth history. However, most systems span fewer years. Some systems are subdivided into two subsystems (i.e., Tertiary System is subdivided into Neogene and Paleogene Subsystems and Carboniferous System is subdivided into Pennsylvanian and Mississippian Subsystems). A list of the main Phanerozoic systems (with approximate age ranges in millions of years) are: Quaternary (0 to 1.64 millions of years); Tertiary (1.64 to 65 millions of years); Cretaceous (65 to 145.8 millions of years); Jurassic (145.8 to 208 millions of years); Triassic (208 to 245 millions of years); Permian (245 to 290 millions of years); Carboniferous (290 to 362.5 millions of years); Devonian (362.5 to 408.5 millions of years); Silurian (408.5 to 439 millions of years);Ordovician (439 to 505 millions of years); Cambrian (505 to 570 millions of years).

The boundary ages are determined by radiometric age-bracketing and biostratigraphic relationships.

Names of systems are of diverse origin arising from workers of the eighteenth and nineteenth centuries. System names indicate either (1) chronostratigraphic position (e.g.,

Tertiary and Quaternary), (2) geologic characteristics, e.g., Carboniferous and Cretaceous (*Creta* is Latin for "chalk"), (3) geographic locations, e.g., Devonian and Permian (named for the Perm Province of Czarist Russi), and (4) native people's tribal names, e.g., Ordovician and Silurian (named for Celtic tribes of southern England). Proper names of systems have no common spelling for their endings, despite some attempts in the past to standardize them. Systems and corresponding periods have the same proper name.

Names of relatively new and all future series, stages, and substages come from local geographic features in the vicinity of their designated stratotype (i.e., the place where the unit is defined for reference purposes) or their GSSP (i.e., the place where the base of the unit is defined for reference purposes). However, some older names (pre-1970s) have come from other sources before the geographic convention was established. Within some systems, names of series are formed from the system plus a positional adjective (lower, middle, or upper). Most names have an "-ian" or "-an" ending. Epochs, ages, and subages have the same name as the corresponding chronostratigraphic unit (i.e., series, stage, and substage). The only exception is where a series bears a positional adjective. In these instances, the positional adjective for the series is replaced by a temporal adjective to form the corresponding epoch. For example, the Lower Devonian Series was formed during the Early Devonian Epoch, Middle Devonian Series was formed during the Middle Devonian Epoch, and Upper Devonian Series was formed during the Late Devonian Epoch.

Names of Erathems and Eonothems reflect major changes in existing life on Earth. Regarding the Erathems, Paleozoic means "old life," Mesozoic means "middle life,'" and Cenozoic means "recent life." For Eonothems, Phanerozoic (which encompasses Paleozoic, Mesozoic, and Cenozoic) means "evident life." Older Eonothems are Hadean (in reference to the fiery beginning of Earth), **Archean** (in reference to ancient times), and Proterozoic (in reference to primitive life). Erathems and Eonothems (with their corresponding approximate ages in millions of years) age span chronostratigraphic units; Phanerozoic—Cenozoic (0 to 65 millions of years), Mesozoic (65to 245 millions of years), and Paleozoic (245 to 570 millions of years)—Proterozoic (570 to 2450 millions of years); Archean (2450 to 3800 millions of years) and Hadean (3800 to 4560 millions of years).

See also Dating methods; Phanerozoic Eon; Stratigraphy

CIRQUE · *see* ARETES

CIRRIFORM · *see* CLOUDS AND CLOUD TYPES

CIRROCUMULOUS · *see* CLOUDS AND CLOUD TYPES

CIRRUS · *see* CLOUDS AND CLOUD TYPES

CLAY

Clay is a fine-grained (small particle size) sedimentary **rock**. Clay is so fine-grained it is rarely possible to see the individual mineral particles with the naked eye. The definition of clays describes rocks with particle sizes of less than 4 μm in diameter. Most **sedimentary rocks** are described using both mineral content and particle size. While this is also true for clays, the particle size description is most reliable and most often used.

The majority of common types of **minerals** found in clays are kaolinite (a soapy-feeling and lightweight mineral), talc, pyrophyllite, all types of micas, minerals from the chlorite group, feldspars, and a lesser amount of **tectosilicates** (including **quartz**).

The mineral content of clays is less variable than other types of sedimentary rock. This is a direct result of the way clays are formed. **Water** carries the bulk of sediments to their resting place where they are cemented together. The transport of sediments is directly related to the force or velocity of water carrying them. The stronger the velocity of water, the larger and heavier the particle it can move. Conversely, the weaker the flow, the smaller the particle that is carried by the water. As a result, water acts as a winnowing filter for certain types of minerals. The heavier minerals are not carried as far by water currents as are the lighter ones. When water finally comes to rest, it deposits its load of minerals. The last to be released are the lighter and smaller particles, the clay minerals.

Where **rivers** meet **oceans**, the clay minerals are so light they are usually carried far out to sea where they fall gently to the bottom forming a fine-grained sediment. These deposits cover organic materials and trap them at the edges of deltas and continental slopes. Over millions of years, the organic materials convert to **petroleum** and remain trapped by the clays. This relationship makes the study of clays extremely important for petroleum geologists. In addition to this important economic consideration, clays provide important economic resources for a wide variety of other industries.

See also Petroleum detection; Sedimentation

CLIMATE · *see* WEATHER AND CLIMATE

CLOUD SEEDING

Mark Twain once said that everyone talks about the **weather**, but no one ever does anything about it. Although he may have been correct in his day, since the 1940s, researchers have been at least partially successful in modifying one aspect of the weather—precipitation.

After about three years of investigative work at the General Electric Research Laboratory in Schenectady, New York, researchers Irving Langmuir and his assistant, Vincent Joseph Schaefer, created the first human-made rainfall. Their work had originated as war-influenced research on airplane wing icing. On November 13, 1946, Schaefer sprinkled several pounds of dry **ice** (frozen **carbon dioxide**) from an airplane into a supercooled cloud, a cloud in which the **water** droplets remain liquid in sub-zero temperatures. He then flew under the cloud to experience a self-induced snowfall. The snow changed to rain by the time it reached Langmuir, who was observing the experiment on the ground.

Langmuir and Schaefer selected dry ice as cloud "seed" for its quick cooling ability. As the dry ice travels through the cloud, the water vapor behind it condenses into rain-producing **crystals**. As the crystals gain weight, they begin to fall and grow larger as they collide with other droplets.

Another General Electric (GE) scientist who had worked with Langmuir and Schaefer, Bernard Vonnegut, developed a different cloud-seeding strategy. The formation of water droplets requires microscopic nuclei. Under natural conditions, these nuclei can consist of dust, smoke, or sea salt particles. Instead of using dry ice as a catalyst, Vonnegut decided to use substitute nuclei around which the water droplets in the cloud could condense. He chose silver iodide as this substitute because the shape of its crystals resembled the shape of the ice crystals he was attempting to create.

The silver iodide was not only successful, it had practical advantages over dry ice. It could be distributed from the ground through the use of cannons, smoke generators, and natural cumulonimbus cloud updrafts. Also, it could be stored indefinitely at room **temperature**.

There is general disagreement over the success and practicality of cloud seeding. Opponents of cloud seeding contend that there is no real proof that the **precipitation** experienced by the seeders is actually of their own making. Proponents, on the other hand, declare that the effect of seeding may be more than local.

Over the years, cloud seeding has become an accepted part of the strategy to combat **drought**. It may indeed bring crop-saving relief to a dry field or may help reinforce subsurface water tables. However, the practice has not begun to eliminate deserts or devastating droughts, for researchers have yet to reproduce the general ground-soaking effects of a well-organized natural storm system so necessary for agriculture and replenishment of water reserves. And today there are environmental concerns over any activity that threatens to change or destroy a bio-community such as the **desert**.

As researchers collect and analyze more information about the weather, other attempts to modify it are bound to be developed.

See also Air masses and fronts; Weather forecasting; Weather forecasting methods

CLOUDS AND CLOUD TYPES

Clouds are condensations of **water** and other particles in the atmosphere. Cloud shapes—and the dynamics of their formation—are accurate indicators of important atmospheric properties, including air stability, moisture content, and motion.

Clouds are divided into families of high level, middle level, low level, and vertically developing clouds, and are classified again, in accord with their general shape (e.g., cumuliform or stratoform)

High level clouds include cirrus, cirrostratus, and cirrocumulus clouds that occur at altitudes between 16,000 and 45,000 feet. Middle level clouds include altostratus, altocumulus, and nimbostratus clouds that occur between 6500 and 22,000 feet. Low-level clouds include stratus and stratocumulus clouds that occur between the surface and 6,500 feet. Vertical development clouds include cumulus and cumulonimbus clouds, and range in their development from the surface to 45,000 feet. The heights of the bases of the clouds used to designate cloud families can vary with **latitude**. At extreme northern or southern latitudes, high altitude family clouds can be observed at much lower altitudes.

In general, cloud shape is determined by the method of cooling to reach **condensation** and the forces of winds that can shear or tear the cloud. Cloud opacity (i.e., whether it is light or dark) is a function of cloud thickness.

Cirrus clouds occur at high levels and are generally wispy and elongate in form. Vertically rising air is unstable and gives rise to cumulus cloud formation. Cumulus clouds are billowy. Stratus clouds (i.e., stratified clouds) are heavily layered and often appear in sheet-like formations. With regard to cloud nomenclature, nimbus clouds (e.g., clouds with the prefix nimbo or the suffix nimbus) are rain-producing clouds. The use of "fracto" designates broken cloud formations.

High clouds—cirrus, cirrostratus, and cirrocumulus—are composed of **ice crystals** and dust or pollution particles. The particles often serve as centers of crystallization or condensation nuclei. Cirrus clouds often produce "mares' tails" that are tail-like wisps of ice crystals. Cirrostratus clouds, because they are thin and the ice crystals act to both reflect and refract sunlight, are often associated with halos of ice crystals that appear to encircle the **Sun** or **Moon**. Cirrocumulus clouds often appear as patch-like thin clouds.

Middle level cloudsmdash; altostratus, altocumulus, and nimbostratus—are composed of water with some ice crystal formation near cloud tops. Both middle level and low level clouds may be composed of super-cooled water (water below **freezing**) that has not yet crystallized around a condensation nucleus. Altostratus clouds often present a bluish-layered appearance. Depending on thickness, altocumulus clouds often have white or gray layers that appear in washboard or wave-like formations. Atmospheric instability and convective air currents can result in the formation of altocumulus castellanus clouds, a form of altocumulus that often appear as isolated cumulous clouds with billowing tops. Another form of altocumulus cloud, a standing lenticular altocumulus clouds, is formed by turbulent updrafts of air uplifted by terrain barriers (e.g., mountains, ridges, etc.). Although dynamic, the standing lenticular altocumulus cloud formations appear static or "standing" over the terrain feature leading to their formation. Nimbostratus clouds often appear as heavy, gray, moisture-laden cloud layers

Low-level stratus clouds are usually gray clouds associated with **precipitation** and **fog**. Stratocumulus clouds present the familiar, cotton ball-like cumulus shapes in an elongate form (a cumulus shape drawn out by shearing winds).

Clouds with extensive vertical development—cumulus and cumulonimbus clouds—often present a gradient of ice and water. Rapid updrafts and downdrafts allow ice crystals to appear at much lower levels than would be expected by atmospheric **temperature**. Although arising from convective currents, cumulus clouds often form in fair **weather** and do not show extensive vertical development. Cumulus clouds present flat bases and curved or domed tops. More extensive vertical development occurs as atmospheric instability increases. Highly developed cumulus clouds often present mushroomed or cauliflower-like tops, and can ultimately produce rain. Under the most unstable of atmospheric conditions, cumulonimbus clouds form. Cumulonimbus clouds are dark clouds with anvil like tops sheared by very high altitude winds. Heavy turbulence, violent rains, **lightning**, and **thunder** often accompany cumulonimbus clouds. Particularly unstable and violent clouds can occur in cells capable of spawning tornadoes.

The identification of cloud types is an important skill for aviators and aviation meteorologists because clouds present variable icing hazards. Ice formation can drastically reduce the effectiveness of airfoils (wings, flaps, rudder, ailerons, elevators) and destroy lift and/or interfere with the ability to control aircraft.

See also Atmospheric circulation; Atmospheric composition and structure; Atmospheric inversion layers; Atmospheric lapse rate; Atmospheric pressure; Phase state changes; Troposphere and tropopause; Weather forecasting methods; Weather forecasting; Weather radar; Wind shear

COAL

Coal is a naturally occurring combustible material consisting primarily of the element **carbon**, but with low percentages of solid, liquid, and gaseous **hydrocarbons** and other materials, such as compounds of nitrogen and sulfur. Coal is usually classified into the sub-groups known as anthracite, bituminous, lignite, and peat. The physical, chemical, and other properties of coal vary considerably from sample to sample.

Coal forms primarily from ancient plant material that accumulated in surface environments where the complete decay of organic matter was prevented. For example, a plant that died in a swampy **area** would quickly be covered with **water**, silt, **sand**, and other sediments. These materials prevented the plant debris from reacting with **oxygen** and decomposing to **carbon dioxide** and water, as would occur under normal circumstances. Instead, anaerobic bacteria (bacteria that do not require oxygen to live) attacked the plant debris and converted it to simpler forms: primarily pure carbon and simple compounds of carbon and hydrogen (hydrocarbons). Because of the way it is formed, coal (along with **petroleum** and **natural gas**) is often referred to as a fossil fuel.

The initial stage of the decay of a dead plant is a soft, woody material known as peat. In some parts of the world, peat is still collected from boggy areas and used as a fuel. It is not a good fuel, however, as it burns poorly and with a great deal of smoke.

If peat is allowed to remain in the ground for long periods of time, it eventually becomes compacted as layers of sediment, known as overburden, collect above it. The additional pressure and heat of the overburden gradually converts peat into another form of coal known as lignite or brown coal. Continued compaction by overburden then converts lignite into bituminous (or soft) coal and finally, anthracite (or hard) coal. Coal has been formed at many times in the past, but most abundantly during the Carboniferous Age (about 300 million years ago) and again during the Upper Cretaceous Age (about 100 million years ago).

Today, coal formed by these processes is often found in layers between layers of sedimentary **rock**. In some cases, the coal layers may lie at or very near the earth's surface. In other cases, they may be buried thousands of feet or meters under ground. Coal seams range from no more than 3–197 ft (1–60 m) or more in thickness. The location and configuration of a coal seam determines the method by which the coal will be mined.

Coal is classified according to its heating value and according to its relative content of elemental carbon. For example, anthracite contains the highest proportion of pure carbon (about 86%–98%), and has the highest heat value (13,500–15,600 Btu/lb [British thermal units per pound]) of all forms of coal. Bituminous coal generally has lower concentrations of pure carbon (from 46%–86%) and lower heat values (8,300–15,600 Btu/lb). Bituminous coals are often sub-divided on the basis of their heat value, and are classified as low, medium, and high volatile bituminous and sub-bituminous. Lignite, the poorest of the true coals in terms of heat value (5,500–8,300 Btu/lb) generally contains about 46%–60% pure carbon. All forms of coal also contain other elements present in living organisms, such as sulfur and nitrogen, that are very low in absolute numbers, but that have important environmental consequences when coals are used as **fuels**.

By far the most important property of coal is that it combusts. When the pure carbon and hydrocarbons found in coal burn completely, only two products are formed, carbon dioxide and water. During this chemical reaction, a relatively large amount of energy is released. The release of heat when coal is burned explains the fact that the material has long been used by humans as a source of energy, for the heating of homes and other buildings, to run ships and trains, and in many industrial processes.

The complete combustion of carbon and hydrocarbons described above rarely occurs in nature. If the **temperature** is not high enough or sufficient oxygen is not provided to the fuel, combustion of these materials is usually incomplete. During the incomplete combustion of carbon and hydrocarbons, other products besides carbon dioxide and water are formed, primarily carbon monoxide, hydrogen, and other forms of pure carbon, such as soot.

During the combustion of coal, minor constituents are also oxidized. Sulfur is converted to sulfur dioxide and sulfur trioxide, and nitrogen compounds are converted to nitrogen oxides. The incomplete combustion of coal and the combustion of these minor constituents result in a number of environmental problems. For example, soot formed during incomplete combustion may settle out of the air and deposit an unattractive coating on homes, cars, buildings, and other structures. Carbon monoxide formed during incomplete combustion is a toxic gas and may cause illness or death in humans and other animals. Oxides of sulfur and nitrogen react with water vapor in the atmosphere and then are precipitated out as **acid rain**. Acid rain is thought to be responsible for the destruction of certain forms of plant and animal (especially fish) life.

In addition to these compounds, coal often contains a few percent of mineral matter: **quartz**, calcite, or perhaps **clay minerals**. These do not readily combust and so become parts of the ash. The ash then either escapes into the atmosphere or is left in the combustion vessel and must be discarded. Sometimes coal ash also contains significant amounts of **lead**, barium, arsenic, or other compounds. Whether air borne or in bulk, coal ash can therefore be a serious environmental hazard.

Coal is extracted using one of two major techniques, sub-surface or surface (strip) mining. The former method is used when seams of coal are located at significant depths below Earth's surface. The first step in sub-surface mining is to dig vertical tunnels into the earth until the coal seam is reached. Horizontal tunnels are then constructed laterally off the vertical tunnel. In many cases, the preferred method of mining coal by this method is called room-and-pillar mining. In this method, vertical columns of coal (the pillars) are left in place as coal around them is removed. The pillars hold up the ceiling of the seam, preventing it from collapsing on miners working around them. After the mine has been abandoned, however, those pillars may often collapse, bringing down the ceiling of the seam and causing subsidence in land above the old mine.

Surface mining can be used when a coal seam is close enough to the earth's surface to allow the overburden to be removed economically. In such a case, the first step is to strip off all of the overburden in order to reach the coal itself. The coal is then scraped out by huge power shovels, some capable of removing up to 100 cubic meters at a time. Strip mining is a far safer form of coal mining, but it presents a number of environmental problems. In most instances, an area that has been strip-mined is scarred, and restoring the area to its original state is a long and expensive procedure. In addition, any water that comes in contact with the exposed coal or overburden may become polluted and require treatment.

Coal is regarded as a non-renewable resource, meaning that it was formed at times during Earth's history, but significant amounts are no longer forming. Therefore, the amount of coal that now exists below the earth's surface is, for all practical purposes, all the coal that humans have available to them for the foreseeable future. When this supply of coal is used up, humans will find it necessary to find some other substitute to meet their energy needs.

Large supplies of coal are known to exist (proven reserves) or thought to be available (estimated resources) in **North America**, the former Soviet Union, and parts of **Asia**, especially China and India. According to the most recent data available, China produces the largest amount of coal each year, about 22% of the world's total. China is also thought to have the world's largest estimated resources of coal, as much as 46% of all that exists.

For many centuries, coal was burned in small stoves to produce heat in homes and factories. Today, the most important use of coal, both directly and indirectly, is still as a fuel. The largest single consumer of coal as a fuel is the electrical power industry. The combustion of coal in power generating plants is used to make steam, which in turn, operates turbines and generators. For a period of more than 40 years, beginning in 1940, the amount of coal used in the United States for this purpose doubled in every decade. Coal is no longer widely used to heat homes and buildings, as was the case a half century ago, but it is still used in industries such as paper production, cement and ceramic manufacture, **iron** and steel production, and chemical manufacture for heating and for steam generation.

See also Environmental pollution

COBBLE · *see* ROCK

COCCOLITHOPHORIDS · *see* CALCAREOUS OOZE

COMETS

Comets are objects—relatively small compared to planets—that are composed of dust and ices of various compounds. Comets orbit the **Sun** in elongated elliptical (eccentric, elongated circle) or parabolic orbits. Accordingly, these objects spend the majority of time in the outer regions of the **solar system**, in some cases well beyond the orbits of Neptune and Pluto. Short-period comets are those with less exaggerated elliptical orbits that carry them out only as far as the region of **space** between the orbits of Jupiter and Neptune. Comets make periodic, brief, but sometimes-spectacular transits through the inner solar system as they approach the Sun. Comet orbits may be prograde, in the same direction as the planets, or retrograde, in the opposite direction. With the aid of a **telescope**, a comet is usually visible from Earth.

The term "comet" derives from the Greek *aster kmetes* (translated literally as "hairy" or long-haired star)—a reference to a sometimes-visible comet tail. If a comet's path takes it close enough to the Sun, the heating causes **melting** and emission of gases (out gassing) and dust that are then swept behind the comet's orbital path (away from the Sun) by the solar **wind** to form the characteristic tail.

Fascination with objects in the night sky dates to the dawn of human civilization. Etchings on **clay** tablets unearthed in the ancient city of Babylon dating to at least 3000 B.C. and **rock** carvings found in prehistoric sites in Scotland dating to 2000 B.C. depict unexplained astronomical phenomena that may have been comets. Until the Arabic astronomers of the eleventh century, the Chinese were by far the most astute skywatchers in the ancient and medieval world. By 400 B.C., their intricate cometary classification system included sketches of

29 comet forms, each associated with a past event and predicting a future one. Comet type 9 was named Pu-Hui, meaning "Calamity in the state, many deaths." In fact, until the seventeenth century when English Astronomer Edmund Halley (1656–1742) predicted the return of a the comet in 1758 (thereafter known as Halley's comet) based, in part, upon calculations derived from English physicist and mathematician Sir Isaac Newton's (1642–1727) work, comets were widely viewed with superstition, as omens portending human disasters and terrestrial catastrophes.

Recorded observation of comets is evident in the records of the Ancient Chinese culture who termed comets "guest stars," a general term also applied to other apparent temporary solar system transients that were later, of course, identified to be much more distant stellar novae. Chinese records clearly indicate the transit of a guest star in approximately 240 B.C. that we now identify as Halley's comet.

In accord with Anasazi Native American accounts, Chinese astronomers also noted the difference in what is now regarded as comets and the 1054 supernova explosion in the Taurus constellation (i.e., a region of the sky associated with the Taurus constellation) that created the Crab Nebula.

Of all the Classical Greek and Roman theories on comets, the most influential, though entirely incorrect, was that of the Greek philosopher Aristotle (384–322 B.C.). His geocentric view of the solar system put Earth at the center circled by the Sun, **Moon**, and visible planets. Stars were stationary and the bodies existed on celestial spheres. Aristotle argued that comets were fires in the dry, sublunar "fiery sphere," a combustible atmosphere "exhaled" from Earth, which accumulated between Earth and the Moon. Comets were therefore considered terrestrial—originating from Earth, rather than celestial—heavenly bodies. Moreover, they were seen as a portent of future events controlled by the gods.

Aristotle's writings formed the basis of later Greek-Alexandrian astronomer Ptolemy's (A.D. 87–150) model of the universe that became strongly supported by the Christian church in Western **Europe**. Because the Ptolemy's model provided accurate results with regard to celestial prediction, it was the most influential astronomical model until the acceptance of the Sun-centered (heliocentric) model put forth by Polish astronomer Nicolaus Copernicus (1473–1543).

In conjunction with the German astronomer and mathematician **Johannes Kepler** (1571–1630), Danish astronomer Tycho Brahe's (1546–1601) observation of the "Great Comet" of 1577 provided evidence that the comet was at least four times further away from Earth than the Moon—a crushing refutation of Aristotle's sublunar positioning.

The study of the Great Comet by Brahe and his contemporaries was the turning point for astronomical science. Throughout the seventeenth and eighteenth centuries, mathematicians and astronomers refined conflicting ideas on the origin, formation, movement, shape of orbit, and meaning of comets. Polish-born scientist Johannes Hevelius (1611–1687), who suggested comets move on a parabola (U-shape) around the Sun; and English scientist Robert Hooke (1635–1703), independent of Newton, introduced the theory of universal gravitational influence based, in part on the periodic behavior

of comets. Newton, however, developed an astounding mathematical model for the parabolic motion of comets, published in his seminal and influential 1687 book, *Philosophiae Naturalis Principia Mathematica* (Mathematical principles of natural philosophy). Until English naturalist Charles Darwin's (1809–1882) writings on **evolution** and German-American physicist Albert Einstein's (1879–1955) twentieth century writings on **relativity theory**, *Principia* remained the single most influential scientific work in the history of science."

By the end of the eighteenth century, comets were understood to be astronomical bodies, the movement of which could be calculated using Newton's laws of planetary motion.

The comet Biela, with a periodic orbit of 6.75 years, split in two during its 1846 appearance. Twin comets reappeared in 1852—but then failed to appear for its next pass. The disappearance fostered scientific speculation regarding comet impacts and their relationships to meteor showers. When Biela's twin offspring should have returned, the meteor shower predicted by some astronomers did indeed appear, strengthening the connection between meteors and dying comets.

The first observation of a comet through a telescope was made in 1618. Until the twentieth century, comets were discovered and observed with the naked eye or through telescopes. Today, most new discoveries are made from photographs of our galaxy and electronic detectors, although many discoveries are still made by amateur astronomers with a passion for careful observation.

The long focal-length refracting telescope, the primary astronomical observation tool of the 1800s, worked well for viewing bright objects but did not collect sufficient light to allow detailed astronomical photography. In 1858, an English artist named Usherwood used a short focal-length lens to produce the first photograph of a comet. In 1864, by using a spectroscope, an instrument that separates the wavelengths of light into spectral bands, Italian astronomer Giovanni Donati (1826–1873) first identified a chemical component in a comet's atmosphere. The first cometary spectrogram (spectral photograph) was taken by amateur astronomer William Huggins of London in 1881.

The early twentieth century saw the development of short focal-length spectrographs that, by the 1950s, allowed identification of several different chemical components in a comet's tail. Infrared spectrography was introduced in the 1960s and, in 1983, the Infrared Astronomy Satellite (IRAS) gathered information on cometary dust particles unattainable from ground-based technology. Observations of comets are now also made by radio wave detection and ultraviolet spectrography.

Spectroscopic evidence indicates that most comets contain a solid nucleus (core) surrounded by a gigantic, glowing mass (coma). Together, the nucleus and coma comprise the comet head. It should be noted that although the tail (when apparent) seems dense with dust and gas, it is still a vacuum that is far less dense than the interplanetary space near the earth (e.g., between the earth and Moon).

Perhaps among the most primitive bodies in the solar system, comets are probably debris from the formation of our Sun and planets some 4.5 billion years ago. One hypothesis

Comet Hale-Bopp. *Jack Newton. Archive Photos, Inc. Reproduced by permission.*

concerning their origin involves the Oort cloud—named for Dutch astronomer Jan Van Oort—a dense shell of debris (dense by interstellar standards) at the frigid, outer edge of the solar system (i.e., the distance at which our Sun's gravitational pull is so weak that beyond this point other stellar bodies exert a greater net attraction). Occasionally, disruptive gravitational forces (perturbations) hurl a piece of debris from the cloud into the gravitational pull of one of the large planets, (e.g. Jupiter or Saturn) that then pull the comet into an elliptical orbit around the Sun. The Kuiper belt, associated with Jupiter's gravitational pull, is more likely the source of the well-known comets, including Halley's comet. Regardless, evidence indicates that comets formed from solar system formation debris.

Short lived comets have orbital durations of less than 200 years. Long-period, having enormous elliptical, nearly parabolic orbital durations of more than 200 years, often traveling far beyond the outer planets. Of the 710 individual comets recorded from 1680 to mid-1992, 121 were short-period comets and 589 were classified as long-period comets.

Two major theories on the composition of the nucleus have developed over time. The "flying sandbank" model, first proposed by Richard Proctor in the mid-1800s and again in the mid-1900s by Raymond Lyttleton, conjectured swarms of tiny solid particles bound together by mutual gravitational attraction. In 1950, Fred Whipple introduced the "icy-conglomerate" model, which described a comet as a solid nucleus of meteoric rock particles and dust frozen in **ice**. Observations of Halley's comet by spacecraft in 1986 strongly support this model.

Evidence to date indicates that within the comet head or nucleus, rocks and dust are held together with ices from **water**, methane, ammonia, and **carbon** monoxide, as well as other ices containing carbon and sulphur. The 1986 studies of Halley's comet revealed the nucleus to be peanut or potato-shaped, 9 mi (15 km) long, and 5.5 mi (8 km) wide. However, visual observation beneath the comet's dark, solid surface proved impossible.

The nuclei of comets are among the smallest bodies in the Solar system, too small, in fact, for observation even through a telescope. As they approach the Sun, however, they produce one of the largest, most spectacular sights in the solar system, a magnificent, glowing coma often visible even to the naked eye. Comet nuclei have been seen to produce sudden, bright flares and some even split into two, three, four, or five refions.

As the nucleus of a comet approaches the Sun, beginning at about the distance of the asteroid belt, its ices begin to vaporize and sublimate (change directly from ice to gas). This off-gassing releases gases of hydrogen, carbon, **oxygen**, nitrogen and other molecules, as well as dust particles. Streaming away at several hundred meters per second, they create an enormous coma hundreds of thousands of kilometers long, completely hiding the nucleus. The Sun's ultraviolet light electrically charges the gaseous molecules, ionizing and exciting them, causing them to fluoresce (emit light) much like a fluorescent light emits light following electrical stimulation. Microscopic mineral particles in the dust reflect and scatter the Sun's light. Only in 1970, during the first spacecraft observation of a comet, was a gigantic hydrogen cloud discovered surrounding the Coma. Depending on the size of the nucleus and its proximity to the Sun, this cloud can be much larger than the comet itself.

As the comet swings around the Sun on its elliptical orbit, gas and dust particles stream from the coma to create two types of tails: the gaseous ion tail, or Type I; and the dust tail, or Type II. In Type I, ionized gases form a thin, usually straight tail, sometimes millions of kilometers long. (The tail of the Great Comet of 1843 stretched out more than 136 million mi [220 million km].) In fact, the tails of comets are the largest measured entities in the solar system. The ion tail, glowing with incredible brightness, does not trail behind but is blown away from the head in a direction almost opposite the Sun by the "solar wind," a continual flow of magnetized plasma emitted by the Sun. The head collides with this plasma, which wraps around the nucleus, pulling the ionized particles with it. Depending on its position to the Sun, the tail may even be traveling almost ahead of the nucleus. A Type II tail is usually shorter and wider, and curves slightly because the heavier particles are carried away at a slower rate. The Great Comet of 1744 actually displayed six brilliant tails fanning above the horizon like peacock feathers.

Comet Hale-Bopp, which streamed across the skies in 1997, boasted a new feature: a third tail composed of electrically neutral sodium atoms. When completely observed using instruments with spectral filters that eliminated all but the yellow light emitted by fluorescing sodium atoms, the tail was more than 370,000 miles wide (600,000 km) and 31 million miles long (50 million km), streaming in a direction close but slightly different to that of the ion tail. Although the exact mechanism is not understood, the tail is thought to be formed of sodium atoms released by dust particles in the coma.

Comets may strike planets without leaving an **impact crater**. The atmospheric energy released by comet vaporization in the atmosphere can, however, be more powerful than a nuclear explosion. The Tunguska event in Siberia in 1908 is thought to have been the result of a comet or stony meteoroid explosion above the ground. In 1979, a United States Air Force space-test satellite took the first photograph of a comet colliding

with the Sun. Late in 1994, the fragmented comet Shoemaker-Levy made spectacular serial collisions with Jupiter.

Some scientists argue that molecules released by comets' vaporized gases may have supplied important molecules in Earth's early atmosphere. When exposed to the Sun's radiation, these molecules began the formation of biochemical compounds that actually began the process of life on Earth—or gave a huge "jump-start" to the evolution of biomolcules. During the recent passage of Hale-Bopp, for example, scientists discovered a variety of complex organic chemicals in the comet.

Some aspects of this theory gained evidence from data gathered by the Polar spacecraft, launched by NASA in 1996. According to some interpretations of observations by the probe, comet-like objects up to 30–40 ft (9–12 m) in diameter may be hitting the atmosphere at the astounding rate of up to 43,000 per day. These cosmic snowballs usually disintegrate in the upper atmosphere, their content liquids and gases entering the **weather** cycle and eventually reaching the terrestrial surface as **precipitation**. Other scientists argue that the evidence of "snowballs" from space is an artifact of instrument background noise or interference.

In a pair of space missions planned for the early part of the twenty-first century, space probes will rendezvous with a pair of short-period comets, hopefully to help scientists reach a better understanding of the **physics** of comets. In 2004, NASA's Stardust mission plans to capture dust from the tail of Comet Wild 2, returning the samples to Earth for analysis. In 2011, the European Space Agency's Rosetta mission will rendezvous with Comet Wirtanen on its trip around the Sun. The Rosetta spacecraft will orbit the comet and send a probe to the surface.

See also Astronomy; Big Bang theory; Impact crater

COMPACTION • *see* LITHIFICATION

CONDENSATION

Condensation occurs when one of the three states of matter in which a gaseous (vapor) substance transforms into a liquid. It is the reverse of vaporization. As a vapor cools, it gives off energy in the form of latent heat. The release of heat causes each vapor molecule to shrink and move more slowly. Strong intermolecular forces (kinetic-molecular theory of gases) push the smaller gas molecules together; the bonding initiates the transformation into a denser, liquid form called condensate.

Condensation can occur from cooling processes such as distillation and steam engine production, or by exerting pressure in a manner that reduces the volume of the gas. Evidence of condensate is often found on a bathroom mirror after a hot shower, or on the outside of a "sweating" soda can as it warms to room **temperature**. Meteorological phenomena such as clouds, **fog**, and dew are also a result of condensation. Clouds form when rising hot air collides with air in cooler elevations.

In **chemistry**, condensation is defined as a reaction involving the joining of atoms in the same or different molecules. Chemists often use the process of condensation to eliminate simple molecules, such as **water** or alcohol, to form a heavier, more complex compound.

See also Evaporation; Hydrologic cycle

CONGLOMERATE ROCK • *see* ROCK

CONSTELLATIONS • *see* CELESTIAL SPHERE: THE APPARENT MOVEMENTS OF THE SUN, MOON, PLANETS, AND STARS

CONTACT AUREOLE • *see* METAMORPHISM

CONTACT METAMORPHISM • *see* METAMORPHISM

CONTINENTAL CRUST • *see* CRUST

CONTINENTAL DIVIDE

A continental divide is a topographic feature separating streams that flow towards opposite sides of a continent. It is a continental scale version of the topographic divides that separate **drainage basins** of all scales.

In the conterminous United States and Canada, the continental divide follows an irregular course from the Basin and Range and Colorado Plateau physiographic provinces of New Mexico north through the Rocky Mountains, the Yellowstone region, and the Canadian Rockies. **Water** in streams to the west of the continental divide flows toward the Pacific Ocean, whereas that to the east of the continental divide flows toward the Atlantic Ocean. In Alaska, however, the continental divide marks the boundary between **rivers** flowing north and west to the Arctic Ocean and those flowing south and west into the Bering Sea.

Continental divides are often associated with mountainous terrain. Elevations along the continental divide through the conterminous United States, however, range from approximately 1400 meters above sea level in the **Basin and Range topography** of southern New Mexico to more than 4000 meters above sea level in the Rocky Mountains of Colorado and Wyoming.

A common, but inaccurate, notion is that **precipitation** falling on different sides of the divide necessarily travels to different **oceans**. Some precipitation that falls as snow, however, may be sublimated. Snowmelt or water that falls as rain may either evaporate or be transpired by vegetation after percolating into the **soil**. In either of those cases, the water may not travel to any ocean until it falls again as precipitation. In the extreme case of internally drained basins common to arid

Aerial view of the continental divide in Colorado. © Dean Conger/Corbis. Reproduced by permission.

regions such as the American Southwest, virtually all water flows towards the center of the basins and is removed through **evaporation**, transpiration, and infiltration. Water that infiltrates deep enough to recharge underlying aquifers may ultimately be discharged on the opposite side of the continental divide because, although **groundwater** divides do exist, they do not necessarily correspond exactly to topographic divides. Humans can also play a role, most notably by constructing diversion tunnels through which water is carried from one side of the continental divide to the other as part of water supply projects. Therefore, it is best to restrict the usage of the term continental divide to a topographic divide that separates streams flowing towards opposite sides of the continent than to include speculations about the ultimate fate of individual drops of water.

See also Drainage basins and drainage patterns; Freshwater; Hydrologic cycle; Landscape evolution; Precipitation; Runoff

CONTINENTAL DRIFT THEORY

Continental drift, in the context of the modern theory of **plate tectonics**, is explained by the movement of **lithospheric plates** over the **asthenosphere** (the molten, ductile, upper portion of the earth's mantle). Precisely used, the term continental drift is actually rooted in antiquated concepts regarding the structure of the earth. Modern geophyicists and geologists explain the movement or drift of the continents within the context of plate tectonic theory. The visible continents, a part of the lithospheric plates upon which they ride, shift slowly over time as a result of the forces driving plate tectonics. Moreover, plate tectonic theory is so robust in its ability to explain and predict geological processes that it is equivalent in many regards to the fundamental and unifying principles of **evolution** in biology, and nucleosynthesis in **physics** and **chemistry**.

The original theory of continental drift made the improbable assertion that the continents moved through and across an underlying oceanic **crust** much as **ice** floats and drifts through **water**. Eventually multiple lines of evidence allowed modern tectonic theory to replace continental drift theory.

Based upon centuries of cartographic depictions that allowed a good fit between the Western coast of **Africa** and the Eastern coast of South America—in 1858, French geographer Antonio Snider- Pellegrini, published a work asserting that the two continents had once been part of larger single continent ruptured by the creation and intervention of the Atlantic Ocean.

In the 1920s, German geophysicist Alfred Wegener's writings advanced the hypothesis of continental drift depicting the movement of continents through an underlying oceanic crust.

Wegener's hypothesis met with wide skepticism but found support and development in the work and writings of South African geologist Alexander Du Toit who discovered a similarity in the **fossils** found on the coasts of Africa and **South America** that were seemingly derived from a common source. Other scientists also attempted to explain **orogeny** (mountain building) as resulting from Wegener's continental drift.

Wegener's initial continental drift assertions were based upon the geometric fit of the displaced continents and the similarity of **rock** ages and Paleozoic fossils in corresponding bands or zones in adjacent or corresponding geographic areas. Wegener also argued that the evidence of Paleozoic **glaciation** in South Africa, South America, India and Australia—sites far removed from estimates of the geographical extent of glaciation—argued strongly for continental drift

The technological advances necessitated by the Second World War made possible the accumulation of significant evidence regarding Wegener's hypothesis, eventually refining and supplanting Wegener's theory of continental drift with modern plate tectonic theory. Although Wegener's theory accounted for much of the then existing geological evidence, Wegener's hypothesis was specifically unable to provide a verifiable or satisfying mechanism by which continents—with all of their bulk and drag—could move over an underlying mantle that was solid enough in composition to be able to reflect seismic S- waves.

In his 1960 publication, *History of Ocean Basins*, geologist and U.S. Navy Admiral Harry Hess (1906–1969) asserted that thermal convection currents in the athenosphere provided the driving force behind plate tectonics. The degree with which the earlier geological community resisted acceptance of Wegener's theory of continental drift is clearly demonstrated by the fact that Hess's assertion of thermal currents was drawn from work done by Author Holmes in the 1930s.

See also Earth, interior structure; Hotspots; Sea-floor spreading; Seismology

CONTINENTAL GLACIER • *see* GLACIERS

CONTINENTAL SHELF

The continental shelf is a gently sloping and relatively flat extension of a continent that is covered by the **oceans**. Seaward, the shelf ends abruptly at the shelf break, the boundary that separates the shelf from the continental slope.

The shelf occupies only 7% of the total ocean floor. The average slope of the shelf is about 10 ft per mile (1.9 m per km). That is, for every one kilometer of distance, the shelf drops 1.9 m in elevation until the shelf break is reached. The average depth of the shelf break is 440 ft (135 m). The greatest depth is found off **Antarctica** (1,150 ft [350 m]), where the great weight of the **ice** on the Antarctic continent pushes the **crust** downward. The average width of the shelf is 43 mi (70 km) and varies from tens of meters to approximately 800 mi (1,300 km) depending on location. The widest shelves are in the Arctic Ocean off the northern coasts of Siberia and **North America**. Some of the narrowest shelves are found off the tectonically active western coasts of North and **South America**.

The shelf's gentle slope and relatively flat terrain are the result of **erosion** and sediment deposition during the periodic fall and rise of the sea over the shelf in the last 1.6 million years. The changes in sea level were caused by the advance and retreat of **glaciers** on land over the same time period. During the last glacial period (approximately 18,000 years ago), sea level was 300–400 ft (90–120 m) lower than present and the shoreline was much farther offshore, exposing the shelf to the atmosphere. During lowered sea level, land plants and animals, including humans and their ancestors, lived on the shelf. Their remains are often found at the bottom of the ocean. For example, 12,000 year old bones of mastodons, extinct relatives of the elephant, have been recovered off the coast of the northeastern United States.

Continental shelves contain valuable resources, such as oil and gas and **minerals**. Oil and gas are formed from organic material that accumulates on the continental shelf. Over time the material is buried and transformed to oil and gas by heat and pressure. The oil and gas moves upward and is concentrated beneath geologic traps. Oil and gas is found on the continental shelf off the coasts of California and Louisiana, for example. Minerals come from rocks on land and are carried to the ocean by **rivers**. The minerals were deposited in river channels and beaches on the exposed continental shelf and sorted (concentrated) by waves and river currents, due to their different densities. Over time as the sea level rose, these minerals were again sorted by waves and ocean currents and finally deposited. The different colored bands of **sand** that one can see on a beach are an example of density sorting by waves. The concentrated minerals are often in sufficient enough quantities to be mined. Examples of important minerals on the shelf are diamonds, chromite (chromium ore), ilmenite (titanium ore), magnetite (**iron** ore), platinum, and gold.

See also Glaciation; Ice ages; Sedimentation; Wave motions

CONVECTION (UPDRAFTS AND DOWN-DRAFTS)

Convection is the vertical transfer of mass, heat, or other properties in a fluid or substance that undergoes fluid-like dynamics. Convection takes place in the atmosphere, in the **oceans**, and in Earth's molten subcrustal **asthenosphere**. Convective currents of air in the atmosphere are referred to as updrafts and downdrafts.

In addition to heat transfer, convention can be driven by other properties (e.g., salinity, density, etc.).

Convection in the mantle drives motion of the **lithospheric plates**. This convection is, in part, caused by **temperature** differences caused by the radioactive decay of the naturally radioactive elements uranium, thorium, potassium.

The temperature differences in **water** cause ocean currents that vertically mix masses of water at different temperatures. In the atmosphere, convection drives the vertical transport of air both upward and downward. In both cases, convection acts toward equilibrium and the lowest energy state by allowing the properties of the differential air or water masses to mix.

Thermal convection is one of the major forces in atmospheric dynamics and greatly contributes to, and directly influences, the development of **clouds** and storm systems. Convective air currents of rising warm and moist air allow a transfer of sensible and latent heat energy from the surface to the upper atmosphere.

One meteorological hypothesis, the convection theory of cyclones, asserts that convection resulting from very high levels of surface heating can be so strong that the current of air can attain cyclonic velocities and **rotation**.

Convection with the earth's mantle results from differential temperatures in mantle materials. In part, these differences can manifest as hot spots or convective currents where less dense and warmer mantle materials form slow moving vertical currents in the plastic (viscous or thick fluid-like) mantle. Phase change differences in materials also change their density and buoyancy.

Convective currents in the mantle move slowly (at a maximum, inches per year), but may last millions of years.

See also Adiabatic heating; Atmospheric circulation; Atmospheric composition and structure; Atmospheric inversion layers; Atmospheric lapse rate; Atmospheric pressure; Insolation and total solar irradiation

CONVERGENT PLATE BOUNDARY

In terms of **plate tectonics**, collision boundaries are sites where **lithospheric plates** move together and the resulting compression causes either subduction (where one or both lithospheric plates are driven down and destroyed in the molten mantle) or crustal uplifting that results in **orogeny** (mountain building).

Colliding plates create tremendous force. Although lithospheric plates move very slowly (low velocities of inches per each), the plates have tremendous mass. Accordingly, at collision, each lithospheric plate carries tremendous momentum (the mathematical product of velocity and mass) that provides the energy to cause subduction or uplifting. In addition, the buoyancy properties of the colliding lithospheric plates determine the outcome of the particular collision. Oceanic **crust** is denser than continental crust and is subductable. Continental crust, composed of lighter, less dense materials, is too light to undergo subduction and so overrides oceanic crust or uplifts.

Earth's crust is fractured into approximately 20 lithospheric plates. Each lithospheric plate is composed of a layer of oceanic crust or continental crust superficial to an outer layer of the mantle. Oceanic crust comprises the outer layer of the **lithosphere** lying beneath the **oceans**. Oceanic crust is composed of high-density rocks, such as **olivine** and **basalt**. Continental crust comprises the outer layer of the lithospheric plates containing the existing continents and some undersea features near the continents. Continental crust is composed of lower density rocks such as **granite** and **andesite**. Containing both crust and the upper region of the mantle, lithospheric plates are approximately 60 miles (approximately 100 km) thick. Lithospheric plates may contain various combinations of oceanic and continental crust in mutually exclusive sections (i.e., the outermost layer is either continental or oceanic crust, but not both except at convergent boundaries where subducting oceanic crust can make material contributions of lighter crustal materials to the overriding continental crust). Lithospheric plates move on top of the **asthenosphere** (the outer plastically deforming region of Earth's mantle).

At convergent boundaries, lithospheric plates move together in collision zones where crust is either destroyed by subduction or uplifted to form **mountain chains**. In zones of convergence, compressional forces (i.e., compression of lithospheric plate material) dominates.

When oceanic crust collides with oceanic crust, both subduct to form an oceanic trench (e.g., Marianas trench). When oceanic crust collides with continental crust, the oceanic crust subducts under the lighter continental crust and both pushes the continental crust upward into mountain chains (e.g., the Andes). The contribution of molten material from the subduction crust contributes to the volcanic arcs found along the Pacific Rim. Because continental crust does not subduct, when continental crust collides with continental crust, there is a uplift of both crusts (e.g., the ongoing collision of India with **Asia** that continues to push the Himalayas upward by about a centimeter a year. Given the expanse of **geologic time**, even modest geomorphologic changes—measured in inches or centimeters a year—can result in substantial changes over millions of years.

At triple points where three plates converge (e.g., where the Philippine sea plate merges into the North American and Pacific plate **subduction zone**), the situation becomes more complex.

Convergent plate boundaries are, of course, three-dimensional. Because Earth is an oblate sphere, lithospheric plates are not flat, but are curved and fractured into curved sections akin to the peeled sections of an orange. Convergent movement of lithospheric plates can best be conceptualized by the movement together of those peeled sections over a curved surface.

Because Earth's diameter remains constant, there is no net creation or destruction of lithospheric plates and so the amount of crust created at divergent boundaries is balanced by an equal destruction or uplifting of crust at convergent lithospheric plate boundaries.

See also Divergent plate boundary; Earth, interior structure; Earthquakes; Geologic time; Mohorovicic discontinuity (Moho); Subduction zone

COOK, JAMES (1728-1779)
English explorer

James Cook was one of the foremost figures of the Age of Exploration. During his career, Cook circumnavigated the globe twice, and captained three voyages of discovery for England. Cook made significant contributions to the fields of surveying, **cartography**, advanced mathematics, **astronomy**, and navigation. The detailed records of his voyages and contacts with various native peoples are considered the first anthropological survey of the Pacific islands, **Australia**, and New Zealand. Cook's voyages sparked European and American interest in Pacific colonization.

James Cook was born in Marton-in-Cleveland, Yorkshire, England. As a youth, he received a modest education, but was a dedicated self-study of mathematics, surveying, and cartography. Cook was apprenticed to a small shop owner, but later left his apprenticeship to join a merchant collier fleet at Whitby. Cook earned his mate's certificate, but his merchant career was cut short by his decision to enlist with the Royal Navy in 1755 at the outbreak of the Seven Year's War (also known as the French and Indian War, 1756–1763).

Cook was sent to America in 1756 as not only seaman, but as a cartographer. His first charge was to conduct soundings and draw charts of the St. Lawrence River. Cook's charts were later used by British forces for their attack on Quebec. He was next named surveyor of New Foundland and carried out that project until 1767. Cook's maps were so precise that many were used for a century.

As the Cook gained renown for his cartography, he also submitted papers to the Royal Academy on astronomical observation and navigation. His work on determining location using the **moon** commanded the attention of not only scholars, but also the British government. In 1766, Cook was appointed to command an expedition to the Pacific, the first of three great voyages. The stated purpose of Cook's Pacific expedition was to observe and document the transit of Venus across the **Sun** during an **eclipse** on June 3, 1779, as part of a scientific endeavor to calculate the distance from Earth to the Sun. At the completion of that task, Cook continued to record significant discoveries. In the South Pacific, he discovered and named the Society Islands. Cook then sailed to New Zealand, which he reported upon favorably as a potential site for British colonization despite the lack of domesticated animals. Venturing from New Zealand, Cook sailed to the eastern coast of Australia and charted the coastline before claiming the land for Britain. On the return voyage, Cook's crew was stricken with disease, a common occurrence at sea then. One-third of his crew died from malarial fever, scurvy, and dysentery.

Cook was scarcely back in Britain for a year before he received his next appointment. He was granted two ships, the *Adventure* and the *Resolution*, and sent back to the Pacific to further complete the exploration of the Southern Hemisphere. Cook was charged with finding a southern continent, which was thought to exist in the extreme South Pacific; the mysterious continent was supposed to be temperate with fertile land. Cook left Britain in 1772 and sailed for the extreme southern Atlantic. Pushing his way through **freezing** temperatures and **ice** flows, cook sailed along the edge of **Antarctica**. The frozen Antarctic was certainly not the fabled southern continent. Cook's circumnavigation of the southern Pole put an end to the legend. Cook again stopped in New Zealand, this time introducing some European plants and domestic animals into the indigenous landscape. He discovered, charted, and named several more islands as he finished his journey.

On his second voyage, Cook also made pioneering provisions for his crew. To avoid the scourge of disease that had plagued the second half of his first voyage, Cook brought an ample supply of lemons aboard and served sauerkraut to the crew in an attempt to ward off scurvy and fevers. The experiment worked; Cook lost only one crewman to disease.

Cook embarked on his third and final voyage in 1776. Instead of returning to the South Pacific, Cook turned his efforts to the Pacific coast of **North America** in search of a northern passage that connected the Atlantic and Pacific **Oceans**. He created detailed maps of the Pacific Coast that were used on later expeditions, including the Lewis and Clark expedition. However, Cook failed to locate the Columbia River and thought that Victoria Island was part of the mainland. Despite these flaws in his cartography, Cook's expedition, and his records of contact with various native peoples who possessed great natural resources, created a new interest in trade and settlement in the Pacific Northwest.

As Cook ventured to the North American Coast, he discovered present-day Hawaii, which he dubbed the Sandwich Islands, in March of 1778. Cook enjoyed a record of very amicable relationships with the native peoples he encountered on his expeditions. His initial contact with the peoples of the Sandwich Islands were no exception; after a time however, Cook felt that relations were beginning to sour so he pulled up anchor and sailed away. Two days later, the foremast of the *Resolution* snapped and Cook returned to the Sandwich

Islands. The native population grew increasingly hostile and stole one of Cook's cutters. In retaliation, Cook took the tribal chief hostage in order to facilitate an exchange. In the ensuing commotion, a shot was fired and the natives threw stones, attacking Cook and his crew. Cook died in the altercation at the age of 51.

COPERNICUS, NICOLAS (1473-1543)

Polish astronomer and mathematician

Nicolas Copernicus was born into a well-to-do family on February 19, 1473. His father, a copper merchant, died when Copernicus was 10, and the boy was taken in by an uncle who was a prince and bishop.

Copernicus was able to afford an excellent education. He entered the University of Cracow in 1491 and studied mathematics and painting. In 1496, he went to Italy for 10 years where he studied medicine and religious law. Two things happened in the year 1500 that influenced Copernicus; he attended a conference in Rome dealing with calendar reform and, on November 6, 1500, he witnessed a lunar **eclipse**.

The tables of planetary positions that were in use at the time were complex and inaccurate. Predicting the positions of the planets over long periods of time was haphazard at best, and the **seasons** were out of step with the position of the **Sun**. Copernicus realized that tables of planetary positions could be calculated much more easily, and accurately, if he made the assumption that the Sun, not Earth, was the center of the **solar system** and that the planets, including Earth, orbited the Sun. He first proposed this theory in 1507.

Copernicus was not the first person to introduce such a radical concept. Aristarchus had come up with the idea in ancient Greece long before, but the teachings of **Ptolemy** had been dominant for 1,300 years. Ptolemy claimed the earth was at the center of the universe, and all the planets (including the Sun and **Moon**) were attached to invisible celestial spheres that rotated around the earth.

Copernicus not only wished to refute Ptolemy's universe, he claimed that Earth itself was very small and unimportant compared to the vast vault of the stars. This marked the beginning of the end of the influence of the ancient Greek scientists.

Copernicus made an incorrect assumption about planetary orbits; he decided they were perfectly circular. Because of this, he found it necessary to use some of Ptolemy's cumbersome epicycles (smaller orbits centered on the larger ones) to reduce the discrepancy between his predicted orbits and those observed. It wasn't until Johannes Kepler's time that this was corrected and the true nature of planetary orbits was understood.

Even so, the heliocentric model developed by Copernicus fit the observed data better than the ancient Greek concept. For example, the periodic "backward" motion in the sky of the planets Mars, Jupiter, and Saturn and the lack of such motion for Mercury and Venus was more readily explained by the fact that the former planets' orbits were outside of Earth's. Thus, the earth "overtook" them as it circled

Nicolas Copernicus. *Library of Congress.*

the Sun. Planetary positions could also be predicted much more accurately using Copernicus' model.

Copernicus was reluctant to make his ideas public. He realized his theory not only contradicted the Greek scientists, it went against the teachings of the Church, the consequences of which could be severe. In 1530, he allowed a summary of his ideas to circulate among scholars, who received it with great enthusiasm, but it was not until just shortly before his death in 1543 that his entire book was published. It took the efforts of the mathematician Rheticus to convince Copernicus to grant him permission to print it. Unfortunately, Rheticus had fallen afoul of official doctrine himself, and found it wise to leave town. Overseeing the publication for Copernicus's book was transferred to the hands of a Lutheran minister named Andreas Osiander (1498–1552).

Osiander now found he was in a tight spot; Martin Luther (1483–1546) had come out firmly against Copernicus' new theory, and Osiander was obligated to follow him. "This Fool wants to turn the whole Art of Astronomy upside down," Luther had said. Copernicus had dedicated his book to Pope Paul III, perhaps to gain favor, but Osiander went one step further; he wrote a preface in which he stated the heliocentric theory was not being presented as actual fact, but just as a concept to allow for better calculations of planetary positions. He did not sign his name to the preface, making it appear that

Copernicus had written it and was debunking his own theory. Copernicus, suffering from a stroke and close to death, could do nothing to defend himself. It is said he died only hours after seeing the first copy of the book. Kepler discovered the truth about the preface in 1609 and exonerated Copernicus.

The immediate reaction to the book, *De Revolutionibus Orbium Coelestium* (Revolution of the heavenly spheres), was subdued. This was primarily due to Osiander's preface, which weakened Copernicus' reputation. In addition, only a limited number of books were printed, they were very expensive and, consequently, had limited circulation. The book did achieve a number of converts, but one had to be a mathematician to fully understand the theories. Still, it was placed on the Roman Catholic Church's list of prohibited books where it remained until 1835.

Almost as significant as proving the heliocentric solar system was possible, was Copernicus's questioning of the ancient Greek scientists. Ptolemy had bent the facts to fit his preconceived theory and his teachings had been accepted, without question, for centuries. Copernicus, on the other hand, did his best to develop his theory to match observed facts, foreshadowing the dawning of modern **scientific method**.

CORAL REEFS AND CORALS

Reefs are found in the **fossil record** and are thought to be about two billion years old. These reefs were built by calcareous algae and not corals. The first true corals did not appear until about 300–500 million years ago. They apparently flourished until a devastating extinction occurred killing many groups of corals. Most of the corals that compose extant (still living) reefs were found around 65 million years ago. They are still a vital part of the living environment. In addition to their ecological role as a foundation for a wide variety of life forms, the coral reefs have become a haven for tourists and scuba divers.

Corals are animals that belong to the monophyletic group called the Cnidaria (formerly called a Phylum). They are named for specialized stinging structures that emit long, venomous barbs when stimulated by the presence of prey or danger. Corals are further classified into a smaller subdivision, the Anthozoa. For geologists, the most important characteristic of many corals is their ability to remove calcium from the **water** and redeposit it as a hardened outer casing in the form of calcium carbonate. Because many corals species are colonial, the chambers of the animals grow together forming a larger hardened structure. Over many years, these constructions may merge with others of the same species or with those of different species. The resulting large and diverse colony makes an impressive and rigid structure that spreads across the sea bottom. At this point, it is identified as a reef. Within the reef, there are hiding places for fish and other marine organisms. Organic material is abundant and many species live their entire lives within the confines of the protection the reef provides.

Coral reefs as ecosystems are being endangered by human activities. © *Stephen Frink/Corbis. Reproduced by permission.*

Corals are believed to live in a symbiotic relationship with species of microscopic algae. As a result, reef corals are rarely found below the photic zone or about 150 feet (46 m). The photic zone is a narrow strip of surface water (about 200 feet) through which sunlight permeates. Below this depth particles in the water prevent light waves from penetrating.

Corals require clear water to flourish. They need as much light energy as possible so the algae in their tissues can thrive. In addition, they prefer areas where wave energy is high. The waves bring floating organic materials, **oxygen**, and nutrients to the corals. Reef-building corals require warm ocean temperatures (68 to 82°F, or 20 to 28°C). Because of this requirement, corals are primarily confined to areas within 30°N and 30°S latitudes. This region includes the tropical and subtropical Western Atlantic and Indo-Pacific **Oceans**. Western Atlantic reefs are found in Bermuda, the Bahamas, the Caribbean Islands, Belize, Florida, and the **Gulf of Mexico**. The Indo-Pacific ocean region extends from the Red Sea and the Persian Gulf through the Indian and Pacific oceans. Reefs have been found as far as the western coast of Panama. The rocky outcrops in some areas of the Gulf of California are also favorable regions for corals. These warm shallow **seas** provide ideal conditions where water temperatures and salinity are high and **carbon dioxide** concentrations are low. In this type of habitat the corals are able to precipitate calcium out of the water.

The colors of the reefs are impressive and add to their beauty. The United States National Oceanographic and Atmospheric Administration describe natural pigments in coral tissue that produce a range of colors including white, red, orange, yellow, green, blue, and purple, along with algae that live within the tissues of some corals that may make the coral appear brown, green, or orange. Unfortunately, this characteristic of some corals has made them prized for their use in jewelry. Removal of living corals for such uses has put many species in danger of extinction or severe harm.

The massive structure of some reefs is especially important to geologists because the colonies can reshape coastal **sedimentation** and deposition regimes. **Barrier islands** provide some of the more spectacular examples of this. As seamounts grow in warm oceans corals land on the emerging **rock** shore and begin their colonies. Over millions of years, the islands may move and sink because of **plate tectonics**. The corals continue to grow on top of one another as they attempt to stay near the surface. The reef expands until large areas of coastline are bordered by the reefs. Waves are slowed by the reefs and any sediment they carry is dropped on the ocean side of the reef. The water that flows over the coral and toward the beach has little energy and is relatively clear. Any sediment it carries is fine-grained and deposited in a space called a lagoon. Lagoons are warm bodies of water that lie between the reef and shore. The waters are calm and warm. Numerous species of plants and animals live in a lagoon. They are favored locations of tourists for swimming and snorkeling and provide ideal spots for recreation.

The continuing cycle of growth of the corals and deposition of sediments are well documented in the fossil record. Reefs sometimes grow to a massive size, and are often identified in rocks such as those found in the Big Bend National Park in Texas. The El Capitan Reef is an exceptional example of such a structure. This is one of the types of evidence geologists use to reconstruct **climate** in various regions of the globe.

Coral reefs are disappearing in many places around Earth. Projects, such as those in Florida where old ships are sunk to provide new surface **area** for coral colonies to grow, are helping save the reefs from destruction by divers and fishing boats. **Water pollution** and disease still threaten many species. Without coral reefs, entire marine ecosystems may vanish, and an ancient geological and biological system might disappear from Earth.

See also Great Barrier Reef; Oceanography

CORIOLIS EFFECT

The Coriolis effect (sometimes called the Coriolis force) is the apparent deflection of air masses and fluids caused by Earth's **rotation**. Named after the French mathematician Gustave-Gaspard Coriolis, (1792-1843), who developed the concept in 1835, the Coriolis force is a pseudoforce (false force) and

should properly be termed the Coriolis effect. As a result of the Coriolis effect, there is an apparent deflection of all matter in motion to the right of their path in the Northern Hemisphere, and to the left in the Southern Hemisphere. In the Northern Hemisphere, air is deflected counterclockwise (to right of its established path of motion) as it moves inward toward a low-pressure **area** (zone of convergence). In the Northern Hemisphere, air is deflected clockwise (again, to the right of its established path of motion) as it moves outward toward a low-pressure area (zone of convergence). These deflections and rotations are reversed in the Southern Hemisphere.

The Coriolis effect is a mechanical principle demonstrating that, on a rotating solid body, an inertial force acts on the body at right angles to its direction of motion. The Coriolis effect is based on the classic laws of motion introduced by English physicist and mathematician **Sir Isaac Newton** (1642-1727) in his work, *Philosophiae Naturalis Principia Mathematica* (Mathematical principles of natural philosophy).

Within its rotating coordinate system, the object acted on by the Coriolis effect appears to deflect off of its path of motion. This deflection is not real. It only appears to happen because the coordinate system that establishes a frame of reference for the observer is also rotating. The Coriolis effect is due to the motion of a rotating frame of reference (e.g., Earth's rotation).

For example, if a missile is launched northward from the equator. The missile will land to the right of a directly northward target because, when launched, the missile moving along with the ground at the equator moves faster to the east than its direct northward target. Conversely, if a missile were fired from the North Pole to a directly southward target (a target on a great circle that also passed through the South Pole) will also land to the right of its intended target because during the missile's flight the target area has moved farther to the east faster. In the Southern Hemisphere these deflections are reversed (i.e., objects are deflected to the left).

The Coriolis effect is important to virtually all sciences that relate to Earth and planetary motions. It is critical to the dynamics of the atmosphere including the motions of winds and storms. In **oceanography**, it helps explains the motions of oceanic currents. Accounting for the Coriolis effect is critical in planning the motions of aircraft and the launch and recovery of spacecraft. In **astronomy** and astrophysics the Coriolis effect explains the rotation of sunspots.

A popular canard (a popular, widely accepted, but false premise) is that **water** in sinks and toilet bowls drains away in counterclockwise or clockwise motion depending on whether the drain is located in the northern or Southern Hemisphere. The fact is that the Coriolis effect acts only on fluids over great distances or long lengths of time, but is not great enough to produce these defections. These deflections are caused by other factors (drain shape, initial water velocity, etc.)

See also Air masses and fronts; Atmospheric circulation; Ocean circulation and currents; Weather and climate; Wind

False color imaging of a solar flare. *U.S. National Aeronautics and Space Administration (NASA).*

CORONAL EJECTIONS AND MAGNETIC STORMS

Coronal mass ejections (CME) are explosive and violent eruptions of charged, magnetic field-inducing particles and gas from the Sun's outer coronal layer. The ejection from the Sun's corona can be massive (e.g., estimates of CME mass often range in the billions of tons. Ejections propel particles in specific directions, some directly crossing Earth's orbital position, at velocities up to 1200 miles per second (1,931 km per second) or 4,320,000 miles per hour in an ionized plasma (also known as the solar **wind**). Solar CMEs that are Earth directed disrupt and distort Earth's **magnetic field** and result in geomagnetic storms.

Although the solar wind is continuous, CMEs reflect large-scale increases in wind (i.e., particle) mass and velocity that are capable of generating geomagnetic storms.

Solar coronal ejections and magnetic storms interact with Earth's magnetosphere to produce spectacular auroral displays. Intense storms may interfere with communications and preclude data transfer from Earth orbiting satellites.

Solar coronal ejections and magnetic storms provide the charged particles that result in the northern and southern lights—Aurora Borealis and Aurora Australialis—electromag-netic phenomena that usually occur near Earth's polar regions. The auroras result from the interaction of Earth's magnetic field with ionic gas particles, protons, and electrons streaming outward in the solar wind.

The rate of solar coronal ejections is correlated to solar sunspot activity that cycles between maximum levels of activity (i.e., the solar maximum) approximately every 11 years. During solar maximums, it is not uncommon to observe multiple coronal ejections per day. At solar minimum, one solar coronal ejection per day is normal.

Earth's core structure provides it with a relatively strong internal magnetic field (oriented about 10–12 degrees from the **polar axis**). Earth's magnetosphere protects the earth from bombardment by Comes by deflecting and modifying the solar wind. At the interface of Earth's magentosphere and the solar wind there is a "bow wave" or magnetic shock wave to form a magnetosheath protecting the underlying magnetosphere that extends into Earth's **ionosphere**.

Coronal mass ejections (CMEs) not only interact with Earth's magnetic field, they also interact with each other. Stronger or faster ejections may subsume prior weaker ejections directed at the same region of **space** in a process known as CME cannibalization. Accordingly, the strength of magnetic storms on Earth may not directly correlate to observed

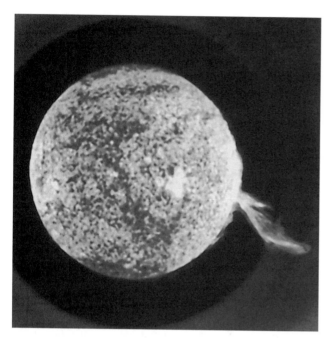

Skylab image of the Sun showing a solar flare. *U.S. National Aeronautics and Space Administration (NASA).*

coronal ejections. In addition, CME cannibalization can alter predicted arrival time of geomagnetic storms because the interacting CMEs can change the eruption velocity.

See also Atmospheric chemistry; Atmospheric composition and structure; Atomic theory; Atoms; Bohr model; Chemical elements; Electricity and magnetism; Electromagnetic spectrum; Quantum electrodynamics (QED); Solar sunspot cycles

CORRELATION (GEOLOGY)

In **geology**, the term correlation refers to the methods by which the age relationship between various strata of Earth's **crust** is established. Such relationships can be established, in general, in one of two ways: by comparing the physical characteristics of strata with each other (physical correlation); and by comparing the type of **fossils** found in various strata (fossil correlation).

Correlation is an important geological technique because it provides information with regard to changes that have taken place at various times in Earth history. It also provides clues as to the times at which such changes have occurred. One result of correlational studies has been the development of a **geologic time** scale that separates Earth history into a number of discrete time blocks known as eras, periods, and epochs.

Sedimentary rocks provide information about Earth history that is generally not available from igneous or metamorphic rocks. For example, suppose that for many millions of years a river has emptied into an ocean, laying down, or

depositing, sediments eroded from the land. During that period of time, layers of sediments would have collected one on top of the other at the mouth of the river. These layers of sediments are likely to be very different from each other, depending on a number of factors, such as the course followed by the river, the **climate** of the **area**, the **rock** types exposed along the river course, and many other geological factors in the region. One of the most obvious differences in layers is thickness. Layers of sedimentary rock may range in thickness from less than an inch to many feet.

Sedimentary layers that are identifiably different from each other are called beds or strata. In many places on Earth's surface, dozens of strata are stacked one on top of each other. Strata are often separated from each other by relatively well-defined surfaces known as **bedding** planes.

In 1669, the Danish physician and theologian Nicolaus Steno (1638–1686) made a seemingly obvious assertion about the nature of sedimentary strata. Steno stated that in any sequence of sedimentary rocks, any one layer (stratum) is younger than the layer below it and older than the layer above it. Steno's discovery is now known as the law of **superposition**.

The law of superposition applies only to sedimentary rocks that have not been overturned by geologic forces. **Igneous rocks**, by comparison, may form in any horizontal sequence whatsoever. A flow of **magma** may force itself, for example, underneath, in the middle or, or on top of an existing rock stratum. It is very difficult to look back millions of years later, then, and determine the age of the igneous rock compared to rock layers around it.

Using sedimentary rock strata it should be possible, at least in theory, to write the geological history of the continents for the last billion or so years. Some important practical problems, however, prevent the full realization of this goal. For example, in many areas, **erosion** has removed much or most of the sedimentary rock that once existed there. In other places, strata are not clearly exposed to view but, instead, are buried hundreds or thousands of feet beneath the thin layer of **soil** that covers most of Earth's surface.

A few remarkable exceptions exist. A familiar example is the Grand **Canyon**, where the Colorado River has cut through dozens of strata, exposing them to view and making them available for study by geologists. Within the Grand Canyon, a geologist can follow a particular stratum for many miles, noting changes within the stratum and changes between that stratum and its neighbors above and below. One of the characteristics observable in such a case is that a stratum often changes in thickness from one edge to another. At the edge where the thickness approaches zero, the stratum may merge into another stratum. This phenomenon is understandable when one considers the way the sediment in the rocks was laid down. At the mouth of a river, for example, the accumulation of sediments is likely to be greatest at the mouth itself, with decreasing thickness at greater distances into the lake or ocean. The principle of lateral continuity describes this phenomenon, namely that strata are three-dimensional features that extend outward in all directions, merging with adjacent deposits at their edges.

Human activity also exposes strata to view. When a highway is constructed through a mountainous (or hilly) area, for example, parts of a mountainside may be excavated, revealing various sedimentary rock strata. These strata can then be studied to discover the correlation among them and with strata in other areas.

Another problem is that strata are sometimes disrupted by earth movements. For example, an **earthquake** may lift one block of Earth's crust over an adjacent block or may shift it horizontally in comparison to the second block. The correlation between adjacent strata may then be difficult to determine.

Physical correlation is accomplished by using a number of criteria. For example, the color, grain size, and type of **minerals** contained within a stratum make it possible for geologists to classify a particular stratum quite specifically. This allows them to match up portions of that stratum in regions that are physically separated from each other. In the American West, for example, some strata have been found to cover large parts of two or more states although they are physically exposed in only a few specific regions.

The stratum tends to have one set of characteristics in one region, which gradually changes into another set of characteristics farther along in the stratum. Those characteristics also change, at some distance farther along, into yet another set of characteristics. Rocks with a particular set of characteristics are called a facies. Facies changes, changes in the characteristics of a stratum or series of strata, are important clues to Earth history. If, for example, a geologist finds that the facies in a particular stratum change from a **limestone** to a shale to a **sandstone** over a distance of a few miles, the geologist knows that limestone is laid down on a sea bottom, shale is formed from compacted mud, and sandstone is formed when **sand** is compressed. The limestone to shale to sandstone facies pattern may allow an astute geologist to reconstruct what Earth's surface looked like when this particular stratum was formed. For example, knowing these rocks were laid down in adjacent environments, the geologist might consider that the limestone was deposited on a coral reef, the shale in a quiet lagoon or coastal swamp, and the sandstone in a nearby beach. So facies changes indicate differences in the environments in which adjacent facies were deposited.

One of the most important discoveries in the science of correlation was made by the English surveyor **William Smith** (1769–1839) in the 1810s. One of Smith's jobs involved the excavation of land for canals being constructed outside of London. As sedimentary rocks were exposed during this work, Smith found that any given stratum always contained the same set of fossils. Even if the stratum were physically separated by a relatively great distance, the same fossils could always be found in all parts of the stratum.

In 1815, Smith published a map of England and Wales showing the geologic history of the region based on his discovery. The map was based on what Smith called his law of faunal succession. That law says simply that it is possible to identify the sequence in which strata are laid down by examining the fossils they contain. The simplest fossils are the oldest and, therefore, strata that contain simple fossils are older than strata that contain more complex fossils.

The remarkable feature of Smith's discovery is that it appears to be valid over very great distances. That is, suppose that a geologist discovers a stratum of rock in southwestern California that contains fossils A, B, and C. If another stratum of rock in eastern Texas is also discovered that contains the same fossils, the geologist can conclude that it is probably the same stratum—or at least of the same age—as the southwestern California stratum.

The correlational studies described so far allow scientists to estimate the relative ages of strata. If stratum B lies above stratum A, B is the younger of the two. However determining the actual, or absolute, age of strata (for example, 3.5 million years old) is often difficult because the age of a fossil cannot be determined directly. The most useful tool in dating strata is radiometric dating of materials. A radioactive isotope such as uranium-238 decays at a very regular and well-known rate. That rate is known as its **half-life**, the time it takes for one-half of a sample of the isotope to decay. The half-life of uranium-238, for example, is 4.5 billion years. By measuring the concentration of uranium-238 in comparison with the products of its decay (especially lead-206), a scientist can estimate the age of the rock in which the uranium was found. This kind of radioactive dating has made it possible to place specific dates on the ages of strata that have been studied and correlated by other means.

See also Cross cutting; Dating methods; Field methods in geology; Landscape evolution; Strike and dip

CORROSION

Corrosion is the deterioration of a material, or of its properties, as a consequence of reaction with the environment.

In addition to corrosion of metals—the effects of **soil**, atmosphere, chemicals, and **temperature** serve as agents of corrosion for a number of materials. The need to understand and control corrosion has given rise to the new sciences of corrosion technology and corrosion control, both of which are solidly based upon **chemistry** and **geochemistry**.

Perhaps the earliest recognition of corrosion was the effect of seawater and sea atmospheres on ships. Salt **water**, continual dampness, and the growth of marine life such as marine borers, led to the decay of wooden hulls. Because of its toxicity, copper cladding of the hulls was widely used to discourage marine growth. In 1824, to protect the copper from deterioration, the team of English scientists Humphrey Davy (1778-1829) and **Michael Faraday** (1791-1867) applied zinc protector plates to the copper sheathing. This was the first successful application of cathodic protection, in which a more readily oxidized metal is attached to the metal to be protected. This procedure was widely used until hulls were replaced by steel or newer materials.

With the development of the industrial age, and the increased use of **iron**, the oxidation of iron, or rust, forced the development of steels and the search for new **metals** and metal coatings to protect surfaces. This gave birth to the science of corrosion control that involves measures of material selection, inhibition, painting, and novel design.

The corrosion of metals is caused by the electrochemical transfer of electrons from one substance (**oxygen** for example) to another. This may occur from the surfaces of metals in contact, or between a metal and another substance when a moist conductor or electrolyte is present. Depending upon the conditions, various types of corrosion may occur (e.g., general corrosion, intergranular and pitting corrosion, stress corrosion cracking, corrosion fatigue, galvanic and cavitation corrosion, etc.)

Throughout the world, the direct and hidden costs of deterioration due to environmental corrosion amount to billions of dollars per year.

See also Atmospheric chemistry; Atmospheric pollution; Chemical bonds and physical properties; Chemical elements; Weathering and weathering series

COSMIC MICROWAVE BACKGROUND RADIATION

In 1965, American physicists **Arno Penzias** (1933–) and Robert Wilson (1936–) announced the discovery of microwave radiation, which uniformly filled the sky and had a blackbody **temperature** of about 3.5K. The pair had been testing a new radio amplifier that was supposed to be exceptionally quiet. After many attempts to account for all extraneous sources of radio noise, they concluded that there was a general background of radiation at the radio frequency they were using. After discussions with a group led by Robert Dicke at nearby Princeton University it became clear that they had in fact detected remnant radiation from the origin of the universe.

Although neither Dicke's group nor Penzias and Wilson realized it at the time, they had confirmed a prediction made by scientists 17 years earlier. Although the temperature that characterized the detected radiation was somewhat different than predicted, the difference could be accounted for by changes to the accepted structure of the universe discovered between 1948 and 1965. The detection of this radiation and its subsequent verification at other frequencies was taken as confirmation of a central prediction of a **cosmology** known as the Big Bang.

The interpretation of the red-shifts of spectral lines in distant galaxies by American astronomer **Edwin Hubble** (1889–1953) 40 years earlier suggested a Universe that was expanding. One interpretation of that expansion was that the universe had a specific origin in **space** and time. Such a universe would have a very different early structure from the present one.

It was the Russian-American physicist George Gamow (1904–1968) and colleagues who suggested that the early phases of the universe would have been hot and dense enough to sustain nuclear reactions. Following these initial phases, the expanding universe would eventually cool to the point at which the dominant material, hydrogen, would become relatively transparent to light and radio waves. We know that for hydrogen, this occurs when the gas reaches a temperature of between

5,000K–10,000K. From that point on in the **evolution** of the universe, the light and matter would go their separate ways.

As every point in the universe expands away from every other point, any observer in the universe sees all objects receding from him or her. The faster moving objects will appear at greater distances by virtue of their greater speed. Indeed, their speed will be directly proportional to their distance, which is what one expects for material ejected from a particular point in space and time. However, this expansion results from the expansion of space itself, and should not be viewed simply as galaxies rushing headlong away from one another through some absolute space. The space itself expands.

As it does, light traveling through it is stretched, becoming redder and appearing cooler. If one samples that radiation at a later date, it will be characteristic of radiation from a much cooler source. From the rate of expansion of the universe, it is possible to predict what that temperature ought to be. Current values of the expansion rate are completely consistent with the current measured temperature of about 2.7K. The very existence of this radiation is strong evidence supporting the expanding model of the universe championed by Gamow and colleagues and disparagingly named the "Big Bang" cosmology by English astronomer **Sir Fred Hoyle** (1915–2001).

Since its discovery in 1965, the radiation has been carefully studied and found to be a perfect blackbody as expected from theory. Since this radiation represents fossil radiation from the initial Big Bang, any additional motion of the earth around the **Sun**, the Sun around the galactic center, and the galaxy through space should be reflected in a slight asymmetry in the background radiation. The net motion of Earth in some specific direction should be reflected by a slight Doppler shift of the background radiation coming from that direction toward shorter wavelengths.

Doppler shift is the same effect that the police use to ascertain the velocity of approaching vehicles. Of course, there will be a similar shift toward longer wavelengths for light coming from the direction from which we are receding. This effect has been observed indicating a combined peculiar motion of Earth, the Sun and galaxy on the order of 600 km/sec.

Finally, small fluctuations in the background radiation are predicted which eventually led to the formation galaxies and clusters of galaxies. Such fluctuations have been found by the CO(smic) B(ackground) E(xplorer) **Satellite**, launched by NASA in 1989. COBE detected these fluctuations at about 1 part in 105, which was right near the detection limit of the satellite. The details of these fluctuations are crucial to deciding between more refined models of the expanding universe. COBE was decommissioned in 1993, but scientists are still unraveling the information contained in its data.

It is perhaps not too much of an exaggeration to suggest that cosmic background radiation has elevated cosmology from enlightened speculative metaphysics to an actual science. We may expect developments of this emerging science to lead to a definitive description of the evolutionary history of the universe in the near future.

See also Big Bang theory; Cosmic ray; Solar system

COSMIC RAY

The term cosmic ray refers to highly-energetic atomic particles (mostly single protons, some proton-neutron pairs, and occasionally subatomic particles and electrons) that travel through **space** near the speed of light. Physicists divide cosmic rays into two categories: primary and secondary. Primary cosmic rays originate far outside Earth's atmosphere. Secondary cosmic rays are particles produced within Earth's atmosphere as a result of collisions between primary cosmic rays and molecules in the atmosphere.

The existence of cosmic radiation was first discovered in 1912, in experiments performed by the Austrian-American physicist Victor Hess (1883–1964). His experiments were sparked by a desire to better understand phenomena of electric charge. A common instrument of the day for demonstrating such phenomena was the electroscope. An electroscope contains thin metal leaves or wires that separate from one another when they become charged, due to the fact that like charges repel. Eventually the leaves (or wires) lose their charge and collapse back together. It was known that this loss of charge had to be due to the attraction by the leaves of charged particles (ions) in the surrounding air. The leaves would attract those ions having a charge opposite to that of the leaves, due to the fact that opposite charges attract; eventually the accumulation of ions in this way would neutralize the charge that had been acquired by the leaves, and they would cease to repel each other. Scientists wanted to know where these ions came from. It was thought that they must be the result of radiation emanating from Earth's **crust**, since it was known that radiation could produce ions in the air. This led scientists to predict that fewer ions would be present the further one traveled away from Earth's surface. Hess's experiments, in which he took electroscopes high above Earth's surface in a balloon, showed that this was not the case. At high altitudes, the electroscopes lost their charge even faster than they had on the ground, showing that there were more ions in the air and thus, that the radiation responsible for the presence of the ions was stronger at higher altitudes. Hess concluded that there was a radiation coming into our atmosphere from outer space.

As physicists became interested in cosmic radiation, they developed new ways of studying it. The Geiger-Muller counter consists of a wire attached to an electric circuit and suspended in a gaseous chamber. The passage of a cosmic ray through the chamber produces ions in the gas, causing the counter to discharge an electric pulse. Another instrument, the cloud chamber, contains a gas that condenses into vapor droplets around ions when these are produced by the passage of a cosmic ray. In the decades following Hess's discovery, physicists used instruments such as these to learn more about the nature of cosmic radiation.

An **atom** of a particular element consists of a nucleus surrounded by a cloud of electrons, which are negatively charged particles. The nucleus is made up of protons, which have a positive charge, and neutrons, which have no charge. These particles can be further broken down into smaller constituents; all of these particles are known as subatomic particles. Cosmic rays consist of nuclei and of various subatomic particles. Almost all of the primary cosmic rays are nuclei of various atoms. The great majority of these are single protons, which are nuclei of hydrogen atoms. The next most common primary cosmic ray is the nucleus of the helium atom, made up of a proton and a neutron. Hydrogen and helium nuclei make up about 99% of the primary cosmic radiation. The rest consists of nuclei of other elements and of electrons.

When primary cosmic rays enter Earth's atmosphere, they collide with molecules of gases present there. These collisions result in the production of more high-energy subatomic particles of different types; these are the secondary cosmic rays. These include photons, neutrinos, electrons, positrons, and other particles. These particles may in turn collide with other particles, producing still more secondary radiation. If the energy of the primary particle that initiates this process is very high, this cascade of collisions and particle production can become quite extensive. This is known as a shower, air shower, or cascade shower.

The energy of cosmic rays is measured in units called electron volts (abbreviated eV). Primary cosmic rays typically have energies on the order of billions of electron volts. Some are vastly more energetic than this; a few particles have been measured at energies in excess of 10^{19} eV. This is in the neighborhood of the amount of energy required to lift a weight of 2.2 lb (1 kg) to a height of 3.3 ft (1 m). Energy is lost in collisions with other particles, so secondary cosmic rays are typically less energetic than primary ones. The showers of particles described above diminish as the energies of the particles produced decrease. The energy of cosmic rays was first determined by measuring their ability to penetrate substances such as gold or **lead**.

Because cosmic rays are mostly charged particles (some secondary rays such as photons have no charge), they are affected by magnetic fields. The paths of incoming primary cosmic rays are deflected by the earth's **magnetic field**, somewhat in the way that **iron** filings will arrange themselves along the lines of force emitted by a magnet. More energetic particles are deflected less than those having less energy. In the 1930s, it was discovered that more particles come to the earth from the West than from the East. Because of the nature of Earth's magnetic field, this led scientists to the conclusion that most of the incoming cosmic radiation consists of positively charged particles. This was an important step towards the discovery that the primary cosmic rays are mostly bare atomic nuclei, since atomic nuclei carry a positive charge.

The ultimate origin of cosmic radiation is still not completely understood. Some of the radiation is thought to have been produced in the "Big Bang" at the origin of the universe. Other cosmic rays are produced by the **Sun**, particularly during solar disturbances such as solar flares. Exploding stars, called supernovas, are also a source of cosmic rays.

The fact that cosmic ray collisions produce smaller subatomic particles has provided a great deal of insight into the fundamental structure of matter. The construction of experimental equipment such as particle accelerators has been inspired by a desire to reproduce the conditions under which high-energy radiation is produced, in order to gain better experimental control of collisions and the production of particles.

See also Astronomy; Big Bang theory; Cosmic microwave background radiation; Quantum theory and mechanics

COSMOLOGY

Cosmology is the study of the origin, structure and **evolution** of the universe.

The origins of cosmology predate the human written record. The earliest civilizations constructed elaborate myths and folk tales to explain the wanderings of the **Sun, Moon,** and stars through the heavens. Ancient Egyptians tied their religious beliefs to celestial objects and Ancient Greek and Roman philosophers debated the composition and shape of the Earth and the Cosmos. For more than 13 centuries, until the Scientific Revolution of the sixteenth and seventeenth centuries, the Greek astronomer Ptolemy's model of an Earth-centered Cosmos composed of concentric crystalline spheres dominated the Western intellectual tradition.

Polish astronomer Nicolaus Copernicus' (1473–1543) reassertion of the once discarded heliocentric (Sun-centered) theory sparked a revival of cosmological thought and work among the astronomers of the time. The advances in empiricism during the early part of the Scientific Revolution, embraced and embodied in the careful observations of Danish astronomer **Tycho Brahe** (1546–1601), found full expression in the mathematical genius of the German astronomer **Johannes Kepler** (1571–1630) whose laws of planetary motion swept away the need for the errant but practically useful Ptolemaic models. Finally, the patient observations of the Italian astronomer and physicist Galileo, in particular his observations of moons circling Jupiter and of the phases of Venus, empirically laid to rest cosmologies that placed Earth at the center of the Cosmos.

English physicist and mathematician Sir Isaac Newton's (1642–1727), important *Philosophiae Naturalis Principia Mathematica* (Mathematical principles of natural philosophy) quantified the laws of motion and **gravity** and thereby enabled cosmologists to envision a clockwork-like universe governed by knowable and testable natural laws. Within a century of Newton's *Principia*, the rise of concept of a mechanistic universe led to the quantification of celestial dynamics, that, in turn, led to a dramatic increase in the observation, cataloging, and quantification of celestial phenomena. In accordance with the development of natural theology, scientists and philosophers debated conflicting cosmologies that argued the existence and need for a supernatural God who acted as "prime mover" and guiding force behind a clockwork universe. In particular, French mathematician Pierre Simon de Laplace (1749–1827) argued for a completely deterministic universe, without a need for the intervention of God. Most importantly to the development of modern cosmology, Laplace asserted explanations for celestial phenomena as the inevitable result of time and statistical probability.

By the dawn of the twentieth century, advances in mathematics allowed the development of increasingly sophisticated cosmological models. Many advances in mathematics pointed toward a universe not necessarily limited to three dimensions and not necessarily absolute in time. These intriguing ideas found expression in the intricacies of relativity and theory that, for the first time, allowed cosmologists a theoretical framework upon which they could attempt to explain the innermost workings and structure of the universe both on the scale of the subatomic world and on the grandest of galactic scales.

As direct consequence of German-American physicist Albert Einstein's (1879–1955) **relativity theory**, cosmologists advanced the concept that space-time was a creation of the universe itself. This insight set the stage for the development of modern cosmological theory and provided insight into the evolutionary stages of stars (e.g., neutron stars, pulsars, black holes, etc.) that carried with it an understanding of nucleosythesis (the formation of elements) that forever linked the physical composition of matter on Earth to the lives of the stars.

Twentieth-century progress in cosmology has been marked by corresponding and mutually beneficial advances in technology and theory. American astronomer Edwin Hubble's (1889–1953) discovery that the universe is expanding, Arno A. Penzias and Robert W. Wilson's observation of cosmic background radiation, and the detection of the elementary particles that populated the very early universe all proved important confirmations of the **Big Bang theory**. The Big Bang theory asserts that all matter and energy in the universe, and the four dimensions of time and **space** were created from the primordial explosion of a singularity of enormous density, **temperature**, and pressure.

During the 1940s Russian-born American cosmologist and nuclear physicist George Gamow (1904–1968) developed the modern version of the big bang model based upon earlier concepts advanced by Russian physicist Alexander (Aleksandr Aleksandrovich) Friedmann (also spelled as Fridman, 1888–1925) and Belgian astrophysicist and cosmologist Abbé Georges Lemaître (1894–1966). Big bang based models replaced static models of the universe that described a homogeneous universe that was the same in all directions (when averaged over a large span of space) and at all times. Big bang and static cosmological models competed with each other for scientific and philosophical favor. Although many astrophysicists rejected the steady state model because it would violate the law of mass-energy conservation, the model had many eloquent and capable defenders. Moreover, the steady model was interpreted by many to be more compatible with many philosophical, social and religious concepts centered on the concept of an unchanging universe. The discovery of **quasars** and of a permeating cosmic background radiation eventually tilted the cosmological argument in favor of big bang-based models.

Technology continues to expand the frontiers of cosmology. The **Hubble Space Telescope** has revealed gas **clouds** in the cosmic voids and beautiful images of fledgling galaxies formed when the universe was less than a billion years old. Analysis of these pictures and advances in the understanding of the fundamental constituents of nature continue to keep cosmology a dynamic discipline of **physics** and the ultimate fusion of human scientific knowledge and philosophy.

See also Cosmic microwave background radiation; Stellar life cycle

COUNTRY ROCK

The term country **rock** refers to a body of rock that receives or hosts an intrusion of a viscous geologic material. Intrusions into country rock are most commonly magmatic, but may also consist of unconsolidated sediments or salt horizons. Country rock may consist any other kind of rock that was present before the intrusion: sedimentary, igneous, or metamorphic.

In most cases, country rock is intruded by an igneous body of rock that formed when **magma** was forced upward through fractures or melted its way up through the overlying rock. The magma then cooled into solid rock forming a mass distinct from the enveloping country rock. Occasionally, a fragment of country rock will break off and become incorporated into the intrusion, and is called a **xenolith**.

The country rock is usually altered by the heat of the intrusion. The change that takes place in country rock as a result of an intrusion cooling off is called contact **metamorphism**. The extent and intensity of contact metamorphism depends on the heat of the magma, the **temperature** of the country rock, the amount of fluids present, the **permeability** of the country rock, and the depth of intrusion (which determines to a great extent, the pressure). The metamorphism is strongest at the contact of the country rock and the intrusion and diminishes outward from the intrusion. A discernable halo of contact metamorphism that extends into the country rock is often produced and is called the contact aureole.

When the country rock has been contact metamorphosed, it often experiences mineralogical alterations that result in a rock quite different from the original. One common rock type produced by contact metamorphism is called **hornfels**. It is a very fine-grained rock with little recognizable texture. Another is called skarn, a rock rich in calc-silicate **minerals** and often the product of a **limestone** or **dolomite** country rock.

The other, less common form of intrusion into country rock consists of geologic material that is able to flow, but is not molten rock. This material can be unconsolidated sediment that has sufficient **water** content to act as a fluid. These are termed soft sediments and if there is sufficient pressure from the overlying rocks, they can be forced into fractures into country rock. The resulting intrusion is called a diatreme. Another type of material that forms diatremes is salt. Salt has a lower density than most other rocks and when buried, salt horizons can become viscous and will flow upward. In both soft sediment intrusions and salt diatremes, the county rock is not metamorphosed. However, diatremes do disrupt the country rock, sometimes producing visible bulges.

See also Intrusive cooling

COUSTEAU, JACQUES-YVES (1910-1997)

French oceanographer

Jacques Cousteau was known as the co-inventor of the aqualung, along with his television programs, feature-length films, and books, all of which have showcased his research on the wonders of the marine world. Cousteau helped demystify undersea life, documenting its remarkable variety, its interdependence, and its fragility. Through the Cousteau Society, which he founded, Cousteau led efforts to call attention to environmental problems and to reduce marine pollution.

Jacques-Yves Cousteau was born in St. André-de-Cubzac, France, on June 11, 1910 to Elizabeth Duranthon and Daniel Cousteau. Jacques, for the first seven years of his life, suffered from chronic enteritis, a painful intestinal condition. In 1918, after the Treaty of Versailles, Daniel found work as legal adviser to Eugene Higgins, a wealthy New York expatriate. Higgins traveled extensively throughout **Europe**, with the Cousteau family in tow. Cousteau recorded few memories from his childhood; his earliest impressions, however, involved **water** and ships. His health greatly improved around this time, thanks in part to Higgins, who encouraged young Cousteau to learn how to swim.

In 1920, the Cousteaus accompanied Higgins to New York City. Here, Jacques attended Holy Name School in Manhattan, learning the intricacies of stickball and roller-skating. He spent his summers at a camp on Vermont's Lake Harvey, where he first learned to dive underwater. At age 13, after a trip south of the American border, he authored a handbound book he called "An Adventure in Mexico." That same year, he purchased a Pathé movie camera, filmed his cousin's marriage, and began making short melodramatic films.

During his teens, Cousteau was expelled from a French high school for "experimenting" on the school's windows with different-sized stones. As punishment, he was sent to a military-style academy near the French-German border, where he became a dedicated student. He graduated in 1929, unsure of which career path to follow. The military won out over filmmaking simply because it offered the opportunity for extended travel. After passing a rigorous entrance examination, he was accepted by the Ecole Navale, the French naval academy. His class embarked on a one-year world cruise, which he documented, filming everything and everyone—from Douglas Fairbanks, the famous actor, to the Sultan of Oman. After graduating second in his class in 1933, he was promoted to second lieutenant and sent to a naval base in Shanghai, China. His assigned duty was to survey and map the countryside, but in his free time he filmed the locals in China and Siberia.

In the mid–1930s, Cousteau returned to France and entered the aviation academy. Shortly before graduation, in 1936, he was involved in a near-fatal automobile accident that mangled his left forearm. His doctors recommended amputation but he steadfastly refused. Instead, he chose rehabilitation, using a regimen of his own design. He began taking daily swims around Le Mourillon Bay to rehabilitate his injured arm. He fell in love with goggle diving, marveling at the variety and beauty of undersea life. He later wrote in his book *The Silent World*: "One Sunday morning...I waded into the Mediterranean and looked into it through Fernez goggles...I was astonished by what I saw in the shallow shingle at Le Mourillon, rocks covered with green, brown and silver **forests** of algae and fishes unknown to me, swimming in crystalline water...Sometimes we are lucky enough to know that our lives have been changed, to discard the old, embrace the new, and run headlong down an immutable course. It happened to me at Le Mourillon on that summer's day, when my eyes were opened on the sea."

During his convalescence he met 17-year-old Simone Melchior, a wealthy high-school student who was living in Paris. After a one-year courtship, the couple married and moved into a house near Le Mourillon Bay. The Cousteaus' first son, Jean-Michel, was born in March of 1938. A second son, Philippe, was born in 1939. Around this time, the new family's tranquil life on the edge of the sea was threatened by world events. In 1939, France began preparing for war, and Cousteau was promoted to gunnery officer aboard the *Dupleix*. The war was largely limited to ground action, however, and Germany quickly overran the ill-prepared French Army. Living in the unoccupied section of France enabled Cousteau to continue his experiments and allowed him to spend many hours with his family. In his free time, he experimented with underwater photography devices and tried to develop improved diving apparatuses. German patrols often questioned Cousteau about his use of diving and photographic equipment. Although he was able to convince authorities that the equipment was harmless, Cousteau was, in fact, using these devices on behalf of the French resistance movement. For his efforts, he was later awarded the Croix de Guerre with palm.

Cousteau regretted the limitations of goggle diving; he simply could not spend enough time under water. The standard helmet and heavy suit apparatus had similar limitations; the diver was helplessly tethered to the ship, and the heavy suit and helmet made Cousteau feel awkward in his movements. A number of experiments with other diving equipment followed, but all the existing systems proved unsatisfactory. He designed his own "oxygen re-breathing outfit," which was less physically constrictive but which ultimately proved ineffective and dangerous. Also during this period he began his initial experiments with underwater filmmaking. Working with two colleagues, Philippe Talliez, a naval officer, and Frédéric Dumas, a renowned spearfisherman, Cousteau filmed his first underwater movie, *Sixty Feet Down*, in 1942. The 18-minute film reflects the technical limitations of underwater photography but was quite advanced for its time. Cousteau entered the film in the Cannes Film Festival, where it received critical praise and was purchased by a film distributor.

As pleased as he was with his initial efforts at underwater photography, Cousteau realized that he needed to spend more time underwater to accurately portray the ocean's mysteries. In 1937, he began collaboration with Emile Gagnan, an engineer with a talent for solving technical problems. In 1942, Cousteau again turned to Gagnan for answers. The two spent approximately three weeks developing an automatic regulator that supplied compressed air on demand. This regulator, along with two tanks of compressed air, a mouthpiece, and hoses, was the prototype Aqualung, which Gagnan and Cousteau patented in 1943.

That summer, Cousteau, Talliez, and Dumas tested the Aqualung off the French Riveria, making as many as five hundred separate dives. This device was put to use on the group's next project, an exploration of the *Dalton*, a sunken British steamer. This expedition provided material for Cousteau's sec-

ond movie, *Wreck*. The film deeply impressed French naval authorities, who recruited Cousteau to assist with the dangerous task of clearing mines from French harbors. When the war ended, Cousteau received a commission to continue his research as part of the Underwater Research Group, which included both Talliez and Dumas. With increased funding and ready access to scientists and engineers, the group expanded its research and developed a number of innovations, including an underwater sled.

In 1947, Cousteau, using the Aqualung, set a world's record for free diving, reaching a depth of 300 feet. The following year, Dumas broke the record with a 306-foot dive. The team developed and perfected many of the techniques of deep-sea diving, working out rigorous decompression schedules that enabled the body to adjust to pressure changes. This physically demanding, dangerous work took its toll; one member of the research team was killed during underwater testing.

On July 19, 1950, Cousteau bought *Calypso*, a converted U.S. minesweeper. The next year, after undergoing significant renovations, *Calypso* sailed for the Red Sea. The *Calypso* Red Sea Expedition (1951–52) yielded numerous discoveries, including the identification of previously unknown plant and animal species and the discovery of volcanic basins beneath the Red Sea. In February of 1952, *Calypso* sailed toward Toulon. On the way home, the crew investigated an uncharted wreck near the southern coast of Grand Congloué and discovered a large Roman ship filled with treasures. The discovery helped spread Cousteau's fame in France. In 1953, with the publication of *The Silent World*, Cousteau achieved international notice. The book, drawn from Cousteau's daily logs, was written originally in English with the help of U.S. journalist James Dugan and later translated into French. Released in more than 20 languages, *The Silent World* eventually sold more than five million copies worldwide.

In 1953, Cousteau began collaborating with Harold Edgerton, a pioneer in high-speed photography who had invented the strobe light and other photographic devices. Edgerton and his son, William, spent several summers aboard *Calypso*, outfitting the ship with an innovative camera that skimmed along the ocean floor, sending back blurry but intriguing photos of deep-sea creatures. The death of William Edgerton in an unrelated diving accident effectively ended the experiments, but Cousteau had already realized the limitations of such a method of exploring the ocean depths. Instead, he and his team began work on a small, easily maneuverable submarine, which he called the diving saucer, or DS–2. The sub has made more than one thousand dives and has been part of countless undersea discoveries.

In 1955, *Calypso* embarked on a 13,800-mile journey that was recorded by Cousteau for a film version of *The Silent World*. The ninety-minute film premiered at the 1956 Cannes International Film Festival, where it received the coveted Palme d'Or. The following year, the film won an Oscar from the American Academy of Motion Picture Arts and Sciences. In 1957, in part due to his film's success, Cousteau was named director of the Oceanographic Institute and Museum of Monaco. He filled the museum's aquariums with rare and unusual species garnered from his ocean expeditions.

Cousteau addressed the first World Oceanic Congress in 1959, an event that received widespread coverage and led to his appearance on the cover of *Time* magazine on March 28, 1960. The highly favorable story painted Cousteau as a poet of the deep. In April of 1961, Cousteau received the National Geographic Society's Gold Medal at a White house ceremony hosted by President John F. Kennedy. The medal's inscription reads: "To earthbound man he gave the key to the silent world."

During the early 1960s, Cousteau and his crew participated in the Conshelf Saturation Dive program, which was intended to prove the feasibility of extended underwater living. The success of the first mission led to Conshelf II, a month-long project involving five divers. The Conshelf program and the DS–2 project provided material for the 53-minute film *World without Sun*, which debuted in the United States in December of 1964.

Cousteau's first hour-long television special, "The World of Jacques-Yves Cousteau," was broadcast in 1966. The program's high ratings and critical acclaim helped Cousteau land a lucrative contract with the American Broadcasting Company (ABC). The *Undersea World of Jacques Cousteau* premiered in 1968, and has since been rebroadcast in hundreds of countries. The program starred Cousteau and his sons, Philippe and Jean-Michel. The show ran for eight **seasons**, with the last episode airing in May of 1976. In 1977, the *Cousteau Odyssey* series premiered on the Public Broadcasting System. The new show reflected Cousteau's growing concern about environmental destruction and tended not to focus on specific animal species.

In the 1970s, the Cousteau Society, a nonprofit environmental group that also focuses on peace issues, opened its doors in Bridgeport, Connecticut. By 1975, the society had more than 120,000 members and had opened branch offices in Los Angeles, New York, and Norfolk, Virginia. Eventually, Cousteau decided to make Norfolk the home base for *Calypso*.

On June 28, 1979, Philippe Cousteau was killed when the seaplane he was piloting crashed on the Tagus River near Lisbon, Portugal. Philippe's death deeply affected Cousteau, who was to his death unable to talk about the accident or the loss of his son. Philippe was expected to eventually take command of his father's empire; instead, Jean-Michel was given increased responsibility for overseeing the Cousteau Society and his father's other ventures.

In 1980, Cousteau signed a one-million-dollar contract with the National Office of Canadian Film to produce two programs on the greater St. Lawrence waterway. In 1984, the *Cousteau Amazon* series premiered on the Turner Broadcasting System. The four shows were enthusiastically reviewed, and called attention to the threatened native South American cultures, Amazon rain forest, and creatures that lived in one of the world's great **rivers**. The final show of the series, "Snowstorm in the Jungle," explored the frightening world of cocaine trafficking. In the mid–1980s "Cousteau/Mississippi: The Reluctant Ally" received an Emmy award for outstanding informational special. In all, Cousteau's television programs have earned more than 40 Emmy nominations.

In addition to his television programs, Cousteau continued to produce new inventions. The Sea Spider, a many-armed diagnostic device, was developed to analyze the biochemistry of the ocean's surface. In 1980, Cousteau and his team began work on the Turbosail, which uses high-tech **wind** sails to cut fuel consumption in large, ocean-going vessels. In spring of 1985, he launched a new wind ship, the *Alcyone*, which was outfitted with two 33-foot-high Turbosails.

In honor of his achievements, Cousteau received the Grand Croix dans l'Ordre National du Mérite from the French government in 1985. That same year, he also received the U.S. Presidential Medal of Freedom. In November of 1987, he was inducted into the Television Academy's Hall of Fame and later received the founder's award from the International Council of the National Academy of Television Arts and Sciences. In 1988, the National Geographic Society honored him with its Centennial Award for "special contributions to mankind throughout the years."

While some critics challenged his scientific credentials, Cousteau never claimed "expert" status in any discipline. His talents appeared as poetic as scientific; his films and books— which include the eight-volume *Undersea Discovery* series and the 21-volume *Ocean World* encyclopedia series—have a lyrical quality that conveys the captain's great love of nature. This optimism was tempered by his concerns about the environment. He emphatically demonstrated, perhaps to a greater degree than any of his contemporaries, how the quality of both the land and sea is deteriorating and how such environmental destruction is irreversible.

Cousteau continued to speak publicly about environmental issues until he was well into his eighties, although he had given up diving in cold water. In the years before his death, he had been planning for the construction of the *Calypso 2* to replace the original *Calypso*, which had sunk in a Singapore shipyard in 1994. The $20 million vessel was to be powered by **solar energy** and include equipment for a television studio, marine laboratory, and **satellite** transmission facility. The oceanographer died of a heart attack in 1997, at his home in Paris, after suffering from a respiratory ailment. He was 87.

COVALENT BONDS • *see* CHEMICAL BONDS AND PHYSICAL PROPERTIES

CRATER, VOLCANIC

A crater is a steep-sided roughly circular to elliptical depression in the earth caused either by volcanic activity or by the impact of an extraterrestrial body. Volcanic craters are formed by explosive events, and/or by the collapse of part of a **volcano** following withdrawal of **magma**. Impact craters are the result of collisions between Earth and extraterrestrial bodies such as meteors or **comets**.

Large volcanic craters are known as calderas among vulcanologists. There are two often-complementary processes involved in their formation; violent eruptions of ash and magma, and/or the collapse of a volcanic surface following withdrawal of a large body of magma from the subsurface. An example of the first type may be Crater Lake in Oregon, thought to have been produced by a violent explosion that destroyed a volcano the size of Mount St. Helens. The **caldera** at Kilauea, in contrast, is thought to be the result of magma drainage from beneath the summit. There is still significant discussion about whether volcanic calderas are formed directly by explosion, indirectly by collapse of the surface following magma ejection or withdrawal, or by both.

Impact craters are the result of collisions of extraterrestrial bodies with the earth. Only recently have scientists begun to understand the importance of impact processes in shaping the planet and life on it. Exploration of our **solar system** has revealed that essentially all planetary bodies are cratered. The density of craters on the older surfaces of the **Moon** indicates an intense bombardment from approximately 4.6 to 3.9 billion years ago. The Moon itself is likely the result of a collision of a Mars size object with a young Earth. The earth experienced the same bombardment as the other planetary bodies. In fact, Earth is subject to about twice as many impacts as the moon because of the difference in **gravity**. This is not obvious because tectonic and **erosion** activity on the earth have removed evidence of most of the impacts that have occurred. Nevertheless, approximately 150 craters have been identified, with more recognized every year.

Perhaps the most well-known **impact crater** on Earth is Chicxulub, a buried crater in the Yucatan, Mexico, that is 110 miles (180 kilometers) in diameter. Most geoscientists now believe that this impact event was responsible for the great mass extinction of the dinosaurs and many other species at the Cretaceous/Tertiary (K-T) boundary, 65 million years ago. Impacts this size occur infrequently, on the order of one every 100 million years. However, impacts that could cause damage similar to a **nuclear winter**, occur at time scales estimated as two or three every million years. This estimate is significant because the most recent known event, Zhamanshin in Kazaksthan, occurred about a million years ago.

See also Meteoroids and meteorites; Volcanic eruptions

CRATON

Cratons are large regions of continental **crust** that have remained tectonically stable for a prolonged period of time, often a billion years or more. **Precambrian** cratons are commonly cored by Archaean granite-greenstone terrains and may be partly covered by sedimentary platform sequences. The North American Craton, Laurentia, which constitutes much of **North America**, formed by the assembly of smaller cratons in the **Archean** and Paleoproterozoic. Cratons are surrounded by orogenic (mountain building) or mobile belts, within which deformation has been localized. For example, cratons comprised of Archean granite-greenstone terrains and Paleoproterozoic sedimentary sequences in **Africa**, central India and Western **Australia** are rimmed by Mesoproterozoic and Neoproterozoic orogenic belts, many of which have been subsequently reactivated during **rifting** and the formation of Paleozoic sedimentary basins. Only minor reactivation of older structural weaknesses occurs in craton interiors during deformation on their margins.

Cratons have thick lithospheric roots or keels. Lithospheric thicknesses for Archean cratons show a bimodal distribution, with thicknesses of approximately 137 mi (220 km) and 218 mi (350 km) predominating. Larger cratons generally have thicker lithospheres. In contrast, post-Archean **lithosphere** is generally 62–124 mi (100–200 km) thick. The physical and/or chemical properties of the deep roots of cratons enable them to resist recycling into the underlying asthenopsheric mantle. This may be responsible for the stability of cratons. Isotopic signatures obtained from mantle lithosphere-derived, **peridotite** xenoliths and inclusions in diamonds imply that roots to Precambrian cratons have been isolated from the convecting mantle for billions of years. Archean subcontinental lithospheric mantle is buoyant relative to the underlying **asthenosphere**. It is therefore not easily delaminated and assimilated into the asthenosphere and will tend to be preserved. Geochemical changes may also impart stability to cratonic roots. **Mantle plumes** are generally unable to break and induce rifting of thick, cratonic lithosphere and may be deflected around cratons. Despite their buoyancy, the margins of cratons can be deformed at subduction zones, however, due to the development of detachments at the interface between crust and mantle lithospheric root. Numerical modeling suggests that the dominant stabilizing factor for the preservation of cratons is the relatively high brittle yield stress of cratonic lithosphere. As the strength of the continental lithosphere resides primarily within the crust, the physical properties of the crust may also play a role in the longevity of cratonic lithosphere.

Favorable pressures and temperatures for the formation and preservation of **diamond** are found beneath cratons. Diamond crystallizes from liquid **carbon** between 1652°–2192°F (900–1200°C) at pressures above 50 kbar. At this pressure, equating to a depth of 93 mi (150 km) or more, temperatures are generally too high for the formation of diamond except in the roots beneath cratons. Most of the world's diamonds come from deep, mantle-sourced intrusive bodies such as kimberlites or lamproites that intrude Archaean cratons. Kimberlite and lamproite magmas intrude extremely rapidly up deep fractures and may bring diamonds to shallow levels in the crust. Near the earth's surface, they erupt explosively due to their high gas pressures, creating **breccia** pipes (called diatremes) and craters.

See also Plate tectonics

CREEP

The slow, often imperceptible downslope movement of **soil** or other debris is called creep. Because creep moves materials so slowly, it is difficult to discern directly. Observation of the effects of creep, such as bent trees, tilted fences, and cracked walls, usually leads to identification of the problem.

Creep is caused by the interaction of multiple factors, but heaving is likely the most important process. Heaving involves the expansion and contraction of **rock** fragments, and occurs during cycles of wetting and drying, as well as **freezing** and thawing. As expansion occurs, particles move outward, perpendicular to the hillside. During contraction, the particles move back toward the hillside, vertically, and end up slightly downslope of where they began. The repeated motion of individual particles results in net downslope movement of the material. Areas that undergo wet/dry or freeze/thaw cycles are most susceptible to creep.

Solifluction is a special type of creep that occurs in cold regions underlain by **permafrost**. During the winter, the ground freezes right up to the surface. When the surface layer thaws, during the spring and early summer, the meltwater cannot percolate downward into the frozen layers beneath. This causes the surface layer of soil to become waterlogged, facilitating downslope movement as the layer becomes saturated. In this case the surface layer flows, riding above the frozen ground beneath. Although most common in permafrost areas, solifluction can occur anywhere that the surface soil layer becomes saturated.

Although movement associated with creep is slow, it causes significant economic damage because it is a widespread phenomenon that is probably occurring to some extent on virtually all soil-covered slopes. Some of the problem relates to the difficulty of detection. Unless trees, walls, or other built structures are deformed, it is difficult to impossible to determine whether or not creep is occurring. Unfortunately, where creep has been identified, it is also difficult to control. The best response to the problem is to avoid building in areas undergoing creep. Where construction is necessary, buildings should be anchored to **bedrock** beneath the creeping soil and debris layer.

See also Mass wasting; Regolith; Weathering and weathering series

CRETACEOUS PERIOD

Cretaceous is the name given to a period of time in Earth's history (i.e., Cretaceous Period) from 145.6 to 65 million years ago. Also, all the rocks that formed during that time have the same proper name of Cretaceous (i.e., they are referred to as the Cretaceous System). Said differently, the Cretaceous System is the **rock** record of events that occurred—and organisms that lived—during a span of geological time that is called Cretaceous Period. Cretaceous was the third and final period of the **Mesozoic Era**.

Cretaceous is a name derived from the Latin word for chalk, *creta*. Chalk is a common type of sedimentary rock formed during this interval of Earth history. The term Cretaceous was first used in 1822 by d'Omalius d'Halloy (1707–1789), a Belgian geologist who was engaged in pioneering efforts at geological mapping of parts of France. He mapped the *Terrain Cretac* (Cretaceous System) within the Paris Basin and later in Belgium. D'Halloy's strata were eas-

ily correlated with chalks mapped earlier, but not formally named, by **William Smith** (1769–1839) on his revolutionary 1815 geological map. Other geologists of the day rapidly correlated the Cretaceous System with chalks they found in northern France, Holland, Denmark, northern Germany, Poland, and Sweden. Since their recognition and definition during the nineteenth century, Cretaceous strata have been mapped on all the world's continents.

During Cretaceous, the breakup of the supercontinent of Pangaea became nearly complete. The Atlantic Ocean opened sufficiently so that a substantial body of **water** existed between **North America** and **Europe** and **South America** and **Africa** were widely separated. While the Atlantic was opening, the Pacific Ocean continued to close rapidly and an episode of major tectonic change occurred as a result in western North and South America. Specifically, a major **subduction zone** developed along the western coast of the Americas and substantial tectonic uplift and volcanic activity occurred along this western margin.

During most of Cretaceous, global sea levels were at some of their highest elevations in the last 500 million years of Earth history. Much of the low-lying areas of the world's continents were covered by shallow **seas**, also known as epicontinental seas. During part of Late Cretaceous, sea level was so high (approximately 275 m above present level) that an epicontinental sea (i.e., Western Interior Seaway) connected the **Gulf of Mexico** with the Arctic Ocean through the center of North America. (The Rocky Mountains were not formed at this time, and thus this **area** was rather low-lying and flat).

Cretaceous was a time of elevated global temperatures and there were essentially no polar or high-altitude **glaciers**. This contributed to elevated sea levels as did the vast development of volcanic activity along Earth's **mid-ocean ridges**. Such volcanic activity and accompanying swelling of these undersea ridges displaced a considerable volume of seawater (strongly exacerbating sea-level rise). Also, there was rapid **sea-floor spreading** during Cretaceous. Further, effusive volcanism from mantle hot spots caused flood basalts to erupt on land, the most noteworthy of these produced the massive Deccan Traps of India.

Cretaceous life in the **oceans** was very plentiful, with numerous species of fish, sharks, rays, ammonites, turtles, mosasaurs, plesiosaurs, and other creatures plying ocean waters. Numerous massive reef systems developed, including one rimming the Gulf of Mexico during Early Cretaceous, which was dominated by a specialized type of large clams called rudists. Large oysters were common at this time. The main plankon, the golden-brown alga, produced massive amounts of calcareous platelets that settled to the sea floor along with abundant foraminifers thus forming an ooze that eventually became the famous Cretaceous chalk (as seen at the White Cliffs of Dover, England, and many other areas).

Cretaceous life on land was dominated by the reptiles, which dwelt for the most part in heavily vegetated terrains. Lush Cretaceous cycad-ginko-conifer **forests** and swamps are well preserved in vast tracts of **coal** and lignite deposits of this age. Late Cretaceous was the time of development of flowing plants and co-evolution of a very diverse insect population. Land animals such as snakes, lizards, crocodiles, dinosaurs, pterosaurs, birds, and small mammals were common. Among the dinosaurs, Cretaceous ecosystems saw the rise of many diverse types, which were indigenous to the rapidly separating continental land masses. The rising groups of dinosaurs during this time included pachycephalosarus, ceratopsians, iguanodonts, hadrosaurs, coelurosaurs, and carnosaurs (e.g., tyrannosaurids).

The end of Cretaceous came abruptly with global ecosystem collapse and mass extinction. The end-Cretaceous catastrophe resulted in rapid death of nearly 50% of all living species of organisms. Known since the nineteenth century as an inexplicable sudden mass death marker, the Cretaceous-Tertiary (K-T) boundary has attracted considerable interest from researchers in recent years because the boundary **clay** layer contains a substantial enrichment in certain **chemical elements** (e.g., iridium), which are much more common in **asteroids** and **comets** than on Earth. Discovery in the 1980s of a 180-km diameter **impact crater** in the state of Yucatan, Mexico, which has the same age as the K-T boundary, indicates a strong connection between the mass death and cosmic impact at this time in Earth history. Subsequent studies have shown that impact of a 10-km diameter asteroid at Yucatan could have resulted in such a global ecosystem collapse and accompanying mass death. The impact crater is known as Chicxulub, which is a Mayan word meaning "tail of the Devil."

See also Chronostratigraphy; K-T event; Impact crater; Mesozoic Era; Stratigraphy; Supercontinents

CRETACEOUS-TERTIARY MASS EXTINCTION EVENT • *see* K-T EVENT

CROSS BEDDING • *see* BEDDING

CROSS CUTTING

Cross-cutting relationships among geological features have been recognized for many years as one of the fundamental ways of determining relative age relationships between adjacent geological features. The principle of cross-cutting relationships, explained by **James Hutton** (1726–1797) in *Theory of the Earth* (1795) and embellished upon by **Charles Lyell** (1797–1875) in his *Principles of Geology* (1830), holds that the geological feature which cuts another is the younger of the two features. For example, in the instance of an igneous **dike** cutting through a layer of **sandstone**, the dike must be younger than the sandstone.

Cross-cutting relationships are of several basic types. There are structural cross-cutting relationships wherein a fault or fracture cuts through older **rock**. Stratigraphic cross-cutting relationships occur where an erosional surface (or unconformity) cuts across older rock layers, geological structures, or other geological features. Sedimentologic cross-cutting relationships occur where currents have eroded or scoured older sediment in a local **area** to produce, for example, a channel

filled with **sand**. Paleontologic cross-cutting relationships occur where animal activity or plant growth produce truncation. This happens, for example, where animal burrows penetrate into pre-existing sedimentary deposits. Geomorphic cross-cutting relationships occur where a surficial feature, such as a river, flows through a gap in a ridge of rock. In a similar example, an **impact crater** excavates into a subsurface layer of rock.

Cross-cutting relationships may be seen cartographically, megascopically, and microscopically. In other words, these relationships have various scales. A cartographic cross-cutting relationship might look like, for example, a large fault dissecting the landscape on a large map. Megascopic cross-cutting relationships are features like igneous dikes, as mentioned above, which would be seen on an outcrop or in a limited geographic area. Microscopic cross-cutting relationships are those that require study by magnification or other close scrutiny. For example, penetration of a fossil shell by the drilling action of a boring organism is an example of such a relationship.

Cross-cutting relationships may be compound in nature. For example, if a fault were truncated by an unconformity, and that unconformity cut by a dike, we can say, based upon compound cross-cutting relationships that the fault is older than the unconformity and that the unconformity is older than the dike. Using such rationale, the sequence of geological events can be better understood.

Cross-cutting relationships can also be used in conjunction with radiometric age dating to effect an age bracket for geological materials that cannot be directly dated by radiometric techniques. For example, if a layer of sediment containing a fossil of interest is bounded on the top and bottom by **unconformities**, where the lower unconformity truncates dike A and the upper unconformity truncates dike B (which penetrates the layer in question), this method can be used. A radiometric age date from **crystals** in dike A will give the maximum age date for the layer in question and likewise, crystals from dike B will give us the minimum age date. This provides an age bracket, or range of possible ages, for the layer in question.

The principle of cross-cutting relationships, like the principles of **superposition** and inclusions, is one of the most basic tools used by geologists to understand relative age relationships on Earth and on planetary and **satellite** surfaces in our **solar system**.

CRUST

Earth's mass is divided into an inner core, outer core, mantle, and crust. The crust is outermost layer of the earth, 3–44 miles (5–70 km) thick and representing less than 1% of the earth's total volume. Thin compared Earth's diameter, the outermost crustal layer is further subdivided into two basic types of crust—each unique in composition, origin and fate. Although the earth is dynamic, with new crust constantly being created and destroyed, the fact that size of the earth remains constant argues that there is no net creation or destruction of force and that these two processes are in equilibrium.

Although there are thousands of **minerals**, about 40 minerals represent more than 99% of the mass of Earth's crust. In terms of percentage of Earth's crust by weight, **oxygen** and **silicon** account for nearly 75% of Earth's crust. Oxygen is the most abundant element (approximately 46.5% followed by silicon, approximately 28%). In order of percentage by weight, other important elements include **aluminum**, **iron**, calcium, sodium, potassium, and magnesium. All other elements (e.g., gold, silver, copper, etc.) compose the remaining one to two percent of the crust.

Crust is classified as oceanic crust or continental crust.

Oceanic crust is thin (3–4.3 mi [5–7 km]), basaltic (<50% SiO_2) and dense. Compositional chemical studies also establish that oceanic crust is substantially younger than the continental crust. No **rock** specimen dated to more than 250 million years old have ever been identified in oceanic crust.

Continental crust is thick (18.6–40 mi [30–65 km]), granitic (>60% SiO_2), light, and old (250–3,700 million years old).

The outer crust is further subdivided into **lithospheric plates**, that contain varying sections of oceanic and continental crust.

At the deepest crustal border there exists a compositional change from crust material to mantle pyriditite called the **Mohorovicic discontinuity**, and the lithospheric plates carrying both oceanic and continental crust move on top of mantle's **asthenosphere**.

See also Dating methods; Earth, interior structure; Geologic time; Hawaiian Island formation; Isostasy; Lithospheric plates; Mid-ocean ridges and rifts; Mohorovicic discontinuity (Moho); Ocean trenches; Plate tectonics; Rifting and rift valleys; Soil and soil horizons; Subduction zone

CRUTZEN, PAUL J. (1933-)
Dutch meteorologist

Paul Crutzen is one of the world's leading researchers in mapping the chemical mechanisms that affect the ozone layer. He has pioneered research on the formation and depletion of the **ozone** layer and the potential threats placed upon it by industrial society. Crutzen has discovered, for example, that nitrogen oxides accelerate the rate of ozone depletion. He has also found that chemicals released by bacteria in the **soil** affect the thickness of the ozone layer. For these discoveries, he has received the 1995 Nobel Prize in Chemistry, along with **Mario Molina** and Sherwood Rowland for their separate discoveries related to ozone and how chlorofluorocarbons (CFCs) deplete the ozone layer. According to Royal Swedish Academy of Science, "by explaining the chemical mechanisms that affect the thickness of the ozone layer, the three researchers have contributed to our salvation from a global environmental problem that could have catastrophic consequences."

Paul Josef Crutzen was born December 3, 1933, to Josef C. Crutzen and Anna Gurek in Amsterdam. Despite growing up in a poor family in Nazi-occupied Holland during 1940–1945, he was nominated to attend high school at a time

when not all children were accepted into high school. He liked to play soccer in the warm months and **ice** skate 50–60 miles (80–97 km) a day in the winter. Because he was unable to afford an education at a university, he attended a two-year college in Amsterdam. After graduating with a civil engineering degree in 1954, he designed bridges and homes.

Crutzen met his wife, Tertu Soininen, while on vacation in Switzerland in 1954. They later moved to Sweden where he worked as a computer programmer for the Institute of **Meteorology** and the University of Stockholm. He started to focus on **atmospheric chemistry** rather than mathematics because he had lost interest in math and did not want to spend long hours in a lab, especially after the birth of his two daughters. Despite his busy schedule, Crutzen obtained his doctoral degree in Meteorology at Stockholm University at the age of 35.

Crutzen's main research focused on ozone, a bluish, irritating gas with a strong odor. Ozone is a molecule made up of three **oxygen** atoms (O_3) and is formed naturally in the atmosphere by a photochemical reaction. The ozone layer begins approximately 10 miles (16 km) above Earth's surface, reaching 20–30 miles (32–48 km) in height, and acts as a protective layer that absorbs high-energy ultraviolet radiation given off by the **sun**.

In 1970, Crutzen found that soil microbes were excreting nitrous oxide gas, which rises to the **stratosphere** and is converted by sunlight to nitric oxide and nitrogen dioxide. He determined that these two gases were part of what caused the depletion of the ozone. This discovery revolutionized the study of ozone and encouraged a surge of research on global **biogeochemical cycles**.

In 1977, while he was the director of the National Center for Atmospheric Research (NCAR) in Boulder, Colorado, Crutzen studied the effects of the mass burning of trees and brush in the fields of Brazil. The theory at the time was that this burning caused more **carbon** compounds—trace gases and carbon monoxide—to enter the atmosphere. These gases were assumed to cause the **greenhouse effect**, or a warming of the atmosphere. Crutzen collected and examined this smoke in Brazil and discovered that the complete opposite was occurring. He stated in *Discover* magazine, "Before the industry got started, the tropical burning was actually decreasing the amount of **carbon dioxide** in the atmosphere." The study of smoke in Brazil led Crutzen to further examine what effects larger amounts of different kinds of smoke might have on the environment, such as smoke from a nuclear war.

The journal *Ambio* commissioned Crutzen and John Birks, his colleague from the University of Colorado, to investigate what effects nuclear war might have on the planet. Crutzen and Birks studied a simulated worldwide nuclear war. They theorized that the black carbon soot from the raging fires would absorb as much as 99% of the sunlight. This lack of sunlight, coined "nuclear winter," would be devastating to all forms of life. For this theory Crutzen was named "Scientist of the Year" by *Discover* magazine in 1984, and awarded the prestigious Tyler Award four years later.

Because of the discoveries by Crutzen and other environmental scientists, a crucial international treaty was established in 1987. The Montreal Protocol was negotiated under the auspices of the United Nations and signed by 70 countries to slowly phase out the production of chlorofluorocarbons and other ozone-damaging chemicals by the year 2000. However, the United States had ended the production of CFCs five years earlier, in 1995. According to the *New York Times*, "the National Oceanic and Atmospheric Administration reported in 1994, while ozone over the South Pole is still decreasing, the depletion appears to be leveling off." Even though the ban has been established, existing CFCs will continue to reach the ozone, so the depletion will continue for some years. The full recovery of the ozone is not expected for at least 100 years.

From 1977 to 1980, Crutzen was director of the Air Quality Division, National Center for Atmospheric Research (NCAR), located in Boulder, Colorado. While at NCAR, he taught classes at Colorado State University in the department of Atmospheric Sciences. Since 1980, he has been a member of the Max Planck Society for the Advancement of Science, and he is the director of the Atmospheric Chemistry division at Max Planck Institute for Chemistry. In addition to Crutzen's position at the institute, he is a part-time professor at Scripps Institution of **Oceanography** at the University of California. In 1995, he was the recipient of the United Nations Environmental Ozone Award for outstanding contribution to the protection of the ozone layer. Crutzen has co-authored and edited several books, as well as having published several hundred articles in specialized publications.

See also Global warming; Greenhouse gases and greenhouse effect; Ozone layer and hole dynamics; Ozone layer depletion

CRYSTALS AND CRYSTALLOGRAPHY

A crystal is a patterned three-dimensional assembly of atoms that is a repetitive (periodic) array of atoms. Crystals contain repeating arrays or atoms arranged in unit cells. Crystallography is the study of the formation processes that produce crystals, and of the structural and identifying details of crystals.

In the ancient and medieval world, crystals were considered a strange union of the animal and mineral kingdoms, growing into predetermined shapes like living things but seemingly without life. Many mineralogists hypothesized that their growth was the result of astrological forces. It was not until Robert Boyle and Robert Hooke began experimenting with microscopes that the true nature of crystals began to be understood. During the course of the last three centuries an entire field of study, crystallography, developed to further the understanding of crystals.

All solid matter is either **amorphous** (without definite shape) or crystalline (from the Greek word for clear **ice**). Crystals are defined by a regular, well-ordered molecular structure called a lattice, consisting of stacked planes of molecules. Because the molecules of the crystal fit together and contain strong electrical attractions between the atoms, a crystal is typically very strong.

There are many shapes in which crystals may be found, depending upon the type of atomic bond that is most dominant

Microscopic view of granite porphyry crystals. © *Lester V. Bergman/Corbis. Reproduced by permission.*

within their molecules (e.g., ionic, covalent, or metallic). Crystals high in ionic bonds are often cubic; these include salt and sugar, as well as **iron** pyrite. Covalent bonds are very strong, producing an extremely durable crystal such as a **diamond**. Metallic bonds are typified by a cloud of free-roaming electrons, giving the compound's shape less definition but allowing for great electrical and thermal conductivity; most **metals** are technically crystals very high in metallic bonds.

All crystals are formed from tiny atomic building blocks called unit cells. By changing the way the unit cells are stacked together, seven different crystal structures can be formed: triclinic, monoclinic, orthorhombic, trigonal, tetragonal, hexagonal, and cubic. In addition to their structure, crystals are classified by their symmetry—that is, the ability of the crystal to look the same when rotated. Some crystals are symmetrical along two axes, some along three axes, and some along four axes; some display no symmetry at all.

Throughout history, many crystalline materials, including most **gemstones**, have been prized for their ability to be cut along flat planes called cleavage faces. This is accomplished by separating one lattice plane from the next, producing a surface that is almost perfectly flat. However, not all crystals allow such clean cuts. Many substances such as metal, stone, and brick behave like crystals but are very difficult to cut along a cleavage face.

Because of the many industrial and scientific applications of crystals, the demand for clear, perfectly formed samples is very high. Unfortunately, nature rarely produces such crystals; more often the crystal has faults or impurities; and it is for this reason that large, perfect gemstones are valued so highly. In order to meet the demand for pure crystals, scientists have developed methods for "growing" crystals. One common method is simply to melt a large supply of unrefined crystal and allow it to reform; while in a liquid state, the molten crystal is often sifted of impurities, in order to yield crystals of higher quality. Another method for growing crystals is called seeding. Here, a small sample of crystal is placed in a vapor or solution; material is allowed to accumulate on the seed until the system reaches equilibrium. Often in crystal seeding a seed of a different material than that of the crystal is used. This is the case in the natural crystal formation called **cloud seeding**, wherein seeds of silver iodide are dropped into **clouds**; the silver iodide accumulates ice crystals which eventually fall in the form of rain or snow.

Scientists began to investigate the nature of crystals as early as the seventeenth century, when the Danish geologist Nicolaus Steno (1638-1686) began his experiments with common crystals. He found that all crystals of a certain compound have characteristic angles at which the faces will meet. This means that every piece of salt will be cubic in shape, and that smashing a piece of salt will yield smaller and smaller cubes. Thus began the science of crystallography, and Steno's observation became its first law.

The next great crystallographer was the French mineralogist René-Just Haüy (1743–1822), who became involved in the science quite accidentally. While browsing through a friend's mineral collection, he dropped a large sample of calcite. He was surprised to note that the sample shattered along straight planes. Although Steno had pointed this out more than a century before, it was Haüy who hypothesized the existence of unit cells, showing how basic cells could be combined to create the different crystal shapes.

By the early 1800s many physicists were experimenting with crystals; in particular, they were fascinated by their ability to bend light and separate it into its component colors. Because of their varying molecular structures, different crystal types would affect light differently. Among the most influential member of the emerging field of optical **mineralogy** was the British scientist David Brewster; by 1819, Brewster had succeeded in classifying most known crystals according to their optical properties.

During the mid-1800s the preeminent French chemist Louis Pasteur examined tartrate crystals under a microscope; these crystals were known to twist the path of light sometimes one direction and sometimes the opposite. He found that the tartrate crystals were not all identical and that some were mirror images of the others. When combined, the two shapes within the whole tartrate would bend light in two possible directions. By using tweezers, Pasteur painstakingly separated the crystals into two piles, which were melted and then reformed into two distinct crystals. Once separated, each new crystal would twist light in only one direction, one clockwise, the other counterclockwise.

Pasteur's work became the foundation for crystal polarimetry, a method by which light is polarized, or aligned to a single plane. It was soon discovered that other crystals were also capable of polarizing light. Today, crystal polarimetry is used extensively in **physics** and optics.

Another phenomenon displayed by certain crystals is piezoelectricity. From a Greek work meaning "to press," piezoelectricity is the creation of an electrical potential by squeezing certain crystals. This strange effect was first discovered by **Pierre Curie** and his brother, Jacques, in 1880, who were surprised to detect a voltage across the face of compressed Rochelle salt.

The piezoelectric effect also works in reverse: when an electrical current is applied to a crystal such as **quartz**, it will contract; if the direction of the current is reversed, the crystal will expand. If an alternating current is used, the piezoelectric crystal will expand and contract rapidly, producing a vibration whose frequency can be regulated. Because of their precise vibrations, piezoelectric crystals are used in radio transmitters and quartz timepieces.

Perhaps the most important application of crystals is in the science of x-ray crystallography. Experiments in this field were first conducted by the German physicist Max von Laue. While an instructor at the University of Munich, Laue had done extensive work with diffraction gratings (metal meshes used to separate light into its component colors). His goal was to apply these gratings to the study of x-ray radiation because, at this time the true nature of x rays was yet to be fully understood. However, the wavelengths of x rays were far too short for diffraction gratings to be used, the x rays would pass through the holes unaffected. What was needed was a grating with microscopically tiny holes; unfortunately, the technology did not exist to construct such a grating.

In 1912, Laue perceived that the regular stacked-plane structure of a crystal would act like a very small diffraction grating; this hypothesis was successfully tested with a crystal of zinc sulfide, and x-ray crystallography was born. Using a crystal, scientists could now measure the wavelength of any x ray as long as they knew the internal structure of their crystal. Also, if an x ray of a known wavelength was used, the molecular structure of unknown crystals could be determined.

X-ray crystallography was perfected just a few years later by the father-son team of William Henry Bragg (1862–1942) and William Lawrence Bragg (1890–1971), who were awarded the 1915 Nobel Prize in physics for their work. Since that time, x-ray crystallography has been used to examine the molecular structure of thousands of crystalline substances and was instrumental in the analysis of DNA. Crystallography remains an important branch of **earth science** because the analysis and study of crystals often yields important information concerning the type and rate of geological processes.

See also Atomic theory; Cave minerals and speleothems; Chemical bonds and physical properties; Minerals; Mineralogy

CUMULIFORM CLOUD • *see* CLOUDS AND CLOUD
TYPES

CUMULONIMBUS CLOUD • *see* CLOUDS AND
CLOUD TYPES

CUMULUS CLOUD • *see* CLOUDS AND CLOUD TYPES

CURIE, MARIE (1867-1934)
Polish-born French physicist

Marie Curie was the first woman to win a Nobel Prize, and one of very few scientists ever to win that award twice. In collaboration with her physicist-husband **Pierre Curie**, Marie Curie developed and introduced the concept of **radioactivity** to the world. Working in primitive laboratory conditions, Curie investigated the nature of high-energy rays spontaneously produced by certain elements, and isolated two new radioactive elements, polonium and radium. Her scientific efforts also included the application of x rays and radioactivity to medical treatments.

Curie was born to her two-schoolteacher parents in Warsaw, Poland. Christened Maria Sklodowska, she was the fourth daughter and fifth child in the family. By the age of five, she had already begun to suffer deprivation. Her mother Bronislawa had contracted tuberculosis and assiduously avoided kissing or even touching her children. By the time Curie was 11, both her mother and her eldest sister Zosia died, leaving Marie an avowed atheist. Curie was also an avowed nationalist (like the other members in her family), and when she completed her elementary schooling, she entered Warsaw's "Floating University," an underground, revolutionary Polish school that prepared young Polish students to become teachers.

Curie left Warsaw at the age of 17, not for her own sake but for that of her older sister Bronya. Both sisters desired to acquire additional education abroad, but the family could not afford to send either of them, so Marie took a job as a governess to fund her sister's medical education in Paris. At first, she accepted a post near her home in Warsaw, then signed on with the Zorawskis, a family who lived some distance from Warsaw. Curie supplemented her formal teaching duties there with the organization of a free school for the local peasant children. Casimir Zorawski, the family's eldest son, eventually fell in love with Curie and she agreed to marry him, but his parents objected vehemently. Stunned by her employers' rejection, Curie finished her term with the Zorawskis and sought another position. She spent a year in a third governess job before her sister Bronya finished medical school and summoned her to Paris.

In 1891, at the age of 24, Curie enrolled at the Sorbonne and became one of the few women in attendance at the university. Although Bronya and her family back home were helping Curie pay for her studies, living in Paris was quite expensive. Too proud to ask for additional assistance, she sub-

Marie Curie. *AP/Wide World. Reproduced by permission.*

sisted on a diet of buttered bread and tea, which she augmented sometimes with fruit or an egg. Because she often went without heat, she would study at a nearby library until it closed. Not surprisingly, on this regimen she became anemic and on at least one occasion fainted during class.

In 1893, Curie received a degree in **physics**, finishing first in her class. The following year, she received a master's degree, this time graduating second in her class. Shortly thereafter, she discovered she had received the Alexandrovitch Scholarship, which enabled her to continue her education free of monetary worries. Many years later, Curie became the first recipient ever to pay back the prize. She reasoned that with that money, yet another student might be given the same opportunities she had.

Friends introduced Marie to Pierre Curie in 1894. The son and grandson of doctors, Pierre had studied physics at the Sorbonne; at the time he met Marie, he was the director of the École Municipale de Physique et Chimie Industrielles. The two became friends, and eventually she accepted Pierre's proposal of marriage. Their Paris home was scantily furnished, as neither had much interest in housekeeping. Rather, they concentrated on their work. Pierre Curie accepted a job at the School of Industrial Physics and Chemistry of the City of Paris, known as the EPCI. Given lab space there, Marie Curie spent eight hours a day on her investigations into the magnetic qualities of steel until she became pregnant with her first child, Irene, who was born in 1897.

Curie then began work in earnest on her doctorate. Like many scientists, she was fascinated by French physicist Antoine-Henri Becquerel's discovery that the element uranium emitted rays that contained vast amounts of energy. Unlike Wilhelm Röntgen's x rays, which resulted from the excitation of atoms from an outside energy source, the "Becquerel rays" seemed to be a naturally occurring part of the uranium ore. Using the piezoelectric **quartz** electrometer developed by Pierre and his brother Jacques, Marie tested all the elements then known to see if any of them, like uranium, caused the nearby air to conduct **electricity**. In the first year of her research, Curie coined the term "radioactivity" to describe this mysterious force. She later concluded that only thorium and uranium and their compounds were radioactive.

While other scientists had also investigated the radioactive properties of uranium and thorium, Curie noted that the **minerals** pitchblende and chalcolite emitted more rays than could be accounted for by either element. Curie concluded that some other radioactive element must be causing the greater radioactivity. To separate this element, however, would require a great deal of effort, progressively separating pitchblende by chemical analysis and then measuring the radioactivity of the separate components. In July, 1898, she and Pierre successfully extracted an element from this ore that was even more radioactive than uranium; they called it polonium in honor of Marie's homeland. Six months later, the pair discovered another radioactive substance—radium—embedded in the pitchblende.

Although the Curies had speculated that these elements existed, to prove their existence they still needed to describe them fully and calculate their atomic weight. In order to do so, Curie needed an abundant supply of pitchblende and a better laboratory. She arranged to get hundreds of kilograms of waste scraps from a pitchblende mining firm in her native Poland, and Pierre Curie's EPCI supervisor offered the couple the use of a laboratory space. The couple worked together, with Marie performing the physically arduous job of chemically separating the pitchblende and Pierre analyzing the physical properties of the substances that Marie's separations produced. In 1902, the Curies announced that they had succeeded in preparing a decigram of pure radium chloride and had made an initial determination of radium's atomic weight. They had proven the chemical individuality of radium.

Pierre Curie's father had moved in with the family and assumed the care of their daughter, Irene, so the couple could devote more than eight hours a day to their work. Pierre Curie's salary, however, was not enough to support the family, so Marie took a position as a lecturer in physics at the École Normal Supérieure; she was the first woman to teach there. In the years between 1900 and 1903, Curie published more than she had or would in any other three-year period, with much of this work being co-authored by Pierre Curie. In 1903, Curie became the first woman to complete her doctorate in France, summa cum laude.

The year Curie received her doctorate was also the year she and her husband began to achieve international recognition for their research. In November, the couple received England's prestigious Humphry Davy Medal, and the follow-ing month Marie and Pierre Curie—along with Becquerel—received the Nobel Prize in physics for their efforts in expanding scientific knowledge about radioactivity. Although Curie was the first woman ever to receive the prize, she and Pierre declined to attend the award ceremonies, pleading they were too tired to travel to Stockholm. The prize money from the Nobel, combined with that of the Daniel Osiris Prize—which she received soon after—allowed the couple to expand their research efforts. In addition, the Nobel bestowed upon the couple an international reputation that furthered their academic success. The year after he received the Nobel, Pierre Curie was named professor of physics of the Faculty of Sciences at the Sorbonne. Along with his post came funds for three paid workers, two laboratory assistants and a laboratory chief, stipulated to be Marie. This was Marie's first paid research position.

In 1904, Marie gave birth to another daughter. Despite the fact that both Pierre and Marie frequently suffered adverse effects from the radioactive materials with which they were in constant contact, their infant daughter was born healthy. The Curies continued their work regimen, taking sporadic vacations in the French countryside with their two children. They had just returned from one such vacation when in April 1906, tragedy struck; while walking in the congested street traffic of Paris, Pierre was run over by a heavy wagon and killed.

A month after the accident, the University of Paris invited Curie to take over her husband's teaching position. Upon acceptance she became the first woman to ever receive a post in higher education in France, although she was not named to a full professorship for two more years. During this time, Curie came to accept the theory of English physicists Ernest Rutherford and Frederick Soddy that radioactivity was caused by atomic nuclei losing particles, and that these disintegrations caused the transmutation of an atomic nucleus into a different element. It was Curie, in fact, who coined the terms disintegration and transmutation.

In 1909, Curie received an academic reward that she had greatly desired: the University of Paris drew up plans for an Institut du Radium that would consist of two branches, a laboratory to study radioactivity—which Curie would run—and a laboratory for biological research on radium therapy, to be overseen by a physician. It took five years for the plans to come to fruition. In 1910, however, with her assistant André Debierne, Curie finally achieved the isolation of pure radium metal, and later prepared the first international standard of that element.

Curie was awarded the Nobel Prize again in 1911, this time "for her services to the advancement of chemistry by the discovery of the elements radium and polonium," according to the award committee. The first scientist to win the Nobel twice, Curie devoted most of the money to her scientific studies. During World War I, Curie volunteered at the National Aid Society, then brought her technology to the war front and instructed army medical personnel in the practical applications of radiology. With the installation of radiological equipment in ambulances, for instance, wounded soldiers would not have to be transported far to be x-rayed. When the war ended, Curie returned to research and devoted much of her time to her work.

By the 1920s, Curie was an international figure; the Curie Foundation had been established in 1920 to accept private donations for research, and two years later the scientist was invited to participate on the League of Nations International Commission for Intellectual Cooperation. Her health was failing, however, and she was troubled by fatigue and cataracts. Despite her discomfort, Curie made a highly publicized tour of the United States in 1921. The previous year, she had met Missy Meloney, editor of the *Delineator*, a woman's magazine. Horrified at the conditions in which Curie lived and worked (the Curies had made no money from their process for producing radium, having refused to patent it), Meloney proposed that a national subscription be held to finance a gram of radium for the institute to use in research. The tour proved grueling for Curie; by the end of her stay in New York, she had her right arm in a sling, the result of too many too strong handshakes. However, with Meloney's assistance, Curie left America with a valuable gram of radium.

Curie continued her work in the laboratory throughout the decade, joined by her daughter, Irene Joliot-Curie, who was pursuing a doctoral degree just as her mother had done. In 1925, Irene successfully defended her doctoral thesis on alpha rays of polonium, although Curie did not attend the defense lest her presence detract from her daughter's performance. Meanwhile, Curie's health still continued to fail and she was forced to spend more time away from her work in the laboratory. The result of prolonged exposure to radium, Curie contracted leukemia and died in 1934, in a nursing home in the French Alps. She was buried next to Pierre Curie in Sceaux, France.

See also History of exploration III (Modern era)

CURIE, PIERRE (1859-1906)

French physicist

Pierre Curie was a physicist who became famous for his collaboration with his wife **Marie Curie** in the study of **radioactivity**. Before joining his wife in her research, Pierre Curie was already widely known and respected in the world of **physics**. He discovered (with his brother Jacques) the phenomenon of piezoelectricity—in which a crystal can become electrically polarized—and invented the **quartz** balance. His papers on crystal symmetry and his findings on the relation between **magnetism** and **temperature** also earned praise in the scientific community.

Pierre Curie was born in Paris, the son of Sophie-Claire Depouilly, daughter of a formerly prominent manufacturer, and Eugène Curie, a free-thinking physician who was also a physician's son. Dr. Curie supported the family with his modest medical practice while pursuing his love for the natural sciences on the side. He was also an idealist and an ardent republican who set up a hospital for the wounded during the Commune of 1871. Pierre was a dreamer whose style of learning was not well adapted to formal schooling. He received his

Pierre Curie. *Library of Congress.*

pre-university education entirely at home, taught first by his mother and then by his father as well as his older brother, Jacques. He especially enjoyed excursions into the countryside to observe and study plants and animals, developing a love of nature that endured throughout his life and that provided his only recreation and relief from work during his later scientific career. At the age of 14, Curie studied with a mathematics professor who helped him develop his gift in the subject, especially spatial concepts. Curie's knowledge of physics and mathematics earned him his Bachelor of Science degree in 1875 at the age of 16. He then enrolled in the Faculty of Sciences at the Sorbonne in Paris and earned his *licence* (the equivalent of a master's degree) in physical sciences in 1877.

Curie became a laboratory assistant to Paul Desains at the Sorbonne in 1878, in charge of the physics students' lab work. His brother Jacques was working in the **mineralogy** laboratory at the Sorbonne at that time, and the two began a productive five-year scientific collaboration. They investigated pyroelectricity, the acquisition of electric charges by different faces of certain types of **crystals** when heated. Led by their knowledge of symmetry in crystals, the brothers experimentally discovered the previously unknown phenomenon of piezoelectricity, an electric polarization caused by force applied to the crystal. In 1880, the Curies published the first in a series of papers about their discovery. They then studied the opposite effect—the compression of a piezoelectric crystal by an electric field. In order to measure the very small amounts of **electricity** involved, the brothers invented a new laboratory instrument: a piezoelectric quartz electrometer, or balance. This device became very useful for electrical researchers and would prove highly valuable to Marie Curie in her studies of radioactivity. Much later, piezoelectricity had important practical applications. Paul Langevin, a student of Pierre Curie's, found that inverse piezoelectricity causes piezoelectric quartz in alternating fields to emit high-frequency sound waves, which were used to detect submarines and explore the ocean's floor. Piezoelectric crystals were also used in radio broadcasting and stereo equipment.

In 1882, Pierre Curie was appointed head of the laboratory at Paris' new Municipal School of Industrial Physics and Chemistry, a poorly paid position; he remained at the school for 22 years, until 1904. In 1883, Jacques Curie left Paris to become a lecturer in mineralogy at the University of Montpelier, and the brothers' collaboration ended. After Jacques's departure, Pierre delved into theoretical and experimental research on crystal symmetry, although the time available to him for such work was limited by the demands of organizing the school's laboratory from scratch and directing the laboratory work of up to 30 students, with only one assistant. He began publishing works on crystal symmetry in 1884, including in 1885, a theory on the formation of crystals and in 1894, an enunciation of the general principle of symmetry. Curie's writings on symmetry were of fundamental importance to later crystallographers, and, as Marie Curie later wrote in *Pierre Curie*, "he always retained a passionate interest in the physics of crystals" even though he turned his attention to other areas.

From 1890 to 1895, Pierre Curie performed a series of investigations that formed the basis of his doctoral thesis: a study of the magnetic properties of substances at different temperatures. He was, as always, hampered in his work by his obligations to his students, by the lack of funds to support his experiments, and by the lack of a laboratory or even a room for his own personal use. His **magnetism** research was conducted mostly in a corridor. In spite of these limitations, Curie's work on magnetism, like his papers on symmetry, was of fundamental importance. His expression of the results of his findings about the relation between temperature and magnetization became known as Curie's Law, and the temperature above which magnetic properties disappear is called the Curie point. Curie successfully defended his thesis before the Faculty of Sciences at the University of Paris (the Sorbonne) in March 1895, thus earning his doctorate. Also during this period, he constructed a periodic precision balance, with direct reading, that was a great advance over older balance systems and was especially valuable for chemical analysis. Curie was now becoming well known among physicists; he attracted the attention and esteem of, among others, the noted Scottish mathematician and physicist William Thomson (Lord Kelvin). It was partly due to Kelvin's influence that Curie was named to a newly created chair of physics at the School of Physics and Chemistry, which improved his status somewhat but still did not bring him a laboratory.

In the spring of 1894, at the age of 35, Curie met Maria (later Marie) Sklodowska, a poor young Polish student who had received her *licence* in physics from the Sorbonne and was then studying for her *licence* in mathematics. They immediately formed a rapport, and Curie soon proposed marriage. Sklodowska returned to Poland that summer, not certain that she would be willing to separate herself permanently from her family and her country. Curie's persuasive correspondence

convinced her to return to Paris that autumn, and the couple married in July 1895, in a simple civil ceremony. Marie used a wedding gift of cash to purchase two bicycles, which took the newlyweds on their honeymoon in the French countryside and provided their main source of recreation for years to come. Their daughter Irene was born in 1897, and a few days later Pierre's mother died; Dr. Curie then came to live with the young couple and helped care for his granddaughter.

The Curies' attention was caught by Henri Becquerel's discovery in 1896 that uranium compounds emit rays. Marie decided to make a study of this phenomenon the subject of her doctor's thesis, and Pierre secured the use of a ground-floor storeroom/machine shop at the School for her laboratory work. Using the Curie brothers' piezoelectric quartz electrometer, Marie tested all the elements then known to see if any of them, like uranium, emitted "Becquerel rays," which she christened "radioactivity." Only thorium and uranium and their compounds, she found, were radioactive. She was startled to discover that the ores pitchblende and chalcolite had much greater levels of radioactivity than the amounts of uranium and thorium they contained could account for. She guessed that a new, highly radioactive element must be responsible and, as she wrote in *Pierre Curie*, was seized with "a passionate desire to verify this hypothesis as rapidly as possible."

Pierre Curie too saw the significance of his wife's findings and set aside his much-loved work on crystals (only for the time being, he thought) to join Marie in the search for the new element. They devised a new method of chemical research, progressively separating pitchblende by chemical analysis and then measuring the radioactivity of the separate constituents. In July 1898, in a joint paper, they announced their discovery of a new element they named polonium, in honor of Marie Curie's native country. In December 1898, they announced, in a paper issued with their collaborator G. Bémont, the discovery of another new element, radium. Both elements were more radioactive than uranium or thorium.

The Curies had discovered radium and polonium, but in order to prove the existence of these new substances chemically, they had to isolate the elements so the atomic weight of each could be determined. This was a daunting task, as they would have to process two tons of pitchblende ore to obtain a few centigrams of pure radium. Their laboratory facilities were woefully inadequate: an abandoned wooden shed in the School's yard, with no hoods to carry off the poisonous gases their work produced. They found the pitchblende at a reasonable price in the form of waste from a uranium mine run by the Austrian government. The Curies now divided their labor. Marie acted as the chemist, performing the physically arduous job of chemically separating the pitchblende; the bulkiest part of this work she did in the yard adjoining the shed/laboratory. Pierre was the physicist, analyzing the physical properties of the substances that Marie's separations produced. In 1902, the Curies announced that they had succeeded in preparing a decigram of pure radium chloride and had made an initial determination of radium's atomic weight. They had proven the chemical individuality of radium.

The Curies' research also yielded a wealth of information about radioactivity, which they shared with the world in a series of papers published between 1898 and 1904. They announced their discovery of induced radioactivity in 1899. They wrote about the luminous and chemical effects of radioactive rays and their electric charge. Pierre studied the action of a **magnetic field** on radium rays, he investigated the persistence of induced radioactivity, and he developed a standard for measuring time on the basis of radioactivity, an important basis for geologic and archaeological dating techniques. Pierre Curie also used himself as a human guinea pig, deliberately exposing his arm to radium for several hours and recording the progressive, slowly healing burn that resulted. He collaborated with physicians in animal experiments that led to the use of radium therapy—often called "Curie-therapie" then—to treat cancer and lupus. In 1904, he published a paper on the liberation of heat by radium salts.

Through all this intensive research, the Curies struggled to keep up with their teaching, household, and financial obligations. Pierre Curie was a kind, gentle, and reserved man, entirely devoted to his work—science conducted purely for the sake of science. He rejected honorary distinctions; in 1903, he declined the prestigious decoration of the Legion of Honor. He also, with his wife's agreement, refused to patent their radium-preparation process, which formed the basis of the lucrative radium industry; instead, they shared all their information about the process with whoever asked for it. Curie found it almost impossible to advance professionally within the French university system; seeking a position was an "ugly necessity" and "demoralizing" for him (*Pierre Curie*), so posts he might have been considered for went instead to others. He was turned down for the Chair of Physical Chemistry at the Sorbonne in 1898; instead, he was appointed assistant professor at the Polytechnic School in March 1900, an inferior position.

Appreciated outside France, Curie received an excellent offer of a professorship at the University of Geneva in the spring of 1900, but he turned it down so as not to interrupt his research on radium. Shortly afterward, Curie was appointed to a physics chair at the Sorbonne, thanks to the efforts of Jules Henri Poincaré. Still, he did not have a laboratory, and his teaching load was now doubled, as he still held his post at the School of Physics and Chemistry. He began to suffer from extreme fatigue and sharp pains through his body, which he and his wife attributed to overwork, although the symptoms were almost certainly a sign of radiation poisoning, an unrecognized illness at that time. In 1902, Curie's candidacy for election to the French Academy of Sciences failed, and in 1903, his application for the chair of mineralogy at the Sorbonne was rejected, both of which added to his bitterness toward the French academic establishment.

Recognition at home finally came for Curie because of international awards. In 1903, London's Royal Society conferred the Davy medal on the Curies, and shortly thereafter they were awarded the 1903 Nobel Prize in physics—along with Becquerel—for their work on radioactivity. Curie presciently concluded his Nobel lecture (delivered in 1905 because the Curies were unable to attend the 1903 award ceremony) by wondering whether the knowledge of radium and radioactivity would be harmful for humanity. He added that he himself felt that more good than harm would result from the

new discoveries. The Nobel award changed the Curies' reclusive work-absorbed life. They were inundated by journalists, photographers, curiosity-seekers, eminent and little known visitors, correspondence, and requests for articles and lectures. Still, the cash from the award was a relief, and the award's prestige finally prompted the French parliament to create a new professorship for Curie at the Sorbonne in 1904. Curie declared he would remain at the School of Physics unless the new chair included a fully funded laboratory, complete with assistants. His demand was met, and Marie was named his laboratory chief. Late in 1904, the Curies' second daughter was born. By early 1906, Pierre Curie was poised to begin work—at last and for the first time—in an adequate laboratory, although he was increasingly ill and tired. On April 19, 1906, leaving a lunchtime meeting in Paris with colleagues from the Sorbonne, Curie slipped in front of a horse-drawn cart while crossing a rain-slicked rue Dauphine. He was killed instantly when the rear wheel of the cart crushed his skull. True to the way he had conducted his life, he was interred in a small suburban cemetery in a simple, private ceremony attended only by his family and a few close friends. In his memory, the Faculty of Sciences at the Sorbonne appointed Curie's widow Marie to his chair.

CUTBANKS • *see* CHANNEL PATTERNS

CUVIER, GEORGES (1769-1832)
French naturalist

Georges Léopold Chrétien Frédéric Dagobert, Baron Cuvier was a French naturalist who is known as the founder of the field of paleontology, as well as the founder of comparative anatomy.

Cuvier was born in Montbeliard, near Basel. Although a French town, Cuvier's birthplace at that time belonged to the Duchy of Wërttemberg. Cuvier was an academically inclined young man and, because his family lived in near-poverty, he accepted the offer to study for free at the Karlsschule in Stuttgart, Germany. He graduated at eighteen, returned home, and then found employment as a tutor in Normandy. While working in Normandy, he familiarized himself with the marine creatures he found on the beach, which he dissected and drew in detail. While doing so, he referred to Aristotle's ideas of comparing different animal structures, Carl Linnaeus's *System of Nature* and Buffon's *Natural History of Animals*. His impressive marine animal drawings came to the attention of Geoffroy Saint-Hilaire, and eventually led to Cuvier's appointment as assistant professor of comparative anatomy at the Museum of Natural History in Paris. Under Napoleon's regime, Cuvier became inspector General in the Department of Education and contributed to significant education reform in France. After Napoleon's fall, Cuvier retained his position and became an accepted authority in science and education, and earned several promotions which include a professorship at the Collège de France and permanent secretary for the

Academy of Sciences. Cuvier died in 1832 of cholera, during the first major epidemic of that disease in **Europe**.

Prior to Cuvier, anatomists such as Louis Daubenton, Johann Friedrich Blumenbach, and Petrus Camper posited the human being as the fundamental form to which all other living creatures were compared. Cuvier, however, decided to create an objective system of comparative anatomy based on observation. His initial field of research was marine animals, particularly mollusks, worms, and various fishes. Later, he extended his investigations to vertebrates in general. The conceptual framework of Cuvier's research was a systematic method of comparative anatomy. According to Cuvier, living beings exhibit certain distinctive anatomic features that enable the scientist to place an individual specimen in the larger context of a general anatomic system. For example, one can make significant generalizations by observing individual features such as dental structure, foot structure, skull shape, etc. Cuvier's comparative research, which expanded from the study of vertebrates to include the entire animal kingdom, was presented in his work *The Animal Kingdom, Distributed According to Its Organization* (1817). While Cuvier's work did not contribute any new facts to the science of anatomy, his method earned him high praise and esteem in the scientific community.

An important element of Cuvier's methodology is his **correlation** theory, which posits the functional interdependence of particular organs within an in individual organism. For example, as Cuvier observed, carnivorous animals possess certain distinctive features that clearly separate them from, say, herbivores. These features include sharp teeth, a certain jaw structure, a digestive system adapted to meat, acute eyesight, sharp claws, powerful and swift locomotion, etc.

In Paris, which is in a calcareous **area**, Cuvier applied his comparative method to study **fossils**. In his carefully organized excavations, particular attention was paid to the specific location, position, and placement of the discovered fossils. In addition, using his correlation theory, he developed a reconstruction method that enabled researchers to identify incomplete skeletons. Furthermore, in order to validate a particular hypothesis concerning the identity of an incomplete skeleton, Cuvier would compare the extinct animal to its closest living relative, in an effort to complete the puzzle. These investigations were described in his seminal *Investigations on Fossil Bones* (1812), establishing Cuvier as the founder of modern paleontology. Using his comparative method, with particular emphasis on dentition and bone structure Cuvier was able to demonstrate that the two types of elephant, Indian and African, classified as examples of one species, in reality constituted two distinct species. In fact, Cuvier found that the extinct mammoth is closer to the Indian elephant than the two existing elephant species are to each other. Extending his research on elephants to Pachydermata in general, Cuvier studied both existing and extinct forms, identifying several new genera, including Palaeotherium and Dinotherium. In addition, he provided the first scientific description of the American giant sloth and named the pterodactyl.

Cuvier, like many of his colleagues, puzzled over the seemingly mysterious fact that animal forms changed through history. However, unlike some his colleagues, who approached the issue with extreme circumspection, Cuvier decided that species do not change. "The immutability of species," wrote Nordenskiöld, "is to Cuvier's mind an absolute fact." In order to explain why certain species were extinct and why fossils of some extinct creatures were unrecognizable from modern creatures, Cuvier invoked the **catastrophism** theory, which posits that a "new" species appear after the extinction, due to a violent upheaval (such as an **earthquake**) of its "old" counterpart. Thus, for example, Cuvier denied the existence of human fossils, asserting that, for example, lion fossils and lions in their present form represent two distinct species. Realizing the absurdity of the idea that species emerged out of nothing following a catastrophe, Cuvier attempted the explain the continuity of life by positing a type of near-extinction, which would allow the survival of small populations of a particular species, positing, as Cassirer has remarked, an **evolution** by analogy, whereby a particular species would be replaced by its new analogue, which to his mind seemed more reasonable than the notion of gradual evolution.

Cuvier's views of classification and evolution were vigorously opposed by several of his prominent contemporaries, who found his systematic philosophy, particularly his adamant insistence of four ground-plans, dogmatic. For example, Geoffroy Saint-Hilaire, who engaged Cuvier in a lengthy polemic, maintained that, because life manifests itself on the basis of a fundamental, indivisible impulse, Cuvier's claim that creatures emerging from different ground plans that cannot be compared does not reflect the true nature of the animal world. Accused by his critics for speculative dogmatism, Cuvier nevertheless, as Cassirer has written, defended his views on the basis of empirical research. As scholars have observed, the polemic between Cuvier and Saint-Hilaire was never resolved owing to the both antagonist defended points of view, which, while seemingly opposed, contributed, as complementary views, to the progress of life sciences and Earth sciences.

See also Evolution, evidence of; Evolutionary mechanisms; Fossil record; Fossils and fossilization

CYCLOSILICATES

The most abundant rock-forming **minerals** in the **crust** of the Earth are the silicates. They are formed primarily of **silicon** and **oxygen**, together with various **metals**. The fundamental unit of these minerals is the silicon-oxygen tetrahedron. These tetrahedra have a pyramidal shape, with a relatively small, positively charged silicon cation (Si^{+4}) in the center and four larger, negatively charged oxygen anions (O^{-2}) at the corners, producing a net charge of –4. **Aluminum** cations (Al^{+3}) may substitute for silicon, and various anions such as hydroxyl (OH^-) or fluorine (F^-) may substitute for oxygen. In order to form stable minerals, the charges that exist between tetrahedra must be neutralized. This can be accomplished by the sharing of oxygen cations between tetrahedra, or by the binding together adjacent tetrahedra with various metal cations. This in turn creates characteristic silicate structures that can be used to classify silicate minerals into cyclosilicates, **inosilicates**, **nesosilicates**, **phyllosilicates**, **sorosilicates**, and **tectosilicates**.

In cyclosilicates, the tetrahedra form rings of 3, 4, or 6 tetrahedra. However, most cyclosilicates are formed from a framework of six tetrahedra, giving them the formula $(Si_6O_{18})^{-12}$, and examples of this type of mineral includes the **gemstones** beryl (including the varieties emerald and aquamarine, and tourmaline. Beryl, found in **granite**, pegmatites, and mica **schist**, has the chemical formula $Be_3Al_2(Si_6O_{18}$.

Deposits of beryl that are not of gem quality are an important ore of the metal beryllium. Minerals in the tourmaline group, found in granite pegmatites and as accessory minerals in igneous and metamorphic rocks, can also be present in veins and as a detrital mineral in sediments and sedimentary **rock**. Varieties of tourmaline include schorlite, also called black tourmaline or **iron** tourmaline, $NaFe_3B_3Al_3(OH)_4(Al_3Si_6O_{27})$; dravite, also called brown tourmaline or magnesium tourmaline, $NaMg_3B_3Al_3(OH)_4(Al_3Si_6O_{27})$, and elbaite, a lighter-colored, lithium-bearing variety also called alkali tourmaline, $Na_2Li_3B_6Al_9(OH)_8(Al_3Si_6O_{27})_2$.

D

DARCEY, HENRY PHILIBERT GASPARD
(1803-1858)
French engineer

Henry Philibert Gaspard Darcy (1803–1858) was an accomplished French engineer, researcher, and civil servant, who is credited with building roads, **water** systems, and railroads. He is best known for a number of major contributions and advances in the understanding of fluid flow in pipes, open channels, and porous media. Darcy was the first researcher to suspect the existence of the boundary layer in fluid flow. His work contributed in the development of the Darcy-Weisbach equation, recognized as the best empirical relation for pipe flow resistance. Darcy also made major contributions to open channel flow research and provided the first quantitative measurements of **artesian** well flow. An enduring legacy of his work, Darcy's Law for flow in porous media, a cornerstone for several fields of study including ground-water hydrology, **soil physics**, and **petroleum** engineering remains widely used.

Darcy was born on June 10, 1803 in Dijon, France, the son of a civil servant. In 1821, he entered L'Ecole Polytechnique in Paris. In 1823 he was admitted to the prestigious L'Ecole des Ponts et Chaussee's (School of Bridges and Roads) in Paris, where he graduated with a degree in Civil Engineering in 1826. After graduation, he took a position with the Corps of Bridges and Roads in Dijon. There, Darcy began work on a project to develop a system for a safe and adequate water supply for the city. That effort eventually resulted in completion in 1844 of a model, completely enclosed, **gravity** driven system that provided water to major buildings and street hydrants throughout the city. Darcy's work was so advanced that it was 20 years before Paris had similar service. Despite his impressive achievements, Darcy refused to accept payment for his work as designer and manager of the water system project. The amount of money Darcy refused would translate to over a million dollars at modern exchange rates.

Darcy completed numerous other civil works in and near Dijon including roadways, bridges, sewers, and a railroad, passing through Dijon, linking Paris with Lyon, the largest industrial city to the south. His design for the railroad included a 2.5-mile-long (4 km) tunnel through the mountains at a time when tunnels of any significant length were considered unacceptable. However, Darcy had enlisted the aid of a geologist and mining engineer for a detailed survey of the site that indicated ideal conditions for constructing a tunnel. He had conceived the best engineering solution by rejecting an old generalization (tunnel length) and analyzed the problem on a site-specific basis. Because of Darcy's plan, the railroad passed through Dijon, ensuring the city's economic future. In 1844, he was awarded the Legion of Honor.

The St. Pierre Basin Fountain was considered a technological marvel for the time. Darcy designed separate valves for controlling an inner and outer ring of jets that allowed variations in the height of the fountain display. In an 1856 report, Darcy provided a theoretical analysis and experimental verification of the jet flow in this artesian system as a function of the height of the water system's two source reservoirs.

In 1848, Darcy was appointed Chief Director for Water and Pavements in Paris. There he began his systematic study of turbulent flow in pipes. He made significant advances in the design of the Pitot tube, used to measure the flow velocity in pipes. Those improvements made possible accurate, detailed measurements of point velocity distributions in pipes, leading to advances in pipe hydraulics and to his recognition of the existence of the boundary layer.

In 1855, Darcy returned to Dijon to carry out his famous **sand** column experiments that ultimately resulted in Darcy's law for flow in porous media. For one-dimensional flow, the law relates the volumetric flow rate to the cross-sectional **area** of a tube or column to the drop in hydraulic head over the length of the flow path, and introduces a proportionality constant for hydraulic conductivity. Subsequent to his work, engineers and scientists have demonstrated the theoretical basis and applicability of Darcy's law in several fields. The law has since been generalized to allow for differential solutions, vector analysis, and unsaturated and multiphase flow.

Darcy died unexpectedly of pneumonia, on January 3, 1858. He is buried in Dijon.

See also Hydrogeology; Hydrostatic pressure

DARWIN, CHARLES ROBERT (1809-1882)

English naturalist

Charles Robert Darwin is credited with popularizing the concept of organic **evolution** by means of natural selection. Though Darwin was not the first naturalist to propose a model of biological evolution, his introduction of the mechanism of the "survival of the fittest" and discussion of the evolution of humans, marked a revolution in both science and natural philosophy.

Darwin was born in Shrewsbury, England and showed an early interest in the natural sciences, especially **geology**. His father, Robert Darwin, a wealthy physician, encouraged Charles to pursue studies in medicine at the University of Edinburg. Darwin soon tired of the subject, and his father sent him to Cambridge to prepare for a career in the clergy. At Cambridge, Darwin rekindled his passion for the natural sciences, often devoting more time to socializing with Cambridge scientists than to his clerical studies. With guidance from his cousin, entomologist William Darwin Fox (1805–1880), Darwin became increasingly involved in the growing circle of natural scientists at Cambridge. Fox introduced Darwin to clergyman and biologist John Stevens Henslow (1796–1861). Henslow became Darwin's tutor in mathematics and theology, as well as his mentor in his personal studies of botany, geology, and zoology. Henslow profoundly influenced Darwin, and it was he who encouraged Darwin to delay seeking an appointment in the Church of England in favor of joining an expedition team and venturing overseas. After graduation, Darwin agreed to an unpaid position as naturalist aboard the H.M.S. *Beagle*. The expedition team was initially chartered for a three year voyage and survey of South America's Pacific coastline, but the ship pursued other ventures after their work was complete and Darwin remained part of H.M.S. *Beagle's* crew for five years.

Darwin used his years aboard *The Beagle* to further his study of the natural sciences. In **South America**, Darwin became fascinated with geology. He paid close attention to changes in the land brought about by earthquakes and volcanoes. His observations led him to reject **catastrophism** (a theory that land forms are the result of single, catastrophic events), and instead espoused the geological theories of gradual development proposed by English geologist **Charles Lyell** (1797–1875) in his 1830 work, *Principles of Geology*. Yet, some of his observations in South America did not fit with Lyell's theories. Darwin disagreed with Lyell's assertion that **coral reefs** grew atop oceanic volcanoes and rises, and concluded that coral reefs built upon themselves. When Darwin returned to England in 1836, he and Lyell became good

Charles Darwin. *Library of Congress.*

friends. Lyell welcomed Darwin's new research on coral reefs, and encouraged him to publish other studies from his voyages.

Darwin was elected a fellow of the Geological Society in 1836, and became a member of the Royal Society in 1839. That same year, he published his *Journal of Researches into the Geology and Natural History of the Various Countries Visited by H.M.S. Beagle*. Though his achievements in geology largely prompted his welcoming into Britain's scientific community, his research interests began to diverge from the discipline in the early 1840s. Discussions with other naturalists prompted Darwin's increasing interest in population diversity of fauna, extinct animals, and the presumed static nature of species. Again, he turned to notes of his observations and various specimens he gathered while on his prior expedition. The focus of his new studies was the Galápagos Islands off the Pacific coast of Ecuador. While there, Darwin was struck by the uniqueness of the island's tortoises and birds. Some neighboring islands had animal populations that were largely similar to that of the continent, while others had seemingly different variety of species. After analyzing finch specimens from the Galapagos, Darwin concluded that species must have some means of transmutation, or ability of a species to alter over time. Darwin thus proposed that as species modified, and as old species disappeared, new varieties could be introduced. Thus, Darwin proposed an evolutionary model of animal populations.

The idea of organic evolution was not novel. French naturalist, Georges Buffon (1707–1788) had theorized that species were prone to development and change. Darwin's own grandfather, Erasmus Darwin, also published research regarding the evolution of species. Although the theoretical concept of evolution was not new, it remained undeveloped prior to Charles Darwin. Just as he had done with Lyell's geological theory, Darwin set about to further the understanding of evolution not merely as a philosophical concept, but as a practical scientific model for explaining the diversity of species and populations. His major contribution to the field was the introduction of a mechanism by which evolution was accomplished. Darwin believed that evolution was the product of an ongoing struggle of species to better adapt to their environment, with those that were best adapted surviving to reproduce and replace less-suited individuals. He called this phenomenon "survival of the fittest," or natural selection. In this way, Darwin believed that traits of maximum adaptability were transferred to future generations of the animal population, eventually resulting in new species.

Darwin finished an extensive draft of his theories in 1844, but lacked confidence in his abilities to convince others of the merits of his discoveries. Years later, prompted by rumors that a colleague was about to publish a theory similar to his own, Darwin decided to release his research. *On the Origin of Species by Means of Natural Selection, or The Preservation of Favoured Races in the Struggle for Life* was published November 1859, and became an instant bestseller.

A common misconception is that *On the Origin of Species* was the introduction of the concept of human evolution. In fact, a discussion of human antiquity is relatively absent from the book. Darwin did not directly address the relationship between animal and human evolution until he published *The Descent of Man, and Selection in Relation to Sex* in 1871. Darwin introduced not only a model for the biological evolution of man, but also attempted to chart the process of man's psychological evolution. He further tried to break down the barriers between man and animals in 1872, with his work *The Expression of the Emotions in Man and Animals.* By observing facial features and voice sounds, Darwin asserted that man and non-human animals exhibited signs of emotion in similar ways. In the last years of his career, Darwin took the concept of organic evolution to its logical end by applying natural selection and specialization to the plant kingdom.

Darwin's works on evolution met with both debate from the scientific societies, and criticism from some members of the clergy. *On the Origin of Species* and *The Descent of Man* were both published at a time of heightened religious evangelicalism in England. Though willing to discuss his theories with colleagues in the sciences, Darwin refrained from participating in public debates concerning his research. In the last decade of his life, Darwin was disturbed about the application of his evolutionary models to social theory. By most accounts, he considered the emerging concept of the social and cultural evolution of men and civilizations, which later became known as Social Darwinism, to be a grievous misinterpretation of his works. Regardless of his opposition, he remained publicly taciturn about the impact his scientific theories on theology, scientific

methodology, and social theory. Closely guarding his privacy, Darwin retired to his estate in Down. He died at Down House in 1882. Though his wishes were to receive an informal burial, Parliament immediately ordered a state burial for the famous naturalist at Westminster Abby. By the time of his death, the scientific community had largely accepted the arguments favoring his theories of evolution. Although the later discoveries in genetics and molecular biology radically refined and reinterpreted Darwin's evolutionary mechanisms, evolutionary theory is the key and unifying theory in all biological science.

See also Evolution, evidence of; Evolutionary mechanisms

DATING METHODS

Dating techniques are procedures used by scientists to determine the age of a specimen. Relative dating methods tell only if one sample is older or younger than another sample; absolute dating methods provide a date in years. The latter have generally been available only since 1947. Many absolute dating techniques take advantage of radioactive decay, whereby a radioactive form of an element is converted into another radioactive isotope or non-radioactive product at a regular rate. Others, such as amino acid racimization and cation-ratio dating, are based on chemical changes in the organic or inorganic composition of a sample. In recent years, a few of these methods have undergone continual refinement as scientists strive to develop the most accurate dating techniques possible.

Relative dating methods determine whether one sample is older or younger than another. They do not provide an age in years. Before the advent of absolute dating methods, nearly all dating was relative. The main relative dating method is **stratigraphy**.

Stratigraphy is the study of layers of rocks or the objects embedded within those layers. It is based on the assumption (which, except at **unconformities**, nearly always holds true) that deeper layers were deposited earlier, and thus are older than more shallow layers. The sequential layers of **rock** represent sequential intervals of time. Although these units may be sequential, they are not necessarily continuous due to erosional removal of some intervening units. The smallest of these rock units that can be matched to a specific time interval is called a bed. Beds that are related are grouped together into members, and members are grouped into formations.

Seriation is the ordering of objects according to their age. It is a relative dating method. In a landmark study, archaeologist James Ford used seriation to determine the chronological order of American Indian pottery styles in the Mississippi Valley. Artifact styles such as pottery types are seriated by analyzing their abundances through time. This is done by counting the number of pieces of each style of the artifact in each stratigraphic layer and then graphing the data. A layer with many pieces of a particular style will be represented by a wide band on the graph, and a layer with only a few pieces will be represented by a narrow band. The bands are arranged into battleship-shaped curves, with each style getting its own curve.

The curves are then compared with one another, and from this the relative ages of the styles are determined. A limitation to this method is that it assumes all differences in artifact styles are the result of different periods of time, and are not due to the immigration of new cultures into the **area** of study.

The term faunal dating refers to the use of animal bones to determine the age of sedimentary layers or objects such as cultural artifacts embedded within those layers. Scientists can determine an approximate age for a layer by examining which species or genera of animals are buried in it. The technique works best if the animals belonged to species that evolved quickly, expanded rapidly over a large area, or suffered a mass extinction. In addition to providing rough absolute dates for specimens buried in the same stratigraphic unit as the bones, faunal analysis can also provide relative ages for objects buried above or below the fauna-encasing layers.

Each year seed-bearing plants release large numbers of pollen grains. This process results in a "rain" of pollen that falls over many types of environments. Pollen that ends up in lakebeds or peat bogs is the most likely to be preserved, but pollen may also become fossilized in arid conditions if the **soil** is acidic or cool. Scientists can develop a pollen chronology, or calendar, by noting which species of pollen were deposited earlier in time, that is, residue in deeper sediment or rock layers, than others. A pollen zone is a period of time in which a particular species is much more abundant than any other species of the time. In most cases, this also reveals much about the **climate** of the period, because most plants only thrive in specific climatic conditions. Changes in pollen zones can also indicate changes in human activities such as massive deforestation or new types of farming. Pastures for grazing livestock are distinguishable from fields of grain, so changes in the use of the land over time are recorded in the pollen history. The dates when areas of **North America** were first settled by immigrants can be determined to within a few years by looking for the introduction of ragweed pollen.

Pollen zones are translated into absolute dates by the use of radiocarbon dating. In addition, pollen dating provides relative dates beyond the limits of radiocarbon (40,000 years), and can be used in some places where radiocarbon dates are unobtainable.

Fluorine is found naturally in ground **water**. This water comes in contact with skeletal remains under ground. When this occurs, the fluorine in the water saturates the bone, changing the mineral composition. Over time, more and more fluorine incorporates itself into the bone. By comparing the relative amounts of fluorine composition of skeletal remains, one can determine whether the remains were buried at the same time. A bone with a higher fluorine composition has been buried for a longer period of time.

Absolute dating is the term used to describe any dating technique that tells how old a specimen is in years. These are generally analytical methods, and are carried out in a laboratory. Absolute dates are also relative dates, in that they tell which specimens are older or younger than others. Absolute dates must agree with dates from other relative methods in order to be valid.

This dating technique of amino acid racimization was first conducted by Hare and Mitterer in 1967, and was popular in the 1970s. It requires a much smaller sample than radiocarbon dating, and has a longer range, extending up to a few hundred thousand years. It has been used to date coprolites (fossilized feces) as well as fossil bones and shells. These types of specimens contain proteins embedded in a network of **minerals** such as calcium.

Amino acid racimization is based on the principle that amino acids (except glycine, a very simple amino acid) exist in two mirror image forms called stereoisomers. Living organisms (with the exception of some microbes) synthesize and incorporate only the L-form into proteins. This means that the ratio of the D-form to the L-form is zero (D/L=0). When these organisms die, the L-amino acids are slowly converted into D-amino acids in a process called racimization. This occurs because protons (H+) are removed from the amino acids by acids or bases present in the burial environment. The protons are quickly replaced, but will return to either side of the amino acid, not necessarily to the side from which they came. This may form a D-amino acid instead of an L– amino acid. The reversible reaction eventually creates equal amounts of L– and D-forms (D/L=1.0).

The rate at which the reaction occurs is different for each amino acid; in addition, it depends upon the moisture, **temperature**, and **pH** of the postmortem conditions. The higher the temperature, the faster the reaction occurs, so the cooler the burial environment, the greater the dating range. The burial conditions are not always known, however, and can be difficult to estimate. For this reason, and because some of the amino acid racimization dates have disagreed with dates achieved by other methods, the technique is no longer widely used.

Cation-ratio dating is used to date rock surfaces such as stone artifacts and cliff and ground drawings. It can be used to obtain dates that would be unobtainable by more conventional methods such as radiocarbon dating. Scientists use cation-ratio dating to determine how long rock surfaces have been exposed. They do this by chemically analyzing the varnish that forms on these surfaces. The varnish contains cations, which are positively charged atoms or molecules. Different cations move throughout the environment at different rates, so the ratio of different cations to each other changes over time. Cation ratio dating relies on the principle that the cation ratio $(K^+ + Ca^{2+})/Ti^{4+}$ decreases with increasing age of a sample. By calibrating these ratios with dates obtained from rocks from a similar microenvironment, a minimum age for the varnish can be determined. This technique can only be applied to rocks from **desert** areas, where the varnish is most stable.

Although cation-ratio dating has been widely used, recent studies suggest it has potential errors. Many of the dates obtained with this method are inaccurate due to improper chemical analyses. In addition, the varnish may not actually be stable over long periods of time.

Thermoluminescence dating is very useful for determining the age of pottery. Electrons from **quartz** and other minerals in the pottery **clay** are bumped out of their normal positions (ground state) when the clay is exposed to radiation. This radiation may come from radioactive substances such as uranium,

Scientist using a spectrograph to determine the age of a sample. © Roger Ressmeyer/Corbis. Reproduced by permission.

present in the clay or burial medium, or from cosmic radiation. When the ceramic is heated to a very high temperature (over 932°F [500°C]), these electrons fall back to the ground state, emitting light in the process and resetting the "clock" to zero. The longer the radiation exposure, the more electrons get bumped into an excited state. With more electrons in an excited state, more light is emitted upon heating. The process of displacing electrons begins again after the object cools. Scientists can determine how many years have passed since a ceramic was fired by heating it in the laboratory and measuring how much light is given off. Thermoluminescence dating has the advantage of covering the time interval between radiocarbon and potassium-argon dating, or 40,000–200,000 years. In addition, it can be used to date materials that cannot be dated with these other two methods.

Optically stimulated luminescence (OSL) has only been used since 1984. It is very similar to thermoluminescence dating, both of which are considered "clock setting" techniques. Minerals found in sediments are sensitive to light. Electrons found in the sediment grains leave the ground state when exposed to light, called recombination. To determine the age of sediment, scientists expose grains to a known amount of light and compare these grains with the unknown sediment. This technique can be used to determine the age of unheated

sediments less than 500,000 years old. A disadvantage to this technique is that in order to get accurate results, the sediment to be tested cannot be exposed to light (which would reset the "clock"), making sampling difficult.

The absolute dating method utilizing tree ring growth is known as dendrochronology. It is based on the fact that trees produce one growth ring each year. Narrow rings grow in cold and/or dry years, and wide rings grow in warm years with plenty of moisture. The rings form a distinctive pattern, which is the same for all members in a given species and geographical area. The patterns from trees of different ages (including ancient wood) are overlapped, forming a master pattern that can be used to date timbers thousands of years old with a resolution of one year. Timbers can be used to date buildings and archaeological sites. In addition, tree rings are used to date changes in the climate such as sudden cool or dry periods. Dendrochronology has a range of one to 10,000 years or more.

As previously mentioned, radioactive decay refers to the process in which a radioactive form of an element is converted into a decay product at a regular rate. Radioactive decay dating is not a single method of absolute dating but instead a group of related methods for absolute dating of samples.

Potassium-argon dating relies on the fact that when volcanic rocks are heated to extremely high temperatures, they release any argon gas trapped in them. As the rocks cool, argon-40 (^{40}Ar) begins to accumulate. Argon-40 is formed in the rocks by the radioactive decay of potassium-40 (^{40}K). The amount of ^{40}Ar formed is proportional to the decay rate (**half-life**) of ^{40}K, which is 1.3 billion years. In other words, it takes 1.3 billions years for half of the ^{40}K originally present to be converted into ^{40}Ar. This method is generally only applicable to rocks greater than three million years old, although with sensitive instruments, rocks several hundred thousand years old may be dated. The reason such old material is required is that it takes a very long time to accumulate enough ^{40}Ar to be measured accurately. Potassium-argon dating has been used to date volcanic layers above and below **fossils** and artifacts in east **Africa**.

Radiocarbon dating is used to date charcoal, wood, and other biological materials. The range of conventional radiocarbon dating is 30,000–40,000 years, but with sensitive instrumentation, this range can be extended to 70,000 years. Radiocarbon (^{14}C) is a radioactive form of the element **carbon**. It decays spontaneously into nitrogen-14 (^{14}N). Plants get most of their carbon from the air in the form of **carbon dioxide**, and animals get most of their carbon from plants (or from animals that eat plants). Relative to their atmospheric proportions, atoms of ^{14}C and of a non-radioactive form of carbon, ^{12}C, are equally likely to be incorporated into living organisms. While a plant or animal is alive, the ratio of ^{14}C/^{12}C in its body will be nearly the same as the ^{14}C/^{12}C ratio in the atmosphere. When the organism dies, however, its body stops incorporating new carbon. The ratio will then begin to change as the ^{14}C in the dead organism decays into ^{14}N. The rate at which this process occurs is called the half-life. This is the time required for half of the ^{14}C to decay into ^{14}N. The half-life of ^{14}C is 5,730 years. Scientists can estimate how many years have elapsed since an organism died by comparing the ^{14}C/^{12}C ratio in the remains with the ratio in the atmosphere. This allows them to determine how much ^{14}C has formed since the death of the organism.

One of the most familiar applications of radioactive dating is determining the age of fossilized remains, such as dinosaur bones. Radioactive dating is also used to authenticate the age of rare archaeological artifacts. Because items such as paper documents and cotton garments are produced from plants, they can be dated using radiocarbon dating. Without radioactive dating, a clever forgery might be indistinguishable from a real artifact. There are some limitations, however, to the use of this technique. Samples that were heated or irradiated at some time may yield by radioactive dating an age less than the true age of the object. Because of this limitation, other dating techniques are often used along with radioactive dating to ensure accuracy.

Accurate radiocarbon dating is that diagenic (after death) demands consideration regarding potential contamination of a specimen and a proper application of changes in the ^{14}C/^{12}C ratio in the atmosphere over time. ^{14}C levels can be measured in tree rings and used to correct for the ^{14}C/^{12}C ratio in the atmosphere at the time the organism died, and can even be used to calibrate some dates directly. Although the magnitude of change of the ^{14}C/^{12}C ratio sometimes stirs controversy, with proper calibration and correction, radiocarbon dating correlates well with other dating techniques and con-

sistently proves to be an accurate dating technique—especially for Pleistocene and Holocene period analysis.

Uranium series dating techniques rely on the fact that radioactive uranium and thorium isotopes decay into a series of unstable, radioactive "daughter" isotopes; this process continues until a stable (non-radioactive) **lead** isotope is formed. The daughters have relatively short half-lives ranging from a few hundred thousand years down to only a few years. The "parent" isotopes have half-lives of several billion years. This provides a dating range for the different uranium series of a few thousand years to 500,000 years. Uranium series have been used to date uranium-rich rocks, deep-sea sediments, shells, bones, and teeth, and to calculate the ages of ancient lakebeds. The two types of uranium series dating techniques are daughter deficiency methods and daughter excess methods.

In daughter deficiency situations, the parent radioisotope is initially deposited by itself, without its daughter (the isotope into which it decays) present. Through time, the parent decays to the daughter until the two are in equilibrium (equal amounts of each). The age of the deposit may be determined by measuring how much of the daughter has formed, providing that neither isotope has entered or exited the deposit after its initial formation. Carbonates may be dated this way using, for example, the daughter/parent isotope pair protactinium-231/uranium-235 (^{231}Pa/^{235}U). Living mollusks and corals will only take up dissolved compounds such as isotopes of uranium, so they will contain no protactinium, which is insoluble. Protactinium-231 begins to accumulate via the decay of ^{235}U after the organism dies. Scientists can determine the age of the sample by measuring how much ^{231}Pa is present and calculating how long it would have taken that amount to form.

In the case of daughter excess, a larger amount of the daughter is initially deposited than the parent. Non-uranium daughters such as protactinium and thorium are insoluble, and precipitate out on the bottoms of bodies of water, forming daughter excesses in these sediments. Over time, the excess daughter disappears as it is converted back into the parent, and by measuring the extent to which this has occurred, scientists can date the sample. If the radioactive daughter is an isotope of uranium, it will dissolve in water, but to a different extent than the parent; the two are said to have different solubilities. For example, ^{234}U dissolves more readily in water than its parent, ^{238}U, so **lakes** and **oceans** contain an excess of this daughter isotope. This excess is transferred to organisms such as mollusks or corals, and is the basis of ^{234}U/^{238}U dating.

Some volcanic minerals and glasses, such as **obsidian**, contain uranium-238 (^{238}U). Over time, these substances become "scratched." The marks, called tracks, are the damage caused by the fission (splitting) of the uranium atoms. When an **atom** of ^{238}U splits, two "daughter" atoms rocket away from each other, leaving in their wake tracks in the material in which they are embedded. The rate at which this process occurs is proportional to the decay rate of ^{238}U. The decay rate is measured in terms of the half-life of the element, or the time it takes for half of the element to split into its daughter atoms. The half-life of ^{238}U is 4.47x10^9 years.

When the mineral or **glass** is heated, the tracks are erased in much the same way cut marks fade away from hard candy that is heated. This process sets the fission track clock to zero, and the number of tracks that then form are a measure of the amount of time that has passed since the heating event. Scientists are able to count the tracks in the sample with the aid of a powerful microscope. The sample must contain enough ^{238}U to create enough tracks to be counted, but not contain too much of the isotope, or there will be a jumble of tracks that cannot be distinguished for counting. One of the advantages of fission track dating is that it has an enormous dating range. Objects heated only a few decades ago may be dated if they contain relatively high levels of ^{238}U; conversely, some meteorites have been dated to over a billion years old with this method.

Although certain dating techniques are accurate only within certain age ranges, whenever possible, scientists attempt to use multiple methods to date specimens. **Correlation** of dates via different dating methods provides a highest degree of confidence in dating.

See also Evolution, evidence of; Fossil record; Fossils and fossilization; Geologic time; Historical geology

Davis, William Morris (1850-1934)
American geologist

William Morris Davis was a geographer, meteorologist, and geologist who devised a relative method of determining the age of a river system. Davis' method of landscape analysis considered the cyclical nature of **erosion** and the subsequent uplift of the surrounding land in order to determine the age of the river in relation to its surroundings.

Davis was born in Philadelphia, Pennsylvania into the city's social elite. His grandmother was Lucretia Mott, the famous abolitionist. William Morris Davis bore the name of his uncle, a congressman. Davis spent many of his childhood summers in the farmlands of Pennsylvania, which instilled in him a deep interest in natural history. This interest spurred Davis to study at the Lawrence Scientific School of Harvard University. After his graduation in 1869, he pursued a master's degree in mining engineering, also at Harvard. Davis embarked on a tour of the mining districts of the Lake Superior region with Raphael Pumpelly during the summer of 1869. Later in the same summer, Davis helped Josiah Dwight Whitney conduct fieldwork in the Rocky Mountains. In 1870, Davis accompanied one of his former teachers, Benjamin Gould, to Argentina for the purpose of organizing an astronomical observatory. Davis remained in Cordoba for two and a half years assisting Gould with the observatory and undertaking meteorological work. Davis then returned to Philadelphia after experiencing differences with Gould.

Davis later became an instructor at Harvard, but initially struggled to interest his students. He commenced a lifelong career in research and writing in 1882. In the 1880s, Davis received notice for his work in **geology** and **meteorology**, but was internationally known for his research in **physical geography**. Davis turned out many articles on the Triassic formation of the Connecticut River Valley and on meteorological topics. In 1889, Davis wrote a paper on "The Rivers and Valleys of Pennsylvania." In this paper, Davis introduced the cycle for river system formations that was reiterated in many more of his works. Davis believed that running **water** is the single most important agent in creating landscapes. At the beginning of his erosion cycle, rivers are small, shallow streams, the result of imperfect drainage of the surrounding **topography**. With time, the streams carve out deeper channels, widen, and contributing tributaries form. These tributaries, in turn, bring in more and more water creating larger, more powerful waterways. Waterfalls are caused by the contrast of hardness of the rocks as they are worn back. Side-streams then form their own valleys and the valley slopes increase as more **soil** is carried downstream. At maturity, the river has a system of headwater branches that gnaw at the uplands, which in turn widen the rivers. The surrounding mountains then slowly erode over time. This erosion deposits a large amount of sediment into the rivers. This causes the rivers to become increasingly sluggish and the tributaries dwindle as the flow of water slows. The cycle then begins again when another episode of uplift rejuvenates the river systems. Davis, who was influenced by the English scientist Charles Darwin's organic evolutionary theories, determined the relative age of the river system by discerning its place in the erosion cycle and thus, proposed a cyclical nature to the **evolution** of the landscape.

Davis published the textbook, *Elementary Meteorology* (1894), which was widely used in colleges. Other relevant scientific literature published by Davis includes: *Elementary Physical Geography* (1902), *Geographical Essays* (1909), *The Lesser Antilles* (1926), and *The Coral Reef Problem* (1928). In *The Coral Reef Problem*, Davis endorsed both James Dwight Dana's and Darwin's belief that barriers such as atolls and reefs are the result from the slow subsidence of the ocean bottom under the upward growing formations of islands.

Davis was appointed Sturgis-Hooper Professor of geology at Harvard in 1898. He retained this position until his retirement in 1912. Davis founded the Association of American Geographers in 1904. He also played a major role in the Geological Society of America. In 1928, Davis, a widower, married his third wife and settled in Pasadena, California where he peacefully lived out his final years until his death in 1934.

See also Landscape evolution

Debris flow

Debris flow is a process in which water-saturated masses of material ranging from **sand** grains to boulders move across low slopes. These flows range from gently flowing sand and **water** slurries to violently surging bouldery masses, and include events described as debris slides, debris torrents, mudflows, mudslides, earthflows, and lahars.

Observations have shown that debris flows often move in waves or surges, each wave consisting of a coarse-grained snout followed by a finer grained and more fluid tail. The consistency of flowing debris has been described as being similar to wet concrete, although water accounts for less than half of the debris flow volume. Debris flows typically have bulk densities almost identical to the water-saturated **regolith** or sediment from which they are derived. Clay- and silt-sized grains are generally very minor constituents.

Most debris flows begin as landslides or slumps. In order for a **landslide** to be mobilized into a debris flow, two conditions (in addition to initial landsliding) must be met. First, the debris mass must contain enough water to flow when agitated during sliding. Second, the gravitational potential energy possessed by the debris must be converted into enough internal kinetic energy to change the mode of movement from rigid block sliding to fluid flow throughout the debris mass. Landslides that mobilize into debris flows often occur along topographic concavities or hollows, which concentrate **groundwater** flow and contain thicker accumulations of regolith than surrounding ridges. Concentrated groundwater flow increases the wetness of regolith in hollows, making it particularly susceptible to destabilizing groundwater pressure increases during and immediately after rainstorms. Debris stops flowing when the internal kinetic energy drops below the level necessary to maintain fluid flow, commonly because the channel through which the debris flows flattens or widens.

Debris flows are also common after intense wildfires. Fires that spawn debris flows create a layer of water repellent (hydrophobic) **soil** a few millimeters below the ground surface. Hydrophobic soil impedes the infiltration of rainwater, mobilizing the overlying soil into small debris flows and forming a drainage network throughout the burned **area**. The small debris flows contribute sediment to nearby stream channels, which can then be mobilized into larger debris flows during heavy rainstorms. Debris flows can also begin when hot volcanic ash flows melt snow and **ice** or when **floods** incorporate large amounts of sediment, but these are rare occurrences relative to debris flows mobilized from landslides.

See also Catastrophic mass movements; Drainage basins and drainage patterns; Erosion; Mass movement; Mass wasting; Mud flow

DEEP SEA EXPLORATION

Deep ocean basins cover almost 70% of Earth's surface, and they contain 96% of the planet's life-sustaining **water**. The **oceans** support the **biosphere** by modulating global **climate** and hydrology, and are home to the marine organisms that form the diverse base of the global ecosystem. Geology's unifying paradigm, the theory of **plate tectonics**, arose from deepsea discoveries of the 1950s and 1960s. The title of marine geologist Philip Kuenen's 1958 paper, "No Geology without Marine Geology," rings especially true with hindsight of late twentieth century advancements in the field of marine geology.

The nature of the seafloor was an unrevealed mystery until the mid-nineteenth century; scientists and artists alike envisioned the deep sea as a lifeless soup of placid water, contained in a bowl of static **rock**. By the late 1860s, however, controversial theories of the **origin of life** by **evolution** and the vastness of **geologic time** had created a climate of scientific

curiosity and piqued a general interest in marine science. The Royal Society of London thus mounted an ambitious oceanographic expedition to augment a sparse collection of existing marine data that included Charles Darwin's observations during the voyage of the *Beagle* (1831–1836), a bathymetric map created by U.S. Navy Lt. Matthew Maury to aid installation of the first trans-Atlantic telegraph cables in 1858, and a few examples of deep marine life. The HMS *Challenger* expedition (1872–1876) covered almost 70,000 miles, and shipboard scientists collected hundreds of sediment samples, hydrographic measurements, and specimens of marine life. They also dredged **mafic basalt** from the seafloor, critical evidence that oceanic **crust** is compositionally different from continental crust, and took almost 500 soundings that revealed the depth and basic physiography of the ocean basins.

While the *Challenger* expedition provided critical data that led to rapid advances in marine biology and **oceanography**, the Victorian concept of a geologically inert seafloor persisted for another seventy years. German meteorologist, **Alfred Wegener**, proposed his **continental drift theory** in 1912, but the idea was generally discarded because of inadequate knowledge of seafloor geology. During the 1920s and 1930s, development of sonar echosounding and detection of seafloor **gravity** anomalies rapidly improved the accuracy of bathymetric maps. The technology and government funding that accompanied naval and submarine warfare during World War II, however, precipitated marine geology's age of discovery. Princeton geologist and naval officer, Harry H. Hess, collected bathymetric, gravity, and magnetic polarity data during transatlantic troop transfers, and went on to become a founder of modern marine geology.

The scientific infrastructure of oceanographic institutions, instruments, and vessels that arose following WWII led directly to the theory of plate tectonics. Lamont Geological Observatory of Columbia University geologists surveyed the ocean basins with newly-refined depth recorders in the early 1950s. They discovered the globe-encircling mid-ocean ridge system, and suggested that oceanic crust is created at these chains of submarine volcanoes. Geophysical surveys of the deep **ocean trenches**, and exploration in manned submersibles, including the U.S. Navy's *Trieste* and *Alvin*, suggested the complementary process of seafloor consumption at subduction zones, and the theory of plate tectonics was born. Plate tectonics became widely accepted in the late 1960s as further geophysical surveys of the ocean, seafloor sampling and drilling by the Deep Sea Drilling Project (DSDP), and continental geology all corroborated the theory.

The technology of deep sea exploration has advanced from twine and cannon ball soundings, to ocean surveys from **space** and robotic exploration of the deep ocean floor. Modern sonar instruments provide high-resolution, three-dimensional images of the seafloor. Seafloor sample collections compiled by the DSPD, its successor the international Ocean Drilling Program (ODP), and many other oceanographic institutions, provide rock and sediment data to augment geophysical images. Seismic reflection surveys allow marine stratigraphers and **petroleum** geologists to investigate strata and potential source rocks beneath the seabed.

Technology, fueled by scientific curiosity, has revealed the deep ocean as a dynamic geological environment. The discoveries of intricate ecosystems at mid-ocean volcanic vents and the unexpected diversity of marine life have revolutionized biological science. Just as marine geology held the keys to understanding earth history, marine science may also be the path to understanding Earth's future. **Ocean circulation and currents** control Earth's climate and hydrology, the continental margins are home to most of Earth's human population, and the techniques developed for deep sea exploration are often applicable to space exploration.

See also Bathymetric mapping; Mid-ocean ridges and rifts; RADAR and SONAR; Remote sensing; Seismology

DEGREES, ARCMINUTES, AND ARCSECONDS • *see* LATITUDE AND LONGITUDE

DELTA

Deltas are complex depositional **landforms** that develop at the mouths of **rivers**. They are composed of sediment that is deposited as a river enters a standing body of **water** and loses forward momentum. Famous deltas include the Mississippi delta in Louisiana and the Nile delta in Egypt.

Every river flows, under the force of **gravity**, from its headwaters to its mouth. The mouth of a river is the location at which the river enters a standing body of water, such as a lake, sea, or the ocean. As the river enters standing water and the current is no longer confined to a channel, it spreads out, slows down, and eventually stops. The reduction in speed of the current causes the river to become unable to continue carrying suspended sediment. As sediment is deposited a series of smaller channels, called distributary channels, forms causing the shoreline to build out, or prograde. The landform created is the delta. In smaller rivers with weaker currents, forward momentum may cease almost immediately upon reaching the lake or ocean. This is especially true where the river empties into an **area** of strong wave action. In this case, no significant delta will be formed. Larger rivers, such as the Amazon, may be able to maintain some current for several miles out to sea, creating an extensive delta.

As a river reaches and enters a standing body of water, sediment is deposited according to grain size. The coarsest sediment, such as **sand**, is dropped first, closest to the mouth of the river. With progressive distance from the mouth, finer sediment including fine sand, silt, and **clay** is deposited. This results in a distinct sequence of layers, known as topsets, foresets, and bottomsets. The topsets, as the name implies, are the uppermost layer. They are comprised of the coarse sediment forming the area of the delta that is above sea level. The foresets include fine sand grading into silt and clay deposited in seaward sloping layers beyond the mouth. Bottomsets are made up of clay particles, carried furthest out to sea where they settle into horizontal layers. Although this sequence is

The Nile River delta, as seen from space. *Corbis/NASA. Reproduced by permission.*

deposited laterally with increasing distance from land, as a delta progrades the bottomsets are covered by new foresets, which are then covered by topsets as sediment builds up, and so on. The resultant coarsening up sequence is a distinguishing feature of deltaic deposits.

The sequence of topsets, foresets, and bottomsets provides an accurate picture of a simple delta system. Large marine deltas are often more complex, depending on whether the river, wave action, or the **tides** play the most important role. In stream-dominated deltas, fluvial deposition processes remain strongest, and distributary channels build far out to sea. These deltas are known as bird's foot deltas because of the appearance of the collection of channels extending into the sea. The Mississippi delta is probably the most famous example of a bird's foot delta. In wave-dominated deltas, distributary channels are not maintained for any great distance out to sea; rather, wave action reforms their sediment into **barrier islands** oriented perpendicular to the direction of flow. This type of delta is more compact, and shaped like a triangle. The Nile delta in Egypt is an example of a wave-dominated delta. Lastly, tide-dominated deltas are also compact, but broad tidal channels and sand bars form parallel to the tide direction. The Mekong delta in Vietnam is an example of a tide-dominated delta.

See also Alluvial system; Estuary; Landforms; Sedimentation

DEPOSITIONAL ENVIRONMENTS

Landscapes form and constantly change due to **weathering** and **sedimentation**. The **area** where sediment accumulates and is later buried by other sediment is known as its depositional environment. There are many large-scale, or regional, environments of deposition, as well as hundreds of smaller sub-environments within these regions. For example, **rivers** are regional depositional environments. Some span distances of hundreds of miles and contain a large number of sub-environments, such as channels, backswamps, **floodplains**, abandoned channels, and **sand** bars. These depositional sub-environments can also be thought of as depositional **landforms**, that is, landforms produced by deposition rather than **erosion**.

Depositional environments are often separated into three general types, or settings: terrestrial (on land), marginal marine (coastal), and marine (open ocean). Examples of each of these three regional depositional settings are as follows: terrestrial-alluvial fans, glacial valleys, **lakes**; marginal marine-beaches, deltas, estuaries, tidal mud and sand flats; marine-coral reefs, **abyssal plains**, and continental slope.

During deposition of sediments, physical structures form that are indicative of the conditions that created them. These are known as sedimentary structures. They may provide information about **water** depth, current speed, environmental setting (for example, marine versus fresh water) or a variety of other factors. Among the more common of these are: **bedding** planes, beds, channels, cross-beds, ripples, and mud cracks.

Bedding planes are the surfaces separating layers of sediment, or beds, in an outcrop of sediment or **rock**. The beds represent episodes of sedimentation, while the bedding planes usually represent interruptions in sedimentation, either erosion or simply a lack of deposition. Beds and bedding planes are the most common sedimentary structures.

Rivers flow in elongated depressions called channels. When river deposits are preserved in the sediment record (for example as part of a **delta** system), channels also are preserved. These channels appear in rock outcrops as narrow to broad, v- or u-shaped, "bellies" or depressions at the base of otherwise flat beds. Preserved channels are sometimes called cut-outs, because they "cut-out" part of the underlying bed.

Submerged bars along a coast or in a river form when water currents or waves transport large volumes of sand or gravel along the bottom. Similarly, **wind** currents form **dunes** from sand on a beach or a **desert**. While these depositional surface features, or **bedforms**, build up in size, they also migrate in the direction of water or wind flow. This is known as bar or dune migration. **Suspended load** or bedload material moves up the shallowly inclined, upwind or upcurrent (stoss) side and falls over the crest of the bedform to the steep, downwind or downcurrent (lee) side. If the bedform is cut perpendicular to its long axis (from the stoss to the lee side) one would observe inclined beds of sediment, called cross-beds, which are the preserved leeward faces of the bedform. In an outcrop, these cross-beds can often be seen stacked one atop another; some may be oriented in opposing directions, indicating a change in current or wind direction.

When a current or wave passes over sand or silt in shallow water, it forms ripples on the bottom. Ripples are actually just smaller scale versions of dunes or bars. Rows of ripples form perpendicular to the flow direction of the water. When formed by a current, these ripples are asymmetrical in cross-

section and move downstream by erosion of sediment from the stoss side of the ripple, and deposition on the lee side. Wave-formed ripples on the ocean floor have a more symmetrical profile, because waves move sediments back and forth, not just in one direction. In an outcrop, ripples appear as very small cross-beds, known as cross-laminations, or simply as undulating bedding planes.

When water is trapped in a muddy pool that slowly dries up, the slow sedimentation of the **clay** particles forms a mud layer on the bottom of the pool. As the last of the water evaporates, the moist clay begins to dry up and crack, producing mud cracks as well as variably shaped mud chips known as mud crack polygons. Interpreting the character of any of the sedimentary structures discussed above (for example, ripples) would primarily provide information concerning the nature of the medium of transport. Mud cracks, preserved on the surface of a bed, give some idea of the nature of the depositional environment, specifically that it experienced alternating periods of wet and dry.

All clastic and organic sediments suffer one of two fates. Either they accumulate in a depositional environment, then get buried and lithified (turned to rock by compaction and cementation) to produce sedimentary rock, or they are re-exposed by erosion after burial, but before **lithification**, and go through one or more new cycles of weathering-erosion-transport-deposition-burial.

DESALINATION

Approximately 97% of Earth's **water** is either sea water or brackish (salt water contained in inland bodies), both of which are undrinkable by humans. Desalination is the process of removing salt from seawater. Natural desalination occurs as a part of the **hydrologic cycle** as seawater evaporates. Manipulated desalinization—desalting, or saline water reclamation—is an energy expensive alternative to natural desalination.

Sea water contains 35,000 parts per million (ppm) (3.5% by weight) of dissolved solids, mostly sodium chloride, calcium and magnesium salts. Brackish water typically contains 5,000-10,000 ppm dissolved solids. To be consumable, or potable, water must contain less than 500 ppm dissolved solids. The method used to reach this level depends on the local water supply, the water needs of the community, and economics. Growing populations in arid or **desert** lands, contaminated **groundwater**, and sailors at sea all created the need for desalting techniques.

In the fourth century B.C., Aristotle related tales of Greek sailors desalting water using **evaporation** techniques. **Sand** filters were also used. Another technique used a wool wick to siphon the water. The salts were trapped in the wool. During the first century A.D., the Romans employed **clay** filters to trap salt. Distillation was widely used from the fourth century on—salt water was boiled and the steam collected in sponges. The first scientific paper on desalting was published by Arab chemists in the eighth century. By the 1500s, methods included filtering water through sand, distillation, and the use of white wax bowls to absorb the salt. The techniques have become more sophisticated, but distillation and filtering are still the primary methods of desalination for most of the world. The first desalination patent was granted in 1869, and in that same year, the first land-based steam distillation plant was established in Britain, to replenish the fresh water supplies of the ships at anchor in the harbor. A constant problem in such a process is scaling. When the water is heated over 160°F (71°C), the dissolved solids in water will precipitate as a crusty residue known as scale. The scale interferes with the transfer of heat in desalting machinery, greatly reducing the effectiveness. Today, the majority of desalting plants use a procedure known as multistage flash distillation to avoid scale. Lowering the pressure on the sea water allows it to boil at temperatures below 160°F (71°C), avoiding scaling. Some of the water evaporates, or flashes, during this low pressure boiling. The remaining water is now at a lower **temperature**, having lost some energy during the flashing. It is passed to the next stage at a lower temperature and pressure, where it flashes again. The condensate of the previous stage is piped through the water at the following stage to heat the water. The process is repeated many times. The water vapor is filtered to remove any remaining brine, then condensed and stored. Over 80% of land-based desalting plants are multistage flash distillation facilities.

A host of other desalinization processes have been developed. An increasingly popular process, reverse osmosis, essentially filters water at the molecular level, by forcing it through a membrane. The pressures required for brackish water range from 250 to 400 pounds per square inch (psi), while those for seawater are between 800 to 1,200 psi. The pressure required depends on the type of membrane used. Membranes have been steadily improving with the introduction of polymers. Membranes were formerly made of cellulose acetate, but today they are made from polyamide plastics. The polyamide membranes are more durable than those of cellulose acetate and require about half the pressure. Solar distillation is used in the subtropical regions of the world. Seawater is placed in a black tray and covered by a sloping sheet of **glass** or plastic. Sunlight passes through the cover. Water evaporates and then condenses on the cover. It runs down the cover and is collected. The salts are left behind in the trays.

Modern desalination technology allows use of desalinated water to supplement regular drinking water. The state of Florida, for example, is using dozens of reverse-osmotic plants to treat undrinkable brackish water and then mixing the treated water with the regular water supply. The intent is to extend the local water supply. Another approach is to make traditional methods, like distillation, more economically feasible.

See also Hydrologic cycle; Saltwater encroachment

DESERT AND DESERTIFICATION

Areas that receive less than 10 inches (25.4 cm) of rain a year are generally classified as deserts. Dry (arid) regions are usually found in **area** of high pressure (subtropical highs, leeward sides of mountains, etc.) associated with descending divergent

air masses that are common between 30 degrees N and 30 degrees S **latitude**.

As a consequence of low moisture, desert vegetation is sparse and specifically adapted to conserve **water**. Deserts are areas of high **relief** (e.g., mesas, buttes, etc). Desert regions typically feature well-sorted sands, often found in various dune formations shaped by **sand** type, moisture content, and eoilian processes.

In desert areas, change usually occurs by some form of physical **weathering**. The wide diurnal **temperature** can make the modest amounts of moisture present powerful **weather** factors through continual **freezing** and thawing cycles that can result in micro-fracturing of **rock**. Winds often allow high levels of physical or frictional abrasion. Oxidation and other forms of chemical weathering produce familiar reddish dessert "varnish."

Desertification refers to the gradual degradation of productive arid or semi-arid land into biologically unproductive land (e.g., a change of grassland to desert). The term desertification was first used by the French botanist Aubreville in 1949, to refer to the transformation of productive agricultural land into a desert-like condition.

However, the processes whereby arid lands are stripped of their productivity do not always result in the development of a desert. In some cases, desertification has been successfully reversed through careful land stewardship, and areas degraded by this process have been restored to a more productive condition. In the worst cases, however, semi-desert and desert lands can lose their sparse complement of plants and animals and become barren, gullied wasteland.

Desertification is sometimes caused by natural influences. This process has been ongoing for eons in some regions, in conjunction with long-term changes in climatic conditions, especially decreased **precipitation**. Until the twentieth century, humans were able to simply move their agricultural activity away from land rendered unusable by desertification. However, this strategy has been rendered less tenable by the immense population increase of humans during the past century, a change that has increased the attention paid to the degradation of once-productive drylands.

Desertification claimed major international attention in the 1970s. This resulted from an extended period of severe **drought** in the Sahel region during 1968 to 1973, affecting six African countries on the southern border of the Sahara Desert. Although international relief measures were undertaken, millions of livestock died during that prolonged drought, and thousands of people suffered or died of starvation.

Arid lands of parts of **North America** are among those severely affected by desertification; almost 90% of such habitats are considered to be moderately to severely desertified. The arid and semi-arid lands of the western and southwestern United States are highly vulnerable to this kind of damage. The perennial grasses and forbs that dominate arid-land vegetation can provide good forage for cattle, but if these animals are kept at too high a stocking density, they will overgraze and degrade the natural vegetation cover, contributing to **erosion** and desertification. In addition, excessive withdrawals of **groundwater** to irrigate crops and supply cities is exceeding the ability of the aquifers to replenish, resulting in a rapid decline in height of the **water table**. Moreover, the salts left behind on the **soil** surface after irrigation water evaporates results in land degradation through salinization, creating toxic conditions for crops.

Desertification is best regarded as a process of continuous ecosystem degradation, including damage to plants and animals, as well as to geophysical resources such as water and soil. Desertification is usually discussed in the context of dry regions and ecosystems, but it can also affect prairies, savannas, rain forest, and mountainous habitats. Such effects can range from minor to severe.

The physical characteristics of land undergoing desertification include the progressive loss of the natural, mature vegetation from the ecosystem; the loss of topsoil; increasing salinity of the soil that reduces crop yields and may produce a salty surface **crust** that hinders the seepage of water into the deeper soil; and an increasing number of gullies or sand **dunes** as the soil is eroded by **wind** action.

Among the natural forces of desertification are wind and water erosion of soil, long-term changes in rainfall patterns, and other changes in climatic conditions. The role of drought is variable and related in part to its duration; a prolonged drought accompanied by poor land management may be devastating, while a shorter drought might not be. As such, drought stresses the ecosystem without necessarily degrading it permanently. Rainfall similarly plays a variable role that depends on its duration, the seasonal pattern of its occurrence, and its spatial distribution. The list of human or cultural influences on desertification includes vegetation loss by overgrazing, the depletion of groundwater, surface **runoff** of rainwater, frequent burning, deforestation, the influence of invasive nonnative species, physical compaction of the soil by livestock and vehicles, and damage by strip-mining.

Land management measures to combat desertification focus on improving sustainability and long-term productivity. It is not always possible to return a desertified area to its predesertified condition. As such, mitigating the effects of desertification is best achieved by converting the degraded ecosystem into a new state that can withstand cultural and climatic land-use pressures. Specific measures include developing a resilient vegetation cover of mixed trees, shrubs, and grasses suitable to local conditions. The soil must be protected against wind and water erosion, compaction, and salinization. Water diversions that excessively lower the water table must be reversed, and if possible new sources of water found for human and animal populations.

See also Adiabatic heating; Atmospheric circulation; Basin and range topography; Depositional environments; Desalination; Dune fields; Dust storms; Eolian processes; Erosion; Evaporation; Global warming; Hydrogeology; Hydrologic cycle; Landforms; Landscape evolution; Rate factors in geologic processes; Seasonal winds; Soil and soil horizons; Water table; Weathering and weathering series

DEVONIAN PERIOD

In **geologic time**, the Devonian Period, the fourth period of the **Paleozoic Era**, covers the time roughly 410 million years ago (mya) until 360 mya.

The Devonian Period spans three epochs. The Early Devonian Epoch is the most ancient, followed in sequence by the Middle Devonian Epoch, and the Late Devonian Epoch.

The Early Devonian Epoch is divided chronologically (from the most ancient to the most recent) into the Gedinnian, Siegenian, and Emsian stages. The Middle Devonian Epoch is divided chronologically (from the most ancient to the most recent) into the Eifelian and Givetian stages. The Late Devonian Epoch is divided chronologically (from the most ancient to the most recent) into the Frasnian and Famennian stages.

In terms of paleogeography (the study of the evolution of the continents from **supercontinents** and the establishment of geologic features), the Devonian Period featured continued cleavage of supercontinent landmass and fusion of plates into the supercontinent Laurasia and eventually the supercontinent Pangaea.

Differentiated by fossil remains and continental movements, the **Silurian Period** preceded the Devonian Period. The Devonian is followed in geologic time by the Carboniferous Period (360 mya to 286 mya). In many modern geological texts, especially those in the United States, the time of Carboniferous Period covered by two alternate geologic periods, the **Mississippian Period** (360 mya to 325 mya) and the **Pennsylvanian Period** (325 mya to 286 mya). A mass extinction marks the end of the Devonian Period. In accord with a mass extinction, many **fossils** dated to the Devonian Period are not found in Carboniferous Period (i.e., alternatively, Mississippian Period and Pennsylvanian Period) formations.

The Devonian Period marked a geologically active period. The North American and European continents—with more tropical climates due to more equatorial positions—drifted together. As a result, the two continents share a similar **fossil record** for the Devonian Period. Similar fossil finds dating to the Devonian Period are found in Germany, Canada, and the United States.

The fossil record indicates that it was during the Devonian Period (also termed the "Age of Fishes" because of the appearance of sharks and bony fishes) that amphibians and more terrestrial (land based) vertebrates evolved. Seed plants also appeared, continuing a diversification and development of botanical species, especially vascular plants. By the end of the Devonian Period, the first **forests** appeared.

There were a number of major impacts from large meteorites that date to the Devonian Period. Similar to the **K-T event**, many scientists argue that these impacts could have provided the environmental stresses that eliminated approximately 25% of Devonian Period species. Impact craters dating to the Devonian Period have been identified in modern China, Canada, Russia, and Sweden.

See also Archean; Cambrian Period; Cenozoic Era; Cretaceous Period; Dating methods; Eocene Epoch; Evolution, evidence of;

Fossils and fossilization; Historical geology; Holocene Epoch; Jurassic Period; Mesozoic Era; Miocene Epoch; Oligocene Epoch; Ordovician Period; Paleocene Epoch; Phanerozoic Eon; Pleistocene Epoch; Pliocene Epoch; Precambrian; Proterozoic Era; Quaternary Period; Tertiary Period; Triassic Period

DEW POINT

The dew point is that **temperature** below which the **water** vapor in a body of air cannot all remain vapor. When a body of air is cooled to its dew point or below, some fraction of its water vapor shifts from gaseous to liquid phase to form **fog** or cloud droplets. If a smooth surface is available, vapor condenses directly onto it as drops of water (dew).

The dew point of a body of air depends on its water vapor content and pressure. Increasing the fraction of water vapor in air (i.e., its relative **humidity**) raises its dew point; the water molecules are more crowded in humid air and thus more likely to coalesce into a liquid even at a relatively warm temperature. Decreasing the pressure of air lowers its dew point; lowering pressure (at constant temperature) increases the average distance between molecules and makes water vapor less likely to coalesce.

Air at ground level often deposits dew on objects at night as it cools. In this case, the dew point of the air remains approximately constant while its temperature drops. When the dew point is reached, dew forms. Ground mist and fog may also form under these conditions.

The dew point can be measured using a dew-point hygrometer. This instrument, invented in 1751, consists essentially of a **glass** with a thermometer inserted. The glass is filled with **ice** water and stirred. As the temperature of the glass drops, the air in contact with it is chilled; when it reaches its dew point, water condenses on the glass. The temperature at which **condensation** occurs is recorded as the dew point of the surrounding air.

If the dew point of a body of air is below 32°F (0°C), its water vapor will precipitate not as liquid water but as ice. In this case, the dew point is termed the frost point.

See also Atmospheric inversion layers; Atmospheric lapse rate; Cloud seeding; Clouds and cloud types; Evaporation; Precipitation; Weather forecasting methods; Weather forecasting

DIAMOND

Diamond is cubic native **carbon** with the same composition as **graphite**, but with different structure. It is the hardest mineral (10 on the **Mohs' scale**), with the highest refractive index of 2.417 among all transparent **minerals**, and has a high dispersion of 0.044. Diamonds are brittle. Under UV light, the diamond frequently exhibits luminescence with different colors. It has a density of 3.52 g/cm³. The mass of diamonds is measured in carats; 1 carat=0.2 grams. Diamonds rarely exceed 15

The Hope Diamond is one of the largest diamonds in the world. © *Richard T. Nowitz/Corbis. Reproduced by permission.*

carats. Diamonds are insoluble in acids and alkalis, and may burn in **oxygen** at high temperatures.

Nitrogen is the main impurity found in diamonds, and influences its physical properties. Diamonds are divided into two types, with type I containing 0.001–0.23% nitrogen, and type II containing no nitrogen. If nitrogen exists as clusters in type I diamonds, it does not affect the color of the stone (type Ia), but if nitrogen substitutes carbon in the crystal lattice, it causes a yellow color (Ib). Stones of type II may not contain impurities (IIa), or may contain boron substituting carbon, producing a blue color and semiconductivity of the diamond.

Diamonds form only at extremely high pressure (over 45,000 atmospheres) and temperatures over 1100°C (2012°F) from liquid ultrabasic magmas or peridotites. Diamonds, therefore, form at great depths in the earth's **crust**. They are delivered to the surface by explosive volcanic phenomena with rapid cooling rates, which preserve the diamonds from transformation. This process happens in kimberlites (a peridotitic type of **breccia**), which constitutes the infill of diamond-bearing pipes. Also found with diamonds are **olivine**, serpentine, carbonates, pyroxenes, pyrope garnet, magnetite, hematite, graphite and ilmenite. Near the surface, kimberlite weathers, producing yellow loose mass called yellow ground, while deeper in the earth, it changes to more dense blue ground. Diamonds are extremely resistive to **corrosion**, so they can be fond in a variety of secondary deposits where they arrived after several cycles of **erosion** and **sedimentation** (alluvial diamond deposits, for example). Even in diamond-bearing **rock**, the diamond concentration is one gram in 8–30 tons of rock.

Most diamonds are used for technical purposes due to their hardness. Gem quality diamonds are found in over 20 counties, mainly in **Africa**. The biggest diamond producer is South Africa, followed by Russia. Usually, diamonds appear as isolated octahedron **crystals**. Sometimes they may have rounded corners and slightly curved faces. Microcrystalline diamonds with irregular or globular appearance are called Bort (or boart), while carbonado are roughly octahedral, cubic or rhombic dodecahedral, blackish, irregular microcrystalline aggregates. Both are valued for industrial applications because they are not as brittle as diamond crystals. Frequently, diamonds have inclusions of olivine, sulfides, chrome-diopside, chrome-spinels, zircon, rutile, disthene, biotite, pyrope garnet and ilmenite. Transparent crystals are usually colorless, but sometimes may have various yellowish tints. Rarely, diamonds may be bright yellow, blue, pale green, pink, violet, and even reddish. Some diamonds are covered by translucent skin with a stronger color. Diamonds become green and radioactive after neutron irradiation, and yellow after further heating. They become blue after irradiation with fast electrons. Diamonds have different hardnesses along their different faces. Diamonds from different deposits also have different hardnesses. This quality allows for the polishing of faceted diamonds by diamond powder.

Most diamond gems are faceted into brilliant cuts. Due to the high reflective index, all light passing through the face of such facetted diamonds is reflected back from the back facets, so light is not passing through the stone. This can be used as a diagnostic property, because most simulants (except cubic zirconia) do not have this property. Diamonds do have many simulants, including zircon, corundum, phenakite, tourmaline, topaz, beryl, **quartz**, scheelite, sphalerite, and also synthetic **gemstones** such as cubic zirconia, Yttrium-aluminum garnet, strontium titanate, rutile, spinel, and litium niobate. Diamonds have high thermal conductivity, which allows it to be readily and positively distinguished from all simulated gemstones. The most expensive diamonds are those with perfect structure and absolutely colorless or slightly bluish-white color. Yellow tint reduces the price of the diamond significantly. Bright colored diamonds are extremely rare, and have exceptionally high prices.

See also Gemstones

DIATOMS

Algae are a diverse group of simple, nucleated, plant-like aquatic organisms that are primary producers. Primary producers are able to utilize photosynthesis to create organic molecules from sunlight, **water**, and **carbon dioxide**. Ecologically vital, algae account for roughly half of photosynthetic produc-

tion of organic material on Earth in both **freshwater** and marine environments. Algae exist either as single cells or as multicellular organizations. Diatoms are microscopic, single-celled algae that have intricate glass-like outer cell walls partially composed of **silicon**. Different species of diatom can be identified based upon the structure of these walls. Many diatom species are planktonic, suspended in the water column moving at the mercy of water currents. Others remain attached to submerged surfaces. One bucketful of water may contain millions of diatoms. Their abundance makes them important food sources in aquatic ecosystems. When diatoms die, their cell walls are left behind and sink to the bottom of bodies of water. Massive accumulations of diatom-rich sediments compact and solidify over long periods of time to form **rock** rich in fossilized diatoms that is mined for use in abrasives and filters.

Diatoms belong to the taxonomic phylum Bacillariophyta. There are approximately 10,000 known diatom species. Of all algae phyla, diatom species are the most numerous. The diatoms are single-celled, eukaryotic organisms, having genetic information sequestered into subcellular compartments called nuclei. This characteristic distinguishes the group from other single-celled photosynthetic aquatic organisms, like the blue-green algae that do not possess nuclei and are more closely related to bacteria. Diatoms also are distinct because they secrete complex outer cell walls, sometimes called skeletons. The skeleton of a diatom is properly referred to as a frustule.

Diatom frustules are composed of pure hydrated silica within a layer of organic, **carbon** containing material. Frustules are really comprised of two parts: an upper and lower frustule. The larger upper portion of the frustule is called the epitheca. The smaller lower piece is the hypotheca. The epitheca fits over the hypotheca as the lid fits over a shoebox. The singular algal diatom cell lives protected inside the frustule halves like a pair of shoes snuggled within a shoebox.

Frustules are ornate, having intricate designs delineated by patterns of holes or pores. The pores that perforate the frustules allow gases, nutrients, and metabolic waste products to be exchanged between the watery environment and the algal cell. The frustules themselves may exhibit bilateral symmetry or radial symmetry. Bilaterally symmetric diatoms are like human beings, having a single plane through which halves are mirror images of one another. Bilaterally symmetric diatoms are elongated. Radially symmetric diatom frustules have many mirror image planes. No matter which diameter is used to divide the cell into two halves, each half is a mirror image of the other. The combination of symmetry and perforation patterns of diatom frustules make them beautiful biological structures that also are useful in identifying different species. Because they are composed of silica, an inert material, diatom frustules remain well preserved over vast periods of time within geologic sediments.

Diatom frustules found in sedimentary rock are microfossils. Because they are so easily preserved, diatoms have an extensive **fossil record**. Specimens of diatom algae extend back to the **Cretaceous Period**, over 135 million years ago. Some kinds of rock are formed nearly entirely of fossilized diatom frustules. Considering the fact that they are microscopic organisms, the sheer numbers of diatoms required to produce rock of any thickness is staggering. Rock that has rich concentrations of diatom **fossils** is known as diatomaceous earth, or diatomite. Diatomaceous earth, existing today as large deposits of chalky white material, is mined for commercial use in abrasives and in filters. The fine abrasive quality of diatomite is useful in cleansers, like bathtub scrubbing powder. Also, many toothpaste products contain fossil diatoms. The fine **porosity** of frustules also makes refined diatomaceous earth useful in fine water filters, acting like microscopic sieves that catch very tiny particles suspended in solution.

Fossilized diatom collections also tell scientists a lot about the environmental conditions of past eras. It is known that diatom deposits can occur in layers that correspond to environmental cycles. Certain conditions favor mass deaths of diatoms. Over many years, changes in diatom deposition rates in sediments, then, are preserved as diatomite, providing clues about prehistoric climates.

Diatom cells within frustules contain chloroplasts, the organelles in which photosynthesis occurs. Chloroplasts contain chlorophyll, the pigment molecule that allows plants and other photosynthetic organisms to capture **solar energy** and convert it into usable chemical energy in the form of simple sugars. Because of this, and because they are extremely abundant occupants of freshwater and **saltwater** habitats, diatoms are among the most important microorganisms on Earth. Some estimates calculate diatoms as contributing 20–25% of all carbon fixation on Earth. Carbon fixation is a term describing the photosynthetic process of removing atmospheric carbon in the form of carbon dioxide and converting it to organic carbon in the form of sugar. Due to this, diatoms are essential components of aquatic food chains. They are a major food source for many microorganisms, aquatic animal larvae, and grazing animals like mollusks (snails). Diatoms are even found living on land. Some species can be found in moist **soil** or on mosses. Contributing to the abundance of diatoms is their primary mode of reproduction, simple asexual cell division. Diatoms divide asexually by mitosis. During division, diatoms construct new frustule cell walls. After a cell divides, the epitheca and hypotheca separate, one remaining with each new daughter cell. The two cells then produce a new hypotheca. Diatoms do reproduce sexually, but not with the same frequency.

See also Atmospheric chemistry; Depositional environments; Fossil record; Fossils and fossilization; Soil and soil horizons

DIKE

A dike is a formation of igneous **rock** that can form exposed vertical or linear ridges. Dikes are formed underground and are an intrusive plutonic rock formation. Intrusive formations form when upwelling **magma** cools and solidifies beneath the surface. As the magma rises it intrudes into the overlying **country rock** (older rock that is also termed "host" rock).

Dikes are vertical formations and thus form at a steep or often near right angle to the surface. Dikes are planar intrusions that, in contrast to horizontal sills, have a discordant form of contact with the host rock into which they intrude. A discordant contact is one that transverses or cuts across the established **bedding** planes of the country or host rock (e.g., at right angles to surrounding sedimentary bedding planes).

Dike texture varies from **aphanitic** (no visible mineral **crystals**) to phaneritic (visible mineral crystals). The texture is determined by the time needed for the upwelling magma to cool and solidify. Because dikes are vertical, gradients of textures can be established within the same dike (e.g., a change form aphanitic to phaneritic texture within the same dike. The longer the magma cooling time, the greater the extent and size of mineral crystal formation in the igneous rock comprising the dike. If, for example a dike—or a region of a dike—cools rapidly, the texture becomes uniform, smooth and without mineral crystals that are discernable upon visual inspection. If the magma in the dike cools over a long period of time, visible crystals form and the texture is described as phaneritic. Variation in the texture of dikes can also result from multiple intrusions of magma.

When exposed, dikes may form visible cliffs. In addition, dikes can form a network of underground passages for magma and when resistance to upward magma flow is encountered, a dike formation may give way to a horizontal **sill** intrusion.

Dikes can vary greatly in thickness across a range from a few inches to hundreds of yards. At the extreme, the great Dike of Zimbabwe extends more than 350 miles and has an average width of about six miles.

The formation of dikes often follows or reflects the fracturing of surrounding country rock. Accordingly, dikes often form in clumps or "swarms" and can occur in radial distributions about a deeper upwelling.

By definition, dikes form and cool underground. However, if the same form of magma upwelling reaches the surface, it results in a usually low viscosity volcanic **fissure** formation. Such formations are common in the Hawaiian islands (formed from an upwelling of a magmatic "hot spot") and Iceland (formed as part of the magma upwelling associated with the Mid-Atlantic Ridge and in other areas of volcanic activity.

The search for such cliff formations is an important part of extraterrestrial studies conducted by the Mars Global Surveyor and probes sent to explore the moons of Jupiter. Identification of dike formations provide easily visible evidence of past volcanic activity, **mantle plumes**, and other forms of plate tectonic activity.

See also Igneous rocks; Pluton and plutonic bodies; Stratigraphy; Volcanic eruptions; Volcano

DIP • *see* STRIKE AND DIP

DISINTEGRATION, AGENTS OF • *see*
WEATHERING AND WEATHERING SERIES

DIURNAL CYCLES

Diurnal cycles refer to patterns within about a 24-hour period that typically reoccur each day. Most daily cycles are caused by the **rotation** of the earth, which spins once around its axis about every 24 hours. The term diurnal comes from the Latin word *diurnus,* meaning daily. Diurnal cycles such as **temperature** diurnal cycles, diurnal **tides**, and solar diurnal cycles affect global processes.

A temperature diurnal cycle is composed of the daily rise and fall of temperatures, called the daily march of temperature. The daily rotation of the earth causes the progression of daytime and nighttime, thus controlling the air temperature. The daily maximum temperature occurs between the hours of 2 P.M. and 5 P.M. and then continually decreases until sunrise the next day. The angle of the **Sun** to the surface of the earth increases until around noon when the angle is the largest. The intensity of the Sun increases with the Sun's angle, so that the Sun is most intense around noon. However, there is a time difference between the daily maximum temperature and the maximum intensity of the Sun, called the lag of the maximum. This discrepancy occurs because air is heated predominantly by reradiating energy from Earth's surface. Although the Sun's intensity decreases after 12 P.M., the energy trapped within the earth's surface continues to increase into the afternoon and supplies heat to the earth's atmosphere. The reradiating energy lost from the earth must surpass the incoming **solar energy** in order for the air temperature to cool.

Diurnal tides are the product of one low tide and one high tide occurring roughly within a 24-hour period. Tides in general result from the relationship of the earth's gravitational attraction to the **Moon** and Sun. Additionally, the motion of the earth and Moon as well as the geometric relationship of the earth's location to both the Moon and Sun affects the tides. More specifically, the tilted axis of the earth in relation to the plane of the Moon plays an important role in creating diurnal tides. Typically, diurnal tides are weak tides that occur when the Moon is furthest from the equator. Diurnal tides occur in the northern part of the **Gulf of Mexico** and Southeast **Asia**.

The earth experiences varying hours of daylight due to the solar diurnal cycle. Solar diurnal cycles occur because the earth's axis is tilted 23.5 degrees and is always pointed towards the North Star, Polaris. The tilt of the earth in conjunction with the earth's rotation around the Sun affects the amount of sunlight the Earth receives at any location on Earth.

On March 21 or 22, the vernal equinox occurs where every location on Earth experiences exactly 12 hours of daylight. On this first day of spring, the Sun rises in the east and sets in the west. After the vernal equinox occurs, the Sun shifts north with each day until the day of the summer solstice on June 21 or 22. On this first day of summer, the Sun is as far north as possible and the Northern Hemisphere experiences its longest day of the year, while the Southern Hemisphere experiences its shortest day of the year. Over the next three months, the Sun moves progressively south and the days begin to get shorter in the Northern Hemisphere. On the day of the autumnal equinox, for the second time a year, the Sun rises due east and sets due west. On this first day of autumn, September 21 or 22, there are exactly 12 hours of daylight everywhere on Earth. After the autumnal equinox, the Sun continues to move south with each day and there continues to be less daylight with each passing day in the Northern Hemisphere until the winter solstice. On this first day of winter, December 21 or 22, the Sun has shifted as far south as possible and the Northern Hemisphere experiences its shortest day of the year while the Southern Hemisphere experiences its longest day of the year.

See also Earth (planet)

DIVERGENT PLATE BOUNDARY

In terms of **plate tectonics**, divergent boundaries are areas under tension where **lithospheric plates** are pushed apart by **magma** upwelling from the mantle. Lithospheric plates are regions of Earth's **crust** and upper mantle that are fractured into plates that move across a deeper plasticine mantle. At divergent boundaries, lithospheric plates move apart and crust is created.

Earth's crust is fractured into approximately 20 lithospheric plates. Each lithospheric plate is composed of a layer of oceanic crust or continental crust superficial to an outer layer of the mantle. Oceanic crust comprises the outer layer of the **lithosphere** lying beneath the **oceans**. Oceanic crust is composed of high-density rocks, such as **olivine** and **basalt**. Continental crust comprises the outer layer of the lithospheric plates containing the existing continents and some undersea features near the continents. Continental crust is composed of lower density rocks such as **granite** and **andesite**. New oceanic crust is created at divergent boundaries that are sites of **seafloor spreading**.

At divergent boundaries, upwelling of magma along **mid-ocean ridges** (e.g., Mid-Atlantic Ridge) creates the tensional forces that drive the lithospheric plates apart.

Although initially formed from convection hot spots in the **asthenosphere**, rift valleys (e.g., Rift Valley of **Africa**) can fuse or interconnect to form zones of divergence that ultimately can fracture the lithospheric plate.

Containing both crust and the upper region of the mantle, lithospheric plates are approximately 60 miles (approximately 100 km) thick. Lithospheric plates may contain various combinations of oceanic and continental crust in mutually exclusive sections (i.e., the outermost layer is either continental or oceanic crust, but not both). Lithospheric plates move on top of the asthenosphere (the outer plastically deforming region of Earth's mantle).

Divergent plate boundaries are, of course, three-dimensional. Because Earth is an oblate sphere, lithospheric plates are not flat, but are curved and fractured into curved sections akin to the peeled sections of an orange. Divergent movement of lithospheric plates can best be conceptualized by the movement apart of those peeled sections over a curved surface (e.g., over a ball).

At divergent boundaries, tensional forces dominate the interaction between plates.

Because Earth's diameter remains constant, there is no net creation or destruction of lithospheric plates and so the amount of crust created at divergent boundaries is balanced by an equal destruction or uplifting of crust at convergent lithospheric plate boundaries.

Evidence of symmetrical bands of **rock** with similar ages located on either side of divergent boundaries offer important evidence in support of plate tectonic theory. In addition, similar fossil and magnetic bands also exist in equidistant bands on either side of a divergent boundary.

See also Convergent plate boundary; Dating methods; Earth, interior structure; Fossil record; Fossils and fossilization; Geologic time; Hawaiian island formation; Mapping techniques; Mid-ocean ridges and rifts; Mohorovicic discontinuity (Moho); Rifting and rift valleys; Subduction zone

DOLOMITE

The term dolomite is used both for the mineral dolomite (calcium magnesium carbonate [$CaMg(CO_3)_2$]) and for the **rock** dolomite, which consists mostly of the mineral dolomite. Dolomite rock is sometimes termed dolostone to distinguish it from the mineral dolomite, but the more confusing terminology is the more prevalent. Dolomite rock is formed from **limestone** (which is mostly calcite, i.e., calcium carbonate [$CaCO_3$]) by the replacement of about half of the limestone's calcium ions by magnesium ions. Because of its close relationship to limestone, dolomite is sometimes categorized as a type of limestone.

Limestone forms primarily in shallow **seas** and coastal waters where shelled marine organisms—crustaceans, mollusks, bivalves, and the like—proliferate. The shells of such creatures consist essentially of calcite. They accumulate on the sea floor in thick beds and are transformed into limestone over time. Some limestone is further transformed to dolomite by processes only partly understood. These various processes are lumped under the term dolomitization. The essential feature of all dolomitization processes is the importation of magnesium ions by **water**. These take up residence in the crystal structure of the limestone and convert it to dolomite.

Dolomites often occur in association with limestone, **gypsum**, and other rocks formed by shallow seas. Dolomite beds one or more meters thick are often sandwiched between similarly thick limestone beds. Dolomite and limestone are difficult to tell apart visually; a common field technique for distinguishing them is to drip hydrochloric acid (a hydrous solution of HCl) onto a hand sample. In response, limestone froths vigorously and dolomite weakly.

Metamorphosed limestone becomes calcite **marble**; metamorphosed dolomite becomes dolomitic marble. Dolomitic marble can be converted to calcite marble by dedolomitization, that is, the **leaching** out of magnesium.

Dolomites are used as magnesium ores, as a source of pharmaceutical magnesia (MgO), and as a flux—aid to the removal of impurities—in metal refining.

See also Fossils and fossilization; Field methods in geology; Industrial minerals

DOUGLAS SEA SCALE

The Douglas Sea Scale was devised by the English Admiral H.P. Douglas in 1917, while he was head of the British Meteorological Navy Service. Its purpose is to estimate the sea's roughness for navigation. The Douglas Scale consists of two codes, one for estimating the state of the sea (fresh waves attributable to local **wind** conditions), the other for describing sea swell (large rolling waves attributable to previous or distant winds).

The Douglas Sea Scale is expressed in one of 10 degrees.
- Degree 0—no measurable wave height, calm sea
- Degree 1—waves >10 cm., rippled sea
- Degree 2—waves 10–50 cm., smooth sea
- Degree 3—waves 0.5–1.25 m., slight sea
- Degree 4—waves 1.25–2.5 m., moderate sea
- Degree 5—waves 2.5–4 m., rough sea
- Degree 6—waves 4–6 m., very rough sea
- Degree 7—waves 6–9 m., high sea
- Degree 8—waves 9–14 m., very high sea
- Degree 9—waves >14 m., phenomenal sea

It was difficult to relate the existing wind scale designed by Sir Frances Beaufort in 1805 to a ship's features, especially as sails were replaced with the rigid structures of powered ships. The Douglas Sea Scale standardized the many variations being used by ship captains from many nations.

See also Beaufort wind scale; Wave motions

DOUGLASS, ANDREW ELLICOTT (1867-1962)

American astronomer and archaeologist

Andrew E. Douglass invented and named dendrochronology, the technique of counting and studying the rings in tree trunks to determine not only the ages of trees, but also the past climatological, geological, agricultural, social, and economic conditions of the local **area**.

Born in Windsor, Vermont, on July 5, 1867, Douglass received his bachelor's degree with honors in 1889 from Trinity College in Hartford, Connecticut. After working five years for the Harvard University Observatory, including an expedition to Arequippa, Peru, from 1891 to 1893 to establish Harvard's Southern Hemisphere Observatory, he accepted the offer of astronomer Percival Lowell (1855–1916) to build an observatory in the American Southwest. They founded Lowell Observatory in 1894 by erecting an 18-inch **telescope** on a mesa outside Flagstaff, Arizona. Lowell was preoccupied with Mars, and some historians argue that Lowell may have skewed Douglass' data in order to support his theories of Martian life and civilization. The two scientists' increasingly hostile disagreements about the proper use of data led Lowell to fire Douglass in 1901.

Douglass then taught school in the Flagstaff area, won an election for probate judge, and around 1904, began to note the connection between tree rings and solar cycles. In 1906, he

This scene from the movie "The Perfect Storm" shows an example of Degree 9 waves, according to the Douglas sea scale. Fortunately, such waves are rare. *The Kobal Collection. Reproduced by permission.*

moved to Tucson and joined the **astronomy** faculty of the University of Arizona. Increasingly interested in the possibility of using tree rings for archaeological dating, he concentrated his research on the ponderosa pine, the Douglas fir, and, in collaboration with Ellsworth Huntington (1876–1947), the giant sequoia. Beginning in 1909 his work received support from Clark Wissler (1870–1947) of the American Museum of Natural History in New York City and philanthropist Archer Milton Huntington (1870–1955). By the second decade of the twentieth century, dendrochronology was widely recognized as an important scientific insight.

After convincing Lavinia Steward (d. 1917) to found a new observatory at the University of Arizona with a bequest of $60,000 from the estate of her late husband, Henry B. Steward (d. 1902), Douglass became the first director of Steward Observatory in 1916. Its 36-inch reflecting telescope, one of the first in the nation, became operational in 1922 and the facility was dedicated in 1923.

Douglass retired from the observatory in 1937, but served the University of Arizona from 1937 until 1958 as the founding director of the Laboratory of Tree-Ring Research. He continued actively engaging in dendrochronological studies until within two years of his death in Tucson on March 20, 1962. His manuscripts and notes, held by the University of Arizona Library Special Collections, reveal an extraordinarily precise, flexible, and meticulous scientist.

See also Archeological mapping; Dating methods; Precipitation; Solar energy; Sun; Weathering and weathering series

DRAAS • *see* DUNE FIELDS

DRAINAGE BASINS AND DRAINAGE PATTERNS

A drainage basin is the **area** that encompasses all the land from which **water** flows into a particular stream or river. Stream is a synonym of river, and although typically something called a stream is smaller than a river, here, any flowing body of water in a clearly defined channel will be called a stream. The size of a drainage basin can vary from being as small as a few square miles or kilometers to as large as part of a continent. An example of a divide is the **continental divide** of **North America**, which separates streams that ultimately empty into one ocean (the Pacific Ocean) from those that ultimately empty into another (the **Gulf of Mexico**). The smallest streams in any particular area are called first order streams, and the land from which water flows into a particular first order stream is called a first order drainage basin. First order streams flow into second order streams, and each second order stream has its own second order drainage basin. There is no limit to how high an order a stream may be.

The drainage pattern that streams in a drainage basin trace out, visible in aerial photographs or even from the window of an airliner, can provide a lot of information about the type of terrain that the streams flow over. The dendritic drainage pattern of streams resembles the veins of a leaf, or the structure of a tree. It typically develops in areas with homogenous or flat-lying rocks that provide no preferred direction to the development of stream channels. Streams that flow over the flat-lying **rock** units of the American Midwest often display this type of drainage pattern. An annular drainage pattern forms when layers of rock are uplifted into a dome or down-warped into a basin, and the stream channels preferentially follow the weakest concentric beds of rock. A radial drainage pattern develops where there is a central highpoint, such as an isolated volcanic peak. The streams all flow away from the highest point. Fractures in massive rock such as **granite** can produce a drainage pattern in which the streams have many right-angle turns, and this is called rectangular drainage. When layered rock units are folded or tilted up, lower-order streams that flow into larger streams tend to be straight and follow weaker beds of rock. This trellis drainage pattern is common in the Appalachian Mountains of the United States. Centripetal drainage is found where streams flow into the center of a depression such as a basin or crater. Deranged drainage forms on terrain that is freshly exposed, and where the streams have not had a chance to develop in response to underlying geologic structure or **bedrock**. Finally, parallel drainage tends to develop in areas of massive rock with a uniform slope, where all the streams tend to flow in the same direction.

See also Avalanche; Delta; Drainage calculations and engineering; Hydrogeology; Runoff

DRAINAGE CALCULATIONS AND ENGINEERING

The design of hydraulic structures from small culverts to large dams requires engineers to calculate the amount of **water** that will flow through the channel along which the structure is built. The rate of flow through a stream channel, or discharge, is measured using units of cubic feet per second for engineer-

Engineers are often challenged when attempting to calculate how much drainage a particular area will need. Nature, however, has proven difficult to predict. *AP/Wide World. Reproduced by permission.*

ing projects in the United States, and in units of cubic meters per second in other countries.

In many cases, knowledge of the maximum rate of flow that is likely to occur in a channel is required. For example, an engineer may wish to ensure that a bridge will be built high enough to allow passage of the largest flood likely to occur during the useful life of the bridge. If the maximum discharge is known or can be estimated, then it is a simple matter to calculate the height to which water will rise in a given channel. This is can be accomplished using Manning's equation, which relates discharge to the channel cross-sectional **area** and perimeter, the channel slope, and channel roughness.

The relationship between **precipitation** and the discharge of nearby streams is controlled by many factors. These include rainfall intensity and duration, drainage basin area, **topography**, **soil** and **bedrock** type, land use, vegetative ground cover, and the amount of precipitation in the days or weeks before a storm. Because it is difficult to incorporate this degree of complexity into mathematical models reflecting the **physics** of precipitation and stream discharges, let alone forecast the **weather**, probabilistic approaches based on the historical frequency of peak discharges are commonly used in engineering calculations. For **drainage basins** in which a long record of maximum annual discharges is available, discharges can be ranked from largest to smallest. Standard formulae are then used to estimate the probability that a given discharge will be exceeded over a specified period of time. A 100-year flood, for example, is one with a discharge that is inferred to occur on average once every 100 years. The term "on average" is an important qualifier because it means that a 100-year flood may occur more or less often than once every 100 years.

The probability that a discharge of given magnitude will occur over a specified time period is estimated using a binomial probability distribution. The binomial distribution can be used to show that there is a 37% chance that no 100-year flood will occur during any given 100-year period. Similarly, there is an 18% probability that two 100-year **floods** will occur during any given 100-year period. This logic can be extended to determine with a specified level of certainty the largest discharge that is likely to occur during the useful life of a hydraulic structure. An engineer designing a flood conveyance channel large enough to handle the discharge with a 90% likelihood of not being exceeded during the 50 year useful life of the channel would not use the discharge of the 50-year flood, but rather the discharge of the 475-year flood.

In some cases, particularly in small or remote drainage basins, flood discharge records may not be available and other methods must be used. One of the simplest techniques is the rational method, which relates peak stream discharge to the product of drainage basin area, rainfall intensity, and a coefficient representing the type of land use or ground cover in the drainage basin. In the United States, the rational method uses units of inches per hour for rainfall intensity and acres for drainage basin area. Values for the coefficient are tabulated in engineering reference books, and range from 0.05 for grassy areas with sandy soil to 0.95 for paved areas. The values can be averaged in cases where there are several different land uses within a drainage basin.

More sophisticated techniques can be used when the change in discharge as a function of time, as opposed to simply the peak discharge, is an important factor in design.

See also Drainage basins and drainage patterns; Hydrologic cycle; Stream valleys, channels, and floodplains

DROUGHT

Drought is a temporary hazard of nature occurring from a lack of **precipitation** over an extended period of time. Drought differs from aridity, a permanent feature of **climate** restricted to regions of low rainfall. Rainfall deficiencies caused by a drought create a severe hydrologic imbalance resulting in considerable **water** shortages.

The beginning of a drought is typically determined by comparing the current meteorological situation to an average based on a 30-year period of record. This "operational" definition of drought allows meteorologists to analyze the frequency, severity, and duration of the aberration for any given historical period and aides in the development of response and mitigation strategies.

The agricultural sector is usually the first to be affected by dryness, since crops are heavily dependent on stored soil water. *Jim Sugar. Jim Sugar Photgraphy/Corbis-Bettmann. Reproduced by permission.*

Characteristics of drought are highly variable from region to region, depending on atmospheric factors such as **temperature**, **wind**, relative **humidity**, and amount of sunshine and cloud cover. High temperatures and lots of sunshine can increase **evaporation** and transpiration to such an extreme that frequent rainfall is incapable of restoring the loss. Meteorological definitions of drought, therefore, may deviate from operational definitions and are usually based on the length of the dry period and the degree of dryness in comparison to the daily average.

Drought is more than a physical phenomena; an extended period of dryness can have a significant socioeconomic impact. Drought presents the most serious physical hazard to crops in nearly all regions of the world. The agricultural sector is usually the first to be affected by dryness, since crops are heavily dependent on stored **soil** water. In addition to a decline in agricultural products, a shortfall in the water supply can disrupt availability of other economic goods such as hydroelectric power. The 1988–89 Uruguay drought resulted in a significant decline of hydroelectric power because the dryness disrupted the streamflows needed for production.

See also Hydrologic cycle

DRUMLINS · *see* GLACIAL LANDFORMS

DRY AIR · *see* HUMIDITY

DUNE FIELDS

Dune fields are large features of eolian or arid environments. They are associated with hot **climate** deserts such as the Sahara. Dune fields are not, however, exclusively restricted to these types of environments. Many dune fields are found in temperate climates where the processes of aridity in an arid climate combine to form **dunes**, but at a much slower rate than hot, arid climates.

The basic processes that contribute to dune formation are straightforward. It is the range of minute geological processes that generates controversy in dune research. When a geographical region experiences prolonged **drought** accompanied by high **evaporation** rates, the soils lose vegetation. Plant roots secure loose particles that make up soils. When **wind** sweeps across the barren **soil**, the first stages of dune and **desert** formation occur.

Wind is an excellent agent for separating grains sizes and weights in soils. Although not as effective as **water** for transporting sediments, it is responsible for a tremendous amount of silt and **sand** relocation around the planet. Wind picks up lighter and smaller particles, such as the types of **clay minerals**, and moves them far away from their source. Atmospheric dust can circulate the globe and may stay in the air for days, weeks, and even years. It later settles in all types of places, even the polar areas.

Unlike these lightweight minerals, heavier minerals such as **quartz** remain in their original spot and begin to accumulate. They are rolled around by wind but not removed. The rolling makes the grains smooth and of the type more commonly associated with dunes than with water deposits. The grains are typically light in color and are what make dune fields light beige to white in hue.

Dune fields are active regions of moving sands that form characteristic shapes including the well-known crescent dune. The macro and microscopic movement of sand particles is an **area** still being intensively studied. One method of particle movement is called **saltation**. This is a process in which the wind is not strong enough to pick up the grains, but, instead, moves the grains along the ground in a hopping and rolling fashion. As the grains climb up the faces of dunes facing the wind, they reach the crest where they bounce off the top. Movies of this action have been carefully studied. This airborne motion of grain movement can be seen as wisps of sand curling from the tops of dunes much like the snows blowing from mountain tops.

The sediments that accumulate on the windward slope are called topset deposits. When they reach the crest, they form an unstable and temporary surface called the brink. When enough sediments are captured on the brink they eventually tumble over the edge onto the slipface. This motion provides the advancement of the dune as it migrates in the direction of the wind. A temporary halt in dune movement can make a thin layer of sediments that become slightly bonded to one another. This layer becomes visible in side view and is even more recognizable in ancient deposits.

The sequence of dune formation has been widely studied. Increased or decreased wind strength is the force that makes the wide variety of dune shapes. This has led to a wide variety of terms used to describe dunes, dune fields, and other structures. Barchan dunes are the traditional crescent shaped dunes where the tips lie pointing away from the direction of the wind. Parabolic dunes are also crescent-shaped, but in this case, the tips lie facing the wind. Star dunes represent a dune formed from wind that blows from a variety of directions. Draas is an antiquated term used to describe huge dune fields that are often only observable from **space**. An erg is a region in the Sahara that is occupied by deep and complex sand dunes. Size is one of the factors that distinguish ergs from draas.

Dune fields themselves are complex environments. Within the field, there are many microenvironments that lie between the dunes and at the bottom of dune valleys. Moisture may even accumulate and form small ponds. Scientists continue to study dunes and dune fields. They are one of the least understood structures in **geology** because of the difficulty in studying them. However, dune fields occur over a significant portion of Earth's surface and certainly command more attention.

See also Beach and shoreline dynamics; Desert and desertification; Seawalls and beach erosion

DUNES

Dunes are well-sorted deposits of materials by **wind** or **water** that take on a characteristic shape and that retain that general shape as material is further transported by wind or water. **Desert** dunes classifications are based upon shape include barchan dunes, relic dunes, transverse dunes, lineal dunes, and blount (parabolic) dunes. Dunes formed by wind are common in desert areas and dunes formed by water are common in coastal areas. Dunes can also form on the bottom of flowing water (e.g., stream and river beds).

When water is the depositing and shaping agent, dunes are a bedform that are created by **saltation** and deposition of particles unable to be carried in suspension. Similar in shape to ripples—but much larger in size—dunes erode on the upstream side and extend via deposition the downstream or downslope side.

Regardless of whether deposited by wind or water, dunes themselves move or migrate much more slowly than any individual deposition particle.

In desert regions, dune shape is dependent upon a number of factors including the type of **sand**, the moisture content of the sand, and the direction and strength of the prevailing wind pattern. Barchan dunes are crescent-shaped small dunes with the terminal points of the crescent pointed downwind (on the lee side of the prevailing wind). Transverse dunes are long narrow dunes (a dune line) formed at right angles to the prevailing wind pattern. Transverse dunes may form from the fusion of individual barchan dunes.

Blount or parabolic dunes may form in regions of higher moisture content where there is sufficient vegetation to retard the migration of sand. Blount dunes take the mirror image shape of barchan dunes—they are crescent-shaped, but the terminal points of the crescent point windward (into the direction of the prevailing winds). Lineal dunes form parallel to prevailing wind patterns. Lineal dunes may be become the dominant **relief** feature and dunes may measure several hundred yards or meters high and extend for more than 50 miles (80 km).

Desert dunes migrate downwind from prevailing winds. Relic dunes form as migration slows and vegetation forms on a dune.

Ergs are "dune seas" ("erg" derives from Arabic) or large complexes of dunes. Very large (generally over 100 meters high and at least a kilometer long) complexes of dunes form a drass. Globally, **dune fields** and **seas** are common between 20° to 40° N, and 20° to 40° S latitudes.

In contrast to well-sorted dunes, a loess is another form of sedimentary, wind-driven deposit usually associated with glacier movements. Loess formations, however, represent layers of settling dust and are not well-sorted.

The formation and movements of dune fields are also of great interest to extraterrestrial or planetary geologists. Analysis of **satellite** images of Mars, for example, allows calculation of the strength and direction of the Martian winds and provides insight into Martian atmospheric dynamics. Dunes fields are a significant Martian landform and many have high rates of migration.

See also Beach and shoreline dynamics; Bed or traction load; Bedforms (ripples and dunes); Desert and desertification; Eolian processes; Glacial landforms; Landscape evolution

DUST STORMS

Dust storms are windstorms that severely blow dust **clouds** across a large **area** in arid or semi-arid regions. Dust storms are different from dust devils, which are small atmospheric dust-filled vortices created by differences in surface heating during fair, hot **weather**. Dust storms can cause poor air quality, decrease visibility, can be hazardous to human and animal health, can interfere with telecommunications, erode away the topsoil, block sunlight, and even can greatly influence not only local, but regional and global weather patterns by accumulating and transporting dust in the atmosphere. For example, after a dust storm in the Sahara, dust can move up to high altitudes, and can be carried hundreds or even thousands of miles away by air streams, causing an illness destroying Caribbean **coral reefs**, resulting in asthma outbreaks in the

Poor planting methods, extended periods of drought, and high winds were all factors contributing to the generation of dust storms during the Dust Bowl in the 1930s. *AP/Wide World. Reproduced by permission.*

United States, or providing good nutrients to the Amazonian rain **forests**. Another example is the dust storm in 2001 that began in Mongolia and gathered industrial pollution from China, then caused a haze in a quarter of the United States mainland.

Although dust storms occur naturally, some anthropogenic activity such as removal of vegetation or overgrazing can increase the amount of sediment available for dust storm events. An example of a prolonged impact of dust storms is the historical event called Dust Bowl in the 1930s, which was a disaster both with ecological and societal consequences. The Dust Bowl took place in the southern Great Plains of the United States, including parts of Kansas, Oklahoma, Texas, New Mexico, and Colorado, and was caused by the combination of poor agricultural practices and years of sustained **drought**. Extreme weather and artificially eroded soils resulted in terrible dust storms alternating with drought, heat, **blizzards** and **floods**. The land dried up because the original grasslands holding the **soil** in place were either plowed, then planted with wheat for many years, or because of overgrazing. Consequently, great clouds of dust and **sand** carried by the **wind** covered the area, sometimes even reaching as far as the Atlantic coast. In many places 8–10 cm (3–4 in) of topsoil was blown away. In 1935, programs for soil conservation and for rehabilitation of the Dust Bowl started, including seeding large areas in grass, crop **rotation**, contour plowing, terracing, and strip planting. Accordingly, subsequent droughts in the region had a much less impact, because the available dust for dust storms was greatly reduced by improved agricultural practices.

See also Desert and desertification; Erosion

Hubble image of the Eagle Nebula, "the Pillars of Creation." See entry, "Astronomy," page 34. *U.S. National Aeronautics and Space Administration (NASA).*

Aurora Borealis, one of the great wonders of the natural world. See entry, "Aurora Borealis and Aurora Australialis," page 46. *FMA. Reproduced by permission.*

Postojna Cave, Slovakia. See entry, "Cave," page 99. *Embassy of the Republic of Slovenia. Reproduced by permission.*

Phosphorescence of calcite cave pearls. See entry, "Cave Minerals and speleothems," page 101. © *Y. Shopov. Reproduced by permission.*

Characteristic red luminenescence of hydrothermal calcite. See entry, "Cave Minerals and speleothems," page 101. © Y. Shopov. *Reproduced by permission.*

Stalactites in Cueva Nerja Cave, Spain. See entry, "Cave Minerals and speleothems," page 101. © *Y. Shopov. Reproduced by permission.*

Large cave ice skeleton crystal in Crow's Nest Pass. See entry, "Cave Minerals and speleothems," page 101. © *Y. Shopov.*
Reproduced by permission.

Channel pattern of the Niger River, Timbuktu, Mali. See entry, "Channel patterns," page 107. © *Wolfgang Kaehler. UPI/Corbis-Bettmann.* *Reproduced by permission.*

Coral reefs as ecosystems are being endangered by human activities. See entry, "Coral reefs and corals," page 128. © *Stephen Frink/Corbis.*
Reproduced by permission.

TOP: False color imaging of a solar flare. See entry, "Coronal ejections and magnetic storms," page 130. *U.S. National Aeronautics and Space Administration (NASA).*

BOTTOM: Microscopic view of granite porphyry crystals. See entry, "Crystals and crystallography," page 143. *© Lester V. Bergman/Corbis. Reproduced by permission.*

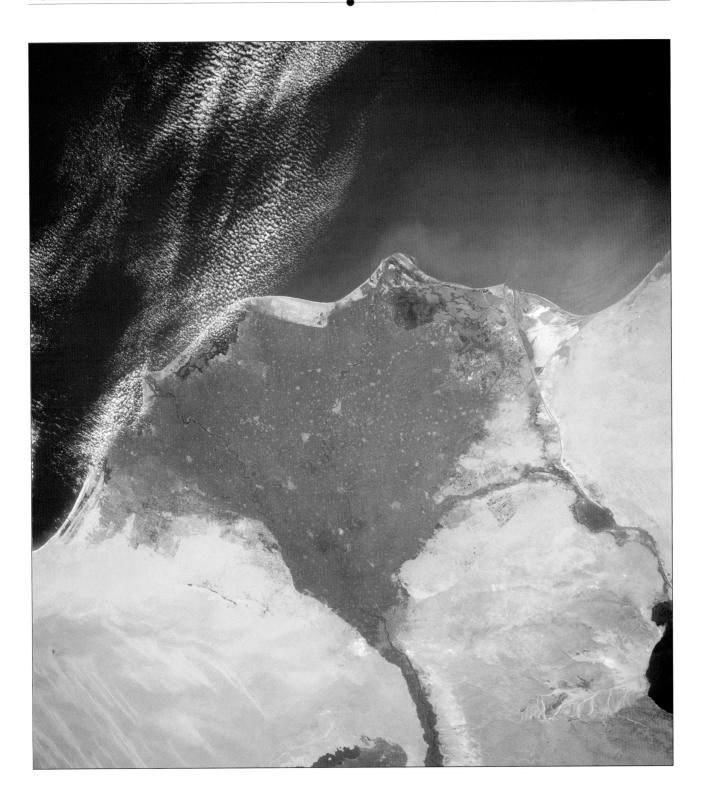

The Nile River delta, as seen from space. See entry, "Delta," page 161. *Corbis/NASA. Reproduced by permission.*

TOP: Far solar corona as seen during a total eclipse. See entry, "Eclipse," page 185. © *Y. Shopov. Reproduced by permission.*

BOTTOM: Giant coronal streamers, visible only from space or the upper stratosphere during an eclipse. See entry, "Eclipse," page 185. © *Y. Shopov.*
Reproduced by permission.

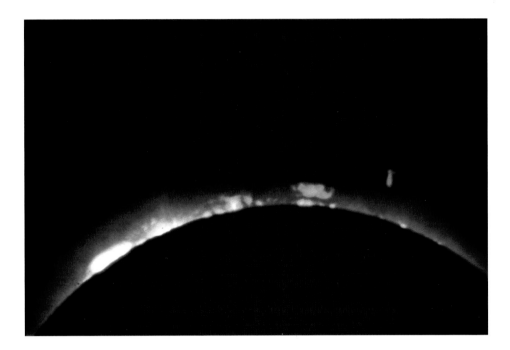

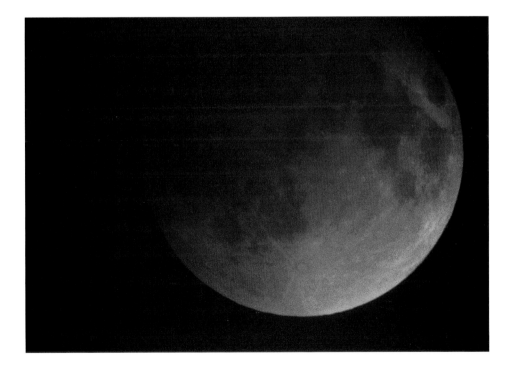

TOP: Solar protuberances at high magnification, as seen during a total eclipse. See entry, "Eclipse," page 185. © *Y. Shopov.*
Reproduced by permission.

BOTTOM: The Moon appears to turn red during a lunar eclipse due to dust and other pollutants in the atmosphere. See entry, "Eclipse,"
page 185. © *Y. Shopov. Reproduced by permission.*

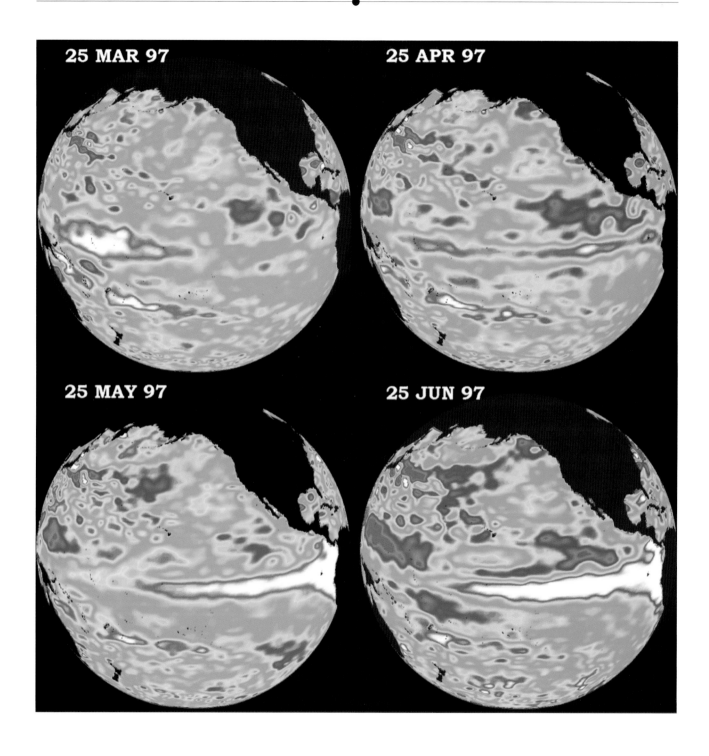

Four false color views of ocean surface height of the Pacific Ocean, with red and white areas showing increased heat storage. See entry, "El Niño and La Niña phenomenon," page 191. *TOPEX/Poseidon, NASA JPL.*

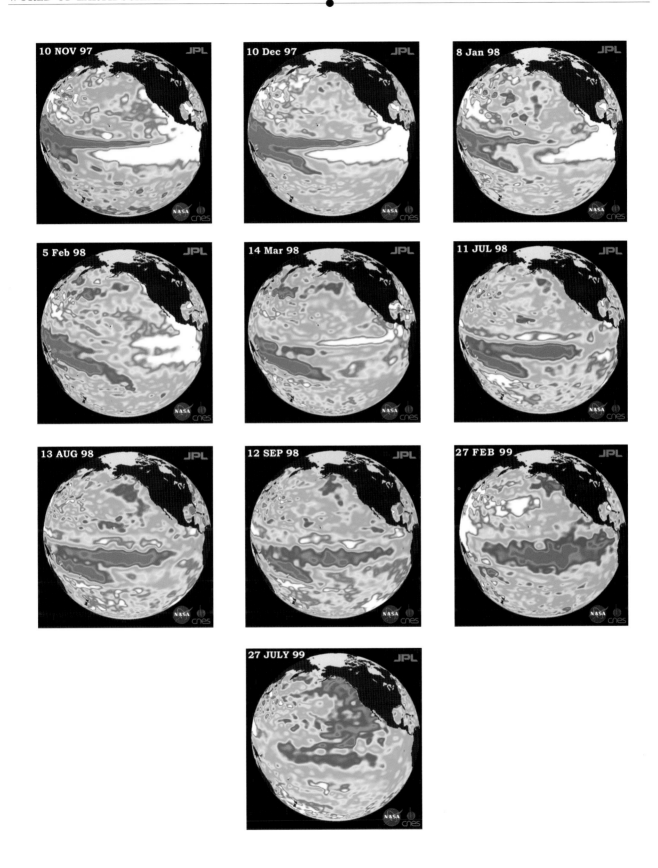

False color image of ocean surface height of the Pacific Ocean, with red and white showing increased heat storage, purple showing low sea level, and green showing normal conditions. See entry, "El Niño and La Niña phenomenon," page 191. *TOPEX/Poseidon, NASA JPL.*

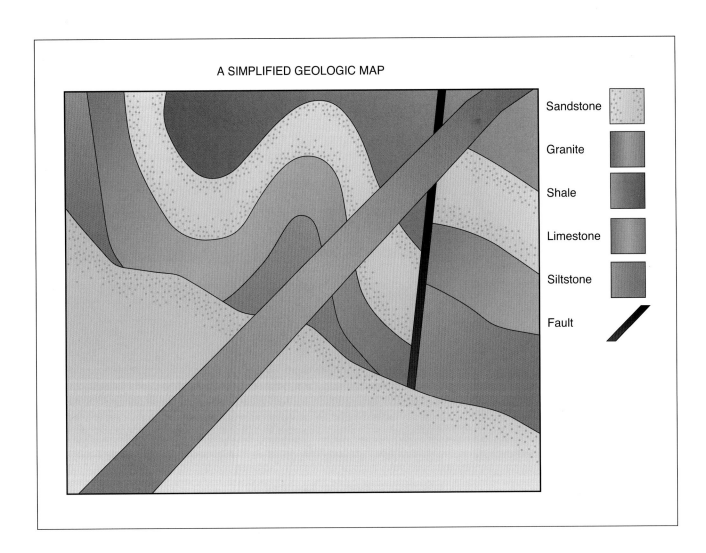

A SIMPLIFIED GEOLOGIC MAP

Sandstone
Granite
Shale
Limestone
Siltstone
Fault

Geologic map. See entry, "Geologic map," page 248.

E

EARLE, SYLVIA A. (1935-)
American oceanographer

Sylvia A. Earle is a former chief scientist of the National Oceanic and Atmospheric Administration (NOAA) and a leading American oceanographer. She was among the first underwater explorers to make use of modern self-contained underwater breathing apparatus (SCUBA) gear, and identified many new species of marine life. With her former husband, Graham Hawkes, Earle designed and built a submersible craft that could dive to unprecedented depths of 3,000 feet.

Sylvia Alice (Reade) Earle was born in Gibbstown, New Jersey, the daughter of Lewis Reade and Alice Freas (Richie) Earle. Both parents had an affinity for the outdoors and encouraged her love of nature after the family moved to the west coast of Florida. As Earle explained to *Scientific American*, "I wasn't shown frogs with the attitude 'yuk,' but rather my mother would show my brothers and me how beautiful they are and how fascinating it was to look at their gorgeous golden eyes." However, Earle pointed out, while her parents totally supported her interest in biology, they also wanted her to get her teaching credentials and learn to type, "just in case."

She enrolled at Florida State University and received her Bachelor of Science degree in the spring of 1955. That fall she entered the graduate program at Duke University and obtained her master's degree in botany the following year. The **Gulf of Mexico** became a natural laboratory for Earle's work. Her master's dissertation, a detailed study of algae in the Gulf, is a project she still follows. She has collected more than 20,000 samples. "When I began making collections in the Gulf, it was a very different body of **water** than it is now—the habitats have changed. So I have a very interesting baseline," she noted in *Scientific American*.

In 1966, Earle received her Ph.D. from Duke University and immediately accepted a position as resident director of the Cape Haze Marine Laboratories in Sarasota, Florida. The following year, she moved to Massachusetts to accept dual roles as research scholar at the Radcliffe Institute and research fellow at the Farlow Herbarium, Harvard University, where she was named researcher in 1975. Earle moved to San Francisco in 1976 to become a research biologist at and curator of the California Academy of Sciences. That same year, she also was named a fellow in botany at the Natural History Museum, University of California, Berkeley.

Although her academic career could have kept her totally involved, her first love was the sea and the life within it. In 1970, Earle and four other oceanographers lived in an underwater chamber for 14 days as part of the government-funded Tektite II Project, designed to study undersea habitats. Fortunately, technology played a major role in Earle's future. A self-contained underwater breathing apparatus had been developed in part by Jacques Cousteau as early as 1943, and refined during the time Earle was involved in her scholarly research. SCUBA equipment was not only a boon to recreational divers, but it also dramatically changed the study of marine biology. Earle was one of the first researchers to don a mask and **oxygen** tank and observe the various forms of plant and animal habitats beneath the sea, identifying many new species of each. She called her discovery of undersea **dunes** off the Bahama Islands "a simple Lewis and Clark kind of observation." But, she said in *Scientific American*, "the presence of dunes was a significant insight into the formation of the area."

Though Earle set the unbelievable record of freely diving to a depth of 1,250 feet, there were serious depth limitations to SCUBA diving. To study deep-sea marine life would require the assistance of a submersible craft that could dive far deeper. Earle and her former husband, British-born engineer Graham Hawkes, founded Deep Ocean Technology, Inc., and Deep Ocean Engineering, Inc., in 1981, to design and build submersibles. Using a paper napkin, Earle and Hawkes roughsketched the design for a submersible they called *Deep Rover*, which would serve as a viable tool for biologists. "In those days we were dreaming of going to thirty-five thousand feet," she told *Discover* magazine. "The idea has always been that scientists couldn't be trusted to drive a submersible by themselves because they'd get so involved in their work they'd run into things." *Deep Rover* was built and continues to operate as a mid-water machine in ocean depths ranging 3,000 feet.

In 1990, Earle was named the first woman to serve as chief scientist at the National Oceanic and Atmospheric Administration (NOAA), the agency that conducts underwater research, manages fisheries, and monitors marine spills. She left the position after eighteen months because she felt that she could accomplish more working independently of the government.

Earle, who has logged more than 6,000 hours under water, is the first to decry America's lack of research money being spent on deep-sea studies, noting that of the world's five deep-sea manned submersibles (those capable of diving to 20,000 feet or more), the United States has only one, the *Sea Cliff*. "That's like having one Jeep for all of North America," she said in *Scientific American*. In 1993, Earle worked with a team of Japanese scientists to develop the equipment to send first a remote, then a manned submersible to 36,000 feet. "They have money from their government," she told *Scientific American*. "They do what we do not: they really make a substantial commitment to ocean technology and science." Earle also plans to lead the $10 million deep ocean engineering project, Ocean Everest, that would take her to a similar depth.

In addition to publishing numerous scientific papers on marine life, Earle is a devout advocate of public education regarding the importance of the **oceans** as an essential environmental habitat. She is currently the president and chief executive officer of Deep Ocean Technology and Deep Ocean Engineering in Oakland, California, as well as the co-author of *Exploring the Deep Frontier: The Adventure of Man in the Sea*.

See also Deep sea exploration; Ocean circulation and currents

EARTH, INTERIOR STRUCTURE

It is 3,950 miles (6,370 km) from the earth's surface to its center. The **rock** units and layers near the surface are understood from direct observation, core samples, and drilling projects. However, the depth of drill holes, and therefore, the direct observation of Earth materials at depth, is severely limited. Even the deepest drill holes (7.5 mi, 12 km) penetrate less than 0.2% of the distance to the earth's center. Thus, far more is known about the layers near the earth's surface, and scientists can only investigate the conditions within the earth's interior (density, **temperature**, composition, solid versus liquid phase, etc.) through more indirect means.

Geologists collect information about Earth's remote interior from several different sources. Some rocks found at the earth's surface, known as kimberlite and ophiolite, originate deep in the **crust** and mantle. Some meteorites are also believed to be representative of the rocks of the earth's mantle and core. These rocks provide geologists with some idea of the composition of the interior.

Another source of information, while more indirect, is perhaps more important. That source is **earthquake**, or seismic waves. When an earthquake occurs anywhere on Earth, seismic waves travel outward from the earthquake's center. The speed, motion, and direction of seismic waves changes dramatically at different levels within Earth, known as seismic transition zones. Therefore, scientists can make various assumptions about the earth's character above and below these transition zones through careful analysis of seismic data. This information reveals that Earth is composed of three basic sections, the crust (the thin outer layer), the mantle, and the core.

The outermost layer of Earth is the crust, or the thin "shell" of rock that covers the globe. There are two types of crust: the continental crust, which consists mostly of light-colored rock of granitic composition that underlies the earth's continents; and the oceanic crust, which is a dark-colored rock of basaltic composition that underlies the **oceans**. One of the most important differences between continental and oceanic crust is their difference in density. The lighter-colored continental crust is also lighter in weight, with an average density of 2.6 g/cm^3 (grams per cubic centimeter), compared to the darker and heavier basaltic oceanic crust, which has an average density of 3.0 g/cm^3. It is this difference in density that causes the continents to have an average elevation of about 2,000 ft (600 m) above sea level, while the average elevation (depth) of the ocean bottom is 10,000 ft (3,000 m) below sea level. The heavier oceanic crust sits lower on the earth's surface, creating the topographic depressions for the ocean basins, while the lighter continental crust rests higher on the earth's surface, causing the elevated and exposed continental land masses.

Another difference between the oceanic crust and continental crust is the difference in thickness. The heavier oceanic crust forms a relatively thin layer of 3–6 mi (5–10 km), while the continental crust is lighter, and the underlying material can support a thicker layer. The continental crust averages about 20 mi (35 km) thick, but can reach up to 40 mi (70 km) in certain sections, particularly those found under newly elevated and exposed mountain ranges such as the Himalayas.

The base of the crust (both the oceanic and continental varieties) is determined by a distinct seismic transition zone called the Mohorovičic discontinuity. The Mohorovičic discontinuity, commonly referred to as "the Moho" or the "M-discontinuity," is the transition or boundary zone between the bottom of the earth's crust and the underlying unit, which is the uppermost section of the mantle called the lithospheric mantle. Like the crust, the lithospheric mantle is solid, but it is considerably more dense. Because the thickness of the earth's crust varies, the depth to the Moho also varies from 3–6 mi (5–10 km) under the oceans to 20–40 mi (35-70 km) under the continents.

This transition between the crust and the mantle was first discovered by the Croatian seismologist Andrija Mohorovičic in 1908. On October 8, 1908, Andrija Mohorovičic observed seismic waves that emitted from an earthquake in Croatia. He noticed that both the compressional, or primary (P), waves and the shear, or secondary (S), waves, at one point in their journey, picked up speed as they traveled

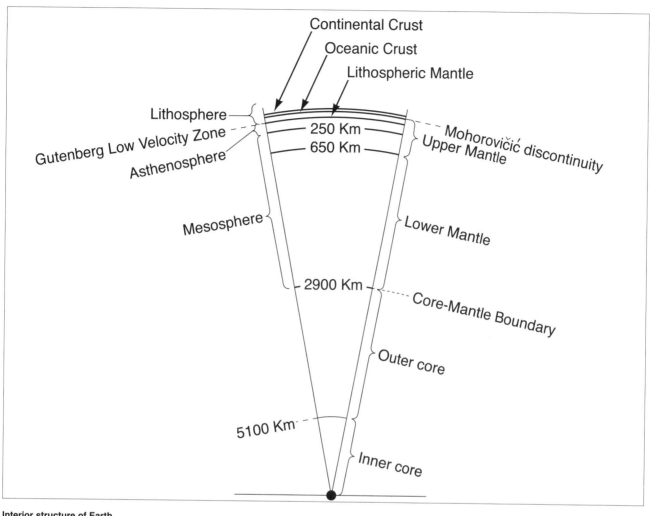

Interior structure of Earth.

farther from the earthquake. This suggested that the waves had been deflected, as if they had encountered something that had affected their energy. He noted that this increase in speed seemed to occur at a depth of about 30 mi (50 km). Since seismic waves travel faster through denser material, he reasoned that there was an abrupt transition from the rocky material in the earth's crust, to denser rocks below. In honor of Andrija Mohorovicic's discovery, this transition zone marking the base of the earth's crust was named after him.

The Moho is a relatively narrow transition zone estimated to be somewhere between 0.1–1.9 mi (0.2–3 km) thick. Currently, the Moho is defined by the level within Earth where P wave velocity increases abruptly from an average speed of about 4.3 mi/second (6.9 km/second) to about 5.0 mi/second (8.1 km/second).

Underlying the crust is the mantle. The uppermost section of the mantle, which is a rigid layer, is called the lithospheric mantle. This section extends to an average depth of about 40 mi (70 km), although it fluctuates between 30–60 mi (50–100 km). The density of this layer is greater than that of the crust, and averages 3.3 g/cm³. But like the crust, this sec-

tion is solid and brittle, and relatively cool compared to the material below. This rigid uppermost section of the mantle (the lithospheric mantle), combined with the overlying solid crust, is called the **lithosphere**, which is derived from the Greek word *lithos*, meaning rock. At the base of the lithosphere, a depth of about 40 mi (70 km), there is another distinct seismic transition called the Gutenberg low velocity zone. At this level, the velocity of S waves decreases dramatically, and all seismic waves appear to be absorbed more strongly than elsewhere within the earth. Scientists interpret this to mean that the layer below the lithosphere is a softer zone of partially melted material (with between 1–10% molten material). This "soft" zone is called the *asthenosphere*, from the Greek word *asthenes* meaning weak. This transition zone between the lithosphere and the **asthenosphere** is named after Beno Gutenberg, a mid-twentieth century geologist who made several important contributions to the study and understanding of the earth's interior. It is at this level that some important Earth dynamics occur, affecting those at the surface. At the Gutenberg low velocity zone, the lithosphere is carried "piggyback" on top of the weaker, less rigid asthenosphere, which seems to be in continual motion. This motion creates stress in the rigid rock layers above it, and the slabs or plates of the lithosphere are forced to jostle against each other, much like **ice** cubes floating in a bowl of swirling **water**. This motion of the **lithospheric plates** is known as **plate tectonics**, and it is responsible for many of the earth's activities that we experience at the surface today, including earthquakes, certain types of volcanic activity, and continental drift.

The asthenosphere extends to a depth of about 155 mi (250 km). Below that depth, seismic wave velocity increases, suggesting an underlying denser, but solid phase.

The rest of the mantle, from the base of the asthenosphere at 155 mi (250) km to the core at 1,800 mi (2,900 km), is called the **mesosphere** (or middle sphere). There are mineralogical and compositional changes suggested by sharp velocity increases within the mesosphere. Notably, there is a thin zone at about the 250 mi (400 km) depth attributed to a possible mineralogical change (presumably from an abundance of the mineral **olivine** to the mineral spinel), and there is another sharp velocity increase at about the 400 mi (650 km) level, attributed to a possible increase in the ratio of **iron** to magnesium in mantle rocks. Except for these variations, down to the 560 mi (900 km) level the mesosphere seems to contain predominantly solid material that displays a relatively consistent pattern of gradually increasing density and seismic wave velocity with increasing depth and pressure. Below the 560 mi (900 km) depth, the P and S wave velocities continue to increase, but the rate of increase declines with depth.

At a depth of 1,800 mi (2,900 km) there is another abrupt change in the seismic wave patterns, known as the Gutenberg discontinuity, or more often referred to as the core-mantle boundary (CMB). At this level, P waves decrease while S waves disappear completely. Since S waves cannot be transmitted through liquids, it is believed that the CMB denotes a phase change from the solid mantle above, to a liquid outer core below. This phase change is believed to be accompanied by an abrupt temperature increase of 1,300° F (704° C). This hot, liquid outer core material is denser than the cooler, solid mantle, probably due to a greater percentage of iron. It is believed that the outer core consists of a liquid of 80–92% iron, alloyed with a lighter element. The composition of the remaining 8–20% is not well understood, but it must be a compressible element that can mix with liquid iron at these immense pressures. Various candidates proposed for this element include **silicon**, sulfur, or **oxygen**.

The actual boundary between the mantle and the outer core is a narrow, uneven zone that contains undulations that may be 3–6 mi (5–8 km) high. These undulations are affected by heat-driven convection activity within the overlying mantle, which may be the driving force for plate tectonics. The interaction between the solid mantle and the liquid outer core is very important to Earth dynamics for another reason. It is the eddies and currents in the core's iron-rich fluids that are ultimately responsible for the earth's **magnetic field**.

Although the core-mantle boundary is currently situated at a depth of about 1,800 mi (2,900 km), this depth has not been constant through **geologic time**. As the heat of the earth's interior is constantly but slowly dissipated, the molten core

within the earth gradually solidifies and shrinks, causing the core-mantle boundary to slowly move deeper and deeper within the earth's core.

There is one final, even deeper transition evident from seismic wave data. Within Earth's core, at the 3,150 mi (5,100 km) level, P waves speed up and are reflected from yet another seismic transition zone. This indicates that the material in the inner core below 3,150 mi (5,100 km) is solid. The phase change from liquid to solid is probably due to the immense pressures present at this depth.

In addition to this phase change in the inner core from liquid to solid, seismic wave velocities, as well as the earth's total weight, suggest that the inner core has a different composition than the outer core. This could be accounted for by a relatively pure iron-nickel composition for the inner core. Although no direct terrestrial evidence for a solid iron-nickel inner core exists, comparative evidence from meteorites supports this theory. Numerous meteorites, fragments presumably from the interior of shattered extraterrestrial bodies within our **solar system**, often contain relatively pure iron or iron-nickel compositions. It is likely that the composition of the core of our own planet is very similar to the composition of these extraterrestrial travelers.

See also Earth (planet); Earth science; Seismograph; Seismology; Volcanic eruptions; Volcanic vent

EARTH OBSERVING SYSTEMS (EOS) · *see* TERRA SATELLITE AND EARTH OBSERVING SYSTEMS (EOS)

EARTH, PLANET

Earth is the third of nine planets in our **solar system**. Its surface is mostly **water** (about 70%), and it has a moderately dense nitrogen and **oxygen** atmosphere that supports life. Rich in **iron** and nickel, Earth is a dense, molten oblate sphere with a solid core and a thin outer **crust**. Earth rotates about its **polar axis** as it revolves around the **Sun**. Earth has one natural **satellite**, the **Moon**. A complete **revolution** of Earth around the Sun takes about one year, while a **rotation** on its axis takes one day. The surface of Earth is constantly changing, as tectonic plates slowly move about on the turbulent foundation of partially molten **rock** beneath them. Collisions between landmasses build mountains; **erosion** wears them down. Slow changes in the **climate** cause equally slow changes in vegetation and animals.

Earth orbits the Sun at a distance of about 93,000,000 mi (150,000,000 km), taking 365.25 days to complete one revolution. Earth is small by planetary standards, only one-tenth the size of Jupiter. The equatorial radius of the earth is about 6,378 km. The polar radius of the earth is about 6,357 km. The difference is due to centrifugal flattening. Earth's mass is estimated at approximately 6.0×10^{24} kg. The difference is due to centrifugal flattening. Unlike the outer planets, which are composed mainly of light gases, Earth is made of heavy elements such as iron and nickel, and is, therefore, much more

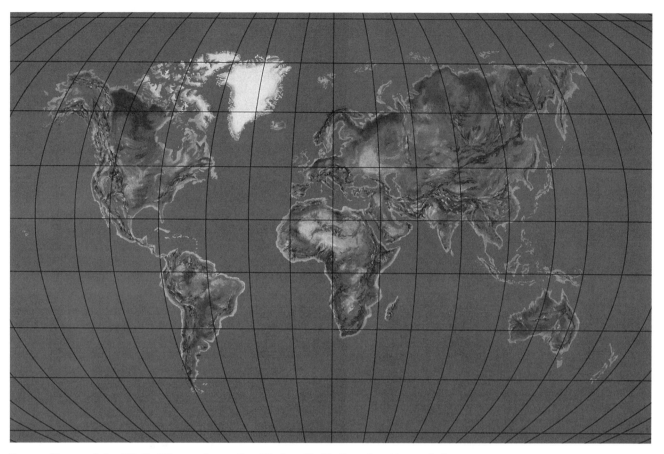

Topographic map of planet Earth. © *Tecmap Corporation, Eric Curry/Corbis. Reproduced by permission.*

dense. These characteristics—small and dense—are typical of the inner four terrestrial planets.

About 4.5 to 5.1 billion years ago, the solar system formed from a contracting cloud of interstellar gas. The cloud underwent compression and heating as it shrank, until its central part blazed forth as the mature, stable star. As the Sun formed, the surrounding gas cloud flattened into a disk. In this disk, the first solid particles formed and then grew as they accreted additional matter from the surrounding gas. Soon sub-planetary bodies, called planetesimals, built up, and then they collided and merged, forming the planets. The high temperatures in the inner solar system ensured that only the heavy elements, those that form rock and metal, could survive in solid form.

Thus were formed the small, dense terrestrial planets. Hot at first due to the collisions that formed it, Earth began to cool. Its components began to differentiate, or separate themselves according to their density. To its core went the heavy abundant elements, iron and nickel. Outside the core were numerous elements compressed into a dense but pliable substance called the mantle. Finally, a thin shell of cool, oxygen- and silicon-rich rock formed at Earth's surface: the crust, or **lithosphere**. Formation of the crust from the initial molten mass took half a billion years.

Earth's atmosphere formed as a result of outgassing of **carbon dioxide** from its interior, and accretion of gases from **space**, including elements brought to Earth by **comets**. The lightest elements, such as helium and most of the hydrogen, escaped to space, leaving behind an early atmosphere consisting of hydrogen compounds such as methane and ammonia as well as water vapor and nitrogen- and sulfur-bearing compounds released by volcanoes. **Carbon** dioxide was also plentiful, but was soon dissolved in ocean waters and deposited in carbonate rocks. As the gases cooled, they condensed, and rains inundated the planet. The lithosphere was uneven, containing highlands made of buoyant rock such as **granite**, and basins of heavy, denser **basalt**. Into these giant basins the rains flowed, forming the **oceans**. Eventually life forms appeared, and over the course of a billion years, plants enriched the atmosphere with oxygen, finally producing the nitrogen-oxygen atmosphere we have today.

Earth's crust is in a constant, though slow, state of change. Landmasses move, collide, and break apart according to a process called **plate tectonics**. The lithosphere (the outer crust and a portion of the upper mantle) is not one huge shell of rock; it is composed of several large pieces called plates. These pieces are constantly in motion, because Earth's interior is dynamic, with its core still molten and with large-scale convective currents in the upper mantle. The thermal forces move the plates a few centimeters a year, but this is enough to have profound consequences over the long expanse of **geologic time**.

For instance, the center of the North American continent is the wide open expanse of the Great Plains and the Canadian Prairies. On the eastern edge, the rolling **folds** of the Appalachian Mountains grace western North Carolina, Virginia, and Pennsylvania. In the west, the jagged, crumpled Rockies thrust skyward, tall, stark, and snow-capped.

These two great ranges represent one of the two basic land-altering processes: mountain building. Two hundred million years ago, **North America** was moving east, driven by plate tectonics. In a shattering, slow-motion collision, it rammed into what is now **Europe** and North **Africa**. The land crumpled, and the ancient Appalachians rose. At that time, they were the mightiest mountains on Earth. A hundred million years later, North America was driven back west. Now the western edge of the continent rumbled along over the Pacific plate, and about 80 million years ago, a massive spate of mountain building formed the Rockies.

During the time since the Appalachians rose, the other land-altering process, erosion, has reshaped the landscape. Battered by **wind** and water, their once sheer mountain flanks have been worn into the low, rolling hills of today.

Mountain building can be seen today in the Himalayas, which are still rising as India moves northward into **Asia**, crumpling parts of Nepal and Tibet. Erosion rules in Arizona's Grand **Canyon**, which gradually is deepening and widening as the Colorado River slices now into ancient granite two billion years old.

This unending cycle of mountain building (caused by movement of the crustal plates) and erosion (by wind and water) has formed every part of Earth's surface today. Where there are mountains, as in the long ranks of the Andes or the Urals, there is subterranean conflict. Where a crustal plate rides over another one, burying and **melting** it in the hot regions below the lithosphere, volcanoes rise, dramatically illustrated by Mt. St. Helens in Washington and the other volcanoes that line the Pacific rim. Where lands lie wide and arid, they are sculpted into long, scalloped cliffs, as one sees in the deserts of New Mexico, Arizona, and Utah.

Earth is mostly covered with water. The Pacific Ocean covers nearly half of Earth; from the proper vantage point in space above the middle of the Pacific Ocean, one would see nothing but water, dotted here and there with tiny islands, with only **Australia** and the coasts of Asia and the Americas rimming the edge of the globe.

The lithosphere rides on a pliable layer of rock in the upper mantle called the **asthenosphere**. Parts of the lithosphere are made of relatively light rocks (continental crust), while others (oceanic crust) are made of heavier, denser rocks. Just as corks float mostly above water while **ice** cubes float nearly submerged, the less dense parts of the lithosphere ride higher on the asthenosphere than the more dense ones. Earth therefore has huge basins, and early in the planet's history, these basins filled with water condensing and raining out of the primordial atmosphere. Additional water was brought to Earth by the impacts of comets, whose nuclei are made of water ice.

The atmosphere has large circulation patterns, and so do the oceans. Massive streams of warm and cold water flow through them. One of the most familiar is the **Gulf Stream**, which brings warm water up the eastern coast of the United States.

Circulation patterns in the oceans and in the atmosphere are driven by **temperature** differences between adjacent areas and by the rotation of Earth, which helps create circular, or rotary, flows. Oceans play a critical role in the overall energy balance and **weather** patterns. Storms are ultimately generated by moisture in the atmosphere, and **evaporation** from the oceans is the prime source of such moisture. Oceans respond less dramatically to changes in energy input than land does, so the temperature over a given patch of ocean is far more stable than one on land.

Earth's atmosphere is the gaseous region above its outer crust, composed of nitrogen (78% by number), oxygen (21%), and other gases (1%). It is only about 50 mi (80 km) from the ground to space: on a typical, 12-in (30 cm) globe the atmosphere would be less than 2 mm thick. The atmosphere has several layers. The most dense and significant of these is the **troposphere**; all weather occurs in this layer, and commercial jets cruise near its upper boundary, 6 mi (10 km) above Earth's surface. The **stratosphere** lies between 6 and 31 mi (10 and 50 km) above, and it is here that the ozone layer lies. In the **mesosphere** and the **thermosphere** one finds auroras occurring after eruptions on the Sun; radio communications "bounce" off the **ionosphere** back to Earth.

The atmosphere is an insulator of almost miraculous stability. Only 50 mi (80 km) away is the cold of outer space, but the surface remains temperate. Heat is stored by the land and the atmosphere during the day, but the resulting heat radiation (infrared) from the surface is prevented from radiating away by gases in the atmosphere that trap infrared radiation. This is the well-known **greenhouse effect**, and it plays an important role in the atmospheric energy budget. According to some models, a global temperature decrease of two degrees could trigger the next advance of an ice age, while an increase of three degrees could melt the **polar ice** caps, submerging nearly every coastal city in the world.

Despite this overall stability, the troposphere is nevertheless a turbulent place. It is in a state of constant circulation, driven by Earth's rotation as well as the constant heating and cooling that occurs during each 24-hour period.

The largest circulation patterns in the troposphere are the Hadley cells. There are three of them in each hemisphere, with the middle, or Ferrel cell, lying over the latitudes spanned by the continental United States. Northward-flowing surface air in the Ferrel cell is deflected toward the east by the Coriolis force, with the result that winds—and weather systems—move from west to east in the middle latitudes of the northern hemisphere.

Near the top of the troposphere are the jet streams, fast-flowing currents of air that circle Earth in sinuous paths.

Circulation on a smaller but violent scale appears in the cyclones and anticyclones, commonly called low and high pressure cells. Lows typically bring unsettled or stormy weather, while highs mean sunny skies. Weather in most areas follows a basic pattern of alternating calm weather and storms, as the endless progression of highs and lows, generated by Earth's rotation and temperature variation, passes by. This is a great simplification, however, and weather in any given place may be affected, or even dominated, by local features. The climate in Los Angeles, California is entirely different from that in Las Vegas, Nevada though the two cities are not very far apart. Here, local features—specifically, the mountains between them—are as important as the larger circulation patterns.

Earth has a **magnetic field** that extends tens of thousands of kilometers into space and shields Earth from most of the solar wind, a stream of particles emitted by the Sun. Sudden enhancements in the solar wind, such as a surge of particles ejected by an eruption in the Sun's atmosphere, may disrupt the magnetic field, temporarily interrupting long-range radio communications and creating brilliant displays of auroras near the poles, where the magnetic field lines bring the charged particles close to the earth's surface.

Farther out, at a mean distance of about 248,400 mi (400,000 km), is Earth's only natural satellite, the Moon. Some astronomers assert that the Earth and the Moon should properly be considered a "double planet," since the Moon is larger relative to our planet than the satellites of most other planets.

The presence of life on Earth is, as far as we know, unique. Men have walked on the Moon, and there is no life on our barren, airless satellite. Unmanned spacecraft have landed on Venus and Mars and have flown close to every other planet in the solar system except Pluto. The most promising possibility, Mars, yielded nothing to the automated experiments performed by the Viking and Mars Surveyor spacecraft that searched for signs of extraterrestrial life.

See also Atmospheric composition and structure; Biogeochemical cycles; Continental drift theory; Cosmology; Earth science; Earth, interior structure; Evolution, evidence of; Evolutionary mechanisms; Gaia hypothesis; Geochemistry; Geologic time; Global warming; Greenhouse gases and greenhouse effect; History of exploration I (Ancient and classical); History of exploration II (Age of exploration); History of exploration III (Modern era); Hydrologic cycle; Landforms; Landscape evolution; Latitude and longitude; Marine transgression and marine regression; Miller-Urey experiment; Oceanography; Origin of life; Orogeny; Revolution and rotation; Space and planetary geology; Supercontinents; Terra satellite and Earth Observing Systems (EOS); Uniformitarianism; Weather and climate

EARTH SCIENCE

Befitting a dynamic Earth, the study of Earth science embraces a multitude of subdisciplines. To understand the complexities of Earth, one must see the patterns of complex interaction through the eyes of the physicist, chemist, geologist, meterologist, and explorer. Just as no single agency has the responsibility to collect geophysical data—in fact there are hundreds worldwide—no single discipline provides the insight to understand or explain all of Earth's complexities.

Moreover, just a cooperation and collaboration are essential for gathering data, an interdisciplinary approach to Earth science is essential to proper evaluation of that data.

At the heart of Earth science is the study of **geology**. Literally meaning "to study the Earth," traditional geological studies of rocks, **minerals**, and local formations have within the last century, especially in the light of the development of plate tectonic theory, broadened to include studies of geophysics and **geochemistry** that offer sweeping and powerful explanations of how continents move, to explanations of the geochemical mechanisms by which **magma** cools and hardens into a multitude of **igneous rocks**.

Earth's formation and the **evolution** of life upon its fragile outer **crust** was dependent upon the conditions established during the formation of the **solar system**. The **Sun** provides the energy for life and drives the turbulent atmosphere. A study of Earth science must, therefore, not ignore a treatment of Earth as an astronomical body in **space**.

At the opposite extreme, deep within Earth's interior, radioactive decay adds to the heat left over from the **condensation** of Earth from cosmic dust. This heat drives the forces of **plate tectonics** and results in the tremendous variety of features that distinguish Earth. To understand Earth's interior structure and dynamics, seismologists probe the interior structure with seismic shock waves.

It does not require the spectacular hurricane, **tornado**, **landslide**, or volcanic eruption to prove that Earth's atmosphere and **seas** are dynamic entities. Forces that change and shape the Earth appear on a daily basis in the form of **wind** and **tides**. What Earth scientists, including meteorologists and oceanographers seek to explain—and ultimately to quantify—are the physical mechanisms of change and the consequences of those changes. Only by understanding the mechanisms of change can predictions of **weather** or climatic change hope to achieve greater accuracy.

The fusion of disciplines under the umbrella of Earth science allows a multidisciplinary approach to solving complex problems or multi-faceted issues of resource management. In a addition to hydrogeologists and cartographers, a study of ground **water** resources could, for example, draw upon a wide diversity of Earth science specialists.

Although modern Earth Science is a vibrant field with research in a number of important and topical areas (e.g., identification of energy resources, **waste disposal** sites, etc), the span of geological process and the enormous expanse of **geologic time** make critical the study of ancient processes (e.g., paleogeological studies). Only by understanding how processes have shaped Earth in the past—and through a detailed examination of the geological record—can modern science construct meaningful predictions of the potential changes and challenges that lie ahead.

See also Astronomy; Chemistry; Earth (planet); Field methods in geology; Geochemistry; Historical geology; History of exploration I (Ancient and classical); History of exploration II (Age of exploration); History of exploration III (Modern era); History of geoscience: Women in the history of geoscience; History of manned space exploration; Oceanography; Physics;

Scientific data management in Earth Sciences; Space and planetary geology

EARTHQUAKE

An earthquake is a geological event inside the earth that generates strong vibrations. When the vibrations reach the surface, the earth shakes, often causing damage to natural and manmade objects, and sometimes killing and injuring people and destroying their property. Earthquakes can occur for a variety of reasons; however, the most common source of earthquakes is movement along a fault.

Some earthquakes occur when tectonic plates, large sections of Earth's **crust** and upper mantle, move past each other. Earthquakes along the San Andreas and Hayward faults in California occur because of this. Earthquakes also occur if one plate overruns another, as on the western coast of **South America**, the northwest coast of **North America**, and in Japan. If plates collide but neither is overrun, as they do crossing **Europe** and **Asia** from Spain to Vietnam, earthquakes result as the rocks at the abutting plates compress into high mountain ranges. In all three of these settings, earthquakes result from movement along faults.

A fault block may also move due to **gravity**, sinking between other fault blocks that surround and support it. Sinking fault blocks and the mountains that surround them form a distinctive **topography** of basins and mountain ranges. This type of fault block configuration is typified by the North American Basin and Range topographic province. In such places, elevation losses by the valleys as they sink between the mountains are accompanied by tremors or earthquakes. Another kind of mountain range rises because of an active thrust fault. Tectonic compression (tectonic, meaning having to do with the forces that deform the rocks of planets) shoves the range up the active thrust fault, which acts like a natural ramp.

Molten **rock** called **magma** moves beneath but relatively close to the earth's surface in volcanically active regions. Earthquakes sometimes accompany **volcanic eruptions** as huge masses of magma move underground.

Nuclear bombs exploding underground cause small local earthquakes, which can be felt by people standing within a few miles of the test site. The earthquakes caused by nuclear bombs are tiny compared to natural earthquakes; but they have a distinctive "sound," and their location can be pinpointed. This is how nuclear weapons testing in one country can be monitored by other countries around the world.

Earth is covered by a crust of solid rock, which is broken into numerous plates that move around on the surface, bumping, overrunning, and pulling away from each other. One kind of boundary between rocks within a plate, as well as at the edges of the plates, is a fault. Faults are large-scale breaks in Earth's crust, in which the rock on one side of the fault has been moved relative to the rock on the other side of the fault by tectonic forces. Fault blocks are giant pieces of crust that are separated from the rocks around them by faults.

When the forces pushing on fault blocks cannot move one block past the other, potential energy is stored up in the fault zone. This is the same potential energy that resides in a giant boulder when it is poised, motionless, at the top of a steep slope. If something happens to overcome the friction holding the boulder in place, its potential energy will convert into kinetic energy as it thunders down the slope. In the fault zone, the potential energy builds up until the friction that sticks the fault blocks together is overcome. Then, in seconds, all the potential energy built up over the years turns to kinetic energy as the rocks surge past each other.

The vibrations of a fault block on the move can be detected by delicate instruments (seismometers and seismographs) in rocks on the other side of the world. Although this happens on a grand scale, it is remarkably like pushing on a stuck window or sliding door. Friction holds the window or door tight in its tracks. After enough force is applied to overcome the friction, the window or door jerks open.

Some fault blocks are stable and no longer experience the forces that moved them in the first place. The fault blocks that face each other across an active fault, however, are still influenced by tectonic forces in the ever-moving crust. They grind past each other along the fault as they move in different directions.

Fault blocks can move in a variety of ways, and these movements define the different types of faults. In a vertical fault, one block moves upward relative to the other. At the surface of the earth, a vertical fault forms a cliff, known as a fault scarp. The sheer eastern face of the Sierra Nevada mountain range is a fault scarp. In most vertical faults, the fault scarp is not truly vertical, and one of the fault blocks "hangs" over the other. This upper block is called the hanging wall and the lower block, the foot wall.

In horizontal faults, the blocks slide past one another without either block being lifted. In this case, the objects on the two sides of the fault simply slide past one another; for example, a road that straddles the fault might be offset by a number of feet. Complex faults display movements with both vertical and horizontal displacements.

Any one of the following fault types can generate an earthquake:

- Normal fault—A vertical fault in which the hanging wall moves down compared to the foot wall.

- Reverse fault—A vertical fault in which the hanging wall moves up in elevation relative to the foot wall.

- Thrust fault—A low-angle (less than 30°) reverse fault, similar to an inclined floor or ramp. The lower fault block is the ramp itself, and the upper fault block is gradually shoved up the ramp. The "ramp" may be shallow, steep, or even curved, but the motion of the upper fault block is always in an upward direction. A thrust fault caused the January 1994 Northridge earthquake near Los Angeles, California.

- Strike-slip (or transform) fault—A fault along which one fault block moves horizontally (sideways), past another fault block, like opposing lanes of traffic. The San Andreas fault in Northern California is one of the best known of this type.

When a falling rock splashes into a motionless pool of **water**, waves move out from the point of impact. These waves appear at the interface of water and air as circular ripples. However, the waves occur below the surface, too, traveling down into the water in a spherical pattern. In rock, as in water, a wave-causing event makes not one wave, but a number of waves, moving out from their source, one after another, like an expanding bubble.

Tectonic forces shift bodies of rock inside the earth, perhaps displacing a mountain range several feet in a few seconds, and they generate tremendous vibrations called seismic waves. The earthquake's focus (also called the hypocenter) is the point (usually deep in the subsurface) where the sudden sliding of one rock mass along a fault releases the stored potential energy of the fault zone. The first shock wave emerges at the surface at a point typically directly above the focus; this surface point is called the epicenter. Seismometers detect seismic waves that reach the surface. Seismographs (devices that record seismic phenomena) record the times of arrival for each group of vibrations on a seismogram (either a paper document or digital data).

Like surfaces in an echoing room that reflect or absorb sound, the boundaries of rock types within the earth change or block the direction of movement of seismic waves. Waves moving out from the earthquake's focus in an ever-expanding sphere become distorted, bent, and reflected. Seismologists (geologists who study seismic phenomena) analyze the distorted patterns made by seismic waves and search through the data for clues about the earth's internal structure.

Different kinds of earthquake-generated waves, moving at their own speeds, arrive at the surface in a particular order. The successive waves that arrive at a single site are called a wave train. Seismologists compare information about wave trains that are recorded as they pass through a number of data-collecting sites after an earthquake. By comparing data from three recording stations, they can pinpoint the map location (epicenter) and depth within the earth's surface (focus or hypocenter) of the earthquake.

These are the most important types of seismic waves:

- P-waves—The fastest waves, these compress or stretch the rock in their path through Earth, moving at about 4 mi (6.4 km) per second.

- S-waves—As they move through Earth, these waves shift the rock in their path up and down and side to side, moving at about 2 mi (3.2 km) per second.

- Rayleigh waves and Love waves—These two types of "surface waves" are named after seismologists. Moving at less than 2 mi (3.2 km) per second, they lag behind P-waves and S-waves but cause the most damage. Rayleigh waves cause the ground surface in their path to ripple with little waves. Love waves move in a zigzag along the ground and can wrench buildings from side to side.

The relative size of earthquakes is measured by the **Richter scale**, which measures the energy an earthquake releases. Each whole number increase in value on the Richter scale indicates a 10-fold increase in the energy released and a

Earthquakes are among the most devastating natural disasters because of their unpredictablility, the large area affected, and the irresistible forces they cause. *AP/Wide World. Reproduced by permission.*

thirty-fold increase in ground motion. An earthquake measuring 8 on the Richter scale is ten times more powerful, therefore, than an earthquake with a Richter Magnitude of 7, which is ten times more powerful than an earthquake with a magnitude of 6. Another scale—the Modified Mercalli Scale uses observations of damage (like fallen chimneys) or people's assessments of effects (like mild or severe ground shaking) to describe the intensity of a quake.

Violent shaking changes water bearing **sand** into a liquid-like mass that will not support heavy loads, such as buildings. This phenomenon, called liquefaction, causes much of the destruction associated with an earthquake in liquefaction-prone areas. Downtown Mexico City rests on the old lakebed of Lake Texcoco, which is a large basin filled with liquefiable sand and ground water. In the Mexico City earthquake of 1985, the wet sand beneath tall buildings turned to slurry, as if the buildings stood on the surface of vibrating gelatin in a huge bowl. Most of the 10,000 people who died as a result of that earthquake were in buildings that collapsed as their foundations sank into liquefied sand.

In the sudden rearrangement of fault blocks in the earth's crust that cause an earthquake, the land surface on the dropped-down side of the fault can fall or subside in elevation by several feet. On a populated coastline, this can wipe out a city. Port Royal, on the south shore of Jamaica, subsided several feet in an earthquake in 1692 and suddenly disappeared as the sea rushed into the new depression. Eyewitnesses recounted the seismic destruction of the infamous pirate anchorage, as follows: "...in the space of three minutes, Port-Royall, the fairest town of all the English plantations, exceeding of its riches,...was shaken and shattered to pieces, sunk into and covered, for the greater part by the sea...The earth heaved and swelled like the rolling billows, and in many places the earth crack'd, open'd and shut, with a motion quick and fast...in some of these people were swallowed up, in others they were caught by the middle, and pressed to death...The whole was attended with...the noise of falling mountains at a distance, while the sky...was turned dull and reddish, like a glowing oven." Ships arriving later in the day found a small shattered remnant of the city that was still above the water. Charts of the Jamaican coast soon appeared printed with the words *Port Royall Sunk.*

In the New Madrid (Missouri) earthquake of 1811, a large **area** of land subsided around the bed of the Mississippi River in west Tennessee and Kentucky. The Mississippi was observed to flow backwards as it filled the new depression, to create what is now known as Reelfoot Lake.

Cities depend on networks of so-called "lifeline structures" to distribute water, power, and food and to remove sewage and waste. These networks, whether power lines, water mains, or roads, are easily damaged by earthquakes. Elevated freeways collapse readily, as demonstrated by a section of the San Francisco Bay Bridge in 1989 and the National Highway Number 2 in Kobe, Japan, in 1995. The combination of several networks breaking down at once multiplies the hazard to lives and property. Live power lines fall into water from broken water mains, creating a deadly electric shock hazard. Fires may start at ruptured gas mains or chemical storage tanks. Although emergency services are needed more than ever, many areas may not be accessible to fire trucks and other emergency vehicles. If the water mains are broken, there will be no pressure at the fire hydrants, and the firefighters' hoses are useless. The great fire that swept San Francisco in 1906 could not be stopped by regular firefighting methods. Only dynamiting entire blocks of buildings halted the fire's progress. Both Tokyo and Yokohama burned after the Kwanto earthquake struck Japan in 1923, and 143,000 people died, mostly in the fire.

Popular doomsayers excite uncomprehending fear by saying that earthquakes happen more frequently now than in earlier times. It is true that more people than ever are at risk from earthquakes, but this is because the world's population grows larger every year, and more people are living in earthquake-prone areas.

Today, sensitive seismometers "hear" every noteworthy earth-shaking event, recording it on a seismogram. Seismometers detect earthquake activity around the world, and data from all these instruments are available on the Internet within minutes of the earthquake. News agencies can report the event the same day. People have ready access to information about every earthquake that happens anywhere on Earth. And the earth experiences a lot of earthquakes—the planet never ceases to vibrate with tectonic forces, although the majority of them are not strong enough to be detected except with instruments. Earth has been resounding with earthquakes for more than 4 billion years. Earthquakes are a way of knowing that the planet beneath us is still experiencing normal operating conditions, full of heat and kinetic energy.

Ultrasensitive instruments placed across faults at the surface can measure the slow, almost imperceptible movement of fault blocks, which tell of great potential energy stored at the fault boundary. In some areas, foreshocks (small earthquakes that precede a larger event) may help seismologists predict the larger event. In other areas, where seismologists believe seismic activity should be occurring but is not, this seismic gap may be used instead to predict an inevitable large-scale earthquake.

Other instruments measure additional fault-zone phenomena that seem to be related to earthquakes. The rate at which radon gas issues from rocks near faults has been observed to change before an earthquake. The properties of the rocks themselves (such as their ability to conduct **electricity**) have been observed to change, as the tectonic force exerted on them slowly alters the rocks of the fault zone between earthquakes. Peculiar animal behavior has been reported before many earthquakes, and research into this phenomenon is a legitimate area of scientific inquiry, even though no definite answers have been found.

Techniques of studying earthquakes from space are also being explored. Scientists have found that ground displacements cause waves in the air that travel into the **ionosphere** and disturb electron densities. By using the network of satellites and ground stations that are part of the global positioning system (**GPS**), data about the ionosphere that is already being collected by these satellites can be used to understand the energy releases from earthquakes, which may help in their prediction.

Scientists have presumed that **tides** do not have any influence on or direct relationship to earthquakes. New studies show that tides may sometimes trigger earthquakes on faults where strain has been accumulating; tidal pull during new or full moons has been discounted by studies of over 13,000 earthquakes of which only 95 occurred during these episodes of tidal stress. Attention is also being directed toward the types of rock underlying areas of earthquake activity to see if rock types dampen (lessen the effects) or magnify earthquake motions.

Seismologists must make a hard choice when their data interpretations suggest an earthquake is about to happen. If they fail to warn people of danger they strongly suspect is imminent, many might die needlessly. But, if people are evacuated from a potentially dangerous area and no earthquake occurs, the public will lose confidence in such warnings and might not heed them the next time.

As more is discovered about how and why earthquakes occur, that knowledge can be used to prevent the conditions that allow earthquakes to cause harm. The most effective way to minimize the hazards of earthquakes is to build new buildings or retrofit old ones to withstand the short, high-speed acceleration of earthquake shocks.

The Moon appears to turn red during a lunar eclipse due to dust and other pollutants in the atmosphere. © Y. Shopov. Reproduced by permission.

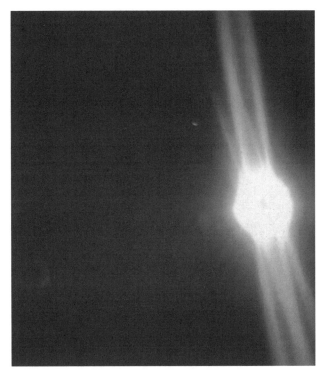

Giant coronal streamers, visible only from space or the upper stratosphere during an eclipse. © Y. Shopov. Reproduced by permission.

See also Convergent plate boundary; Earth, interior structure; Faults and fractures; Mid-plate earthquakes; Plate tectonics; Tsunami

EARTHQUAKE PREDICTION, ADVANCES IN RESEARCH • *see* SEISMOLOGY

ECLIPSE

An eclipse is a phenomenon in which the light from a celestial body is temporarily obscured by the presence of another.

A solar eclipse occurs when the **Moon** is aligned between the **Sun** and Earth. The trace of the lunar shadow (where the solar eclipse is visible) is less than 270 km (168 mi) wide. A partial eclipse is visible over a much wider region. When the Moon is further away from Earth, the lunar disc has a smaller visible diameter than the solar disc, so a narrow ring of the Sun remains uncovered, even when the three bodies are aligned. This produces an annular solar eclipse. The ratio between the visible lunar and solar diameters is called the magnitude of the eclipse. At the beginning of the solar eclipse, the Moon progressively covers the solar disk. Illumination of Earth's surface rapidly diminishes. The air **temperature** falls a few degrees. Seconds before the totality of the eclipse, shadow bands appear. Shadow bands are irregular bands of shadow, a few centimeters wide and up to

a meter apart, moving over the ground. The **diamond** ring phase of the eclipse then shines for few seconds and later, Bailey's beads appear on the solar limb. Bailey's beads are a string of bright beads of light produced by the uneven shape of the lunar limb.

In the first two to three seconds of the total phase of the eclipse (totality), the chromosphere is visible as a pink halo around half of the limb. Maximal duration of the totality varies from eclipse to eclipse, up to 7.5 minutes. The brightest stars and planets are observable on the sky during the totality. Prominences are the brightest objects visible continuously during the totality. They are **clouds** of relatively cold (10,000 K) and dense matter with the same properties as that of the chromosphere matter. They emit in lines of hydrogen, helium and calcuim, which produce the pink color of prominences and the chromosphere, and can always be observed in monochromatic light.

White corona can be observed from Earth only during total solar eclipses, because its intensity is much lower than the brightness of the sky. It has several components emitting in the entire visible region of spectra. The K- (Electron or continuum) corona is due to scattering of sunlight on free high-energy electrons, which are at a temperature of 1 million degrees, and contain continuous spectra and linear polarization of the light. The K-corona dominates in the corona, have distinct 11-year cycles, and have variable structures depending on the level of solar activity. During the solar maximum, it is circular. During the solar minimum, it is symmetrical and elongated in the equatorial region, while

Solar protuberances at high magnification, as seen during a total eclipse. © Y. Shopov. Reproduced by permission.

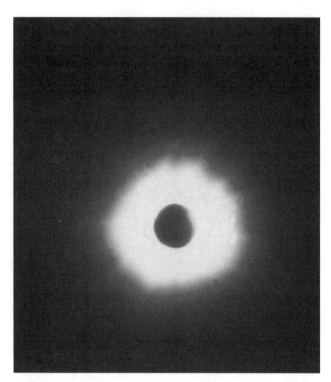

Far solar corona as seen during a total eclipse. © Y. Shopov. Reproduced by permission.

in the polar regions, it has bunches of short rays or plumes. During intermediate phases, it has asymmetric structure with many streamers of different lengths. The F- (Fraunhofer or Dust) corona is due to scattering of sunlight on dust particles. An F-corona has Fraunhofer spectra with absorption lines. Due to heating of dust particles close to the Sun, the F-corona evaporates, producing a large cavity in the dust distribution. An F-corona has oval shape. Its intensity decreases slowly with the distance from the Sun, and it predominates over the K-corona at long distances. The F-corona reaches near-Earth **space**, producing Zodiacal light (a faint conical glow extending along the ecliptic, visible after sunset or before sunrise in a dark, clear sky). The Thermal (T) corona is due to thermal emission of dust particles heated by the Sun.

Solar corona also have components emitting linear spectrum. The E- (Emission) corona is due to emission lines of highly ionized atoms of **iron**, nickel, and calcium. The E-corona intensity decreases rapidly with its distance from the Sun and is visible up to a 2-solar radius in monochromatic light. The S- (Sublimation) corona, was recently found, but as of 2002, its existence is still debatable. It consists of emission of low ionized atoms of Ca(II) produced by sublimation of dust particles in relatively cold parts of the corona. All these components are visible together in the corona during total eclipses.

The last and most mysterious component of the corona is giant coronal streamers observed only from the orbital coronagraph LASCO and from stratospheric flights during total eclipses. The giant coronal streamer shape and properties are different from those of any other component of the corona. Animations of their timed development look similar to visualizations of gusts of solar **wind**. In the last few years, evidence has arisen demonstrating that its nature is the same as that of plasma tails of **comets**, fluorescence of ionized gas molecules (originated by **evaporation** of comets near the Sun), and is due to interaction with the solar wind and sunlight. This component of the corona is called Fluorescent (Fl) corona, but this hypothesis needs further scientific verification. The corona is divided arbitrarily to Internal corona (up to 1.3 radius), which can be observed any time by coronagraph, Medium (1.3- 2.3 radius), and External corona (over 2.3 radius) where F-corona dominates. Edges of the corona gradually disappear in the background of the sky. Therefore, the size of the corona greatly depends on the spectral region of observations and clearness of the sky.

Lunar eclipses occur when the Moon passes into Earth's shadow. The Moon does not normally disappear completely; its disc is illuminated by light scattered by the Earth's atmosphere. Color of the lunar eclipse depends highly on the composition of the atmosphere (amount of **ozone** and dust). The full shadow (umbra) cast by Earth is surrounded by a region of partial shadow, called the penumbra. Some lunar eclipses are visible only as penumbral, other as partial. The length of the Moon's path through the umbra, divided by the Moon's diameter, defines the magnitude of a lunar eclipse.

See also Coronal ejections and magnetic storms

EINSTEIN, ALBERT (1879-1955)

German-born American physicist

Albert Einstein ranks as one of the most remarkable theoreticians in the history of science. He was also a heartfelt pacifist dedicated to world peace. During a single year, 1905, he produced three papers that are among the most important in twentieth-century **physics**, and perhaps in all of the recorded history of science, for they revolutionized the way scientists looked at the nature of **space**, time, and matter. These papers dealt with the nature of particle movement known as Brownian motion, the quantum nature of electromagnetic radiation as demonstrated by the photoelectric effect, and the special theory of relativity. Although Einstein is probably best known for the last of these works, it was for his quantum explanation of the photoelectric effect that he was awarded the 1921 Nobel Prize in physics. In 1915, Einstein extended his special theory of relativity to include certain cases of accelerated motion, resulting in the more general theory of relativity.

Einstein was born in Ulm, Germany, the only son of Hermann and Pauline Koch Einstein. Both sides of his family had long-established roots in southern Germany, and, at the time of Einstein's birth, his father and uncle Jakob owned a small electrical equipment plant. When that business failed around 1880, Hermann Einstein moved his family to Munich to make a new beginning. A year after their arrival in Munich, Einstein's only sister, Maja, was born.

Although his family was Jewish, Einstein was sent to a Catholic elementary school from 1884 to 1889. He was then enrolled at the Luitpold Gymnasium in Munich. During these years, Einstein began to develop some of his earliest interests in science and mathematics, but he gave little outward indication of any special aptitude in these fields. Indeed, he did not begin to talk until the age of three and, by the age of nine, was still not fluent in his native language.

In 1894, Hermann Einstein's business failed again, and the family moved once more, this time to Pavia, near Milan, Italy. Einstein was left behind in Munich to allow him to finish school. Such was not to be the case, however, since he left the *gymnasium* after only six more months. Einstein's biographer, Philipp Frank, explains that Einstein so thoroughly despised formal schooling that he devised a scheme by which he received a medical excuse from school on the basis of a potential nervous breakdown. He then convinced a mathematics teacher to certify that he was adequately prepared to begin his college studies without a high school diploma. Other biographies, however, say that Einstein was expelled from the *gymnasium* on the grounds that he was a disruptive influence at the school.

In any case, Einstein then rejoined his family in Italy. One of his first acts upon reaching Pavia was to give up his German citizenship. He was so unhappy with his native land that he wanted to sever all formal connections with it; in addition, by renouncing his citizenship, he could later return to Germany without being arrested as a draft dodger. As a result, Einstein remained without an official citizenship until he became a Swiss citizen at the age of 21. For most of his first year in Italy, Einstein spent his time traveling, relaxing, and

teaching himself calculus and higher mathematics. In 1895, he thought himself ready to take the entrance examination for the Eidgenössische Technische Hochschule (the ETH, Swiss Federal Polytechnic School, or Swiss Federal Institute of Technology), where he planned to major in electrical engineering. When he failed that examination, Einstein enrolled at a Swiss cantonal high school in Aarau. He found the more democratic style of instruction at Aarau much more enjoyable than his experience in Munich and soon began to make rapid progress. He took the entrance examination for the ETH a second time in 1896, passed, and was admitted to the school. (In *Einstein*, however, Jeremy Bernstein writes that Einstein was admitted without examination on the basis of his diploma from Aarau.)

The program at ETH had nearly as little appeal for Einstein as had his schooling in Munich, however. He apparently hated studying for examinations and was not especially interested in attending classes on a regular basis. He devoted much of this time to reading on his own, specializing in the works of Gustav Kirchhoff, Heinrich Hertz, **James Clerk Maxwell**, Ernst Mach, and other classical physicists. When Einstein graduated with a teaching degree in 1900, he was unable to find a regular teaching job. Instead, he supported himself as a tutor in a private school in Schaffhausen. In 1901, Einstein also published his first scientific paper, "Consequences of Capillary Phenomena."

In February, 1902, Einstein moved to Bern and applied for a job with the Swiss Patent Office. He was given a probationary appointment to begin in June of that year and was promoted to the position of technical expert, third class, a few months later. The seven years Einstein spent at the Patent Office were the most productive years of his life. The demands of his work were relatively modest and he was able to devote a great deal of time to his own research.

The promise of a steady income at the Patent Office also made it possible for Einstein to marry. Mileva Maric (also given as Maritsch) was a fellow student in physics at ETH, and Einstein had fallen in love with her even though his parents strongly objected to the match. Maric had originally come from Hungary and was of Serbian and Greek Orthodox heritage. The couple married in 1903, and later had two sons, Hans Albert and Edward.

In 1905, Einstein published a series of papers, any one of which would have assured his fame in history. One, "On the Movement of Small Particles Suspended in a Stationary Liquid Demanded by the Molecular-Kinetic Theory of Heat," dealt with a phenomenon first observed by the Scottish botanist Robert Brown in 1827. Brown had reported that tiny particles, such as dust particles, move about with a rapid and random zigzag motion when suspended in a liquid.

Einstein hypothesized that the visible motion of particles was caused by the random movement of molecules that make up the liquid. He derived a mathematical formula that predicted the distance traveled by particles and their relative speed. This formula was confirmed experimentally by the French physicist Jean Baptiste Perrin in 1908. Einstein's work on the Brownian movement is generally regarded as the first direct experimental evidence of the existence of molecules.

Albert Einstein. *Library of Congress.*

A second paper, "On a Heuristic Viewpoint concerning the Production and Transformation of Light," dealt with another puzzle in physics, the photoelectric effect. First observed by Heinrich Hertz in 1888, the photoelectric effect involves the release of electrons from a metal that occurs when light is shined on the metal. The puzzling aspect of the photoelectric effect was that the number of electrons released is not a function of the light's intensity, but of the color (that is, the wavelength) of the light.

To solve this problem, Einstein made use of a concept developed only a few years before, in 1900, by the German physicist **Max Planck**, the quantum hypothesis. Einstein assumed that light travels in tiny discrete bundles, or "quanta," of energy. The energy of any given light quantum (later renamed the photon), Einstein said, is a function of its wavelength. Thus, when light falls on a metal, electrons in the metal absorb specific quanta of energy, giving them enough energy to escape from the surface of the metal. But the number of electrons released will be determined not by the number of quanta (that is, the intensity) of the light, but by its energy (that is, its wavelength). Einstein's hypothesis was confirmed by several experiments and laid the foundation for the fields of quantitative photoelectric **chemistry** and quantum mechanics. As recognition for this work, Einstein was awarded the 1921 Nobel Prize in physics.

A third 1905 paper by Einstein, almost certainly the one for which he became best known, details his special theory of relativity. In essence, "On the Electrodynamics of Moving Bodies" discusses the relationship between measurements made by observers in two separate systems moving at constant velocity with respect to each other.

Einstein's work on relativity was by no means the first in the field. The French physicist Jules Henri Poincaré, the Irish physicist George Francis FitzGerald, and the Dutch physicist Hendrik Lorentz had already analyzed in some detail the problem attacked by Einstein in his 1905 paper. Each had developed mathematical formulas that described the effect of motion on various types of measurement. Indeed, the record of pre-Einstein thought on relativity is so extensive that one historian of science once wrote a two-volume work on the subject that devoted only a single sentence to Einstein's work. Still, there is little question that Einstein provided the most complete analysis of this subject. He began by making two assumptions. First, he said that the laws of physics are the same in all frames of reference. Second, he declared that the velocity of light is always the same, regardless of the conditions under which it is measured.

Using only these two assumptions, Einstein proceeded to uncover an unexpectedly extensive description of the properties of bodies that are in uniform motion. For example, he showed that the length and mass of an object are dependent upon their movement relative to an observer. He derived a mathematical relationship between the length of an object and its velocity that had previously been suggested by both FitzGerald and Lorentz. Einstein's theory was revolutionary, for previously scientists had believed that basic quantities of measurement such as time, mass, and length were absolute and unchanging. Einstein's work established the opposite—that these measurements could change, depending on the relative motion of the observer.

In addition to his masterpieces on the photoelectric effect, Brownian movement, and relativity, Einstein wrote two more papers in 1905. One, "Does the Inertia of a Body Depend on Its Energy Content?" dealt with an extension of his earlier work on relativity. He came to the conclusion in this paper that the energy and mass of a body are closely interrelated. Two years later he specifically stated that relationship in a formula, $E=mc^2$ (energy equals mass times the speed of light squared), that is now familiar to both scientists and non-scientists alike. His final paper, the most modest of the five, was "A New De Determination of Molecular Dimensions." It was this paper that Einstein submitted as his doctoral dissertation, for which the University of Zurich awarded him a Ph.D. in 1905.

Fame did not come to Einstein immediately as a result of his five 1905 papers. Indeed, he submitted his paper on relativity to the University of Bern in support of his application to become a *privatdozent*, or unsalaried instructor, but the paper and application were rejected. His work was too important to be long ignored, however, and a second application three years later was accepted. Einstein spent only a year at Bern, however, before taking a job as professor of physics at the University of Zurich in 1909. He then went on to the German University of Prague for a year and a half before returning to Zurich and a position at ETH in 1912. A year later Einstein was made director of scientific research at the Kaiser Wilhelm Institute for Physics in Berlin, a post he held from 1914 to 1933.

Einstein was increasingly occupied with his career and his wife with managing their household; upon moving to Berlin in 1914, the couple grew distant. With the outbreak of World War I, Einstein's wife and two children returned to Zurich. The two were never reconciled; in 1919, they were formally divorced. With the outbreak of the war, Einstein's pacifist views became public knowledge. When 93 leading German intellectuals signed a manifesto supporting the German war effort, Einstein and three others published an antiwar counter-manifesto. He also helped form a coalition aimed at fighting for a just peace and for a worldwide organization to prevent future wars. Towards the end of the war, Einstein became very ill and was nursed back to health by his cousin Elsa. Not long after Einstein's divorce from Maric, he was married to Elsa, a widow. The two had no children of their own, although Elsa brought two daughters to the marriage.

The war years also marked the culmination of Einstein's attempt to extend his 1905 theory of relativity to a broader context, specifically to systems with non-zero acceleration. Under the general theory of relativity, motions no longer had to be uniform and relative velocities no longer constant. Einstein was able to write mathematical expressions that describe the relationships between measurements made in any two systems in motion relative to each other, even if the motion is accelerated in one or both. One of the fundamental features of the general theory is the concept of a space-time continuum in which space is curved. That concept means that a body affects the shape of the space that surrounds it so that a second body moving near the first body will travel in a curved path.

Einstein's new theory was too radical to be immediately accepted, for not only were the mathematics behind it extremely complex, it replaced Newton's theory of gravitation that had been accepted for two centuries. So, Einstein offered three proofs for his theory that could be tested: first, that relativity would cause Mercury's perihelion, or point of orbit closest to the **sun**, to advance slightly more than was predicted by Newton's laws. Second, Einstein predicted that light from a star would be bent as it passes close to a massive body, such as the sun. Last, the physicist suggested that relativity would also affect light by changing its wavelength, a phenomenon known as the redshift effect. Observations of the planet Mercury bore out Einstein's hypothesis and calculations, but astronomers and physicists had yet to test the other two proofs.

Einstein had calculated that the amount of light bent by the sun would amount to 1.7 seconds of an arc, a small but detectable effect. In 1919, during an **eclipse** of the sun, English astronomer Arthur Eddington measured the deflection of starlight and found it to be 1.61 seconds of an arc, well within experimental error. The publication of this proof made Einstein an instant celebrity and made "relativity" a household word, although it was not until 1924 that Eddington proved the final hypothesis concerning redshift with a spectral analysis of the star Sirius B. Thus, it was proved that light would be shifted to a longer wavelength in the presence of a strong gravitational field.

Einstein's publication of his general theory in 1916, the *Foundation of the General Theory of Relativity*, essentially brought to a close the revolutionary period of his scientific career. In many ways, Einstein had begun to fall out of phase with the rapid changes taking place in physics during the 1920s. Even though Einstein's own work on the photoelectric effect helped set the stage for the development of **quantum theory**, he was never able to accept some of its concepts, particularly the uncertainty principle. In one of the most-quoted comments in the history of science, he claimed that quantum mechanics, which could only calculate the probabilities of physical events, could not be correct because "God does not play dice." Instead, Einstein devoted his efforts for the remaining years of his life to the search for a unified field theory, a single theory that would encompass all physical fields, particularly gravitation and electromagnetism.

In early 1933, Einstein made a decision. He was out of Germany when Hitler rose to power, and he decided not to return. In March 1933, he again renounced his German citizenship. His remaining property in German was confiscated and his name appeared on the first Nazi list of those who were stripped of citizenship. He accepted an appointment at the Institute for Advanced Studies in Princeton, New Jersey, where he spent the rest of his life. In addition to his continued work on unified field theory, Einstein was in demand as a speaker and wrote extensively on many topics, especially peace.

The growing fascism and anti-Semitism of Hitler's regime, however, convinced him in 1939 to sign his name to a letter written by American physicists warning President Franklin D. Roosevelt that the Germans were nearing the possibility of an atomic bomb, and that Americans must develop the technology first. This letter led to the formation of the Manhattan Project for the construction of the world's first nuclear weapons. Although Einstein's work on relativity, particularly his formulation of the equation $E=mc^2$, was essential to the development of the atomic bomb, Einstein himself did not participate in the project. He was considered a security risk, although he had renounced his German citizenship and become a U.S. citizen in 1940, while retaining his Swiss citizenship.

In 1944, he contributed to the war effort by hand writing his 1905 paper on special relativity, and putting it up for auction. The manuscript, which raised $6 million, is today the property of the U.S. Library of Congress.

After World War II and the bombing of Japan, Einstein became an ardent supporter of nuclear disarmament. He continued to support the efforts to establish a world government and the Zionist movement to establish a Jewish state. In 1952, after the death of Israel's first president, Chaim Weizmann, Einstein was invited to succeed him as president; he declined the offer.

Among the many other honors given to Einstein were the Barnard Medal of Columbia University in 1920, the Copley Medal of the Royal Society in 1925, the Gold Medal of the Royal Astronomical Society in 1926, the Max Planck Medal of the German Physical Society in 1929, and the Franklin Medal of the Franklin Institute in 1935. He also received honorary doctorates in science, medicine and philosophy from many

European and American universities and was elected to memberships in all of the leading scientific academies in the world. In December 1999, *Time* magazine named Einstein "Person of the Century," stating: "In a hundred years, as we turn to another new century—nay, ten times a hundred years, when we turn to another new millennium—the name that will prove most enduring from our own amazing era will be that of Albert Einstein: genius, political refugee, humanitarian, locksmith of the mysteries of the **atom** and the universe."

A week before he died, Einstein agreed to include his name on a manifesto urging all nations to give up nuclear weapons. Einstein died in his sleep at his home in Princeton at the age of 76, after suffering an aortic aneurysm. At the time of his death, he was the world's most widely admired scientist and his name was synonymous with genius. Yet Einstein declined to become enamored of the admiration of others. He wrote in his book *The World as I See It*: "Let every man be respected as an individual and no man idolized. It is an irony of fate that I myself have been the recipient of excessive admiration and respect from my fellows through no fault, and no merit, of my own. The cause of this may well be the desire, unattainable for many, to understand the one or two ideas to which I have with my feeble powers attained through ceaseless struggle."

See also History of exploration III (Modern era); Relativity theory

EIS (ENVIRONMENTAL IMPACT STATEMENT)

Societies are attempting to develop means of coping with the environmental stressors caused by human activities. These include the activities of individual people as they go about their daily lives, as well as the larger enterprises of corporations, governments, and society at large. One of the procedures that is increasingly becoming a routine component of planning for potential damages is known as environmental impact assessment.

The environmental impact assessment (EIA), and the subsequently prepared statement (EIS), is a process that can be used to identify and estimate the potential environmental consequences of proposed developments and policies. Environmental impact assessment is a highly interdisciplinary process, involving inputs from many fields in the sciences and social sciences. Environmental impact assessments commonly examine ecological, physical/chemical, sociological, economic, and other environmental effects.

Environmental assessments may be conducted to review the potential effects of: (1) individual projects, such as the construction of a particular power plant, incinerator, airport, or housing development; (2) integrated development schemes, or proposals to develop numerous projects in some **area**. Examples include an industrial park, or an integrated venture to harvest, manage, and process a natural resource, such as a pulp mill with its associated wood-supply and forest-management plans; or (3) government policies that carry a risk

of having substantial environmental effects. Examples include decisions to give national priority to the generation of **electricity** using nuclear reactors, or to clear large areas of natural forest to develop new lands for agricultural use.

An extraordinary variety of environmental and ecological changes can potentially be caused by any project, scheme, or policy. Consequently, it is never practical in an environmental impact assessment to consider all of the potential effects of a proposal. Usually certain indicators, called "valued ecosystem components" (VECs), are carefully selected for study, on the basis of their importance to society. The valued ecosystem components are often identified through consultations with representatives of regulatory agencies, scientists, non-governmental organizations, and the public.

Examples of commonly examined VECs include: (1) resources that are economically important, such as agricultural or forest productivity, and populations of hunted fish or game; (2) rare or endangered species and natural ecosystems; (3) particular species, communities, or landscapes that are of cultural or aesthetic importance; and (4) simple indicators of a complex of ecological values. The spotted owl, for example, is an indicator of the integrity of certain types of old-growth conifer **forests** in western **North America**. Any proposed activity, likely associated with forestry, that threatens a population of these birds also indicates a challenge to the larger, old-growth forest ecosystem.

The initial phase of an environmental impact assessment attempts to screen for potentially important conflicts between proposed activities and valued ecosystem components. Essentially, this is done by predicting the dimensions in **space** and time of stressors associated with the proposed development, and then comparing these with the known boundaries of VECs. This is a preliminary, scoping exercise, which may require environmental scientists to make professional judgments about the severity and importance of potential interactions between stressors and VECs. The best-available information is used to guide these interpretations, although the existing knowledge is almost always incomplete. To better determine the risks of interactions identified during the preliminary screening, it is highly desirable to undertake field, laboratory, or simulation research. However, time and funds are not always available to support new research, and this can constrain some impact assessments.

Once potentially important risks to VECs are identified and studied, it is possible to consider various planning options. During this stage of the impact assessment, environmental specialists provide objective information and their professional opinions to decision makers, who must then make choices that deal with the conflicts. There are three broad types of choices that can be made:

(1) The predicted damages can be avoided, by not proceeding with the development, or by modifying its structure. However, avoidance is often considered an undesirable option by proponents of a development. This is because irreconcilable conflicts with environmental quality can result in substantial costs, including canceled projects. Regulators and politicians also tend to resent this option, because socioeconomic opportunities can be lost, and there is often intense controversy.

False color image of ocean surface height of the Pacific Ocean, with red and white showing increased heat storage, purple showing low sea level, and green showing normal conditions. *TOPEX/Poseidon, NASA JPL.*

False color image of ocean surface height of the Pacific Ocean, with red and white showing increased heat storage, purple showing low sea level, and green showing normal conditions. *TOPEX/Poseidon, NASA JPL.*

(2) Often, mitigations can be designed and implemented to prevent or significantly reduce damages to the VEC. For example: (a) if an industrial activity is predicted to acidify a lake, an appropriate mitigation might be liming to reduce the acidification; (b) if habitat of an endangered species is threatened, the risk may be mitigated by moving the population to another site; or (c) the emissions of **carbon dioxide** from a proposed coal-fired generating station could be offset by planting trees to achieve no net change in atmospheric concentrations of this gas. Mitigations are common ways of resolving conflicts between project-related stressors and VECs. However, there are always substantial risks with the use of mitigations, because ecological and environmental knowledge are incomplete, and it is not necessarily known if mitigations will actually work properly to protect the VEC.

(3) Another option that is often selected is to allow the damages to the VEC to occur, and to accept the degradation as an unfortunate cost of achieving the perceived socioeconomic benefits of proceeding with the development. This choice is common, because not all environmental damages can be avoided or mitigated, and many damaging activities can yield large, short-term profits.

It is not possible to carry out large industrial or economic developments without causing some environmental damages. However, a properly done impact assessment can help decision makers to understand the dimensions and importance of those damages, and to decide whether they are acceptable, or whether they must be reduced or avoided.

See also Earth (planet); Earth science

El Niño and La Niña phenomena

El Niño and La Niña are the names given to changes in the winds, **atmospheric pressure**, and seawater that occur in the Pacific Ocean near the equator. El Niño and La Niña are opposite phases of a back and forth cycle in the Pacific Ocean and the atmosphere above it. Unlike winter and summer, however, El Niño and La Niña do not change with the regularity of the **seasons**; instead, they repeat on average about every three or four years. They are the extremes in a vast repeating cycle called the Southern Oscillation, El Niño being the warm extreme and La Niña the cold extreme.

Although El Niño and La Niña take place in a small portion of the Pacific, the changes caused by Southern Oscillation can affect the **weather** in large parts of **Asia**, **Africa**, Indonesia and North and **South America**. Scientists have only recently become aware of the far-reaching effects of the Southern Oscillation on the world's weather. An El Niño during 1982–83 was associated with record snowfall in parts of the Rocky Mountains, flooding in the southern United States, and heavy rain storms in southern California, which brought about **floods** and mud slides.

The name El Niño comes from Peruvian fishermen. They noticed that near the end of each year, the seawater off the South American coast became warmer, which made fishing much poorer. Because the change appeared each year close to Christmas, the fishermen dubbed it El Niño, Spanish for "the boy child" referring to the Christ child. Every few years, the changes brought with El Niño were particularly strong or long lasting. During these strong El Niños, the warmer sea waters nearly wiped out fishing and brought sig-

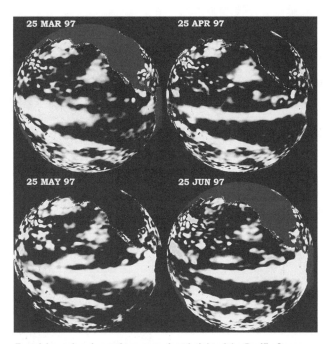

Four false color views of ocean surface height of the Pacific Ocean, with red and white areas showing increased heat storage. *TOPEX/Poseidon, NASA JPL.*

False color image of ocean surface height of the Pacific Ocean, with red and white showing increased heat storage, purple showing low sea level, and green showing normal conditions. *TOPEX/Poseidon, NASA JPL.*

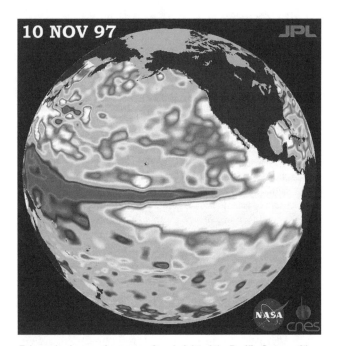

False color image of ocean surface height of the Pacific Ocean, with red and white showing increased heat storage, purple showing low sea level, and green showing normal conditions. *TOPEX/Poseidon, NASA JPL.*

False color image of ocean surface height of the Pacific Ocean, with red and white showing increased heat storage, purple showing low sea level, and green showing normal conditions. *TOPEX/Poseidon, NASA JPL.*

False color image of ocean surface height of the Pacific Ocean, with red and white showing increased heat storage, purple showing low sea level, and green showing normal conditions. *TOPEX/Poseidon, NASA JPL.*

False color image of ocean surface height of the Pacific Ocean, with red and white showing increased heat storage, purple showing low sea level, and green showing normal conditions. *TOPEX/Poseidon, NASA JPL.*

nificant changes in weather. For example, normally dry areas on shore could receive abundant rain, turning deserts into lush grasslands for as long as these strong El Niños lasted. In the 1950s and 60s it was found that strong El Niños were associated with increased sea surface temperatures throughout the eastern tropical Pacific. In recent years, these strong El Niños have been recognized as not just a local change in the sea, but as one half of a vast atmospheric-oceanic cycle.

The other half of the repeating cycle has been named La Niña, or the girl child. This phase of the Southern Oscillation is also sometimes called El Viejo, or the old man.

The Southern Oscillation was detected in the early 1920s by Sir Gilbert Walker. He was trying to understand the variations in the summer monsoons (rainy seasons) of India by studying the way atmospheric pressure changed over the Pacific Ocean. Based on meteorologists' previous pressure observations from many stations in the southern Pacific and Indian **oceans**, Walker established that over the years, atmospheric pressure seesawed back and forth across the ocean. In some years, pressure was highest over northern **Australia** and lowest over the southeastern Pacific, near the island of Tahiti. In other years, the pattern was reversed. The two pressure patterns had specific weather patterns associated with each, and the change from one phase to the other could mean the shift from rainfall to **drought**, or from good harvests to famine. In the late 1960s, Jacob Bjerknes, a professor at the University of California, first proposed that the Southern Oscillation and the strong El Niño sea warming were two aspects of the same vast atmosphere-ocean cycle.

El Niño and the Southern Oscillation (often referred to as ENSO) take place in the tropics, a part of the world dominated by prevailing winds, called the trade winds. Near the equator in the tropical Pacific, these easterly (east to west) winds blow day in and day out and tend to pull the surface **water** of the ocean along with them. This pulls the warm surface water westward, where it collects on the western edge of the ocean basin, the **area** that includes Indonesia, eastern Australia and many Pacific Islands. The warm waters literally pile up in these areas, where the sea level is about 16 inches (40 cm) higher than in the eastern Pacific.

Meanwhile, along the coast of South America colder water from the ocean depths rises to the top, since the warmer water has been blown westward. The result is called upwelling, and it occurs along much of the coasts of South and **North America**. Upwelling has two important consequences. The cold deep waters tend to have more nutrients than surface water; these nutrients are essential to phytoplankton, the tiny plants of the sea that provide food for many other types of sea life. Thus, upwelling zones are very productive for fish and the animals (and people) that depend on fish for food. The second result of upwelling cold water is that it cools the air above it. Cool air is denser than warm air, and cool air in the atmosphere cannot begin rising to form **clouds** and thunderstorms. As a result, the areas near upwelling zones tend to be arid (desert-like) because rain clouds rarely form.

The warmer water that builds up in the western Pacific warms the air above it. This warm moist air frequently rises to form clouds, which eventually produce rainfall. When the trade winds are blowing the warm water their way, the lands along the western Pacific enjoy abundant rainfall. Many rain

False color image of ocean surface height of the Pacific Ocean, with red and white showing increased heat storage, purple showing low sea level, and green showing normal conditions. *TOPEX/Poseidon, NASA JPL.*

False color image of ocean surface height of the Pacific Ocean, with red and white showing increased heat storage, purple showing low sea level, and green showing normal conditions. *TOPEX/Poseidon, NASA JPL.*

forests are found in these areas, such as those of Borneo and New Guinea.

The pattern of winds described above is the La Niña phase of the Southern Oscillation. It sets up the areas of high and low atmospheric pressure observed by Walker and others: in the west, warm air rising produces low pressure, while farther east the cooler, denser air leads to areas of high pressure.

The atmosphere and the ocean form a system that is coupled, that is, they respond to each other. Changes in the ocean will cause a response in the winds above it, and vice versa. For reasons not yet fully understood, the coupled atmosphere-ocean system of the La Niña phase begins to change, slowly developing the characteristics of El Niño phase. The trade winds weaken somewhat, so that they pull less warm water to the western edge of the Pacific. This causes far-reaching changes. Fewer rain clouds form over the lands along the western Pacific. The lush rain forests dry out and become fuel for forest fires. The area of heavy rain shifts to the mid-southern Pacific, where formerly **desert** island are soaked day after day. In the eastern Pacific, the surface water becomes warmer, since it is no longer being driven westward. Ocean upwelling is weakened, so the water near the surface soon runs low on nutrients, which support the ocean food chain. Many species of fish are driven elsewhere to find food; in severe El Niño years fish populations may be almost completely wiped out. Bird species that depend on fish must look elsewhere, and the human fishing population face economic hardship. At the same time, the warmer waters offshore encourage the development of clouds and thunderstorms. Normally dry areas in western South America, such as Peru

and Ecuador, may experience torrential rains and flooding during the El Niño phase.

While its effects have long been noted in the tropical Pacific, El Niño is now being studied for its impact on weather around the world. The altered pattern of winds and ocean temperatures during an El Niño is believed to change the high-level winds, called the jet streams, that steer storms over North and South America. El Niños have been linked with milder winters in western Canada and the northern United States, as most severe storms are steered northward to Alaska. As Californians saw in 1982-83, El Niño can cause extremely wet winters along the west coast, bringing torrential rains to the lowlands and heavy snow packs to the mountains. The jet streams altered by El Niño can also contribute to storm development over the **Gulf of Mexico**, which bring heavy rains to the southeastern United States. Similar changes occur in countries of South America, such as Chile and Argentina, while droughts may affect Bolivia and parts of Central America.

El Niño also appears to affect monsoons, which are annual shifts in the prevailing winds that bring on rainy seasons. The rains of the monsoon are critical for agriculture in India, Southeast Asia and portions of Africa; when the monsoon fails, millions of people are at risk of starvation. At present it appears that while El Niños do not always determine monsoons, they are associated with weakened monsoons in India and southeastern Africa, while tending to strengthen those in eastern Africa.

In general, the effects of El Niño are reversed during the La Niña extremes of the Southern Oscillation cycle. During the 1999 La Niña episode, for example, the central and north-

False color image of ocean surface height of the Pacific Ocean, with red and white showing increased heat storage, purple showing low sea level, and green showing normal conditions. *TOPEX/Poseidon, NASA JPL.*

eastern United States experienced record snowfall and sub-zero temperatures, while rainfall increased in the Pacific Northwest and a record number of tornadoes plagued the southern states. Not all El Niños and La Niñas have equally strong effects on the global **climate** because every El Niño and La Niña event is of a different magnitude and duration.

The widespread weather impacts of the two extreme phases of the Southern Oscillation make their understanding and prediction a high priority for atmospheric scientists. Researchers have developed computer models of the Southern Oscillation that mimic the behavior of the real atmosphere-ocean system. These computer simulations require the input of mountains of data about sea and **wind** conditions in the equatorial Pacific. The measurements are provided by a large and growing network of instruments. Ocean buoys, permanently moored in place across the Pacific, constantly relay information on water **temperature**, wind, and air pressure to weather prediction stations around the world. The buoys are augmented by surface ships, island weather stations, and Earth observing satellites.

Even with mounting data and improving computer models, El Niño, La Niña and the Southern Oscillation remain difficult to predict. However, the Southern Oscillation models are now being used in several countries to help prepare for the next El Niño. Countries most affected by the variations in El Niño, such as Peru, Australia and India, have begun to use El Niño prediction to improve agricultural planning.

See also Air masses and fronts; Atmospheric circulation; Ocean circulation and currents

ELECTRICITY AND MAGNETISM

Electricity and **magnetism** are manifestations of a single underlying electromagnetic force. Electromagnetism is a branch of physical science that describes the interactions of electricity and magnetism, both as separate phenomena and as a singular electromagnetic force. A **magnetic field** is created by a moving electric current and a magnetic field can induce movement of charges (electric current). The rules of electromagnetism also explain geomagnetic and electromagnetic phenomena by explaining how charged particles of atoms interact.

Before the advent of technology, electromagnetism was perhaps most strongly experienced in the form of **lightning**, and electromagnetic radiation in the form of light. Ancient man kindled fires that he thought were kept alive in trees struck by lightning. Magnetism has long been employed for navigation in the compass. In fact, it is known that Earth's magnetic poles have exchanged positions in the past.

Some of the rules of *electrostatics*, the study of electric charges at rest, were first noted by the ancient Romans, who observed the way a brushed comb would attract particles. It is now known that electric charges occur in two different forms, positive charges and negative charges. Like charges repel each other, and differing types attract.

The force that attract positive charges to negative charges weakens with distance, but is intrinsically very strong—up to 40 times stronger than the pull of **gravity** at the surface of the earth. This fact can easily be demonstrated by a small magnet that can hold or suspend an object. The small magnet exerts a force at least equal to the pull of gravity from the entire Earth.

The fact that unlike charges attract means that most of this force is normally neutralized and not seen in full strength. The negative charge is generally carried by the atom's electrons, while the positive resides with the protons inside the atomic nucleus. Other less known particles can also carry charge. When the electrons of a material are not tightly bound to the atom's nucleus, they can move from **atom** to atom and the substance, called a conductor, can conduct electricity. Conversely, when the electron binding is strong, the material resists electron flow and is an insulator.

When electrons are weakly bound to the atomic nucleus, the result is a semiconductor, often used in the electronics industry. It was not initially known if the electric current carriers were positive or negative, and this initial ignorance gave rise to the convention that current flows from the positive terminal to the negative. In reality we now know that the electrons actually flow from the negative to the positive.

Electromagnetism is the theory of a unified expression of an underlying force, the electromagnetic force. This is seen in the movement of electric charge, that gives rise to magnetism (the electric current in a wire being found to deflect a compass needle), and it was Scottish physicist **James Clerk Maxwell** (1831–1879), who published a unifying theory of electricity and magnetism in 1865. The theory arose from former specialized work by German mathematician Carl Fredrich Gauss (1777–1855), French physicist Charles Augustin de Coulomb (1736–1806), French scientist André Marie Ampère

•

Electricity turns this large piece of metal into a powerful magnet, allowing scrap metal to be easily moved. © Corbis/Bettmann. Reproduced by permission.

(1775–1836), English physicist **Michael Faraday** (1791–1867), American scientist and statesman **Benjamin Franklin** (1706–1790), and German physicist and mathematician Georg Simon Ohm (1789–1854). However, one factor that did not contradict the experiments was added to the equations by Maxwell to ensure the conservation of charge. This was done on the theoretical grounds that charge should be a conserved quantity, and this addition led to the prediction of a wave phenomena with a certain anticipated velocity. Light, with the expected velocity, was found to be an example of this electromagnetic radiation.

Light had formerly been thought of as consisting of particles (photons) by Newton, but the theory of light as particles was unable to explain the wave nature of light (diffraction and the like). In reality, light displays both wave *and* particle properties. The resolution to this duality lies in **quantum theory**, where light is neither particles nor wave, but both. It propagates as a wave without the need of a medium and interacts in the manner of a particle. This is the basic nature of quantum theory.

Classical electromagnetism, useful as it is, contains contradictions (acausality) that make it incomplete and drive one to consider its extension to the **area** of quantum **physics**, where electromagnetism, of all the fundamental forces of nature, it is perhaps the best understood.

There is much symmetry between electricity and magnetism. It is possible for electricity to give rise to magnetism, and symmetrically for magnetism to give rise to electricity (as in the exchanges within an electric transformer). It is an exchange of just this kind that constitutes electromagnetic waves. These waves, although they don't need a medium of propagation, are slowed when traveling through a transparent substance.

Electromagnetic waves differ from each other only in amplitude, frequency, and orientation (polarization). Laser beams are particular in being very coherent, that is, the radiation is of one frequency, and the waves coordinated in motion and direction. This permits a highly concentrated beam that is used not only for its cutting abilities, but also in electronic data storage, such as in CD-ROMs.

The differing frequency forms are given a variety of names, from radio waves at very low frequencies through light itself, to the high frequency x rays and gamma rays.

The unification of electricity and magnetism allows a deeper understanding of physical science, and much effort has been put into further unifying the four forces of nature (e.g., the electromagnetic, weak, strong, and gravitational forces. The weak force has now been unified with electromagnetism, called the electroweak force. There are research programs attempting to collect data that may lead to a unification of the strong force with the electroweak force in a grand unified theory, but the inclusion of gravity remains an open problem.

Maxwell's theory is in fact in contradiction with Newtonian mechanics, and in trying to find the resolution to this conflict, Einstein was lead to his theory of special relativity. Maxwell's equations withstood the conflict, but it was Newtonian mechanics that were corrected by relativistic mechanics. These corrections are most necessary at velocities, close to the speed of light.

Paradoxically, magnetism is a counter example to the frequent claims that relativistic effects are not noticeable for low velocities. The moving charges that compose an electric current in a wire might typically only be traveling at several feet per second (walking speed), and the resulting Lorentz contraction of special relativity is indeed minute. However, the electrostatic forces at balance in the wire are of such great magnitude, that this small contraction of the moving (negative) charges exposes a residue force of real world magnitude, namely the magnetic force. It is in exactly this way that the magnetic force derives from the electric. Special relativity is indeed hidden in Maxwell's equations, which were known before special relativity was understood or separately formulated by Einstein.

Electricity at high voltages can carry energy across extended distances with little loss. Magnetism derived from that electricity can then power vast motors. But electromagnetism can also be employed in a more delicate fashion as a means of communication, either with wires (as in the telephone), or without them (as in radio communication). It also drives motors and provides current for electronic and computing devices.

See also Aurora Borealis and Aurora Australialis; Earth, interior structure; Electromagnetic spectrum; Ferromagnetic;

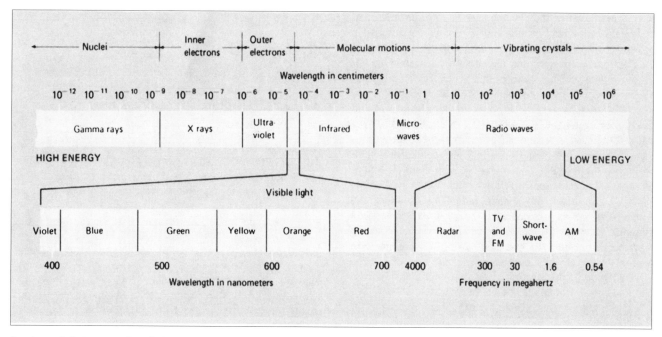

Spectrum of electromagnetic radiation. *Illustration by Robert L. Wolke. Reproduced by permission.*

Quantum electrodynamics (QED); Quantum theory and mechanics

ELECTROMAGNETIC SPECTRUM

The electromagnetic spectrum encompasses a continuous range of frequencies or wavelengths of electromagnetic radiation, ranging from long wavelength, low energy radio waves, to short wavelength, high frequency, high-energy gamma rays. The electromagnetic spectrum is traditionally divided into regions of radio waves, microwaves, infrared radiation, visible light, **ultraviolet rays**, x rays, and gamma rays.

Scottish physicist James Clerk Maxwell's (1831–1879) development of a set of equations that accurately described electromagnetic phenomena allowed the mathematical and theoretical unification of electrical and magnetic phenomena. When Maxwell's calculated speed of light fit well with experimental determinations of the speed of light, Maxwell and other physicists realized that visible light should be a part of a broader electromagnetic spectrum containing forms of electromagnetic radiation that varied from visible light only in terms of wavelength and wave frequency. Frequency is defined as the number of wave cycles that pass a particular point per unit time, and is commonly measured in Hertz (cycles per second). Wavelength defines the distance between adjacent points of the electromagnetic wave that are in equal phase (e.g., wavecrests).

Exploration of the electromagnetic spectrum quickly resulted practical advances. German physicist Henrich Rudolph Hertz regarded Maxwell's equations as a path to a "kingdom" or "great domain" of electromagnetic waves. Based on this insight, in 1888, Hertz demonstrated the exis-

tence of radio waves. A decade later, Wilhelm Röentgen's discovery of high-energy electromagnetic radiation in the form of x rays quickly found practical medical use.

At the beginning of the twentieth century, German physicist, Maxwell Planck, proposed that atoms absorb or emit electromagnetic radiation only in certain bundles termed quanta. In his work on the photoelectric effect, German-born American physicist **Albert Einstein** used the term photon to describe these electromagnetic quanta. Planck determined that energy of light was proportional to its frequency (i.e., as the frequency of light increases, so does the energy of the light). Planck's constant, $h=6.626 \times 10^{-34}$ joule-second in the meter-kilogram-second system, relates the energy of a photon to the frequency of the electromagnetic wave and allows a precise calculation of the energy of electromagnetic radiation in all portions of the electromagnetic spectrum.

Although electromagnetic radiation is now understood as having both photon (particle) and wave-like properties, descriptions of the electromagnetic spectrum generally utilize traditional wave-related terminology (i.e., frequency and wavelength).

Electromagnetic fields and photons exert forces that can excite electrons. As electrons transition between allowed orbitals, energy must be conserved. This conservation is achieved by the emission of photons when an electron moves from a higher potential orbital energy to a lower potential orbital energy. Accordingly, light is emitted only at certain frequencies characteristic of every **atom** and molecule. Correspondingly, atoms and molecules absorb only a limited range of frequencies and wavelengths of the electromagnetic spectrum, and reflect all the other frequencies and wavelengths of light. These reflected frequencies and wavelengths are often the actual observed light or colors associated with an object.

The region of the electromagnetic spectrum that contains light at frequencies and wavelengths that stimulate the rod and cones in the human eye is termed the visible region of the electromagnetic spectrum. Color is the association the eye makes with selected portions of that visible region (i.e., particular colors are associated with specific wavelengths of visible light). A nanometer (10^{-9} m) is the most common unit used for characterizing the wavelength of visible light. Using this unit, the visible portion of the electromagnetic spectrum is located between 380nm-750nm and the component color regions of the visible spectrum are: Red (670–770 nm), Orange (592–620 nm), Yellow (578–592 nm), Green (500–578 nm), Blue (464–500 nm), Indigo (444–464 nm), and Violet (400–446 nm). Because the energy of electromagnetic radiation (i.e., the photon) is inversely proportional to the wavelength, red light (longest in wavelength) is the lowest in energy. As wavelengths contract toward the blue end of the visible region of the electromagnetic spectrum, the frequencies and energies of colors steadily increase.

Like colors in the visible spectrum, other regions in the electromagnetic spectrum have distinct and important components. Radio waves, with wavelengths that range from hundreds of meters to less than a centimeter, transmit radio and television signals. Within the radio band, FM radio waves have a shorter wavelength and higher frequency than AM radio waves. Still higher frequency radio waves with wavelengths of a few centimeters can be utilized for **RADAR** imaging.

Microwaves range from approximately a foot in length to the thickness of a piece of paper. The atoms in food placed in a microwave oven become agitated (heated) by exposure to microwave radiation. Infrared radiation comprises the region of the electromagnetic spectrum where the wavelength of light is measured region from one millimeter (in wavelength) down to 400 nm. Infrared waves are discernible to humans as thermal radiation (heat). Just above the visible spectrum in terms of higher energy, higher frequency and shorter wavelengths is the ultraviolet region of the spectrum with light ranging in wavelength from 400 to 10 billionths of a meter. Ultraviolet radiation is a common cause of sunburn even when visible light is obscured or blocked by **clouds**. X rays are a highly energetic region of electromagnetic radiation with wavelengths ranging from about ten billionths of a meter to 10 trillionths of a meter. The ability of x rays to penetrate skin and other substances renders them useful in both medical and industrial radiography. Gamma rays, the most energetic form of electromagnetic radiation, are comprised of light with wavelengths of less than about ten trillionths of a meter and include waves with wavelengths smaller than the radius of an atomic nucleus (10^{15} m). Gamma rays are generated by nuclear reactions (e.g., radioactive decay, nuclear explosions, etc.).

Cosmic rays are not a part of the electromagnetic spectrum. Cosmic rays are not a form of electromagnetic radiation, but are actually high-energy charged particles with energies similar to, or higher than, observed gamma electromagnetic radiation energies.

ELECTRONS • *see* ATOMIC THEORY

ELEMENTS • *see* CHEMICAL ELEMENTS

ELLES, GERTRUDE (1872-1960)
English geologist

Throughout her life, Gertrude Elles made significant contributions to both the status of women in science, especially in the field of Earth sciences, and to the understanding of graptolites as zone **fossils** and their place within wider fossil communities.

Gertrude Lilian Elles was born in Wimbledon, Surrey near London, on October 8, 1872. At the age of 19, she attended Cambridge University studying the natural science Tripos, gaining a first class honors degree in 1895 and continuing on to become the first female to be awarded a Cambridge University Readership, 30 years later. She never married, but spent the majority of her life in Cambridge at Newnham College and was recognized as an excellent and enthusiastic teacher. Her name, however, was made not in the field of teaching, but in that of research. Elles' contribution to the study and classification of graptolites has not been surpassed to date. She spent 12 years compiling the Treatise on British Graptolites (with her colleague Ethel Wood) under the guidance of Charles Lapworth, (who named the **Ordovician Period**). Their names are inextricably linked with graptolite research. Her work on the genera of graptolites from North Wales and the Skiddaw Slates of the Lake District, England and from the Wenlock Shales of the Welsh borders eventually led to Elles' receiving the prestigious Lyell Fund from the Geological Society of London. She was not able to receive it in person as women were at that time (1900) barred from attending the meetings. She was one of the first scientists to look at not individual specimens of fossils, but at the concept of communities of organisms. In 1919, she became one of the first women to become a Fellow of the Geological Society of London and the same year she received the Murchison medal from the Society in recognition of her work. In 1922, she published a seminal work on the evolution and classification of graptolites from her long study of the group. However, her work also concentrated on **stratigraphy** and she published over 10 papers on lower Palaeozoic stratigraphy.

Among Elles' other accolades, she received the Medal of Member of the British Empire for her work with the British Red Cross in Britain during the First World War. She was an active worker with them for many years. She was also President of the British Association in 1923. Elles considered fieldwork the key to good **geology** and an understanding of paleontology and stratigraphy. Eventually her love of her homeland Scotland called her back permanently (she had always spent considerable time there fishing and researching metamorphic rocks), and Gertrude Elles died in Scotland in 1960 at the age of 88.

See also Fossil record; Fossils and fossilization

EMPEDOCLES OF ACRAGAS, (CA. 492 B.C.-CA. 432 B.C.)

Greek philosopher, poet, and politician

A philosopher, poet, politician, and visionary, Empedocles of Acragas developed radical new ideas about the nature of the universe. His philosophy of the four elements in the universe and the definition of matter as the various ratios of these elements foreshadowed later developments in **atomic theory** by philosophers such as Democritus of Abdera (c. 460–c. 370 B.C.).

Empedocles was born in Acragas, Sicily. His father, Meto, was wealthy, and his grandfather, also named Empedocles, was renowned for winning a horse race in the Olympia. Empedocles is believed to have travelled to Thourioi shortly after it was established approximately 444 B.C. Empedocles's keen intellect enabled him to combine talents in philosophy, natural history, poetry, and politics, and to achieve superstar status in his day. According to the Greek philosopher Aristotle (384–322 B.C.), Empedocles was the inventor of rhetoric, a talent Empedocles often utilized as a statesman. He became popular among his fellow citizens through his support of democracy.

Empedocles's scientific inquiries usually included mysticism. However, his philosophies contained early insight into basic laws of **physics**, including atomic theory. Although sometimes labeled a Pythagorean, Empedocles followed the Greek philosopher Parmenides (c. 515–c. 445 B.C.) in the belief that matter (or, "what is") is indestructible. Empedocles claimed that matter was the only principle of all things and that four elements in the universe—air, fire, earth, and water—made up all things according to various ratios of these elements. Empedocles further stated that two forces, which he called love and hate, or eros and strife, controlled how the four elements come together or move apart. In addition to creating a philosophy that closely resembles modern atomic theory, Empedocles also studied the nature of change in the universe. Empedocles asserted that the cyclical nature of the universe introduces the possibility of reincarnation because nothing that comes into being can be destroyed but only transformed. Empedocles later wrote a poetic treatise *On Nature* containing the ideas of **evolution**, the circulation of the blood, and **atmospheric pressure**. He stated that the **Moon** shone by reflected light and estimated that the Moon was one-third the distance from the earth to the **Sun**.

The object of admiration, Empedocles, according to Aristotle, was offered a kingship but refused to be considered king. Nevertheless, some scholars claim that Empedocles assumed royal status and went so far as to claim himself a deity. Viewed by some as a demi-god and by others as a charlatan, Empedocles made important contributions to the philosophy of science in his day. Galen (c. 130–c. 200), the physician to several Roman emperors, also credits Empedocles with founding the Italian school of medicine. In addition, Empedocles was an accomplished poet. However, little remains of his writings except for segments of his poems *On Nature* (*Peri Physeos*) and *Purifications* (*Katharmoi*).

That Empedocles had a flair for self-promotion and public relations is evident in scholarly writings. Empedocles was reported to have leapt into the crater of the Mt. Etna **volcano** so he would have a death befitting a god. The English poet, Mathew Arnold, wrote a poem about the episode entitled *Empedocles on Etna*. Some scholars dispute the story of Empedocles' fiery death. According to the writings of Aristotle, Empedocles died at the age of 60.

ENERGY TRANSFORMATIONS

Energy is a state function that is best defined as the capacity to do work or to produce heat. There are many forms of energy (e.g., radiant energy, kinetic energy, potential energy, etc) each of which can be converted into other forms of energy. The fundamental law of thermodynamics states that the total energy of the universe is fixed and that energy can not be created or destroyed—only converted from one form to another.

Energy can be changed, or transformed, from one form into another. Energy transformation is also called energy conversion. The Système International d'Unités (SI) unit for energy is the joule (J), named after James Joule, who demonstrated that work can be converted into heat. The Joule is the fundamental unit of energy for both work and heat and is the work done by a force of one Newton acting through a distance of one meter. The joule is also equal to 1/4.184 of a calorie. Energy is often expressed as the calorie (cal), which is the amount of heat needed to raise the **temperature** of one gram of **water** by one degree Celsius at a pressure of one atmosphere. One calorie is equal to 4.184 joules. The Calorie (Cal; also called the kilocalorie) that is used to express the energy in food, is equal to 1,000 calories.

Kinetic energy is the energy of an object in motion and is related by the Newtonian formula $1/2mv^2$ (where m equals mass in Kg and v equals velocity in meters/second). An object in motion can cause another object to do work by colliding with it, causing it to move a particular distance. The colliding objects can be a hammer swinging down on a nail, or two atoms colliding in a chemical reaction. Examples of kinetic energy include mechanical energy (caused by motion of parts) and thermal energy (caused by the random motion of particles of matter). An object that has potential energy has energy by virtue of position and is related by the Newtonian equation PE=mgh (where m equals mass in Kg, g the acceleration due to gravity—approximately 9.8 m/s^2 near Earth's surface—and h equals height in meters). At some point, that object had work performed on it, which resulted in energy storage. One example of performing work on an object to give it potential energy is the lifting of a body against the gravitational force of the Earth. As it is lifted, the body gains potential energy that is converted into kinetic energy as the body falls.

Another example of potential energy is water in an elevated tank. If water is allowed to fall on a wheel and the wheel turns, the turning wheel can be used to produce **electricity**. The water in the tank has gravitational potential energy. The potential energy of the water is transformed into mechanical energy of the wheel, that is then further transformed into electrical energy.

In 1845, Joule performed an experiment that demonstrated energy transformation both qualitatively and quantitatively. The experiment was not complicated—he placed a paddle wheel in a tank of water and measured the temperature of the water. He cranked the wheel in the water for a period of time, then read the temperature again. He found that the temperature of the water rose as he cranked the paddle wheel. Joule quantified this observation and discovered that an equal amount of energy was always required to raise the temperature of the water by one degree. He also discovered that it did not have to be mechanical energy; it could be energy in any form. He obtained the same results with electrical or magnetic energy as he did with mechanical energy. Joule's experiments showed that different forms of energy are equivalent and can be converted from one form to another.

Interestingly, as Joule expressed it, the energy required to increase one pound of water by one degree on the Fahrenheit scale is equal to the amount of energy obtained by a weight of 890 pounds after falling one foot in Earth's gravitational field.

These observations led to what is now called the "Law of conservation of energy." This law states that any time energy is transferred between two objects, or converted from one form into another, no energy is created and none is destroyed. The total amount of energy involved in the process remains the same.

Most chemical reactions involve transformations in energy. A chemical reaction is simply the process whereby bonds are broken between atoms and new ones are made.

The ultimate example of energy transformation is that of the radiant energy of the **Sun**. All of the energy on Earth originates from the Sun, energy left over from the formation of Earth (usually thermal energy caused by the gravitational collapse of matter), or energy derived from nuclear decay in Earth's interior.) The thermal energy in Earth's interior drives **plate tectonics** and, at the surface, the Sun's radiant energy is converted by plants into chemical energy through the process of photosynthesis. This chemical energy is stored in the form of sugars and starches. When these plants are eaten by animals (i.e., as part of the food chain), this chemical energy is either transformed into another form of chemical energy (fats or muscle) or used for mechanical or thermal energy. With regard to fossil **fuels**, the fuels used in the modern era derive from the transformations of **solar energy** over millions of years.

See also Earth (planet); Earth, interior structure; Earthquakes; K-T event; Landslide; Mass movement; Radioactivity

ENVIRONMENTAL POLLUTION

Scottish-American naturalist and Sierra Club founder, **John Muir** (1838–1914), wrote, "When we try to pick out any-

thing by itself, we find it hitched to everything else in the Universe." Our rapidly growing, ever more industrialized human population exists within a carefully balanced global system of physical processes that circulates **chemical elements** through the solid earth, hydrosphere, atmosphere, and **biosphere**. From agricultural land and **water** management, to extraction and combustion of fossil **fuels**, to industrial and municipal disposal of waste products, modern human activity has overprinted natural Earth cycles with synthetic ones. In many cases, these man-made alterations to the natural environment negatively impact the very Earth systems that sustain human life. Contamination of the hydrosphere and atmosphere, depletion of radiation-shielding stratospheric **ozone**, and anthropogenic global **climate** change are examples of changes induced by human environmental pollution.

Accessible, uncontaminated water is essential to all human activities, and **water pollution** is a persistent environmental issue. Contamination of surface, ground, ocean, and atmospheric water occurs when chemical, radioactive, and organic waste is washed, spilled, or dumped into water reservoirs at point and non-point sources. Point sources of water pollution introduce concentrated waste products into **rivers**, aquifers, and **oceans** at focused entry points. Point sources such as oil spills, chemical leaks, and sewage discharges can often be easily corrected; the inflow of hazardous waste can be stopped, and the contaminated water reservoir can sometimes be cleansed. However, the immediate damage to ecosystems and water quality by highly concentrated chemicals at the spill site or pipe outlet may be irreversible, and cleanup is usually costly and difficult. The 1989 Exxon Valdez oil spill in Prince William Sound, Alaska was a dramatic example of a point source of marine water pollution.

Damaging materials also flow into streams and aquifers from diffuse, non-point sources like agricultural lands, logging tracts, mines, residential leach fields, and urban pavement. While non-point pollution is usually less concentrated, it is also more difficult to control, contain, and regulate. Furthermore, the environmental effects of non-point pollutants like fertilizers, pesticides, animal manure, and mining leachates often manifest themselves as systemic changes to aquatic environments that, in turn, reduce water quality. For example, addition of organic materials, fertilizers, and detergents to streams and **lakes** enhances the natural process of eutrophication, in which aquatic vegetation chokes a stream or lake, and eventually kills the reservoir's aquatic fauna. Even very low concentrations of toxic heavy **metals** like those found in leachates from mine tailings, or **lead** plumbing, can result in toxic contamination of fish and mammals in an aquatic ecosystem. Untreated sewage and agricultural **runoff** may introduce viral and bacterial pathogens that cause an array of human illnesses from typhoid to dysentery.

Groundwater pollution occurs when contaminants enter an **aquifer** from a point or non-point source in a recharge zone, contaminated surface water infiltrates, or buried tanks and landfills leak into the groundwater. Groundwater flow paths are complex, and the ultimate site of contamination is often difficult to predict. In karst aquifers, groundwater flows fairly rapidly through interconnected **limestone** dissolution cavities with little filtration of dissolved materials. Pollutants may thus be flushed from the groundwater in months or weeks, but contaminants often take unexpected paths through limestone aquifers, and eventually discharge, undiluted, at unexpected locations. In homogenous, porous aquifers, like the **sandstone** Ogalla aquifer in the south central United States, pollutants flow slowly from their points of entry, and are naturally filtered over time. However, it is difficult to flush contamination from a sandstone aquifer, and recharge with fresh water is extremely slow. Groundwater contamination is of particular concern for sitting buried landfills, **petroleum** tanks, and particularly nuclear waste repositories. Groundwater contamination by harmful radioactive waste buried at nuclear weapons laboratories in Hanford, Washington, and Oak Ridge, Tennessee has cast doubt on nuclear **waste disposal** schemes.

Though contamination is often introduced into the atmosphere at point sources like smokestacks and exhaust pipes, air pollution is usually diffuse because **atmospheric circulation** is unconfined. Sulfur dioxide emitted by coal-burning electrical generators disperses widely into the atmosphere before chemically combining with water vapor to form sulfuric acid. The resulting corrosive **acid rain** falls on widespread areas far downwind of the original point source. Nitrogen oxides released from automobile engines are a main component of the brown **smog** that blankets many cities. Nitrogen oxide and sulfur dioxide combine with other atmospheric chemicals in strong sunlight to form ozone, the component of smog that affects respiration and irritates humans' eyes. Ironically, ozone is harmful to humans in the lower atmosphere, but ozone in the outer atmosphere shields us from harmful, carcinogenic ultra-violet radiation. Another class of man-made chemicals, called chlorofluorocarbons (CFCs), has chemically destroyed the shielding ozone in the **stratosphere** over **Antarctica**, creating the "Ozone Hole." CFCs are common industrial chemicals used in air conditioners, aerosol spray cans, refrigerators, and foam packaging.

The dramatic decrease in air and water quality during the twentieth century has spurred the scientific community to better understand the types of environmental pollution described above, and to devise solutions that reduce contamination. Many governments have enacted legislation that encourages these solutions. In the United States, the Environmental Protection Agency's Clean Water Act of 1972, and Clean Air Act of 1990 strictly regulate industrial, agricultural and municipal sources of air and water pollution. Improved understanding of such complex processes as groundwater and atmospheric flow has led to safer methods of waste disposal, from properly-sited, lined landfills, to air filters on smoke stacks, to carburetors on automobiles. Countries that have enacted these relatively inexpensive measures now enjoy much cleaner air and water than existed in the 1970s. In 1987, the international community signed the Montreal Protocol that eventually bans production of CFCs. However, a handful of the thorniest environmental problems facing the Earth's human population have consequences we

have yet to understand, let alone reverse. Solutions to the most threatening and highly-politicized environmental issues, including global climate change, overpopulation, and loss of biological diversity may require significant international socioeconomic changes.

See also Atmospheric pollution; Global warming; Greenhouse gases and greenhouse effect; Ozone layer and hole dynamics; Ozone layer depletion

EOCENE EPOCH

The Eocene Epoch, second of the five epochs into which the **Tertiary Period** is divided, lasted from 54 to 38 million years ago. Mammals became the dominant land animals during this epoch.

The Eocene Epoch (meaning dawn of the recent period, from the Greek *eos*, dawn, and *koinos*, recent), like the other epochs of the Tertiary Period, was originally defined in 1833 by the English geologist **Charles Lyell** (1797–1875) on the basis of how many modern species are found among its **fossils**. The Eocene Epoch was defined by Lyell as that time where 1–5% of the species were modern (i.e., are still alive today). The Eocene Epoch's boundaries are therefore arbitrary, not set by mass extinctions or other clear-cut events.

For most of the Eocene Epoch, the global climate was warm and rainy. **Ice** caps were small or nonexistent. Early Eocene Epoch sea levels were low, creating land bridges between **Asia** and **North America** via the Bering Strait, North America and **Europe** via Greenland, and **Australia** and **Antarctica**. Late in the epoch Antarctica drifted south, opening a deep-water channel between it and Australia that caused a global cooling trend by allowing the formation of the circum-Antarctic current.

The Eocene Epoch saw the replacement of older mammalian orders by modern ones. Hoofed animals first appeared, including the famous Eohippus (dawn horse) and ancestral rhinoceroses and tapirs. Early bats, rabbits, beavers, rats, mice, carnivorous mammals, and whales also evolved during the Eocene Epoch. The earliest Eocene Epoch mammals were all small, but larger species, including the elephant-sized titanothere, evolved toward the end of the epoch.

Many flowering plants evolved in the Eocene Epoch. Especially important are the grasses, which had first appeared in the late **Cretaceous Period** but did not become diverse and ubiquitous until the Eocene Epoch. Abundant grass encouraged the **evolution** of early grazing animals, including Eohippus. Familiar tree species such as birch, cedar, chestnut, elm, and beech flourished during the Eocene Epoch; aquatic and insect life were much the same as today.

See also Archean; Cambrian Period; Carbon dating; Cenozoic Era; Dating methods; Devonian Period; Fossil record; Geologic time; Historical geology; Holocene Epoch; Jurassic Period; Mesozoic Era; Miocene Epoch; Mississippian Period;

Oligocene Epoch; Ordovician Period; Paleozoic Era; Pennsylvanian Period; Phanerozoic Eon; Pleistocene Epoch; Pliocene Epoch; Precambrian; Proterozoic Era; Quaternary Period; Silurian Period

EOLIAN PROCESSES

Eolian processes are processes of **relief** formation resulting from the action of **wind**. The term comes from the name of the Greek god of winds, Aeolus, and is sometimes referred to as Aeolian processes.

The effectiveness of Eolian processes depend on several factors: the average wind speed in a given **area**, the availability of transportable material, the factors hindering this transportation (such factors are mainly rich vegetation coverage on the surface and high moisture of **soil** and sediments). Eolian processes are inherent mostly in deserts and in areas with arid climates, but they occur also on beaches, glacial outwash valleys, snow surfaces, and in several other kinds of environment.

There are various modes of Eolian transport. **Creep** refers to a rolling and sliding transport. **Saltation** involves short hops ranging from centimeters to a meter. During reptation, numerous particles are displaced as splash close to the surface by impact bombardment of higher energy saltating grains. In suspension, short (up to hundreds of meters) and long term (up to thousands of kilometers) of fine-grained silts and **clay** sized sediment are transported.

Spatial and temporal variations in the Eolian transport processes and in the conditions of their development give rise to various erosional and depositional **landforms**. Ventifacts, rocks abraded and fluted by constant impact of **sand** grains, are an example of erosional landforms. Yardangs are abrasion ridges aligned in the direction of transporting winds. Other erosional landforms include **desert** pavement and deflation lag deposits. Depositional landforms are massive fine-grained deposits of windblown loess (silt), giving rise to sand sheets, ripples, and **dunes**. **Superposition** of forms of different orders is characteristic of Eeolian landforms.

Eolian processes are of interest to scientists of applied **geology**. For example, they can influence the formation of gold placers in several regions.

See also Dune fields; Erosion

EON • *see* GEOLOGIC TIME

EPOCH • *see* GEOLOGIC TIME

EQUATOR • *see* LATITUDE AND LONGITUDE

EQUINOX • *see* LATITUDE AND LONGITUDE

ERA • *see* GEOLOGIC TIME

ERATOSTHENES OF CYRENE
(276 B.C.-194 B.C.)
Greek astronomer

Using elegant mathematical reasoning and limited empirical measurement, in approximately 240 B.C., Eratosthenes of Cyrene (in current-day Libya) made an accurate measurement of the circumference of Earth. In addition to providing evidence of scientific empiricism in the ancient world, this and other contributions to geodesy (the study of the shape and size of the earth) spurred subsequent exploration and expansion. Ironically, centuries later the Greek mathematician and astronomer Claudius Ptolemy's erroneous rejection of Eratosthenes' mathematical calculations, along with other mathematical errors, resulted in the mathematical estimation of a smaller Earth that, however erroneous, made extended seagoing journeys and exploration seem more tactically achievable.

Eratosthenes served as the third librarian at the Great Library in Alexandria. Serving under **Ptolemy** III and tutor to Ptolemy IV, the head librarian post was of considerable importance because the library was the central seat of learning and study in the ancient world. Ships coming into the port of Alexandria, for example, had their written documents copied for inclusion in the library and, over the years, the library's collection grew to encompass hundreds of thousands of papyri and vellum scrolls containing much of the intellectual wealth of the ancient world.

In addition to managing the collection, reading, and transcription of documents, Eratosthenes' own scholarly work concentrated on the study of the mathematics related to Platonic philosophy. Although Eratosthenes actual writings and calculations have not survived, it is known from the writings of other Greek scholars that Eratosthenes' writings and work treated the fundamental concepts and definitions of arithmetic and geometry. Eratosthenes' work with prime numbers, for example, resulted in a prime number sieve still used in modern number theory. Eratosthenes' also contributed to the advancement of scientific knowledge and reasoning, including the compilation of a large catalogue of stars and the preparation of influential chronologies, calendars and maps. So diverse were Eratosthenes' scholarly abilities that he was apparently referred to by his contemporaries as "Beta," a reference to the fact that although Eratosthenes was well-grounded in many scholarly disciplines, he was seemingly second-best in all.

Eratosthenes is, however, best known for his accurate and ingenious calculation of the circumference of the earth. Although Eratosthenes' own notes regarding the methodology of calculation are lost to modern historians, there are tantalizing references to the calculations in the works of Strabo and other scholars. There are extensive references to Eratosthenes' work in Pappus's Collection, an A.D. third century compilation and summary of work in mathematics, **physics**, **astronomy**, and geography. Beyond making an accurate estimate of the earth's circumference, based on observations of shifts in the zenith position of the **Sun**, Eratosthenes also made very accurate measurements of the tilt of the earth's axis.

Apparently inspired by observations contained in the scrolls he reviewed as librarian, Eratosthenes noticed subtle differences in accounts of shadows cast by the midday summer sun. In particular, Eratosthenes read of an observation made near Syene (near modern Aswan, Egypt) that at noon on the summer solstice, the Sun shone directly on the bottom of a deep well and that upright pillars were observed to cast no shadow. In contrast, Eratosthenes noticed that in Alexandria on the same day the noon Sun cast a shadow on a stick thrust into the ground and upon pillars.

Based upon his studies of astronomy and geometry, Eratosthenes assumed that the Sun was at such a great distance that it could be assumed that its rays were essentially parallel by the time they reached the Earth. Although the calculated distances of the Sun and the **Moon**, supported by measurements and estimates made during lunar eclipses, were far too low, the assumptions made by Eratosthenes proved essentially correct. Utilizing such an assumption regarding the parallel incidence of light rays, it remained to determine the angular variance between the shadows cast at Syene and Alexandria at the same time of day. In addition, Eratosthenes needed to calculate the distance between Syene and Alexandria.

Given modern methodologies, it seems intuitive that Eratosthenes would set out to establish the values of angle and distance needed to complete his calculations. In the ancient world, however, Eratosthenes' empiricism, reflected in his actual collection of data, reflected a significant break from a scholarly tradition that relied upon a more philosophical or mathematical approach to problems. Moreover, Eratosthenes' solution relied upon a world-view, especially reflecting the spheroid shape of the world, that itself was subject to philosophical debate.

Eratosthenes ultimately determined the angular difference between the shadows at Syene and Alexandria to be about seven degrees. Regardless of how he obtained the distance to Syene (some legends hold that he paid a messenger runner to pace it off), Eratosthenes reasoned that the ratio of the angular difference in the shadows to the number of degrees in a circle (360 degrees) must equal the ratio of the distance (about 500 mi, or 805 km) to the circumference of the earth. The resulting estimate, about 25,000 mi (40,200 km), is astonishingly accurate.

In making his calculations, Eratosthenes measured distance in units termed stadia. Although the exact value of the stadia is not known, modern estimates place it in the range of 525 ft (160 m). Depending on the exact value of the stadia, Eratosthenes' estimate varied only a few percent from the modern value of 24,902 mi (40,075 km) at the equator. It is necessary to specify that this is the circumference at the equator because the earth is actually an oblate sphere (a slightly compressed sphere with a bulge in the middle) where the circumference at the equator is greater than the distance of a great circle passing through the poles.

Although Eratosthenes' calculations were disputed in his own time, they allowed the development of maps and globes that remained among the most accurate produced for

Erosive forces are pushing this cliff further back, endangering the fence and gatehouse, and eventually the main house itself. *© Vince Streano/Corbis. Reproduced by permission.*

more than a thousand years. The interest in geography and geodesy emboldened regional seafaring exploration using only the most primitive navigational instruments. Moreover, Eratosthenes' calculations fostered the persistence of belief in the sphericity of the earth that ultimately allowed for the development of the concept of antipodes and an early theory of climatic zones dependent on distance from the earth's equator. Eratosthenes' work, encapsulating many of his theories, was reported to have been the first to use the term geography to describe the study of Earth.

ERGS · *see* DUNE FIELDS

EROSION

Erosion is the reduction or breakdown of **landforms** exposed to the forces of **weathering** (disintegration and decomposition). Weathering and subsequent erosion may be caused by both chemical or mechanical forces. Mechanical weathering agents include **wind**, **water**, and **ice**. Chemical weathering leading to erosion results from bio-organic breakdown, hydration, hydrolysis, and oxidation processes. The process of transportation describes the movements of eroded materials.

Erosion requires a transport mechanism (e.g., **gravity**, wind, water, or ice). Wind, water, and ice are also agents of erosion that cause the physical breakdown of **rock** and landforms.

A special form of erosion, **mass wasting**, describes the transport of material downslope under the influence of gravity. Landslides are a common example of mass wasting.

Erosion processes can also cause indirect landform alteration by breaking down overburden of rock and precipitating a pressure release that can crack and shift rock layers. The cracking process results in peels, **exfoliation**, or spalling. For example, the erosion of overburden can expose batholiths and these exposed formations can form exfoliation domes.

Organic materials can frequently contribute to erosion by pressure that results in structural cracking or in the formation of acidic compounds that **weather** rock.

Rapid **temperature** changes or large diurnal temperature changes (the difference between the highest daytime temperature and the coolest nighttime temperature) can accelerate erosional exfoliation, jointing, and ice wedging.

See also Acid rain; Catastrophic mass movements; Depositional environments; Dunes; Eolian processes; Faults and fractures; Freezing and melting; Glacial landforms; Glaciation; Hydrothermal processes; Ice heaving and ice wedging; Impact crater; Landforms; Landscape evolution; Leaching; Oxidation-reduction reaction; Precipitation; Rapids and waterfalls; Rate factors in geologic processes; Rock; Rockfall; Salt wedging; Seawalls and beach erosion; Soil and soil horizons; Talus pile or talus slope.

Escape velocity · *see* Gravity and the gravitational field

Eskers · *see* Glacial landforms

Estuary

Estuaries are unique and complex environments located between **oceans** and river mouths. As **freshwater** flows into the sea from land, it dilutes the salty **water** in a small **area** around the shore. This relatively small space is the site of sediment build-up resulting from fluvial (stream or river) **erosion** along the riverbanks. The organic sediments and brackish (slightly salty, but undrinkable) water make a unique environment that supports a diverse community of plants and animals. The sediments themselves also form characteristic types of deposits and bed forms (the appearance of the horizontal layers of **sand**) that are easily seen in cross-section.

It is well known that **rivers** often carry tremendous amounts of sediment which, when emptied into the oceans,

construct distinctive patterns in the underlying sediments. In an estuary, the deposition of sediments is greatly influenced by tidal currents and ocean waves. Even **climate** is a factor in how the sands and muds settle into distinct patterns. During seasonal storms, erosion is increased and the waters become heavily laden with a wide variety of sediments. Unlike deltas, in which the finer sediments are often carried far out to sea, the estuary is bordered on the deeper ocean side by heavy sands while clays and muds are dropped at river mouths. As tidal forces work the sediments by tumbling and rolling them, the lighter and finer particles are left near the river mouth. The build-up of coarse-grained (larger particle size) sands at the estuary edge often makes a barrier at the outer edge of the estuary that contains the bulk of the fine sediment and diluted water. The sediment structures in these ridges are defined as longitudinal or oblique bars. The structures in the upper reaches of the estuary are described as asymmetric and longitudinal bars become point bars similar to those observed in rivers. A dendritic (tree-shaped) pattern of channels occurs in these finer, flat lying sediments.

The greatest force at work shaping and changing the estuary comes from moving water. Daily tidal fluctuation brings saltier water into the estuary and pushes medium-grained sands into the main body of the tidal flats. On the ocean side of the estuary, the sand bars are penetrated by channels through which the flow of water into and out of the estuary is restrained. These containment structures close the general water and sediment circulation paths around the main body of the estuary. Water and sediment flow is greatly restricted and additional build-up of medium and fine-grained sediments occurs. Water is forced to leave the estuary by these well-defined channels.

The tidal or exit channels of the estuary can be dangerous places for some life forms. As low tide occurs, the ebb of general sea level reverses the oceanic flow into the estuary. Water laden sediments release their burden and contribute to the general volume of water leaving the flats. The force of water exiting through the channels becomes great. The velocity of the water can reach dangerously high levels. In well-established estuaries, large animals can be swept to sea. However, for many marine animals this is a benefit.

The estuaries are safe places for many creatures such as crabs and other crustaceans to lay their eggs. The hatched larvae live in the estuary until they are ready to join the zooplankton community of the larger sea. The swift release of water from the tidal flats helps the floating larvae to jet far out into the ocean where they will spend the next phase of their lives. If not for the tidal currents, the larvae would have to live dangerously exposed to shore birds and other marine carnivores. The estuarine water channels help them grow and gain a slight advantage for survival that would not be found in typical shore lines.

When high tide returns, the encroaching water brings **oxygen** to the anaerobic (without oxygen) sediments. The water also brings marine organic food for mud dwelling inhabitants. The sediments are refreshed with salt water and another cycle of replenishment occurs. During stormy **seasons**, this effect can be exaggerated and actually quite harmful to inhabitants as sediments are churned and redeposited. However, this continuous recycling of sediments and resources keeps the estuary healthy and flourishing.

Because the water plays such a physical and active role near the outer borders of the estuary, the inner regions of the estuary are more protected. By containing the general flow of water to the channels, the finer sediments, such as silt and **clay**, are left relatively undisturbed near the river mouths. They build up into areas of fine muds and contribute to the distinctive tidal flats. Organic debris is carried along by the rivers as they carve through valleys and plains of the terrestrial environment. This lightweight material comes to rest in the tidal flat as the velocity of the water is drastically reduced in the tidal flats. The decay and spreading of organic material throughout the flats makes them rich in nutrients. Subsequently, clams and other burrowing animals thrive in the rich sediments of the upper estuary. In turn, birds are lured to this feast where they are able to rear young on the nearby shore. These life forms are relatively protected because the muds make it difficult for heavier predators to walk out into the estuary with any stealth. There are even places where the muds act as a sort of quicksand and can be very dangerous.

Estuaries are fragile environments that are becoming increasingly threatened. They are being geologically altered as sediments are trapped upstream by dams. Diversion of water and sediments by agriculture is also reducing flow to the estuary. As a direct result, the life forms that rely on the dynamics of the estuary are decreasing in numbers. Many people are realizing the importance of estuarine environments and the important role they play in both marine and terrestrial ecosystems.

See also Oceanography; Sedimentation; Tides

EUROPE

The continent of Europe is a landmass bounded on the east by the Ural Mountains, on the south by the Mediterranean Sea, and on the north and west by the Arctic and Atlantic **Oceans**. Numerous islands around this landmass are considered a part of Europe. Europe is also the westernmost part of the Eurasian supercontinent (several continental masses joined together). Europe is a collection of different kinds of geologic regions located side by side. Europe holds a unique place among the continents; much of it is new, in geologic terms.

Plate tectonics is the main force of nature responsible for the geologic history of Europe. European geologic history, like that of all the continents, involves the formation of features as a result of plate tectonics.

When the edge of a plate of Earth's **lithosphere** runs over another plate, forcing the lower plate deep into the elastic interior, a long, curved chain of volcanic mountains usually erupts on the forward-moving edge of the upper plate. When this border between two plates forms in an ocean, the volcanic mountains constitute a string of islands (or archipelago). This is called an island arc. Italy's Appenine Mountains originally formed as an island arc, then later became connected into a single peninsula.

A continental arc is exactly like an island arc except that the volcanos erupt on a continent, instead of in the middle of an ocean. The chemical composition of the erupted **rock** is changed, because old continental rocks at the bottom of the lithosphere have melted and mixed with the **magma**. A clear-cut example of this kind of mountain chain no longer exists in Europe, but ancient continental arcs once played an important part in Europe's geologic past. Sicily's Mt. Aetna and Mt. Vesuvius on the Bay of Naples are good examples of the type of **volcano** that commonly make up a continental arc.

A suture describes the place where two parts of a surgery patient's tissue are sewed or rejoined; it also describes the belts of mountains that form when two continents are shoved into each other, over tens of millions of years, to become one. The Alps and other ranges in southern Europe stand tall because of a continental collision between Europe and **Africa**. The Alps, and other European ranges, are the forerunners of what may be a fully developed suture uniting Europe and Africa. This suture would be a tall mountain range that stretches continuously from Iberia to easternmost Europe.

The collision of Africa with Europe is a continuous process that stretches tens of millions of years into the past and into the future. All the generations of humanity together have seen only a tiny increment of the continental movement in this collision. However, throughout history, people have felt the collision profoundly, in earthquakes and volcanos, with all the calamities that attend them.

Sometimes a continent is torn in pieces by forces moving in opposite directions beneath it. On the surface, this tearing at first makes a deep valley, which experiences both volcanic and **earthquake** activity. Eventually the valley becomes wide and deep enough that its floor drops below sea level, and ocean **water** moves in. This process, called **rifting**, is the way ocean basins are born on Earth; the valley that makes a place for the ocean is called a rift valley. Pieces of lithosphere, rifted away from Africa and elsewhere, have journeyed across the Earth's surface and joined with the edge of Europe. These pieces of lithosphere lie under southern England, Germany, France, and Greece, among other places.

When a continent-sized "layer cake" of rock is pushed, the upper layers move more readily than the lower layers. The upper layers of rock are heavy but easier to move than those beneath it (much like a full filing cabinet is heavy but, when pushed, moves more easily than the floor beneath it). Between near surface rocks and the deeper, more ancient crustal rocks, a flat-lying fault forms, also called a detachment fault. This horizontal crack, called a thrust fault, contains fluid (water, mostly). The same hydraulic force that makes hydraulic machines lift huge weights functions in this crack as well. The fluid is so nearly incompressible that a sizeable piece of a continent can slide on it when pushed. The fault block floats on fluid pressure between the upper and lower sections of the lithosphere like a fully loaded tractor trailer gliding effortlessly along a rain-slicked road. The mountains that are heaved up where the thrust fault reaches the surface are one kind of fault block mountains. Both the Jura Mountains and the Carpathians are fault block mountains.

The white cliffs of Dover, composed almost entirely of chalk, are a familiar landmark of the British Isles. *JLM Visuals. Reproduced by permission.*

Europe was not formed in one piece, or at one time. Various parts of it were formed all over the ancient world, over a period of four billion years, and were slowly brought together and assembled into one continent by the processes of plate tectonics. What is now called Europe began to form more than 3 billion years ago, during the **Archean** Eon.

Most geologists feel that prior to and during the creation of the oldest parts of Europe, Earth only superficially resembled the planet we live on today. Active volcanos and rifts abounded. The planet had cooled enough to have a solid **crust** and oceans of liquid water. The crust may have included hundreds of small tectonic plates, moving perhaps ten times faster than plates move today. These small plates, carrying what are now the earth's most ancient crustal rocks, moved across the surface of a frantic crazy-quilt planet. Whatever it truly was like, the oldest regions in Europe were formed in this remote world. These regions are in Finland, Norway (Lofoten Islands), Scotland, Russia, and Bulgaria.

The piece of Europe that has been in its present form for the longest time is the lithospheric crust underneath Scandinavia, the Baltic states, and parts of Russia, Belarus, and Ukraine. This region moved around independently for a long time, and is referred to as Baltica. It is the continental core, or **craton**, to which other parts were attached to form Europe.

At the opening of the **Mesozoic Era**, 245 million years ago, a sizeable part of western and southern Europe was squeezed up into the Central Pangean mountain system that sutured Laurasia and Gondwana together. Pangea was the huge landmass from which all continents drifted; Laurasia and Gondwana were the two **supercontinents** that Pangea separated into about 200 million years ago. Europe was almost completely landlocked, its southern regions part of a mountain chain that stretched from Kazakhstan to the west coast of **North America**.

The birth of a new ocean basin, the Atlantic, signaled the end of Pangea. The Central Pangean Mountains, after tens of millions of years, had worn down to sea level and below. A new ocean basin, not the Mediterranean, but rather the Ligurean Ocean, began to open up between Africa and Europe. This formed a seaway between the modern North Atlantic and the Neo-Tethys Ocean (which no longer exists). Sea water began to leave deposits where high mountains had stood, and a layer cake of sediment—laid down on the shallow sea bottom—began to accumulate throughout Europe.

Beginning at the close of the Mesozoic Era (66 million years ago), and continuing through the **Cenozoic Era** to the present day, a complex **orogeny** (mountain building) has taken place in Europe. The ocean basin of Tethys was entirely destroyed, or if remnants still exist, they are indistinguishable from the ocean crust of the Mediterranean Sea. Africa has shifted from west of Europe (and up against the United States' East Coast) to directly south of Europe, and their respective tectonic plates are now colliding.

As in the collision that made the Variscan mountain belt, a couple of dozen little blocks are being pushed sideways into southern Europe. The tectonic arrangement can be compared with a traffic jam in Rome or Paris, where numerous moving objects attempt to wedge into a **space** in which they cannot all fit.

When sea level fell below the level of the Straits of Gibraltar around six million years ago, the western seawater passage from the Atlantic Ocean to the Mediterranean Sea closed, and water ceased to flow through this passage. At about the same time, northward-moving Arabia closed the eastern ocean passage out of the Mediterranean Sea and the completely landlocked ocean basin began to dry up. Not once, but perhaps as many as 30 times, all the water in the ancestral Mediterranean, Black, and Caspian **Seas** completely evaporated, leaving a thick crust of crystallized sea **minerals** such as **gypsum**, sylvite, and halite (salt). It must have been a lifeless place, filled with dense, hot air like modern below-sea-level deserts such as Death Valley and the coast of the Dead Sea. The **rivers** of Europe, **Asia**, and Africa carved deep valleys in their respective continental slopes as they dropped down to disappear into the burning salt wasteland.

Many times, too, the entire basin flooded with water. A rise in global sea level would lift water from the Atlantic Ocean over the barrier mountains at Gibraltar. Then the waters of the Atlantic Ocean would cascade 2.4 mi (4 km) down the mountainside into the western Mediterranean basin. From Gibraltar to central Asia, the bone-dry basin filled catastrophically in a geological instant—a few hundred years. This "instant ocean" laid deep-sea sediment directly on top of the layers of salt. The widespread extent and repetition of this series of salt and deep-sea sediment layers is the basis for the theory of numerous catastrophic **floods** in the Mediterranean basin.

For reasons not yet fully understood, Earth periodically experiences episodes of planet-wide climatic cooling, the most recent of which is known as the **Pleistocene Epoch**. Large areas of the land and seas become covered with **ice** sheets thousands of feet thick that remain unmelted for thousands or hundreds of thousands of years. Since the end of the last ice age about 8,000–12,000 years ago, only Greenland and **Antarctica** remain covered with continent-sized **glaciers**. But during the last two million years, Europe's northern regions and its mountain ranges were ground and polished by masses of water frozen into miles-thick continental glaciers.

Ice in glaciers is not frozen in the sense of being motionless. It is in constant motion—imperceptibly slow, but irresistible. Glaciers subject the earth materials beneath them to the most intense kind of scraping and scouring. An alpine glacier has the power to tear **bedrock** apart and move the shattered pieces miles away. These are the forces that shaped the sharp mountain peaks and u-shaped mountain valleys of modern Europe. Many European mountain ranges bear obvious scars from alpine **glaciation**, and the flat areas of the continent show the features of a formerly glaciated plain.

Humans have lived in Europe for much of the Pleistocene Epoch and the entire **Holocene Epoch** (beginning at the end of the last ice age, about 10,000 years ago). During the past few thousand years, humans have been significantly altering the European landscape. Wetlands across Europe have been drained for agricultural use from the Bronze Age onward. The Netherlands is famous for its polders, below-sea-level lands made by holding back the sea with dikes. Entire volcanoes (cinder cones) have been excavated to produce frost-resistant road fill.

The northwest fringe of Europe is made up of the two very old islands, Great Britain and Ireland, and numerous smaller islands associated with them. Geologically, these islands are a part of the European continent, although culturally separate from it. Unlike many islands of comparable size, the British Isles do not result from a single group of related tectonic events. They are as complex as continents themselves, which in the last two centuries has provided plenty of subject matter for the new science of **geology**.

Scotland and Ireland are each made of three or four slices of continental crust. These slices came together around 400 million years ago like a deck of cards being put back together after shuffling.

The Iberian Peninsula, occupied today by Portugal and Spain, is one of the pieces of lithosphere that was welded to Europe during the Variscan mountain-building event. Like Britain, it is an unusual "micro-continent" with a complex geologic history.

Since the **Paleozoic Era**, southern Europe has continued to acquire a jumbled mass of continental fragments from Africa. Even today, the rocks of Europe from the Carpathian Mountains southwestward to the Adriatic and Italy are made up of "tectonic driftwood," and are not resting on the type of solid, crystalline basement that underlies Scandinavia and Ukraine.

Since the late Mesozoic Era, the widening Atlantic Ocean has been pushing Africa counterclockwise. All the blocks of lithosphere between Africa and Europe, including parts of the Mediterranean seafloor, will in all likelihood eventually become a part of Europe.

The Alps resulted from Europe's southern border being pushed by the northern edge of Africa. In Central Europe, freshly-made **sedimentary rocks** of early Mesozoic age, along with the older, metamorphosed, Variscan rocks below, were pushed into the continent until they had no other way to go but up. Following the path of least resistance, these rocks were shaped by powerful forces into complex **folds** called nappes, which means tablecloth in French. The highly deformed rocks in these mountains were later carved into jagged peaks by glaciers during the Pleistocene Epoch.

The Jura Mountains, the Carpathians, and the Transylvanian Alps are made of stacks of flat-lying sedimentary rock layers. These mountain ranges were thrust forward in giant sheets out in front of the rising Alps.

A complex story of tectonic movement is recorded in the sea-floor rocks of the western Mediterranean. Corsica, Sardinia, Iberia, and two pieces of Africa called the "Kabylies"—formerly parts of Europe—moved in various directions at various speeds throughout the Cenozoic Era.

On the western Mediterranean floor, new oceanic lithosphere was created. A **subduction zone** formed as an oceanic plate to the east sank below the western Mediterranean floor. The magma generated by this event gave rise to the Appenine Mountains, which formed as an island arc on the eastern edge of the western Mediterranean oceanic plate. The Appenines began to rotate counterclockwise into their present position. The Tyrrhenian Sea formed as the crust stretched behind this forward-moving island arc. In the Balkans, blocks of lithosphere have piled into each other over tens of millions of years.

The Dinarides and Hellenides, mountains that run down the east coast of the Adriatic Sea, form the scar left after an old ocean basin closed. The compressed and deformed rocks in these mountain ranges contain pieces of ocean floor. Just east of these seacoast mountains is a clearly-recognized plate boundary, where the European and African plates meet. The boundary runs from the Pannonian Basin (in Hungary, Romania, and Yugoslavia), cuts the territory of the former Yugoslavia in half, and winds up in Greece's Attica, near Athens.

Further inland, the Pannonian Basin results from the lithosphere being stretched as the Carpathian Mountains move eastward and northward.

The Aegean Sea seems to have formed as continental crust has been stretched in an east-west direction. It is a submerged basin-and-range province, such as in the western United States. The Pelagonian Massif, a body of igneous and **metamorphic rock** that lies under Attica, Euboea, and Mount Olympus, forms part of the Aegean sea floor. The Rhodopian Massif, in northern Greece, Bulgaria, and Macedonia, also extends beneath the Aegean Sea. Faults divide the ridges from the troughs that lie between them. The faults indicate that the troughs have dropped into the crust between the high ridges.

The Balkan Range in Bulgaria is thought to mark the crumpled edge of the European craton—the Proterozoic-age rocks extending north into Russia.

Europe is also host to isolated volcanoes related to structural troughs within the continent. The Rhine River flows in a trough known as the Rhine Graben. Geologists believe the Rhine once flowed southward to join the Rhone River in France, but was diverted by upwarping of the crust around the Vogelsberg Volcano. The Rhine then changed its course, flowing out to meet England's Thames River in the low-sea-level Ice Age.

Europe continues to change today. From the Atlantic coast of Iberia to the Caucasus, Europe's southern border is geologically active, and will remain so effectively forever, from a human frame of reference. Africa, Arabia, and the Iranian Plateau all continue to move northward, which will insure continued mountain-building in southern Europe.

Geologists are concerned about volcanic hazards, particularly under the Bay of Naples and in the Caucasus. In historic times, in the Aegean Sea and at Pompeii, Herculaneum, and Lisbon, entire cities have been devastated or destroyed by volcanoes, earthquakes, and seismic sea waves. These larger-scale natural disasters can and will continue to happen in Europe on an unpredictable schedule with predictable results.

See also Continental drift theory; Earth (planet); Historical geology

EVAPORATION

Evaporation is a geologic process that concentrates the ion solute residues in the ocean basins. At a fundamental level, evaporation is the transition of the molecule of a liquid from the liquid state to the gaseous state by diffusion from the surface of the liquid.

Driven by **solar energy**, the only significant loss of **water** from the ocean basin occurs via evaporation. As the ocean surface and atmospheric interface is small compared to the total volume of the ocean, estimates of the time a particular molecule remains in the liquid phase range in the order of thousands to tens of thousands of years before once again entering the atmosphere as part of the **hydrologic cycle**.

Because solutes (e.g., dissolved salts) from **weathering** and **erosion** are not as volatile (i.e., as easy to move into the gas or vapor phase as the water molecules, evaporation plays a significant role in the formation of many geologic features (e.g., Great Salt Lake, Dead Sea, etc.).

Evaporation is usually also responsible for the majority of the loss of water from **precipitation** and results in a high cycling of water molecules during the hydrologic cycle.

Evaporation may be driven by solar energy or be a directed process used to concentrate an aqueous solution of nonvolatile solutes and a volatile solvent. In evaporation, a portion of the solvent is vaporized or boiled away, leaving a thick liquid or solid precipitate as the final product. The vapor is condensed to recover the solvent or it can simply be discarded. A typical example is the evaporation of sea water to produce salt.

Evaporation may also be used as a method to produce a liquid or gaseous product obtained from the condensed vapor. For instance, in desalinization processes, sea water is vaporized and condensed in a water-cooled heat exchanger and forms the fresh water product.

Although evaporation can be driven by the random motion of molecules near the liquid-gas interface, the addition of heat to a system speeds the evaporative process.

See also Caliche; Condensation; Drainage calculations and engineering; Leaching; Oceans and seas; Phase state changes; Runoff

EVOLUTION

Evolution is the gradual, cumulative change over time of the characteristics of groups of organisms in a heritable manner. Eventually, these minute changes add up to produce an individual that is markedly different from its distant ancestors, but almost indistinguishable from its most immediate ancestors. These changes are brought about by the organism's genetic response to the environment, and, over the entire course of history, evolution has given rise to all different forms of life on Earth.

Evolution does not occur rapidly on the individual unit of life; changes are too small and slow to be effective at the individual level. In fact, evolution is more efficient at the pop-

The Bonneville Salt Flats in Utah were at one time covered by a sea of salt water. Evaporation removed the water, leaving the salt behind. © *Buddy Mays/Corbis. Reproduced by permission.*

ulation level among groups of organisms that are capable of successfully breeding with each other. With organisms that do not breed with other individuals, the rate of evolutionary change is slower than it is among outbreeding organisms.

Evolution leads to increasing complexity and, eventually, to the production of new species, which survive or become extinct depending upon their reaction to the environment and its continuing changes. Evidence for evolution comes from the **fossil record**, genetics, and comparative studies.

The mechanism behind evolution is natural selection. Small, individual changes that arise by chance can confer an advantage to those possessing them; this group then has better success at breeding, and successful genes are consequently spread further throughout the population. The theory of evolution is now widely accepted, but when it was first put forward in the nineteenth century by English naturalist Charles Darwin there was much opposition, particularly from religious quarters. Opponents to the theory of evolution often argue for special creation, which states that each type of species was created in the form in which it currently exists, and that no two species are related, by descent, to any other. Most scientists now accept the theory of evolution, as the concept of evolution fits available evidence. There exist some gaps in scientific knowledge of evolution, such as the discovery of the common ancestor for both apes and humans, often referred to as the missing link, but, with time, these knowledge gaps have become smaller.

Evolution does not proceed at a constant rate. At times, a gradual change occurs that allows for a good reconstruction of the process from the fossil record. This is known as phyletic gradualism. The other method of evolution, which can leave gaps in the fossil record is the quicker and more explosive form, called punctuated equilibrium.

See also Cosmology; Evolution, evidence of; Evolutionary mechanisms; Fossils and fossilization

EVOLUTION, EVIDENCE OF

Evidence of **evolution** can be observed in a number of different ways, including distribution of **fossils** of species both geographically and through **geologic time**. Evolution is a major scientific theory. As such, it has a tremendous amount of supporting evidence and no clearly contradicting evidence. If new evidence appears to refute it, then a new theory must be formulated. Any evidence requiring a totally new theory, however, would have to be staggering in its scope and strength. The new evidence that has been forthcoming in recent scientific studies supports the theory of evolution and merely fine tunes scientific understanding of the mechanisms involved.

A clear and strong argument in favor of evolutionary theory is found in the **fossil record**. Paleontology (the study of fossils) provides an unarguable record that many species no longer exist. By such techniques as **carbon dating** and studying the placement of fossils within the ground, an age can be given for fossils. By placing fossils together based on their ages, a gradual change in form can be surmised which can be followed and extrapolated to the species that exists today. Although the fossil record is incomplete, and many intermediate species are missing, the weight of evidence from those that do exist favors the theories of evolution and natural selection.

Extremely strong evidence supporting evolutionary theory is found in the strata of fossils. No fossils more ancient than those found in underlying layers have ever been found. This **correlation** of the geologic record with the biological evolutionary record is profound.

English naturalist Charles Darwin (1809–1882) formulated the theory of organic evolution through natural selection in his groundbreaking publication *The Origin of Species by Means of Natural Selection*, published in 1859. One of the first pieces of evidence that started the young scientist thinking along evolutionary lines was given to him on his journeys aboard the H.M.S. *Beagle*. Darwin made extensive collections of plants and animals that he came across wherever the ship stopped, and soon he started to notice patterns within the organisms he studied.

There were similarities between organisms collected from widely differing areas. As well as the similarities, there were also striking differences. For example, mammals are present on all of the major landmasses, but Darwin did not find the same mammals even in similar habitats. One explanation of this is that in the past when the landmasses were joined, mammals spread over all of the available land. Subsequently this land moved apart, and the animals became isolated. As time passed, random variation within the populations was acted upon by natural selection. This process is known as adaptive radiation—from the same basic stock, many different forms have evolved. Each environment is slightly different, and slightly different forms are better suited to survive there. An example of this, which is seen at a formative stage, is the case of the finches on the Galápagos Islands. All of the Galápagos finches bear similarities to the mainland finches, but each species has evolved to fill a particular **niche**, which is not already filled by an animal on the islands even though there are species filling these ecological openings on the mainland.

If it is true that widely separated groups of organisms have ancestors in common, then logic dictates that they would have certain basic structures in common as well. The more structures they have in common then the more closely related they must be. The study of evolutionary relationships based on commonalties and structural differences is termed comparative anatomy. What scientists look for are structures that may serve entirely different functions but are basically similar. Such homologous structures suggest a common ancestor. A classic example of this is the pentadactyl limb, which in suitably modified forms can be seen in all mammals. A greater modified version of this can also be seen amongst birds. This limb has been used by different groups for slightly different purposes and so provides an example of divergent evolution.

These evolutionary relationships are reflected in taxonomy. Taxonomy is an artificial, hierarchical system showing relationships between species. Each level progressed within the taxonomic system denotes a greater degree of relatedness for the organism in that group to the level above.

In embryology, the developing fetus is studied, and similarities with other organisms are observed. This adds evidence to a past recent common ancestor. It is not, however, true that a developing organism replays its evolutionary stages as an embryo; there are some similarities with the more conserved regions, but embryonic development is subjected to evolutionary pressures as much as other areas of the life cycle.

Cell biology is an area where many similarities can be seen between organisms. Many structures and pathways within the cell are vital for the continuance of life. The more important and basic to the whole structure of life a pathway or organelle is, the more likely it is to be observed. For example, the DNA code is the same in virtually all living organisms, as are such structures as mitochondria. These are virtually ubiquitous throughout known life. Most scientists hold that it is inconceivable that each of these things arose separately for each species of living organism. The conclusion advanced by science is that all life forms arose from the same basic source many millennia ago. The examples visible today have survived, and the organisms carrying these processes on have adapted, yielding the diversity of forms seen today.

See also Cosmology; Dating methods; Fossils and fossilization; Stratigraphy

EVOLUTIONARY MECHANISMS

Evolution is the process of biological change over time. Such changes, especially at the genetic level, are accomplished by a complex set of evolutionary mechanisms that act to increase or decrease genetic variation. Changes at the genetic level are often directly caused by physical phenomena (e.g., cosmic rays). Genetic changes are also indirectly acted upon by the physical environment in such a manner that the number of genetic changes (e.g., mutations) within a population can either increase or decrease in subsequent generations.

Evolutionary theory is the cornerstone of modern biology, and unites all the fields of biology under one theoretical umbrella to explain the changes in any given gene pool of a population over time. Fundamental to the concept of evolutionary mechanism is the concept of the syngameon, the set of all genes. By definition, a gene is a hereditary unit in the syngameon that carries information that can be used to construct proteins via the processes of transcription and translation. A gene pool is the set of all genes in a species or population.

Another essential concept, important to understanding evolutionary mechanisms, is an understanding that there are no existing (extant) primitive organisms that can be used to study evolutionary mechanism. For example, all eukaryotes (organisms having a true nucleus) derived from a primitive, common prokaryotic (organisms such as bacteria that lack a true nucleus) ancestral bacterium. Accordingly, all living eukaryotes have evolved as eukaryotes for the same amount of time. Additionally, no eukaryote plant or animal cell is more primitive with regard to the amount of time they have been subjected to evolutionary mechanisms. Seemingly primitive characteristics are simply highly efficient and conserved characteristics that have changed little over time.

Evolution requires genetic variation, and these variations or changes (mutations) can be beneficial, neutral, or deleterious. In general, there are two major types of evolutionary mechanisms—those that act to increase genetic variation, and mechanisms that operate to decrease genetic mechanisms.

Mechanisms that increase genetic variation include mutation, recombination and gene flow.

Mutations generally occur via chromosomal mutations, point mutations, frame shifts, and breakdowns in DNA repair mechanisms. Chromosomal mutations include translocations, inversions, deletions, and chromosome non-disjunction. Point mutations may be nonsense mutations leading to the early termination of protein synthesis, missense mutations (a mutation that results in a substitution of one amino acid for another in a protein), or silent mutations that cause no detectable change. Point mutations may result from natural phenomena such as cosmic radiation.

Recombination involves the re-assortment of genes through new chromosome combinations. Recombination occurs via an exchange of DNA between homologous chromosomes (crossing over) during meiosis. Recombination also includes linkage disequilibrium. With linkage disequilibrium, more variations of the same gene (alleles) occur in combinations in the gametes (sexual reproductive cells) than should occur according to the rules of probability.

Gene flow occurs when individuals change their local genetic group by moving from one place to another. These migrations allow the introduction of new variations of the same gene (alleles) when they mate and produce offspring with members of their new group. In effect, gene flow acts to increase the gene pool in the new group. Because genes are usually carried by many members of a large population that has undergone random mating for several generations, random migrations of individuals away from the population or group usually do not significantly decrease the gene pool of the group left behind.

In contrast to mechanisms that operate to increase genetic variation, there are fewer mechanisms that operate to decrease genetic variation. Mechanisms that decrease genetic variation include genetic drift and natural selection.

Genetic drift results from the changes in the numbers of different forms of a gene (allelic frequency) that result from sexual reproduction. Genetic drift can occur as a result of random mating (random genetic drift) or be profoundly affected by geographical barriers, catastrophic events (e.g., natural disasters or wars that significantly affect the reproductive availability of selected members of a population), and other political-social factors.

Natural selection is based upon the differences in the viability and reproductive success of different genotypes with a population (differential reproductive success). Natural selection can only act on those differences in genotype (type of genes present) that appear as phenotypic (visible characteristics) differences that affect the ability to attract a mate and produce viable offspring that are, in turn, able to live, mate and continue the species. Evolutionary fitness is the success of an entity in reproducing (i.e., contributing alleles to the next generation).

There are three basic types of natural selection. With directional selection, an extreme phenotype is favored (e.g., for height or length of neck in giraffe). Stabilizing selection occurs when intermediate phenotype is fittest (e.g., neither too high nor too low a body weight) and for this reason it is often referred to a normalizing selection. Disruptive selection occurs when two extreme phenotypes are fitter than an intermediate phenotype.

Natural selection does not act with foresight. Rapidly changing environmental conditions can, and often do, impose new challenges for a species that result in extinction. In addition, evolutionary mechanisms, including natural selection, do not always act to favor the fittest in any population, but instead may act to favor the more numerous but tolerably fit. Thus, the modern understanding of evolutionary mechanisms does not support the concepts of social Darwinism.

The operation of natural evolutionary mechanisms is complicated by geographic, ethnic, religious, and social groups and customs. Accordingly, the effects of various evolution mechanisms on human populations are not as easy to predict. Increasingly sophisticated statistical studies are carried out by population geneticists to characterize changes in the human genome.

See also Cosmic ray; Evolution, evidence of; Fossil record; Fossils and fossilization; Geologic time; Radioactivity; Uniformitarianism

EXFOLIATION

Exfoliation is the term used to describe the peeling away of sheets of **rock** millimeters to meters in thickness from a rock's surface due a range of physical and chemical processes during exhumation and **weathering**. Exfoliation can occur due to several processes.

Unloading or release of stress in a rock that produces expansion joints can cause exfoliation. A reduction in stress occurs when rocks previously buried deeply are exposed due to **erosion** of overlying rocks, or when **ice** sheets that bury rocks melt. During a combination of physical and chemical weathering, exfoliation may occur parallel to a rock's outer surface due to a combination of chemical breakdown of **minerals**, especially in the presence of **water**. Such 'onion-skin' style weathering occurs especially in **igneous rocks** (e.g., **granite**) as micas, amphiboles and pyroxenes, common minerals in many igneous rocks, break down to **clay**. Clays swell in the presence of water, so alternating wetting and drying of a rock may lead to consecutive expansion and shrinking that can result in disintegration and exfoliation.

Stresses induced in a rock due to the expansion of water trapped between grains or in fractures in a rock during **freezing** may result in fracturing. Shattering of rock into small fragments by the expansion of water during the formation of ice is common in arctic environments (causing a problem for field geologists looking for rock relationships and structures). Likewise, changes in **temperature** of a rock may cause exfoliation. Stresses due to variability in the rates and amounts of expansion of different minerals in a rock, or due to alternating expansion and shrinkage from day to night in **desert** areas, may result in exfoliation. Rapid temperature changes may also occur due to lightning strikes followed by cooling in the ensuing rain. Although generally a naturally occurring process, exfoliation was also induced by man to obtain rock sheets several centimeters in thickness to thin, sharp shards of some fine-grained rocks for use as scrapers and knives by heating the rock with fire, then pouring water on the rock's surface.

See also Weathering and weathering series

EXTRUSIVE COOLING

Igneous rocks formed at ground level are termed extrusive. A body of extruded **magma** (molten **lava**) cools more rapidly than an equal amount of magma intruded into preexisting **rock** because all extruded magma is bathed, at least on its upper surface, in a coolant (e.g., air or **water**). Gobs of lava blown high into the air by a volcanic eruption may even solidify before reaching the ground, producing the streamlined, glassy rocks termed volcanic bombs. At the opposite extreme, a lava flow many yards thick may take days or weeks to crystallize all the way through and years to cool to ambient **temperature**. Even a thick lava flow, however, cools very rapidly compared to an intrusion of comparable dimensions, which may take hundreds or thousands of years to crystallize.

Fast cooling does not permit the formation of large **crystals**, so extrusive cooling produces either very fine-grained crystalline rock or volcanic **glass**, which contains no crystals at all. Because glasses are inherently unstable and spontaneously reorganize into fine-grained crystalline rocks over millions of years, truly old (pre-Cenozoic) glasses are rare.

Another feature of extrusive cooling is that **atmospheric pressure** is much lower than the pressures under which magmas form. Magma's volatile components, that is, those substances that tend to separate out at high temperature and low pressure (especially water), are therefore quickly lost by extruded magma, and are not present during crystallization. Reduced water content in a magma permits many **minerals** to crystallize at higher temperatures, further speeding the rapid crystallization caused by fast cooling.

See also Amorphous; Intrusive cooling; Lava; Volcanic eruptions; Volcano

F

FAHRENHEIT, DANIEL GABRIEL
(1686-1736)
Dutch physicist

Daniel Fahrenheit invented the first truly accurate thermometer using mercury instead of alcohol and **water** mixtures. In the laboratory, he used his invention to develop the first **temperature** scale precise enough to become a worldwide standard.

The eldest of five children born to a wealthy merchant, Fahrenheit was born in Danzig (Gdansk), Poland. When he was fifteen his parents died suddenly, and he was sent to Amsterdam to study business. Instead of pursuing this trade, Fahrenheit became interested in the growing field of scientific instruments and their construction. Sometime around 1707, Fahrenheit began to wander the European countryside, visiting instrument makers in Germany, Denmark, and elsewhere, learning their skills. He began constructing his own thermometers in 1714, and it was in these that he used mercury for the first time.

Previous thermometers, such as those constructed by Galileo and Guillaume Amontons, used combinations of alcohol and water; as the temperature rose, the alcohol would expand and the level within the thermometer would increase. These thermometers were not particularly accurate, however, since they were too easily thrown off by changing air pressure. The key to Fahrenheit's thermometer was a new method for cleaning mercury that enabled it to rise and fall within the tube without sticking to the sides. Mercury was an ideal substance for reading temperatures since it expanded at a more constant rate than alcohol and is able to be read at much higher and lower temperatures.

The next important step in the development of a standard temperature scale was the choosing of fixed high and low points. It was common in the early eighteenth century to choose as the high point the temperature of the body, and as the low point the **freezing** temperature of an ice-and-salt mixture, then believed to be the coldest temperature achievable in the laboratory. These were the points chosen by Claus Roemer,

a German scientist whom Fahrenheit visited in 1701. Roemer's scale placed blood temperature at 22.5°F and the freezing point of pure water at 7.5°F. When Fahrenheit graduated his own scale he emulated Roemer's fixed points; however, with the improved accuracy of a mercury thermometer, he was able to split each degree into four, making the freezing point of water 30°F and the temperature of the human body 90°F. In 1717, he moved his points to 32°F and 96°F in order to eliminate fractions.

These points remained fixed for several years, during which time Fahrenheit performed extensive research on the freezing and boiling points of water. He found that the boiling point was constant, but that it could be changed as **atmospheric pressure** was decreased (such as by increasing elevation to many thousand feet above sea level). He placed the boiling point of water at 212°F, a figure that was actually several degrees too low. After Fahrenheit's death scientists chose to adopt this temperature as the boiling point of water and to shift the scale slightly to accommodate the change. With 212°F as the boiling point of water and 32°F as the freezing point, the new normal temperature for the human body became 98.6°F.

In 1742, Fahrenheit was admitted to the British Royal Society despite having had no formal scientific training and having published just one collection of research papers.

See also Temperature and temperature scales

FAHRENHEIT SCALE • *see* TEMPERATURE AND TEMPERATURE SCALES

FARADAY, MICHAEL (1791-1867)
English physicist and chemist

Michael Faraday's early life had a remarkable resemblance to that of **Benjamin Franklin**. Faraday was born in Newington,

Surrey, England. Like Franklin, Michael Faraday was part of a large family. His father was a blacksmith who lacked the resources to obtain a formal education for his son.

Franklin had been apprenticed to a printer; young Faraday went a similar route, becoming apprenticed to a bookbinder. In each case, this led to a voracious love of books. Michael was especially interested in **chemistry** and **electricity**. He studied the articles about electricity in the *Encyclopaedia Britannica*. His employer not only allowed him to read all that he wanted, he encouraged the boy to attend scientific lectures.

A turning point occurred in 1812. In that year, Faraday obtained tickets to hear the lectures of Humphry Davy at the Royal Institution. Faraday took careful notes extending to 386 pages, which he had bound in leather. He sent a copy to Sir Joseph Banks (1743–1820), president of the Royal Society of London, who wielded great influence over European scientific investigation. Faraday hoped to make a favorable impression, but if Banks ever looked through the book with its carefully drawn and colored diagrams, Faraday never knew it.

Determined not to be ignored, Faraday sent a copy of his notes to Davy and included an application for a job as Davy's assistant. Davy was impressed, but did not offer Faraday work as he already had an assistant. Later, however, after firing the assistant in 1813, Davy contacted Faraday. The job description was not quite what Faraday had in mind. A trustee of the Royal Institution had said, "Let him wash bottles. If he is any good, he will accept the work; if he refuses, he is not good for anything." Faraday accepted, even though it meant he would be paid less than what he was making as a bookbinder.

Shortly thereafter, Davy resigned his post at the Royal Institution, married a wealthy widow, and decided to travel through **Europe**. Faraday accompanied the couple and met such illustrious men as Italian physicist Alessandro Volta and French chemist Louis-Nicholas Vauquelin.

In 1820, Danish Physicist Hans Christian Oersted amazed scientists with the discovery that electric current produced a **magnetic field**. Faraday had a greater goal in sight: Oersted had converted electric current into a magnetic force; Faraday intended to reverse the process and create electricity from **magnetism**. Within a year Faraday, now back in England, constructed a device that essentially consisted of a hinged wire, a magnet and a chemical battery. When the current was turned on, a magnetic field was set up in the wire, and it began to spin around the magnet. Faraday had just invented the electric motor. Faraday's motor was certainly an interesting device, but it was treated as a toy.

At this point, Davy, realizing that Faraday had the potential to eclipse him, jealously claimed that Faraday had taken his own idea for the experiment.

Faraday's first major contribution to chemistry came a few years later. In 1823 he unknowingly became the "father" of cryogenics by producing laboratory temperatures that were below **freezing**. He discovered how to liquefy **carbon dioxide**, chlorine, hydrogen sulfide, and hydrogen bromide gases by placing them under pressure, but again Davy took the credit. Two years later Faraday discovered the compound benzene, which became his greatest contribution in chemistry. In 1865

German chemist Friedrich Kekulé was able to determine the structure of benzene, leading to the understanding of molecular structure in general.

The studies of **magnetism** and electricity were still main interests of Faraday, so he elaborated on Davy's pioneering work in electrochemistry. Davy had passed an electric current through a variety of molten **metals** and created new metals in the process. Faraday named this process electrolysis. He also bestowed names that were suggested by the British scholar Whewell and are still in use today: electrolyte, for the compound or solution that conducts electricity; electrode, for the metal rod inserted in the object; and anode and cathode for the positively and negatively charged electrodes, respectively.

In 1832, Faraday devised what became known as Faraday's laws of electrolysis, which hold that the mass of the substance liberated at an electrode during electrolysis is proportional to the amount of electricity going through the solution; and the mass liberated by a given amount of electricity is proportional to the atomic weight of the element liberated and inversely proportional to the "combining power" of the element liberated. The two laws showed there was a connection between electricity and chemistry. They also supported the suggestion that Franklin had made nearly 100 years earlier when he claimed electricity was composed of particles, a theory that would be another 50 years in the making.

In another experiment, Faraday sprinkled **iron** filings on a paper which was held over a magnet and noticed the filings had arranged themselves along what he called "lines of force." The connections along the lines showed where the strength of the field was equal. With the magnetic field now "visible," scientists began to wonder if **space** itself was filled with interacting fields of various types and this helped establish a new way of thinking about the universe. Up to this point most scientists had believed in the mechanical nature of the universe as established by Galileo and Isaac Newton.

Taking the concept of his lines of force one step farther, Faraday realized that when an electric current began to flow it caused lines of force to expand outward. When the current stopped, the lines collapsed. If the lines expanded and collapsed across an intervening wire, an electric current would be induced to flow through it, first forward then in reverse.

By now Faraday was giving public lectures, which were very popular, at the Royal Institute, just as Davy had. Faraday reasoned that if electricity could induce a magnetic field, then it should be possible for the reverse to be true. Taking an iron ring during one demonstration, Faraday wrapped half of it with a coil of wire that was attached to a battery and switch. André Ampère (1775–1836) had shown that electricity would set up a magnetic field in the coil. The other half of the ring was wrapped with a wire that led to a galvanometer. In theory, the first coil would set up a magnetic field that the second coil would intercept and convert back to electric current, which the galvanometer would register. Faraday threw the switch: the experiment worked. He had just invented the transformer. However, the result was not exactly what he expected. Instead of registering a continuous current, the galvanometer moved only when the circuit was opened or closed. Ampère had observed the same effect a decade earlier but ignored it because it did not fit his theories. Deciding to make the theory fit the observation, instead of the other way around, Faraday concluded that when the current was turned on or off, it caused magnetic "lines of force" from the first coil to expand or contract across the second coil, inducing a momentary flow of current in the second coil. Faraday had now discovered electrical induction. Meanwhile, in the United States, physicist Joseph Henry had independently made the same discovery.

In 1839, at the age of 48, Faraday had a nervous break-down. Faraday never completely recovered. It is also possible that he was afflicted with a low-grade chemical poisoning. This was a common ailment that affected chemists at the time; Davy had suffered from it as well. In any event, Faraday's failing memory forced him to leave the laboratory.

Faraday published a book describing his lines of force in 1844, but because he lacked a formal education it was written without mathematical equations. Consequently the book was not taken seriously. When **James Clerk Maxwell** investigated the subject, he essentially came to the same conclusion as Faraday had, but used mathematics to prove his theory.

Although out of the laboratory, Faraday was by no means inactive. He investigated the effect of weak magnetic fields on nonmetallic substances and coined the words para-magnetic and diamagnetic to differentiate between the force of attraction and repulsion. Development of a theory to explain the two opposing forces, however, had to wait more than 50 years for the work of Paul Langevin. In the 1850s, during the Crimean War, the British government sought his opinion on the feasibility of using poisonous chemical weapons, and asked him to oversee their development. Faraday immediately said the project was very feasible, but he refused to have anything to do with its initiation.

Faraday died at Hampton Court, Middlesex, England. The word "farad," which is a unit of capacitance, was named in his honor.

See also Electricity and magnetism; Electromagnetic spectrum

FAULTS AND FRACTURES

Fractures and faults are planes of tensile or shear failure at microscopic to regional scales in brittle rocks. Faults may constitute a single plane or comprise zones of parallel or oblique shear planes, fault **breccia** or gouge (finely ground **rock**) across which there has been relative displacement of rocks on either side. Faults and fractures dominate approximately the upper 9 mi (15 km) of the earth's **crust**. Earthquakes are the expression of rapid displacement along faults. Most upper crustal rocks deform in a brittle manner at rapid deformation (strain) rates due to low **temperature** and confining pressure. Faults and fractures also develop in competent rocks at deeper crustal levels and in some dry rocks in the lower crust.

Fractures develop when the applied stress exceeds a rock's elastic limit. Regional stresses in plate interiors responsible for fracturing and faulting may represent far-field effects of tectonic forces acting on plate margins. They can also result

The San Andreas fault is a famous example of a strike-slip fault. *Photograph by Robert E. Wallace. U.S. Geological Survey.*

from gravitational instabilities (e.g., when a less dense, ductile rock such as salt is overlain by brittle, denser rocks), unloading (e.g., in response to removal of an **ice** sheet) or **mantle plumes**. Rocks may also fracture at regional stresses below their elastic limit if pore fluid pressure increases sufficiently. This phenomenon (called hydrofracturing or hydraulic fracturing) is used commercially in the **petroleum** industry to increase **permeability** of petroleum reservoirs. It may also be induced inadvertently, such as by building a large dam over a previously inactive fault zone, triggering fracturing that may potentially cause the dam to fail.

Tensile (extensional) fractures develop normal to the maximum extension direction. Tensile fractures may dilate and infill with **minerals** such as **quartz** and calcite precipitated from fluids within the rock to form extensional veins. The direction of vein opening (deduced from the orientation of quartz or calcite fibers that track the incremental extension direction) is orthogonal to fracture margins. Shear fractures (fractures along which lateral displacement occurs) develop oblique to the principal stresses. Shear fractures commonly develop in two preferred orientations where fractures with the opposite sense of displacement form at an angle of approximately 60° to each other, constituting a conjugate set. Conjugate fractures are bisected acutely by the maximum shortening direction and obtusely by the maximum extension direction. Their intersection parallels the intermediate principal stress. The acute angle between conjugate shear fractures and faults may be less than 60° if the rock is very brittle or greater than 60° if the rock is more ductile. One or more of the following types of minor fractures commonly develop prior to the formation of through-going shear fractures or faults:

- tensile fractures that **strike** parallel to each other in both conjugate zones,
- shear fractures along each zone that trend parallel to, and have the same sense of displacement as the conjugate zone,
- shear fractures (called Riedel shears) that make an angle of approximately 15° to each zone.

All three may step en échelon along incipient **shear zones**. The sense of stepping may be used to determine the sense of displacement along a fault where there is no clear offset of marker layers.

There are three end-member types of faults, with each type forming under different orientations of principal stresses:

- Normal or extensional faults are inclined structures along which rocks above the fault plane (i.e. in the hangingwall) are displaced down the fault with respect to rocks beneath the fault plane (i.e. in the footwall). Such faults were called normal as they were the most commonly observed faults by Welsh **coal** miners who termed the name. They form when the maximum extension direction is horizontal during vertical shortening (due to the gravitation loading of overlying rock). Normal faults develop in sedimentary basins during **rifting** and in areas of localized horizontal extension, such as above salt diapirs or during collapse or slumping.
- Reverse or contractional faults are moderately inclined structures along which the hanging wall is displaced up the footwall. The name reverse fault also comes from Welsh miners as they showed the opposite sense of displacement to the normal faults. Shallowly dipping faults with reverse displacement are called thrusts. Reverse and thrust faults imply horizontal shortening and vertical extension, and are commonly formed in convergent plate margins. Reverse faults in sedimentary basins may also form in the toes of deltas because of local shortening due to sediment loading or slumping or on the margins of laterally expanding salt diapirs.
- Faults along which there has been lateral, sub-horizontal displacement are called strike-slip, transcurrent or wrench faults. They are generally steeply dipping structures. They also form during regional horizontal shortening, but where the maximum extension direction is horizontal. When the left side (observed in map view) is displaced towards the observer, the fault is said to show a sinistral or left lateral sense of displacement. When the right side is displaced towards the observer, the displacement is dextral or right-lateral.

Oblique-slip faults have components of both transcurrent and either normal or reverse displacement. The direction of displacement along a fault is indicated by fine scratch or gouge marks (called slickenlines) and/or mineral fibers that infill spaces created by displacement of irregular, stepped surfaces. Fault planes that contain striations and/or fibers are called slickensides. Where the sense of displacement cannot be seen by the offset of markers, it can be determined from the sense of stepping of irregularities along the fault surface, the location of dilatational sites in which mineral fibers have grown, gouge marks formed by the incutting of rigid bodies in the rock, and the en échelon stepping of minor fractures (as described above).

Faults are important in mineral and petroleum exploration as they may either seal and act as a barrier to fluid flow (e.g., due to smearing of mud or shale along them), or may be important conduits for the migration of petroleum or mineralizing fluids. Many mineral deposits are fault and fracture controlled. Recognition of faults is also important in hydrogeological studies as fracturing along faults may produce hard-rock aquifers.

See also Petroleum detection; Plate tectonics

FAUNAL SUCCESSION • *see* FOSSIL RECORD

FELDSPAR

Feldspar is the most common mineral on Earth, constituting approximately 60% of the **crust**. It forms directly from cooling **magma** and is a major component of **granite** and most other **igneous rocks**.

The term feldspar actually covers a whole family of **minerals**, all of which consist of a framework of **aluminum**, **oxygen**, and **silicon** atoms plus an additive, usually potassium, sodium, or calcium. Feldspars vary in color from pink to gray, and are categorized by the additives they contain. Pure potassium feldspar is orthoclase ($KAlSi_3O_8$), pure sodium feldspar is albite ($NaAlSi_3O_8$), and pure calcium feldspar is anorthite ($CaAl_2Si_3O_8$). A feldspar may contain both sodium and calcium or sodium and potassium. The sodium–calcium feldspars form a continuum from albite to anorthite, the **plagioclase** feldspar series, which corresponds to the continuous branch of **Bowen's reaction series**. The sodium–potassium feldspars form a continuum from albite to orthoclase that is termed the orthoclase or alkali feldspar series. Feldspars containing significant quantities of both calcium and potassium are not found, as such mixtures are not chemically stable in cooling magma and react to form other minerals.

Orthoclase feldspars cleave along two planes that are at right angles, and plagioclase feldspars cleave along two planes that are not quite at right angles. Feldspar nomenclature is based on these mechanical properties: *ortho*, *plagio*, and *clase* are the Greek for right, slanted, and breaking, respectively.

Feldspar is less chemically stable when exposed to **water** than **quartz**, the other major ingredient of granite. Granite exposed to **weather** therefore becomes crumbly as its feldspar decays, and mechanical forces (e.g., **wind**, running water) break the granite up into **sand**. Rough, rapid fragmentation liberates some feldspar before it has had time to decay chemically, so a sand's ratio of feldspar to quartz records the rate at which its source granite was fragmented. This information is used by geologists to deduce ancient patterns of mountain-building and **erosion**.

See also Weathering and weathering series

FELSIC

Geologists sometimes find it useful to classify **igneous rocks** based on color. Because color is sensitive to minor chemical differences it is not a very reliable index to the history or composition of any given **rock**; however, it has the merit of being obvious at a glance, making color classification an indispensable aid to describing rocks in the field. **Minerals** are classed in two general color groups: felsic (light) and **mafic** (dark). Rocks may contain a mixture of mafic and felsic minerals, and are termed felsic if felsic minerals predominate, mafic otherwise. Alternatively, a numerical color index can be assigned to a rock based on visual estimation of the percentage of mafic or felsic minerals it contains.

Felsic minerals are usually higher in silica (SO_2) and **aluminum** and of lower density than mafic minerals. Common felsic minerals are **quartz, feldspar**, and the feldspathoids, and common felsic rocks (i.e., rocks high in felsic minerals) are **granite** and **rhyolite**. Mafic minerals are usually higher high in **iron** and magnesium than felsic minerals; common mafic minerals are pyroxene, amphibole, **olivine**, mica, and biotite, and common mafic rocks are **basalt** and gabbro.

The term mafic is also used in a precise chemical sense, that is, to denote rocks consisting of 45–52% silica regardless of color. Since the non-silica fraction of a rock often consists largely of iron and magnesium compounds, rocks that are mafic in the chemical sense are usually also mafic in the color sense.

See also Silicic

FERREL'S LAW

Ferrel's law, named after American meteorologist W. Ferrel (1817–1891), is the rule that air or **water** moving horizontally in the Northern Hemisphere is deflected or pushed to the right of its line of motion while air or water moving horizontally in the Southern Hemisphere is deflected to the left of its line of motion. Ferrel's law, which predicts the directions of the large-scale circulations of the earth's atmosphere and **oceans**, is a restatement in global terms of the action of the Coriolis force.

The Coriolis force is a consequence of the conservation of angular momentum and arises as follows. Consider a spinning disk with an ant perched on its outer edge. The ant's angular momentum (P) is given by its mass (m) times the square of its distance from the center of the disk (r^2) times the radial velocity of the disk (ω, how fast it is turning): $P = mr^2\omega$. If the ant crawls along a straight radial line toward the center

of the disk, its radial velocity ω—the number of turns around the axis it makes per second—remains constant. Its mass m also remains constant. However, its distance from the center r decreases. By the above formula, therefore, its angular momentum P also decreases. Yet a force is required to change the angular momentum of an object. Therefore, as it walks straight toward the center of the disk the ant must experience a force (a sideways push on its feet) that decelerates its rate of spin. From the perspective of the ant, the result is straight-line relative to the disk, and this sideways push seems required to balance a force tending to accelerate the ant sideways in the direction of the disk's **rotation**. This apparent force, which only seems to act while the ant is in motion toward or away from the disk's axis of rotation (i.e., is changing its angular momentum), is termed the Coriolis force. If the surface of the disk is too slippery to enable the moving ant to completely resist this apparent force, the ant will be deflected by it, relative to the disk's surface, in the direction of the disk's spin: that is, it will retain some or all of its original angular momentum by rotating more rapidly as it nears the disk's axis of rotation, just as a spinning ice skater's limbs rotate more rapidly as she or he retracts them toward her or his axis of rotation.

An ant trying to crawl axisward on a slippery disk is like a body of the air trying to drift northward or southward on the spinning (rotating) earth. Because Earth rotates eastward, air moving toward the axis of rotation (i.e., toward either the North or South Pole) tends to preserve its angular momentum by accelerating eastward; that is, it experiences an eastward Coriolis force that deflects it in the direction described by Ferrel's law. Eastward acceleration of a north-moving object in the Northern Hemisphere is to the right, as viewed along the object's line of motion; of a south-moving object in the Southern Hemisphere, to the left. Movements *away* from the axis of rotation are deflected *westward* in both hemispheres—again, to the right of the line of motion in the north, to the left in the south. The result is that high pressure systems, as seen from **space** tend to spin clockwise in the Northern Hemisphere and counterclockwise in the Southern Hemisphere. Low pressure systems spin in the respective reverse direction.

Ferrel's law applies equally to air and ocean movements, so the oceans circulate in the same sense as the air in both hemispheres. However, though it is often said to do so, Ferrel's law does not govern the direction of whirlpool spin in draining sinks and toilets. The Coriolis force is too weak to determine the behavior of fluids in such small basins.

See also Silicic

FERROMAGNETIC

Iron, cobalt, nickel, and various alloys of these materials are called ferromagnetic. Ferromagnetic materials can be permanently magnetized through exposure to an external **magnetic field**. They are strongly drawn towards a magnetic field. Their magnetic susceptibility, which is a material specific constant that relates applied field and magnetic response linearly, is orders of magnitude stronger than the susceptibility of paramagnetic or diamagnetic materials.

Paramagnetic materials are drawn towards magnets, while diamagnetic materials are repelled. Neither material can become permanently magnetized—or carry a remanent magnetization—and this is independent of **temperature** for all practical purposes. Their magnetic susceptibility is weakly positive and negative, respectively. The strength of a material's magnetic susceptibility is solely dependent on crystal structure. Paramagnetism usually dominates over diamagnetism. Most **rock** forming **minerals** are diamagnetic (e.g., **quartz**, **limestone**) or paramagnetic (e.g., micas, amphiboles).

Ferromagnetic behavior is different from diamagnetism and paramagnetism in several respects. First, it is strongly dependent on temperature. A ferromagnet looses its ability to carry a remanent magnetization and simply become paramagnetic if heated above its specific Curie temperature. A second fundamental property of ferromagnets is hysteresis. Hysteresis means that the application of an external field changes a ferromagnet irreversibly. The magnetic state of a ferromagnet depends not only on the strength of an applied field, but also on the history of the magnet. Any applied field can produce four different magnetic answers in a ferromagnet, once it has been magnetized initially. Additional to their susceptibility, ferromagnets are characterized by their coercivity, which is proportional to the field strength necessary to remagnetize it, and by their saturation remanence.

The most important variants of ferromagnetism are ferromagnetism and antiferromagnetism. Magnetite is the most abundant representative of the first family. It is a product of abiotic geochemical processes. Pure magnetite can be grown inter- and extra-cellularly by bacteria. These iron-oxide minerals have different crystal lattices resulting in dramatically differing magnetic properties. An important antiferromagnetic mineral is goethite, which is a product of **weathering** processes.

The properties of ferromagnets are not only determined by their crystalline structure, but also depend strongly on the grain-size of a particle. Ferromagnets develop magnetic domains above a critical volume. They are not capable of having a remanent magnetization below this volume, in which case they are called superparamagnetic. Magnetization carried by particles just above the critical threshold are extremely stable, because these single domain particles can only be magnetized parallel to their long, easy, axis. A further increase of the particle volume leads to the development of an increasing number of domains that destabilizes the remanent magnetization.

Both palaeomagnetic and rock magnetic research use the ferromagnetic properties of rocks. Palaeomagnetism uses the fact that minute amounts of ferromagnets acquire a magnetization parallel to the magnetic field of the earth at the time of the rocks' formation. This naturally occurring magnetization of rocks can be used in **plate tectonics** and magnetostratigraphy to reconstruct the former distribution of tectonic plates and continents, and to date sedimentary sequences.

Rock magnetic research uses the fact that it is relatively easy to measure the ferromagnetic properties of rocks. Additionally, rock **magnetism** is a fast and non-destructive method. Because composition and grain size distribution of any assemblage of iron-oxide minerals is a highly sensitive indicator of past environmental change, rock **magnetism** has become ever more important in environmental and palaeoclimatic research. Today, environmental magnetism is routinely incorporated in research projects designed to understand the environmental history of a site, material or region.

See also Paleomagnetics

FERSMAN, ALEKSANDR EVGENIEVICH (1883-1945)

Russian geochemist

Aleksandr Evgenievich Fersman was a Russian geochemist and mineralogist. He made major contributions to Russian **geology**, both in theory and exploration, advancing scientific understanding of crystallography and the distribution of elements in the earth's **crust**, as well as founding a popular scientific journal and writing biographical sketches of eminent scientists. He was known as a synthesizer of ideas from different subdisciplines.

Fersman was born in St. Petersburg on November 8, 1883, to a family that valued both art and science. His father, Evgeny Aleksandrovich Fersman, was an architect and his mother, Maria Eduardovna Kessler, a pianist and painter. Fersman's maternal uncle, A. E. Kessler, had studied **chemistry** under Russian chemist Aleksandr Mikhailovich Butlerov.

At the family's summer estate in the Crimea, Fersman first discovered **minerals** and began to collect them. When his mother became ill, the family traveled to Karlovy Vary (Carlsbad) in Czechoslovakia. There the young Fersman explored abandoned mines and added to his collection of **crystals** and druses (crystal-lined rocks).

Fersman graduated from the Odessa Classical Gymnasium in 1901 with a gold medal and entered Novorossisk University. He found the **mineralogy** course so dull that he decided to study art history instead. He was dissuaded by family friends (the chemist A. I. Gorbov and others) who encouraged him to delve into molecular chemistry. He subsequently studied physical chemistry with B. P. Veynberg, who had been a student of Russian chemist Dmitri Ivanovich Mendeleev. Veynberg taught Fersman about the properties of crystals.

The Fersman family moved to Moscow in 1903 because Aleksandr's father became commander of the First Moscow Cadet Corps. Fersman transferred to Moscow University, where his interest in the structure of crystals continued. Studying with mineralogist V. I. Vernadsky, he became an expert in goniometry (calculation of angles in crystal) and published seven scientific papers on crystallography and mineralogy as a student. When Fersman graduated in 1907, Vernadsky encouraged him to become a professor.

By 1908, Fersman conducted postgraduate work with **Victor Goldschmidt** at Heidelberg University in Germany. Goldschmidt sent him on a tour of Western **Europe** to examine the most interesting examples of natural **diamond** crystals in the hands of the region's jewelers. This work formed the basis of an important monograph on diamond crystallography Fersman and Goldschmidt published in 1911.

While a student in Heidelberg, Fersman also visited French mineralogist François Lacroix's laboratory in Paris and encountered pegmatites for the first time during a trip to some islands in the Elbe River that were strewn with the rocks. Pegmatites are granitic rocks that often contain rare elements such as uranium, tungsten, and tantalum. Fersman was to devote years to their study later in his career.

In 1912, Fersman returned to Russia, where he began his administrative and teaching career. He became curator of mineralogy at the Russian Academy of Science's Geological Museum. He would be elected to the Academy and become the museum's director in 1919. During this period Fersman also taught **geochemistry** at Shanyavsky University and helped found *Priroda,* a popular scientific journal to which he contributed throughout his life.

Fersman participated in an Academy of Science project to catalogue Russia's natural resources starting in 1915, traveling to all of Russia's far-flung regions to assess mineral deposits. After the Russian Revolution, Lenin consulted Fersman for advice on exploiting the country's mineral resources. During World War I Fersman consulted with the military, advising on strategic matters involving geology, as he would also later do in World War II.

In the early 1920s, Fersman devoted himself to one of geochemistry's major theoretical questions regarding the distribution of the **chemical elements** in the earth's crust. Fersman worked out the percentages for most of the elements and proposed that these quantities be called "clarkes" in honor of Frank W. Clarke, an American chemist who had pioneered their study. Clarkes had traditionally been expressed in terms of weight percentages; Fersman calculated them in terms of atomic percentages. His work showed different reasons for the terrestrial and cosmic distribution of the elements. He was interested in the ways in which elements are combined and redistributed in the earth's crust. He coined the term "techno-genesis" for the role of humans in this process, concentrating some elements and dispersing others through extraction and industrial activities.

Over the next twenty years, Fersman was responsible for a reassessment of the U.S.S.R.'s mineral resources. There were many areas, such as Soviet Central **Asia** and Siberia, which were thought to be resource-poor. Fersman showed otherwise, traveling from the Khibiny Mountains north of the Arctic Circle near Finland to the Karakum **Desert** north of Iran. He found rich deposits of apatite (a phosphorus-bearing mineral useful in fertilizers) in the former and a lode of elemental sulfur in the latter.

Fersman was acutely aware of the history of his profession and of science in general, passing on to his students his respect for his predecessors, especially Mendeleev and Vernadsky. He wrote many biographical sketches of distinguished scientists and published a number of popular works on mineral collecting. He was active in the Academy of Science of the U.S.S.R., serving in five different administrative posts,

and received a number of honors, including the Lenin Prize. He died in the Soviet Georgian city of Sochi on May 20, 1945.

See also Earth, interior structure; Mineralogy

FEYNMAN, RICHARD (1918-1988)
American physicist

Richard Feynman's career spanned some of the greatest discoveries of twentieth century **physics**, from developing the atomic bomb and studying **quantum electrodynamics (QED)** to solving the riddle of the **space shuttle** *Challenger* disaster.

Feynman received the 1965 Nobel Prize for his work regarding the interaction of light and matter, which he shared with Shin'ichio Tomonaga and Julian Schwinger. Other honors he received include the **Albert Einstein** Award (1954), the **Niels Bohr** International Gold Medal (1973) and membership in the National Academy of Sciences (1954).

Richard Phillips Feynman was born in Queens, New York. His parents were Lucille Phillips and Melville Feynman, a clothing salesman originally from Minsk. Feynman was interested in science from an early age, when he tinkered with crystal radio sets. His father had predicted that his first child, if a boy, would be a scientist; Mr. Feynman got more than he bargained for, for Richard's younger sister Joan also became a physicist

Feynman attended New York public schools, and after high school graduation went on to study physics at the Massachusetts Institute of Technology. After receiving his bachelor's degree in 1939, he went on to do his doctoral work at Princeton, where he served as a research assistant to John A. Wheeler, a Nobel-Prize-winning physicist. As a graduate student under Wheeler, he concerned himself with the knotty problem of how electrons interact, a question that would occupy him for years.

During the 1920s, Paul Dirac had introduced the theory that described the behavior of electrons in such a way that satisfied both quantum mechanics and Albert Einstein's theory of relativity. However, many problems arose with Dirac's equation when the known principles governing electromagnetic interactions were brought to bear on it; then Dirac's equation involved dividing by zero, which resulted in infinite answers, which were for all practical purposes useless.

Feynman—who from his undergraduate days had a well-deserved reputation for finding better ways to complete calculations—found a way to circumvent these useless answers. By "renormalizing" or redefining, the existing value of the electron's mass and charge, he was able to make irrelevant the parts of Dirac's theory that led to the troublesome answers.

But Feynman's work did more than just clean up some messy mathematics. It also provided physicists a new way to work with electrons. It opened the way for a new examination and description of the hydrogen **atom**. It gave scientists a look at what really happens when electrons, anti-electrons (or positrons) and photons (light particles) collide.

Richard Feynman. *Library of Congress.*

After receiving his doctorate, Feynman and his wife, Arlene Greenbaum, moved to Los Alamos, N.M., where he went to work with the Manhattan Project. There he worked, and held his intellectual own, with noted scientists such as Enrico Fermi and Hans Bethe, who headed Feynman's division. Feynman's wife Arlene, who had been his high-school sweetheart, died in 1945 after battling lymphatic tuberculosis. Less than a month after Arlene's death, Feynman became one of the first people in the world to witness the explosion of an atomic bomb.

After World War II, Bethe offered Feynman a position at Cornell University. While teaching physics, he also studied the question of the interaction of light and matter. In his description of the problem, he discarded the effect of the electromagnetic field and concentrated on the interactions of the particles themselves, as ruled by least action. In 1945, he also created a visual way to keep track of the interactions of the particles within time and **space**. Read from the bottom up (which indicates the passage of time), the diagrams show incoming particles (electrons) as straight lines. Their interactions (when they meet) are illustrated with wavy lines, indicating photons, which transmit the interactions. The straight lines then resume, indicating the departing particles after the interaction. Known as Feynman diagrams, they are still in use

by theoretical physicists in such diverse areas as acousto-optics, QED, and studies of electroweak interactions.

Feynman left Cornell in 1950 to join the California Institute of Technology, or CalTech, with which he would be affiliated for the rest of his life. During his time at CalTech he turned his prodigious mind and imagination to a staggering variety of problems, including the superfluidity of helium, superconductivity, and quark theory.

He married again in 1952, to Mary Louise Bell; they divorced four years later. In 1960, he married again, for the last time, to Gweneth Howarth; the couple had two children.

In 1979, Feynman was diagnosed with cancer, which he would battle for the next decade, before his death at age 69. During his last decade, Feynman became one of the world's most popular scientists, with the publication of two autobiographies, *"Surely You're Joking, Mr. Feynman!" Adventures of a Curious Character* (1984) and *"What Do You Care What Other People Think?" Further Adventures of a Curious Character* (1988). He also published *QED: The Strange Theory of Light and Matter* in 1985. During the 1960s, he published *Quantum Electrodynamics* (1961), *The Character of Physical Law* (1965), and *The Feynman Lectures*, three volumes of transcribed physics lectures he gave at CalTech that beautifully explain everything from the fall of **water** to QED.

Feynman was part of the group that investigated the explosion of the space shuttle *Challenger* in 1986. Feynman memorably demonstrated the failure of the shuttle's O-ring gaskets by placing a piece of gasket in a clamp and dropping it into **ice** water. The material became brittle, and the simple demonstration showed that the gaskets' failure had caused the explosion.

See also Bohr Model; Electromagnetic spectrum; Quantum theory and mechanics

FIELD METHODS IN GEOLOGY

Geology is, at heart, a field science. Even though much work is done in the laboratory and at the computer, geological samples and information must initially be obtained from the context in which they occur in nature. This natural setting would typically be a field locale chosen by the investigating geologist, or by his employers.

The main field instruments used by geologists include the Brunton compass (and/or Silva compass), tape measures, and plane table and alidade. The Brunton compass is a compact device that permits compass bearings to be made upon linear features (including strike lines) and lines connecting any two points. The Brunton may also function as a protractor (when placed upon a map) and as a device for measuring structural **dip** and vertical angles (using its internal clinometer). The Silva compass is somewhat similar, except that it does not have a bubble level or adjustable clinometer, so a task like measuring a vertical angle is not possible, and **strike and dip** measurements may not be as accurately made as with a Brunton. This Silva does not come in a rugged case, as does the Brunton, but its design as a flat, blade shape allows it to be used for map work more easily than a Brunton. The geologist's tape measure is usually the reel-in variety, which is marked in meters and feet. A typical tape length is 100 ft (30.5 m). The plane table and alidade are surveying devices used to measure distance and relative elevation. The plane table sits atop a tripod and a geological or topographic map under construction would be taped to its top. The alidade is a telescopic device that can be moved over the map surface as sightings are made. This device allows measure of horizontal distance and elevation. Horizontal distance and elevation of a point on the earth's surface is obtained by viewing through the alidade sight, a rod with a printed scale upon it (called a *stadia*). Data recorded during this observation are used for recording distance and elevation of the surveyed point upon a map.

The field instruments above would accompany most fully equipped field geologists on an expedition of mapping and sample collection. The geologist would typically also carry along a field notebook, hand lens, hammer, acid bottle, knife, shovels or trowels, sample bags, pens and pencils, aerial photographs and **satellite** imagery, maps and literature, camping equipment, and a camera. In the modern era, these materials could also be supplemented by a global positioning satellite system (**GPS**) receiver (for determining location and retracing routes), laptop computers, digital cameras, and portable geophysical equipment (including a gravimeter, altimeter, magnetic susceptibility meter, etc.). Occasionally, a geologist will bring along power tools for cutting or drilling **rock** or plaster, and burlap for wrapping delicate samples such as fossil bones.

Field methods in geology may be broken down into four main groups: (1) obtaining and marking samples and describing and measuring where they came from in an outcrop; (2) measuring and recording orientation (i.e., altitude) of strata or other planar features; (3) measuring dimensions (height and width); and (4) constructing geologic and topographic maps.

Obtaining and marking samples and describing and measuring where they originate in an outcrop requires observational skills and patience to record all information that might be obtained at one outcrop. Typically, the thickness of strata at an outcrop is recorded in a notebook where the layers are drawn to scale and described as to rock type, grain size, fossil content, color, sedimentary structures, and other attributes. Thickness of strata is measured using a tape measure or a Jacob's staff, which is a long stick made for sighting intervals of equal stratigraphic thickness (usually 5 ft, or 1.5 m). In the field notebook, detail is given about sampling locations and where photographs of the rocks are made. Samples are marked with an arrow indicating 'up' direction and labeled with a number which relates to the notebook number for the outcrop plus a number relating to feet or meters above the base of the stratigraphic section at that location. The same process is followed at each locale. Later, this information is compiled into a measured and described section for each outcrop, which may be used for **correlation** between outcrops. In terrains where igneous and metamorphic rocks occur, it is usually not so important for sample information to be recorded about the up direction and elevation above the base of outcrop.

A geologist scoops pahoehoe lava from a Hawaiian flow. © *Roger Ressmeyer/Corbis. Reproduced by permission.*

Measuring and recording orientation (i.e., attitude) of strata or other planar features is another important field activity that relates to understanding geological structures and to the making of geological maps. Strike, dip direction, and dip magnitude of rock layers and other planar geological features (e.g., foliation) are obtained in as many places as possible within a study **area** in order to understand completely all the geological structures (i.e., **folds** and fault patterns) of an area. Analysis of geological structures can help geologists interpret the conditions of deformation of rocks in an area. Generally, the geologist tries to obtain as many orientation or attitude measurements as possible in the field area being studied.

Measuring dimensions (height and width) of an area or of features in an area, is an important aspect of many geological studies. This may be done as an estimate by using the moveable clinometer in a Brunton compass and employing trigonometric relationships to compute the height or width. For example, if one uses a tape measure (or number of foot paces, if the average foot pace of the observer is known) to measure distance to a cliff wall, and then uses the clinometer in his Brunton to measure angle between his eye level and the top of the cliff, a computation of cliff height can be made. In this instance, the cliff height is equal to the person's eye height plus the product of the horizontal distance to the cliff times the tangent of the sighted angle. Geologists sometimes make simple maps, called "pace and compass maps," using the Brunton compass to take bearings and his measured pace length as a distance measure.

Constructing geologic and topographic maps is another field activity that occupies geologists. Geological maps are made by using a base map or set of aerial photographs to record the observed rock type (preferably a measured and described section, as noted above) and rock attitude at numerous locales in the study area. It is the task of the geologist to ultimately fashion a **geologic map** that is the simplest interpretation of all the surficial data about rock type and rock attitude in the area. Topographic maps are made by plane table and alidade, as noted above, and these may form the base map for geological mapping studies because surficial elevation is important in interpreting physical relationships between rock formations.

Other types of geological field work include reconnaissance studies of areas where detailed mapping is yet to be done, geological sample analysis conducted on-site at drilling operations, geophysical studies (where the objective is to collect data such as **gravity** strength, magnetic characteristics, etc.), surface- and ground-water studies (where the emphasis is upon **water** distribution, quality, and its relationship to geologic features), economic geology studies (where mines and excavations are studied and areas explored for the value of potential new mining), engineering geology field work (where studies assess the impact of human disturbance upon rock and **soil** stability), and many others.

See also Cartography; Topography and topographic maps

FISSION • *see* NUCLEAR FISSION

FISSURE

Any extensive crack in the earth is a fissure. When a small or medium-size fissure is filled with **magma** it is termed a **dike**. A large, magma-filled fissure that breaches the surface may erupt along its whole length or manifest as a chain of craters, each connected by a short central pipe to the magma-filled fissure below.

Small fissures are a common feature of volcanoes built by central activity (i.e., fed by a single pipelike conduit at their core). Indeed, most central volcanoes begin as eruptions from fissures and later localize to a single, central vent. High pressure in the central pipe may cause cracks in the surrounding cone or shield; if such a fissure breaches the surface, it may become a secondary point of eruption or even take over for the central crater. A fissure of this type typically appears at the surface as a hairline crack that gradually increases in width. Sulfurous fumes and steam emerge first, followed by small, glowing crumbs of red-hot **rock**. Later, viscous **lava** begins to bulge and ooze from the fissure, followed by increasingly fluid and voluminous flow. Not all fissures open so gradually;

where magma meets subsurface **water**, steam explosions can open or widen a fissure suddenly.

Small fissures around central volcanoes are a parasitic phenomenon. In contrast, eruptions along large, independent fissures are a distinct type of volcanism. Such eruptions may be pyroclastic (i.e., explosive eruptions of solid fragments), such as that which covered the Valley of Ten Thousand Smokes in Alaska with some 1.7 mi³ (7 km³) of ash and **pumice** in 1912, or those which covered Nevada and western Utah with 12,000 mi³ (50,000 km³) of welded **tuff** in the early Oligocene and late Pliocene Epochs. Fissure eruptions may also be gradual, such as the Great Tolbachik Fissure Eruption on the Kamchatka Peninsula in Russia, that in 1975 vented lava from a fissure 19 mi (30 km) long for 450 days and covered more than 15 mi² (40 km²) with lava flows.

Iceland is widening by about .5–1 in (1–2 cm) per year because it sits astride the Mid-Atlantic Ridge, and so is infiltrated by stretching-induced fissures that yield numerous fissure eruptions. Although not all independent fissure eruptions are on the largest scale, the most voluminous **volcanic eruptions** have all been fissure eruptions.

See also Crater, volcanic; Pipe, volcanic; Sea-floor spreading; Volcanic eruptions; Volcanic vent

FJORDS

Fjords (sometimes spelled fiords) are drowned glacier valleys. The depth of a fjord may exceed 1 mi (1.6 km) while the length sometimes exceeds 60 mi (97 km).

Geologic evidence indicates that some fjords form when a glacier cuts a deep, U-shaped valley through a river valley and advances into the sea. Given enough height and mass, the glacier may cut a valley into the sea floor as well. As it advances further into the sea, the glacier melts. The resulting reduction in size and mass causes the leading tongue of the glacier to float. A steep ridge is formed at the farthest reach of the glacier where the cutting stopped; this ridge is called a **sill**. In general, fjords can be identified by their steep sided, narrow channel, and sills.

Other evidence regarding the formation of fjords suggests that some of the deepest and most dramatic (such as those of Alaska, Norway, British Columbia, New Zealand, and other locations in the high latitudes) were formed during the most recent **ice** age. Evidence supports two scenarios to explain this process.

The first scenario suggests that some fjords resulted from landlocked **glaciers**. During the Ice Age, vast amounts of **water** were locked in enormous ice sheets. This reduced the availability of liquid water to Earth's **oceans** and exposed miles of coastline. In the coldest latitudes, glaciers excavated valleys as they moved onto the newly exposed coastline without ever reaching the sea. With the coming of warmer climatic conditions, ice sheets began to melt, causing the oceans to rise. At the same time, glaciers on the dry coastal shelves retreated. As they retreated, the rising **seas** filled the U-shaped valleys left behind.

This water treatment plant in Des Moines, Iowa, was flooded by the Raccoon River in 1993. Flooding was a serious problem all over the United States that year due to rainfall amounts far above average. *AP/Wide World. Reproduced by permission.*

The second scenario poses a combination of events that might account for the existence of some of the largest fjords. This explanation suggests that some glaciers did not remain land-locked on the coastal shelves during the ice age, but advanced for great distances and then continued their cutting on the sea floor. As the **climate** warmed, the rising seas occupied the deeply cut sea floor and advanced up the glacial valleys as the glaciers retreated.

It should be noted that fjords are classified as estuaries. However, due to their great depth and length, plus the height of their sills, they are semi-enclosed environments, similar to that of the open ocean. This characteristic sets fjords apart from all other estuarial environments.

See also Glacial landforms

FLOODPLAIN • *see* STREAM VALLEYS, CHANNELS, AND FLOODPLAINS

FLOODS

Floods can be defined as an overflow or downpour of **water** accumulating in an **area** where water is normally absent. Floods usually occur within a short period of time due to the soil's inability to absorb the water fast enough. According to the United States Geological Survey, floods were the natural disaster that caused the highest number of deaths and the most property damage in the United States during the twentieth century. Of all the natural disasters, floods are the most common and occur in the most places, with the only exception being fire. Flooding results in heavy currents that have the capacity to loosen structures and collapse foundations, destroying even the toughest of buildings.

The most common type of flood is the regional flood. Regional floods typically occur during the winter and spring months when the snow melts too rapidly or an excessive amount of water falls too quickly during spring rains or thunderstorms. Additionally, regional floods can result from tropical storms or hurricanes occurring along the coast or even far inland due to drastic changes in **weather** patterns. Floods can occur with no warning, but often occur over a period of days. If cold temperatures keep the ground frozen or the ground is already immersed with water, the water will run off into **rivers**. However, all too quickly the water rises above the banks of the rivers and flows onto dry land. Other types of floods include flash floods, ice-jam floods, storm-surge floods, dam-failure floods, and debris, **landslide**, and mudflow floods.

Regional floods across the U.S. that have occurred since 1990, include the Trinity, Arkansas, and Red Rivers in Texas, Arkansas, and Oklahoma in April of 1990, each caused by recurring thunderstorms. The number of reported deaths was 17 and the approximate cost of damage was one billion dollars. In January of 1993, the Gila, Salt, and Santa Cruz Rivers in Arizona flooded due to persistent winter **precipitation**, causing 400 million dollars in damages with the number of deaths unknown. From May through September of 1993, the Mississippi River Basin in the central U.S. flooded due to excessive rainfall, causing 48 deaths and 20 billion dollars in damages. In May 1995, flooding occurred in the south central U.S. from recurring thunderstorms causing 32 deaths and over five billion dollars in damages. Winter storms in California killed 27 people and caused three billion dollars in damages between January and May of 1995. Torrential rains and snowmelt caused flooding in the Pacific Northwest and western Montana in February of 1996 and again between December 1996, through January 1997. This caused nine deaths and one billion dollars in damages and 36 deaths and over two billion dollars in damages, respectively. The Ohio River and its tributaries flooded in March of 1997, causing more than 50 deaths and 500 million dollars in damages from a slow-moving frontal system. Snow **melting** caused the Red River of the North in North Dakota and Minnesota to flood between April and May of 1997, causing eight deaths and two billion dollars in damages. In September 1999, Hurricane Floyd destroyed eastern North Carolina, causing 42 deaths and six billion dollars in damages.

Flooding does not always prove destructive. Occasionally, floods can be beneficial, leaving **soil** laden with **minerals** and organic matter from the debris carried by the flood. Annual flooding of the Nile River enabled agriculture to be the foundation for Egyptian civilization. At the same time each year, the Nile River would flood, providing enough water to the soil to make lands fertile. With little rainfall, the Egyptians were dependent on the annual flooding to sustain their agriculture. Accordingly, Egypt's soil containing minerals and organic debris was a result of river sediment brought by the yearly floods.

See also Debris flow; Mud flow; Sedimentation

FOCUSED ION BEAM (FIB)

Focused ion beams have been used since the 1960s to investigate the chemical and isotopic composition of **minerals**. A focused ion beam blasts atoms and molecules free from the surface of a small sample of material; some of these free particles are also ions, and these are guided by electric fields to a mass spectrometer which identifies them with great precision.

An ion is an **atom** or molecule with a net electric charge. Electric fields subject electric charges to forces; therefore, electric fields can be used to move and steer ions. A continuous stream of ions moving together is termed an ion beam; a focused ion beam (FIB) is produced by using electric fields to guide a beam of ions.

In a typical FIB analysis, a narrow beam of argon, gallium, or **oxygen** ions traveling about 500,000 mph (800,000 kph) is directed at a polished flake of the material to be analyzed. Some of the atoms and molecules in the sample are kicked loose by the beam, a process termed sputtering. Some of these sputtered particles are themselves ions and so can be collected and focused by electric fields. The sputtered ions are directed to a mass spectrometer, which sorts them by mass. Even the very slight mass differences between isotopes of a single element can be distinguished by mass spectrometry; thus, not only the chemical but the isotopic composition of a sample can be determined with great precision. Very small, even microscopic, samples can be analyzed by FIB techniques.

The abundance of trace elements in a mineral can reveal information about the processes that formed it, helping petrologists and geochemists unravel geological history. Further, the decay of radioactive elements into isotopes of other elements acts as a built-in clock recording when the host mineral was formed. The hands of this clock are the relative isotope abundances in the mineral, and these can be determined by FIB analysis. **Carbon** isotope ratios also reveal whether a carbon-containing mineral was assembled by a living organism or by a nonliving process. Using FIB analysis, scientists have exploited this property of carbon isotopes to show that life existed on earth at least 3.85 billion years ago and that certain rocks originating on Mars and recovered as meteorites lying on the Antarctic **ice** probably, despite appearances, do not contain **fossils** of Martian microbes.

FIB facilities are complex and expensive. Accordingly, only about 15 facilities devoted to Earth sciences exist worldwide. FIBs are also used extensively in the manufacture of electronic microchips.

See also Carbon dating; Dating methods; Radioactivity

FOG

If the atmospheric visibility near the earth's surface is reduced to 0.62 mi (1 km) or less due to floating **water** droplets in the air, it is called fog. Fog can form in two ways: either by cooling the air to its **dew point** (e.g., radiation fog, **advection** fog, upslope fog), or by **evaporation** and mixing, when moisture is added to the air by evaporation, and then it is mixed with drier air (e.g., evaporation fog, frontal fog). Other types of fog include **ice** fog (a fog of suspended ice **crystals**, frequently forming in Arctic locations), acid fog (fog forming in polluted air, and turning acidic due to oxides of sulfur or nitrogen), or **smog** (fog consisting of water and smoke particles). While any type of fog can be hazardous because of its effects on atmospheric visibility for ground and air transportation, acid fog and smog can pose additional risk to human health, causing eye irritations or respiratory problems.

Radiation fog (or ground fog) occurs at night, when radiational cooling of the earth's surface cools the shallow moist air layer near the ground to its dew point or below, so the moisture in the air condenses into fog droplets. It occurs under calm **weather** conditions, when light **wind**, or no wind at all is present, since a strong wind would mix the lower-level cold air with the higher-level dry air, thus preventing the air at the bottom from becoming saturated enough to create fog. The presence of **clouds** at night can also prevent fog formation of this type, because they trap the earth's heat, not allowing the cooling of the air for **condensation**. Radiation fog often forms in late fall and winter nights, especially in lower areas, because cold and heavy air moves downhill, and gathers in valleys. Accordingly, radiation fog is also called valley fog. In the morning it usually dissipates or "burns off" when the Sun's heat warms the ground and air.

Advection fog forms when warm, moist air horizontally moves (which is called advection) over a cold surface, which cools the air to its dew point. Advection fog can form any time, and can be very persistent. It is common along coastlines where moist air moves from over the water to over the land, or when an air mass moves over a cold surface (e.g., snow), and the moisture in the air condenses into fog as the surface cools it. Advection-radiation fog forms when warm, moist air moves over a cold surface, which is cold as a result of radiation cooling. When warm, humid air moves over cold water, it is called sea fog.

Upslope fog forms in higher areas, where a moist air mass is forced to move up along a mountain. While the air mass is moving up the slope, it is cooled beyond its dew point and produces fog. It requires a fast wind, and warm and humid conditions at the surface. Unlike radiation fog, this type of fog dissipates when no more wind is available, and it can also form under cloudy skies. Upslope fog is usually dense, and extends to high altitudes.

Evaporation fog forms by the mixing of two unsaturated air masses. Steam fog is a type of evaporation fog, which appears when cold, dry air moves over warm water or warm, moist land. When some of the water evaporates into low air layers, and the warm water warms the air, the air rises, mixes with colder air, cools, and condenses some of its water vapor. Over **oceans**, it is referred to as sea smoke. Examples of cold air over warm water occur over swimming pools or hot tubs, where steam fog easily forms. It is common, especially in the fall season, when winds are getting colder but the water is only slowly turning colder.

Precipitation fog is a type of evaporation fog that happens when relatively warm rain or snow falls through cool, almost saturated air, and evaporation from the precipitation saturates the cool air. It can turn dense, persist for a long time, and may extend over large areas. Although it is mostly associated with warm fronts, it can occur with slow cold fronts or stationary fronts as well, hence the name frontal fog is also used.

See also Hydrologic cycle

FOHN · *see* SEASONAL WINDS

FOLDS

Compositional or metamorphic layers of rocks may bend during ductile deformation to produce folds. Folds commonly form during regional horizontal shortening in orogenic (mountain building) belts at microscopic to regional scales in all **rock** types (given suitable deformation conditions). Even rocks that at Earth's surface may be brittle and shatter when rapidly deformed, may fold during the application of regional, tectonic stresses over a long period of time at depth. Such a change in rock rheology is due to elevated **temperature** and confining pressure and the presence of fluids at deeper levels of the **crust**.

Upright layers (where young beds overlie older beds) that are arched upward are called anticlines. If the direction of younging (facing) is not known, such folds are called antiforms. Layers that are bent downward are called synclines (where beds are upright) or synforms where facing is not known. Cylindrical folds show the same profile in sections normal to their axes at any position along the axis. Folds where profiles vary from section to section and layers describe part of a cone are called conical folds. Folds are also classified according to the orientation of their hinge line or fold axis (the axis of curvature) and of their axial surface (the surface that bisects fold limbs and passes through the fold axis). The angle the fold hinge makes with the horizontal is called the plunge of a fold. Folds plunge gently when this angle is 10–30°, moderately between 30–60°, steeply between 60–90°, and are vertical when axes plunge 90°. Folds are upright where the axial surface is steeply dipping, inclined where the axial surface is moderately dipping, overturned where the axial surface is shallowly dipping and one limb is inverted, and recumbent where the axial surface is horizontal. In parallel folds, the layer thickness measured normal to the layer is constant around the fold. In similar folds, layer thicknesses measured parallel to the axial plane are constant. In describing folds, it is also important to note the inter-limb angle and whether fold hinges are rounded or angular.

Strong (competent) layers interlayered with more ductile (incompetent) layers buckle during layer-parallel shortening. The wavelength of the resulting folds depends on both the layer thickness and the viscosity (competence) contrast between layers. Larger wavelength folds develop in thick or competent layers. Folds may also develop during ductile flow in high-grade metamorphic rocks and in incompetent, lower-grade rocks. Irregular and often highly contorted syn-sedimentary folds can form during deposition of **sedimentary rocks** within slumps (which may be triggered by earthquakes).

When rocks that have already been folded are subjected to further shortening, early-formed folds may be refolded. Different fold interference patterns develop depending on the relative orientations of axes and axial surfaces for both generations of folds. A "dome and basin" (or, "egg carton") pattern results from the interference between two sets of upright folds whose axial surfaces are at a large angle to each other. A mushroom-shaped interference pattern results where folds with horizontal or shallowly dipping axial surfaces are folded by upright folds. A "hook" interference pattern occurs where fold axes are of similar orientation, but where axial surfaces are at a high angle to each other.

Folds may also form during regional crustal extension, such as in sedimentary basins. Roll-over antiforms develop over curved extensional (normal) faults in the upper, brittle crust or ductile **shear zones** in the middle to lower crust. Synforms are formed above areas where the underlying fault or ductile shear zone changes from shallowly to steeply dipping. Folds may also form during back-rotation of layers between two extensional faults or ductile shear zones. In high-grade rocks, folds may also form in surrounding layers when a competent layer pinches and swells or separates into barrel-shaped fragments (boudins) during layer-parallel extension.

Folds control the formation and localization of some **petroleum** and mineral deposits. Many oil and gas traps are created by regional-scale antiforms or domes formed by fold **superposition**, in wrench zones, or on the margins of salt diapirs. Some gold deposits are also controlled by folds. Differences in fold style of adjacent beds may lead to parting of beds along fold hinges. **Quartz** and, if chemical conditions are favorable, gold, may be deposited from fluids that migrate to such dilatational sites forming saddle reefs. In higher-grade rocks, rare metal pegmatites may intrude dilatational sites along fold hinges. Folds also provide geologists with valuable information about the orientation of stresses in Earth's crust at the time of their formation, helping them to unravel regional geological history.

See also Industrial minerals; Orogeny; Plate tectonics

FORAMINIFERAL OOZE • *see* CALCAREOUS OOZE

FORESTS AND DEFORESTATION

A forest is any ecological community that is structurally dominated by tree-sized woody plants. Forests occur anywhere that the **climate** is suitable in terms of length of the growing season, air and **soil temperature**, and sufficiency of soil moisture. Forests can be classified into broad types on the basis of their geographic range and dominant types of trees. The most extensive of these types are boreal coniferous, temperate angiosperm, and tropical angiosperm forests. However, there are regional and local variants of all of these kinds of forests. Old-growth tropical rainforests support an enormous diversity of species under relatively benign climatic conditions, and this ecosystem is considered to represent the acme of Earth's ecological development. Within the constraints of their regional climate, temperate and boreal forests also represent peaks of ecological development.

Many countries have developed national schemes for an ecological classification of their forests. Typically, these schemes are based on biophysical information and reflect the natural, large-scale patterns of species composition, soil type, **topography**, and climate. However, these classifications may vary greatly among countries, even for similar forest types.

An international system of ecosystem classification has been proposed by a scientific working group under the auspices of the United Nations Educational, Scientific and Cultural Organization (UNESCO). This scheme lists 24 forest types, divided into two broad classes: closed-canopy forests with a canopy at least 16.5 ft (5 m) high and with interlocking tree crowns, and open woodlands with a relatively sparse, shorter canopy.

Forests are among the most productive of Earth's natural ecosystems. Mature forests store more **carbon** (in biomass) than any other kind of ecosystem. This is especially true of old-growth forests, which typically contain large trees and, in temperate regions, a great deal of dead organic matter. Because all of the organic carbon stored in forests was absorbed from the atmosphere as **carbon dioxide** (CO_2), these ecosystems are clearly important in removing this greenhouse gas from the atmosphere. Conversely, the conversion of forests to any other type of ecosystem, such as agricultural or urbanized lands, results in a large difference in the amount of carbon stored on the site. That difference is made up by a large flux of CO_2 to the atmosphere. In fact, deforestation has been responsible for about one-half of the CO_2 emitted to the atmosphere as a result of human activities since the beginning of the industrial revolution.

Because they sustain a large biomass of foliage, forests evaporate large quantities of **water** to the atmosphere, in a hydrologic process called evapotranspiration. Averaged over the year, temperate forests typically evapotranspire 10–40% of their input of water by **precipitation**. However, this process is most vigorous during the growing season, when air temperature is highest and the amount of plant foliage is at a maximum. In fact, in many temperate forests evapotranspiration rates during the summer are larger than precipitation inputs, so that the ground is mined of its water content, and in some cases streams dry up.

Intact forests are important in retaining soil on the land, and they have much smaller rates of **erosion** than recently harvested forests or deforested landscapes. Soil eroded from disturbed forests is typically deposited into surface waters such as streams and **lakes**, in a process called **sedimentation**. The resulting shallower water depths makes flowing waters more prone to spilling over the banks of **rivers** and streams, causing flooding.

Forests are also important in moderating the peaks of water flow from landscapes, both seasonally and during extreme precipitation events. When this function is degraded by deforestation, the risk of flooding is further increased.

Although trees are the largest, most productive organisms in forests, the forest ecosystem is much more than a population of trees growing on the land. Forests also provide habitat for a host of other species of plants, along with numerous animals and microorganisms. Most of these associated species cannot live anywhere else; they have an absolute requirement of forested habitat. Often that need is very specific, as when a bird species needs a particular type of forest, in terms of tree species, age, and other conditions.

Generally, forests provide the essential habitat for most of Earth's species of plants, animals, and microorganisms. This is especially true of tropical rain forests. Recent reductions of forest **area**, which since the 1950s have mostly been associated with the conversion of tropical forest into agricultural land-use, are a critical environmental problem in terms of losses of biodiversity. Deforestation also has important implications for climate change and access to natural resources.

Forests are an extremely important natural resource that can potentially be repeatedly harvested and managed to yield a diversity of commodities of economic importance. Wood is by far the most important product harvested from forests. The wood is commonly manufactured into paper, lumber, plywood, and other products. In addition, in most of the forested regions of the less-developed world firewood is the most important source of energy used for cooking and other purposes. Potentially, all of these forest products can be sustained indefinitely. Unfortunately, in most cases forests have been irresponsibly over-harvested, resulting in the "mining" of the forest resource and widespread ecological degradation. It is critical that in the future all forest harvesting is conducted in a manner that is more responsible in terms of sustaining the resource.

Many other plant products can also be collected from forests, such as fruits, nuts, mushrooms, and latex for manufacturing rubber. In addition, many species of animals are hunted in forests, for recreation or for subsistence. Forests provide additional goods and services that are important to both human welfare and to ecological integrity, including the control of erosion and water flows, and the cleansing of air and water of pollutants. These are all important forest values, although their importance is not necessarily assessed in terms of dollars. Moreover, many of these values are provided especially well by old-growth forests, which in general are not very compatible with industrial forestry practices. This is one of the reasons why the conservation of old-growth forest is such a controversial topic in many regions of **North America** and elsewhere. In any event, it is clear that when forests are lost or degraded, so are these important goods and services that they can provide.

The global area of forest of all kinds was about 8.4 billion acres (3.4 billion ha) in 1990, of which 4.3 billion acres (1.76 billion ha) were tropical forest and the rest temperate and boreal forest. That global forest area is at least one-third smaller than it was prior to extensive deforestation caused by human activities. Most of the deforested land has been converted to permanent agricultural use, but some has been ecologically degraded into semi-desert or **desert**. This global deforestation, which is continuing apace, is one of the most serious aspects of the environmental crisis.

Deforestation refers to a longer-term conversion of forest to some other kind of ecosystem, such as agricultural or urbanized land. Sometimes, however, the term is used in reference to any situation in which forests are disturbed, for example by clear-cut harvesting, even if another forest subsequently regenerates on the site. Various human activities result in net losses of forest area and therefore contribute to defor-

estation. The most important causes of deforestation are the creation of new agricultural land and unsustainable harvesting of trees. In recent decades, deforestation has been proceeding most rapidly in underdeveloped countries of the tropics and subtropics.

The most important ecological consequences of deforestation are: the depletion of the economically important forest resource; losses of biodiversity through the clearing of tropical forests; and emissions of carbon dioxide with potential effects on global climate through an enhancement of Earth's **greenhouse effect**. In some cases, indigenous cultures living in the original forest may be displaced by the destruction of their habitat.

There are numerous references in historical, religious, and anthropological literature to forests that became degraded and were then lost through over harvesting and conversion. For example, the cedars of Lebanon were renowned for their abundance, size, and quality for the construction of buildings and ships, but today they only survive in a few endangered groves of small trees. Much of the deforestation of the Middle East occurred thousands of years ago. However, even during the Crusades of the eleventh century through the thirteenth century, extensive pine forests stretched between Jerusalem and Bethlehem, and some parts of Lebanon had cedar-dominated forests into the nineteenth century. These are all now gone.

Similar patterns of deforestation have occurred in many regions of the world, including most of the Mediterranean area, much of **Europe**, south **Asia**, much of temperate North and **South America**, and, increasingly, many parts of the subtropical and tropical world.

In recent decades, the dynamics of deforestation have changed greatly. The forest cover in wealthier countries of higher latitudes has been relatively stable. In fact, regions of Western Europe, the United States, and Canada have experienced an increase in their forest cover as large areas of poorer-quality agricultural land have been abandoned and then regenerated to forest. Although these temperate regions support large forest industries, post-harvest regeneration generally results in new forests, so that ecological conversions to agriculture and other non-forested ecosystems do not generally occur.

In contrast, the rate of deforestation in tropical regions of Latin America, **Africa**, and Asia has increased alarmingly in recent decades. This deforestation is driven by the rapid growth in size of the human population of these regions, with the attendant needs to create more agricultural land to provide additional food, and to harvest forest biomass as fuel. In addition, increasing globalization of the trading economy has caused large areas of tropical forest to be converted to agriculture to grow crops for an export market in wealthier countries, often to the detriment of local people.

In 1990, the global area of forest was 4.23 billion acres (1.71 billion ha), equivalent to 91% of the forest area existing in 1980. This represents an annual rate of change of about -0.9% per year, which if projected into the future would result in the loss of another one-half of Earth's remaining forest in only 78 years. During this period of time, deforestation (indicated as % loss per year) has been most rapid in tropical regions, especially West Africa (2.1%), Central America and Mexico (1.8%), and Southeast Asia (1.6%). Among nations, the most rapid rates of deforestation are: Côte d'Ivoire (5.2%/year), Nepal (4.0%), Haiti (3.7%), Costa Rica (3.6%), Sri Lanka (3.5%), Malawi (3.5%), El Salvador (3.2%), Jamaica (3.0%), Nicaragua (2.7%), Nigeria (2.7%), and Ecuador (2.3%).

These are extremely rapid rates of national deforestation. A rate of forest loss of 2% per year translates into a loss of one-half of the woodland area in only 35 years, while at 3% per year, the **half-life** is 23 years, and at 4%, it is 18 years.

Potentially, forests are a renewable natural resource that can be continually harvested to gain a number of economically important products, including lumber, pulp for the manufacture of paper, and fuel wood to produce energy. Forests also provide a habitat for game species and also for the much greater diversity of animals that are not hunted for sport or food. In addition, forests sustain important ecological services related to clean air and water and the control of erosion.

Any loss of forest area detracts from these important benefits and represents the depletion of an important natural resource. Forest harvesting and management can be conducted in ways that encourage the regeneration of another forest after a period of recovery. However, this does not happen in the cases of agricultural conversion and some types of unsustainable forest harvesting. In such cases, the forest is "mined" rather than treated as a renewable natural resource, and its area is diminished.

At the present time, most of Earth's deforestation involves the loss of tropical forests, which are extremely rich in species. Many of the species known to occur in tropical forests have local (or endemic) distributions, so they are vulnerable to extinction if their habitat is lost. In addition, tropical forests are thought to contain millions of additional species of plants, animals, and microorganisms as yet undiscovered by scientists.

Tropical deforestation is mostly caused by various sorts of conversions, especially to subsistence agriculture, and to market agriculture for the production of export commodities. Tropical deforestation is also caused by unsustainable logging and fuel wood harvesting (about two thirds of tropical people use wood **fuels** as their major source of energy. Less important causes of tropical deforestation include hydroelectric developments that flood large reservoirs and the production of charcoal as an industrial fuel. Because these extensive conversions cause the extinction of innumerable species, tropical deforestation is the major cause of global biodiversity concern.

Mature forests contain large quantities of organic carbon, present in the living and dead biomass of plants, and in organic matter of the forest floor and soil. The quantity of carbon in mature forests is much larger than in younger, successional forests, or in any other type of ecosystem, including human agroecosystems. Therefore, whenever a mature forest is disturbed or cleared for any purpose, it is replaced by an ecosystem containing a much smaller quantity of carbon. The difference in carbon content of the ecosystem is balanced by an emission of carbon dioxide (CO_2) to the atmosphere. This

CO_2 emission always occurs, but its rate can vary. The CO_2 emission is relatively rapid, for example, if the biomass is burned, or much slower if resulting timber is used for many years and then disposed into an anaerobic landfill, where biological decomposition is very slow.

Prior to any substantial deforestation caused by human activities, Earth's vegetation stored an estimated 990 billion tons (900 billion metric tons) of carbon, of which 90% occurred in forests. Mostly because of deforestation, only about 616 billion tons (560 billion metric tons) of carbon are presently stored in Earth's vegetation, and that quantity is diminishing further with time. It has been estimated that between 1850 and 1980, CO_2 emissions associated with deforestation were approximately equal to emissions associated with the combustion of fossil fuels. Although CO_2 emissions from the use of fossil fuels has been predominant in recent decades, continuing deforestation is an important source of releases of CO_2 to the atmosphere.

The CO_2 concentration in Earth's atmosphere has increased from about 270 ppm prior to about 1850, to about 360 ppm in 1999, and it continues to increase. Many atmospheric scientists hypothesize that these larger concentrations of atmospheric CO_2 will cause an increasing intensity of an important process, known as the greenhouse effect, that interferes with the rate at which Earth cools itself of absorbed solar radiation. If this theory proves to be correct, then a climatic warming could result, which would have enormous implications for agriculture, natural ecosystems, and human civilization.

See also Acid rain; Atmospheric circulation; Atmospheric composition and structure; Atmospheric inversion layers; Exfoliation; Floods; Global warming; Greenhouse gases and greenhouse effect; History of exploration I (Ancient and classical); History of exploration II (Age of exploration)

FOSSIL FUELS • *see* PETROLEUM

FOSSIL RECORD

The fossil record is the record of life on Earth as it is preserved in **rock** as **fossils**. The fossil record provides evidence of when and how life began on the planet, what types of organisms existed and how long they persisted, how they lived, died, and evolved, and what the **climate** was and how it changed. The fossil record also has allowed scientists to correlate rocks on a worldwide basis and to determine the relative ages of rock formations.

Fossils record life by preserving remains of organisms. A fossil is a rare thing. Most organisms decay and disappear quickly after dying. Of the tiny minority of organisms that do become preserved as fossils, an even smaller fraction survives the geologic cycle to become exposed and visible. As a result the fossil record is incomplete; there is no record of most organisms that probably lived and died.

Trilobite fossils. © James L. Amos/Corbis. Reproduced by permission.

The interpretation of the fossil record requires describing fossils, classifying them to place them in a biological context, and determining their age to give them chronological context. Fossil classification follows the same system of taxonomy as modern biology. Fossil organisms are placed in a genus, species, etc. Owing to the incompleteness of the fossil record, the classification of fossil organisms includes only about 250,000 species, a small number when compared to the over 2 million species of modern organisms that have been identified.

The most direct information the fossil record provides is of an organism's physical structure and what it may have looked like, thereby enabling it to be classified. Other information such as its environment, its diet, and its life cycle is deduced from its physical attributes, from other fossils found in association, and from the types of rocks containing the fossils. Trace fossils, or fossilized marks left as a result of the activities of creatures such as trails, footprints, and burrows also provide important information.

Of critical importance to the fossil record is the age of fossils. Many theories about how Earth and life on it evolved would not be possible without knowing the time sequence of the fossil record. The age of fossils is determined by two methods: relative dating and absolute dating. Relative dating involves comparing one rock formation with another and deciding the relative ages of the two formations. For example, when one formation is found above another, the lower formation and the fossils it contains must have been deposited prior to the overlying formation and so must be older. This rule, known as the principle of **superposition**, holds as long as the rocks have not been overturned by faulting or folding. In determining an absolute age, radiometric age dating is used. This method measures the abundance of a radioactive element in a fossil or an associated rock. An absolute age is then reverse-calculated based on the rate of decay, or **half-life** of the element.

Often, certain fossils are found in a limited vertical sequence of rock and are assumed to represent a limited time period. These fossils, known as index fossils, are useful for determining relative ages and for correlating rock formations on a worldwide basis. Early workers used index fossils and rock **correlation** to develop the geologic scale. Originally, the geologic scale was relative, based largely on the fossil record. Subsequently, absolute ages have been applied to the geologic scale.

By synthesizing the fossil record, classifying fossils, aging them, and placing them in the context of the geologic scale, scientists have revealed the sequence of life on Earth. In many cases, the scale shows how some organisms evolved systematically over time, each subsequent version of an organism displaying modifications over the earlier. In other cases, there are large gaps in the fossil record and the developmental process for some organisms is not as clear. Often, the **evolution** of organisms leads to a dead end. The fossil record shows that throughout **geologic time**, life often evolved slowly, punctuated by explosions of life when a large number new organisms appeared. For example, the beginning of the **Cambrian Period** of the geologic scale contains a phenomenal number of new organisms. It also shows that, periodically, mass extinctions occurred, such as at the end of the **Cretaceous Period**, when a majority of species came to an end over a relatively short amount of time.

The fossil record begins with 3.5 to 3.0 billion-year-old rocks from **Australia** and South **Africa**, which preserve the remains of blue-green algae. These fossils resemble the modern stromatolites that grow in oceanic tidal areas. The fossil record shows a steady increase in the complexity of marine organisms over the next three billion years. Eventually, about 435 million years ago, terrestrial organisms appeared. The subsequent rise and fall of different creatures, from insects to fish, dinosaurs to mammals, has all been deduced from the fossil record.

In addition to outlining the history of life on Earth, the fossil record provides clues to climatic and tectonic evolution of the planet. Plant fossils and microscopic fossils such as pollen are particularly useful for the evidence they provide about the climate of the earth. For example, the Carboniferous period must have been very warm and moist because of the presence of abundant fossils of ferns and other tropical plants from that time. Also, the lack of fossilized remains of **oxygen** breathing organisms and the dominance of photosynthetic algae fossils from the very early Earth suggest that the primordial atmosphere was devoid of oxygen. The concept of **plate tectonics** was greatly aided by the observation that fossils now found widely spaced across the globe must have actually lived on the same original landmass that subsequently split apart.

See also Evolution, evidence of; Fossils and fossilization; Uniformitarianism

FOSSILIZATION OF BACTERIA

Studies of **fossilization** of bacteria provide an indication of the age of ancient bacteria and of the rate of geological and geochemical processes on ancient Earth. **Fossils** of cyanobacteria or "blue-green algae" have been recovered from rocks that are nearly 3.5 million years old. Bacteria known as magnetobacteria form very small **crystals** of a magnetic compound inside the cells. These crystals have been found inside **rock** that is two billion years old.

The fossilization process in cyanobacteria and other bacteria appears to depend on the ability of the bacteria to trap sediment and **metals** from the surrounding solution. Cyanobacteria tend to grow as mats in their aquatic environ-

ment. The mats can retain sediment. Over time and under pressure the sediment entraps the bacteria in rock. As with other living organisms, the internal structure of such bacteria is replaced by **minerals**, notably pyrite or siderite (**iron** carbonate). The result, after thousands to millions of years, is a replica of the once-living cell.

Other bacteria that elaborate a carbohydrate network around themselves also can become fossilized. The evidence for this type of fossilization rests with laboratory experiments where bacteria are incubated in a metal-containing solution under conditions of **temperature** and pressure that attempt to mimic the forces found in geological formations. Experiments with *Bacillus subtilis* demonstrated that the bacteria act as a site of **precipitation** for silica, the ferric form of iron, and of elemental gold. The binding of some of the metal ions to available sites within the carbohydrate network then acts to drive the precipitation of unstable metals out of solution and onto the previously deposited metal. The resulting cascade of precipitation can encase the entire bacterium in metallic species. On primordial Earth, this metal binding may have been the beginning of the fossilization process.

The deposition of metals inside carbohydrate networks like the capsule or exopolysaccharide surrounding bacteria is a normal feature of bacterial growth. Indeed, metal deposition can change the three-dimensional arrangement of the carbohydrate strands so as to make the penetration of antibacterial agents through the matrix more difficult. In an environment—such as occurs in the lungs of a cystic fibrosis patient—this micro-fossilization of bacteria confers a survival advantage to the cells.

In contrast to fossils of organisms such as dinosaurs, the preservation of internal detail of microorganisms seldom occurs. Prokaryotes have little internal structure to preserve. However, the mere presence of the microfossils is valuable, as they can indicate the presence of microbial life at that point in geological time.

Bacteria have been fossilized in amber, which is fossilized tree resin. Several reports have described the resuscitation of bacteria recovered from amber as well as bacteria recovered from a crystal in rock that is millions of years old. Although these claims have been disputed, a number of microbiologists assert that the care exercised by the experimenters lends increases the validity of their studies.

In the late 1990s a meteorite from the planet Mars was shown to contain bodies that appeared very similar to bacterial fossils that have been found in rocks on Earth. Since then, further studies have indicated that the bodies may have arisen by inorganic (non-living) processes. Nonetheless, the possibility that these bodies are the first extraterrestrial bacterial fossils has not been definitively ruled out.

See also Atmospheric chemistry; Carbon dating; Dating methods; Evolution, evidence of; Fossil record; Fossils and fossilization; Geochemistry; Miller-Urey experiment; Murchison meteorite; Paleoclimate; Petroleum microbiology; Phanerozoic Eon; Precambrian; Rate factors in geologic processes

Fossils and fossilization

A fossil is the remains of an ancient life form—plant or animal—or its traces, such as nesting grounds, footprints, worm trails, or the impressions left by leaves, preserved in **rock**. Fossil traces are called ichnofossils. Fossilization refers to the series of postmortem changes that lead to replacement of **minerals** in the original hard parts (shell, skeleton, teeth, horn, scale) with different minerals, a process known as remineralization. Infrequently, soft parts may also be mineralized and preserved as fossils. A new category of subfossil—a fossil that has not yet begun to mineralize—is increasingly recognized in the scientific literature. Many subfossils originated in the Holocene or Recent, the period that we now live in, and cannot be dated with any greater accuracy. In addition, the term "fossil" is applied in other ways, for example, to preserved soils and landscapes such as fossil **dunes**. Through the study of fossils, it is possible to reconstruct ancient communities of living organisms and to trace the **evolution** of species.

Fossils occur on every continent and on the sea floors. The bulk of them are invertebrates with hard parts (for example, mussels). Vertebrates, the class that includes reptiles (for example, dinosaurs) and mammals (mastodons, humans), are a relatively late development, and the finding of a large, complete vertebrate fossil, with all its parts close together, is rare. Microfossils, on the other hand, are extremely common. The microfossils include very early bacteria and algae; the unicellular organisms called foraminiferans, which were common in the Tertiary Periods; and fossil pollen. The study of micro fossils is a specialized field called micropaleontology.

Fossils of single-celled organisms have been recovered from rocks as old as 3.5 billion years. Animal fossils first appear in late **Precambrian** rocks dating back about a billion years. The occurrence of fossils in unusual places, such as dinosaur fossils in **Antarctica** and fish fossils on the Siberian steppes, reflects both shifting of continental position by **plate tectonics** and environmental changes over time. The breakup of the supercontinent Pangaea in the **Triassic Period** pulled apart areas that were once contiguous and shared the same flora and fauna. In particular, the plates carrying the southern hemisphere continents—South America, southern **Africa**, the Indian subcontinent, **Australia**, and Antarctica—moved in different directions, isolating these areas. Terrestrial vertebrates were effectively marooned on large islands. Thus, the best explanation for dinosaurs on Antarctica is not that they evolved there, but that Antarctica was once part of a much larger land mass with which it shared many life forms.

An important environmental factor influencing the kinds of fossils deposited has been radical and episodic alteration in sea levels. During episodes of high sea level, the interiors of continents such as **North America** and Australia are flooded with seawater. These periods are known as marine transgressions. The converse, periods of low sea level when the waters drain from the continents, are known as marine regressions. During transgressions, fossils of marine animals may be laid down over older beds of terrestrial animal fossils. When sea level falls, exposing more land at the edges of continents, fossils of terrestrial animals may accumulate over older marine animals. In this way plate tectonics and the occasional marine flooding of inland areas could result in unusual collections of fossil flora and fauna where the living plants or animals could not exist today—such as fishes on the Siberian steppes.

Changes in sea level over the past million years or so have been related to episodes of **glaciation**. During glaciation, proportionately more **water** is bound up in the **polar ice** caps and less is available in the **seas**, making the sea levels lower. It is speculated, but not certain, that the link between glaciation and lower sea levels holds true for much of Earth's history. The periods of glaciation in turn are related to broad climatic changes that affect the entire Earth, with cooler **weather** increasing glaciation and with warmer temperatures causing glacial **melting** and a rise in sea levels. A change in **temperature** would also affect the availability of plants for herbivores to eat, and the availability of small animals for carnivores to eat. Thus, even modest temperature changes, if long-lasting enough, could produce large changes in the flora and fauna available to enter the **fossil record** in any given locale.

The principal use of fossils by geologists has been to date rock strata (layers) that have been deposited over millions of years. As different episodes in Earth's history are marked by different temperature, aridity, and other climatic factors, as well as different sea levels, different life forms were able to survive in one locale or period but not in another. Distinctive fossilized life forms that are typically associated with given intervals of **geologic time** are known as index fossils, or indicator species. The concepts that different fossil species correlate with different strata and that, in the absence of upheaval, older strata underlie younger ones are attributed to the English geologist **William Smith**, who worked in the early nineteenth century.

The temporal relationship of the strata is relative: it is more important to know whether one event occurred before, during, or after another event than to know exactly when it occurred. Recently geologists have been able to subdivide time periods into smaller episodes called zones, based on the occurrence of characteristic zonal indicator species, with the smallest time slices about one-half million years. Radiometric dating measures that measure the decay of radioactive isotopes have also been used to derive the actual rather than relative dates of geological periods; the dates shown on the time scale were determined by radiometry. The relative dating of the fossil clock and the quantitative dating of the radiometric clock are used in combination to date strata and geological events with good accuracy.

The fossil clock is divided into units by index fossils. Certain characteristics favor the use of one species over another as an index fossil. For example, the ammonoids (ammonites), an extinct mollusk, function as index fossils from the lower Devonian through the upper Cretaceous—a period of about 350 million years. The ammonoids, marine animals with coiled, partitioned shells, in the same class (Cephalopoda) as the present-day Nautilus, were particularly long lasting and plentiful. They evolved quickly and colonized most of the seas on the planet. Different species preferred warmer or colder water, evolved characteristically sculpted shells, and exhibited more or less coiling. With thousands of variations on a few basic, easily visible features—variations unique to each species in its own time and place—the ammonoids were obvious candidates to become index fossils. For unknown reasons, this group of immense longevity became extinct during the Cretaceous-Triassic mass extinction. The fossils are still quite plentiful; some are polished and sold as jewelry or paperweights.

Index fossils are used for relative dating, and the geologic scale of time is not fixed to any one system of fossils. Multiple systems may coexist side-by-side and be used for different purposes. For example, because macrofossils such as the ammonoids may break during the extraction of a core sample or may not be frequent enough to lie within the exact **area** sampled, a geologist may choose to use the extremely common microfossils as the indicator species. Workers in the oil industry may use conodonts, fossils commonly found in oil-bearing rocks. Regardless of which system of index fossils is used, the idea of relative dating by means of a fossil clock remains the same.

The likelihood that any living organism will become a fossil is quite low. The path from **biosphere** to lithosphere—from the organic, living world to the world of rock and mineral—is long and indirect. Individuals and even entire species may be snatched from the record at any point. If an individual is successfully fossilized and enters the **lithosphere**, ongoing tectonic activity may stretch, abrade, or pulverize the fossil to a fine dust, or the sedimentary layer housing the fossil may eventually be subjected to high temperatures in Earth's interior and melt, or be weathered away at Earth's surface. A fossil that has survived or avoided these events may succumb to improper collection techniques at the hands of a human.

Successful fossilization begins with the conditions of death in the biosphere. Fossils occur in sedimentary rock, and are incorporated as an integral part of the rock during rock formation. Unconsolidated sediments such as **sand** or mud, which will later become the fossiliferous (fossil-bearing) **sandstone** or **limestone**, or shale, are an ideal matrix for burial. The organism should also remain undisturbed in the initial phase of burial. Organisms exposed in upland habitats are scavenged and weathered before they have an opportunity for preservation, so a low-lying habitat is the best. Often this means a watery habitat. The fossil record is highly skewed in favor of organisms that died and were preserved in calm seas, estuaries, tidal flats, or the deep ocean floor (where there are few scavengers and little disruption of layers). Organisms that died at altitude, such as on a plateau or mountainside, and are swept by **rivers** into a **delta** or **estuary** may be added to this death assemblage, but are usually fragmented.

A second factor contributing to successful fossilization is the presence of hard parts. Soft-bodied organisms rarely make it into the fossil record, which is highly biased in favor of organisms with hard parts—skeletons, shells, woody parts, and the like. An exception is the Burgess Shale, in British Columbia, where a number of soft-bodied creatures were fossilized under highly favorable conditions. These creatures have few relatives that have been recorded in the fossil record; this is due to the unlikelihood of the soft animals being fossilized.

From the time of burial on, an organism is technically a fossil. Anything that happens to the organism after burial, or anything that happens to the sediments that contain it, is encompassed by the term diagenesis. What is commonly called fossilization is simply a postmortem alteration in the **mineralogy** and **chemistry** of the original living organism.

Fossilization involves replacement of minerals and chemicals by predictable chemical means. For example, the shells of mollusks are made of calcium carbonate, which typically remineralizes to calcite or aragonite. The bones of most vertebrates are made of calcium phosphate, which undergoes subtle changes that increase the phosphate content, while cement fills in the pores in the bones. These bones may also be replaced by silica.

Because of the nature of fossilization, fossils are often said to exist in communities. A fossil community is defined by **space**, not time. Previously fossilized specimens of great age may be swept by river action or carried by scavengers into young sediments that are just forming, there to join the fossil mix. For this reason, it may be very difficult to date a fossil with precision based on a presumed association with nearby fossils. Nevertheless, geologists do hope to confirm relationships among once living communities by comparing the makeup of fossil communities.

One of the larger goals of paleontologists is to reconstruct the prehistoric world, using the fossil record. Inferring an accurate life assemblage from a death assemblage is insufficient and usually wrong. The fossil record is known for its extreme biases. For example, in certain sea environments over 95% of species in life may be organisms that lack hard parts. Because such animals rarely fossilize, they may never show up in the fossil record for that locale. The species diversity that existed in life will therefore be much reduced in the fossil record, and the proportional representation of life forms greatly altered.

In some cases, however, a greater than usual proportion of preservable individuals in a community has fossilized in place. The result is a bed of fossils, named after the predominant fossil component, "bone bed" or "mussel bed," for example. Geologists are divided over whether high-density fossil deposits are due to reworking and **condensation** of fossiliferous sediments or to mass mortality events. Mass mortality—the contemporaneous death of few to millions of individuals in a given area—usually is attributed to a natural catastrophe. In North America, natural catastrophe is thought to have caused the sudden death of the dinosaurs in the bone beds at Dinosaur National Park, Colorado, and of the fossil fishes in the Green River Formation, Wyoming. These are examples of local mass mortality. When mass mortality occurs on a global scale and terminates numerous species, it is known as a mass extinction. The greatest mass extinctions have been used to separate the geological eras: the Permian-Triassic extinction separates the Palaeozoic Era from the Mesozoic; the Cretaceous-Tertiary extinction, which saw the demise of the dinosaurs and the rise of large mammalian species to fill newly available biological niches, separates the Mesozoic from the Tertiary. Thus, mass extinctions are recorded not only in the high-density fossil beds but in the complete disappearance of many species from the fossil record.

The fossil record—the sum of all known fossils—has been extremely important in developing the phylogeny, or evolutionary relations, of ancient and living organisms. The contemporary understanding of a systematic, phylogenetic hierarchy descending through each of the five kingdoms of living organisms has replaced earlier concepts that grouped organisms by such features as similar appearance. It is now known that unrelated organisms can look alike and closely related organisms can look different; thus, terms like "similar" have no analytical power in biology. Charles Darwin, working in the mid-1800s, was the chief contributor to the systematic approach to biological relationships of organisms.

In addition to providing important information about the history of Earth, fossils have industrial uses. Fossil **fuels** (oil, **coal**, **petroleum**, bitumen, **natural gas**) drive industrialized economies. Fossil aggregates such as limestone provide building material. Fossils are also used for decorative purposes. This category of functional use should be distinguished from the tremendous impact fossils have had in supporting evolutionary theory.

See also Evolution, evidence of; Geologic time; Marine transgression and marine regression

FRACTIONAL CRYSTALIZATION • *see* CRYSTALS
AND CRYSTALLOGRAPHY

FRANKLIN, BENJAMIN (1706-1790)

American scientist and statesman

Before serving his fledgling country during time of revolution, Benjamin Franklin also achieved international recognition for his scientific acumen, especially in his experimentation with **electricity**.

Born in the British colony of Boston, Massachusetts, Franklin was the fifteenth of seventeen children. His father was an impoverished candlemaker, unable to afford to send young Benjamin to school. As a result, he received only two years of formal education. Franklin was working in his father's shop at the age of ten, and later was apprenticed to his brother, a printer, where he developed a love for books. In 1724, he went to London where he became skilled at printing, returning to Philadelphia two years later. In Philadelphia he made a name for himself, as well as a small fortune, publishing the *Pennsylvania Gazette* and *Poor Richard's Almanack*.

In addition to his pursuit of printing, Franklin became interested in the study of electricity in 1746. During this period, scientists around the globe, many of whom had advanced degrees, were investigating the phenomena of static electricity. A less confident man might have felt inadequate to compete, but Franklin, who was essentially self-educated, obtained a Leyden jar and began his own research.

Benjamin Franklin. *Library of Congress.*

The Leyden jar, invented by Musschenbroek, was a water-filled bottle with a stopper in the end. Through the stopper was a metal rod that extended into the **water**. A machine was used to create a static electric charge, which could be stored in the jar. A person who touched the end of the charged rod received an electrical jolt. Public demonstrations, in which many people joined hands and received a simultaneous shock, were very popular. Franklin saw such a demonstration, and that initiated his interest in electricity.

It was Franklin's originality and tenacity that earned him the reputation as a leading scientist. He was the first person to wonder how the Leyden jar actually worked, and performed a series of experiments to find the answer. He poured the "charged" water out of the jar into another bottle, and discovered the water had lost its charge. This indicated that it was the **glass** itself, the material that insulated the conductor, which produced the shock. To verify this, Franklin took a windowpane and placed a sheet of **lead** on each side. He "electrified" the lead, removed each sheet one at a time, and tested for a charge. Neither sheet gave so much as a single spark, but the windowpane had been charged. Franklin had unknowingly invented the electrical condenser. The condenser, also known as a capacitor, was destined to be one of the most important elements in electric circuits. Today the condenser, which received its name from Alessandro Volta, is used in radios, televisions, telephones, radar systems, and many other devices.

Drawing a parallel between the sparking and crackling of the charged Leyden jar and **lightning** and **thunder**, Franklin wondered if there was an electrical charge in the sky. He planned to erect a long metal rod atop Christ Church in Philadelphia to conduct electricity to a sentry box in which a man, standing on an insulated platform, would be able to collect an electric charge. Because he was a proponent in the free exchange of ideas, Franklin had written a book outlining his theories, which received wide circulation in **Europe**. A French scientist named D'Alibard used Franklin's idea and performed the experiment himself on May 10, 1752, charging a Leyden jar with lightning. Franklin generously gave D'Alibard credit for being the first to "draw lightning from the skies." If nothing else, Franklin did receive credit for the invention of the lightning rod.

While waiting for the rod to be installed atop Christ Church, Franklin had come up with an idea of a faster way to get a conductor into the sky. He tied a large silk handkerchief to two crossed wooden sticks, attached a long silken thread with a metal key at the end, and waited for a thunderstorm. The rain made the thread an excellent conductor, and the static charge traveled down to the key. When Franklin brought his knuckle to the key, a spark jumped from the key to his hand, proving the existence of electricity in the sky.

Franklin had been wise enough to connect a ground wire to his key; two other scientists, attempting to duplicate the experiment but neglecting the ground wire, were killed when they were actually struck by lightning. Still, Franklin was lucky he was not hit by lightning himself. Franklin invented the lightning rod from his work with electricity. The lightning rod became indispensable for protecting buildings from the destructive force of lightning. Because he had discovered he could get the Leyden jar to spark over a greater distance with a sharply pointed rod, Franklin's lightning rods had very sharp points. (In 1776, after the conflict between the Colonies and King George III had erupted, the king ordered that lightning rods with blunt ends be installed on his palace.) By 1782, there were four hundred lightning rods in Philadelphia.

His discovery of sky-borne electricity led Franklin to speculate on the nature of the **aurora borealis**, the "northern lights" that illuminate the sky. Franklin thought they might be electrical in nature, and suggested that conditions in the upper atmosphere might be responsible.

His work on electricity led to a plethora of new words (battery, condenser, conductor, armature, charge, and discharge to name a few) and concepts. He suggested that electrical charge was due to the abundance or lack of "something" that resulted in attraction and repulsion, and he established the concept of positive and negative charges, believing (incorrectly) that electrical flow went from positive to negative. In fact, the opposite is true.

Continuing his observations of the **weather**, he noticed there was a prevailing pattern as it moved from west to east and suggested the circulation of air masses was responsible, establishing the concept of high and low pressure. He went on to show that the boiling point of water was affected by air pressure; as he created a vacuum in a sealed water bottle, the **temperature** needed to boil the water dropped. He also charted the flow on the **Gulf Stream** in the Atlantic Ocean.

Volumes have been written about Franklin's life as a statesman. He founded service organizations, became Postmaster of Philadelphia, and established a college that eventually became the University of Pennsylvania. He returned to London in 1757 as an Agent of the Pennsylvania Assembly and remained there until 1775. After warning that the "Stamp Tax" was not a good way to obtain revenue from the American Colonies, he returned and joined the committee drafting the Declaration of Independence.

During Franklin's long life he developed many inventions (such as bifocal lenses and the Franklin stove), received numerous honors and achieved an international reputation, becoming one of few Americans of colonial days to do so. He died in Philadelphia, at the age of eighty-four.

See also Atmospheric pressure; Electricity and magnetism

FRASCH, HERMAN (1851-1914)
German-born American chemist

Herman Frasch, the son of a prosperous apothecary, was born in Gaildorf, Württemberg (now part of Germany) on Christmas Day 1851. He studied at the gymnasium in Halle but rather than attend the university, he decided to immigrate to the United States in 1868. Frasch taught at the Philadelphia College of Pharmacy and continued to study **chemistry** with an eye to becoming an expert in a newly-emerging field, **petroleum**.

The oil industry in the United States began with the opening of the Titusville, Pennsylvania, oil field in 1859. In 1870, John D. Rockefeller formed Standard Oil—which refined a majority of the oil in the country—in Cleveland, Ohio. Frasch sold his patent for an improved process for refining paraffin wax to a subsidiary of Standard Oil in 1877 and moved to Cleveland to open a laboratory and consulting office. Soon he became the city's outstanding chemical consultant. In 1882, he sold to the Imperial Oil Company in Ontario, Canada, a process for reducing the high sulfur content of petroleum, which gave it a disagreeable odor and caused the kerosene refined from it to burn poorly. When Standard Oil discovered a field of "sour oil" in Indiana and Ohio, the company hired Frasch as a full time consultant, bought his process and the Empire Oil Company he had recently purchased in Ontario, and gave him charge of the American petroleum industry's first experimental research program. Frasch's process for removing sulfur, patented in 1887, was to treat the petroleum with a variety of metallic oxides to precipitate the sulfur and recover the oxides for further use. He continued with Standard Oil as special consultant for the development of new petroleum by-products and became wealthy. He refused to join Standard Oil as an executive, choosing instead to be a lifetime consultant.

Frasch turned his attention to sulfur, the substance his process removed from petroleum. The island of Sicily held a virtual monopoly on this valuable mineral from which sulfuric acid, industry's most vital chemical, was made. While Sicilian sulfur deposits were near the earth's surface and more easily mined, sulfur deposits in Texas and Louisiana were deeper, and American laborers were unwilling to go into sulfur mines. Frasch believed that sulfur could be melted and pumped from the ground in much the same manner petroleum was, but boiling **water** was not hot enough to liquefy the sulfur. He organized the Union Sulfur Company in 1892, and two years later began employing the method he had patented a year earlier. His process required three concentric pipes to be sunk into the sulfur deposit. Water, superheated under pressure to above 241°F (116°C), was pumped into the sulfur deposit through the outside pipe. Compressed air was forced down the center pipe, and through the center pipe the melted sulfur flowed to the surface where it was pumped into bins to solidify. The major problem with this method was the cost of heating the water, but the discovery of the East Texas oil fields in the early twentieth century provided an inexpensive, readily available fuel supply. Frasch expanded his research into the use of sulfur as an insecticide and a fungicide. Other companies infringed on his patent rights, and his company disappeared, but the use of the Frasch process enabled the United States to become self-sufficient in the production of sulfur needed to supply its growing chemical industry.

Frasch died in Paris on May 1, 1914. Among his honors was the Perkin Medal in 1912. His greatest honor was the distinction of having two chemical processes, one for producing sulfur and the other for removing sulfur from petroleum, carry his name.

See also Petroleum detection; Petroleum, economic uses of; Petroleum extraction; Petroleum, history of exploration

FREEZING AND MELTING

Freezing is the change that occurs when a liquid changes into a solid as the **temperature** decreases. Melting is the opposite change, from a solid to a liquid as the temperature increases. These are both examples of changes in the states of matter of substances.

Substances freeze at exactly the same temperature as they melt. As a consequence, the temperature at which—under a specified pressure—liquid and solid exist in equilibrium is defined as the melting or freezing point. When the pressure is one atmosphere, this temperature is known as the normal freezing (or melting) point. A change in pressure will change the temperature at which the change in the state of matter occurs. A decrease in pressure will decrease the temperature at which this occurs and an increase in pressure will increase the temperature required.

At a fundamental level freezing and melting represent changes in the energy levels of the molecules of the substance under consideration. Freezing is a change from a high energy state to one of lower energy, the molecules are moving less as their temperature falls. They become more ordered and fixed in shape. When a substance melts the average energy level of the constituent molecules increases. The molecules are moving more rapidly and in a less ordered manner in a liquid than in a solid. It is this greater freedom of movement that allows a liquid to flow to touch the walls of its container whereas a solid is fixed in a rigid shape. This consideration of the energy of the molecules is known as the kinetic molecular theory.

The temperature at which substances freeze and melt is different for different chemicals. The chemical formula of a substance is not necessarily a true indicator of what the freezing or melting point may be. Isomers of substances can have different physical properties including freezing and melting points. Similarly the presence of hydrogen bonds and other attractive forces such as van der Waals forces can influence the bonding within the substance and hence the freezing and melting points. If any intermolecular forces are present more energy must be added to the system to change from a solid to a liquid. This is because the intermolecular bonds have to be overcome to allow the molecules to move more freely. This is less of a change than occurs from the change from liquid to gas, because the molecules are still touching each other in both liquids and solids.

The purity of the compound can influence the temperature at which the solid-liquid change takes place. For example adding sodium chloride (common salt) or another salt to **water** depresses the freezing point, which is why salt is put on roads to stop their icing over. A pure substance has a definite melting or freezing point, the addition of an impurity lowers this temperature as well as spreads it so that there is a less definite, more diffuse melting or freezing point. This means that we can use the freezing or melting point as an indicator of the purity of a substance. When a solid is melted by heating or a liquid frozen while cooled, the temperature remains constant. Thus, if a graph of temperature is plotted against heat added a shoulder or plateau will be seen which represents the freezing or melting point. With an impure substance, this shoulder will not be so precise. A graph of this nature is known as a heating curve. The conversion between solid and liquid occurs at a constant temperature.

With most substances the solid is denser than the liquid phase. As a result of this when freezing the solid will sink to the bottom of the liquid. Water does not behave in this manner. **Ice** is less dense than water and consequently ice will float on water. Water has its maximum density at 39°F (4°C). This is caused by hydrogen bonding, which in the liquid phase is unordered. When the water freezes to form ice, the molecules assume an open ordered pattern that allows the maximum amount of hydrogen bonding. This characteristic has had a profound effect on life on Earth (e.g., it allows **lakes** and streams to freeze at the surface and provide insulation to life underneath the ice during frigid winter months) and results in an active agent of geological change. Because water expands when freezing it is able to crack **rock**; the cyclic freezing and refreezing of water is an important **weathering** agent.

Normally, when we talk about a substance being a solid or a liquid we are referring to its appearance at standard temperature and pressure, this is a pressure of one atmosphere and a temperature of 68°F (20°C). If the melting point is below this temperature and the boiling point is above it then the chemical is a liquid at standard temperature and pressure.

It is possible to cool a liquid below its freezing point and still have it remain as a liquid. This is known as a super-cooled liquid. This represents an unstable equilibrium and in time the liquid freezes. It is very easy to supercool water down to 12°F (−11.1°C) and still have it remain a liquid. The super-cooled liquid will not start to freeze until there is a point for the ice to start to form. This may be a single piece of dust, which acts as a nucleation point for the ice to start forming. Supercooled water is not encountered in nature because there is too much particulate material in the atmosphere. If any of these particles lands in a supercooled liquid it will instantly turn into the solid form.

Some chemicals do not have a point at which they turn from solid to liquid—they can change directly from solid to gas, a property called sublimation. Dry ice, solid **carbon dioxide**, exhibits this. Like melting and freezing this also happens at one specific temperature.

Solids and liquids are both densely packed at a molecular level. One difference in terms of the molecules is that with a liquid the molecules are more readily capable of slipping over each other. It is this property that makes it easier to pour a liquid. The molecules in a liquid are still touching each adjacent molecule (as they do in a solid), although they are less freely held.

Ionic compounds generally have a higher melting point than covalent compounds. This is because the intermolecular forces in an ionic compound are much stronger. If the pressure is increased the molecules are forced closer together and this means that the intermolecular forces are holding the particles closer together and more tightly, so a higher temperature is required to make the material melt.

Melting is also called fusion, and the energy required to bring about this change of state is called the heat of fusion or the enthalpy of fusion. For ice to turn into liquid water the heat of fusion is 6.01 kJ/mol. Melting and sublimation are both endothermic processes and freezing is an exothermic process. Whenever a material changes from one state to another there is an energy change within the system. For melting the order of the system is decreasing, so energy must be supplied to increase the randomness of the molecules. For freezing the molecules are becoming more ordered, so energy is lost from the system.

Freezing and melting are the change of state from liquid to solid and from solid to liquid. For any given pure chemical they happen at a specific temperature, which is the same for freezing and melting.

See also Chemical bonds and physical properties; Chemical elements; Evaporation; Faults and fractures; Glacial landforms; Glaciation; Glaciers; Glass; Ice heaving and ice wedging; Ice

FRESHWATER

Freshwater is chemically defined as containing a concentration of less than two parts per thousand (<0.2%) of dissolved salts.

Freshwater can occur in many parts of the environment. Surface freshwaters occur in **lakes**, ponds, **rivers**, and streams. Subsurface freshwater occurs in pores in **soil** and in subterranean aquifers in deep geological formations. Freshwater also occurs in snow and glacial **ice**, and in atmospheric vapors, **clouds**, and **precipitation**.

Most of the dissolved, inorganic chemicals in freshwater occur as ions. The most important of the positively charged ions (or cations) in typical freshwaters are calcium (Ca^{2+}), magnesium (Mg^{2+}), sodium (Na^+), ammonium (NH_4^+), and hydrogen ion (H^+). This hydrogen ion is only present if the solution is acidic; otherwise a hydroxy ion (OH^-) occurs. The most important of the negatively charged ions (or anions) are sulfate (SO_4^{2-}), chloride (Cl^-), and nitrate (NO_3^-). Other ions are also present, but in relatively small concentrations. Some freshwaters can have large concentrations of dissolved organic compounds, known as humic substances. These can stain the **water** a deep-brown, in contrast to the transparent color of most freshwaters.

At the dilute end of the chemical spectrum of surface waters are lakes in watersheds with hard, slowly **weathering bedrock** and soils. Such lakes can have a total concentration of salts of less than 0.002% (equivalent to 20 mg/L, or parts per million, ppm). For example, Beaverskin Lake in Nova Scotia has very clear, dilute water, with the most important dissolved chemicals being: chloride (4.4 mg/L), sodium (2.9 mg/L), sulfate (2.8 mg/L), calcium (0.41 mg/L), magnesium (0.39 mg/L), and potassium (0.30 mg/L). A nearby body of water, Big Red Lake, has similar concentrations of these inorganic ions. However, this lake also receives drainage from a nearby bog, and its **chemistry** includes a large concentration of dissolved organic compounds (23 mg/L), which stain the water the color of dark tea.

More typical concentrations of major inorganic ions in freshwater are somewhat larger: calcium 15 mg/L; sulfate 11 mg/L; chloride 7 mg/L; silica 7 mg/L; sodium 6 mg/L; magnesium 4 mg/L; and potassium 3 mg/L.

The freshwater of precipitation is considerably more dilute than that of surface waters. For example, precipitation falling on the Nova Scotia lakes is dominated by sulfate (1.6 mg/L), chloride (1.3 mg/L), sodium (0.8 mg/L), nitrate (0.7 mg/L), calcium (0.13 mg/L), ammonium (0.08 mg/L), magnesium (0.08 mg/L), and potassium (0.08 mg/L). Because the sampling site is within 31 mi (50 km) of the Atlantic Ocean, its precipitation is significantly influenced by sodium and chloride originating with sea sprays. More continental locations have much smaller concentrations of these ions in their precipitation water. For example, precipitation at a remote place in northern Ontario has a sodium concentration of 0.09 mg/L and chloride 0.15 mg/L, compared with 0.75 mg/L and 1.3 mg/L, respectively, at the maritime Nova Scotia site.

See also Clouds and cloud types; Drought; Estuary; Floods; Glaciers; Groundwater; Humidity; Hydrologic cycle; Rapids and waterfalls; Stream capacity and competence; Stream valleys, channels, and floodplains

FUELS AND FUEL CHEMISTRY

A fuel is any compound that has stored energy. This energy is captured in **chemical bonds** through processes such as photosynthesis and respiration. Energy is released during oxidation. The most common form of oxidation is the direct reaction of a fuel with **oxygen** through combustion. Wood, gasoline, **coal**, and any number of other fuels have energy-rich chemical bonds created using the energy from the **Sun**, which is released when the fuel is burned (i.e., the release of chemical energy). Chemical fuels or the fossil fuels are useful reserve of fuels and are therefore used extensively to satisfy the demands of an energy-dependent civilization.

Fossil fuels are principally **hydrocarbons** with minor impurities. They are so named because they originate from the decayed and fossilized remains of plants and animals that lived millions of years ago.

Fossil fuels can be separated into three categories. The first is **petroleum** or oil. This is a mixture of light, simple hydrocarbons dominated by the fractions with 6 to 12 carbons but also containing some light hydrocarbons (e.g., methane and ethane). Fully half of the energy consumed in the United States is from petroleum used to produce fuels for automobiles, recreational vehicles, home heating, or industrial production.

The principal use of petroleum is the production of gasoline. Over 40% all of all production ends up consumed in automobiles and such. Smaller fractions are turned into fuel oil (27%), jet fuel (7.4%), and other miscellaneous fuels, while the small fraction (about 10%) is used for the synthesis of the thousands of petrochemicals used in our daily lives. Indeed, many food compounds and pharmaceuticals owe their synthesis to a petrochemical precursor.

The second most prominent and naturally most abundant fossil fuel is coal. Coal also originates from decayed vegetative material buried eons ago, but the process is slightly different, being less oxidizing. The resulting material still has some of the original lignin-like structure exhibiting many fused rings and a large fraction of aromatic compounds. Consequently, coal is more of a polymeric substance than petroleum and is found as a solid not a liquid. The **carbon** to hydrogen ratio in coal is close to 1:1 (depending upon the type of coal), whereas the carbon to hydrogen ration in petroleum is closer to the 1:2 value expected for a hydrocarbon chain.

Minable coal is defined as 50% of the coal in a seam of at least 12 in thickness. The proven reserves of minable coal are sufficient to supply the industrial needs of modern society for the next four to five hundred years. Unfortunately, as a fuel source, coal has many disadvantages. It is a very dirty fuel that produces a large amount of unburned hydrocarbon, particulate, and—most damaging of all—significant quantities of sulfur dioxide byproducts. Indeed, it is the coal burning power plants of the eastern United States that are responsible for much of the **acid rain** and environmental damage observed in upstate New York and eastern Canada. The other significant disadvantage of coal is that it is not liquid, making it awkward to transport and store and limiting its use in applications like automobiles. A great deal of research has been done on the liquefaction of coal but with little economically viable success.

The third major fossil fuel is **natural gas**. This is a generic term for the light hydrocarbon fractions found associated with most oil deposits. Natural gas is mostly methane with small quantities of ethane and other gases mixed in. It is hydrogen- rich, since methane has a carbon to hydrogen ratio of 1:4. It is also an excellent fuel, burning with a high heat output and little in the way of unwanted pollution. It does produce **carbon dioxide**, which is a greenhouse gas, but all organic compounds also generate carbon dioxide on combustion. Natural gas is also easy to transport through pressurized pipelines.

All of this would appear to make natural gas the perfect fuel. However, it is not without its drawbacks. This includes the presence of hydrogen sulfide in some gas fields, leading to the term "sour gas." Hydrogen sulfide is the smell of rotten eggs, but if smell were the only problem, this would be of little concern. However, hydrogen sulfide is extremely corrosive to the pipes used to transport natural gas and is a very toxic compound being lethal at levels around 1,500 ppm.

In addition, natural gas is potentially explosive as the gas must be maintained under pressure, and any hydrocarbon, in a gaseous state, can explode. This is in contrast to the use of gasoline, which is a much safer fuel. Nevertheless, both the gaseous and liquid forms of hydrocarbons are much more volatile and represent a hazard compared to coal.

It has been suggested that because all of the oxygen in the atmosphere came from the splitting of carbon dioxide via photosynthesis, the total oxygen content may lead to an estimate of the total carbon reserves. However, it is not the total abundance of fuels that is critical—it is the accessibility from both an engineering and economics standpoint that makes near-term global fuel shortages appear probable.

In addition there are serious flaws in the gross estimate of fuels as a significant amount of the world's carbon reserves are tied up in calcium carbonate **rock** formations. Loosely speaking, calcium carbonate is **limestone** and there is abundance of limestone present throughout the world. Accordingly, most industry analysts place available fossil fuel reserves at much lower levels. It is estimated by a wide variety of sources that we will reach maximum oil production in the next twenty years. After that, production will decline worldwide and we will be forced to wean ourselves from an oil-based society. The estimates for coal provide a slightly better prognosis giving a window of about 500 years for consumption of all known reserves.

Natural gas also has one significant advantage over the other fossil fuels: it is "renewable." Natural gas is found as a side product of any decaying material. Methanogenic bacteria—literally, methane-making bacteria—exist in the garbage dumps and **waste disposal** sites of the industrial world, busily producing methane from garbage. Enough that the Fresh Kills garbage dump on Staten Island, New York is capable of heating 16,000 homes.

Various companies have been exploring the use of hydrogen as a fuel. When used in simple combustion, hydrogen has some of the problems associated with natural gas. It must be stored under pressure and is extremely explosive upon ignition in air (e.g. the Hindenburg disaster, in which a hydrogen-filled airship explosively burned). The type of explosion—the shape of the detonation—also makes hydrogen unsuitable as an alternative fuel for the conventional automobile. The sharpness of the explosion would quickly rattle pistons to pieces.

However, hydrogen does not need to be burned directly with oxygen to provide energy. Fuel cells combine hydrogen and oxygen at electrodes to produce **electricity**, which can then be used to run an electric motor or a spacecraft. NASA has been employing hydrogen/oxygen fuel cells for years to provide the electricity for both manned and unmanned spacecraft. In addition, fuel cells have the added bonus of providing crew members with fresh drinking **water**, as the only product of the reaction is pure water. Using fuel cells, hydrogen could potentially be used for conventional automobiles. Questions about storing hydrogen and long-term viability of the fuel cells need to be addressed, but the future of this technology looks promising. At present, however, our economically viable supply of hydrogen is obtained from fossil fuels because hydrogen is released from hydrocarbons during the refining process.

Both the use of sunlight and solar panels to create sufficient electricity to electrolyze water and bacteria capable of splitting water to generate hydrogen and oxygen may make hydrogen the chemical fuel of the future. In addition, research is underway into the use of methanol as a potential partner for a fuel cell, eliminating the need for hydrogen and greatly reducing the difficulties with storage and filling the tank.

The United States leads the world in energy consumption—over 90 quadrillion Btu are used annually in the United States alone (year 2001 estimates).

See also Chemical bonds and physical properties; Energy transformations; Environmental pollution; Petroleum detection; Petroleum, economic uses of; Petroleum extraction; Petroleum, history of exploration

FUJITA SCALE · *see* TORNADO

FUJITA, TETSUYA THEODORE
(1920-1998)
Japanese-born American meteorologist

In 1974, Theodore Fujita became the first scientist to identify microburst **wind shear**, a particularly intense and isolated form of **wind** shear later blamed for several devastating airline accidents in the 1980s. The development of Doppler radar allowed Fujita to track and explain the microburst phenomenon, which is now far better understood and avoided by aviators. Fujita also lent his name to the "F Scale" he developed to measure the strength of tornadoes by analyzing the damage they cause on the ground.

Fujita was born in Kitakyushu City, Japan, to Tomojiro, a schoolteacher, and Yoshie (Kanesue) Fujita. Fujita showed an early aptitude for science and obtained the equivalent of a bachelor's degree in mechanical engineering from the Meji College of Technology in 1943. It was while he was working as an assistant professor of **physics** at Meji that U.S. forces dropped the **atom** bomb on the Japanese cities of Hiroshima and Nagasaki. Fujita visited the ruins three weeks after the bombings. By measuring the scorch marks on bamboo vases in a cemetery in Nagasaki, Fujita was able to show that only one bomb had been dropped. Surveying the damage in Hiroshima, Fujita calculated how high above the ground the bombs had exploded in order to create their unique starburst patterns, which would become important to his later work.

Leaving Meji College in 1949, Fujita became an assistant professor at Kyushu Institute of Technology while pursuing his Ph.D. in atmospheric science at Tokyo University. Like others involved in atmospheric science, he had read the published articles of Horace R. Byers of the University of Chicago, who had conducted groundbreaking research on thunderstorms in 1946 and 1947. Fujita translated two of his own articles on the same subject into English and sent them to Byers. Byers was impressed with Fujita's work and the two men began a correspondence. In 1953, the year Fujita received his doctorate, Byers extended an invitation to the Japanese scientist to work at the University of Chicago as a visiting research associate.

Fujita worked at the University as a senior meteorologist until 1962, when he became an associate professor. For two years beginning in 1961, he was the director of the Mesometeorological Research Project, and in 1964, Fujita became the director of the **Satellite** and Mesometeorology Research Project. Fujita was made a full professor in 1965, and held the Charles E. Merriam Distinguished Service Professorship on an active basis and on an emeritus basis. He became a naturalized U.S. citizen in 1968, and adopted the first name Theodore for use in the United States. He married Sumiko Yamamo in June, 1969, and has a son from his first marriage.

From the mid–1960s, Fujita and his graduate students did extensive aerial surveys of tornadoes. Fujita claims to have logged over 40,000 miles flying in small planes under the worst of **weather** conditions. In the late 1960s, Fujita developed his **tornado** "F Scale." Traditionally, meteorologists listed only the total number of tornadoes that occurred, having no objective way to measure storm strength. Fujita constructed a system of measurement that correlates ground damage to windspeed. His six point system operates on an F–0 to F–5 scale and is similar to the **Richter scale** used to measure the strength of earthquakes.

Fujita did not actually witness a tornado until June 12, 1982, so the mainstay of his work was research on the aftermath of tornadoes. While at the National Center for Atmospheric Research in Denver, Colorado, Fujita spotted a tornado in the region early and collected some of the best data on the phenomenon ever.

In 1974, Fujita began analyzing the phenomenon of microbursts. Flying over the devastation wrought by a tornado, he noticed patterns of damage similar to those he had witnessed in Hiroshima and Nagasaki. "If something comes down from the sky and hits the ground it will spread out; it will produce the same kind of outburst effect that was in the back of my mind, from 1945 to 1974," Fujita explained in *The Weather Book: An Easy to Understand Guide to the U.S.A.'s Weather.* Meteorologists knew by the mid–1970s that severe storms produce **downdrafts**, but they assumed those downdrafts lost most of their force before they hit the ground and, therefore, did not cause much damage, so the phenomenon was largely ignored. Encouraged by Byers, Fujita coined the term "downburst" and began research to prove his thesis that downdraft is a significant weather phenomenon. Aided by the National Center for Atmospheric Research, he set up a project near Chicago that detected 52 downbursts in 42 days.

Fujita was eventually able to show that downdrafts cause so-called wind shear, a sudden and dramatic change in wind velocity, which causes damage on the ground and is a particular hazard in aviation, especially to planes taking off, landing, or flying low. Windspeeds up to F–3 are common for downbursts (higher F Scale readings usually indicate tornadoes). Fujita has commented that a lot of damage attributed to tornadoes in the past has really been the work of downbursts. "After I pointed out the existence of downbursts, the number of tornadoes listed in the United States decreased for a number of years," Fujita noted in *The Weather Book.*

Fujita's research finally gained national attention in the 1980s. Wind shear caused by downdraft was cited as a contributing factor in the July 1982, crash of a Pan American 727 in New Orleans, Louisiana, which killed 154 people. During that event, the airliner was observed sinking back to the ground shortly after takeoff—the apparent result of wind shear. Another accident occurred in August 1985, when Delta Flight 191 crashed at Dallas-Ft. Worth Airport, killing 133 people. Again, wind shear was suspected to be the immediate cause of the catastrophe.

Air safety has improved dramatically because of Fujita's work, which led to the development of Doppler radar. Doppler radar is so sensitive it actually picks up particles of debris in the air that are as fine as dust. Movements of these particles are tracked to measure shifts in wind velocity. "This is particularly important in being able to detect the precursor events for severe weather," Frank Lepore, public affairs officer for the National Weather Service, told Joan Oleck in an interview. By 1996, the Weather Service and U.S. Air Force together installed 137 Doppler systems, essentially blanketing the continental United States. Harking back to the airline accidents of the 1980s, Lepore noted "a reduction in those incidents today because there is Doppler radar available. By being able to measure the internal velocity of air moving inside a storm system, [aviators] can see rising and falling volumes of air.... They're now getting 20- and 25-minute warning on the systems that cause tornadoes."

In 1988, Fujita assumed directorship of the Wind Research Lab at the University of Chicago. Among his many awards were the 1989 Medaille de Vermeil from the French National Academy of Air and Science for identifying microbursts; the 1990 Fujiwara Award Medal from the Meteorological Society of Japan for his research on mesometeorology; the 1991 Order of the Sacred Treasure, Gold and

Silver Star from the Government of Japan for his tornado and microburst work; and the 1992 Transportation Cultural Award from the Japanese Government for his contributions to air safety. Fujita died in Chicago at the age of 78.

FUMAROLE

Any opening in the ground that emits hot steam or gas is a fumarole. Fumaroles are common on the flanks of volcanoes as well as in their craters and calderas. Extensive fumarole fields occur in areas where a shallow volcanic heat source is overlaid by water-permeable **rock**, as at Yellowstone National Park in the United States and Rotorua in New Zealand.

All fumaroles require both heat and a source of gas or **water**. They are most often supplied with heat and gas by **magma** or masses of freshly ejected volcanic rock and with water by **precipitation** that seeps into the ground. Subterranean heated water also produces hot **springs** and geysers; hot springs are more common than fumaroles, geysers less common. Geysers are distinguished from both hot springs and fumaroles by their specialized plumbing systems, while the difference between a hot spring and a fumarole is simply the degree of heating. If the heat source is not strong enough to boil water, the result is a hot spring. Even if water is boiled, the resulting steam may be condensed by passing through liquid **groundwater** before reaching the surface, in which case the result is still a hot spring. Only if steam reaches the surface is a fumarole produced. Some vents are hot springs in the wet season and fumaroles in the dry, when there is less groundwater to condense steam rising from below.

A deposit of hot ash and shattered rock laid down by an explosive volcanic eruption may cover many square miles of ground and be hundreds of feet deep. During the years it takes to cool, such a deposit may produce a vast field of fumaroles. This occurred in the Valley of Ten Thousand Smokes in Alaska, where a thick ash-and-rock layer was laid down by a large eruption in 1912. Immediately after deposition, this layer was dotted by tens of thousands of fumaroles, some venting from openings many feet across. Over the next half-century, as the underlying mass cooled, most of these fumaroles became extinct.

Fumaroles whose gases are particularly sulfurous are termed solfataras. (Some geologists use the terms fumarole and solfatara synonymously regardless of sulfur content.) Furthermore, some gas-emitting vents have temperatures below the boiling point of water and emit mostly **carbon dioxide** (CO_2) and other gases with little water vapor; geologists term such dry, cool vents mofettes to distinguish them from fumaroles.

See also Crater, volcanic; Hydrothermal processes; Volcanic vent

FUSION • *see* NUCLEAR FUSION

G

GAGARIN, YURI A. (1934-1968)

Russian cosmonaut

Yuri A. Gagarin was the first human in **space**. In 1961, the boyish-looking Soviet cosmonaut captured the attention of the world with his short flight around the earth. "He invited us all into space," American astronaut **Neil Armstrong** said of him, as quoted in *Aviation Week and Space Technology*.

The third of four children, Yuri Alekseevich Gagarin was born on a collective farm in Klushino, in the Smolensk region of the Russian Federation. His father, Aleksey Ivanovich Gagarin, was a carpenter on the farm and his mother, Anna, a dairymaid. Gagarin grew up helping them with their work. Neither of his parents had much formal education, but they encouraged him in his schooling. During World War II, the family was evicted from their home by invading German troops, and Gagarin's older brother and sister were taken prisoner for slave labor, though they later escaped.

After the war, Gagarin went to vocational school in Moscow, originally intending to become a foundry worker, and then he moved on to the Saratov Industrial Technical School. He was still learning to be a foundryperson, although his favorite subjects were **physics** and mathematics. In 1955, during his fourth and final year of school, he joined a local flying club. His first flight as a passenger, he later wrote in *Road to the Stars*, "gave meaning to my whole life." He quickly mastered flying, consumed by a new determination to become a fighter pilot. He joined the Soviet Air Force after graduation. The launch of Sputnik—the first artificial **satellite** sent into space—occurred on October 4, 1957, while he pursued his military and flight training. He graduated with honors that same year and married medical student Valentina Ivanova Goryacheva. They would have two children, a daughter and a son.

Gagarin volunteered for service in the Northern Air Fleet and joined the Communist Party. He followed closely news of other Sputnik launches; although there had been no official announcement, Gagarin guessed that preparations for manned flights would soon begin and he volunteered for cos-

monaut duty. Gagarin completed the required weeks of physical examinations and testing in 1960, just before his twenty-sixth birthday. He was then told that he had been made a member of the first group of twelve cosmonauts. The assignment was a secret, and he was forbidden to tell even his wife until his family had settled into the new space-program complex called Zvezdniy Gorodok (Star Town), forty miles from Moscow. An outgoing, natural leader, the stocky, smiling Gagarin stood out even among his well-qualified peers. Sergei Korolyov, the head of the Soviet space program and chief designer of its vehicles, thought Gagarin had the makings of a first-rate scientist and engineer, as well as being an excellent pilot. In March of 1961, Korolyov approved the selection of Gagarin to ride Vostok I into orbit.

Senior Lieutenant Gagarin made history on April 12, 1961, when a converted ballistic missile propelled his Vostok capsule into Earth orbit from the remote Baikonur Cosmodrome. The Vostok was controlled automatically, and Gagarin spent his time reporting observations of the Earth and his own condition. He performed such tasks as writing and tapping out a message on a telegraph key, thus establishing that a human being's coordination remained intact even while weightless in space. Proving that people could work in space, he also ate and drank to verify that the body would take nourishment in weightlessness. He commented repeatedly on the beauty of the earth from space and on how pleasant weightlessness felt.

Gagarin rode his spacecraft for 108 minutes, ejecting from the spherical reentry module after the craft reentered the atmosphere just short of one complete orbit. Ejection was standard procedure for all Vostok pilots, although Gagarin dutifully supported the official fiction that he had remained in his craft all the way to the ground—a requirement for international certification of the flight as a record. Cosmonaut and capsule landed safely near the banks of the Volga River.

After doctors proclaimed him unaffected by his flight, Gagarin was presented to the public as an international hero. He received an instant promotion to the rank of major and made appearances around the world. He was named a Hero of the Soviet Union and a Hero of Socialist Labor, and he became an honorary citizen of fourteen cities in six countries. He received the Tsiolkovsky Gold Medal of the Soviet Academy of Sciences, the Gold Medal of the British Interplanetary Society, and two awards from the International Aeronautical Federation. The flight had many implications for international affairs: American leaders extended cautious congratulations and redoubled their own efforts in the space race, while the Soviet media proclaimed that Gagarin's success showed the strength of socialism.

Gagarin became commander of the cosmonaut team. In 1964, he was made deputy director of the cosmonaut training center at the space program headquarters complex—where he oversaw the selection and training of the first women cosmonauts. He served as capsule communicator—the link between cosmonauts and ground controllers—for four later space flights in the Vostok and Voskhod programs. At various times during this period, he also held political duties; he chaired the Soviet-Cuban Friendship Society and served on the Council of the Union and the Supreme Soviet Council of Nationalities.

Gagarin always wanted to venture back to space, and in 1966, he was returned to active status to serve as the backup cosmonaut to Vladimir Komarov for the first flight of the new Soyuz spacecraft. When the Soyuz 1 mission ended and Komarov died due to a parachute malfunction, Gagarin was assigned to command the upcoming Soyuz 3. But Gagarin himself did not live to fly the Soyuz 3 mission. On March 27, 1968, he took off for a routine proficiency flight in a two-seat MiG–15 trainer. He and his flight instructor became engaged in low-level maneuvers with two other jets. Gagarin's plane crossed close behind another jet and was caught in its vortex; he lost control and the jet crashed into the tundra at high speed, killing both occupants instantly.

Gagarin was given a hero's funeral. The Cosmonaut Training Center was renamed in his honor, as were his former hometown, a space tracking ship, and a lunar crater. His wife continued to work as a biomedical laboratory assistant at Zvezdniy Gorodok, and Gagarin's office there was preserved as a museum; a huge statue of him was erected in Moscow. His book *Survival in Space* was published posthumously. Written with space-program physician Vladimir Lebedev, the work outlines Gagarin's views on the problems and requirements for successful long-term space flights. On April 12, 1991, thirty years after Gagarin's flight, his cosmonaut successors, along with eighteen American astronauts, gathered at Baikonur to salute his achievements.

See also History of manned space exploration

GAIA HYPOTHESIS

The Gaia hypothesis is a recent and controversial theory that views Earth as an integrated, pseudo-organismic entity, and not as a mere physical object in **space**. Gaia, Earth, was believed by the ancient Greeks to be a living, fertile ancestor of many of their important gods. The Gaia hypothesis suggests that organisms and ecosystems on Earth cause substantial

changes to occur in the physical and chemical nature of the environment, in a manner that improves the living conditions on the planet. In other words, it is suggested that Earth is an organismic planet, with homeostatic mechanisms that help to maintain its own environments within the ranges of extremes that can be tolerated by life. According to the Gaian hypothesis, **evolution** is the result of cooperative, not competitive processes. The Gaian hypothesis holds that evolution of life on Earth was enhanced by two processes: sexual reproduction, which introduced enormous variety in the gene pool, and the development of consciousness, which enabled genetic methods of evolution to be replaced with more efficient social mechanisms.

Earth is the only planet in the universe that is known to support life. This is one of the reasons why the Gaia hypothesis cannot be tested by rigorous, scientific experimentation—there is only one known replicate in the great, universal experiment. However, some supporting evidence for the Gaia hypothesis can be marshaled from certain observations of the structure and functioning of the planetary ecosystem.

One supporting line of reasoning for the Gaia hypothesis concerns the presence of **oxygen** in Earth's atmosphere. Scientists believe that the primordial atmosphere of Earth did not contain oxygen. The appearance of this gas required the evolution of photosynthetic life forms, which were initially blue-green bacteria and, somewhat later, single-celled algae. Molecular oxygen is a waste product of photosynthesis, and its present atmospheric concentration of about 21% has entirely originated with this biochemical process (which is also the basis of all biologically fixed energy in ecosystems). Of course, the availability of atmospheric oxygen is a critically important environmental factor for most of Earth's species and for many ecological processes. In addition, it appears that the concentration of oxygen in the atmosphere has been relatively stable for an extremely long period of time, perhaps several billions of years. This suggests the existence of a long-term equilibrium between the production of this gas by green plants, and its consumption by biological and non-living processes. If the atmospheric concentration of oxygen were much larger than it actually is, say about 25% instead of the actual 21%, then biomass would be much more readily combustible. These conditions could lead to much more frequent and more extensive forest fires. Such conflagrations would be severely damaging to Earth's ecosystems and species.

Some proponents of the Gaia hypothesis interpret the above information to suggest that there is a planetary, homeostatic control of the concentration of molecular oxygen in the atmosphere. This control is intended to strike a balance between the concentrations of oxygen required to sustain the metabolism of organisms, and the larger concentrations that could result in extremely destructive, uncontrolled wildfires.

Another line of evidence in support of the Gaian theory concerns **carbon dioxide** in Earth's atmosphere. To a substantial degree, the concentration of this gas is regulated by complex biological and physical processes by which **carbon dioxide** is emitted and absorbed. This gas is well known to be important in the planet's **greenhouse effect**, which is critical to maintaining the average **temperature** of the surface within a range that organisms can tolerate. It has been estimated that in the absence of this greenhouse effect, Earth's average surface temperature would be about −176°F (−116°C), much too cold for organisms and ecosystems to tolerate over the longer term. Instead, the existing greenhouse effect, caused in large part by atmospheric carbon dioxide, helps to maintain an average surface temperature of about 59°F (15°C). This is within the range of temperature that life can tolerate.

Again, advocates of the Gaia hypothesis interpret these observations to suggest that there is a homeostatic system for control of atmospheric carbon dioxide, and of climate. This system helps to maintain conditions within a range that is satisfactory for life.

All scientists agree that there is clear evidence that the non-living environment has an important influence on organisms, and that organisms can cause substantial changes in their environment. However, there appears to be little widespread support within the scientific community for the notion that Earth's organisms and ecosystems have somehow integrated in a mutually benevolent symbiosis (or mutualism), aimed at maintaining environmental conditions within a comfortable range.

Still, the Gaia hypothesis is a useful concept, because it emphasizes the diverse connections of ecosystems, and the consequences of human activities that result in environmental and ecological changes. Today, and into the foreseeable future, humans are rapidly becoming a dominant force that is causing large, often degradative changes to Earth's environments and ecosystems. Hopefully, the changes wrought by humans will not exceed the limits of homeostatic tolerance and repair of the planet and its ecological components. If these possibly Gaian limits of tolerance are exceeded, some scientists assert the consequences could be catastrophic for life on Earth.

See also Earth (planet); Evolutionary mechanisms; Evolution, evidence of

GALILEI, GALILEO (1564-1642)

Italian mathematician and astronomer

Galileo Galilei is credited with establishing the modern experimental method. Before Galileo, knowledge of the physical world that was advanced by scientists and thinkers was for the most part a matter of hypothesis and conjecture. In contrast, Galileo introduced the practice of proving or disproving a scientific theory by conducting tests and observing the results. His desire to increase the precision of his observations led him to develop a number of inventions and discovery, particularly in the fields of **physics** and **astronomy**.

The son of Vincenzo Galilei (c.1520–1591), an eminent composer and music theorist, Galileo was born in Pisa. He received his early education at a monastery near Florence, and in 1581, entered the University of Pisa to study medicine. While a student he observed a hanging lamp that was swinging back and forth, and noted that the amount of time it took the lamp to complete an oscillation remained constant, even as the arc of the swing steadily decreased. He later experimented with other suspended objects and discovered that they

behaved in the same way, suggesting to him the principle of the pendulum. From this discovery he was able to invent an instrument that measured time, which doctors found to be useful for measuring a patient's pulse rate, and Christiaan Huygens later adapted the principle of a swinging pendulum to build a pendulum clock.

While at the University of Pisa, Galileo listened in on a geometry lesson and afterward abandoned his medical studies to devote himself to mathematics. However, he was unable to complete a degree at the university due to lack of funds. He returned to Florence in 1585, having studied the works of Euclid and Archimedes. He expanded on Archimedes' work in hydrostatics by creating a hydrostatic balance, a device designed to measure the density of objects. The following year, he published an essay describing his new invention, which determined the specific **gravity** of objects by weighing them in **water**. With the hydrostatic balance, Galileo gained a scientific reputation throughout Italy.

In 1592, Galileo was appointed professor of mathematics at Padua University in Pisa, where he conducted experiments with falling objects. Aristotle had stated that a heavier object should fall faster than a lighter one. It is said that Galileo tested Aristotle's assertion by climbing the leaning tower of Pisa, dropping objects of various weights, and proving conclusively that all objects, regardless of weight, fall at the same rate.

Some of Galileo's experiments did not turn out as expected. He tried to determine the speed of light by stationing an assistant on a hill while he stood on another hill and timed the flash of a lantern between the hills. He failed because the hilltops were much too close together to make a measurement.

In 1593, Galileo invented one of the first measuring devices to be used in science: the thermometer. Galileo's thermometer employed a bulb of air that expanded or contracted as **temperature** changed and in so doing caused the level of a column of water to rise or fall. Though this device was inaccurate because it did not account for changes in air pressure, it was the forerunner of improved instruments.

From 1602 to 1609, Galileo studied the motion of pendulums and other objects along arcs and inclines. Using inclined planes that he built, he concluded that falling objects accelerate at a constant rate. This law of uniform acceleration later helped Isaac Newton derive the law of gravity.

Galileo did not make his first contribution to astronomy until 1604, when a supernova abruptly exploded into view. Galileo postulated that this object was farther away than the planets and pointed out that this meant that Aristotle's "perfect and unchanging heavens" were not unchanging after all. Ironically, Galileo's best-known invention, the **telescope**, was *not* his creation after all. The telescope was actually invented in 1608 by **Hans Lippershey**, a Danish spectacle maker. When Galileo learned of the invention in mid-1609, he quickly built one himself and made several improvements. His altered telescope could magnify objects at nine-power, three times the magnification of Lippershey's model. Galileo's telescope proved to be very valuable for maritime applications, and

Galileo, pleading at his trial before the Inquisition. © *Corbis-Bettmann. Reproduced by permission.*

Galileo was rewarded with a lifetime appointment to the University of Venice.

He continued his work, and by the end of the year he had built a telescope that could magnify at 30-power. The discoveries he made with this instrument revolutionized astronomy. Galileo saw jagged edges on the **Moon**, which he realized were the tops of mountains. He assumed that the Moon's large dark areas were bodies of water, which he called maria ("seas"), though we now know there is no water on the Moon. When he observed the Milky Way, Galileo was amazed to discover Jupiter, which resulted in his discovery of its four moons; he later called them "satellites," a term suggested by the German astronomer **Johannes Kepler**. Galileo named the moons of Jupiter, Sidera Medicea ("Medicean stars") in honor of Cosimo de Medici, the Grand Duke of Tuscany, whom Galileo served as "first philosopher and mathematician" after leaving the University of Pisa in 1610. Also, with repeated observation, he was able to watch the moons as they were being eclipsed by Jupiter and from this he was able to correctly estimate the period of **rotation** of each of the moons.

In 1610, Galileo outlined planetary discoveries in a small book called *Siderus Nuncius* ("The Sidereal Messenger"). Venus, seen through the telescope, exhibited phases like the Moon, and for the same reasons: Venus did not produce its own light but was illuminated by the **Sun**.

Saturn was a mystery: Galileo's 30-power telescope was at the limit of its ability to resolve Saturn, and the planet appeared to have three indistinct parts. When Galileo looked at the Sun, he saw dark spots on its disc. The position of the spots changed from day to day, allowing Galileo to determine the rotational rate of the Sun.

In 1613, Galileo published a book in which for the first time he presented evidence for and openly defended the model of the **solar system** earlier proposed by the Polish astronomer

Nicholas Copernicus, who argued that Earth, rather than being positioned at the center of the universe, as in the Ptolemaic design, was only one of several galactic bodies that orbited the Sun. While there was some support even among ecclesiastical authorities for Galileo's proof of the Copernican theory, the Roman Catholic hierarchy ultimately determined that a revision of the long-held astronomical doctrines of the church was unnecessary. Thus, in 1616, a decree was issued by the church declaring the Copernican system "false and erroneous," and Galileo was ordered not to support this system.

Following this run-in with the Catholic Church and the inquisition that forced his adherence to the Copernican theory of the solar system, Galileo focused on the problem of determining **longitude** at sea, which required a reliable clock. Galileo thought it possible to measure time by observing eclipses of Jupiter's moons. Unfortunately, this idea was not practical for eclipses could not be predicted with enough accuracy and observing celestial bodies from a rocking ship was nearly impossible.

Galileo wanted to have the edict against the Copernican theory revoked, and in 1624, traveled to Rome to make his appeal to the newly elected pope, Urban VIII. The pope would not revoke the edict but did give Galileo permission to write about the Copernican system, with the provision that it would not be given preference to the church-sanctioned Ptolemaic model of the universe.

With Urban's imprimatur, Galileo wrote his *Dialogue Concerning the Two Chief World Systems—Ptolemaic and Copernican*, which was published in 1632. Despite his agreement not to favor the Copernican view, the objections to it in the *Dialogue* are made to sound unconvincing and even ridiculous. Summoned to Rome to stand before the Inquisition, Galileo was accused of violating the original proscription of 1616 forbidding him to promote the Copernican theory. Put on trial for heresy, he was found guilty and ordered to recant his errors. At some point during this ordeal Galileo is supposed to have made his famous statement: "And yet it moves," referring to the Copernican doctrine of Earth's rotation on its axis.

While the judgment against Galileo included a term of imprisonment, the pope commuted this sentence to house arrest at Galileo's home near Florence. Although he was forbidden to publish any further works, he devoted himself to his work on motion and parabolic trajectories, arriving at theories that were later refined by others and made an important impact on gunnery. Galileo died blind at the age of 78.

See also Gravity and the gravitational field

GEIGER COUNTER • *see* FISSION

GELL-MANN, MURRAY (1929-)
American physicist

Prior to the 1940s and 1950s, only a handful of fundamental particles—among them the proton, neutron, electron, and positron—had been discovered in particle **physics** research. The study of cosmic rays and particle accelerator reactions revealed that the composition of matter was much more complex than previously thought. Dozens, and then hundreds of new particles were discovered. Most appeared to meet the criterion of being a basic form of matter, but they often had unexpected properties. For example, some had lifetimes much longer (10^{-9} second) than was predicted for them, based on their mass. Because of these properties, they were collectively referred to as "strange" particles. Before long, physicists aggressively began searching for a way to organize and make sense out of the particle zoo they had discovered. A leading figure in this search was Murray Gell-Mann.

Gell-Mann was born in New York City in 1929. He earned his bachelor of science degree at Yale University at the age of nineteen and his Ph.D. from the Massachusetts Institute of Technology three years later. He worked briefly at the Institute for Advanced Studies and then taught at the University of Chicago from 1952 to 1954. Gell-Mann then moved to the California Institute of Technology, where he became R.A. Millikan professor of theoretical physics in 1966.

Gell-Mann has made a number of contributions to the effort to organize the "particle zoo." In 1953, he suggested that basic particles contain an intrinsic property known as "strangeness," not unlike charge or spin. He showed how the conservation of strangeness in a particle reaction could explain a number of observations made of these new particles. A similar concept was developed independently by the Japanese physicist, Kazuhiko Nishijima.

Gell-Mann next applied himself to the development of a system for placing the known elementary particles into a small number of groups. He observed that particles could be classified into a relatively small number of families of multiplets that have similar properties. Gell-Mann referred to his classification system as *the eight-fold way*, after the eight ways of right living taught by the Buddha.

Gell-Mann's scheme accomplished for elementary particles what Dmitri Mendeleev's **periodic table** had achieved for the elements. Furthermore, like the periodic table, the eight-fold way predicted the existence of new elementary particles. The discovery in 1964 of one such particle, the omega minus ($\Omega-$) provided dramatic confirmation of Gell-Mann's ideas. The Israeli physicist, Yuval Ne'emann, independently proposed a similar system of classification at about the same time.

Finally, Gell-Mann suggested that the hundreds of elementary particles might, in fact, be composed of a very small number of even more basic particles. He called these particles quarks, from James Joyce's *Finnegan's Wake*, "Three quarks for Master Mark!" The first three quarks to be discovered were given the somewhat whimsical names of "up," "down," and "strange." Gell-Mann has also made important contributions to the theory of quantum chromodynamics, which attempts to explain interactions among quarks.

See also Atomic structure; Quantum theory and mechanics

GEMSTONES

Gemstones are **minerals** or other materials that, because of certain outstanding physical properties such as color, clarity, and hardness, have aesthetic value for use in jewelry and other adornments. Of the over 3,000 different mineral varieties known, about 50 are commonly used as gemstones. In general, for a mineral to be used as a gemstone it must be beautiful when polished, cut, or faceted, and it must be hard and durable. Rarity is another characteristic that lends value to a gemstone.

Most gemstones are minerals, but gemstones are given a name based on their appearance, as opposed to the more scientifically strict names of minerals. As a result, a mineral may have a different name for its gem version. For example, sapphire and ruby, two well-known gemstones of distinctly different color, are actually the same mineral: corundum. Emerald and aquamarine are gem forms of beryl. **Quartz** is called amethyst if it is purple, citrine if yellow. Other gemstones are known by their mineral name such as **diamond**, garnet, and topaz.

Although a gemstone may have many properties that make it appealing, the beauty of a gemstone is generally a factor of its color, clarity, and luster. The color of a gem is largely due to its chemical composition. If the color is the result of elements that are an essential part of the mineral structure, it is termed idiochromatic. These minerals usually produce gems of a consistent color, such as peridot (mineral name: **olivine**), which is always green. An allochromatic gem derives its color from elemental impurities that are not integral to mineral. In this case, a mineral can vary in color, based on the varying trace impurities. Corundum, for example, is white in the pure mineral state, but slight amounts of chromium and **iron** will produce the red color of rubies while a combination of iron and titanium will result in sapphire blue. The color variation in diamond and quartz are also due to chemical impurities.

The clarity is the degree to which a gemstone is free of visible impurities, or inclusions. Inclusions may be tiny gas bubbles trapped in the crystal, internal fractures, or microscopic specks of a differing mineral. Inclusions are a very common result of the natural formation processes of minerals and it is the exception to find a mineral free of them and is why the most valued gems are free of inclusions. Some minerals have a greater tendency to contain inclusions, such as emerald.

The luster of a gemstone is the overall appearance as light strikes it. Gemstones are valued for a luster that is very shiny and glasslike and for one that yields a high degree of internal reflections. The latter, termed **adamantine**, is enhanced greatly by faceting, or the grinding of regular, angled surfaces. There are numerous patterns of faceting that are designed to maximize the natural luster of a gemstone. Diamond is a prime example of how faceting brings out its natural brilliance. Chatoyancy in gemstones, commonly known as "cat's eyes" or "stars," occurs when light reflects perpendicularly from mineral channels or mineral fibers inside the gemstone. Parallel fibers will result in a cat's eye effect; when the reflecting fibers extend in different directions, a star effect will result.

Not all gemstones are minerals. Some are naturally occurring organic materials. The popular gemstone amber, for example, is fossilized tree resin. Another, pearl, is produced when an oyster attempts to isolate a foreign particle within its shell by coating it with the same material that lines its shell: mother-of-pearl. In addition, many gemstones are now synthesized and produced in large quantities in factories.

See also Industrial minerals; Minerology

GEOCENTRIC MODEL ● *see* ASTRONOMY

GEOCHEMISTRY

Geochemistry is the study of the chemical processes that form and shape the earth.

Earth is essentially a large mass of crystalline solids that are constantly subject to physical and chemical interaction with a variety of solutions (e.g., **water**) and substances. These interactions allow a multitude of chemical reactions.

It is through geochemical analysis that estimates of the age of Earth are formed. Because radioactive isotopes decay at measurable and constant rates (e.g., **half-life**) that are proportional to the number of radioactive atoms remaining in the sample, analysis of rocks and **minerals** can also provide reasonably accurate determinations of the age of the formations in which they are found. The best measurements obtained via radiometric dating (based on the principles of nuclear reactions) estimate the age of Earth to be four and one half billion years old.

Dating techniques combined with spectroscopic analysis provide clues to unravel Earth's history. Using neutron activation analysis, Nobel Laureate Luis Alvarez discovered the presence of the element iridium when studying samples from the K-T boundary layer (i.e., the layer of sediment laid down at the end of the Cretaceous and beginning of the Tertiary Periods). Fossil evidence shows a mass extinction at the end of the **Cretaceous Period**, including the extinction of the dinosaurs. The uniform iridium layer—and presence of **quartz crystals** with shock damage usually associated only with large asteroid impacts or nuclear explosions—advanced the hypothesis that a large asteroid impact caused catastrophic climatic damage that spelled doom for the dinosaurs.

Although hydrogen and helium comprise 99.9% of the atoms in the universe, Earth's **gravity** is such that these elements readily escape Earth's atmosphere. As a result, the hydrogen found on Earth is found bound to other atoms in molecules.

Geochemistry generally concerns the study of the distribution and cycling of elements in the **crust** of the earth. Just as the biochemistry of life is centered on the properties and reaction of **carbon**, the geochemistry of Earth's crust is centered upon **silicon**. Also important to geochemistry is **oxygen**. Oxygen is the most abundant element on Earth. Together, oxygen and silicon account for 74% of Earth's crust.

The type of **magma** (Basaltic, Andesitic or Ryolytic) extruded by volcanoes and fissures (magma is termed **lava** when at Earth's surface) depends on the percentage of silicon and oxygen present. As the percentage increases, the magma becomes thicker, traps more gas, and is associated with more explosive eruptions.

The eight most common elements found on Earth, by weight, are oxygen (O), silicon (Si), **aluminum** (Al), **iron** (Fe), calcium (Ca), sodium (Na), potassium (K), and magnesium (Mg).

Unlike carbon and biochemical processes where the covalent bond is most common, however, the ionic bond is the most common bond in **geology**. Accordingly, silicon generally becomes a cation and will donate four electrons to achieve a noble gas configuration. In quartz, each silicon **atom** is coordinated to four oxygen atoms. Quartz crystals are silicon atoms surrounded by a tetrahedron of oxygen atoms linked at shared corners.

Rocks are aggregates of minerals and minerals are composed of elements. A mineral has a definite (not unique) formula or composition. Diamonds and **graphite** are minerals that are polymorphs (many forms) of carbon. Although they are both composed only of carbon, diamonds and graphite have very different structures and properties. The types of bonds in minerals can affect the properties and characteristics of minerals.

Pressure and **temperature** affect the structure of minerals. Temperature can determine which ions can form or remain stable enough to enter into chemical reactions. **Olivine**, $((Fe, Mg)_2 SiO_4)$, for example is the only solid that will form at

1,800°C. According to olivine's formula, it must be composed of two atoms of either Fe or Mg. Olivine is built by the ionic substitution of Fe and Mg—the atoms are interchangeable because they the same electrical charge and are of similar size—and thus, olivine exists as a range of elemental compositions termed a **solid solution series**. Olivine can thus be said to be "rich" in iron or rich in magnesium. As magma cools larger atoms such as potassium ions enter into reactions and additional minerals form.

The determination of the chemical composition of rocks involves the crushing and breakdown of rocks until they are in small enough pieces that decomposition by hot acids (hydrofluoric, nitric, hydrochloric, and perchloric acids) allows the elements present to enter into solution for analysis. Other techniques involve the high temperature fusion of powdered inorganic reagent (flux) and the **rock**. After **melting** the sample, techniques such as x-ray fluorescence spectrometry may be used to determine which elements are present.

Chemical and mechanical **weathering** break down rock through natural processes. Chemical weathering of rock requires water and air. The basic chemical reactions in the weathering process include solution (disrupted ionic bonds), hydration, hydrolysis, and oxidation.

The geochemistry involved in many environmental issues has become an increasing important aspect of scientific and political debate. The effects of **acid rain** are of great concern to geologists not only for the potential damage to the **biosphere**, but also because acid rain accelerates the weathering process. Rainwater is made acidic as it passes through the atmosphere. Although rain becomes naturally acidic as it contacts nitrogen, oxygen, and **carbon dioxide** in the atmosphere, many industrial pollutants bring about reactions that bring the acidity of rainwater to dangerous levels. Increased levels of carbon dioxide from industrial pollution can increase the formation of carbonic acid. The rain also becomes more acidic. **Precipitation** of this "acid rain" adversely affects both geological and biological systems.

According to plate tectonic theory, the crust (**lithosphere**) of Earth is divided into shifting plates. Geochemical analysis of Earth's tectonic plates reveals a continental crust that is older, thicker and more granite-like than the younger, thinner oceanic crusts made of basaltic (iron, magnesium) materials.

See also Atomic mass and weight; Atomic number; Atomic structure; Big Bang theory; Chemical bonds and physical properties; Chemical elements; Geologic time

GEOGRAPHIC AND MAGNETIC POLES

Earth's geographic poles are fixed by the axis of Earth's **rotation**. On maps, the north and south geographic poles are located at the congruence of lines of **longitude**. Earth's geographic poles and magnetic poles are not located in the same place—in fact they are hundreds of miles apart. As are all points on Earth, the northern magnetic pole is south of the northern geographic pole (located on the **polar ice** cap) and is

presently located near Bathurst Island in northern Canada, approximately 1,000 mi (1,600 km) from the geographic North Pole. The southern magnetic pole is displaced hundreds of miles away from the southern geographic pole on the Antarctic continent.

Although fixed by the axis of rotation, the geographic poles undergo slight wobble-like displacements in a circular pattern that shift the poles approximately six meters per year. Located on shifting polar **ice**, the North Pole (geographic pole) is technically defined as that point 90° N **latitude**, 0° longitude (although, because all longitude lines converge at the poles, any value of longitude can be substituted to indicate the same geographic point). The South Pole (geographic pole) is technically defined as that point 90° S latitude, 0° longitude. Early explorers used sextants and took celestial readings to determine the geographic poles. Modern explorers reply on **GPS** coordinates to accurately determine the location of the geographic poles.

Earth's **magnetic field** shifts over time, eventually completely reversing its polarity. There is evidence in magnetic mineral orientation that, during the past 10–15 million years, reversals have occurred as frequently as every quarter million years. Although Earth's magnetic field is subject to constant change (periods of strengthening and weakening) and the last magnetic reversal occurred approximately 750,000 years ago, geophysicists assert that the next reversal will not come within the next few thousand years. The present alignment means that at the northern magnetic pole, a **dip** compass (a compass with a vertical swinging needle) points straight down. At the southern magnetic pole, the dip compass needle would point straight up or away from the southern magnetic pole.

The magnetic poles are not stationary and undergo polar wandering. The north magnetic pole migrates about 6.2 mi (10 km) per year. The magnetic reversals mean that as **igneous rocks** cool from a hot **magma**, those that contain magnetic **minerals** will have those minerals align themselves with the magnetic polarity present at the time of cooling. These volcanic rocks preserve a history of magnetic reversals and when found in equidistant banded patterns on either side of sites of sea floor spreading, provide a powerful paleomagnetic proof of **plate tectonics**.

Navigators using magnetic compass readings must make corrections both for the distance between the geographic poles and the magnetic poles, and for the shifting of the magnetic poles. Moreover, the magnetic poles may undergo displacements of 25–37 mi (40–60 km) from their average or predicted position due to **magnetic storms** or other disturbances of the **ionosphere** and/or Earth's magnetic field. Angular corrections for the difference between the geographic poles and their corresponding magnetic pole are expressed as magnetic declination. The values for magnetic declination vary with the observer's position and are entered into navigation calculations to relate magnetic heading to true directional heading.

See also Bowen's reaction series; Cartography; Continental drift theory; Earth, interior structure; Ferromagnetic; GPS; Magnetic field; Magnetism and magnetic properties; Polar axis and tilt

Earth's magnetic fields and the Van Allen radiation belts. *U.S. National Aeronautics and Space Administration (NASA).*

GEOGRAPHY • *see* EARTH SCIENCE

GEOLOGIC MAP

A geologic map shows the types of rocks or loose sediments at or below Earth's surface, along with their distribution. Geologic maps also illustrate the relative ages of, and physical relationships between, Earth's materials. Geologic maps are used for a variety of purposes, including natural resource development, land use planning, and natural hazard studies.

There are three steps to constructing a geologic map. First, the geologist locates natural or man-made exposures of **rock** called outcrops. Second, the geologist records outcrop locations and characteristics on a simple base map. Finally, the geologist prepares a geologic map by interpreting the distribution of and relationships between rock units.

Outcrops provide several kinds of data that are critical to map construction. The geologist records the rock type, for example, **sandstone** or **granite**, and a detailed description of its specific physical characteristics. **Fossils**, if present, may

allow the rock to be dated fairly accurately. If more than one rock type is present, the nature of the contact between them is important. In addition, the geologist records the shape and spatial orientation of each rock unit.

The accuracy of a geologic map is primarily dependent upon the spacing of outcrops. If outcrops are widely spaced, the geologist must use his or her knowledge of local and regional **geology** to fill in, or interpolate, between the outcrops. In most cases, the best interpretation will be the simplest one that fits all the known data.

The spacing of outcrops is dependent upon several factors. If the terrain is steep or vegetation is sparse, outcrops are better exposed. Rock units that are resistant to **erosion** also form good outcrops. Finally, in flat terrain, resistant beds are better exposed if they are steeply inclined (due to deformation).

The fundamental rock unit for mapping is the formation. A formation is a body of rock consisting of one or more rock types, usually at least 10 ft (3 m) in thickness, that is present over a large geographic **area**. These characteristics allow it to be the basis for geologic mapping. Formations are named for a geographic location, such as a town, near where

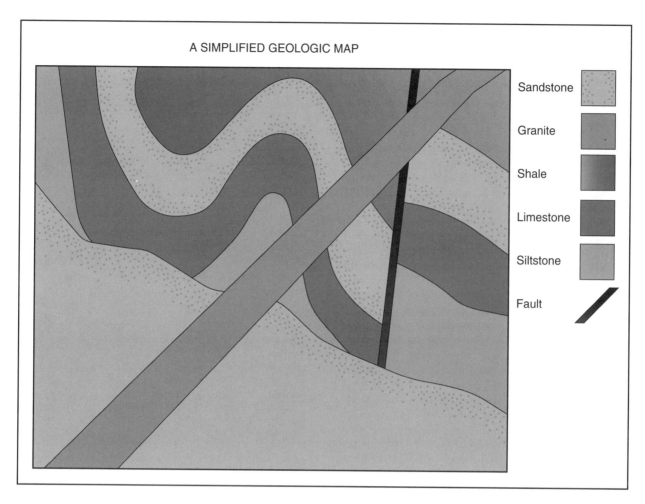

A SIMPLIFIED GEOLOGIC MAP

Sandstone

Granite

Shale

Limestone

Siltstone

Fault

Geologic map.

they were first described. For example, the Miami **Limestone** was named for Miami, Florida.

The distribution of each formation is shown on the map using a separate color, a letter code, or both. A line indicates where the contact is located between adjacent formations. If the contact is exposed at the surface, a solid line is used; if covered, a dashed line is used. Heavier lines indicate the location of faults. With training, one can learn to determine the geologic history of an area from looking at a geologic map.

See also Bathymetric mapping; Cartography

GEOLOGIC TIME

There are several different ways that Earth scientists consider time. Geologic time is generally thought of as the period of time that begins with the initial formative processes of Earth and ends with the onset of recorded human history. The human era begins with the end of geologic time and continues today.

Archeologic time, which begins with early hominid **evolution** and also continues through today, overlaps with geologic time. Encompassing all other measures of time is cosmologic time, which begins with the formation of the Universe and will continue until it ends.

In the early days of the Enlightenment era, geologic and cosmologic time were viewed as one, because there was no perceived difference between the age of the earth and the Universe as a whole. **James Hutton** (1726–1795), a founding father of **geology** during this era, stated his view of geologic/cosmologic time thusly: "We find no vestige of a beginning—no prospect of an end." One of Hutton's contemporaries, John Playfair (1748–1819), is noted for having said of the well-exposed stratigraphic relations at Siccar Point, Scotland: "The mind seemed to grow giddy by looking so far into the abyss of time." Then, as today, vast spans of pre-historic time seemed unimaginable. Now we know that there is a difference between geologic and cosmologic time, and appreciate that there was a beginning for Earth—and for the Universe—and that there is a theoretical end for both.

Notions of archaeologic, geologic, and cosmologic time have been rejected and in some instances suppressed by persons who possessed strong beliefs in the concept of Biblical time. Biblical time was the prevailing view of time espoused as a generally unquestioned belief by Christian people for nearly 2,000 years. In 1650–1654, Archbishop James Ussher (1581–1656) of England attempted to quantify Biblical time by studying Biblical genealogies. Based on his analysis, he reckoned that the Earth was formed on October 23, 4004 B.C. This date was printed after 1658 as a footnote in the *English Bible*, and was accepted as a part of scripture. Until the Enlightenment era (c. 1750–1850), any estimate of Earth antiquity that conflicted with Ussher's age was dismissed as inconsistent with scripture. During and after the Enlightenment (and in some quarters still today), resistance persisted to notions of ages that predate the approximate 6,000-year figure of Ussher.

Enlightenment era notions that geologic time was infinite, indefinite, or inconceivably long, would be replaced in time by finite estimates of ages of geological materials. This occurred not long after the nature of radioactive decay was understood and analytical equipment was developed to a level of precision and accuracy in order to measure minute amounts of radioactive isotopes. The advent of this analytic process lead to radiometric dating of **minerals** within rocks using isotopic ratios. For instance, the American chemist Bertram Boltwood (1870–1927) was obtaining **rock** ages, based on lead isotopes, of 410 to 2.2 billion years in the early years of the twentieth century. In 1931, a carefully prepared and highly reviewed report to the United States National Research Council established the validity of radiometric dating of rocks in a robust way that has not been debated since that time. Not long after radiometric dating of rocks began, the geological time scale, which had been only a relative time scale up to that time, became a geochronometric scale as well (i.e., ages of the geologic eras and periods, and eventually their finer subdivisions, were added and refined). For the first time, it was known that, for example, Cretaceous spanned 146.5 to 65 million years ago.

Geologic time for the geologist is the numerical aspect of change in the history of the earth (i.e., the rock record). This differs from other scientific views of time. For example, for the classical physicist, time is more the numerical aspect of motion, because motion is measured by time. Non-scientists normally do not think of time in either of these ways. To the average person, time is the numerical aspect that allows one to order events within a day, a year, or a lifetime or a way to know when to begin or end a task. Thus, the geologic concept of time is rather foreign to most people. Because of its vast dimensions, geologic (and cosmologic) time has the potential to broaden the perspective of anyone who is willing to consider its implications and compare it with human or historic time. Among the many implications of geologic time is the fact that an unlikely or very rare event will become common. Geologic time is vast enough to accommodate, for example, all stages in organic evolution, mass extinctions and biotic recoveries, crustal plate motions and attendant episodes of crustal deformation, and long-term climatic changes.

See also Chronostratigraphy; Fossil record; Historical geology; Radioactivity; Stratigraphy

Geologists from Hawaiian Volcano Observatory studying the crust of a cooled lava flow from Kilauea, Hawaii, Hawaii. © *Roger Ressmeyer/Corbis. Reproduced by permission.*

GEOLOGY

Geology is the study of the earth. Specifically, geologists may study mountains, valleys, plains, sea floors, **minerals**, rocks, **fossils**, and the processes that create and destroy each of these. Geology consists of two broad categories of study. Physical geology studies Earth's materials (**erosion**, volcanism, sediment deposition, etc.) that create and destroy the materials and **landforms**. **Historical geology** explores the development of life by studying fossils (petrified remains of ancient life) and the changes in land (for example, distribution and **latitude**) via rocks. The two categories overlap in their coverage: for example, to examine a fossil without also examining the **rock** that surrounds it tells only part of the preserved organism's history.

Physical geology further divides into more specific branches, each of which deals with its own part of Earth's materials, landforms, and/or processes. **Mineralogy** and petrology investigate the composition and origin of minerals and rocks, respectively. Sedimentologists look at **sedimentary rocks**, products of the accumulation of rock fragments and other loose Earth materials, to determine how and where they formed. Volcanologists tread on live, dormant, and extinct volcanoes checking **lava**, rocks and gases. Seismologists set up instruments to monitor and to predict earthquakes and **volcanic eruptions**. Structural geologists study the ways rock layers bend and break. **Plate tectonics** unifies most aspects of physical geology by demonstrating how and why plates (sections of Earth's outer **crust**) collide and separate and how that movement influences the entire spectrum of geologic events and products.

Fossils are used in historical geology as evidence of the **evolution** of life on Earth. Plate tectonics adds to the story with details of the changing configuration of the continents and **oceans**. For years paleontologists observed that the older the rock layer, the more primitive the fossil organisms found therein, and from those observations developed evolutionary theory. Fossils not only relate evolution, but also speak of the environment in which the organism lived. Corals in rocks at the top of the Grand **Canyon** in Arizona, for example, show a shallow sea flooded the **area** around 290 million years ago. In addition, by determining the ages and types of rocks around the world, geologists piece together continental and oceanic history over the past few billions of years. For example, by matching fossil and tectonic evidence, geologists reconstructed the history and shape of the 200–300 million year-old supercontinent, Pangaea.

Many other sciences also contribute to geology. The study of the **chemistry** of rocks, minerals, and volcanic gases is known as **geochemistry**. The **physics** of the earth is known as geophysics. Paleobotanists study fossil plants. Paleozoologists reconstruct fossil animals. Paleoclimatologists reconstruct ancient climates.

Much of current geological research focuses on resource utilization. Environmental geologists attempt to minimize human impact on Earth's resources and the impact of natural disasters on human kind. Hydrology and **hydrogeology**, two subdisciplines of environmental geology, deal specifically with **water** resources. Hydrologists study surface water whereas hydrogeologists study ground water. Both disciplines try to minimize the impact of pollution on these resources. Economic geologists focus on finding the minerals and fossil **fuels** (oil, **natural gas**, **coal**) needed to maintain or improve global standards of living. Extraterrestrial geology, a study in its infancy, involves surveying the materials and processes of other planets, trying to unlock the secrets of the universe and even to locate mineral deposits useful to those on Earth.

GEOTHERMAL DEEP-OCEAN VENTS

Geothermal deep-ocean vents are undersea hot **springs** that occur in clusters along the **mid-ocean ridges**. Nutrients and energy supplied by vents support communities of deep-sea organisms found in no other environment.

Most deep-ocean vent action is powered by the heat of the same bodies of **magma** that drive **sea-floor spreading**. A vent forms when seawater seeps downward through cracks in the flanks of a mid-ocean ridge to depths of 1.25–2 mi (2–3 km), about halfway through the thickness of the oceanic **crust**,

Black smoker. *Photograph by P. Rona. OAR/National Undersea Research Program (NURP)/National Oceanic and Atmospheric Administration.*

and is there heated to 750–840°F (400–450°C). This superheated **water** then re-ascends to the center of the mid-ocean ridge and emerges as a fast jet at about 350°C.

As this hot jet mixes with cold (34–36°F [1–2°C]) ocean water, hydrogen sulfide conveyed in solution from the deep rocks precipitates instantly, often coloring the jet black. Such jets are termed "black smokers." Hydrogen sulfide is usually poisonous to life, but the specialized communities around black smokers could not live without it.

A vibrant community of bacteria, tubeworms that are unique to the geothermal vent environment, and other creatures exists around hydrothermal vents. The entire ecosystem is possible because of the activity of the bacteria. These bacteria have been shown, principally through the efforts of the Holger Jannasch (1927–1998) of Woods Hole Oceanographic Institution, to accomplish the conversion of sulfur to energy in a process that does not utilize sunlight called chemosynthesis. The energy is then available for use by the other life forms, which directly utilize the energy, consume the bacteria, or consume the organisms that rely directly on the bacteria for nourishment. For example, the tubeworms have no means with which to take in or process nutrients. Their existence relies entirely on the bacteria that live in their tissues.

Sulfur-oxidizing bacteria pervade the waters around the vents and live symbiotically in the tissues of certain species of animals unique to the vent environment, including 5 ft (2 m) red-blooded worms. These bacteria derive energy by oxidizing the sulfur in hydrogen sulfide, and derive **carbon** from **carbon dioxide** dissolved in the seawater. The bacteria depend directly on the vent and all the other organisms in the vent's vicinity—including over fifty species of clams, mussels, crabs, worms, tube-dwelling worms, and sea anemones not found in any other environment—depend, directly or indirectly, on the bacteria.

Deep-sea vents and their associated fauna were unknown until the late 1970s. They have aroused keen interest for several reasons. First, they demonstrate that life can thrive on energy from purely geothermal sources, isolated from **solar energy**. This suggests the possibility of life in places in the **solar system** where liquid water and heat are available but sunlight is not, such as under the ice-encrusted **oceans** of Jupiter's **moon** Europa. Second, some biologists believe that some vent species are living **fossils** or survivors from earlier periods of **geologic time**. Third, some molecular biologists have theorized that complex chemical reactions in and around ancient deep-sea vents may have synthesized amino acids and other organic molecules key to the spontaneous **origin of life**.

Vents remain an active research topic. In 2001, a new class of deep-sea vents was discovered that derives its energy not from volcanic magma but from a chemical reaction between seawater and the rocks of the upper oceanic crust.

See also Deep sea exploration; Geothermal gradient; Mantle plumes; Ocean trenches

GEOTHERMAL ENERGY

The discovery that the **temperature** in deep mines exceeded the surface temperature implied the existence of a source of deep geothermal energy. Within the continental crusts, the temperature differential gradient averages about one micro calorie per square centimeter (equivalent to an increase of about 95°F per mile or 33°C per kilometer of increasing depth).

In some areas geothermal energy is a viable economic alternative to conventional energy generation. Commercially viable geothermal fields have the same basic structure. The source of heat is generally a magmatic intrusion into Earth's **crust**. The **magma** intrusion generally measures 1110–1650°F (600–900°C), at a depth of 4.3–9.3 mi (7–15 km). The **bedrock** containing the intrusion conducts heat to overlying aquifers (i.e., layers of porous **rock** such as **sandstone** that contain significant amounts of **water**) covered by a dome-shaped layer of impermeable rock such as shale or by an overlying fault thrust that contains the heated water and/or steam. A productive geothermal generally produces about 20 tons (18.1 metric tons) of steam, or several hundred tons of hot water, per hour. Historically, some heavily exploited geothermal fields have had decreasing yields due to a lack of replenishing water in the **aquifer**, rather than to cooling of the bedrock.

There are three general types of geothermal fields: hot water, wet steam, and dry steam. Hot water fields contain reservoirs of water with temperatures between 140–212°F (60–100°C), and are most suitable for space heating and agricultural applications. For hot water fields to be commercially viable, they must contain a large amount of water with a temperature of at least 140°F (60°C) and lie within 2,000 meters of the surface.

Wet steam fields contain water under pressure and usually measure 212°F (100°C). These are the most common commercially exploitable fields. When the water is brought to the surface, some of the water flashes into steam, and the steam may drive turbines that can produce electrical power.

Dry steam fields are geologically similar to wet steam fields, except that superheated steam is extracted from the aquifer. Dry steam fields are relatively uncommon.

Because superheated water explosively transforms into steam when exposed to the atmosphere, it is much safer and generally more economical to use geothermal energy to generate **electricity**, which is much more easily transported. Because of the relatively low temperature of the steam/water, geothermal energy may be converted into electricity with an efficiency of 10–15%, as opposed to 20–25% for **coal** or oil fired generated electricity.

To be commercially viable, geothermal electrical generation plants must be located near a large source of easily accessible geothermal energy. A further complication in the practical utilization of geothermal energy derives from the corrosive properties of most **groundwater** and steam. In fact, prior to 1950, metallurgy was not advanced enough to enable the manufacture of steam turbine blades resistant to **corrosion**. Geothermal energy sources for space heating and agriculture have been used extensively in Iceland, and to some degree Japan, New Zealand, and the former Soviet Union. Other applications include paper manufacturing and water **desalination**.

While geothermal energy is generally presented as non-polluting energy source, water from geothermal fields often contains large amounts of hydrogen sulfide and dissolved **metals**, making its disposal difficult.

See also Earth, interior structure; Energy transformations; Geothermal deep ocean vents; Geothermal gradient; Temperature and temperature scales

GEOTHERMAL GRADIENT

The geothermal gradient is the rate of change of **temperature** (ΔT) with depth (ΔZ), in the earth. Units of measurement are °F/100 ft or °C/km. In the geosciences, the measurement of T is strongly associated with heat flow, Q, by the simple relation: $Q = K\Delta T/\Delta Z$, where K is the thermal conductivity of the **rock**.

Temperatures at the surface of the earth are controlled by the **Sun** and the atmosphere, except for areas such as hot **springs** and **lava** flows. From shallow depths to about 200 ft (61 m) below the surface, the temperature is constant at about 55°F (11°C). In a zone between the near surface and about 400 ft (122 m), the gradient is variable because it is affected by atmospheric changes and circulating ground **water**. Below that zone, temperature almost always increases with depth. However, the rate of increase with depth (geothermal gradient) varies considerably with both tectonic setting and the thermal properties of the rock.

High gradients (up to 11°F/100 ft, or 200°C/km) are observed along the oceanic spreading centers (for example, the Mid-Atlantic Rift) and along **island arcs** (for example, the Aleutian chain). The high rates are due to molten volcanic rock (**magma**) rising to the surface. Low gradients are observed in tectonic subduction zones because of thrusting of cold, water-filled sediments beneath an existing **crust**. The tectonically stable shield areas and sedimentary basins have average gradients that typically vary from 0.82–1.65°F/100 ft (15–30°C/km).

Measurements of thermal gradient data in Japan range widely and over short horizontal distances between to 0.6–4.4°F/100 ft (10–80°C/km). The Japanese Islands are a volcanic island arc that is bordered on the Pacific side by a trench and subduction complex. The distribution of geothermal gradients is consistent with the tectonic settings. In the northeastern part of Japan, the thermal gradient is low on the Pacific side of the arc and high on the back-arc side. The boundary between the outer low thermal gradient and the high thermal gradient regions roughly coincides with the boundary of the volcanic front.

The geothermal gradient is important for the oil, gas, and **geothermal energy** industries. Downhole logging tools must be hardened if they are to function in deep oil and gas wells in areas of high gradient. Calculation of geothermal gradients in the geological past is a critical part of modeling the generation of **hydrocarbons** in sedimentary basins. In Iceland, geothermal energy, the main source of energy, is extracted from those areas with geothermal gradients ≥2.2°F/100 ft (≥40°C/km).

See also Island arcs; Subduction zone; Hydrothermal processes

Gesner, Konrad von (1516-1565)
German physician, zoologist, and naturalist

Konrad von Gesner, a dedicated physician by many accounts, somehow managed to produce approximately 90 manuscripts during his short life span. The topics of his publications were encyclopedic in scope and ranged from zoology to theology, mountains to medicines, and to many other subjects that struck his fancy. Of all his works, the one of most interest to geologists is *Fossils, Gems, and Stones* (the full Latin title of this work is *De Rerum Fossilium, Lapid um et Gemmarum maxime, figuris et similitudinibus Liber: non solum Medicis, sed omnibus rerum Naturae ac Philogiae studiosis, utilis et juncundus futurus*). It was published in the year of his death (1565).

Gesner's *Fossils, Stones, and Gems* is significant primarily for two contributions to the study of **fossils**, **minerals**, rocks, and gems. First, although he did not recognize fossils as the remains of once living things (he labeled them stony con-

cretions), Gesner realized that their unusual appearance deserved recognition. Therefore, he assembled the first extensive collection of fossil illustrations. However, he was not the first to publish fossil illustrations, as some historians have suggested; German naturalist Christophorus Encelius (1517–1583) included illustrations of four fossils in a publication 14 years prior to Gesner's work. Gesner's illustrations went far beyond Encelius' work in scope and even included the four illustrations from Encelius' publication.

Second, like his contemporary, German scientist **Georgius Agricola** (1494–1555), Gesner recognized the inadequacy of past methods of classifying fossils, rocks, minerals, and gems which ranged from alphabetical listings to nonsensical mystical properties. Agricola solved the classification problem by carefully identifying physical and chemical properties of certain minerals, a methodology that remains in effect today. Gesner approached the classification problem from a very different perspective. He constructed a list of 15 classes into which he held that most fossils, rocks, minerals, and gems could be categorized.

To the modern geologist, some of these classes may seem trivial or illogical. In class three, for example, fossil Echinoderms, Neolithic stone axes, and minerals that have a smokey appearance are all grouped together as objects that fell from the sky. This class was derived from the teachings of the Greek philosopher Aristotle (384–322 B.C.) and was believed to be the literal truth. Other classes appear to have little value. Class thirteen bears this out by including only fossils, rocks, minerals, and gems that derive their names from birds. A few of Gesner's classes, however, proved valuable to future fossil research. Despite the fact that neither Gesner nor his contemporaries recognized fossils as the remains of living things, he applied his remarkable knowledge of zoology to a practical classification of many fossils. Class 10 (coral in appearance), Class 11 (coral like sea plants in appearance), Class 14 (appearance of things living in the sea), and Class 15 (appearance of insects and serpents) all make this point.

It was not until 136 years after Gesner's death that English naturalist, **John Ray** (1628–1705), declared that fossils were the remains of ancient life and another hundred years before Ray's views were generally accepted. But it was Gesner's early illustrations and some of his methods of classification that highlighted the remarkable similarities between the **fossil record** and living organisms.

Historians have recorded that Konrad von Gesner refused to leave his patients when the plague struck Zürich, Switzerland in 1565. After contracting the disease himself, he asked to be carried to his study when he felt death was near. It was there, among his voluminous library and eclectic collections, that he died.

See also Fossil record; Fossils and fossilization

Geyser

A geyser is an intermittent spout of geothermally heated **groundwater**. The word *geyser* comes from the name of a single Icelandic geyser, Geysir, written mention of which dates back to A.D. 1294.

Some geysers erupt periodically, others irregularly; a few send jets of **water** and steam hundreds of feet into the air, others only a few feet. There are fewer than 700 geysers in the world, all concentrated in a few dozen fields. More than 60% of the world's geysers are in Yellowstone National Park in the northwestern United States, including the famous geyser, "Old Faithful."

Geysers form only under special conditions. First, a system of underground channels must exist in the form of a vertical neck or series of chambers. The exact arrangement cannot be observed directly, and probably varies from geyser to geyser. This system of channels must vent at the surface. Second, water deep in the system—tens or hundreds of meters underground—must be in contact with or close proximity to **magma**. Third, this water must come in contact with some **rock** rich in silica (**silicon** dioxide, SiO_2), usually **rhyolite**.

Silica dissolves in the hot water and is chemically altered in solution. As this water moves toward the surface, it deposits some of this chemically altered silica on the inner surfaces of the channels through which it flows, coating and sealing them with a form of opal termed *sinter*. Sinter sealing allows water and steam to be forced through the channels at high pressure; otherwise, the pressure would be dissipated through various cracks and side-channels.

The episodic nature of geyser flow also depends on the fact that the boiling point of water is a function of pressure. In a vacuum (zero pressure), liquid water boils at 0°C; under high pressure, water can remain liquid at many hundreds of degrees. Water heated above 100°C but kept liquid by high pressure is said to be *superheated*.

The sequence of events in an erupting geyser follows a repeating sequence. First, groundwater seeps into the geyser's reservoirs (largely emptied by the previous eruption), where it is heated—eventually, superheated—by nearby magma. Steam bubbles then form in the upper part of the system, where the boiling point is lower because the pressure is lower. The steam bubbles eject some water onto the surface and this takes weight off water deeper in the system, rapidly lowering its pressure and therefore its boiling point. Ultimately, the deeper water flashes to steam, forcing a mixed jet of water and steam through the geyser's surface vent.

Many of the world's geysers are endangered by drilling for **geothermal energy** in their vicinity. Drilling draws off water and heat, disrupting the unusual balance of underground conditions that makes a geyser possible.

See also Bedrock; Country rock; Crater, volcanic; Geothermal deep ocean vents; Geothermal gradient; Hotspots; Magma chamber; Pluton and plutonic bodies; Volcanic eruptions; Volcanic vent; Water table

GIS

GIS is the common abbreviation for Geographic Information Systems, a powerful and widely used computer software pro-gram that allows scientists to link geographically referenced information related to any number of variables to a map of a geographical **area**.

GIS programs allow scientists to layer information so that different combinations of data plots can be assigned to the same defined area. GIS also allows scientists to manipulate data plots to predict changes or to interpret the **evolution** of historical data.

GIS maps are able to convey the same information as conventional maps, including the locations of **rivers**, roads, topographical features, and geopolitical information (e.g., location of cites, political boundaries, etc.).

In addition to conventional map features, GIS offers geologists, geographers, and other scholars the opportunity to selectively overlay data tied to geographic position. By overlaying different sets of data, scientists can look for points or patterns of correspondence. For example, rainfall data can be layered over another data layer describing terrain features. Over these layers, another layer data representing **soil** contamination data might be used to identify sources of pollution. In many cases, the identification of data correspondence spurs additional study for potential causal relationships.

GIS software data plots (e.g., sets of data describing roads, elevations, stream beds, etc.) are arranged in layers that be selectively turned on or turned off.

In addition to scientific studies, GIS technology is increasingly used in emergency planning and resource management. When tied in with **GPS** data, GIS provides very accurate mapping. GIS provides, for example, powerful data **correlation** between pollution patterns monitored at specific points and wildlife population changes monitored by GPS tracking tags.

Broad in scope, GIS is becoming more widely used in business and marketing studies.

See also Archeological mapping; Area; Cartography

GLACIAL LANDFORMS

Glacial **landforms** are deposits of sediments produced by the advance and retreat of **glaciers**. As a glacier forms and advances, large amounts of **rock** and **soil** are picked up and incorporated into the base of the **ice**. In alpine glaciers, **erosion** along valley walls may also contribute sediment build-up on the top of the glacier. The continual flow of the glacier carries these materials forward until they reach the end of the glacier, where they are deposited when the toe of the glacier melts away. If the sediments are deposited directly from the ice, they are called **till**. Till consists of a range of unconsolidated and usually unstratified materials in a range of sizes, from **clay** to large boulders. Sediments may also be transported and deposited by glacial melt **water** and are termed glaciofluvial deposits.

A distinct deposit of till is called a moraine. Because till is deposited more or less as it is turned out by a glacier, the resulting **moraines** tend to be irregularly and randomly shaped mounds and hills. A variety of moraine types is distinguished.

A moraine that is deposited at the toe of a glacier is an end moraine. Till deposited along the edge of a glacier is a lateral moraine. The furthest extent of a glacier before it retreats is marked by a specific type of end moraine called a terminal moraine. Intermittent halts in glacial retreat may produce recessional moraines.

If a portion of a glacier melts in place, leaving all the accumulated sediment in place, it produces a till plain. Large isolated boulders that have been transported and deposited by glaciers are called erratics.

A glacier can also produce more regularly shaped landforms as well. Eskers are snakelike deposits of **sand** and gravel deposited in ice tunnels at the base of a glacier formed by flowing streams. Because the sediments in an esker were deposited by running water, they tend to be better sorted than till and can be layered. A kettle is a circular depression, often filled by a lake, that is produced when a large block of ice is detached from a glacier and is subsequently surrounded by till. The ice eventually melts, leaving the depression or lake in its place. Drumlins are elongated, asymmetric mounds of till that resemble a teardrop. They are deposited by a retreating continental glacier with the tapered end indicating the direction of retreat. Kames are conical mounds of sediment deposited where a stream exits a glacier. Outwash plains are large areas of well-sorted glaciofluvial sand deposited beyond the end of a glacier by numerous converging streams of meltwater.

See also Glaciation; Ice ages

GLACIATION

Glaciation is an extended period of time during which **glaciers** are present and active. It also refers to all the processes that form glaciers and that are at work within a glacier. A glacier is a land-based mass of highly compacted **ice** that moves downward and outward under its own weight due to **gravity**. Glaciers may be large enough to cover a continent or small enough to fill a mountain valley and periods of glaciation can last hundreds, thousands, or millions of years.

A glacier is formed by a series of processes that begins with accumulation. Accumulation occurs when the buildup of snow and ice through snowfall, avalanching, or **wind** transport during cold months greatly exceeds the loss through **melting** or sublimation (the direct conversion of a solid to a gas) in warmer months. As snow accumulates and deepens, its weight causes increased pressure that converts snowflakes first into granular snow and eventually into dense ice granules. Continual and sustained accumulation, compaction, melting, and refreezing eventually create a very dense mass of interlocking ice with about 10% void space. As a comparison, freshly fallen snow contains about 80% void space. The formation of new glacial ice takes several decades.

A mass of accumulated snow and ice is not strictly considered a glacier until it begins to move downhill. Glacial ice will begin to move when it becomes too thick and heavy to hold its position against gravity. The instant that glacial ice will begin to move depends on the steepness of the ground, the ice thickness, and the ice **temperature**. In mountainous regions a glacier will start to **creep** when it reaches a thickness of 65.5–131 ft (20–40 m). Glaciers usually move very slowly, less than one meter a day, but can move up to 164 ft (50 m) a day.

Although a glacier is constantly creeping downhill, the front edge of it may appear to remain in the same place, or even retreat uphill. This is because at the same time a glacier is moving, it may be growing or shrinking. A glacier fluctuates depending on the rate of the accumulation of new ice versus the rate of ablation, or the loss of ice due to melting, sublimation, and wind **erosion**. A glacier will appear to advance if accumulation exceeds ablation and it will recede if ablation is greater than accumulation. If the two are equal, the glacier will appear to remain stationary. When a glacier advances far enough to reach a body of **water**, large chunks may break off and fall in into the water. This is called calving and is the source of **icebergs**.

Throughout **geologic time**, Earth has experienced major periods of glaciation when glaciers covered large portions of Earth's surface for up to many millions of years. In the last 500,000 years, four major periods of glaciation have occurred. During these times, ice sheets several kilometers thick covered as much as 30% of the global land surface. This type of glaciation, which extends over vast areas of lowlands and mountains, is known as continental glaciation. The most recent major glaciation ended about 10,000 years ago.

Today, continental glaciation occurs only at the polar regions, mostly in Greenland and **Antarctica**. Active glaciation in other parts of the world exists in mountainous regions at high altitudes and is called alpine glaciation. The causes of glaciation are not completely understood, but major periods of glaciation are attributed to decreases in the amount of sunlight the earth receives due to very long-term cyclic variations in Earth's **rotation** and angle of orbit.

See also Glacial Landforms

GLACIERS

Glaciers are large land-bound bodies of **ice**. To be called a glacier, the ice mass must be moving, or show evidence of having moved in the past. Covering about 10% of Earth's surface, glaciers store a significant amount of Earth's supply of **freshwater**.

Glaciers form from the buildup of snow over time. As snow accumulates, it is compressed under its own weight. Compaction, along with partial thawing and refreezing, converts the original snow to a type of granular ice called firn. As snow continues to accumulate, the firn is buried and further compacted and is eventually converted into glacial ice. Sufficient accumulation of snow and ice is critical to the ability of glacial ice to deform and flow. Pressure from above allows the solid ice to flow at depth.

Glacial ice flows outward from the center of the accumulation and/or downhill under the force of **gravity**. Plastic flow, the chief mechanism of glacial movement, occurs when individual ice **crystals** within the center of the mass move very small distances. The cumulative motion of a large number of

Glacier, Muir Bay Inlet, Glacier Bay National Park, Alaska. *JLM Visuals. Reproduced by permission.*

ice crystals results in movement of the glacier as a whole. Basal slip is another important mechanism of glacial flow. It occurs when the glacier slides along its base, usually aided by the presence of meltwater between the ice and the land surface. Glacial movement is generally so slow as to be imperceptible to the human eye. Rates of movement typically range from a few millimeters to a few meters per day. Occasionally, however, a glacier may surge, moving tens of meters per day for a period of months to years, before slowing down again.

The two main types of glaciers, classified based on size and location, are continental and alpine glaciers. A continental glacier, as the name suggests, is one that covers a large portion of a continent. These glaciers, also known as ice sheets, flow out from one or more centers of accumulation. Such zones of accumulation are typically more than 1.9 mi (3 km) thick, and continental glaciers cover most of the region's topographical features except for the highest mountain peaks. Continental glaciers are presently found only in Greenland and **Antarctica**. Over 695,000 mi^2 (1,800,000 km^2) of Greenland is hidden beneath its ice sheet, while continental glaciers in Antarctica cover 4,885,350 mi^2 (12,653,000 km^2). During the Pleistocene, continental glaciers were widespread, and massive ice sheets covered much of **North America**. Depositional and erosional evidence of their existence can be found throughout Canada and the northern portion of the United States. Smaller versions of continental glaciers, known as ice caps, cover less than 19,300 mi^2 (50,000 km^2). The small continental glacier covering much of Iceland is an example of an ice cap. Ice caps are also found in the polar regions of Canada, including Baffin Island.

Alpine glaciers are those that occur in mountainous regions. Alpine glaciers are much smaller than continental glaciers, and flow from areas of higher elevation to lower areas. Large alpine glaciers in North America occur in Alaska and the Rocky Mountains. As alpine glaciers flow down from mountaintops, they are generally confined to a valley or system of valleys. These valley glaciers are analogous to streams, and often have smaller tributary glaciers that flow into them. Valley glaciers leave behind clear evidence of their existence.

River **erosion** typically produces a valley that is narrow at the bottom, with a V-shaped cross section. Wide, flat-floored valleys with a U-shaped cross section indicate the previous existence of a valley glacier. Matanuska Glacier, located along the highway 90 mi (145 km) northeast of Anchorage, Alaska, is a famous example of a valley glacier.

Glaciers are responsible for reshaping the land by eroding and transporting a great deal of material. Glaciers erode material by plucking and abrasion. Plucking refers to the removal of **rock** by the advancing ice. Material may be plucked from beneath or along the sides of a glacier, and then picked up and transported by the moving ice mass. Abrasion differs from plucking in that it doesn't involve removing large particles of rock; rather it refers to grinding or filing processes. Highly polished rock surfaces are formed by abrasion. Where large rock fragments are embedded into the bottom of the ice, they may grind into the underlying **bedrock** forming glacial striations, or scratches. The Pleistocene glaciers formed in Canada and the northern United States have left behind large areas of polished and striated bedrock, exposed when glaciers scraped off the surface sediment. Other erosional features of glaciers include hanging valleys, cirques, arêtes, and horns.

Glaciers eventually deposit the material they accumulate. Material deposited by glaciers includes glacial erratics, large boulders transported by ice and deposited far from their source; glacial **till**, unsorted deposits of sediment and rocks; and stratified drift, layered, sorted deposits that originated from glacial streams. Extensive deposits of sediment from glacial streams are known as outwash plains. Depositional **landforms** created by glaciers include **moraines**, drumlins, kanes, and eskers.

See also Glacial landforms; Glaciation; Polar Ice

GLASS

Although a glass is a substance that is non-crystalline, it is almost completely undeformable and therefore brittle. A glass exists in a state of matter termed a vitreous state. Vitreous substances, when heated, will transform slowly through stages of decreasing viscosity. As a sample of glass is heated, it becomes increasingly deformable, eventually reaching a point where it resembles a very viscous liquid. **Ice**, on the other hand, does not go through these changes as it is heated.

Excepting sublimation (direct solid to gas transformations) most substances change directly from a solid to a liquid. Ice, therefore, is not a vitreous substance. Glasses are only very slightly deformable. Glasses tend to bend and elongate under their own weight, especially when formed into rods, plates, or sheets. Glasses can be either organic or inorganic materials.

Because solidification is the act of crystallization, the depiction of glass as a non-crystalline solid may not be entirely correct. However, true crystallization occurs when the molecules of a substance arrange themselves in a systematic, periodic fashion. The atoms or molecules of glass do not exhibit this periodicity; this is consistent with the depiction of glass as an extremely viscous, or "supercooled" liquid.

Glass is often referred to as an **amorphous** solid. An amorphous solid has a definite shape without the geometric regularity of crystalline solids. Glass can be molded into any shape. If glass is shattered, the resulting pieces are irregularly shaped. A crystalline solid would exhibit regular geometrical shapes when shattered. Amorphous solids tend to hold their shape, but they also tend to flow very slowly. If left undisturbed for a long period of time, a glass will very slowly crystallize. Once it crystallizes, it is no longer considered to be glass. At this point, it has devitrified. This crystallization process is extremely slow and in many cases may never occur.

The chemical make-up of standard window glass, which will be described in greater detail below, is quite similar to the mineral **quartz**. An x-ray crystallographic picture of quartz would show atoms arranged in an orderly, periodic sequence. X-ray crystallography studies of glass show no such arrangement. The atoms in glass are disordered and show no periodic structure. This irregular arrangement of atoms not only defines a substance as glass, but also determines several of its properties.

The bonds between the molecules or ions in a glass are of varying length, which is why they show no symmetry or periodic structure. Because the bonds are not symmetrical, glass is isotropic and has no definite **melting** point. The melting of glass instead takes place over a wide **temperature** range. Changing the state of a substance with asymmetric bonds requires more energy than a crystalline structure would. The tendency of glass to devitrify is a result of the atoms moving from a higher to a lower energy state.

The most common glasses are **silicon** based. Most glasses are 75% silicate. These glasses are based on the SiO_2 molecule. This molecule creates an asymmetric, aperiodic structure. Some of the **oxygen** atoms are not bridged together, creating ions that need to be neutralized by metal cations. These metal cations are randomly scattered throughout the glass structure, adding to the asymmetry. The oxides of elements other than silicon can also form glasses. These other oxides include Al_2O_3, B_2O_3, P_2O_5, and As_2O_5.

The production of glasses is a complicated process. In general, certain molten materials are cooled in a specific manner so that no crystallization occurs, i.e., they remain amorphous. There are four basic materials that are used in glass production. These materials are the glass-forming substances, fluxes, stabilizers, and secondary components.

A glass-forming substance is any mineral that remains vitreous when cooled. Glass-forming substances are usually silica, boric oxide, phosphorous pentoxide, or feldspars. Sometimes **aluminum** oxide (Al_2O_3) is used. Silica, as the most commonly used material to make glass, is usually obtained from **sand**, which is 99.1-99.7% SiO_2. Occasionally, natural silica deposits are discovered that are pure enough to use in glass manufacturing, but these deposits are rare and the silica found in them is usually expensive to obtain. Even the lowest quality sands can be purified rather economically. Impurities in the natural silica are important because they can dramatically alter the quality of the glass produced. The most common impurities found in natural silica are **iron** sesquioxide

(Fe_2O_3), alumina (Al_2O_3), and calcium compounds. Ferric oxide is sometimes found as an impurity. Even if the amount of ferric oxide in a natural silica sample is only 0.1% of the sample, the glass produced would have a deep yellow-green color and the impurity would have detrimental effects on the thermal and mechanical properties of the glass.

Occasionally impurities are added to the glass-forming substances to give the glass certain qualities such as transparency, fusibility, or stability. Stabilizers also are used to give the finished product particular characteristics. For example, calcium carbonate can be added as a stabilizer that will make the glass produced insoluble in **water**. **Lead** oxide added as a stabilizer gives the glass extreme transparency, brightness, and a high refractive index. Lead oxide also makes glass easier to cut. Zinc oxide can be added to glass to make it more resistant to changes in temperature as well as to increase its refractive index (a measure of the ability to bend light). Aluminum oxide can also be added as a stabilizer to increase the physical strength of the glass. Secondary components are added to determine some of the final properties of the glass and to correct any defects in the glass. The secondary components can be classified as decolorants, opacifiers, colorants, or refiners.

The production of glass includes many steps that can be generalized as follows. First, the fluxes, glass-forming substances, and stabilizers are crushed and milled, then blended and mixed together. They are then re-milled and granulated. At this point, the secondary components are added, if needed. The granules are then fused, refined, homogenized, and corrected, using more secondary components if necessary. Finally, the glass is formed and finished.

The final product is one of many hundreds of different types of glass. One popular type of glass, especially in laboratory settings or for use in the kitchen, is borosilicate glass. Some well-known borosilicate glasses are Jena, Pyrex, Durax, and Thermoglass. These glasses contain 12% or more B_2O_3. The addition of the boron oxide increases the softening temperature of the glass, making it more resistant to high temperatures such as those experienced while cooking or while performing laboratory experiments. Borosilicate glasses are also used in the production of thermometers, television tubes, and other objects that need to have constant dimensions or a high softening point.

See also Chemical bonds and physical properties

GLENN JR., JOHN H. (1921-)
American astronaut and senator

John H. Glenn Jr. was the first American to orbit the earth. In the wake of this 1962 feat, Glenn became a national hero on the order of Trans-Atlantic aviator Charles A. Lindbergh—a status that helped carry him to a second career in the United States Senate. As a 77-year-old, he made history again when he became the oldest American to travel in **space** on Oct. 29, 1998, aboard the **space shuttle** *Discovery*. His mission was a series of experiments on aging.

John Glenn. *U.S. National Aeronautics and Space Administration (NASA).*

John Herschel Glenn Jr. was born in Cambridge, Ohio, and grew up in nearby New Concord. He was the son of plumber John Herschel Glenn and Clara Sproat. Glenn credits his parents with instilling his deep-rooted Presbyterian faith and the accompanying philosophy that everyone is given certain talents and a duty to use them to the fullest. In high school Glenn was a diligent student who earned top grades. He worked hard athletically as well, lettering in three sports. After high school Glenn entered Muskingum College in New Concord, majoring in **chemistry**. His high school sweetheart, Anna Castor, enrolled as well.

After two and a half years of study, Glenn entered a local civilian pilot training program and learned to fly. He then left college to enter the Naval Aviation Cadet Program. In 1943, he was graduated and commissioned as a lieutenant in the Marine Corps. He married Annie before going on to advanced training and assignment to a combat unit, and the couple eventually had children. Glenn flew F4U Corsair fighter-bombers on 59 missions in the Pacific theater during World War II.

When peace came, Glenn remained in the corps, serving as a fighter pilot and then as a flight instructor. In 1952, Major Glenn was sent to Korea. He flew primarily ground-attack missions in that war as well, repeatedly returning in aircraft riddled with bullet and shrapnel holes. Through an interservice exchange program, Glenn transferred to an Air Force squadron

just before the end of the war. Flying the F–86 Sabre, Glenn downed three North Korean MiG fighters in nine days.

Following Korea, Glenn attended the Naval Test Pilot School, part of the Naval Test Center in Patuxent River, Maryland. After graduating as a test pilot, he spent two years as a project officer evaluating new aircraft. Glenn moved on to the Navy Bureau of Aeronautics in Washington, D.C., where he continued to oversee development of new fighters. These included the F8U Crusader, a plane Glenn made famous in 1957. In Project Bullet, a test Glenn conceived himself, he flew a Crusader coast to coast, making the first transcontinental supersonic flight in a record time of three hours and twenty-three minutes.

When Glenn learned of the upcoming astronaut program, he was captivated by the challenge of spaceflight. He immediately began to strengthen his qualifications, improving his physical condition, volunteering for centrifuge tests and other research projects, and pursuing courses at the University of Maryland. (Glenn did not actually receive a college degree until after he had flown in space, when Muskingum College awarded him a bachelor's degree in mathematics.) In April 1959, the newly promoted Lieutenant Colonel Glenn was selected as one of America's seven Mercury astronauts.

Glenn helped design the cockpit layout and instrumentation of the *Mercury* capsule. He became the unofficial spokesperson for the astronaut team, and it was a surprise to the country and to Glenn when fellow astronaut **Alan B. Shepard**, Jr., a lieutenant commander with the U.S. Navy, was chosen to make the first U.S. spaceflight. Shepard and then Gus Grissom, an Air Force captain and astronaut, made suborbital flights, in which the *Mercury* craft was launched by a Redstone rocket. These efforts were eclipsed in the popular imagination by the Soviet Union's successful orbital manned flights, and the pressure was on the National Aeronautics and Space Administration (NASA) to match the Russian feat as soon as possible. Glenn was chosen to make the first orbital effort, officially known as *Mercury-Atlas 6*.

After several frustrating postponements caused by unsuitable **weather** and technical glitches, Glenn's capsule, *Friendship 7*, roared into orbit on February 20, 1962. The astronaut fed ground controllers a constant stream of observations and physiological reports, performing experiments such as pulling on an elastic cord to determine the effects of physical work in weightlessness. Tremendous publicity surrounded the flight, in contrast to the secretive Russian launches. Not publicized at the time, however, was a telemetry signal's indication that Glenn's heat shield, vital for safe reentry to the earth's atmosphere, might not be secured to the capsule. Glenn was directed to change the original plan of jettisoning his retro-rocket package after it had been used to slow the capsule; instead it would be kept in place, strapped over the heat shield, to keep the shield from coming loose.

Glenn was briefed on the problem. (He later argued that NASA policy should be to notify airborne astronauts as soon as any abnormality is detected.) Glenn left the retropack on and took manual control of his craft, guiding the capsule to a perfectly safe reentry after three orbits of the earth. It was later determined that the telemetry signal was false, but the incident solidified Glenn's view that spacecraft needed humans aboard who could respond to the unexpected.

Glenn was bathed in national attention. President John F. Kennedy awarded him the NASA Distinguished Service Medal. He was invited to address a joint session of Congress, an honor normally reserved for top officials and visiting heads of state. Glenn told the assembly that the real benefits of space exploration were "probably not even known to man today. But exploration and the pursuit of knowledge have always paid dividends in the long run—usually far greater than anything expected at the outset."

Glenn received hundreds of thousands of letters, some of which he collected in a book, *Letters to John Glenn*. Glenn also became friends with President Kennedy and his brother, U.S. Attorney General Robert Kennedy. The president urged Glenn to enter politics and, unknown to the astronaut, directed that Glenn's life not be risked by another spaceflight. Glenn worked on the preliminary designs for Project Apollo, which had the goal of putting a man on the **Moon**, then left NASA and applied for military retirement to enter the Ohio Senate race in 1964. He withdrew from that contest after suffering a serious head injury in a bathroom fall.

Colonel Glenn retired from the Marines on January 1, 1965, with six Distinguished Flying Crosses and eighteen Air Medals, among other decorations. He had logged over 5,400 hours of flying time. Glenn's space exploit also garnered him numerous civilian honors, including induction into the Aviation Hall of Fame and the National Space Hall of Fame, and, in 1978, the award of the Congressional Space Medal of Honor. He was granted honorary doctorates in engineering by four universities.

After retiring from the military, Glenn went into business, first with the Royal Crown cola company and later with a management group that operated Holiday Inn hotels. His business ventures made Glenn a millionaire, but his political dreams remained foremost in his mind. In 1970, Glenn again declared his candidacy for the U.S. Senate. He narrowly lost in the Democratic primary to Howard Metzenbaum, who outspent and out-organized his less experienced rival. When another Senate seat opened in 1974, Glenn started earlier, ran harder, and won the election.

Despite being new to Washington politics, Glenn gained a reputation for hard work and effective legislating. His voting record marked him as generally liberal on both domestic and foreign policy. Glenn was considered for the vice presidency by presidential nominee Jimmy Carter, but Walter Mondale, a Minnesota senator, was chosen instead. In the Senate, Glenn became best known for his work against nuclear proliferation. He was willing to oppose President Carter on some issues, most notably the second Strategic Arms Limitation Talks (SALT II) arms accord, which Glenn considered unverifiable. He sought the Democratic presidential nomination in 1984; once again, however, Mondale grasped the prize Glenn sought, eliminating Glenn before the party's convention with a better-run campaign. Glenn then served as chair of the Senate Governmental Affairs Committee from 1987 to 1995.

Glenn continued serving in the Senate, where he supported increased funding for education, space exploration, and basic scientific research. He was a strong advocate of a permanent research station in space. Outside the Senate, Glenn served on the National Space Society's Board of Governors, on Ohio's Democratic Party State Executive Committee, and as a Presbyterian elder, among many other commitments.

But not satisfied with his political career and longing to return to space, Glenn asked NASA if he could fly again, but only if he met the agency's physical and mental requirements. On January 16, 1998, NASA announced that Glenn, who had made history 36 years before as the first American to orbit the earth, would fly in space again as a payload specialist on shuttle mission STS-95 October 29. Glenn took part in experiments to study the connection between the adaptation to weightlessness and the aging process. The highly successful mission concluded with the landing on Saturday, November 7, 1998. While he tolerated space flight surprisingly well, Glenn confirmed it was unlikely that he would ever fly in space again.

See also Spacecraft, manned

GLOBAL WARMING

Global warming, as used in the popular context, is a scientifically controversial phenomenon that attributes an increase in the average annual surface **temperature** of Earth to increased atmospheric concentrations of **carbon dioxide** and other gases. Global warming describes only one of several components involved in **climate** change and specifically refers to a warming of Earth's surface outside of the range of normal fluctuations that have occurred throughout Earth's history.

Climate describes the long-term meteorological conditions or average **weather** for a region. Throughout Earth's history there have been dramatic and cyclic changes in climatic weather patterns corresponding to cycles of glacial advance and retreat that occur on the scale of 100,000 years. Within these larger cycles are shorter duration warming and cooling trends that last from 20,000 to 40,000 years. Scientists estimate that approximately 10,000 years have elapsed since the end of the last **ice** age, and examination of physical and biological processes establishes that since the end of the last ice age there have been fluctuating periods of global warming and cooling.

Measurements made of **weather and climate** trends during the last decades of the twentieth century raised concern that global temperatures are rising not in response to natural cyclic fluctuations, but rather in response to increasing concentrations of atmospheric gases that are critical to the natural and life-enabling **greenhouse effect** (infrared re-radiation, mostly from **water** vapor and **clouds**, that warms the earth's surface).

Observations collected over the last century indicate that the average land surface temperature increased by 0.8–1.0°F (0.45–0.6°C). The effects of temperature increase, however, cannot be fully isolated and many meteorological models suggest that such increases temperatures also result in increased **precipitation** and rising sea levels.

Measurements and estimates of global precipitation indicate that precipitation over the world's landmasses has increased by approximately 1% during the twentieth century. Further, as predicted by many global warming models, the increases in precipitation were not uniform. High **latitude** regions tended to experience greater increases in precipitation while precipitation declined in tropical areas.

Measurements and estimates of sea level show increases of 6–8 in (15–20 cm) during the twentieth century. Geologists and meteorologists estimate that approximately 25% of the sea level rise resulted from the **melting** of mountain **glaciers**. The remainder of the rise can be accounted for by the expansion of ocean water in response to higher atmospheric temperatures.

Many scientists express concern that the measured increases in global temperature are not natural cyclic fluctuations, but rather reflect human alteration of the natural phenomena known as the greenhouse effect by increasing concentrations of greenhouse-related atmospheric gases. Estimates of atmospheric **greenhouse gases** prior to the nineteenth century (extrapolated from measurements involving ice cores) indicate that of the last few million years the concentration of greenhouse gases remained relatively unchanged prior to the European and American industrial revolutions. During the last two centuries, however, increased emissions from internal combustion engines and the use of certain chemicals have measurably increased concentrations of greenhouse gases that might result in an abnormal amount of global warming.

Although most greenhouse gases occur naturally, the **evolution** of an industrial civilization has significantly increased levels of these naturally occurring gases. In addition, new gases have been put into the atmosphere that potentiate (i.e., increase) the greenhouse effect. Important greenhouse gases in the modern Earth atmosphere include water vapor and **carbon** dioxide, methane, nitrous oxides, **ozone**, halogens (bromine, chlorine, and fluorine), halocarbons, and other trace gases.

The sources of the greenhouse gases are both natural and man-made. For example, ozone is a naturally occurring greenhouse gas found in the atmosphere. Ozone is constantly produced and broken down in natural atmospheric processes. In contrast, halocarbons enter the atmosphere primarily as the result of human use of products such as chlorofluorocarbons (CFCs). Water vapor and carbon dioxide are natural components of respiration, transpiration, **evaporation** and decay processes. Carbon dioxide is also a by-product of combustion. Although occurring at lower levels than water vapor or carbon dioxide, methane is also a potent greenhouse gas. Nitrous oxides, enhanced by the use of nitrogen fertilizers, nylon production, and the combustion of organic material, including fossil **fuels** have also been identified as contributing to strong greenhouse effects.

Alterations in the concentrations of greenhouse gases results in a disruption of equilibrium processes. Both increased formation and retardation of destruction cause compensatory mechanisms to fail and result in an increased or potentiated greenhouse effect. For example, the amount of water vapor released through evaporation increases directly with increases in the surface temperature of Earth. Within normal limits, increased levels of water vapor are usually con-

trolled by increased warming and precipitation. Likewise, within normal limits, concentrations of carbon dioxide and methane are usually maintained with specified limits by a variety of physical and chemical processes.

Measurements made late in the twentieth century showed that since 1800, methane concentrations have doubled and carbon dioxide concentrations measured at he highest values estimated to have existed during the last 160,000 years. In fact, increases in carbon dioxide over the last 200 years were exponential up until 1973 (the rate of increase has since slowed).

Although the effects of these increases in global greenhouse gases are debated among scientists, the correlation of the increased levels of greenhouse gases with a measured increase in global temperature during the twentieth century, have strengthened the arguments of models that predict pronounced global warming over the next few centuries. In the alternative, some scientists remain skeptical because the earth has not actually responded to the same extent as predicted by these models. For example, where many models based upon the rate of change of greenhouse gases predicted a global warming of .8°F to 2.5°F (0.44°C to 1.39°C) over the last century, the actual measured increase is significantly less with a mean increase generally measured at .9°F (.5°C) and that this amount of global warming is within the natural variation of global temperatures.

One problem in reaching a scientific consensus regarding global warming is that the data used in many models is neither global nor a result of high-reliance systematic scientific measurement (i.e., that it generally neglects **oceans** and vast uninhabited areas). Other problems involve forming an accurate articulation of the interplay of global surface warming phenomena that include thermal conduction, greenhouse radiation, and convective currents. Most scientists agree, however, that an enhanced greenhouse effect will result in some degree of global warming.

See also Acid rain; Atmospheric pollution

GNEISS

Gneiss (pronounced "nice") is a **metamorphic rock** consisting mostly of **quartz** and **feldspar** and showing distinct layering or banding. The layering of a gneiss may be weak or well-developed and consists of varying concentrations of biotite, garnet, hornblende, mica, and other **minerals**. These structures do not record a layered deposition process but arise from preferential recrystallization along flow or stress lines during metamorphism of the parent **rock** (protolith).

The gneisses are a very varied group, including both **igneous rocks** and metamorphosed **sedimentary rocks**, and may be categorized as quartzofeldspathic, pelitic, calcarous, or hornblende gneiss.

Quartzofeldspathic gneiss forms by metamorphosis of either **silicic** igneous rocks such as **granite**, **rhyolite**, and rhyolitic tuff—or silicic sedimentary rocks such as **sandstone**. Quartzofeldspathic gneiss containing eye-shaped feldspar **crystals** is termed augen gneiss after the German *augen* (eyes).

Pelitic gneiss is formed by metamorphosis of clay-textured sedimentary rocks, particularly those rich in **iron**.

Calcareous gneiss contains calcite ($CaCO_3$). It is formed by metamorphosis of limestones and dolomites containing large fractions of **sand** and **clay**. Calcareous gneisses with large fractions of calcite blur conceptually with the marbles (metamorphosed limestones).

Hornblende gneiss contains a large fraction of hornblende in addition to its quartz and feldspar.

The gneisses can be alternatively categorized simply as orthogneisses and paragneisses. The former are metamorphosed from igneous protoliths and the latter from sedimentary protoliths.

The gneisses and schists are closely related. Both are metamorphosed igneous or sedimentary rocks showing foliation or layering. The difference is primarily one of degree; schists are less coarsely crystallized and more prone to cleave into flakes or slabs. Gneisses represent a higher grade of metamorphosis—more thorough melting—and are distinguished by their coarser texture and their resistance to cleavage.

See also Migmatite

GODDARD, ROBERT H. (1882-1945)

American physicist

Robert H. Goddard was foremost among the first generation of rocket and **space** pioneers. Goddard not only contributed to space flight theory, but also engaged over most of his adult life in the actual development of rockets. As a result, he is credited with launching the world's first liquid-propellant rocket. He developed and patented a large number of innovations in rocket technology that were later used in the much larger rockets and missiles employed by the Germans during World War II and, thereafter, by the United States' and Soviet Union's missile and space programs, among others. Paradoxically, Goddard's influence upon modern rocketry was not as great as it would have been had he been less a solitary inventor and more inclined to publish his findings in scientific journals and elsewhere.

Robert Hutchings Goddard was born in Worcester, Massachusetts, to Nahum Danford Goddard, himself something of an inventor, and Fannie Louise Hoyt Goddard, the daughter of a machine knife manufacturer for whom her husband worked at the time of their marriage. Of modest means but old New England stock, Goddard's parents had a second son who died in infancy. Goddard himself was prone to illness and fell behind in school, compensating with self-education. Encouraged by his father in his early inclinations towards experimentation and invention, Goddard also heeded his father's advice to mind his own business and work for himself rather than someone else. Science fiction proved another early influence upon him, one that apparently led to a transforming experience he had in a cherry tree on October 19, 1899, when he imagined a device that might ascend to Mars. As he stated in an autobiographical memoir, the experience suddenly made life seem purposeful to him. Throughout the rest of his life, he

Robert Goddard. *U.S. National Aeronautics and Space Administration (NASA).*

recorded the date in his diary as "anniversary day," and he revisited the tree on that date whenever he was in Worcester.

Goddard received his early education in the Boston **area**, where his father had been working, and had not done well in algebra during his first year in high school. When the family moved back to Worcester in 1898, after his mother was diagnosed with tuberculosis, his experience in the cherry tree compelled him to excel in math and **physics** at South High School. Because of his own illnesses, Goddard did not graduate from South High until 1904, when he was 21. He went on to earn a bachelor's degree in general science, with a concentration in physics, from Worcester Polytechnic Institute in 1908, and a master's degree from Clark University in 1910. By 1909, Goddard had already begun teaching physics at Worcester Polytechnic and shortly after receiving his doctorate from Clark in 1911, he became an honorary fellow in physics there. Working as a research instructor in physics at Princeton University, Goddard fell dangerously ill in 1913 and, like his mother, was diagnosed with tuberculosis. Initially given only two weeks to live, he recovered sufficiently the following year to become a physics instructor at Clark, where he was promoted to assistant professor in 1915. Goddard would remain at Clark throughout much of his academic career, allowing for leaves of absence to pursue rocket research. Goddard eventually became head of Clark's physics department and director of the physical laboratories, obtaining the rank of full professor in 1934. In 1924, Goddard married Esther Christine Kisk, the secretary to the president of Clark. Although the couple had no children, they became devoted to one another and to Goddard's rocket research, in which Esther became very much a partner.

Goddard apparently did not begin serious work on rocket development until early 1909, while a graduate student at Clark. He had, by 1914, obtained a patent for a two-stage powder rocket, followed by patents for a cartridge-loading rocket and a rocket that burned a mixture of gasoline and liquid nitrous oxide. While he was aware of the greater efficiencies of liquid propellants, Goddard found them hard to obtain, preferring instead, smokeless powder, which offered fewer experimental difficulties. Using a steel combustion chamber and a sleeker exhaust nozzle, named for Swedish engineer Carl de Laval, Goddard was able to achieve higher rates of energy efficiency and exhaust velocities than previous rockets had exhibited. He also developed a device that allowed him to fire a rocket in a vacuum, showing that it could operate in the upper atmosphere where air density was small and also demonstrating that it did not require a reaction against the air, as many knowledgeable people at the time supposed.

Until 1916, Goddard had conducted these experiments using the meager funds and facilities provided by Clark, as well as money from his own pocket. No longer able to support the research required to advance his theories, Goddard applied for funding to the Aero Club of America and the Smithsonian Institution. After several inquiries into his request, Goddard reported to the Smithsonian that he had developed a means of propelling meteorological recording devices to heights previously unattainable by sounding balloons, indicating that altitudes of 100–200 mi (161–322 km) could be reached within a year's time. By January 1917, The Smithsonian had awarded Goddard a grant for $5,000. This proved to be the first of many grants from the Smithsonian, Clark University, the Carnegie Institution of Washington, Daniel Guggenheim, and especially the Guggenheim Foundation.

Before the Smithsonian funds could be put to use, America became embroiled in World War I. Supported by the U.S. Army, Goddard and a number of technicians developed both multiple-charge and single-charge recoilless rockets, the latter serving as a prototype for the bazooka which proved effective against tanks during World War II. While tests proved these weapons successful, the armistice intervened before they could be employed. Once World War I was over, Goddard's department head at Clark prodded him into publishing the results of his solid-propellant rocket researches in a paper entitled "A Method of Reaching Extreme Altitudes," which appeared in the *Smithsonian Miscellaneous Collections*. In it, Goddard not only explained the experiments he had conducted, but laid the foundations for much of the early theory of modern rocketry. While devoted primarily to the solid propellants he had used in his research, the paper did mention the greater efficiencies of propellants such as hydrogen and **oxygen** used in their liquid states. The paper briefly discussed the use of stages (propulsion units coupled together to fire in sequence) in order to reach extreme altitudes, and included numerous calculations of such matters as the reduced resist-

ance a rocket would face as it climbed higher and entered less dense portions of the earth's atmosphere.

The reaction to this paper was shaped by a Smithsonian press release emphasizing a point Goddard had not intended as the focus of the work. It suggested the possibility of using a rocket to send a small quantity of flash powder to the dark side of the **Moon**, where, when ignited, it could be viewed from the earth through telescopes, thereby proving that extreme altitude had been reached. The press played up the idea of a moon rocket, and Goddard was embarrassed by the publicity. His inclination against publicizing his work until rockets were actually capable of reaching such altitudes was reinforced. Nevertheless, he persisted in his rocket development in his native Massachusetts for the next decade. Frustrated at the problems he encountered in using solid propellants, he switched to liquid propellants in 1921, though it was not until March 16, 1926—almost ten years after his initial proposal to the Smithsonian—that he launched the world's first liquid-propellant rocket from a hill in Auburn, Massachusetts. Since this was an important event in the history of rocketry, it is noteworthy that the hill, on his Aunt Effie's farm, had an Indian name meaning "a turning point or place." The small rocket only rose 41 ft (12.5 m)—far short of the altitudes he sought to reach—but it represented a significant beginning to the age of rocket flight, comparable, perhaps, to the Wright brothers' contributions to aviation.

From a number of standpoints, including its **weather** and its population density, Massachusetts was hardly an ideal location for launching noisy, fire-belching rockets. So, when Goddard received a generous $50,000 grant from philanthropist Daniel Guggenheim in mid–1930, he took a two-year leave of absence from Clark University and, with his wife and some technical assistants, rented a farmhouse near Roswell, New Mexico, where he proceeded with his rocket development. Loss of funding after 1932 interrupted his research there, but he returned to Roswell in 1934 to resume his testing. In the process, he invented and patented a large number of innovations, including a gyroscopically-controlled guidance system, and a method for cooling the combustion chamber that used a film of propellant streaming along the sides of the chamber. Parachutes were incorporated for recovery of the rocket and a number of instruments were devised for measuring the rocket's performance. Goddard also searched for ways to make a more lightweight, streamlined rocket casing. But he never succeeded in putting all of these components together to create a vehicle capable of reaching anything close to the 100–200 mi (161–322 km) of altitude he had originally expected to achieve. The greatest height one of his rockets reached was estimated at 8,000–9,000 ft (2,440–2,740 m) in March, 1937.

In 1941, he discontinued his attempt to reach extreme altitudes and began work for the armed forces on defense-related rocket research as he had during World War I. In 1942, he moved his crew of assistants to the Naval Engineering Experimental Station in Annapolis, Maryland, where they worked on developing jet-assisted take-off devices for aircraft, pumps, and a variable-thrust rocket motor that became the basis for the one later used on the Bell X–2 rocket plane, the first aircraft in America to use a throttleable engine. This, like the bazooka, was a very important and tangible result of his research. His many patented inventions were also significant. In June 1960, the Army, Air Force, Navy, and National Aeronautics and Space Administration recognized their importance when they granted Mrs. Esther C. Goddard and the Guggenheim Foundation a settlement of $1,000,000 for the right to use many of Goddard's patents.

Despite his technical achievements, however, Goddard's career remained somewhat flawed by his failure to reach the extreme altitudes he sought, and by his secretive nature and consequent failure to communicate most of the details of his research to other scientists and engineers. In 1936, he did publish another paper entitled "Liquid-propellant Rocket Development." Here, Goddard devoted much more attention to liquid propulsion than he had in 1919, and while he did include pictures of some of his rockets and discussed some of their features, the brevity of his treatment (some seventeen pages in his published papers) made the work of limited utility to other scientists and engineers engaged in rocket development. While some of them were inspired by Goddard's example, for the most part they had to develop their own counterparts to his innovations without the benefit of a detailed knowledge of his pioneering inventions.

Despite this failing, Goddard was a remarkable figure in the history of rocket development. Of the many streets, buildings, and awards named in his honor, perhaps the most significant is NASA's Goddard Space Flight Center, dedicated on March 16, 1961—the 35th anniversary of the first flight of a liquid-propellant rocket. On that occasion, Mrs. Goddard accepted a Congressional Gold Medal presented posthumously to him. A little more than nine years later, Clark University named its new library after Goddard. Since 1958, the National Space Club in Washington, DC, has awarded a Goddard Memorial Trophy for achievement in missiles, rocketry, and space flight. Finally, it might be noted that in 1960, Goddard was the ninth recipient of the Langley Gold Medal, awarded only sparingly since 1910 by the Smithsonian Institution for excellence in aviation.

See also Aerodynamics; Satellite; Spacecraft, manned

GOLDSCHMIDT, VICTOR (1888-1947)
Norwegian geochemist

Victor Goldschmidt, helped lay the foundations for the field of crystal **chemistry**. He was a mineralogist, petrologist, and geochemist who devoted the bulk of his research to the study of the composition of the earth. During his many years as a professor and director of a mineralogical institute in Norway, he also investigated solutions to practical geochemical problems at the request of the Norwegian government.

Victor Moritz Goldschmidt was born on January 27, 1888, in Zürich, Switzerland, to Heinrich Jacob Goldschmidt, a distinguished professor of physical chemistry, and Amelie Kohne. His family left Switzerland in 1900 and moved to Norway, where his father took a post as professor of physical chemists at the University of Christiania (now Oslo). Goldschmidt's family obtained Norwegian citizenship in 1905, the same year he entered the university to study chemistry, **geology**, and **mineralogy**. There he studied under the noted geologist and petrologist Waldemar Brogger, becoming a lecturer in mineralogy and crystallography at the university in 1909.

Goldschmidt obtained his Ph.D. in 1911. His doctoral dissertation on contact metamorphic rocks, which was based on **rock** samples from southern Norway, is considered a classic in the field of **geochemistry**. It served as the starting point for an investigation of the **chemical elements** that Goldschmidt pursued for three decades. In 1914, he became a full professor and director of the University of Christiana's mineralogical institute. In 1917, the Norwegian government asked Goldschmidt to conduct an investigation of the country's mineral resources, as it needed alternatives to chemicals that had been imported prior to World War I and were now in short supply. The government appointed him Chair of the Government Commission for Raw Materials and head of the Raw Materials Laboratory.

This led Goldschmidt into a new **area** of research—the study of the proportions of chemical elements in the earth's **crust**. His work was facilitated by the newly developed science of x-ray crystallography, which allowed Goldschmidt and his colleagues to determine the crystal structures of 200 compounds made up of 75 elements. He also developed the first tables of atomic and ionic radii for many of the elements, and showed how the hardness of **crystals** is based on their structures, ionic charges, and the proximity of their atomic particles.

In 1929, Goldschmidt moved to Gottingen, Germany, to assume the position of full professor at the Faculty of Natural Sciences and director of its mineralogical institute. As part of his investigation of the apportionment of elements outside the earth and its atmosphere, he began studying meteorites to ascertain the amounts of elements they contained. He researched numerous substances, including germanium, gallium, scandium, beryllium, selenium, arsenic, chromium, nickel, and zinc, using materials from both the earth and meteorites to devise a model of Earth. In this model, elements were distributed in different parts of Earth based on their charges and sizes. Goldschmidt stayed at Gottingen until 1935, when Nazi anti-Semitism made it impossible for him to continue his work. Returning to Oslo, he resumed work at the university there and assembled data he had collected at Gottingen on the distribution of chemical elements in Earth and in the cosmos. He also began studying ways to use Norwegian **olivine** rock for use in industry.

When World War II began, Goldschmidt had confrontations with the Nazis that resulted in his imprisonment on several occasions. He narrowly escaped internment in a concentration camp in 1943 when, after the Nazis arrested him, he was rescued by the Norwegian underground. They managed to secretly get him onto a boat to Sweden, where fellow scientists arranged for a flight to Scotland.

In Scotland, Goldschmidt worked at the Macaulay Institute for **Soil** Research in Aberdeen. Later during the war, he worked as a consultant to the Rothamsted Agricultural Experiment Station in England. As reported in *Chemists,* Goldschmidt carried with him a cyanide suicide pill for use in the event the Nazis invaded England. When a colleague asked him for one, he responded, "Cyanide is for chemists; you, being a professor of mechanical engineering, will have to use the rope."

After the war, Goldschmidt returned to Oslo and his job as professor and director of the geological museum. There he worked on a newly equipped raw materials laboratory supplied by the Norwegian Department of Commerce. He continued his work until his death on March 20, 1947.

Goldschmidt was a member of the Royal Society and the Geological Society of London, the latter of which awarded him the Wollaston Medal in 1944. He was also an honorary member of the British Mineralogical Society, the Geological Society of Edinburgh, and the Chemical Society of London. He wrote over 200 papers as well as a treatise, *Geochemistry,* which was published posthumously in 1954.

See also Mineralogy

GONDWANALAND · *see* SUPERCONTINENTS

GPS (GLOBAL POSITIONING SYSTEM)

Long before the **space** age, people used the heavens for navigation. Besides relying on the **Sun, Moon,** and stars, the early travelers invented the magnetic compass, the sextant, and the seagoing chronometer. Eventually, radio navigation in which a position could be determined by receiving radio signals broadcast from multiple transmitters came into existence. Improved high frequency signals gave greater accuracy of position, but they were blocked by mountains and could not bend over the horizon. This limitation was overcome by moving the transmitters into space on Earth-orbiting satellites, where high frequency signals could accurately cover wide areas.

The principle of **satellite** navigation is relatively simple. When a transmitter moves toward an observer, radio waves have a higher frequency, just like a train's horn sounds higher as it approaches a listener. A transmitter's signal will have a lower frequency when it moves away from an observer. If measurements of the amount of shift in frequency of a satellite radiating a fixed frequency signal with an accurately known orbit are carefully made, the observer can determine a correct position on Earth.

The United States Navy developed such a system, called Transit, in the late 1960s and early 1970s. Transit helped submarines update their on-board inertial navigation systems. After nearly ten years of perfecting the system, the Navy released it for civilian use. It is now used in surveying, fishing, private and commercial maritime activities, offshore oil exploration, and drifting buoys. However, a major drawback to Transit was that it was not accurate enough; a user had to wait until the satellite passed overhead, position fixes required some time to be determined, and an accurate fix was difficult to obtain on a moving platform.

As a result of these shortcomings, the United States military developed another system: Navstar (Navigation Satellite for Time and Ranging) Global Positioning System. This system consists of 24 operational satellites equally divided into six different orbital planes (each containing four satellites) spaced at 60° intervals. The new system can measure to within 33 ft (10 m), whereas Transit was accurate only to 528 ft (161 m).

With the new Global Positioning System (GPS), two types of systems are available with different frequencies and levels of accuracy. The Standard Positioning System (SPS) is used primarily by civilians and commercial agencies. As of midnight, May 1, 2000, the SPS system became 30 times more accurate when President Bill Clinton ordered that the Selective Availability (SA) component of SPS be discontinued. SA was the deliberate decrease of accurate positioning information available for commercial or civilian use. The SPS obtains information from a frequency labeled GPS L1. The United States military has access to GPS L1 and a second frequency, L2. The use of L1 and L2 permits the transfer of data with a higher level of security. In addition to heightened security, the United States military also has access to much more accurate positioning by using the Precise Positioning System (PPS). Use of the PPS is usually limited to the U.S. military and other domestic government agencies.

Both Transit and Navstar use instantaneous satellite position data to help users traveling from one place to another. But another satellite system uses positioning data to report where users have been. This system, called Argos, is a little more complicated: an object on the ground sends a signal to a satellite, which then retransmits the signal to the ground. Argos can locate the object to within 0.5 mi (0.8 km). It is used primarily for environmental studies. Ships and buoys can collect and send data on **weather**, currents, winds, and waves. Land-based stations can send weather information, as well as information about hydrologic, volcanic, and seismic activity. Argos can be used with balloons to study weather and the physical and chemical properties of the atmosphere. In addition, the system is being perfected to track animals.

Use of the GPS system in our everyday lives is becoming more frequent. Equipment providing and utilizing GPS is shrinking both in size and cost, while it increases in reliability. The number of people able to use the systems is also increasing. GPS devices are being installed in cars to provide directional, tracking, and emergency information. People who enjoy the outdoors can pack hand held navigational devices that show their position while exploring uncharted areas. Emergency personnel can respond more quickly to 911 calls thanks to tracking signal devices in their vehicles and in the cell phones of the person making the call. As technology continues to advance the accuracy of navigational satellite and without the impedance of Selective Availability, the uses for GPS will continue to develop.

See also Archeological mapping; Weather satellite

GRANITE

Granite, which makes up 70–80% of Earth's **crust**, is an igneous **rock** formed of interlocking **crystals** of **quartz**, **feldspar**, mica, and other **minerals** in lesser quantities. Large masses of granite are a major ingredient of mountain ranges. Granite is a plutonic rock, meaning that it forms deep underground. Slow cooling gives atoms time to migrate to the surfaces of growing crystals, resulting in a coarse or mottled crystalline structure easily visible to the naked eye.

Geologists have debated rival theories of granite's origin for over 150 years. The two theories most favored today are the magmatic theory and the hypermetamorphic theory. Supporters of the magmatic theory observe that granite is strongly associated with mountain ranges, which in turn tend to follow continental edges where one plate is being subducted (wedged under another). Tens of kilometers beneath the continental edge, the pressure and friction caused by subduction are sufficient to melt large amounts of rock. This melted rock or **magma** ascends toward the surface as large globules or plutons, each containing many cubic kilometers of magma. A **pluton** does not emerge suddenly onto the surface but remains trapped underground, where it cools slowly and may be repeatedly injected from beneath with pulses of fresh magma. To become surface rock, a solidified pluton must finally be uplifted to the surface and stripped bare by **erosion**.

The ultrametamorphic theory, in contrast, argues that granite is not formed from raw magma but consists of sedimentary rock thoroughly melted and re-crystallized. Most geologists now argue that granites can be formed by magmatism, ultrametamorphosis, or a combination of both.

Until recently, geologists thought that plutons of granitic magma would require millions of years to ascend to the surface. However, laboratory experiments with melted rock has shown that granitic magma is thin and runny enough (i.e., of low viscosity) to squirt rapidly upward to the surface through small cracks in the crust. Granite plutons may thus be created in 1,000–100,000 years, rather than in the millions of years previously thought. The precise origin and process of granite formation continues to be a subject of active research.

See also Bowen's reaction series; Convergent plate boundary; Plate tectonics

GRANULITE • *see* METAMORPHIC ROCK

GRAPHITE

Graphite is a soft, black, metallic mineral composed of the element **carbon**. It is nontoxic and rubs off easily on rough surfaces, which is why graphite mixed with fine **clay**, rather than actual **lead**, is used to make pencil leads. The word graphite derives from the Greek *gréphein*, to write or draw.

Graphite has the same chemical formula as **diamond** (C), yet the two **minerals** could hardly be more unlike. Diamond is the hardest of minerals, graphite one of the softest; diamond is transparent, graphite opaque; and diamond is almost twice as dense as graphite. These radically different properties arise from the way the atoms are arranged in each substance. In graphite, carbon atoms are linked in hexagonal sheets resembling chicken-wire fencing. These sheets slide over each other easily, which accounts for graphite's slipperiness. In diamond, carbon atoms are linked in a potentially endless matrix of tetrahedra (four-cornered pyramids), an extremely strong arrangement. Surprisingly, however, graphite is stable under ordinary atmospheric conditions and diamond is not; that is, at standard **temperature** and pressure diamond transforms spontaneously to graphite. The rate of transformation is extremely slow because carbon atoms organized into diamond are separated from the lower-energy graphite state by an activation-energy barrier similar to that which keeps an explosive from going off until triggered by a spark.

The carbon in most graphite and diamonds derives from living things. The organic (carbon-containing) remains of organisms may be transformed into **coal** or into impurities in **limestone**; under some conditions, metamorphosis of these rocks purifies these organic materials to produce graphite. Further metamorphosis under extremely high pressures, such as occur many miles underground, is needed to produce diamond.

Because graphite is resistant to heat and slows neutrons, it was used in the early years of nuclear-power research as a matrix to contain radioactive fuel elements and moderate their chain reaction. Early atomic reactors were termed atomic piles because they consisted mostly of large piles of graphite blocks. Graphite is used not only in pencils, but also as a lubricant for locksmiths and for bearings operating in vacuum or at high temperatures. Because it is both conductive and slippery, graphite is used in generator brushes. It is also employed in making metallurgical crucibles and electrical batteries. Most of the graphite used is manufactured from coal in electrical furnaces, not mined.

See also Chemical bonds and physical properties; Chemical elements

GRAVITATIONAL CONSTANT

The gravitational constant is fundamental quantity of the universe. The gravitational constant, G, was the first great universal constant of **physics** (the others subsequently being the speed of light and Planck's constant) and modern physicists still argue its importance and relationships to **cosmology**. Regardless, almost all the major theoretical frameworks dictate that the value for the gravitational constant (G) is in some regard related to the large-scale structure of the cosmos. Ironically, despite centuries of research, the gravitational constant, G, is—by a substantial margin—the least understood, most difficult to determine, and least precisely known fundamental constant value. The quest for "G" provides a continuing challenge to the experimental ingenuity of physicists, and often spurs new generations of physicists to recapture the inventiveness and delicacy of measurement first embodied in the elegant experiments conducted by English physicist Henry Cavendish (1731–1810).

The Cavendish constant "G" must not be confused with the "g" (designated in lowercase) that geophysicists use to designate gravitational acceleration (i.e., a change in the velocity of an object due to the gravitational field (commonly referred to as the gravitational force) of the earth that is due to the mass of the earth. Although the gravitational field of the earth fluctuates with the mass underneath the **area** in question, the overall average "g" is 9.80665 m/s².

In 1798, Cavendish performed an ingenious experiment that led to the determination of the gravitational constant (G). Cavendish used a carefully constructed experiment that utilized a torsion balance to measure the very small gravitational attraction between two masses suspended by a thin fiber support. (Cavendish actually measured the restoring torque of the fiber support). Cavendish's experimental methodology and device design was not novel. Similar equipment had been designed by English physicist John Mitchell (1724–1793), and a similar apparatus had been designed by French physicist Charles Coulomb and others for electrical measurements and calibrations. Cavendish's use, however, of the torsional balance to measure the gravitational constant of Earth, was a triumph of empirical skill.

Cavendish balanced his apparatus by placing balls of identical mass at both ends of a crossbar suspended by a thin wire. By **lead** balls of known mass, Cavendish was able to account for both the masses in the Newtonian calculation and thereby allowing a determination of the gravitational constant (G). The Cavendish experiment worked because not much force was required to twist the wire suspending the balance. In addition, Cavendish brought relatively large masses close to the smaller weights—actually on symmetrically opposite sides of the weights—so as to double the actual force and make the small effects more readily observable. Over time, due to the mutual gravitational attraction of the weights the smaller balls moved toward the larger masses. The smaller balls moved because of their smaller mass and inertia (resistance to movement). Cavendish was able to measure the force of the gravitational attraction as a function of the time it took to produce any given amount of twist in the suspending wire. The value of the gravitational constant determined by this method was not precise by modern standards (only a 7% precision but with 1% accuracy) but was an exceptional value for the eighteenth century given the small forces being measured. Because all objects exert a gravitational "pull," precision in Cavendish type experiments is often hampered by a number of factors, including underlying **geology** or factors as subtle as movements of furniture or objects near the experiment.

The Cavendish experiment was, therefore, a milestone in the advancement of scientific empiricism. In fact, accuracy of the Cavendish determination remained unimproved for almost another century until Charles Vernon Boys (1855–1944) used the Cavendish Balance to make a more accurate determination of the gravitational constant. More importantly, the Cavendish experiment proved that scientists could construct experiments that were able to measure very small forces. Cavendish's work spurred analysis of the funda-

mental force of electromagnetism (a fundamental force far stronger than gravity) and gave confidence to the scientific community that Newton's laws were not only valid, they were also testable on exceedingly small scales.

In modern physics, the speed of light, Planck's constant, and the gravitational constant are among the most important of fundamental constants. According to **relativity theory**, G is related to the amount of space-time curvature caused by a given mass. Modern concepts of gravity and of the ramifications of the value of the gravitational constant are subject to seemingly constant revision as scientists aim to extend the linkage between the gravitational constant (G) and other fundamental constants. Although profoundly influential and powerful on the cosmic scale, the force of gravity is weak in terms of human dimensions. Accordingly, the masses must be very large before gravitational effects can be easily measured. Even using modern methods, different laboratories often report significantly different values for G.

See also Gravity and the gravitational field

GRAVITY AND THE GRAVITATIONAL FIELD

Geophysicists utilize slight variations in gravitational force to characterize the mass of subsurface features. Particularly useful in **petroleum** exploration, subtle gravitational field differences can help identify solid subsurface plutonic bodies or fluid filled reservoirs.

In 1687, English physicist **Sir Isaac Newton** (1642–1727) published a law of universal gravitation in his important and influential work *Philosophiae Naturalis Principia Mathematica (Mathematical Principles of Natural Philosophy)*. In its simplest form, Newton's law of universal gravitation states that bodies with mass attract each other with a force that varies directly as the product of their masses and inversely as the square of the distance between them. This mathematically elegant law, however, offered a remarkably reasoned and profound insight into the mechanics of the natural world because it revealed a cosmos bound together by the mutual gravitational attraction of its constituent particles. Moreover, along with Newton's laws of motion, the law of universal gravitation became the guiding model for the future development of physical law.

Newton's law of universal gravitation was derived from German mathematician and astronomer Johannes Kepler's (1571–1630) laws of planetary motion, the concept of "action-at-a-distance," and Newton's own laws of motion. Building on Italian astronomer and physicist Galileo Galilei's (1564–1642) observations of falling bodies, Newton asserted that gravity is a universal property of all matter. Although the force of gravity can become infinitesimally small at increasing distances between bodies, all bodies of mass exert gravitational force on each other. Newton extrapolated that the force of gravity (later characterized by the gravitational field) extended to infinity and, in so doing, bound the universe together.

Newton's law of gravitation, mathematically expressed as $F = (G)(m_1 \, m_2) / r^2$, stated that the gravitational attraction between two bodies with masses m_1 and m_2 was directly proportional to the masses of the bodies, and inversely proportional to the square of the distance (r) between the centers of the masses. Accordingly, a doubling of one mass resulted in a doubling of the gravitational attraction while a doubling of the distance between masses resulted in a reduction of the gravitational force to a fourth of its former value. Nearly a century passed, however, before English physicist Henry Cavendish (1731–1810) was to determine the missing **gravitational constant** (G) that allowed a reasonably accurate determination of Earth's actual gravitational force.

The force exerted by a gravitational field on a body, such as forces produced by Earth's gravitational field, is called the weight of the body. The weight of a body is equal to the product of its mass m and the acceleration due to gravity g, (w = mg). Weight should not be confused with mass, which is an intrinsic property of matter that is not altered by an change in the gravitational field (i.e., the mass of an object on Earth is the same in the lower gravity environment on the **Moon**).

Newton's second law states that the net force F acting on an object is equal to the mass of the object m multiplied by its acceleration a (F = ma). Freely falling bodies experience acceleration (g) due to Earth's gravitational field. The force of this field is directed towards the center of Earth. By applying Newton's second law to freely falling bodies, with a = g and F = w, the weight of the body is given as w = mg. Because weight depends upon the gravitational field, it varies with geographical location. Because g decreases with increasing distance from the center of Earth, bodies weigh less at higher altitudes than at sea level. Because of this, weight, unlike mass, is not an inherent property of a body.

The value of "g" (9.82 m/s^2) is the average measure of the strength of Earth's gravitational field (i.e., the acceleration produced on a mass regardless of the composition of the mass.) The value of "g" can vary locally depending on subsurface mass (e.g., plutonic bodies) and so the value (9.82 m/s^2) is an average. Using "g" the gravitational field can then be expressed as force per kilogram exerted by Earth's gravitational field. Although Earth's gravitational field extends to infinity (i.e., as do all other objects with mass in the universe, Earth's gravitational field affects all other entities with mass), because the magnitude of the force of gravity declines as the square of the distance between objects, the force drops dramatically as objects move away from Earth.

Astronauts orbiting Earth do not experience weightlessness because of a lack of gravity. Rather, the apparent weightlessness in a decreased Earth gravity environment results from uniform acceleration toward Earth in such a way that the spacecraft and all objects in it are constantly falling toward Earth in a manner akin to the objects inside a free-falling elevator. In order to achieve orbit, rockets must be powerful enough to achieve escape velocity, the velocity that, at a minimum, allows their vertical "fall" to match the falling away of Earth spherical surface beneath them. Earth's escape velocity measures 6.959 mi/s (11.2 km/s)—more than 25,000 mph.

A tremendous amount of thrust is required to overcome Earth's gravitational field and lift the Space Shuttle into orbit. *U.S. National Aeronautics and Space Administration (NASA).*

Weight is usually expressed in pounds or grams. Although weight and mass are not synonymous terms, they are often used interchangeably. One concept associated with weight is Archimedes' principle that states that a body immersed in a fluid is acted upon by a force equal and opposite in direction to the weight of the displaced fluid. This principle explains the buoyancy of ships, as well as the rise of helium filled balloons.

Another important property associated with weight is specific gravity. The specific gravity of a material is the ratio of the weight of a given volume of that substance to the weight of an equal volume of **water**. For example, because of the salt, the specific gravity of a **saltwater** solution is greater than one. This high specific gravity gives saltwater its large buoyancy power because the weight of the volume displaced by an object in the ocean is larger than the weight of the volume displaced by the same object in **freshwater**.

The molecular weight of a substance is usually expressed in **atomic mass** units which is exactly 1/12 the mass of a carbon-12 **atom**.

Although Newton's law of gravitation offered no fundamental explanatory mechanism for gravity, it's usefulness of explanation lies in a higher level of cause and effect. An explanation of gravity continues to elude physicists. The two great theories of modern physics—relativity theory and quantum theory—explain gravity in very different ways. According to **relativity theory**, gravity is a consequence of the fusion of space and time. Quantum theory proposes that graviton particles (as of yet undiscovered) act as bosons (carriers) of gravitational force.

See also Aerodynamics; Astronomy; Atomic mass and weight; Aviation physiology; Big Bang theory; Crust; Earth (planet); Gravitational constant; Mohs' scale; Petroleum detection; Quantum theory and mechanics

GREAT BARRIER REEF

The Great Barrier Reef lies off the northeastern coast of **Australia** and is the largest structure ever made by living organisms including human beings, consisting of the skeletons of tiny coral polyps and hydrocorals bounded together by the soft remains of coralline algae and microorganisms.

The Great Barrier Reef is over 1,250 mi (2,000 km) long and is 80,000 mi² (207,000 km²) in surface **area**, which is larger than the island of Great Britain. It snakes along the coast of the continent of Australia, roughly paralleling the coast of the State of Queensland, at distances ranging 10–100 mi (16–160 km) from the shore. The reef is so prominent a feature on Earth that it has been photographed from satellites. The reef is located on the **continental shelf** that forms the perimeter of the Australian landmass where the ocean **water** is warm and clear. At the edge of the continental shelf and the reef, the shelf becomes a range of steep cliffs that plunge to great depths with much colder water. The coral polyps require a **temperature** of at least 70°F (21°C), and the water temperature often reaches 100°F (38°C).

The tiny coral polyps began building their great reef in the **Miocene Epoch** that began 23.7 million years ago and ended 5.3 million years ago. The continental shelf has subsided almost continually since the Miocene Epoch. In response, the reef has grown upward with living additions in the shallow, warm water near the surface; live coral cannot survive below a depth of about 25 fathoms (150 ft, or 46 m) and also depend on the salt content in seawater. As the hydrocorals and polyps died and became cemented together by algae, the spaces between the skeletons were filled in by wave action that forced in other debris called infill to create a relatively solid mass at depth. The upper reaches of the reef are more open and are riddled with grottoes, canyons, caves, holes bored by mollusks, and many other cavities that provide natural homes and breeding grounds for thousands of other species of sea life. The Great Barrier Reef is, in reality, a string of 2,900 reefs, cays, inlets, 900 islands, lagoons, and shoals, some with beaches of **sand** made of pulverized coral.

The reef is the product of over 350 species of coral and red and green algae. The number of coral species in the northern section of the reef exceeds the number (65) of coral species found in the entire Atlantic Ocean. Polyps are the live organisms inside the coral, and most are less than 0.3 in (8 mm) in diameter. They feed at night by extending frond-like fingers to wave zooplankton toward their mouths. In 1981, marine biologists discovered that the coral polyps spawn at the same time on one or two nights in November. Their eggs and sperm form

an orange and pink cloud that coats hundreds of square miles of the ocean surface. As the polyps attach to the reef, they secrete lime around themselves to build secure turrets or cups that protect the living organisms. The daisy- or feather-like polyps leave **limestone** skeletons when they die. The creation of a 1 in (2.5 cm) thick layer of coral takes five years.

The coral is a laboratory of the living and once-living; scientists have found that coral grows in bands that can be read much like the rings in trees or the icecaps in polar regions. By drilling cores 25 ft (7.6 m) down into the coral, 1,000 years of lifestyles among the coral can be interpreted from the density, skeleton size, band thickness, and chemical makeup of the formation. The drilling program also proved that the reef has died and revived at least a dozen times during its 25-million-year history, but it should be understood that this resiliency predated human activities. The reef as we know it is about 8,000 years old and rests on its ancestors. In the early 1990s, study of the coral cores has yielded data about temperature ranges, rainfall, and other **climate** changes; in fact, rainfall data for design of a dam were extracted from the wealth of information collected from analysis of the coral formation.

Animal life forms flourish on and along the reef, but plants are rare. The Great Barrier Reef has a distinctive purple fringe that is made of the coralline or encrusting algae *Lithothamnion* (also called stony seaweed), and the green algae *Halimeda discodea* that has a creeping form and excretes lime. The algae are microscopic and give the coral its many colors; this is a symbiotic relationship in which both partners, the coral and the algae, benefit. Scientists have found that variations in water temperature stress the coral causing them to evict the resident algae. The loss of color is called coral bleaching, and it may be indicative of **global warming** or other effects like El Niño.

This biodiversity makes the reef a unique ecosystem. Fish shelter in the reef's intricacies, find their food there, and spawn there. Other marine life experience the same benefits. The coastline is protected from waves and the battering of storms, so life on the shore also thrives.

See also El Niño and La Niña phenomena; Greenhouse gases and greenhouse effect; Tropical cyclone

GREAT CIRCLE • *see* LATITUDE AND LONGITUDE

GREAT LAKES

The Great Lakes are a system of five large **freshwater** lakes in central North America—Lake Erie, Lake Huron, Lake Michigan, Lake Ontario, and Lake Superior—that drain into the Atlantic Ocean via the St. Lawrence Seaway. Combined, the Great Lakes constitute the largest surface **area** of unfrozen fresh **water** in the world: 94,850 mi² (245,660 km²), an area larger than the United Kingdom. Except for Lake Michigan, which is wholly contained in the United States, the Great Lakes form a natural segment of the U.S.-Canadian border.

Lake Superior is the largest of the five lakes by almost 10,000 mi^2 (41,682 km^2), and has the greatest average (and maximum) depth. As a result, Lake Superior contains slightly more water than all the other Great Lakes combined—almost 3,000 mi^3 (12,504 km^3). The deepest parts of all the Great Lakes except Lake Erie are below sea level; in Lake Superior's case, over 600 ft (183 m) below.

Lake Superior has an average depth of 487 ft (148 m), a maximum depth of 1,302 ft (397 m) and covers 31,820 mi^2 (82,413 km^2). Lake Huron has an average depth of 195 ft (59 m), a maximum depth of 750 ft (229 m), and covers 23,010 mi^2 (59,596 km^2). Lake Michigan, covering 22,400 mi^2 (58,016 km^2), has an average depth of 276 ft (84 m), and a maximum depth of 923 ft (281 m). Lake Erie has an average depth of 62 ft (19 m), a maximum depth of 210 ft (64 m), and covers approximately 9,930 mi^2 (25,719 km^2). Lake Ontario, the smallest of the Great Lakes in terms of surface area (7,520 mi^2/19,477 km^2), has an average depth of 62 ft (19 m) but reaches a maximum depth of 778 ft (237 m).

The Great Lakes drain 295,800 mi^2 (766,118 km^2) of watershed (counting the surfaces of lakes themselves), or about 3% of the continent. Half the water entering the lakes evaporates; the rest flows from lake to lake, west to east, until it reaches Lake Ontario and then the St. Lawrence River.

By geological standards, the Great Lakes formed very recently. Prior to the beginning of the **ice ages** of the Pleistocene Epoch—about 1 million years ago—river valleys drained through the areas now occupied by the five lakes. As the ice-sheets flowed southward they favored these preexisting channels, scouring them and so increasing their depth. The latest glacial episode was the Wisconsin **Glaciation**, which ended about 18,000 years ago. When **melting** removed the glacier's enormous weight, the land began to rise. (It is still rising, at about .12 in / 3 mm per year.) This rising of the land, along with deposition of glacial sediments (**moraines**), blocked all drainage from the Great Lakes area except eastward via the St. Lawrence. Lake Superior is the only Great Lake not formed by glacial scouring and deposition of moraines. Lake Superior's basin, although somewhat enlarged by glacial scouring, is the trough of a V-shaped fold in the **rock** termed a **syncline**.

But the Wisconsin glaciation did not simply advance to a most southerly limit, then retreat in an orderly way. It advanced and retreated several times over thousands of years in a three-steps-northward, two-steps-southward fashion. These oscillations partially uncovered and recovered the Great Lakes basins, forming a series of lakes corresponding partly to the modern ones. At one point, a superlake submerged what are today the basins of Superior, Michigan, and Huron. The history of these fluctuations can be traced primarily by the many abandoned beaches that are today found far above water level (often hundreds of feet above). Each abandoned beach records a lake stage or period during which the water level was stable long enough to form a beach. From these and other data, it is known that Erie reached its present level about 10,000 years ago; Ontario about 7,000 years ago; and Superior, Michigan, and Huron only about 3,000 years ago.

Human activity has significantly altered the **chemistry** and ecology of the lower four lakes, which are ringed by such cities as Buffalo, Chicago, Cleveland, Detroit, Gary, Milwaukee, Rochester, Toledo, and Toronto. Sewage and industrial effluents have burdened these lakes increasingly for over a century. (Chicago and several other cities, however, now divert their sewage southward, away from the lakes.) Lake Superior has been less affected by pollution, having no major settlements on its shores.

Another detrimental side-effect of human activity is the introduction into the lake ecosystem, both deliberate and accidental, of non-native species. The sea lamprey (1930s), alewife (probably 1940s), and zebra mussel (1980s) have been particularly destructive to the native lake fauna. Alewives are now the most abundant fish species in the lakes. They suffer intermittent mass die-offs, wash up on the beaches by the millions, and must be removed using bulldozers and trucked away.

See also Drainage basins and drainage patterns; Glacial landforms; Syncline and anticline; Water pollution and biological purification

GREENHOUSE GASES AND GREENHOUSE EFFECT

The greenhouse effect is the physical mechanism by which the atmosphere helps to maintain Earth's surface **temperature** within a range comfortable for organisms and ecological processes. The greenhouse effect is largely a natural phenomenon, but its intensity may be changing because of increasing concentrations of **carbon dioxide** and some other gases in the atmosphere. These increased concentrations are occurring as a result of human activities, especially the burning of fossil **fuels** and the clearing of **forests**. A probable consequence of an intensification of Earth's greenhouse effect will be a significant warming of the atmosphere. This could likely result in important secondary changes, such as a rise in sea level, variations in the patterns of **precipitation**, and large and difficult ecological and socio-economic adjustments.

Earth's greenhouse effect is a well-understood physical phenomenon. Scientists believe that in the absence of the greenhouse effect, Earth's surface temperature would average about –0.4°F (–18°C), which is colder than the **freezing** point of **water**, and more frigid than life could tolerate long term. By slowing the rate at which the planet cools itself, the greenhouse effect helps to maintain Earth's surface at an average temperature of about 59°F (15°C). This is about 59.5°F (33°C) warmer than it would otherwise be, and is within the range of temperature that life can tolerate.

An energy budget is a physical analysis of all of the energy coming into a system, all the energy going out, and any difference that might be internally transformed or stored. Almost all of the energy coming to Earth from outer **space** has been radiated by the closest star, the **Sun**. The Sun emits electromagnetic energy at a rate and spectral quality determined by its surface temperature—all bodies do this, as long as they have a temperature greater than absolute zero, or –459°F

(−273°C). Fusion reactions occurring within the Sun maintain an extremely hot surface temperature, about 10,800°F (6,000°C). As a direct consequence of this surface temperature, about one-half of the Sun's emitted energy is so-called "visible" radiation with wavelengths between 0.4 and 0.7 μm (this is called visible radiation because it is the range of electromagnetic energy that the human eye can perceive), and about one-half is in the near-infrared wavelength range between about 0.7 and 2.0 μm. The Sun also emits radiation in other parts of the **electromagnetic spectrum**, such as ultraviolet and cosmic radiation. However, these are relatively insignificant amounts of energy (although even small doses can cause biological damage).

At the average distance of Earth from the Sun, the rate of input of **solar energy** is about 2 cal cm^{-2} min^{-1}, a value referred to as the solar constant. There is a nearly perfect energetic balance between this quantity of electromagnetic energy incoming to Earth, and the amount that is eventually dissipated back to outer space. The myriad ways in which the incoming energy is dispersed, transformed, and stored make up Earth's energy budget.

On average, one-third of incident solar radiation is reflected back to space by the earth's atmosphere or its surface. The planet's reflectivity (or albedo) is strongly dependent on cloud cover, the density of tiny particulates in the atmosphere, and the nature of the surface, especially the cover of vegetation and water, including **ice** and snow.

Another one-third of the incoming radiation is absorbed by certain gases and vapors in Earth's atmosphere, especially water vapor and **carbon** dioxide. Upon absorption, the solar electromagnetic energy is transformed into thermal kinetic energy (that is, heat, or energy of molecular vibration). The warmed atmosphere then re-radiates energy in all directions as longer-wavelength (7–14 μm) infrared radiation. Much of this re-radiated energy escapes to outer space.

The remaining one-third of the incoming energy from the Sun is transformed or dissipated by the following processes:

Absorption and radiation at the surface

Much of the solar radiation that penetrates to Earth's surface is absorbed by living and non-living materials. This results in a transformation to thermal energy, which increases the temperature of the absorbing surfaces. Over the medium term (days) and longer term (years) there is little net storage of energy as heat. This occurs because almost all of the thermal energy is re-radiated by the surface, as electromagnetic radiation of a longer wavelength than that of the original, incident radiation. The wavelength spectrum of typical, re-radiated electromagnetic energy from Earth's surface peaks at about 10 μm, which is within the long-wave infrared range.

Evaporation and melting of water

Some of the electromagnetic energy that penetrates to Earth's surface is absorbed and transformed to heat. Much of this thermal energy subsequently causes water to evaporate from plant and inorganic surfaces, or it causes ice and snow to melt.

Winds, waves, and currents

A small amount (less than 1%) of the absorbed solar radiation causes mass-transport processes to occur in the **oceans** and lower atmosphere, which disperses of some of Earth's unevenly distributed thermal energy. The most important of these physical processes are winds and storms, water currents, and waves on the surface of the oceans and **lakes**.

Photosynthesis

Although small, an ecologically critical quantity of solar energy, averaging less than 1% of the total, is absorbed by plant pigments, especially chlorophyll. This absorbed energy is used to drive photosynthesis, the energetic result of which is a temporary storage of energy in the inter-atomic bonds of biochemical compounds.

If the atmosphere was transparent to the long-wave infrared energy that is re-radiated by Earth's atmosphere and surface, then that energy would travel unobstructed to outer space. However, so-called radiatively active gases (or RAGs; also known as "greenhouse gases") in the atmosphere are efficient absorbers within this range of infrared wavelengths, and these substances thereby slow the radiative cooling of the planet. When these atmospheric gases absorb infrared radiation, they develop a larger content of thermal energy, which is then dissipated by a re-radiation (again, of a longer wavelength than the electromagnetic energy that was absorbed). Some of the secondarily re-radiated energy is directed back to Earth's surface, so the net effect of the RAGs is to slow the rate of cooling of the planet.

This process has been called the "greenhouse effect" because its mechanism is analogous to that by which a glass-enclosed space is heated by solar energy. That is, a greenhouse's **glass** and humid atmosphere are transparent to incoming solar radiation, but absorb much of the re-radiated, long-wave infrared energy, slowing down the rate of cooling of the structure.

Water vapor (H_2O) and carbon dioxide (CO_2) are the most important radiatively active constituents of Earth's atmosphere. Methane (CH_4), nitrous oxide (N_2O), **ozone** (O_3), and chlorofluorocarbons (CFCs) play a more minor role. On a per-molecule basis, these gases differ in their ability to absorb infrared wavelengths. Compared with carbon dioxide, a molecule of methane is 11–25 times more effective at absorbing infrared, nitrous oxide is 200–270 times, ozone 2,000 times, and CFCs 3,000–15,000 times.

Other than water vapor, the atmospheric concentrations of all of these gases have increased in the past century because of emissions associated with human activities. Prior to 1850, the concentration of CO_2 in the atmosphere was about 280 ppm, while in 1994 it was 355 ppm. During the same period CH_4 increased from 0.7 ppm to 1.7 ppm, N_2O from 0.285 ppm to 0.304 ppm; and CFCs from zero to 0.7 ppb. These increased concentrations are believed to contribute to a hypothesized increase in the intensity of Earth's greenhouse effect, an increase attributable to human activities. Overall, CO_2 is estimated to account for about 60% of this enhancement of the greenhouse effect, CH_4 15%, N_2O 5%, O_3 8%, and CFCs 12%.

The physical mechanism of the greenhouse effect is conceptually simple, and this phenomenon is acknowledged by scientists as helping to keep Earth's temperature within the comfort zone for organisms. It is also known that the concentrations of CO_2 and other RAGs have increased in Earth's atmosphere, and will continue to do so. However, it has proven difficult to demonstrate that a warming of Earth's surface or lower atmosphere has been caused by a stronger greenhouse effect.

Since the beginning of instrumental recordings of surface temperature around 1880, it appears that almost all of the warmest years have occurred during the late 1980s and 1990s. Typically, these warm years have averaged about 1.5–2.0°F (0.8–1.0°C) warmer than occurred during the decade of the 1880s. Overall, Earth's surface air temperature has increased by about 0.9°F (0.5°C) since 1850.

However, the temperature data on which these apparent changes are based suffer from some important deficiencies, including: (1) air temperature is variable in time and space, making it difficult to determine statistically significant, longer-term trends; (2) older data are generally less accurate than modern records; (3) many **weather** stations are in urban areas, and are influenced by "heat island" effects; and (4) **climate** can change for reasons other than a greenhouse response to increased concentrations of CO_2 and other RAGs, including albedo-related influences of volcanic emissions of sulfur dioxide, sulfate, and fine particulates into the upper atmosphere. Moreover, it is well known that the interval 1350 to 1850, known as the Little Ice Age, was relatively cool, and that global climate has been generally warming since that time period.

Some studies have provided evidence for linkages between historical variations of atmospheric CO_2 and surface temperature. Important evidence comes from a core of Antarctic glacial ice that represents a 160,000-year time period. Concentrations of CO_2 in the ice were determined by analysis of air bubbles in layers of known age, while changes in air temperature were inferred from ratios of **oxygen** isotopes (because isotopes differ in weight, their rates of diffusion are affected by temperature in predictably different ways, and this affects their relative concentrations in the glacial ice). Because changes in CO_2 and surface temperature were positively correlated, a potential greenhouse mechanism is suggested. However, this study could not determine whether increased CO_2 might have resulted in warming through an intensified greenhouse effect, or whether warming could have increased CO_2 release from ecosystems by increasing the rate of decomposition of biomass, especially in cold regions.

Because of the difficulties in measurement and interpretation of climatic change using real-world data, computer models have been used to predict potential climatic changes caused by increases in atmospheric RAGs. The most sophisticated simulations are the so-called "three-dimensional general circulation models" (GCMs), which are run on supercomputers. GCM models simulate the extremely complex, mass-transport processes involved in **atmospheric circulation**, and the interaction of these with variables that contribute to climate. To perform a simulation experiment with a GCM model,

components are adjusted to reflect the probable physical influence of increased concentrations of CO_2 and other RAGs.

Many simulation experiments have been performed, using a variety of GCM models. Of course, the results vary according to the specifics of the experiment. However, a central tendency of experiments using a common CO_2 scenario (a doubling of CO_2 from its recent concentration of 360 ppm) is for an increase in average surface temperature of 1.8–7.2°F (1–4°C). This warming is predicted to be especially great in polar regions, where temperature increases could be two or three times greater than in the tropics.

One of the best-known models was designed and used by the International Panel on Climate Change (IPCC). This GCM model made assumptions about population and economic growth, resource availability, and management options that resulted in increases or decreases of RAGs in the atmosphere. Scenarios were developed for emissions of CO_2, other RAGs, and sulfate aerosols, which may cool the atmosphere by increasing its albedo and by affecting cloud formation. For a simple doubling of atmospheric CO_2, the IPCC estimate was for a 4.5°F (2.5°C) increase in average surface temperature. The estimates of more advanced IPCC scenarios (with adjustments for other RAGs and sulfate) were similar, and predicted a 2.7–5.4°F (1.5–3°C) increase in temperature by the year 2100, compared with 1990.

It is likely that the direct effects of climate change caused by an intensification of the greenhouse effect would be substantially restricted to plants. The temperature changes might cause large changes in the quantities, distribution, or timing of precipitation, and this would have a large effect on vegetation. There is, however, even more uncertainty about the potential changes in rainfall patterns than of temperature, and effects on **soil** moisture and vegetation are also uncertain. Still, it is reasonable to predict that any large changes in patterns of precipitation would result in fundamental reorganizations of vegetation on the terrestrial landscape.

Studies of changes in vegetation during the warming climate that followed the most recent, Pleistocene, **glaciation**, suggest that plant species responded in unique, individualistic ways. This results from the differing tolerances of species to changes in climate and other aspects of the environment, and their different abilities to colonize newly available habitat. In any event, the species composition of plant communities was different then from what occurs at the present time. Of course, the vegetation was, and is, dynamic, because plant species have not completed their post-glacial movements into suitable habitats.

In any region where the climate becomes drier (for example, because of decreased precipitation), a result could be a decreased **area** of forest, and an expansion of savanna or **prairie**. A landscape change of this character is believed to have occurred in the New World tropics during the Pleistocene glaciations. Because of the relatively dry climate at that time, presently continuous rainforest may have been constricted into relatively small refugia (that is, isolated patches). These forest remnants may have existed within a landscape matrix of savanna and grassland. Such an enormous restructuring of the character of the tropical landscape must have had a tremendous effect on the multitude of rare species that live in that region. Likewise, climate change potentially associated with an intensification of the greenhouse effect would have a devastating effect on Earth's natural ecosystems and the species that they sustain.

There would also be important changes in the ability of the land to support crop plants. This would be particularly true of lands cultivated in regions that are marginal in terms of rainfall, and are vulnerable to **drought** and desertification. For example, important crops such as wheat are grown in regions of the western interior of **North America** that formerly supported natural shortgrass prairie. It has been estimated that about 40% of this semiarid region, measuring 988 million acres (400 million ha), has already been desertified by agricultural activities, and crop-limiting droughts occur there sporadically. This climatic handicap can be partially managed by irrigation. However, there is a shortage of water for irrigation, and this practice can cause its own environmental problems, such as salinization. Clearly, in many areas substantial changes in climate would place the present agricultural systems at great risk.

Patterns of wildfire would also be influenced by changes in precipitation regimes. Based on the predictions of climate models, it has been suggested that there could be a 50% increase in the area of forest annually burned in Canada, presently about 2.5–4.9 million acres (1–2 million ha) in typical years.

Some shallow marine ecosystems might be affected by increases in seawater temperature. Corals are vulnerable to large increases in water temperature, which may deprive them of their symbiotic algae (called zooxanthellae), sometimes resulting in death of the colony. Widespread coral "bleachings" were apparently caused by warm water associated with an **El Niño** event in 1982–83.

Another probable effect of warming could be an increase in sea level. This would be caused by the combination of (1) a thermal expansion of the volume of warmed seawater, and (2) **melting** of polar **glaciers**. The IPCC models predicted that sea level in 2100 could be 10.5–21 in (27–50 cm) higher than today. Depending on the rate of change in sea level, there could be substantial problems for low-lying, coastal agricultural areas and cities.

Most GCM models predict that high latitudes will experience the greatest intensity of climatic warming. Ecologists have suggested that the warming of northern ecosystems could induce a positive feedback to climate change. This could be caused by a change of great expanses of boreal forest and arctic tundra from sinks for atmospheric CO_2, into sources of that greenhouse gas. In this scenario, the climate warming caused by increases in RAGs would increase the depth of annual thawing of frozen soils, exposing large quantities of carbon-rich organic materials in the **permafrost** to microbial decomposition, and thereby increasing the emission of CO_2 to the atmosphere.

It is likely that an intensification of Earth's greenhouse effect would have large climatic and ecological consequences.

Under the auspices of the United Nations Environment Program, various international negotiations have been undertaken to try to get nations to agree to decisive actions to reduce

their emissions of RAGs. One recent major agreement came out of a large meeting held in Kyoto, Japan, in 1997. There, industrial countries, such as those of North America and Western **Europe**, agreed to reduce their CO_2 by as much as 5–7% of their 1990 levels by the year 2012. These reductions will be a huge challenge for those countries to achieve.

One possible complementary way to balance the emissions of RAGs would be to remove some atmospheric CO_2 by increasing its fixation by growing plants, especially through the planting of forests onto agricultural land. Similarly, the prevention of deforestation will avoid large amounts of CO_2 emissions through the conversion of high-carbon forests into low-carbon agro-ecosystems.

See also Atmospheric chemistry; Atmospheric circulation; Atmospheric composition and structure; Desert and desertification; Earth (planet); El Niño and La Niña phenomena; Forests and deforestation; Fuels and fuel chemistry; Global warming; Ozone layer and hole dynamics; Ozone layer depletion; Petroleum, economic uses of

GREENSTONE BELT

Greenstone belts are generally elongate, **Archean** to Proterozoic terrains comprising intrusive and extrusive **mafic** to ultramafic **igneous rocks**, **felsic** volcanics, and inter-flow or cover **sedimentary rocks**. Greenstone belts occur sandwiched between regions dominated by granitoids and **gneiss**. Greenstones are generally of low to moderate metamorphic grade. The term greenstone comes from the green color of many mafic to ultramafic constituents due to an abundance of chlorite. A common igneous **rock** in greenstones is komatiite. Komatiites are rocks with greater than 18 weight percent magnesium oxide and a well-developed spinifex texture of interlocking bladed or acicular (pointed) **crystals** of **olivine** or pyroxene. Spinifex texture (named after similarities in crystal shape and pattern to the pointed spinifex grass that grows in South **Africa** and Western **Australia**) implies rapid cooling or decompression of the **magma**. Komatiites formed as volcanic flows and less commonly as intrusive sills. Sedimentary sequences within greenstone belts comprise both clastic (e.g., conglomerate, **quartz** arenite, shale and graywacke) and chemically precipitated (e.g., banded **iron** formation and **chert**) components. Greenstones may also be intruded by syn- to post-tectonic granitoids. Greenstone belts check host many major mineral deposits, such as gold and nickel. Greenstone belts were previously often thought to continue to large depths in the **crust**. Reflection seismic profiles over the Norseman Wiluna Belt of the Yilgarn **Craton**, Western Australia, however, indicate that this greenstone belt has a relatively shallow (3.7–5.6 mi [6–9 km]) flat-base and overlies a uniformly thick crust.

Contrasting models have been proposed for the origins of greenstone belts. Some geologists believe magmatic and tectonic processes during formation of greenstone belts in Archaean times were different to present-day **plate tectonics**. Earth's mantle would have then been far hotter. They cite differences between greenstone belts and Phanerozoic orogens (such as the abundance of komatiitic lavas) and point out that there are no modern analogues to greenstone belts. Opponents to Archean plate tectonics contend that greenstone belts commonly represent a laterally continuous **volcano** sedimentary sequence (sometimes on a granite-gneiss basement) essentially undeformed prior to late tectonism and may not therefore represent relics of volcanic chains. They consider that Archaean tectonics was dominated by **mantle plumes** and was possibly analogous to the tectonics of Venus. Greenstone belts are interpreted as oceanic plateaus generated by mantle plumes, similar to plume-generated oceanic plateaus in the southern Caribbean. A mantle plume origin is also proposed for neighboring tonalite-trondhjemite-granodiorite sequences.

The alternate view is that tectonic processes comparable to present-day plate tectonics were operative during the Late Archaean, and possibly were similar to plate tectonics since the Hadean-Archean transition (between 4.0 and 4.2 billion years ago). In a plate tectonic context, greenstones may have formed in volcanic arcs or inter-arc or back-arc basins. Greenstone belts are interpreted to represent collages of oceanic crust, **island arcs**, accretionary prisms, and possible plateaus. Recent experimental work on the origin of komatiitic magmas indicates that they were hydrous and that temperatures for their formation do not indicate that the Archean upper mantle was significantly hotter than today. Komatiites and similar rocks have also been found in younger orogens. Komatiites may not therefore require different tectonic processes or conditions for their formation, as previously thought.

In many granitoid-greenstone terrains, greenstone belts constitute synformal keels between circular to elliptical granitoid bodies. This outcrop pattern is generally thought to be due to deformation resulting solely from the greater density of greenstones compared to underlying granitoid and gneiss. Due to gravitational instability, the underlying, less dense granitoid-gneiss basement domed upwards and rose to form mushroom-shaped bodies called diapirs whilst the denser greenstones sank into the basement. **Shear zones** were formed along some granite-greenstone contacts due to differential vertical displacement and upright **folds** developed in the greenstones. This process of either solid-state and/or magmatic diapirism was independent to any tectonic processes that may have acted on margins to granite-greenstone terrains. The formation of granitoid domes in granite-greenstone terrains has also been attributed to crustal extension (producing metamorphic core complexes) or polyphase folding during regional shortening. The more linear form of some greenstone belts is due to subsequent deformation, especially the **superposition** of regional-scale transcurrent shear zones on early-formed structures.

See also Geologic time

GROUND FOG · *see* FOG

GROUND MORAINE · *see* MORAINES

GROUNDWATER

Groundwater occupies the void **space** in geological strata. It is one element in the continuous process of moisture circulation on Earth, termed the **hydrologic cycle**.

Almost all groundwater originates as surface **water**. Some portion of rain hitting the earth runs off into streams and **lakes**, and another portion soaks into the **soil**, where it is available for use by plants and subject to **evaporation** back into the atmosphere. The third portion soaks below the root zone and continues moving downward until it enters the groundwater. **Precipitation** is the major source of groundwater. Other sources include the movement of water from lakes or streams and contributions from such activities as excess irrigation and seepage from canals. Water has also been purposely applied to increase the available supply of groundwater. Water-bearing formations called aquifers act as reservoirs for storage and conduits for transmission back to the surface.

The occurrence of groundwater is usually discussed by distinguishing between a zone of saturation and a zone of aeration. In the zone of saturation, the pores are entirely filled with water, while the zone of aeration has pores that are at least partially filled by air. Suspended water does occur in this zone. This water is called vadose, and the zone of aeration is also known as the vadose zone. In the zone of aeration, water moves downward due to **gravity**, but in the zone of saturation it moves in a direction determined by the relative heights of water at different locations.

Water that occurs in the zone of saturation is termed groundwater. This zone can be thought of as a natural storage **area** or reservoir whose capacity is the total volume of the pores of openings in rocks.

An important exception to the distinction between these zones is the presence of ancient seawater in some sedimentary formations. The pore spaces of materials that have accumulated on an ocean floor, which has then been raised through later geological processes, can sometimes contain salt water. This is called connate water.

Formations or strata within the **saturated zone** from which water can be obtained are called aquifers. Aquifers must yield water through wells or **springs** at a rate that can serve as a practical source of water supply. To be considered an **aquifer** the geological formation must contain pores or open spaces filled with water, and the openings must be large enough to permit water to move through them at a measurable rate. Both the size of pores and the total pore volume depends on the type of material. Individual pores in fine-grained materials such as **clay**, for example, can be extremely small, but the total volume is large. Conversely, in coarse material such as **sand**, individual pores may be quite large but total volume is less. The rate of movement for fine-grained materials, such as clay, will be slow due to the small pore size, and it may not yield sufficient water to wells to be considered an aquifer. However, the sand is considered an aquifer, even though they yield a smaller volume of water, because they will yield water to a well.

The **water table** is not stationary, but moves up or down depending on surface conditions such as excess precipitation, **drought**, or heavy use. Formations where the top of the saturated zone or water table define the upper limit of the aquifer are called unconfined aquifers. The hydraulic pressure at any level with an aquifer is equal to the depth from the water table, and there is a type known as a water-table aquifer, where a well drilled produces a static water level which stands at the same level as the water table.

A local zone of saturation occurring in an aerated zone separated from the main water table is called a perched water table. These most often occur when there is an impervious strata or significant particle-size change in the zone of aeration, which causes the water to accumulate. A confined aquifer is found between impermeable layers. Because of the confining upper layer, the water in the aquifer exists within the pores at pressures greater than the atmosphere. This is termed an **artesian** condition and gives rise to an artesian well.

Groundwater can be pumped from any aquifer that can be reached by modern well-drilling apparatus. Once a well is constructed, hydraulic pumps pull the water up to the surface through pipes. As water from the aquifer is pulled up to the surface, water moves through the aquifer towards the well. Because water is usually pumped out of an aquifer more quickly than new water can flow to replace what has been withdrawn, the level of the aquifer surrounding the well drops, and a cone of depression is formed in the immediate area around the well.

Groundwater can be polluted by the spilling or dumping of contaminants. As surface water percolates downward, contaminants can be carried into the aquifer. The most prevalent sources of contamination are **waste disposal**, the storage, transportation and handling of commercial materials, mining operations, and nonpoint sources such as agricultural activities. Two other forms of groundwater pollution are the result of pumping too much water too quickly, so that the rate of water withdrawal from the aquifer exceeds the rate of aquifer recharge. In coastal areas, salty water may migrate towards the well, replacing the fresh water that has been withdrawn. This is called salt-water intrusion. Eventually, the well will begin pulling this salt water to the surface; once this happens, the well will have to be abandoned. A similar phenomenon, called connate ascension, occurs when a **freshwater** aquifer overlies a layer of **sedimentary rocks** containing connate water. In some cases, over pumping will cause the connate water to migrate out of the sedimentary rocks and into the freshwater aquifer. This results in a brackish, briny contamination similar to the effects of a salt-water intrusion. Unlike salt water intrusion, however, connate ascension is not particularly associated with coastal areas.

Groundwater has always been an important resource, and it will become more so in the future as the need for good quality water increases due to urbanization and agricultural production. It has recently been estimated that 50% of the drinking water in the United States comes from groundwater; 75% of the nation's cities obtain all or part of their supplies from groundwater, and rural areas are 95% dependent upon it. For these reasons every precaution should be taken to protect groundwater purity. Once contaminated, groundwater is difficult, expensive, and sometimes impossible to clean up.

See also Freshwater; Hydrogeology

GUERICKE, OTTO VON (1602-1686)
German politician and physicist

Otto von Guericke, born in Magdeburg, Germany, was a scientific showman during the seventeenth century. He studied mathematics, law, and engineering. Following travels to England and France, Guericke returned to Magdeburg in 1627 and became a politician. Unfortunately, this was during the Thirty Years' War; Guericke and his family had to flee the city in 1631. Following the war, he returned and helped rebuild the city, becoming mayor in 1646. Twenty years later he became a noble and added "von" to his name.

Otto von Guericke spent his leisure time dabbling in science, and he became involved in discussions surrounding the possibility of the existence of a vacuum. Most scientists were inclined to disavow that a vacuum could exist, primarily because of the teachings of Aristotle. Aristotle's theory was a masterpiece of reverse logic; he believed that if the air became less dense, an object would be able to move faster. If there was a vacuum, he erroneously added, an object could move with infinite speed, but because infinite speed was not possible, a vacuum was not possible either.

Unlike scientists who blindly accepted the ancient teachings, Guericke attempted to get a definite answer by experimentation. In 1650, he built the first air pump and proceeded to put on his production.

Guericke's first vacuum experiment was with a bell. Placing it in a vessel from which he had removed the air, thereby creating a vacuum, he showed that the bell could not be heard. This proved one of Aristotle's theories, which stated that sound would not travel through a vacuum. In addition, Guericke showed that lit candles would go out, and animals could not live in a vacuum.

For added drama, Guericke tied a rope to a piston in a cylinder and had fifty men pulling on it as he created a vacuum on the other side of the piston. The piston was drawn down into the cylinder in spite of the men trying to pull it the other direction.

For his next trick, in 1657, Guericke fitted two 12–ft (3.6–m) diameter metal hemispheres together, removed the air and created a vacuum that held the halves in place. Sixteen horses were unable to pull the hemispheres apart, yet when air was returned to the sphere, the halves fell apart. Emperor Ferdinand III, in the audience, was impressed. Guericke had placed the air valve in the bottom of his sphere because he was under the impression that air, like **water**, would seek the lowest level. He discovered that air could be removed no matter where the valve was located. Obviously the air was evenly distributed throughout the sphere. This led him to speculate that air decreased in density as one's altitude above the earth increased. He used this knowledge to build a water barometer in 1672, and used it to forecast the **weather**.

In addition to his experiments with the vacuum, Guericke built a device that created static **electricity**, similar to

Robert Van de Graaff's (1901–1967) generator. A sphere of sulfur was rotated on a shaft. When it was rubbed, it built up a static charge that caused sizeable electric sparks to discharge. Guericke did not realize the electrical effect was a special phenomenon, but he was responsible for instigating a century of investigation by others.

Guericke also was interested in **astronomy**, suggesting that **comets** were members of the **solar system** and made regular returns as they orbited around the **sun**. **Edmond Halley** jumped on this idea and became famous when his observed comet returned precisely when he predicted. Guericke also believed that a magnetic force caused celestial objects to interact with each other across empty **space**. Isaac Newton would show that interaction did occur, but not because of **magnetism**.

After holding the position of mayor of Magdeburg for 35 years, Guericke retired. He died in Hamburg, Germany at the age of 83.

See also Atmospheric pressure; Electricity and magnetism; Gravity and the gravitational field

GULF OF MEXICO

The Gulf of Mexico is a unique, semi-enclosed sea located between the Yucatan and Florida peninsulas, at the southeast shores of the United States. The Gulf of Mexico borders five of the 50 United States (Alabama, Florida, Louisiana, Mississippi, and Texas), and also Cuba and the eastern part of Mexico. Sometimes it is also called America's Sea. The Straits of Florida divides the Gulf from the Atlantic Ocean, while the Yucatan Channel separates it from the Caribbean Sea. The Gulf of Mexico covers more than 600,000 mi² (almost 1.5 million km²), and in some areas its depth reaches 12,000 ft (3660 m), where it is called Sigsbee Deep, or the "Grand **Canyon** under the sea." About two-thirds of the contiguous United States (31 states between the Rocky Mountains and the Appalachian Mountains) belongs to the watershed **area** of the Gulf of Mexico, while it receives **freshwater** from 33 major river systems, and many small **rivers**, creeks, and streams. This watershed area covers a little less than two million mi² (almost 5 million km²).

The currents in the Gulf of Mexico form a complex system. Its dominant feature is the Caribbean Current, coming from the warm Caribbean Sea by the Yucatan Channel, meandering around in the Gulf, then leaving through the Straits of Florida. Together with the Antilles Current, the Caribbean Current forms the **Gulf Stream**. The Gulf of Mexico has **tides** (the ocean waters' response to the Moon's and Sun's gravitational pull) of normally 2 ft (0.6 m) or less.

According to the modified Trewartha climate system, most of the Gulf Coast area is in the subtropical climate region with a summer **precipitation** maximum. The southern tip of the Yucatan Peninsula belongs to the savanna climate, and between the subtropical and the savanna lies a small area of tropical dry savanna. The hurricane season is between June and November, when hurricanes from the Atlantic Ocean, the Caribbean Sea, or the Gulf of Mexico can damage the Gulf

shore, and beyond it. These hurricanes also help to balance the salinity of the **water**, while also moderating the atmosphere. The Gulf of Mexico plays an important role as a fuel injector for hurricanes before landfall, since major hurricanes are rapidly intensified by passing over deep and warm water.

The Gulf of Mexico has several environmental quality problems originating either from natural processes, or from anthropogenic pollution, or their combination. The problems range from **erosion**, and topsoil washing from the land into the Gulf, to oil spills and hazardous material spills, or trash washing ashore. These problems not only affect the estuaries, wetlands, and water quality in the Gulf, but have led to problems such as hypoxia (a zone of oxygen-depleted water), declining fish catch, contaminated fish, fish kills, endangered species, and air and water quality problems.

The role of the Gulf of Mexico is complex. The Gulf hosts important ocean currents (the area where hurricanes can gain strength before hitting land). The Gulf of Mexico and the Caribbean area contain some of the most spectacular wildlife in the world. The Gulf also partially supplies moisture for the North American Monsoon, and is also an important area for recreation and commercial fisheries. Many onshore refineries and offshore drilling platforms operate in the Gulf area, and produce about a quarter of the crude oil and almost one third of the **natural gas** in the United States. The Gulf also links the ports of the five southern states and Mexico with the ocean; about half of all the cargo shipped in and out of the United States travels through the Gulf. The Gulf of Mexico provides food, energy, jobs, recreation, and government revenue, not only benefiting the population on the shoreline of the Gulf, but the whole country.

See also Delta; Dunes; Estuary; Gulf stream; Petroleum extraction; Red tide; Rip current; Seawalls and beach erosion

GULF STREAM

The Gulf Stream is a well-known, fast, intense, and warm ocean current in the North Atlantic Ocean. Its path goes from the **Gulf of Mexico** and the Caribbean Sea, along the eastern coast of the United States, heading to the northeast Atlantic Ocean, to the British Isles, and the Norway coasts. This western boundary current is responsible for the mild **climate** of western **Europe**, which is located at a much higher **latitude** than most of New England, but experiences much milder **weather**.

The origin of the Gulf Stream goes back to the broad, slow, and warm North Equatorial Current under the trade winds, which moves to the west, and when it reaches the Caribbean Sea, its **water** moves through the Yucatan Channel. Here, it becomes not only narrower, but also gains strength, meandering around the Gulf of Mexico (here it is often referred to as the Loop Current), then exiting the Gulf at the Straits of Florida (here, it is called the Florida Current). Along the east coast of Florida, the current meets the Antilles Current, and the flow, now called the Gulf Stream, runs parallel to the coast until reaching Cape Hatteras, North Carolina, where it moves away from the coast. Around 50 degrees West, it splits into different currents, the largest of which is the North Atlantic Current, which also feeds the northbound Norwegian Current. The Canary Current flows towards the equator on the eastern side of the Atlantic Ocean.

The Gulf Stream also has mesoscale eddies or rings, large, concentric cylinders, reaching deep down in the water, which are usually about 62–186 mi (100–300 km) in diameter. They appear on both sides of the Gulf Stream, forming as a meandering loop cut off from the current, and can contain both a warm or a cold core. These rings help to maintain the termohaline (**temperature** and salinity) balance in the ocean basin.

The Gulf Stream not only helps to redistribute heat by carrying warm waters towards the North Pole, but also has a large impact on the climate on land by bringing humid, mild air to the British Isles and Northwest Europe, causing significantly milder winters than at the same latitudes in the West.

See also Ocean circulation and currents

GUYOT, ARNOLD HENRI (1807-1884)
Swiss geologist and geographer

Arnold Henri Guyot's geological field studies advanced the knowledge of **lakes**, **glaciers**, **ice ages**, mountains, erratic boulders, **evolution**, and **weather**.

Guyot was born in Boudevilliers, Switzerland, on September 28, 1807. After graduating from the University of Neuchâtel, Switzerland, in 1825, he went to Germany to continue his studies in botany, zoology, entomology, geography, and theology. While living and studying with botanist Alexander Braun (1805–1877) in Karlsruhe, Germany, he met naturalist **Louis Agassiz** (1807–1873) and botanist Karl Friedrich Schimper (1803–1867), who would later coin the term "ice age." During this time, Guyot considered becoming a minister, but decided instead on a career in science. He received his doctorate in **geology** from the University of Berlin in 1835 with a dissertation on lakes. Among his professors at Berlin was the geographer **Carl Ritter** (1779–1859).

For the next four years, Guyot worked as a private tutor for the family of the Count of Pourtalès-Gorgier in Paris and traveled throughout **Europe**. Reacquainted with Agassiz in Paris in 1838, Guyot became interested in glaciers, though he disagreed with Agassiz on many points of interpretation. They decided to collaborate on a study of Alpine glaciers, but through a misunderstanding between them, the resultant publication appeared under Agassiz's name alone in 1847. Guyot and Agassiz remained friends, but Guyot only received credit for his work on this project in the 1880s, after both his and Agassiz's deaths.

Pursuing research in botany, geology, geography, glaciology, **meteorology**, and **cartography**, Guyot taught history, natural history, and **physical geography** at the Neuchâtel Academy from 1839 until the Revolutions of 1848 closed that institution and deprived him of his livelihood. In consequence, he, Agassiz, and many other first-rate scientists immigrated to America. Lectures he presented at the Lowell Technological Institute in Boston, Massachusetts, published as *Earth and Man* in 1849, quickly established his reputation in the English-speaking world.

In 1854, Guyot became professor of geology and physical geography at Princeton University, where he remained until his death in Princeton, New Jersey, on February 8, 1884. He founded what became the Princeton Department of Geosciences in 1855 and the Princeton Museum of Natural History in 1856. Princeton named him the first John I. Blair professor of geology in 1864. During his summers on vacation from Princeton, Guyot conducted extensive on-site meteorological studies of the Appalachians from Mt. Katahdin, Maine, to Mt. Oglethorpe, Georgia, sponsored by the Smithsonian Institution. These explorations eventually led to the creation of the Appalachian Trail. He encouraged his student, William Berryman Scott (1858–1947), later the second Blair professor, to lead a dangerous expedition to Colorado to gather **fossils** in 1877.

The "guyot," a flat-topped undersea mountain, was named for him by the sixth Blair professor, **Harry Hammond Hess** (1906–1969). Also named in his honor are three mountains, one in New Hampshire, one on the North Carolina-Tennessee border, and one in Colorado, as well as Guyot Hall, the geology building at Princeton.

See also Evolution, evidence of; Glacial landforms; Glaciation; Moraines; Mountain chains; Weather forecasting methods; Rock

GUYOTS AND ATOLLS

A guyot is a flat-topped submarine mountain, or seamount, that once emerged above sea level as a volcanic island, and then resubmerged when volcanic activity ceased. **Erosion** by wave activity during submergence creates the characteristic flat-topped profile of a guyot. In some cases, carbonate reefs fringing an aging volcanic island continue to grow as the island sinks below sea level, leaving a circular island of coral, or an atoll, surrounding a round lagoon where the peak of the extinct **volcano** once stood. The British naturalist, Charles Darwin (1809–1882), observed atolls in the southwest Pacific Ocean during his nineteenth century travels aboard the *Beagle*, and he was the first to suggest that an atoll is a crown of coral on a newly-submerged guyot.

Seamounts are volcanic hills or plateaus formed by extrusion of **lava** onto the seafloor in places where plates of oceanic **lithosphere** override hot areas in the mantle near divergent plate tectonic boundaries called **mid-ocean ridges**, and over localized intraplate mantle upwellings called hot spots. Most seamounts never grow tall enough to become islands, but some very large ocean-floor volcanoes, particularly those above well-established **hotspots**, emerge above sea level before plate motion removes them from their magmatic sources. The Hawaiian Islands and Iceland are examples of oceanic islands created by vigorous hot spot volcanism.

After growth and emergence, large volcanic islands evolve through several stages of decline and submergence. The weight of a cooling volcanic construction depresses the lithosphere into an underlying plastic layer of the mantle called the

asthenosphere. When volcanic activity ceases or slows, the rate of depression, or isostatic subsidence, outpaces the rate of volcanic construction, and the island sinks. Wave erosion cuts a bench encircling the declining island, and carbonate organisms, including corals, construct a ring of shallow-water carbonate rocks around it. As subsidence and erosion continue, the peak of the extinct volcano is planed off, and a carbonate lagoon fills the flat wave-cut surface creating an atoll. Eventually, the atoll sinks below the biologically productive photic zone, and carbonate production ceases. Most tropical guyots have a carbonate cap, while most high-latitude guyots do not. Once the guyot has fully cooled, it reaches a state of isostatic equilibrium and stops subsiding. Plate motion then carries the guyot passively toward a **subduction zone** where it will eventually accrete to a continental margin or subduct into the mantle.

See also Isostasy

GYPSUM

Gypsum, a white mineral soft enough to be scratched with a fingernail, is hydrated calcium sulfate [Ca(SO$_4$)·2H$_2$O]. Gypsum often begins as calcium sulfate dissolved in an isolated body of salt **water**. As the water evaporates, the calcium sulfate becomes so concentrated that it can no longer remain in solution and crystallizes out (precipitates) as gypsum. Many large beds of gypsum have been formed in this way.

Gypsum occurs in a number of distinct forms, including a clear, parallelogram-shaped crystal (selenite); a white, **amorphous** form (alabaster, used for ornamental carving); and a fibrous, lustrous form (satin spar, used in jewelry). When ground up and heated to drive off its water, gypsum becomes a powder termed plaster of Paris. Plaster of Paris has the useful property of hardening in any desired shape when mixed with water, molded, and allowed to dry.

Gypsum is one of the most widely used **minerals** in the world. Some 90 countries mine gypsum, producing more than 100 million tons (91 million metric tons) annually. The construction industry has long been particularly gypsum intensive. In the late nineteenth and early twentieth centuries gypsum was widely used in plastering, which since the 1950s has been displaced by gypsum drywall (sheetrock). The average new U.S. home contains tons of gypsum drywall. Gypsum is also an ingredient in portland cement, which is used in the construction of bridges, buildings, highways, and the like, and millions of tons of gypsum are used annually as fertilizer. Small quantities of pure gypsum are essential in smelting, glassmaking, and other industries.

Low-grade gypsum is manufactured synthetically at coal-fired electric power plants as a by-product of pollution-control processes that remove sulfur from flue gas. Synthetic gypsum production exceeds 110 million tons (100 million metric tons) annually.

See also Mohs' scale

H

HAIL • *see* PRECIPITATION

HALF-LIFE

As defined by geophysicists, the half-life (or half-value period) of a substance is the time required for one-half of the atoms in any size sample to radioactively decay.

Radioactive elements have different isotopes that decay at different rates. As a result, half-life varies with regard to the particular isotope under consideration. Some isotopes have very short half-lives, for example oxygen-14 has a half-life of only 71 seconds, some are even shorter—with values measured in millionths of a second not being uncommon. Other elements' isotopes can have a much longer half-life, thallium-232 has a half-life of 1.4×10^{10} years and carbon-14 has a half-life of 5,730 years. This latter figure is used as the basis of radiocarbon dating.

While living, an organism takes in an amount of carbon-14 at a relatively constant rate. Once the organism dies no more carbon-14 is taken in and the amount of carbon-14 present overall starts to decrease, decreasing by half every 5,730 years. By measuring the ratio of carbon-12 to carbon-14 an estimate of the date when carbon-14 stopped being assimilated can be calculated. This figure can also be obtained by comparing the levels of **radioactivity** of the test material to that of a piece of identical material that is fresh. Other radioactive elements can be used to date older, inorganic materials (e.g., rocks).

Strontium-90 has a half-life of 29 years. If starting with a 2.2 lb (1 kg) mass of strontium-90, then after 29 years there will only be 1.11 lb (0.5 kg) of strontium-90 remaining. After a further 29 years there will only be 0.55 lb (0.25 kg). Strontium-90 decays to give yttrium-90 and one free electron. Half-life is independent of the mass of material present.

The half-life ($t_{1/2}$) of a material can be calculated by dividing 0.693 by the decay constant (which is different for different radionucleotides). The decay constant can be calculated by dividing the number of observed disintegrations per unit time by the number of radioactive nuclei in the sample. The decay constant is usually given the symbol k or λ.

The half-life of a material is a measure of how reactive it is either in terms of radioactive decay or in participation in specific reactions.

See also Atomic mass and weight; Atomic number; Atomic theory; Cosmic microwave background radiation; Dating methods; Geologic time

HALLEY, EDMOND (1656-1743)
English astronomer

The son of a wealthy merchant, Edmond Halley was attracted to **astronomy** after seeing two **comets** as a child. By the age of eighteen, he had found errors in authoritative tables on the positions of Jupiter and Saturn and by nineteen, had published a paper on the laws of **Johannes Kepler**. In 1676, Halley left England for St. Helena, an island west of **Africa**, to map the southern constellations, a task never before undertaken. Although the **climate** of St. Helena proved less than ideal for Halley's purposes, he was able to catalogue 341 stars before returning to England. His pioneering work on the island assured his place in England's scientific community, and Halley was awarded a master's degree from Oxford as well as election to the Royal Society.

In 1684 Halley entered into a conversation with biologist Robert Hooke and architect Christopher Wren (1632–1723) that concerned the force that drove the movement of the planets. Unable to reach a satisfactory conclusion, Halley turned to his friend Isaac Newton. Discovering that Newton had already answered the question using his law of **gravity**, Halley convinced his reticent friend to publish his findings. Using funds bequeathed to him by his father, Halley financed the publication of *Mathematical Principles of Natural Philosophy*, now considered one of the classic texts of modern scientific thought.

At the age of 39 Halley turned his attention to comets, which, as they streaked unexpectedly through the sky, appeared ungoverned by Newton's law. Halley, however, believed that gravity did indeed dictate their path and that the rarity of their appearances was due to the vast length of their orbit, which was elliptical. With the help of Newton, Halley compared the paths of past comets that had appeared in 1531, 1607, and 1682. From this data he was able to determine that these seemingly separate comets were indeed the same comet and accurately predicted its reappearance in 1758. In 1705, Halley published his findings in *A Synopsis of the Astronomy of Comets*. Eventually, the comet that he predicted was named for him.

In addition to his findings concerning comets, Halley undertook a lengthy study of solar eclipses and discovered that the so-called fixed stars actually moved with respect to each other. He also wrote in favor of the theory that the universe is limitless and has no center. Halley's scientific interests, however, extended beyond astronomy. He played a major role in transforming the Royal Society from a social club into a well-respected clearing-house for scientific ideas. He devised the first **weather** map and calculated the amount of salt deposited by **rivers** into seawater over millions of years which allowed him to draw conclusions about the age of Earth. He also invented, developed, and tested one of the first practical diving bells. He served as chief science advisor to Peter the Great when the Russian czar came to England in an attempt to integrate Western advances into his country's society. From 1698 to 1700, Halley commanded the *Paramour*, a Royal Navy ship, for a scientific expedition which studied the effects of the Earth's **magnetic field** on magnetic needle compasses. He became Astronomer Royal in 1720, and continued to make astronomical observations and attend scientific meetings until shortly before his death in Greenwich at the age of 86.

HARRISON, JOHN (1693-1776)

English clockmaker and carpenter

John Harrison solved the so-called "longitude problem," that is, he developed the means to enable navigators to calculate their east-west (longitudinal) positions at sea. A ship's north-south (latitudinal) position is easily computed from the **Sun**, stars, date, and local time, but to calculate longitudinal position a navigator must also know the current time at the home port and compare it with the local time of the ship as determined by observing the Sun and stars. This calculation is based on the fact that every hour represents 15 degrees of **longitude**. The principle was known centuries before Harrison, but using it was not possible in practical navigation until he invented his portable, durable, and extremely accurate clock, which proved reliable under the harsh conditions of the sea.

Born the son of a carpenter on March 24, 1693, in Foulby, Yorkshire, England, Harrison early learned his father's trade as well as surveying, clockmaking, bell tuning, and several other practical skills. He also enjoyed music and was a

singer. The family moved to Barrow-on-Humber, Lincolnshire, while Harrison was still a child. In the 1720s, he began a professional association with his brother James, born in 1704. Together until 1739, they designed and built beautiful, precise, reliable clocks, soon renowned as the most accurate in Britain.

As the world's dominant sea power from the end of the sixteenth century until the middle of the nineteenth, Britain was keenly aware of the longitude problem and was quite serious about solving it. In 1675, King Charles II founded the Royal Observatory at Greenwich, England, mainly to gather data for the longitude problem. In 1714, the British Parliament passed the Queen Anne Act, which offered a prize of £20,000 (about £1,932,917 in 2001 British money or $2,811,860 in 2001 American money) to anyone who could calculate a ship's longitude to within a half-degree throughout its voyage from Britain to the West Indies. Many tried and failed to win that prize.

Fixed on winning the prize, Harrison began working on the longitude problem in 1730 and submitted his first sea clock, known as the H1 chronometer, to the Board of Longitude in London in 1736. On the basis of H1's partial success, the Board gave him financial assistance to continue his research. From 1737 to 1740, he worked on a larger instrument, H2, but it failed. His experiments with H3 from 1740 to 1749 also ended in failure. In 1755, he stumbled across an entirely different design for H4, similar to a pocket watch. The test results of H4 on the voyage of the *Deptford* to Jamaica in 1761–62 and on the voyage of the *Tartar* to Barbados in 1764 exceeded the stipulations of the Queen Anne Act.

Despite the success of H4, the Board of Longitude awarded Harrison only £10,000. The Royal Astronomer Nevel Maskelyne (1732–1811), even though he was aboard the *Tartar* during the trial of H4, remained unconvinced that any timepiece could be a more accurate indicator of longitude than the popular "lunar distance method," by which navigators computed longitude from their observations of the Moon's position relative to selected stars, according to tables prepared by the Royal Observatory. Maskelyne was jealous of Harrison, whom he spurned as a mere "mechanic," and changed the rules of the contest to favor astronomers.

Harrison, with his son William, spent the rest of his life trying to claim the second half of his prize. The Board was adamant about not giving it to him but, in 1773, after Harrison appealed to King George III, Parliament grudgingly recognized his having solved the longitude problem and gave him an additional £8,750. All four of his marine chronometers are now in the National Maritime Museum, London. He died in London on his birthday in 1776.

See also History of exploration II (Age of exploration); Latitude and longitude; Time zones

HAUPTMAN, HERBERT A. (1917-)

American mathematician and biophysicist

In the early 1950s, Herbert A. Hauptman and former classmate, Jerome Karle, developed a mathematical system, usually referred to as the "direct method," for the interpretation of data on atomic structure collected through x-ray crystallography. The system, however, did not come into general use until the l960s, and it was only in 1985 that Hauptman and Karle were jointly awarded the Nobel Prize in chemistry for their accomplishment.

Hauptman and Karle developed a complex series of mathematical formulas, relying heavily on probability theory, which made it possible to correctly infer the phases from the data that was recorded on the photographic film. Their new mathematical system came to be known as the determination of molecular structure by "direct method." They demonstrated the workability of their new technique in 1954 by calculating by hand, in collaboration with researchers at the United States Geological Survey, the atomic structure of the mineral colemanite.

Herbert Aaron Hauptman was born in New York City on February 14, 1917, the son of Israel Hauptman, an Austrian immigrant who worked as a printer, and Leah (Rosenfeld) Hauptman. He grew up in the Bronx and graduated from Townsend Harris High School. At the City College of New York, he majored in mathematics and received a Bachelor of Science degree in 1937. Karle, his later collaborator, also graduated from City College the same year. Hauptman went on to complete a master's degree in mathematics at Columbia University in 1939. He married Edith Citrynell, a schoolteacher, on November 10, 1940; they eventually had two daughters. Hauptman worked for two years as a statistician in the United States Bureau of the Census before serving in the United States Army Air Force from 1942 to 1947. After his period of service ended, Hauptman went to work as a physicist and mathematician at the Naval Research Laboratory in Washington, remaining there until 1970. While working at the laboratory, he enrolled in the doctoral program in mathematics at the University of Maryland and received his Ph.D. in 1955.

At the Naval Research Laboratory, Hauptman renewed his acquaintance with Karle, who had come to the laboratory in 1946. The two men soon began to work together on the problem of determining molecular structures through the methodology of x-ray crystallography. Most of the work that later led to their **joint** Nobel Prize was done between 1950 and 1956. A brief monograph, *Solution of the Phase Problem, 1. The Centrosymmetric Crystal,* was published in 1953 that revealed many of the results of their studies.

The German physicist Max Laue had discovered as far back as 1912 that it was possible to determine the arrangement of atoms within a crystal by studying the patterns formed on a photographic plate by x rays passed through a crystal. Since that time x-ray crystallography had become a standard tool for chemists, physicists, geologists, biologists, and other scientists concerned with determining the atomic structure of substances. X-ray crystallography, for example, had made possible the discovery of the double-helical structure of deoxyribonucleic acid (DNA) by molecular biologists Francis Crick, James Watson, and others in the 1950s. The problem with the technique was that interpreting the patterns on the photographic plates was a difficult, laborious, and time-con-

suming task. The accurate determination of the atomic structure of a single substance could require one or more years of work based upon indirect inferences that often amounted to educated guesswork. The greatest difficulty arose from the fact that while photographic film could record the intensity of the x-ray dots that formed the patterns, it could not record the phases (the minute deviations from straight lines) of the x rays themselves.

Hauptman and Karle's system met with a good deal of skepticism and resistance from the specialists in x-ray crystallography in the 1950s and was largely ignored for about ten years. This was partly due to the fact that most crystallographers of the time lacked the mathematical knowledge and sophistication to make use of the new technique. It also stemmed from the fact that the necessary mathematical calculations themselves were a laborious process. It was the introduction of computers and the development of special programs to deal with the Hauptman-Karle method in the 1960s that finally led to its widespread acceptance and use. The work that originally required months or years to complete could now be done in a matter of hours or, at most, days. By the mid 1980s the atomic structures of approximately 40,000 substances had been determined through use of the direct method, as compared to only some 4,000 determined by other methods in all the years prior to 1970, and some 4,000 to 5,000 new structures were being determined each year.

Hauptman left the Naval Research Laboratory in 1970 to become head of the biophysics laboratory at the Medical Foundation of Buffalo, a small but highly regarded organization specializing in research on endocrinology. He also became professor of biophysical science at the State University of New York at Buffalo. Hauptman served as executive vice president and research director of the Medical Foundation from 1972 to 1985 and president from 1985 onwards. There he worked to perfect the direct method and to extend its use to the study of very large atomic structures. Hauptman has received numerous awards, including the 1985 Nobel Prize in chemistry shared with Karle, the Award in Pure Sciences from the Research Society of America in 1959, and, also with Karle, the A. L. Patterson Memorial Award of the American Crystallography Association in 1984.

See also Atomic theory; Crystals and crystallography; Geochemistry; Industrial minerals; Mineralogy

HAWAIIAN ISLAND FORMATION

The Hawaiian archipelago is a group of 132 islands, reefs, and shoals in the North Pacific Ocean that extends about 1,525 mi (2,454 km) from Kure Atoll (29°N, 178°W) to the big island of Hawaii (19°N, 156°W). This string of geographically remote and geologically unique volcanic islands makes up the U.S. state of Hawaii, and includes the eight main Hawaiian Islands of Ni'ihau, Kauai, Oahu, Molokai, Lanai, Maui, Kahoolawe, and Hawaii. The islands are progressively younger in geologic age toward the southeast; Kauai and Ni'ihau are about 5 million years in age, and the big island of

Hawaii is less than 0.5 million years old. Indeed, new volcanic rocks are being deposited at Mt. Kilauea on Hawaii today.

The Hawaiian Islands are the exposed summits of the southernmost seafloor mountains, or seamounts, in the Hawaiian-Emperor seamount chain. This 3,105 mi (5,750 km) line of 107 volcanoes has formed over the last 70 million years as the Pacific Lithospheric Plate has moved to the northwest over a stationary magmatic hot spot in the mantle. Each individual **volcano** in the seamount chain formed as heat from the Hawaiian hot spot melted the overlying oceanic **crust**, and generated buoyant molten **rock**, or **magma**, which migrated upward and erupted onto the seafloor as **lava**. Many sequential lava flows then amalgamated to form seamounts composed mainly of an **iron** and magnesium-rich, or **mafic**, volcanic rock called **basalt**. Eventually, some of these seamounts grew tall enough to emerge above sea level. The Hawaiian-Emperor seamounts are examples of basaltic volcanoes with low-angle slopes and wide bases, called shield volcanoes. Mauna Loa, the central volcanic peak on Hawaii, is the world's tallest and most massive mountain when measured from its submarine base. It has a total elevation of about 32,000 ft (10 km), and its base covers an **area** about the size of the U.S. state of Connecticut.

Ongoing northwestward migration of the Pacific Plate at 3.4 in/year (9 cm/yr) has carried all but the newest Hawaiian-Emperor seamounts away from the hot spot. As a seamount moves away from the hot spot, volcanic activity ceases, its rock base cools, and it begins to subside into the surrounding ocean crust. An aging oceanic island then sinks below sea level, and wave **erosion** levels off the volcanic peak, creating a flat-topped seamount called a guyot. Sometimes, **coral reefs** fringing a volcanic island continue to grow after the island has subsided below sea level, creating a ring-shaped carbonate island called an atoll, or annular island. English naturalist Charles Darwin (1809–1882) first suggested this explanation for the formation of atolls during the voyage of the HMS *Beagle* from 1831 to 1836.

By this mechanism of sequential island formation and subsidence, the Hawaiian hot spot has perforated the Pacific Plate with a line of volcanoes that are younger and higher toward the southeast. The Hawaiian-Emperor Chain propagated northward, beginning at least 70 million years ago, the age of the Meiji seamount at the Aleutian Trench. About 40 million years ago, a dogleg bend in the chain suggests a shift to northwestward plate motion, possibly due to the collision of the Indian subcontinent with **Asia** that created the Himalayan Mountains at that time. Since 40 million years ago, the Pacific Plate has moved northwest, bringing the hot spot to its present position beneath the southern shore of the island of Hawaii. The newest volcano in the Hawaiian-Emperor Chain, the Lo'ihi seamount, is presently forming on the seafloor about 25 mi (40 km) southeast of Hawaii.

Today, geologists at the Hawaiian Volcano Observatory at Mt. Kilauea, and visitors to Hawaii Volcanoes National Park, can observe active **volcanic eruptions**. Low-viscosity basaltic lava erupts from volcanic vents at about 1,830°F (1,000°C). The surface of fast-flowing lava streams cools to create a ropy-textured skin called *pahoehoe*. (*Pahoehoe* means

"rope" in Hawaiian.) After the surface of a flow has cooled, lava may continue to move beneath the surface in lava tubes. Sometimes, dissolved volatile gases escape during cooling, and the lava forms a jumble of sharp blocks called *aa*. When lava flows into the ocean, it cools very rapidly to form pillow basalt, the most common submarine basaltic texture. The Hawaiian Islands are the world's best natural laboratory for the study of hot spot dynamics, basaltic volcanism and ocean island formation.

See also Volcanic eruptions; Volcanic vent

HAWKING, STEPHEN (1942-   )

English physicist

Stephen Hawking has been called the most insightful theoretical physicist since **Albert Einstein**. His work concentrates on the puzzling cosmic bodies called black holes and extends to such specialized fields as particle **physics**, supersymmetry, and quantum **gravity**. The origin and fate of the universe are a central concern of Hawking's work. Though few people are able to understand the intricacies of these abstruse subjects, Hawking has gained a worldwide following, not only among other scientists, but also among a great many laypeople. As an author and lecturer, he has achieved celebrity status.

Stephen William Hawking was born in Oxford, England. He often refers to the fact that his birth date coincided with the 300th anniversary of Galileo Galilei's death. Hawking was the eldest child of an intellectual and accomplished family. His father, Frank Hawking, was a physician and research biologist who specialized in tropical diseases; his mother, Isobel, the daughter of a Glasgow physician and a well-read, lively woman, was active for many years in Britain's Liberal Party

Stephen Hawking's earliest years were spent in Highgate, a London suburb. In 1950, when he was eight, the family moved to St. Albans, a cathedral town some twenty miles northwest of London. Two years later, his family enrolled him in St. Albans School, a private institution affiliated with the cathedral. As Michael White and John Gribbin describe the young schoolboy in *Stephen Hawking: A Life in Science*, "He was eccentric and awkward, skinny and puny. His school uniform always looked a mess and, according to friends, he jabbered rather than talked clearly, having inherited a slight lisp from his father." Young Hawking's abilities made little impact on his teachers or fellow students. But he already knew he wanted to be a scientist, and by the time he reached his middle teens, he had decided to pursue physics or mathematics.

Gangly and unathletic, Hawking formed close friendships with a small group of other precocious boys at school. Intrigued by subjects that focused on measurable quantities and objective reasoning, Hawking began to show increasing skill at mathematics, and soon he was outdistancing his peers with high grades while spending very little time on homework. In 1958, Hawking and his friends built a primitive computer that actually worked. In the spring of 1959, Hawking won an open scholarship in natural sciences to University College, Oxford—his father's old college—and in October he enrolled there. It was at Oxford that his unusual abilities began to become more obvious. Hawking's ease at handling difficult problems made it seem to others that he didn't need to study. In *Stephen Hawking's Universe*, John Boslough wrote, "He took an independent and freewheeling approach to studies although his tutor, Dr. Robert Berman, recalls that he and other dons were aware that Hawking had a first-rate mind, completely different from his contemporaries."

In 1962, after receiving a first-class honors degree from Oxford, Hawking set off for Cambridge University to begin studying for a Ph.D. in **cosmology**. Now he was beginning to deal with some of the themes that would preoccupy him throughout his life. One of these was the poorly understood question of black holes. As scientists were later to realize, a black hole is a cosmic body that by its very nature can never be seen. One type of black hole is thought to be the remnant of a collapsed star, which possesses such intense gravity that nothing can escape from it, not even light. Hawking was also intrigued by "space-time singularities," those phenomena in the physical universe or moments in its history where physics seems to break down. In attempting to understand a black hole and the space-time singularity at its center, Hawking made pioneering studies, using formulas developed more than half a century earlier by Einstein.

Hawking received his Ph.D. in 1965 and obtained a fellowship in theoretical physics at Gonville and Caius College, Cambridge. He continued his work on black holes, frequently collaborating with Roger Penrose, a mathematician a decade his senior, who like Hawking was deeply interested in theories of space-time. Though still in his twenties, Hawking was beginning to acquire a reputation, and he would often attend conferences where he shocked people by questioning the findings of eminent scientists much older than himself.

In 1968, Hawking joined the staff of the Institute of Astronomy in Cambridge. He and Penrose began using complex mathematics to apply the laws of thermodynamics to black holes. He continued to travel to America, the Soviet Union, and other countries, and in 1973, he published a highly technical book, *The Large Scale Structure of Space-Time*, written with G. F. R. Ellis. Not long afterward, Hawking made a startling discovery: whereas virtually all previous thinking assumed that black holes could not emit anything, Hawking theorized that under certain conditions they could emit subatomic particles. These particles became known as Hawking Radiation.

Early in 1974, at the unusually young age of 32, Hawking was named a fellow of the Royal Society. Soon afterward, he spent a year as Fairchild Distinguished Scholar at the California Institute of Technology in Pasadena. On returning to England, he continued to work toward a theory of the origin of the universe. In this endeavor, he made progress toward linking the theory of relativity, which deals with gravity, with quantum mechanics, which deals with minuscule events inside the **atom**. Such a theoretical linkage, long sought by researchers, is called the Grand Unification Theory. In 1978, Hawking received the Albert Einstein Award of the

Stephen Hawking. *AP/Wide World. Reproduced by permission.*

Lewis and Rose Strauss Memorial Fund, the most prestigious award in theoretical physics. The following year he co-edited a book with Werner Israel, called *General Relativity: An Einstein Centenary Survey*. In 1979, Hawking was named Lucasian Professor of Mathematics at Cambridge—a position held three centuries earlier by **Sir Isaac Newton**. In the 1980s, his work was beginning to lead him to question the **big bang theory**, which most other scientists were accepting as the probable origin of the universe. Hawking now asked whether there really had ever been a beginning to space-time (a big bang), or whether one state of affairs (one universe, to put it loosely) simply gave birth to another without beginning or end. Hawking suggested that new universes might be born frequently through little-understood anomalies in space-time. He also investigated string theory and exploding black holes, and showed mathematically that numerous miniature black holes may have formed early in the history of our universe.

In 1988, Hawking's *A Brief History of Time: From the Big Bang to Black Holes* was published. Intended for a general audience, it leapt onto best-seller lists in both America and Britain and remained there for several years. In that book, Hawking explained in simple language the evolution of his own thinking about the cosmos. Major articles followed in *Time, Popular Science,* and other magazines; films and television programs featured Hawking. He received honorary degrees from many institutions, including the University of Chicago, Princeton University, and the University of Notre Dame. His numerous awards included the Eddington Medal of the Royal Astronomical Society, in 1975; the Pius XI Gold Medal, in 1975; the Maxwell Medal of the Institute of Physics, in 1976; the Franklin Medal of the Franklin Institute, in 1981; the Gold Medal of the Royal Society, in 1985; the Paul Dirac Medal and Prize, in 1987; and the Britannica Award, in 1989.

In 1965, Hawking married Jane Wilde, and they had two sons and a daughter. The couple separated in 1990. Hawking suffers from amyotrophic lateral sclerosis, also called Lou Gehrig's disease, which confines him to a wheelchair and requires him to use a computer and voice synthesizer to speak.

HELIOCENTRIC MODEL • *see* ASTRONOMY

HERSCHEL, CAROLINE LUCRETIA (1750-1848)
German astronomer

Caroline Herschel was the first female astronomer to discover a comet. Herschel grew up in a home where her father encouraged learning, much to the displeasure of her mother, who believed girls should focus their education solely on skills necessary to manage a well-appointed traditional home. After her father's death in 1767, Herschel's formal education in mathematics and science ceased, as she ceded to the wishes of her mother. Finally, in 1772, Herschel left Germany to pursue a musical career in Bath, England, living with her brother, the astronomer William Herschel.

In England, Herschel trained to become a professional singer, but she also began to study mathematics under her brother's tutelage. William Herschel soon involved her in his hobby, **telescope** building. She helped him grind and polish mirrors for his telescopes, while copying catalogs and tables for his reference. After he discovered the planet Uranus in 1781, William Herschel was awarded a yearly stipend by King George III that allowed him and his sister to pursue **astronomy** full time.

As she became more proficient with her own telescope, Herschel made a name for herself in this largely male domain. In 1783, she discovered three new nebulas and from 1786–1797, she discovered eight **comets**. George III awarded her a salary as well, a rare gesture at the time. She also took on the formidable task of making a thorough index of the star catalog created by John Flamsteed (1646–1719), the first Royal Astronomer. This job called for perseverance, accuracy, and attention to detail, all qualities in which Herschel excelled.

Following her brother's death, Herschel returned to Hanover, Germany, but remained in close contact with her brother's son, astronomer John Herschel (1792–1871), for whom she compiled a new catalog of nebulae. Herschel and Scottish scientific writer Mary Somerville (1780–1872) became the first women to be awarded an honorary membership in the Royal Society. In spite of her informal training, Herschel became a well-known figure in her own time and an important figure in the history of astronomy.

HERSCHEL, SIR WILLIAM (1738-1822)
German-born English astronomer

Sir William Herschel was among the preeminent astronomers of the eighteenth century, and is credited with discovering the planet Uranus, binary stars, nebulae, and for correctly describing the form of the Milky Way galaxy.

Herschel was born in Hanover, Germany at a time when the city belonged to England under the rule of George II. As Herschel's father was a musician in the Hanoverian army, Herschel himself was trained in music in order to enter the same profession. The Seven Years' War, however, made military life an unattractive option, and in 1757, Herschel arrived in England where he began working as an organist and music teacher. Herschel learned of **astronomy** through his interest in the theory of music and the scientific basis for musical sounds, which led him to mathematics and then optics.

Newton's treatise on optics inspired Herschel with his desire to study the stars. Unable to find a **telescope** of a high enough resolution, he decided to grind his own lenses and to design his own instruments. With his first telescope, a 6-foot Gregorian reflector that was one of the best of its kind, he decided that its first application would be to conduct a systematic survey of the stars and planets. Herschel was assisted in this endeavor by his sister Caroline, who also discovered eight **comets** and produced two astronomy catalogues in her lifetime. Throughout his life, Herschel built numerous telescopes, each one more sophisticated and more powerful than the last.

Herschel's first major discovery occurred in 1781, during his second survey of the sky when he announced the existence of a new planet to be found in the constellation of Taurus. Herschel's name for the new planet was *Georgium Sidus,* George's star, in honor of King George III, but it eventually came to be known as Uranus, after the mythical father of Saturn. The discovery of Uranus, which effectively doubled the previously accepted size of the **solar system,** caused a popular and scientific sensation, and George III appointed Herschel to the position of King's Astronomer while providing him with a small annuity that allowed him to pursue astronomy full time.

Herschel's most significant achievements were in the **area** of sidereal astronomy, to which he contributed the first systematic body of evidence on the order and nature of the stars and the planets. Whereas plenty of theories had been put forward by prominent philosophers of the time on the systems that might govern the universe, none were supported by any scientific gathering of data. In 1783, Herschel began to search for nebulae in the sky, and raised their known total from little more than 100 to 2,500. Much of eighteenth-century astronomy set out to determine the distances between stars; trigonometrical calculations based on their apparent annual movement, however, had failed. **Galileo Galilei** had proposed the use of double stars, pairs of stars very close together, to calculate stellar distance, where the fainter member of the pair was so far away as to represent a fixed point from which the annual movement of its brighter companion could be measured. In Herschel's second survey, he searched for double stars, producing three catalogs over the next 40 years and listing 848 examples. It was later discovered by another astronomer who had seen Herschel's work that these double stars were in fact companions in **space** held together by gravitational forces and therefore equidistant from the earth; Herschel had assumed that companions in space would have been of equal brightness, and had therefore discounted this possibility. Nonetheless, much of Herschel's work was concerned with producing evidence for the powers of attraction between stars. In three of his papers delivered between 1784 and 1789, he proposed a cosmogony for the universe in which stars, initially randomly scattered throughout the universe, clustered together over time around the regions from which they originally developed.

Herschel was the first to embark upon a scientific study of the Milky Way and half of his work, though less influential, focuses upon the solar system. He studied the **Sun**, observing that what we see is not the Sun itself but the **clouds** of gases that cover its surface, and examined the nature of the infrared section of the spectrum by which some of the Sun's heat is transmitted. Besides calculating the height of lunar mountains, Herschel devoted most of his attention to the other known planets, Venus, Mars, Jupiter, and Saturn, determining their **rotation** period and checking the inclination of their axes, their shape, and the nature of their atmospheres. Herschel devoted most of his attention to examining Saturn and its rings, arguing at one point that the rings were solid, but later conceding that they were in fact composed of floating particles.

Herschel's work on nebulae had led him to conclude that they might well be other solar systems seen only as a luminous cluster of stars around a brighter one. As a result, he saw the Milky Way and Earth as only one rather insignificant part of the universe. In this sense, he changed the status of the solar system within the universe in much the same way as **Nicolas Copernicus** had the earth when he showed that the planets revolved around the Sun rather than Earth.

See also Cosmology

HESS, HARRY HAMMOND (1906-1969)

American geologist

Harry Hammond Hess spent much of his career studying what the ocean floor was made of and where it came from. He was

a renowned geologist whose interests and influence ranged from **oceanography** to **space** science. One of Hess's most important contributions to science was the concept of seafloor spreading, which became a cornerstone in the acceptance of the **continental drift theory** during the 1960s. As an officer in the United States Naval Reserve, he was able to combine military service with scientific investigation; in his later years, he became an important figure in NASA, helping direct the science of lunar exploration.

Hess was born in New York City to Julian S. Hess, a member of the New York Stock Exchange, and Elizabeth Engel Hess. He attended Asbury Park High School in New Jersey before entering Yale University in 1923. At Yale, he intended to study electrical engineering, but changed his mind and graduated in 1927 with a B.S. degree in **geology**. Hess then spent two years in northern Rhodesia (now Zambia) as an exploration geologist. Returning to the United States, Hess received his doctorate from Princeton University in 1932. He taught at Rutgers University for a year, conducted research at the Geophysical Laboratory at the Carnegie Institute of Washington, and then returned to Princeton in 1934. Hess would remain at Princeton for essentially the rest of his career, serving as chair of the university's geology department from 1950 to 1966.

Annette Burns, daughter of a botany professor at the University of Vermont, became Hess's wife in 1934. She was a source of strong support for Hess throughout his life, and accompanied him to conferences and scientific meetings. The couple had two sons.

As a professor at Princeton, Hess continued his work on mountain ranges and **island arcs**, which are arc-shaped chains of islands that usually contain active volcanoes. By 1937, he had developed a unifying hypothesis that tied together the creation of island arcs with the presence of **gravity** anomalies and magnetic belts of serpentine (a **rock** which is formed by the crystallization of **magma**).

Hess's geological research was halted during World War II because he was a reserve officer in the Navy. He was initially assigned to duty in New York City, where he was responsible for estimating the positions of enemy submarines in the North Atlantic. Hess was then assigned to active sea duty and eventually became commander of an attack transport ship. This vessel carried equipment for sounding the ocean floor, and Hess took full advantage of it. He mapped a large part of the Pacific Ocean, discovering in the process the underwater flat-topped seamounts that he named **guyots**, in honor of A.H. Guyot, the first professor of geology at Princeton. The origin of guyots was puzzling, for they were flat on top as if they had been eroded off at the ocean surface, yet were two kilometers below sea level. As commander of the USS *Cape Johnson*, Hess also participated in four major combat landings, including one at Iwo Jima. Remaining a reserve officer after the war, Hess was called on for advice in such emergencies as the Cuban missile crisis in October 1962. By the time of his death he had achieved the rank of rear admiral.

After the war ended, Hess continued to study guyots as well as midocean ridges, which run down the centers of the Atlantic and Pacific **Oceans** like an underwater backbone. He also continued his mineralogical studies on the family of pyroxenes, an important group of rock-forming **minerals**. In 1955, he proposed that the boundary between the **crust** and the mantle of the earth is due to a change in the chemical composition of rocks.

During the 1950s, Hess became an influential backer of the ill-fated "Project Mohole," which proposed to drill a hole through the shallow oceanic crust into the earth's mantle for scientific sampling. In 1961, an experimental hole was bored through 11,600 ft (3,535 m) of **water**, 600 ft (183 m) of sediments, and 44 ft (13 m) of **basalt**. President John F. Kennedy telegraphed his congratulations to the National Science Foundation; John Steinbeck wrote an article for *Life* magazine about it. Despite amassing 25 million dollars in federal funding, Project Mohole foundered in 1966 under rising costs and political intrigue. It did, however, become an important stepping stone for the Deep Sea Drilling Project, successfully begun in the late 1960s.

Hess accepted visiting professorships at South Africa's Capetown University from 1949 to 1950, and at Cambridge University in 1965. Otherwise, he remained at Princeton until his death. He received numerous awards and honors, both at home and abroad, and was a major figure in the American Miscellaneous Society, a loosely-gathered group of scientists from different fields who liked to discuss "miscellaneous" ideas, such as Project Mohole.

From 1962 until his death, Hess chaired the Space Science Board that advised NASA on its lunar exploration program. He lived long enough to see the first person walk on the **Moon** in July 1969. One month later, while attending a space science conference in Woods Hole, Massachusetts, Hess died even as he was consulting a doctor about chest pains that he was experiencing.

Hess made a major contribution to the continental drift theory, which viewed continental and oceanic positions as the result of the break up of a single "supercontinent" (a theory first proposed by **Alfred Wegener** in 1912). Suggesting a mechanism by which continents could move away from each other without tearing up a rigid seafloor, Hess managed to unite several disparate elements: the youth of the ocean floor, the origin of midocean ridges, and the presence of island arcs and deep sea trenches surrounding the Pacific.

Hess's hypothesis gave geologists their first clue that drifting continents are carried passively on the spreading seafloor. In 1963, Fred Vine and **Drummond Matthews** at Cambridge University proposed a corollary to Hess's hypothesis: if the seafloor is created at the midocean ridges and spreads outward—and if the earth's **magnetic field** reverses polarity every few thousands of years—then the seafloor should be made of magnetized strips running parallel to the midocean ridges, alternating between normal and reverse polarity. Their idea, proposed independently by Lawrence Morley of the Geological Survey of Canada, was confirmed a few years later when scientists found the underwater bands of differently-magnetized rocks.

This oceanographic data established that continental drift does in fact occur. Over the next couple of years, geologists eventually accepted the new and revolutionary idea.

Although certain details of Hess's seafloor spreading hypothesis have become outdated, its central idea—that seafloor is created at ridges and destroyed under continents—has become an important foundation of modern **earth science**.

HILDEBRAND, ALAN RUSSELL (1955-   )
Canadian geologist

Alan Russell Hildebrand is part of the Geological Survey in Canada and The Geology and Geophysics Department at the University of Calgary. Hildebrand's greatest accomplishment in the field of geology includes the discovery of the catastrophic Chicxulub Crater in the Yucatan Peninsula of Mexico that is considered the site of the catastrophic event that resulted in the extinction of the dinosaurs. Additionally, Hildebrand is one of the leaders for the **Prairie** Meteorite Search and is a member of the Meteorites and Impacts Advisory Committee.

Alan R. Hildebrand began his career in geology after receiving his bachelor's degree in geology at the University of New Brunswick in 1977. Initially, Hildebrand worked in the mineral exploration industry. Eventually Hildebrand returned to school to seek his Ph.D. in planetary sciences. As part of his dissertation, Hildebrand discovered the Chicxulub crater in the Yucatan Peninsula of Mexico. In 1992, he received his Ph.D. from the University of Arizona. In 1999, he became a member of the faculty in the Department of Geology and Geophysics at the University of Calgary and currently holds a Canada Research Chair in Planetary Sciences. Additionally, Hildebrand is a research scientist for the Geological Survey of Canada, with his research focusing mainly on the **K-T event** as well as meteorite processes and impacts.

It has been theorized that the dinosaurs became extinct 65 million years ago by a disaster that has become known as the K-T event (Cretaceous-Tertiary Mass Extinction event). Presumably an asteroid hit the earth, its impact killing 70% of the species on Earth. In 1990, Alan Hildebrand discovered the Chicxulub Crater in the Yucatan Peninsula of Mexico. The crater has a diameter of 112 mi (180 km) and has been dated at about 65 million years ago. The asteroid that is responsible for creating the Chicxulub Crater is believed to have had a diameter of 6 mi (10 km).

Alan Hildebrand is currently one of the leaders for the Prairie Meteorite Search, a national project that focuses on the discovery of meteorites by prairie farmers. Researchers involved in the project travel to prairie farms to instruct locals on what meteorites look like and examine specimens found by the farmers. Additionally, Hildebrand volunteers for the Meteorites and Impacts Advisory Committee to the Canadian Space Agency. This committee is responsible for discovering meteorites around Canada and investigates possible fireballs.

See also K-T event; Meteoroids and meteorites; Cretaceous Period

HISTORICAL GEOLOGY

All areas of geologic study are subdisciplines of either historical **geology**, which focuses on the chemical, physical, and biological history of Earth, or physical geology, which is the study of Earth materials and processes. Historical geology uses theory, observation, and facts derived from studying rocks and **fossils** to learn about the **evolution** of Earth and its inhabitants.

According to the principle of **uniformitarianism**, most physical and chemical processes occurring today are very similar to those that operated in the geologic past, although their rates may be different. Therefore, by studying modern geologic activities and their products, geologists can understand how these activities produced the ancient **rock** record. In other words, the present is the key to the past. The principle of uniformitarianism has been very useful in deciphering much of the rock record.

Studies in historical geology rely on the rock record for factual information about Earth's past. As geologists collect data, they develop hypotheses to explain phenomena they observe. Geologists test hypotheses by making further observations of rocks and the fossils they contain. If this and other research supports a hypothesis, eventually it will be accepted as a theory explaining how Earth, and the life on it, evolved through time.

Rocks preserve a record of the events that formed them. The trained observer can examine the physical, chemical, and biological characteristics of a rock and interpret its origin. Fossils are an especially useful type of biological evidence preserved in **sedimentary rocks** (they do not occur in igneous or metamorphic rocks). Organisms thrive only in those conditions to which they have become adapted over time. Therefore, the presence of particular fossils in a rock provides paleontologists with very specific insights into the environment that formed that rock.

In addition to body fossils, sediments also preserve a variety of tracks and trails (for example, footprints, burrows, etc.). These biological impressions preserve traces of the daily activities of organisms, rather than their bodies, and so are called trace fossils. These too provide important clues to certain aspects of Earth history.

Through studies of rocks and fossils, geologists have produced what is called the **geologic time** scale. This is a convenient way of representing the vast amounts of time and the numerous details of historical geology in a way that is easily expressed and understood. The geologic time scale consists of the dates of major events in Earth's history, placed in chronological order. These events, primarily major extinctions and episodes of organic evolution, separate the scale into distinct time units. From largest to smallest, these units are the geologic eon, era, period, and epoch. The age of each boundary event is determined by radiometric dating of rocks associated with the time unit boundary. Radiometric dating uses the rates of atomic decay for radioactive elements to determine the age of geologic materials.

See also Big Bang theory; Dating methods; Earth science; Fossil record; Evolution, evidence of; Evolutionary mechanisms; Stratigraphy

HISTORY OF EXPLORATION I (ANCIENT AND CLASSICAL)

As early as the dawn of the world's major civilizations, people developed a long-standing curiosity about their world and universe. Exploration was a means of pushing the boundaries of known lands, as well as creating a new interpretation of the workings of the cosmos. As man wandered farther from home, he found new civilizations, wide **oceans**, and exotic goods. Growing curiosity, the desire to enhance military might, and demand for goods linked exploration and trade.

The Egyptians were the first build sea worthy ships. The earliest expedition recorded in Egyptian hieroglyphics is that of Pharaoh Snefru in about 3200 B.C. In 2750 B.C., Hannu led an expedition to explore the Arabian Peninsula and the Red Sea. After Hannu's voyage, Egyptian exploration declined until the first millennium B.C. In 550 B.C., Egyptian vessels circumnavigated **Africa**. They also constructed a canal between the Red Sea and the Nile River to facilitate trade.

The Phoenicians were perhaps the most prolific seafarers and traders of the ancient world. From their main port of Carthage, the Phoenicians dominated trade in the Mediterranean Sea. The Phoenician monopoly of trade reached from the Straits of Gibraltar to the far reaches of Persia (present-day Iran).

In 510 B.C., Greek explorer Scylax, who served in the Persian Navy, traveled to the Indus River and the mountains of present-day Afghanistan and Pakistan. He searched for new trade routes and a way to break the Phoenician trade monopoly. Pytheas sailed to the coast of modern France and established a Greek port and military garrison at Massalia (Marseilles). He then continued his expedition, later circumnavigating Britain and exploring the North Sea. The invention of a new ship, the bireme, which had two decks and four rows of oarsmen, aided the Greeks in assuming dominance over the Mediterranean.

The Roman Empire, which reached the height of its power from 100 B.C. to A.D. 400, commanded both sea and land. Sea vessels were largely used as battleships, and while the Romans did have a considerable trade fleet, the most ambitious expeditions used large war ships that carried soldiers, slaves, and plundered goods. The **area** that the Phoenicians once controlled with trade, the Romans governed over directly. The continued success of Rome depended on military conquest, territorial expansion, and the growth of the imperial economy. Rome gained dominion over lands from Northern **Europe** to Northern Africa, from Spain to Persia. They developed circular trade routes that insured that various regions of the empire received the goods and raw materials desired. Timber was exported the peripheral regions where trees were scarce. Slaves were transported to regions of production and building. Olive oil and wine was traded throughout the

Empire. These complex trade routes that insured a steady stream of raw materials and luxury goods were the model for the Atlantic triangular trade routes of the 1700s.

The European Old World was not the only venue for world exploration. In the first century A.D., Chinese explorers made rapid technological advancements, inventing the compass and complex sailing vessels, which aided open **water** exploration. Most ships had to remain in sight of land in order to navigate, but the Chinese compass, as well as Phoenician astronomical charts, permitted longer voyages, sometimes beyond the sight of land. Early Chinese sailors explored many of Asia's **rivers** and surrounding **seas**. They ventured as far as India and the eastern coast of Africa. Exploration and trade aided in the creation of a powerful and far-reaching Chinese empire.

In the South Pacific, Polynesian mariners explored the regional islands even before the recorded history. In 100-ft (30.5-m) canoes with minimal sails, Polynesians hopped from island to island, as well as made long open sea voyages. By A.D. 1000, Polynesian explorers had set foot in Hawaii and New Zealand. These Pacific sailors had a deep understanding of ocean currents and prevailing winds that was not achieved in the Atlantic until the sixteenth century.

As exploration pushed the boundaries of the known world, philosophers, astronomers, and mathematicians devised new interpretations for the workings of the world and universe. Some focused on practical challenges, such as navigation, and devised complex charts of stars. Others took a universal approach, mingling religion with exploration and science to devise of theories of how the universe and Earth itself were structured. These structures, or cosmologies, dictated the bounds of scientific reasoning and exploration. The Greek mathematician, Ptolemy, devised a model for the universe that persisted for centuries, most especially through Europe's Dark and Middle Ages (496–1450). Not until the fifteenth century and Copernican Revolution—the reemergence of concepts of a spherical Earth, and a solar system that revolves around the Sun—did scientific exploration of the earth, and beyond, reemerge.

See also History of exploration II (Age of exploration); History of exploration III (Modern era)

HISTORY OF EXPLORATION II (AGE OF EXPLORATION)

In the seventeenth century, the nature of colonialism changed. While daring expeditions at sea and discoveries of new lands still defined exploration, European nations had become dependent on the trade and resources of their New World colonies. This prompted governments to encourage settlers to move to colonial territories to establish trading ports and protect land interests. As more unknown lands were discovered, they were quickly claimed by European nations. The great territorial race began with clamoring for ownership of the vast land and resources of the New World. By the mid-eighteenth century, nations focused their attention to exploring **Africa**, the Pacific, and **Australia**. By the end of the era, European nations fought both each other and existing civilizations in the Far East for shipping and trade strongholds in **Asia**.

Colonialism and maritime discovery were not the only forces that shaped the exploration from 1600 to 1850. Knowledge gained from exploration yielded a new interest in studying the world. The Enlightenment, a resurgence in science, reason, and learning during the late eighteenth century, fostered a **climate** of scientific curiosity. Not only did people sail the **seas** to discover and claim new lands, they carefully catalogued the differing plants, animals, crops, and people in the lands they explored. New "natural sciences" such as biology and **geology** became popular pastimes for individuals. Eventually, the natural sciences gained academic credibility, by 1830, France, Spain, England, and Holland all had national geological and geographical societies.

In 1620, the Pilgrims, English settlers, landed at Plymouth, New England. While the Massachusetts Bay Colony was not the first successful settlement in the New World, it was the first major stronghold in **North America**. Soon after the English settled in Massachusetts, the Dutch sent settlers to New Amsterdam (now New York). Later in the 1600s, French Explorer La Salle claimed the northwestern coast of the **Gulf of Mexico**, naming the land Louisiana in honor of the French king. In the 1650's, the triangular trade route began. Europeans traded slaves for sugar in the West Indies, and sugar for rum, molasses, and timber in New England. This ensured a steady stream of both raw materials for industry and luxury goods for consumers.

A century later, exploration and settlement focused on Asia, the South Pacific, and Australia. In 1776, the British government proposed settlement of New South Wales, the **area** of Australia explored by **James Cook**. Two years later, the first settlers arrived in Australia. They settled around Botany Bay before relocating because of disease and poor **soil** conditions. Settlement in Australia grew quickly, especially as the French and Dutch claimed their own ports in the region, and trade in Asia flourished. The entire coastline of Australia was not fully mapped until 1822.

This mercantile economy, however, was dependent not only on colonists, but also on the procurement of African slaves. Exploration and colonialism had a devastating effect on the native populations of Africa and the Americas. Exploration and settlement brought European diseases to the populations of other continents. Virgin soil epidemics—epidemics that erupt in populations that previously had no exposure to the diseases—killed millions of people in the New World and Africa. These European diseases changed and adapted to their new climates, sometimes producing more virulent and destructive strains of disease. Thinking that certain environments caused disease, colonists became conscious of where they built their homes and towns, avoiding swamps and beaches. They often avoided fresh air and local fruits and fish, opting instead for dried beef and staples from home. Medicine during this era was closely linked to environment, a connection that sparked people's interest in the environment itself.

English explorer James Cooke embarked on perhaps the most ambitious voyage of the era in 1768. Cooke's expedition circumnavigated the globe, spending a great deal of time surveying the South Pacific and Indian **Oceans**. He discovered several island chains and surveyed a large portion of the coast of Australia. He was the first to realize that Australia was a vast continent. In 1778, Cooke voyaged to the northern Pacific in search of a passage through North America. He and his crew wintered in California, comprehensively charted the west coast of North America, and sailed as far north as Alaska and the Arctic Circle. Unable to locate an inland passage, Cooke and his crew sailed to Hawaii, which he had explored several years earlier. Cooke died in the Hawaiian Islands in 1779, but his crew returned home nearly a year later. Cooke's long expeditions fundamentally changed the shape of the known world—maps made after the Cooke voyages are some of the first to resemble modern maps.

Cooke possessed a distinct navigational advantage over his predecessors, which greatly expedited his journey and allowed for greater accuracy in mapping. In 1714, the British Parliament offered a prize to anyone who could devise a method for accurately and reliably determining longitude (position on a vertical grid of the earth). For centuries, mariners had been able to find their latitude (position on a horizontal grid of the earth), thus aiding the determination of position and distance when traveling north or south, and giving a rough position when sailing east or west. However, no means existed for accurately measuring distance and position when sailing west or east, the predominant direction of voyages to the Americas and the Far East. Most scientists thought the best way to solve the longitude problem was through the creation of astronomical charts and complex tables and equations. However, these could only be used in good **weather**, and for a few hours of the day, thus restricting their ultimate usefulness. In 1735, British inventor **John Harrison** determined that longitude is most easily calculable when one knows the exact time. On land, the most precise timekeepers were watches. Harrison thus began constructing chronometers, or clocks, for use at sea. He tested several models on various voyages, fine-tuning his models each time to account for the rocking of the ship, vibrations, and other factors that influenced the reliability of the chronometers. After nearly 30 years, Harrison perfected his chronometer in 1761. Longitudinal navigation was no longer a mystery. The British Parliament was reluctant to award the prize to an amateur scientist however, and Harrison did not receive the promised prize for several years. Realizing its potential, several navigators began to use chronometers, despite their relatively high cost, before Parliament recognized the magnitude of the discovery.

Exploration, settlement, and medicine prompted people to more closely look at nature and the environment. In 1798, Thomas Malthus published his first essay on population. The work classified peoples by region and racial characteristics, but also dealt with the relationship of each group to their environment. In the 1820s, the first rough theories of **evolution** appeared. While none deduced a specific scheme for evolving life, the concept of development began to fascinate a few peo-

ple interested in the natural sciences. Along with the questioning of man's antiquity, natural scientists studied the earth itself. In 1930, the Royal Geographical Society was founded in Britain. That same year, English geologist **Charles Lyell** published his *RMS Principles of Geology*, a work that served as the standard methodology for the discipline for nearly a century. Natural scientists were fascinated with classifying all aspects of the world around them, and probing their interconnectedness. Thirty years later, a theory that proposed to explain those very connections would be the center of scientific debate in the nineteenth century. In 1831, English naturalist Charles Darwin embarked on an expedition to the Galapagos Islands, a remote island chain off the coast of present-day Ecuador. His discoveries made aboard the HMS *Beagle* foreshadowed a new age of scientific exploration, and the modern era.

See also Cartography; Evolution, evidence of; Evolutionary mechanisms; History of exploration I (Ancient and classical); History of exploration III (Modern era); Latitude and longitude

HISTORY OF EXPLORATION III (MODERN ERA)

Until the dawn of the twentieth century, exploration of Earth's surface was limited to the surface itself. The summits of the highest mountains, the depths of the **oceans**, and sky and **space** were unexplored. Technological advances, as well as fundamental shifts in scientific theory, opened the entire world, and beyond, to exploration. The twentieth century was the golden age of discovery, rivaled only by the Copernican Revolution and the European discovery of the New World in the fifteenth century.

In 1859, Charles Darwin published a natural history work that sparked great controversy. Darwin's book, *On the Origin of Species*, was a compilation of his scientific observations from an expedition to the Galapagos Islands. Darwin's observations led him to construct a model of the **evolution** of various animal species, including man. Over the course of his career, Darwin built upon Charles Lyell's earlier works on the antiquity of Earth. Darwin proposed that the dawn of man was not because of spontaneous creation, but was the result of a process of evolution and natural selection, the principle of survival of the fittest, which occurred over hundreds of thousands of years. Darwin thus concluded that man, animals, plants, **geology**, and environment were all connected in their development. During the course of the nineteenth century, some aspects of Darwin's theory fell out of favor, such as the principles of natural selection and inheritance, after the discovery of genetics. However, Darwin's work popularized the idea of evolution, brought it into mainstream science, and opened up the fields of anthropology (the study of man) paleontology (the study of **fossils**) and geology (the study of Earth.)

At the turn of the twentieth century, wider communication was possible through the telegraph and telephone.

Modern mapping techniques allow scientists to explore Earth's least hospitable regions. Only decades ago, the bottom of Lake Tahoe was a mystery. *AP/Wide World. Reproduced by permission.*

Railroads, steamships, and later automobiles allowed people to travel with greater ease. A voyage that took Columbus four months in 1492, took passengers a scant four days in 1915. The Panama Canal connected the Atlantic and Pacific Oceans, eliminating the need to traverse the feared and time-consuming Straits of Magellan. The advent of flight in 1901 provided the final link; within 60 years, even transatlantic passenger ships were rendered artifacts of the past. Flight also allowed man to see the surface of the earth from above, changing the way cartographers created maps. **Electricity** and cameras further revolutionized the perspectives from which people viewed the earth. This connectivity with once distant and remote places made the world seem smaller, while photography made the remote corners of the world accessible to everyone.

Exploration of the earth's surface continued. In the early decades of the twentieth century, several men led expeditions to the South Pole. In 1909, British explorer, Ernst Shackleton's voyage to **Antarctica** was particularly ill-fated, leaving much of the crew stranded on the continent as **ice** floes crushed the ship. The team did not reach magnetic south, but provided some of the first observational clues that Antarctica was a landmass, not only a **polar ice** cap like its northern counterpart. Norwegian explorer Roald Amundsen reached magnetic south in 1911, and attempted to reach the North Pole in 1926, but failed. Not until 1978 did a manned expedition reach the surface of the North Pole. In 1953, British mountaineer Sir Edmund Hillary and his guide Tenzing Norgay climbed Mt. Everest on the border between Nepal and China. The mountain is the highest peak on the earth's surface (the only larger mountains are submerged beneath the sea). Everest continues to be a draw to adventurers and explorers, claiming several lives on its harsh slopes.

By the middle of the twentieth century, most of Earth's surface had been explored. As technology grew more sophisticated and complex, explorers turned to space, the inner layers of the earth, and the unknown depths of the sea as venues for modern exploration and discovery. After World War II, the technology that was developed for the war was not only used for weapons, but also for scientific research. **Remote sensing** equipment such as sonar was used for mapping the ocean floor and locating ridges, reefs, and other underwater features. Jet engines, rockets, and robotics allowed for the launching of satellites by the United States and the Soviet Union in the late 1950s. The "Space Race" followed, with the Soviet Union and the United States each competing to achieve landmark space explorations before the other nation. In 1961, The USSR sent the first man, Yuri Gagarin, into space. The United States soon followed, sending the first astronaut to orbit Earth. Then, the ultimate goal of each nation was to send a team of astronauts to the **Moon**. In 1969, the United States sent the first team of astronauts, aboard the *Apollo 11* craft, to the surface of the moon. Aside from **meteoroids** that crashed into the earth, the Moon mission was man's first contact with an alien geology.

In the 1990s, the United States and several other nations, including Russia, renewed their interest in space research. The **International Space Station** was planned and constructed, providing the first non-earthbound research laboratory. A growing fascination with deep-space exploration and **astronomy** prompted the launch of the **Hubble Space Telescope** to photograph stellar bodies that could not be clearly investigated on the earth's surface. In 2001, missions of unmanned probes to Mars yielded information about Martian geology, giving scientists their first basis of comparison for Earth's geological processes.

Just as exploration in centuries past was largely driven not only by curiosity but also by the need for certain goods and resources, exploration in the twentieth century was similarly motivated. The Industrial Revolution created the need for vast amounts of fuels—at first **coal**, then natural gases, and then nuclear materials. The first coal mines were dug in Britain in the 1200s, and the process of mining changed little from then until the 1800s, when railroads, steam engines, and steel were introduced to expedite production, processing, and increased mine safety. Research about the deep geology of the earth's **crust** not only altered mining processes but also opened new venues to resource acquisition, such as the ocean floor, and led to the discovery of new **fuels**, such as **petroleum** and uranium. Deep-crust geology gave scientists a better understanding of how forces within the earth combine with organic materials to produce these fuels.

Today, only the uppermost layers of the planet have been explored. The technology does not currently exist to drill or explore to the molten sub-crust surfaces. While studying earthquakes and volcanoes has helped researchers understand geological processes, sub-surface exploration remains limited.

With less than ten percent of the ocean floor mapped, explored, or even seen by human eyes, the depths of the oceans are perhaps the last great arena for man's exploration of his planet. Submarines and submersibles first appeared in the 1780s, but not until the invention of the German U-boat (battle submarine) in World War I was there technology for breathing apparatuses, long-term, deep submersion, and propulsion. The first viable and practical self-contained breathing apparatus (or SCUBA) was introduced in the early 1900s. In 1930, William Beebe created a bathysphere, a personal submersible that was lowered from a ship and dragged at a given depth. Beebe dove to 1,427 ft (435 m), deeper than modern personal SCUBA equipment can attain. Beebe thus pioneered the small submersible, which later became the fixture of deep-sea exploration.

By World War II, submarines were standard naval warfare ships, and diving gear such as the Mark V diving helmet, were standard. After the war, their principals of design were applied to scientific research. In the 1970s, scientists and engineers designed a small submersible named ALVIN. The submersible had a mechanical arm, modern surveying and recording equipment, and room for up to three researchers. The submersible was motorized, and once lowered, its crew could drive it for a more free range of exploration. This generation of submersibles could be used miles beneath the surface, and entered realms of the ocean that no human had ever seen. Some of the first notable discoveries were that of deep-water marine life. Without light or heat, scientists thought that no life could survive on the sea floor. Instead, researchers discovered a vibrant community of strange and primitive-looking creatures that thrived at great depths.

In 1973 and 1974, the Mid-Atlantic Ridge, a large underwater mountain range, was explored by divers in small submersibles that were equipped with sensory and photography devices. That same decade, scientists discovered underwater volcanic vents, thus gaining insight into the creation of the ocean floor itself. The discovery of hydrothermal vents provided a new means of understanding ocean **chemistry**, and the circulation of **minerals**, necessary for sustaining marine life, at great depths. Deep-sea exploration was further utilized in archaeological endeavors to locate sunken shipwrecks and other submerged sites. In 1985, Dr. Robert Ballard, who was part of many of the great deep-ocean expeditions of the late twentieth century, discovered the wreck of the RMS *Titanic*, a British luxury liner—the technological marvel of its time—that sank in 1912.

Not all modern exploration is achieved on a grand scale. During the modern era, explorers and scientists have also embarked on projects to explore mankind, both in the past and present. Physical anthropologists unearthed early hominid (human or human-like) fossils in the African Rift Valley as early as the nineteenth century. However, not until the last half of the twentieth century did scientists attempt to establish a chronology of man to match the **desert** fossils. In 1974, anthropologists Donald Johanson and Tom Gray discovered a nearly complete hominid skeleton. The bones, named Lucy, formed the then most complete skeleton of one of the earliest known human ancestors, Australopithecus. The skeleton dates back 3.18 million years—far longer than Darwin predicted over a century before.

Today, ethnographers (people who record other cultures) and other anthropologists study groups of people around the globe, especially those who live in the most remote areas and have little contact with other peoples. Archaeologists work to unearth the material remains of the past as a means of scientifically exploring the course of human history. Thus, the exploration of man is as concerned with re-discovery as it is with novel discovery.

Marine biologists are searching for the most primitive forms of life, often small, wormlike creatures and sponges, in the depths of the ocean. Mathematician and physicist, **Albert Einstein**, paved the way for exploration of atoms, and other fundamental structures of the universe. Physicists continue to break down these elemental structures into sub-atomic particles. Other physicists and astronomers research particles, gases, and forces in space to determine provide clues about the few hundredths of a second in the life of the universe.

See also Bathymetric mapping; Dating methods; Deep sea exploration; Evolution, evidence of; Geographic and magnetic poles; Historical geology; History of exploration I (Ancient and classical); History of exploration II (Age of exploration); History of manned space exploration; Hubble Space Telescope (HST); Petroleum, history of exploration; Physical geography

HISTORY OF GEOSCIENCE: WOMEN IN THE HISTORY OF GEOSCIENCE

Women in science have been conspicuous by their apparent absence throughout history. Yet from Theano, the wife of Pythagoras, in the fifth century B.C. or Hypatia in the last days of the Roman Empire (born in A.D. 370) to **Marie Curie**, women have been practicing and teaching science and mathematics. Their profile has, however, been so low that most female scientists never made an impact on general society. From original research carried out along the western seaboard of **Europe** (United Kingdom, Ireland, France, Spain), it is clear that the population in general can only consistently cite Marie Curie as a memorable female scientist. The second most remembered female scientist is Rosalind Franklin or Dian Fosse in the English speaking countries and Hypatia (the mathematician) in France and Spain. Indeed, for nearly fifteen centuries, Hypatia was considered to be the only female scientist in history.

It is clear from the above introduction that women earth scientists were not recognized at all. Perhaps this is because it is a relatively young science; but that assumption does not stand up to scrutiny if we realize that, in what was then Germany, Agricola published his "de re Metallica," a treatise on mining techniques, in 1556. Until recently, **geology** was divided into a practical science and a theoretical one. In addition, the amateur of the science practiced collecting for its own sake. This also applies to all the descriptive sciences. Also, it is worth noting that the prevalent view between 350

B.C. and A.D. 1600 was that science has been predominantly a male subject **area**. Most female scientists, even during the eighteenth and nineteenth centuries, were denied access to formal education and had to rely on their male relatives for tuition. With the rise of public education in America and Europe at the end of the nineteenth century, the situation did start to change. Before this, women could be considered amateur geologists in the strict sense of the word as they were rarely paid for their work. Within geology, the most popular area for female work was initially paleontology. In a paid capacity, drawing of samples or illustrating books was a favored pastime, as for example, in *Sowerby's Mineral Conchology* in 1813.

Arguably, the first female paleontologist of note was **Etheldred Benett**, an English spinster living in the south of England. Bennett was both a scientific researcher in paleontology and an accomplished artist. The most famous early female geologist was **Mary Anning** from the United Kingdom. Sometimes called "the dinosaur woman" or the "mother of paleontology," Anning supposedly had the tongue twister "She sells sea shells by the sea shore" made up about her. These two remarkable women were unique in that they took geological knowledge forward. However, it was not until the end of the nineteenth century that women in Europe had the chance to become professionally educated and, therefore, become professional geologists.

In America, Florence Bascom influenced female geological education significantly. She was born in 1862 in Williamstown, Massachusetts. Her mother was a suffragette, her father a professor and then president of University of Wisconsin. Her interest in geology was aroused after a visit to Mammoth **Cave**. She received degrees in 1884 and 1887, did graduate work at John Hopkins University in petrology, and was the first woman to receive a doctorate in geology in 1893. After time teaching at Ohio State University and as assistant geologist at the United States Geological Survey working on the Mid-Atlantic piedmont, Bascom subsequently founded the department of geology at Bryn Mawr College where she spent most of her adult life.

The United States Geological Survey employed their first woman geologist, Florence Bascom, in 1896, whereas the British Geological Survey did not appoint a woman until the 1920s when they were invited to apply for appointment. The female candidates were required to be unmarried or widows, and were also required to resign on marriage. Miss Eileen Guppy, the first successful woman to be employed by the British Survey, worked in the petrology department in 1927. During the Second World War, women were employed to look at water-supply boreholes and wells. The first woman was employed as a scientific officer in 1957, and the first woman to research at sea was employed in 1967.

In Canada, Alice Wilson, born in 1881 in Cobourg, Ontario, became the first woman to reach a prominent position within the Geological Survey of Canada. Her interest in Ordovician sediments and **fossils** of the Ottawa valley pushed forward geological knowledge in that country. Alice was helped in her academic pursuits by having two gifted brothers, one in geology and one in mathematics, and parents who supported education of all their children. Her father was Professor of Classics at Victoria University.

When looking at the above women it is obvious that they would not have achieved their status without determination, but also family help. That most of them were single also meant that they did not have the pressure to maintain families and children placed upon them by Victorian society or earlier prejudices, as in the case of Etheldred Benett and Mary Anning. For example, it was considered inappropriate for mothers to have careers and work outside the family sometimes even for pleasure at that time.

The early female geologists appeared to prefer paleontology. They were also single, determined, and intelligent. However, it must not be thought that the twentieth and twenty-first centuries have advanced the position of women within geology. The first fellow of the Geological Society of America was Mary Emilee Holmes in 1889. By 1930, only eleven women had been elected to fellowship of the Geological Society of America and seven of these were paleontologists. Women were only allowed into the Geological Society of London, the oldest geological society in the world, in 1919, but awards had been made to nine women prior to this date. Two late twentieth century reports from prestigious societies in Britain and the United States have highlighted the lack of equality between the genders. In 1977, a report on the status of women professional geoscientists, (something that could not have happened in the previous century) from the American Geological Institute commissioned by the Women Geoscientists Committee showed that women were moving away from the education sector toward industrial employment. However, salary inequalities still existed for more experienced women across all sectors. A report on the status of women in the Geological Society of London 20 years later in 1997 showed that women members of the society had increased to 12% (1,040) and a high percentage of these were less than 40 years old.

In United States higher education in 1977, 17% of geoscience students were female. Twenty years later in the United Kingdom, the figure had risen to 25–30%. A recent report from the Helsinki Group, a European Union research group promoting scientific research to women in general, which was reported in the journal *Science*, again emphasized the lack of women in research science. At the top of the list for female researchers is Portugal, with 48% in the natural sciences including geology. At the bottom of the list is the Netherlands with 8%. In Germany in 1994, higher education figures showed the percentage of first-year university students studying geosciences to be 30%, dropping to 15% awarded PhDs, to less than 1% for professors of geology. Women are still underrepresented in the geological world at the higher levels of expertise, but perhaps as we move through the twenty-first century, the role models from previous eras will act as an incentive for women to participate in sustaining our geological heritage.

See also Historical geology

HISTORY OF MANNED SPACE EXPLORATION

The history of manned **space** exploration is essentially the history of the United States and Soviet/Russian space programs. Although the European Space Agency and China are expected to begin manned exploration of space in the early twenty-first century—manned exploration of space in the twentieth century resulted initially from a hotly contested "space race" that was, perhaps, the most visible of Cold War competitions between the Soviet and American superpowers. Initially driven by national pride and a quest for perceived strategic military advantage, over the last two decades, the exploration of space has become a more scientifically oriented and cooperative enterprise, especially in the ongoing **joint** construction of the **International Space Station** (ISS).

The official Soviet Space Program (SSP) began in May 1946, as it was then that the government made the decision to set up an industrial branch for missile "armamentation" in the Union of Soviet Socialist Republics (USSR).

The decision to create a space program, however, was not made overnight. Since the early 1930s in the USSR, small groups of enthusiasts attempted to create rockets. These groups were made up of engineers who afterwards would play leading roles in the Soviet Space Program; among them was Soviet aeronautical engineer Sergei Pavlovich Korolev (1906–1966). A victim of Soviet purges during the late 1930s, Korolev survived a term in Stalin's Gulag system to become the "Chief Designer" of the Soviet Space Program. Although during his lifetime Korolov's identity was never publicly revealed to western sources, Korolev became the eventual leader of most of the SSP projects.

The success of missile building in Nazi Germany, including the creation of the V-2 rocket, influenced the beginning of rocket system development in the USSR. Immediately after the end of the World War II, a group of Soviet engineers traveled to Germany, where they carefully studied captured German documents and equipment intended for missile creation. The group even worked in Germany for a time before returning to the USSR. By 1948, the P-1 missile was already developed (the analogue of V-2) and was officially accepted as the main armament in the Soviet army. In 1950, an improved version of this missile was created, and in 1951, the new P-2 missile was first created. By 1956, the army accepted the new missile, which could carry a nuclear charge and reach its target from a distance of about 932 mi (1,500 km).

The future United States space program also received an important boost from German scientists who either fled to the United States before the war, or who intentionally fled the advancing Soviet armies to surrender to United States and British forces. Notable among these scientists was **Wernher von Braun** (1912–1977). Greatly advancing the work of early American rocket designer **Robert H. Goddard** (1882–1945), and others, von Braun would become one of the chief architects of the American Space Program and go on to design the Saturn V rocket that ultimately achieved the escape velocities needed to propel America's Apollo program astronauts to the **Moon**.

Skylab was America's first space station. Americans have recently returned to long-term space missions with the International Space Station. *U.S. National Aeronautics and Space Administration (NASA).*

In 1957, the two-stepped intercontinental ballistic missile P-7 was created in the USSR. Using this rocket, the world's first Earth-orbiting **satellite**, *Sputnik,* was launched on September 4, 1957 from the Baikonur cosmodrome (now in Kazakhstan). *Sputnik* weighed 187 lb (83.6 kg) and completed an elliptical orbit around the earth every 98 minutes. *Sputnik* was designed to return data about the composition and density of Earth's upper atmosphere. *Sputnik* transmitted via radio signals for approximately three weeks.

Also in the 1950s, the largest and most famous Soviet space vehicle-launching site, Baikonur, was built. This launching site is located in the Kazakhstan steppe, and the most important launches of Soviet/Russian spacecraft, in particular all spacecraft with men and women (cosmonauts) aboard, were launched from the Baikonur launch complex. Following the collapse of the USSR, the Baikonur launching site was located in an independent Kazakhstan, but Russia continues to rent and use the site for its space program.

The experimental research component of space study began under Korolev's supervision as early as 1949. Gradually, geophysics and meteorological rockets began to be launched, measurements of different geophysical field parameters at different heights were executed, and later, launching of rockets with animals on board were executed. Also in the 1950s, in parallel with military missile projects, work began on sending man into space. The result was the launch on April 12, 1961, of the spacecraft *Vostok* with the first cosmonaut, Yuri Gagarin. Aboard *Vostok,* Gagarin spent about one hour in space and completed one circuit around the earth. Shortly afterward, on August 6, 1961, the second cosmonaut, German Titov, flew to space on board the spacecraft *Vostok 2,* spending approximately one day orbiting the earth (Titov died in Moscow in 2000).

The launch of the first *Sputnik* and Gagarin's flight were triumphs of the Soviet Space Program, and the start of the unofficial space race with the Americans.

The effect on America of the spectacular early success of the Soviet Union in space exploration can rarely be overstated. Despite the design-based successes of Soviet-made MiG jet aircraft used against American forces during the Korean War, since 1947, when Charles E. "Chuck" Yeager (1923–) became the first man to break the sound barrier, America assumed it held a vast technological superiority over the Soviet Union in aeronautics and other science and engineering fields. Americans regarded the Soviet Union as nation with a struggling economy and often politically repressed scientific research programs (e.g., the debilitating effects of Lysenkoism—a Stalin-supported pseudoscientific interpretation of genetics that suppressed early Russian advances in genetics and contributed widespread Soviet agricultural shortages).

The Soviet launch of *Sputnik* and subsequently of launching the first man into orbit inflicted a deep wound to America's pride and assumed technological superiority. Near hysteria swept the American government as it feared—at a time of increasing nuclear tensions—that the demonstrated Soviet capabilities in peaceful space exploration would easily be translated into a destabilizing advantage in nuclear weapons delivery capability. Decimating the manpower and budgets of its successful high altitude aeronautics programs, the United States military and, after its founding in 1958, the National Aeronautics and Space Administration (NASA) accelerated rocket development programs. Moreover, the Soviet successes in space so rocked the American psyche that major reforms were undertaken in the educational system to close an apparent gap in scientific and engineering expertise. Education of scientists became a strategic national priority.

Although Soviet secrecy made direct comparisons difficult, the early record of the American space programs was notable for its very public failures. Rockets continually destroyed themselves in spectacular launch explosions. America's first attempted response to *Sputnik,* the Vanguard TV3 on December 6, 1957 ended in a launch failure. During 1958, four out of five intended American Pioneer probe missions (intended for lunar flybys) ended in launch failures.

Consistently behind the Soviet Union, the Americans finally successfully launched *Explorer 1* on February 1, 1958. Weeks after Gagarin's orbital flight, on May 5, 1961, the United States launched its first astronaut, **Alan B. Shepard** Jr. into a sub-orbital flight from Cape Canaveral. It was not until February 20, 1962, that astronaut John Glenn became the first American to orbit the earth. Shepard's flight so captivated and buoyed Americans, that then U.S. President John F. Kennedy issued a challenge that America would dedicate the resources needed to land a man on the Moon and return him safely before the end of the 1960s.

Before America could complete its one-man Mercury program—a series of 20 unmanned and six manned flights designed principally to test elements of rocketry and whether humans could survive and work in space—the Soviet Union increased its number of "firsts" in space exploration. One of the lone bright spots for the American space program was the success of the *Mariner 2* interplanetary probe. Launched in December 1962, *Mariner* passed within approximately 21,000 mi (33,800 km) of Venus and was able to transmit back to Earth the first useful data from an interplanetary probe.

On July 14–16, 1963, two spacecraft, *Vostok 5* and *Vostok 6,* were launched with cosmonaut V. Bykovskii on board one, and with the first woman cosmonaut **Valentina Tereshkova** on board the second. On March 18, 1965, cosmonaut Aleksei Leonov was the first to go out in open space (spacewalk) from onboard the three-man *Voskhod* spacecraft.

On June 3, 1965, American astronaut Edward H. White II became the first American to "walk in space" (i.e., perform an Extra Vehicular Activity or EVA). White's tethered EVA was part of a methodical two-man Gemini program, designed to test equipment and refine skills in maneuver and rendezvous that would be required on subsequent three-man Apollo lunar missions. Although the Soviets maintained an impressive lead in space "firsts," it was during the Gemini program—consisting of 2 unmanned and 10 manned missions—that America gained the technological ability to move into the ambitious Apollo missions.

In contrast, despite being shrouded in secrecy, by the mid-1960s, the first failures of the Soviet Space Program were apparent. First launches of Soviet N-1 rockets (the rocket intended to take Soviet cosmonauts to the Moon) were not successful, and during tests, the rocket did not reach past an altitude of 70,000 ft (21,335 m). While the "Moon race" between the United States and the USSR continued, the Soviets spent vast amounts of money (equivalent to over 600 million U.S. 1969 dollars), but when it became obvious that the USSR would not win the race—i.e., it would not be able to be the first country which would land a man on the Moon—the Moon project stagnated and the remaining N-1 rockets were dismantled.

A devastating fire during a prelaunch test on January 27, 1967, killed three *Apollo 1* astronauts and put NASA's quest to put a man on the moon by the end of the decade in jeopardy. The Soviet space program also encountered fatalities. During the first test of the *Soyuz* spacecraft in April 1967, cosmonaut V. M. Komarov was killed in a crash resulting from entangled parachute shroud lines. Another *Soyuz* accident occurred in June 1971, when a pressure leak during reentry killed all three *Soyuz* cosmonauts.

After a series of unmanned test flights, Americans returned to space on October 11, 1968, with the launch of *Apollo 7.* The success of the mission, and the stellar performance of the redesigned Apollo spacecraft put NASA on the fast track to a lunar mission. In December 1968, *Apollo 8* astronauts Frank Borman, James A. Lovell Jr., and William A. Anders became the first manned spacecraft to leave Earth orbit when they traveled to the Moon and completed 10 orbits before returning to Earth. In March 1969, the flight of *Apollo 9* remained in Earth orbit to successfully test the Lunar Excursion Module (LEM)—the first true spacecraft never designed to enter Earth's atmosphere. The 2-stage LEM was designed to carry astronauts from lunar orbit to the lunar surface and the upper stage was designed to return them to the Apollo command module that would remain in lunar orbit. In May, 1969, the flight of *Apollo 10* tested the LEM in the lunar gravitational field, as astronauts undocked the LEM from the

Apollo command module and flew within approximately 50,000 ft (15,420 m) of the lunar surface.

On July 16, 1969, the launch of *Apollo 11* propelled astronauts Neil A. Armstrong (Commander), Edwin E. "Buzz" Aldrin, Jr. (Lunar Module Pilot), and Michael Collins (Command Module Pilot) toward the Moon. On July 20, 1969, Armstrong became the first man to set foot on another world. Armstrong and Aldrin left behind an American flag and a plaque that read: "Here Men From Planet Earth First Set Foot Upon the Moon. July 1969 A.D. We Came In Peace For All Mankind."

Having won the "space race," the American public's interest in lunar exploration quickly waned. Other than a renewed concern for the astronauts about the ill-fated *Apollo 13* mission, public interest and the political will to continue to shoulder the financial burdens of lunar exploration brought the Apollo program to a halt after the flight of *Apollo 17* in December, 1972.

Interestingly, it was only on the last flight of Apollo that a trained professional scientist—geologist astronaut Jack Schmitt—was able to conduct observations and conduct experiments on the lunar surface. *Apollo 17's* emphasis on lunar **geology** and science heralded a new age of space exploration.

Also frustrating and unsuccessful for the Soviets was the Soviet **space shuttle** project, begun in 1974. In the middle of the 1980s, an experimental version of the shuttle was created, the *Buran*, but the spacecraft executed only one flight in the automatic mode (without a pilot on board) on November 15, 1988. After the collapse of the USSR, work on this project was cut because of lack of funding.

One of the most successful projects of the Soviet Space Program, however, was the creation and work of the *Mir* orbital space station. Based on extensive experience with earlier *Salyut* space stations, *Mir* became the premier Earth orbiting laboratory. The base module of *Mir*, which weighed 36 tons (32.7 metric tons), was launched on February 20, 1986. In 1987, the *Kvant* module was linked up with the station; in 1990, the *Kristall* module was also linked up with the station, and in 1990, special equipment for docking of American shuttles in the station was placed aboard *Mir*.

For several years, *Soyuz TM* spacecraft with cosmonauts on board were regularly sent to *Mir*, changing crews after periods of about 5–6 months. After launch, the spacecraft docked with the *Mir* station after traveling in space for about two days. During several missions, the crews of the spacecraft and the station conducted joint exercises, and afterwards, the old crew "passed watch" to the new one. Several days after the spacecraft's arrival at the station, it returned to the earth with the crew who had finished the last watch.

Many scientific research projects were carried out on board *Mir*; 73 persons from nine countries visited the station and it was calculated that the total time spent onboard by all the cosmonauts was about 40 years. During 1988, cosmonauts Musa Manrov and Vladimir Titov set records for what was then the longest period of time in space for humans (366 days). Subsequently, the longest flight record was extended to 438 days by cosmonaut Vladimir Polyakov. The duration of *Mir's* active status was much longer then expected. After ten years of utilization, different mechanical drawbacks occurred

more frequently, and eventually the decision was made to end the project. The Soviet/Russian space station *Mir* was taken out of orbit and re-entered Earth's atmosphere in 2001; its fragments sank in the Pacific Ocean.

Following the Apollo lunar program, the United States completed a more modest Skylab program. In a welcome de-escalation of Cold War tensions in 1975, the U.S. and Soviet Space programs cooperated in a joint *Apollo-Soyuz* rendezvous and docking mission. The mission was designed to pave the way for future cooperation in spaceflight.

NASA development of the Space Transportation System (STS)—the Space Shuttle—has made space more accessible to a wider variety of scientists and experts. The space shuttle has become the American workhorse for orbital delivery of wide range of satellites and repair of science instruments (e.g., the **Hubble Space Telescope**). America's first shuttle, *Columbia*, was launched in April, 1981.

In January 1986, a disastrous explosion 73 seconds after liftoff destroyed the space shuttle *Challenger*. The explosion was due to a faulty "O" ring—a ring sealing joints in the segmented solid **rock** boosters—that was made less flexible by the cold **weather** conditions prevailing during the launch sequence. A ring failure allowed hot gasses to escape the right solid rocket booster and then to burn through to the main auxiliary fuel tank used at liftoff. The explosion claimed the life of the seven-member crew—including America's first teacher in space, astronaut Christa McAuliffe—and halted manned U.S. space flight for more than two years.

In general, after the collapse of the USSR, the Russian space program was reduced and only recently began a revival. In spite of many different difficulties, Russia participated in the project of creating and utilizing the International Space Station. The first module (*Zarya*) of the ISS was created in Russia and launched on November 20, 1998, from Baikonur.

On December 4, 1998, the United States shuttle *Endeavor* was launched, carrying the ISS module *Unity*. American and Russian astronauts joined the two modules in open space. This event marked the beginning of the International Space Program (ISP), the official opening of which occurred on December 10, 1998. Construction of and research aboard the ISS continues.

See also History of exploration I (Ancient and classical); History of exploration II (Age of exploration); History of exploration III (Modern era); Hubble Space Telescope (HST); Space and planetary geology; Space physiology; Space probe; Spacecraft, manned

HOLMES, ARTHUR (1890-1965)

English geophysicist

Arthur Holmes, geologist and geophysicist, was born in Gateshead, England. From a modest family background, in 1907 he gained a scholarship to study **physics** at the Royal College of Science (Imperial College), London, where he became interested in the newly emerging science of **radioac-**

tivity and its application to dating **minerals** to solve geological problems.

Throughout his early life, Holmes struggled against financial hardship and frequently sought other work to support his research. Thus in 1911, he accepted a contract to prospect for minerals in Mozambique, where he conceived his vision of building a geological timescale based on radiometric dates. It was also there that he contracted a severe form of malaria.

In 1912, as a demonstrator at Imperial College, Holmes pioneered radiometric dating techniques and wrote the first of three editions of his celebrated booklet *The Age of the Earth*. In it, he estimated the earth to be 1,600 million years old, at that time an immense age and considered by many to be unacceptable. For the next 30 years, he pursued the topic, but it was not until the early 1940s that real progress was made on the geological timescale. By 1947, Holmes had pushed back the earth's age to 3,350 million years. However, it was not until 1956 that it was established at 4,550 million years.

Holmes married Maggie Howe (1885–1938) in 1914 and his first son was born in 1918. He escaped active service in the First World War due to recurring bouts of malaria but, by 1920, he was still only earning 200 pounds a year. Once again financial necessity compelled him to accept a post abroad, this time in Burma as chief geologist to the Yomah Oil Company (1920) Ltd. By 1922, however, the company had collapsed. Six weeks before leaving for home his young son caught dysentery and died. Eighteen months of unemployment followed Holmes' return to England, during which time he opened a Far Eastern craft shop in Newcastle-upon-Tyne.

In 1924, Holmes was offered the headship of a one-man **geology** department at Durham University. With his fortunes revived, and enhanced by the birth of his second son, this period saw an invigorated renewal of his research activities. An immediate supporter of the **continental drift theory**, originally proposed by Wegener in 1912, Holmes saw at once that it explained how identical **fossils** and **rock** formations occurred on either side of the Atlantic. However, the theory was then highly controversial, as no force was considered adequate to move continental slabs over the surface of the globe. It was Holmes' profound understanding of radioactivity—the amount of heat it generated and the enormous time it bestowed on geology for infinitely slow processes—that placed him in a unique position to formulate such a mechanism.

In 1927, Holmes gave a seminal paper which proposed that differential heating of the earth's interior, generated by the decay of radioactive elements, caused convection of the substratum (mantle). He calculated that convection could produce a force sufficient to drag continents apart, allowing the substratum to rise up and form new ocean floor. Evidence to corroborate this theory was not found until 1965, the year of Holmes' death, but by then his ideas were largely forgotten.

During the Second World War, Holmes was commissioned to write *Principles of Physical Geology*. Published in 1944, this famous book with its heretical chapter on continental drift, became an international best-seller that influenced generations of geologists. When Holmes retired in 1956, he set out to update the book, but with failing health it was a mammoth task, completed only months before he died.

Recognition of Holmes' outstanding contributions to geology came when he was elected Fellow of the Royal Society in 1942 and a year later appointed to the Regius chair in geology at Edinburgh University. In 1956, the Geological Societies of London and America awarded him their highest honors. In 1964, he was presented with the Vetlesen Prize, the geologist's equivalent of the Nobel Prize, for his "uniquely distinguished achievement in the sciences resulting in a clearer understanding of the earth, its history, and its relation to the universe."

Holmes was a deep thinker on the broad philosophical aspects of geology, with ideas far ahead of his time. Always of smart appearance, he was a gentleman of quiet charm and unfailing kindness. He had an exceptional talent for playing the piano, was fascinated by history, and loved poetry. In 1931, Holmes met Doris Reynolds (1899–1985), a geologist then working at University College, London. He engineered a lectureship for her in the Durham geology department, but they were unable to marry until 1939, nine months after the death of his first wife from cancer. Holmes died of bronchial pneumonia in 1965, leaving his beloved Doris to succeed him by 20 years.

See also Dating methods; Geologic time

HOLOCENE EPOCH

Earth is currently in the Holocene Epoch. In **geologic time**, the Holocene Epoch represents the second epoch in the current **Quaternary Period** (also termed the Anthropogene Period) of the current **Cenozoic Era** of the ongoing **Phanerozoic Eon**. The Holocene Epoch ranges from approximately 10,000 years ago until present day.

Also termed the Recent Epoch, the Holocene Epoch is thus far notable for the retreat of glaciers—a major force in producing the landscape topographical features evident today—and the geological time during which humans (*Homo sapiens*) became the dominant life form on Earth, increased their societal relationships, and produced major civilizations. The retreat of **glaciers** and the gradual climactic warming in the Northern Hemisphere encouraged migration and biological radiation of species.

Although *Homo sapiens* appeared during the preceding **Pleistocene Epoch**, and were fully differentiated as a species by the beginning of the Holocene Epoch, human societal **evolution** has taken place during the Holocene Epoch. As expected, the most recent and superficial of sedimentary remains were laid down during the Holocene Epoch. The **fossil record** is also dotted with an archaeological record of human activity and civilization.

Human societal and intellectual development during the Holocene Epoch produced the first species capable of significantly and consciously altering geophysical processes. In addition to deliberate reworking of topographical features and use of natural resources, byproducts of human civilization and industrialization have affected **groundwater** reservoirs, the type of abundance of **weathering** agents, the **geochemistry** of atmospheric processes on a local scale (e.g., **acid rain**); and possible atmospheric and/or marine processes on a global scale (e.g., possible **global warming**).

The general retreat of **glaciation** was punctuated by smaller-scale "ice ages"—including the "Little **Ice** Age" that occurred between approximately 1150 and 1700. One of the reasons that it is difficult for modern scientists to quantify the possible extent in global warming is that accurate climatic data extends back, at best, only about a hundred years. Accordingly, it is difficult to determine whether any data indicating global warming is simply a normal variation in a general downtrend, or a normal variation in generalized warming pattern.

See also Archean; Cambrian Period; Cretaceous Period; Dating methods; Devonian Period; Eocene Epoch; Evolution, evidence of; Fossil record; Fossils and fossilization; Geologic time; Historical geology; Holocene Epoch; Jurassic Period; Mesozoic Era; Miocene Epoch; Mississippian Period; Oligocene Epoch; Ordovician Period; Paleocene Epoch; Paleozoic Era; Pennsylvanian Period; Pliocene Epoch; Precambrian; Proterozoic Era; Silurian Period; Supercontinents; Tertiary Period; Triassic Period

HORNFELS

Hornfels is a fine-textured **metamorphic rock** formed by contact **metamorphism**. Contact metamorphism occurs when a mass of hot **magma** intrudes into preexisting **rock**, whether by injecting itself into a crack or by ascending in a large body (e.g., **pluton**). Rock in close proximity to the magma is temporarily softened or melted and recrystallizes with an altered texture, producing a hornfels. The term hornfels is often restricted to rocks produced by contact metamorphism of shale, **slate**, or mudstone.

In contrast to schists and gneisses, hornfelses show little or no foliation or layering. They form under conditions of approximately anisotropic (directionless) stress, so there is no tendency for the **crystals** to align in any particular direction. Traces of **bedding** present in the parent rock may remain in a hornfels but are not caused by metamorphosis.

Because they form by contact metamorphosis, hornfelses occur in shells or layers around bodies of intrusive magmatic rock. When seen in cross-section, as at Earth's surface, these shells or layers appear as rings or bands surrounding areas of magmatic rock. These rings are termed contact aureoles. A contact aureole may be only a few centimeters thick or several kilometers thick, depending on the size of the magmatic intrusion. An aureole of less-metamorphosed rocks, often spotted slates and semihornfels, frequently surrounds the hornfels aureole and blends smoothly with it. As is generally the rule with metamorphic rocks, coarser texture in a hornfels indicates more thorough **melting** and slower recrystallization.

Hornfels may be chemically altered by the magma that metamorphoses them, but generally reflect the chemical composition of their parent rocks; thus, **quartz**, **feldspar**, biotite, muscovite, pyroxenes, garnet, and calcite are common ingredients of hornfelses. However, as hornfelses are defined by process of origin (contact metamorphism), not by composition, one must establish that a rock has originated in a contact aureole to classify it as a hornfels.

See also Country rock; Intrusive cooling; Metamorphic rock

HOTSPOTS

Hotspots are localized areas of volcanism and high heat flow within the earth's **lithosphere** above a mantle plume hot enough to melt portions of the overriding plate. As a plate moves over a mantle plume, volcanoes previously above the plume cease to be active and new volcanoes form, creating an arcuate chain of volcanoes whose ages change progressively along the chain. The approximately 3,790 mi (6,100 km) long chain of Hawaiian volcanic islands and Emperor seamounts (submarine volcanoes) were formed by the displacement of the Pacific plate over a mantle plume. The age of volcanoes, degree of **erosion**, and general maturity of the Hawaiian Islands decreases progressively southeastward. The volcanically active Big Island overlies the present-day hotspot. The chain of volcanoes decreasing in age from west to east across the western United States and ending at Yellowstone National Park formed by the North American plate moving westward over the Yellowstone plume. Iceland lies above a hotspot at a mid-oceanic ridge.

Hotspots were thought to remain in a fixed position with respect to the earth's lower mantle. Hotspots were therefore used to define a unique, absolute reference frame to quantify the displacement of **lithospheric plates**. Maps of absolute velocity vectors for the earth's plates relative to hotspots (first produced in the late 1970s) provide a visual representation of plate motion from which the sense of displacement and resulting style of deformation at plate margins can be deduced. Hotspot tracks define segments of small circles about a fixed pole of **rotation** for the plate (called the Euler pole). Paleomagnetic studies of volcanic rocks formed above hotspots and detailed global plate reconstructions suggest that at least some hotspots may not have remained stationary. Rates of relative hotspot movement have been generally estimated at approximately 0.8–1.2 in (20–30 mm) per year. This implies that plumes may have moved with part of the mantle and displaced relative to each other. If the density distribution of the mantle changed in the past, the whole mantle may have rotated with respect to the earth's rotational axis to a new stable position (a process called true polar wander), systematically rotating all hotspots relative to the rotational axis. Paleomagnetic studies of volcanoes along hot-spot tracks are used to determine if true polar wander is likely to have occurred in the period of time recorded by the volcanoes.

See also Hawaiian Island formation; Paleomagnetics

HOWARD, LUKE (1772-1864)

English pharmacist and meteorologist

Luke Howard classified and named cloud formations. He understood that even though **clouds** have a countless variety of shapes, they have only three basic forms, which he termed cirrus (hair curl), cumulus (heap), and stratus (layer). There can be combinations of any of these three, such as cumulostratus or cirrocumulus. Any of them can also be a "nimbus" (rain) cloud, such as cumulonimbus. High clouds are designated by the prefix "alto-," such as altostratus.

Howard was born in London, England, on November 28, 1772, the eldest son of a prosperous businessman, Robert Howard, and his wife Elizabeth, née Leatham. As devout Quakers, the family enrolled Luke in a prominent Quaker institution, Thomas Huntly's School in Burford, near Oxford, from 1780 to 1787. That was the extent of his formal education. Although not trained as a scientist, he learned enough **chemistry** and pharmacy on his own to become a successful manufacturer, wholesaler, and retailer of pharmaceutical preparations. He began his business in London in 1793, partnered with William Allen in London and Plaistow, Essex, from 1796 until Allen's death in 1803, moved the business to Stratford while continuing to live in Plaistow, and eventually became head of the firm Howards and Sons. Throughout his life he supported himself with this trade.

Howard was fully dedicated to four main concerns: his business, his religion, his family, and his hobby, **meteorology**. On December 7, 1796, he married Mariabella Eliot, who shared his amateur interest in meteorology, helped him to gather data, and encouraged him to disseminate his findings. From an early age he loved clouds, and would spend hours watching them or painting watercolors of them. Gradually his observations and experiments, mostly conducted at home in his garden, became more precise and systematic.

Before Howard's time there was no useful classification of clouds. They were described haphazardly in terms of their color, size, shape, density, persistence, altitude, and moisture content. The eighteenth-century enthusiasm for classifying everything imaginable had not succeeded with clouds, even though scientists as reputable as Jean Baptiste Lamarck (1744–1829) had worked on the problem. Howard solved it with a simple threefold schema, which he presented in a famous lecture to the Askesian Society in London in 1802. This talk was published as "On the Modifications of Clouds" in 1803. Overnight Howard was a sensation. Within a decade, his classification was in general use throughout Western **Europe**. Not only scientists, but also poets such as Percy Bysshe Shelley (1792–1822) and Johann Wolfgang von Goethe (1749–1832) and painters such as Joseph Mallord William Turner (1775–1851) and John Constable (1776–1837) praised him for his meteorological breakthrough and its contribution to their professions.

Besides his work on clouds, Howard published articles and essays on pollen, **atmospheric pressure**, meteorological instrumentation, the **seasons**, **precipitation**, **electricity**, and **evaporation**, plus a three-volume book called *The Climate of London*, as well as anti-slavery pamphlets and several apolo-

gies for Quakerism. For his contributions to meteorology and climatology, he was elected a Fellow of the Royal Society in 1821.

Howard moved from Plaistow to Tottenham, near London, in 1812 and to Ackworth, Yorkshire, in the mid-1820s. After Mariabella died in 1852, he moved in with his son, Robert, at Bruce Grove, Tottenham, where he died on March 21, 1864.

See also Clouds and cloud types; Meteorology

HOYLE, FRED (1915-1999)

English astronomer

A prolific and talented author in both science fact and fiction, Fred Hoyle is best known for publicizing the controversial steady state theory of the creation of the universe. Hoyle also helped develop radar and advance the understanding of the nuclear processes that power the stars. He has taught at both Cambridge and Cornell universities, received numerous awards and honors, and was knighted in 1972.

Born in Bingley, Yorkshire, England, Hoyle was the son of Benjamin Hoyle and Mabel (Picard) Hoyle. He attended Bingley Grammar School and went on to Emmanuel College at Cambridge, where he studied mathematics and **astronomy**, receiving his master of arts degree in 1939. On December 28, 1939, Hoyle married Barbara Clark and the couple eventually had two children.

During World War II, Hoyle served in the Admiralty at London, where he helped the British Navy develop radar (radio detection and ranging) technology. The Royal Air Force's victory in the Battle of Britain has been credited to the navy's improvement of radar during this period. After the war, numerous radar dishes were acquired by fledgling radio astronomers and converted into radio telescopes. These amateurs' discoveries in the 1960s ultimately helped to refute the theories Hoyle developed in the 1940s and 1950s.

During the early 1940s, Hoyle focused his attention on an issue that arose through the work of physicist Hans Bethe: energy production in stars. In 1938, Bethe had suggested a sequence of nuclear reactions that fuel the stars: Four hydrogen atoms were fused into a single **atom** of helium, resulting in a minute amount of mass being converted into energy. While this process of **nuclear fusion** was consistent with the predicted amounts of stellar energy observed, Bethe's theory did not account for the production of elements heavier than helium—heavy elements that exist within other stars and that are also abundant on Earth.

Hoyle expanded Bethe's findings. Elaborating on gravitational, electrical and nuclear fields, he determined what would happen to elements at ever increasing temperatures. He theorized that when a star has nearly exhausted its supply of hydrogen, nuclear fusion halts, and the outward radiation pressure generated by the fusion reaction also comes to a halt. Without this outward flow, the star begins to collapse because of gravitation. This causes the core of the star to heat up and reach a **temperature** great enough to fuse helium into **carbon**.

•

The collapse of the star is then halted by the outward pressure of this new fusion radiation, and the star becomes stable. Hoyle's investigation into the nature of the carbon atom had the added benefit of helping scientists understand the origin of the atoms within the human body.

As the fusion cycle of **stellar evolution** continues, **oxygen**, magnesium, sulfur and heavier elements build up until the element **iron** is formed. At this point, no more fusion reactions can occur, and the star collapses catastrophically, becoming a white dwarf (a star dimmer than the **Sun** but much more dense). During this implosion, the star's outer layers ignite to become a supernova (an explosion whose luminosity is many times greater than the Sun). The supernova explosion creates elements heavier than iron, which are then hurtled into **space** by the explosion's force. It was from stellar debris such as this, Hoyle hypothesized, that the second generation stars with the heavier elements were formed.

Hoyle further proposed that the Sun was once part of a binary (double) star system whose companion became a supernova eons ago. The resulting heavy elements it ejected into space became the material from which the planets were formed. Hoyle's remarkable theory of stellar **evolution** appeared to be correct; it agreed with scientists' observations and accounts for the heavy elements in the **solar system**. However, whether the Sun had a companion star or not is still disputed; some believe a passing star was the culprit.

Following the war, Hoyle returned to Cambridge and became a professor of astronomy and mathematics. The pivotal point in his career came in 1948 when nuclear physicist George Gamow, building upon a theory first suggested by Georges Lemaître, a Jesuit priest and astonomer, and supported by the telescopic observations of the astronomer Edwin Powell Hubble, published what became known as the **big bang theory** of the creation of the universe. The big bang theory states that billions of years ago there was an enormous explosion in which all the matter of the universe was created. Galaxies formed and evolved from this matter and are still moving away from each other at tremendous velocities as a result of the explosion.

The concept that the universe had a specific beginning—and the implication that it will have an end—was abhorred by many scientists and laymen. Consequently, Thomas Gold and Hermann Bondi, an astronomer and mathematician respectively, proposed the steady state theory that theorized that the universe was perpetual, an idea that appeared to agree with scientific observation. Through the steady state theory, Gold and Bondi conceived of a universe in which matter was created continuously. As galaxies drift apart, new matter appears in the void and evolves into new galaxies. Since the universe seemed homogeneous (the same) regardless from which direction it was observed, or how far away (i.e. how far back in time) it was observed, Gold and Bondi suggested the cosmos was the same every "where" and every "when." That is, the physical state of the universe remains the same in the past, the present, and the future. The steady state concept had several virtues, not the least of which was avoiding the troublesome issue of the beginning and end of creation. It was simple, symmetrical,

and attracted as many adherents as did Gamow's big bang theory. Hoyle became one of steady state's most influential and talented supporters.

Gold and Bondi had not based their concept on general field theory, but instead on an intuitive physical principle. To rectify this, Hoyle delved into the complex equations of **Albert Einstein**, modified them, and produced a mathematical model that supported the steady state theory, thereby giving it both respectability and plausibility. He became the official spokesperson for the theory and produced many books, some extremely technical, others geared for popular consumption, that publicized steady state **cosmology**.

The greatest objection to the steady state theory concerned the issue of the continual creation of new matter forming from nothing—an idea that seemed to violate the laws of nature. Hoyle claimed it was easier to accept the idea of matter being created slowly and continuously over the eons than believing that all matter in the universe was created in a single instant from a single blast. For the next fifteen years, proponents of each side interpreted new astronomical discoveries in ways that supported the theory to which each adhered.

In 1952, however, astronomer Walter Baade demonstrated that the accepted cosmological "yardstick" of measurement was seriously flawed. This "yardstick" was derived from the relationship between the brightness and the rate of pulsation of certain stars called Cepheid variable stars. According to Baade's findings, such stars were much farther away than had been previously calculated. This meant that the universe was much older, had been evolving longer, and was more than two times larger than had been believed. If the steady state theory were to hold up, astronomers surveying space would expect to see "old" galaxies created billions of years ago and containing aging stars, as well as "new," recently-created galaxies containing lighter elements and new stars. Yet observed galaxies appeared to be similar in age, supporting the big bang theory. Proving that matter is continuously created was more complicated than it seemed for the steady state theorists. Since space is so vast and the amount of matter that needs to be created at a given moment for the theory to be proven was so small, scientists were not able to detect the instantaneous creation of matter.

The debate between the factions continued. Hoyle acknowledged in his 1962 book *Astronomy* that there are "cosmological theories in which the universe had a finite and 'explosive' origin," but he manages to discuss them without once using the contentious phrase "big bang"; an ironic point since the term "big bang" is attributed to Hoyle. A decade earlier, Gamow's book *The Creation of the Universe* remarked that "Astronomical observations" concerning the brightness of the Milky Way stars in relation to the brightness of neighboring stars suggest "that the theory of [Bondi, Gold, and Hoyle] may not correspond to reality." In order to maintain the relevance of his work, Hoyle made several modifications to the steady state theory throughout the 1950s and 1960s.

The 1963 discovery of **quasars** by Maartin Schmidt created an awkward complication for the steady state theory.

Quasars, distant objects brighter than and emitting more energy than stars, did not fit into the steady state explanation of the universe. This tipped the balance toward the big bang theory, which had no trouble embracing these "quasi-stellar" objects. In the following year, **Arno Penzias** and **Robert W. Wilson** discovered background microwave radiation in outer space by using radio telescopes. Claiming they had discovered the "remnants" of the big bang explosion with their telescopes, which had evolved from Hoyle's work on radar during World War II, Penzias and Wilson sealed the fate of the steady state theory, which was now abandoned in favor of the big bang theory. Subsequently, Hoyle found working with radio astronomers at Cambridge University increasingly difficult. When his proposed grant for a computer was rejected by the Science Research Council in 1972, Hoyle left the university in favor of working elsewhere.

Hoyle stirred up controversy again in 1981, when he proposed that one-celled life could be found in interstellar dust or **comets** and life on Earth may have originated from a close encounter with a comet. He also suggested that the abrupt appearance of global epidemics could be caused by space-borne contaminants, a suggestion not taken seriously by most scientists. In 1985, Hoyle ignited yet another controversy when he claimed that the British Museum's fossil of Archaeopteryx was a fake, but he had not been alone in that contention.

A prodigious amount of information has flowed from Hoyle's pen during his career. With his talent for simplifying complex theories for general audiences, he has produced technical treatises, textbooks, popular science fiction stories, an opera libretto, even a radio and a television play. The radio play, *Rockets in Ursa Major*, and the television play, *A for Andromeda*, were both written in collaboration with his son, Geoffrey, in 1962. His research on the development of stars and their age, including giants and white dwarfs, helped establish some of cosmology's major theories.

During his career, Hoyle was widely recognized for his achievements with many honors. In 1956, he became a member of the staff at Mount Wilson and Palomar observatories. In 1957, he was elected to the Royal Society of London; the following year he became Plumian Professor of Astronomy and Experimental Philosophy at Cambridge, and in 1962, he became the director of the Institute of Theoretical Astronomy. Following his departure from Cambridge in 1972, he became professor-at-large at Cornell University. Hoyle died at the age of 86.

See also Stellar life cycle

HUBBLE, EDWIN (1889-1953)

American astronomer

Edwin Hubble was an American astronomer whose impact on science has been compared to pioneering scientists such as the English physicist Isaac Newton and the Italian astronomer Galileo. Hubble helped to change perceptions of the universe in two very important ways. In an era when the Milky Way

was perceived as the extent of the entire Universe, Hubble confirmed the existence of other galaxies through his observations from the Mount Wilson Observatory in Pasadena, California. Furthermore, with the help of other astronomers of his time, Hubble showed that this newly discovered Universe was expanding, and developed a mathematical concept to quantify this expansion now known as Hubble's law.

Edwin Powell Hubble was born in Marshfield, Missouri to John P. Hubble, an agent in a fire insurance firm, and Virginia Lee James Hubble, a descendant of the American colonist Miles Standish. The third of seven children, Hubble spent his early childhood in Missouri, entering grade school in 1895. In 1898, John Hubble transferred to the Chicago office of his firm, and the Hubble family moved first to Evanston and then to Wheaton, both Chicago suburbs.

Hubble attended Wheaton High School, excelling in both sports and academics. He graduated in 1906 at the age of sixteen, two years earlier than most students. For his efforts, he received an academic scholarship to the University of Chicago, where he studied mathematics, **physics**, **chemistry**, and **astronomy**. In the summer, Hubble tutored and worked to earn money for his college expenses. In his junior year he received a scholarship in physics, and by his senior year he was working as a laboratory assistant to physicist Robert A. Millikan. Hubble graduated in 1910 with a B.S. in mathematics and astronomy. In addition to his academic career, the six-foot-two-inch Hubble was an amateur heavyweight boxer.

In 1910, Hubble was awarded a Rhodes Scholarship, following which he went to attend Queen's College at the University of Oxford in England. There he studied jurisprudence, completing the two-year course in 1912. He began working on a bachelor's degree in law during his third year, but renounced it for Spanish instead. He also continued his athletic endeavors, excelling in the high jump, broad jump, shot put, and running. In 1913, Hubble returned to the United States and began practicing law in Louisville, Kentucky, where his family was now living. Bored with his law career within a year, Hubble returned to the University of Chicago in 1914 to work towards his doctorate in astronomy.

At the time Hubble attended the University of Chicago, Yerkes was a waning institution that did not actually offer formal courses in astronomy. However, working under the supervision of Edwin B. Frost, the observatory's director, Hubble made regular observations on Yerkes' **telescope** and studied on his own. It is believed that Hubble's work at this time was influenced by a lecture he attended at Northwestern University. At the presentation, Lowell Observatory astronomer, Vesto M. Slipher presented evidence that spiral nebulae (in that era, the term nebulae was used to describe anything not obviously identifiable as a star) had high radial velocities—the velocities with which objects appear to be moving toward or away from us in the direct line of sight. Slipher found spiral nebulae that were moving at much higher velocities than stars generally moved—evidence that the nebulae might not be part of the Milky Way.

During his term at Yerkes, Hubble also met astronomer George Ellery Hale, founder of the Yerkes Observatory and then the director of the Mount Wilson Observatory in Pasadena, California. Hale had heard of Hubble, and in 1916, invited him to join the Mount Wilson Staff once he received his doctorate. However, Hubble's acceptance of this offer was delayed by World War I, which he joined in 1917. Hubble attained the rank of major, and after his discharge in 1919, he finally began work at Mount Wilson. The observatory had two telescopes, a 60–in (152–cm) reflector and a newly operational 100–in (254–cm) telescope, the largest in the world at that time. It was here that Hubble began the major portion of his life's work.

Hubble's first notable achievement at Mount Wilson was the confirmation of the existence of galaxies outside the Milky Way. From observations made in October 1923, Hubble was able to identify a type of variable star known as a Cepheid in the Andromeda nebula (known today as the Andromeda galaxy). By using information about the relationship between brightness, luminosity (how much light a star radiates) and the distances of Cepheid stars in our galaxy, Hubble was able to estimate the distance to the Cepheid in the Andromeda nebula to be about one million light years. Hubble also discovered other Cepheids, as well as other objects, and calculated the distances to them. Since scientists knew that the maximum diameter of the Milky Way was only 100,000 light years, Hubble's figures established the existence of galaxies outside our own. Eventually, he determined the distances to nine galaxies. Consistent with scientific terminology of his time, Hubble called these "extragalactic nebulae." The results of Hubble's work were publicly announced at the December 1924 meeting of the American Astronomical Society, settling one of the great scientific debates of that era.

Also in 1924, Hubble married Grace Burke Leib. His personal interests included dry-fly fishing (his favorite fishing haunts were in the Rocky Mountains and in England) and collecting antique books about the history of science. He served as a member of the Board of Trustees of the Huntington Library in San Marino, California from 1938 until he died in 1953.

Hubble's work at Mount Wilson was interrupted during World War II, when he served as chief of exterior ballistics and director of the supersonic **wind** tunnel at the Ballistics Research Laboratory at Aberdeen Proving Ground in Maryland. He worked at Aberdeen from 1942 to 1946 and received a Medal of Merit for his efforts.

Returning to Mount Wilson after the war, Hubble continued his observations of galaxies. In 1925 he introduced a system for classifying them at a meeting of the International Astronomical Union; according to this system, galaxies were either "regular" or "irregular." In addition, regular galaxies were either spiral or elliptical, and each of these classes could be further subdivided. The system used to classify galaxies today is still based on Hubble's structure.

In 1927 Hubble was elected a member of the National Academy of Sciences, but another great achievement was yet to come. By combining his own work on the distances of galaxies with the work of American astronomers Vesto M. Slipher and Milton L. Humason, Hubble proposed a relationship between the high radial velocities of galaxies and distance. He systematically looked at a number of galaxies and

found that except for a few nearby, all of the others were moving away from us at high speed. He discovered a **correlation** between this velocity and distance, and the result was a mathematical concept now known as Hubble's law. Simply put, Hubble's law states that the more distant a galaxy is from us, the faster it's moving away from us. Although Hubble didn't actually discover that the universe is expanding, he put the theory together in a coherent way. Today, the expanding universe is part of the big-bang theory of the creation of the universe.

HUBBLE SPACE TELESCOPE (HST)

The Hubble Space **Telescope** (HST) is a large Earth-orbiting astronomical telescope designed by the United States National Aeronautics and Space Administration (NASA) and the European Space Agency (ESA). Hubble observes the heavens from 380 mi (612 km) above the earth, relaying pictures and data captured above the distortions of Earth's atmosphere. The HST is named after American astronomer Edwin P. Hubble (1889-1953), who early in the twentieth century provided evidence of an expanding universe consisting of many galaxies beyond our Milky Way galaxy. The HST has provided scientists with the clearest views yet obtained of the universe. Moreover, stunning images and spectrographic data sent from the HST provide scientists with critical data relevant to studies regarding the birth of galaxies, the existence of black holes, and the workings of planetary systems around stars.

Deployed from the **space shuttle** *Discovery* on April 25, 1990, the Hubble Space Telescope was the culmination of a 20-year scientific effort to construct one of the largest and most complex satellites ever built. Astronomers first proposed the idea of building an orbiting observatory in the 1940s. The $1.5 billion project to build the Hubble Space Telescope began in earnest in 1977 after the United States Congress passed a resolution granting approval for the HST construction. By 1985, the HST was completed and ready for launch. The explosion of the space shuttle *Challenger* and loss of its crew in January 1986 delayed the Hubble's launch four years. As NASA officials re-evaluated the space shuttle program, the HST was relegated to storage—at a maintenance cost of up to one million dollars a month.

The HST is roughly the size of a school bus, and is modular in design to facilitate in-orbit servicing. Like any reflecting telescope, the Hubble uses a system of mirrors to magnify and focus light. The primary mirror is concave, and a smaller convex secondary mirror is placed in front of the primary mirror to boost the telescope's total effective focal length. The telescope receives its main power from a pair of flexible, lightweight solar arrays. Each array is a large (40 ft by 8 ft, or 12.2 m by 2.4 m) rectangle of light-collecting solar cells. Exterior thermal blanketing protects the HST from the extreme **temperature** changes encountered during each 95-minute orbit of the earth.

Shortly after the 1990 launch of the HST, scientists found the telescope was unable to adequately focus light to provide desired resolutions. Fuzzy halos appeared around objects observed by the HST. The culprit was found to be a defect in the primary mirror. As a result of an incorrect adjustment to a testing device, the mirror was precisely, but inaccurately, ground to a curvature that was too flat at its edge. Although the error measured less than a micron (one ten-thousandth of an inch), the defect caused a spherical aberration when light reflected by the mirror focused across a wider **area** than necessary for a sharp image. The problem was corrected in December 1993 when, following an orbital rendezvous between the space shuttle *Endeavor* and the HST, the crew of *Endeavor* completed the first Hubble servicing mission. During the eleven-day operation, the Corrective Optics Space Telescope Axial Replacement (COSTAR) was installed. COSTAR corrected the spherical aberration of the HST primary mirror with a series of mirrors designed to act as corrective "eyeglasses" able to focus the blurred uncorrected image.

The Hubble Space Telescope carries a variety of onboard, scientific instruments designed to collect and send data to awaiting scientists. As needed, instruments are replaced or added during Hubble servicing missions. In 1977, two spectrographs were replaced with the Near-Infrared Camera and Multi-Object Spectrometer (NICMOS), and the Space Telescope Imaging Spectrograph (STIS). NICMOS allows the telescope to see objects in near-infrared wavelengths. These observations are important in **astronomy**, as well as in the study of the visible-light-obscuring gas and dust nebular clouds where stars are born. The STIS collects light from hundreds of points across a target and spreads it out into a spectrum, creating an image from which scientists can study individual wavelengths of radiation from a distant source. STIS is especially helpful to scientists studying regions of space where black holes are presumed to exist. In 1993, the HST's original Wide Field Planetary Camera was replaced with an updated version complete with relay mirrors spherically aberrated to correct for the spherical aberration on the Hubble's primary mirror. In 1999, the HST received a new high-speed computer.

Once the Hubble gathers data and pictures from celestial objects, its computers send the digitized information to Earth as radio signals. The HST signal is passed through a series of **satellite** relays, then to the Goddard Space Flight Center in Maryland before reaching the Space Telescope Science Institute at Johns Hopkins University. Here, the signal is converted back into pictures and data. Scientists at these institutions are responsible for the daily programming and operations of the HST.

Scheduled to serve until the year 2010, the Hubble Space Telescope continues to provide dramatic observations that stretch the boundaries of the known universe. Among its accomplishments so far, the HST has provided evidence of the existence of massive black holes at the centers of galaxies, captured the first detailed image of the surface of Pluto, detected protogalaxies (structures presently thought to have existed close to the time of the origin of the universe), and captured spectacular images of the comet Shoemaker-Levy as its parts collided with Jupiter.

In order to provide continuous and broader astronomical observations, NASA is expected to launch the Hubble's successor (tentatively named the Next Generation Space Telescope) more fully equipped with cameras and spectrographs sensitive to multiple regions of the **electromagnetic spectrum** prior to the end of the HST's expected service life.

See also Big Bang theory; Cosmology; History of manned space exploration; Quasars; Solar system; Spacecraft, manned; Stellar life cycle

HUMBOLDT, ALEXANDER VON (1769-1859)

German explorer and naturalist

Alexander Humboldt pursued a lifetime of exploration and discovery, and was best known for his expeditions to Central and **South America**. A master of observation and analysis, Humboldt was also a prolific writer and recorder of his observed scientific data.

Humboldt was born in Berlin, the son of a Prussian army officer and a Huguenot (French Protestant) mother. He experienced poor health as a child and was unimpressive as a student. He was raised under his mother's strict Calvinistic beliefs and remained unmarried throughout his life.

Perhaps his most notable accomplishment was his five-year expedition to South and Central America made from 1799 to 1804. Spain had been preoccupied with the pursuit of wealth and conquest in its American colonies, and it was rare for a learned individual like Humboldt to gain permission to visit these areas. Once there, his perseverance took him to the edges of human endurance.

South America was a largely unknown land, and much of what Humboldt observed was new knowledge. Traveling by foot and canoe, he discovered a connection between the Orinoco and Amazon River systems. He climbed volcanoes in Ecuador and observed how they were positioned in a line, as though following a flaw in the earth's **crust**. He collected thousands of plant specimens. He observed ocean currents in the Pacific Ocean including one, now called the Peru Current, which was also named after him. No matter where his location or surroundings, Humboldt tirelessly recorded his observations. This proved to be Humboldt's greatest legacy.

Humboldt resided in Paris from 1805 to 1827, enjoying a cosmopolitan lifestyle that allowed him to associate with many of his fellow professionals. He published more than 30 volumes of his data during this time, proving his excellence as a writer and artist.

Humboldt spent his later years in Berlin, where he had become a notable figure. At the invitation of the Russian government, he traveled for three months in the Urals and Siberia, and brought with him his knowledge of mining techniques.

The ceremonial trappings of this visit only interfered with his ability to observe the region.

Humboldt died while working on the fifth volume of his book *Kosmos*. This work was his attempt to give a unified explanation of all existence, and gathered most of the available scientific knowledge of the time. During his life, Humboldt had been a meteorologist, botanist, geologist, geographer, and oceanographer.

See also Cosmology; Ocean circulation and currents

HUMIDITY

Humidity is a measure of the quantity of **water** vapor in the air. There are different methods for determining this quantity and those methods are reflected in a variety of humidity indexes and readings.

The humidity reading in general use by most meteorologists is relative humidity. The relative humidity of air describes the saturation of air with water vapor. Given in terms of percent humidity (e.g., 50% relative humidity), the measurement allows a comparison of the amount of water vapor in the air with the maximum amount water vapor that—at a given temperature—represents saturation. Saturation exists when the **phase state changes** of **evaporation** and **condensation** are in equilibrium.

Approximately one percent of Earth's total water content is suspended in the atmosphere as water vapor, **precipitation**, or **clouds**. Humidity is a measure only of the vapor content.

Because water vapor exerts a pressure, the presence of water vapor in the air contributes vapor pressure to the overall **atmospheric pressure**. Actual vapor pressures are measured in millibars. One atmosphere of pressure (1 atm) equals 1013.25 mbar.

In contrast to the commonly used relative humidity, the absolute humidity is a measure of the actual mass of water vapor in a defined volume of air. Absolute humidity is usually expressed in terms of grams of water per cubic meter.

Specific humidity is a measure of the mass of water vapor in a defined volume of air relative to the total mass of gas in the defined volume.

The amount of water vapor needed to achieve saturation increases with **temperature**. Correspondingly, as temperature decreases, the amount of water vapor needed to reach saturation decreases. As the temperature of a parcel of air is lowered it will eventual reach saturation without the addition or loss of water mass. At saturation (**dew point**) condensation or precipitation. This is the fundamental mechanism for cloud formation as air moving aloft is cooled. The level of cloud formation is an indication of the humidity of the ascending air because—given the standard temperature lapse rate—a parcel of air with a greater relative humidity will experience condensation (e.g., cloud formation) at a lower altitude than a parcel of air with a lower relative humidity.

The differences in the amount of water vapor in a parcel of air can be dramatic. A parcel of air near saturation may contain 28 grams of water per cubic meter of air at 30°F (–1°C), but only 8 grams of water per cubic meter of air at 10°F (–12°C).

An increasingly popular measure of comfort, especially in the hotter summer months, is the heat index. The heat index is an integrated measurement of relative humidity and dry air temperature. The measurement is useful because higher humidity levels retard evaporation from the skin (perspiration) and lower the effectiveness of physiological cooling mechanisms.

Absolute humidity may be measured with a sling cyclometer. A hydrometer is used to measure water vapor content. Water vapor content can also be can be expressed as grains/cubic ft. A grain, a unit of weight, equals 1/7000 of a pound.

See also Adiabatic heating; Atmospheric composition and structure; Atmospheric inversion layers; Atmospheric lapse rate; Atmospheric pressure; Hydrologic cycle; Hydrostatic pressure; Weather forecasting methods; Weather forecasting; Wind chill

HURRICANE • *see* TROPICAL CYCLONE

HUTTON, JAMES (1726-1797)
Scottish physician

James Hutton, a Scottish physician and farmer, is considered by many to be the father of **geology**. Hutton observed geological changes and theorized that the forces that were changing the landscape of his farm were the same forces that had changed Earth's surface in the past. He built on this theory to form his principle of uniformitariansm in 1785.

The principle states that current geological processes, for example volcanic activity and **erosion**, are the same processes that were at work in the past, and will still be at work in the future. A summary of his theory is the phrase "the present is the key to the past." Hutton watched these slow changes occurring on his own farm and theorized that over time, a stream could carve a valley, rain would erode **rock**, and sediment could accumulate and form new **landforms**. He realized that these forces must be acting very slowly, and therefore, Earth must be older than theologians at the time argued it to be. He published this theory in 1790 in his work *The Theory of the Earth*.

Modern evidence supports the essential elements of Hutton's theory. Earth is approximately 4.6 billion years old, and there is abundant evidence that slow processes have worked to mold and shape the planet. Moreover, the same forces that acted in the past are active now, even though the relative rates may vary over time. When Hutton published his theories, however, they were not met with enthusiasm. **Uniformitarianism** went against both religious beliefs and the theory of **catastrophism**, the accepted theory of the time. Catastrophism states that the earth was formed not by slow processes, but by violent, worldwide disturbances such as earthquakes and **floods**. It was not until the nineteenth century that Sir **Charles Lyell**, in his 1830 work *Principles of Geology*, popularized the theory of uniformitarianism.

James Hutton was not only known for his uniformitarianism theory, but also for developing the concept of the rock cycle. This theory describes the interrelationships between igneous, sedimentary, and metamorphic rocks. The matter that makes up these rocks is neither created nor destroyed, but instead transformed from one rock type to another. He also suggested that the study of the earth be called "geophysiology." Hutton's theories about Earth as an entity that undergoes dynamic cycles are considered by some to be the basis of the **Gaia hypothesis**, the concept of the "living earth."

See also Earth (planet); Geologic time

HYDROCARBONS

Hydrocarbons are compounds composed solely of **carbon** and hydrogen. Despite their simple composition, hydrocarbons include a large number of different compounds with a variety of chemical properties. Hydrocarbons are derived from oil deposits, and are the source of gasoline, heating oil, and other "fossil fuels." Found in a variety of geological settings, hydrocarbons provide the carbon skeletons required for the thousands of chemicals produced by the chemical industry.

Hydrocarbons are classified on the basis of their structure and bonding. The three major classes are aliphatics, alicyclics, and aromatics. Aliphatics have carbon backbones that form straight or branched chains, with no rings. Alicyclics are ring compounds that, while they may have one or more double bonds, do not form conjugated sets of double bonds around the ring like benzene. Aromatics are compounds with at least one benzene ring. Aliphatic means fatty, and aromatic refers to odor, but these terms no longer have significance for the compounds they describe.

Because hydrocarbons are nonpolar, they are generally insoluble in **water**, and dissolve in nonpolar solvents.

The aliphatic hydrocarbons are further divided into alkanes, alkenes, and alkynes. Alkanes (sometimes called paraffins) have only single bonds, while alkenes (sometimes called olefins) have a carbon-carbon double bond, and alkynes have a carbon-carbon triple bond. Compounds with two double bonds are known as dienes. Compounds with double or triple bonds are referred to as "unsaturated," while those without are "saturated," meaning all of their carbons are bonded to the maximum number of hydrogens.

All alkanes have the general formula C_nH_{2n+2}, where n is the number of carbon atoms in the compound. Alkenes have the formula C_nH_{2n}, while alkynes are C_nH_{2n-2}. Because of these regularities, the members of each group are known as a homologous series.

The simplest alkane (and indeed, the simplest hydrocarbon) is methane, CH_4. The four C-H bonds are directed towards the four corners of a tetrahedron, with carbon at its center and hydrogens at each vertex. The C-H bonds are slightly polar, and equivalent in length and strength. The angle formed by any pair of bonds is 109.5 degrees, the tetrahedral angle.

Despite the bond polarity, there is no net dipole because of the symmetry of the molecule, and methane has very weak intermolecular attractions, consisting only of van der Waals attractions. These attractions are proportional to surface **area**, which is small for the compact methane molecule and, as a result, methane has a very low **melting** point and boiling point, –297.4°F (–183°C) and –258.7°F (–161.5°C), respectively.

Methane is found in oil deposits, forming the majority of the "natural gas" fraction. Methane is also formed by certain anaerobic bacteria, especially in swamp bottoms, where it bubbles to the surface as "marsh gas." It is used in large quantities as fuel for heating and cooking because it burns with **oxygen** to produce **carbon dioxide** and water. Methane burns cleanly, with very little soot or smoke.

The next alkane is ethane, C_2H_6. Its higher melting and boiling points (–277.6°F [–172°C] and –127.3°F [–88.5°C], respectively) reflect its larger surface area and consequently greater van der Waals attraction.

Propane, C_3H_8, is an important fuel because it can be liquified at pressures low enough to easily maintain in commercial and consumer apparatus, allowing easy transport and storage, but vaporizes to burn almost as cleanly as methane.

Higher alkanes continue the trend of increased surface area and higher melting and boiling points. However, as the number of carbon atoms increases, structural isomerism becomes possible, allowing the same molecular formula to describe two or more compounds with different structures and different physical properties. For example, Butane (C_4H_{10}) can be either a straight-chain molecule, or a branched one, with three carbons in a straight chain and the fourth branching off from the middle carbon. Compounds with the same molecular formula but with different three-dimensional structures are called isomers. The straight chain isomer is called n-butane (for "normal"), while the branched one is called either iso-butane, or 2-methyl propane. This latter name indicates that the longest straight-chain backbone within the molecule has three carbons (propane), and there is a single-carbon branch (methyl) at the second position in the main chain. The extended structure of n-butane gives it a boiling point of 32°F (0°C), while the more spherical iso-butane boils twelve degrees lower, at 10.4°F (–12°C), due to its smaller surface area. As the number of carbons increase, so too does the number of possible isomers.

The simplest alkene is ethene, C_2H_4, also called ethylene. Ethene is a planar molecule, with the four hydrogens splayed out in a plane and angles between bonds of 120 degrees. Ethene melts at –272.2°F (–169°C), and boils at –151.6°F (–102°C).

Higher alkenes take their names from the corresponding alkanes: propene, butene, pentene, etc. However, beginning with butene, isomerism is possible based on the position of the double bond. If the bond is between C1 and C2, the compound is 1-butene, while if it lies between C2 and C3, it is 2-butene (the compound with a double bond between C3 and C4 is equivalent to 1-butene). In addition, the non-rotation of the double bond means there are two structural isomers of 2-butene, one with the two terminal carbons on the same side of the C=C long axis, termed *cis*-2-butene, and one with them on opposite sides, *trans*-2-butene. As might be expected, the number of possible isomers rises with an increase in number of carbons.

The chemical reactions of the alkenes is more complex and richer than that of the alkanes, due to the presence of the double bond. The most common reaction is of addition across the double bond. For instance, addition of water to ethene creates ethyl alcohol. Alkenes are often prepared by the reverse of this reaction, dehydration across the double bond.

Alkynes have a triple bond. The simplest alkyne is C_2H_2, athyne, also called acetylene. Acetylene is an important high-temperature fuel used especially for metal cutting.

Cyclic alkenes, such as cyclohexene (C_6H_{10}), are common solvents, and also serve as starting points for a number of organic syntheses.

Benzene (C_6H_6) is a cyclohexane (the carbons are bonded in a ring) with three double bonds alternating around the ring. The electrons in benzene's bonds are highly delocalized, to the point that it is no longer accurate to describe them as belonging to individual carbons. Instead, they form a cloud "ring" above and below the plane of the carbons and, as a result, all the carbon-carbon bonds are of equivalent length and strength.

See also Atomic theory; Fuels and fuel chemistry; Geochemistry; Petroleum detection; Petroleum, economic uses of; Petroleum extraction; Petroleum, history of exploration; Petroleum

HYDROGEOLOGY

Hydrogeology is the study of **water** contained in materials of Earth's **crust**, the physical and chemical characteristics of this water, its origin, **evolution**, and ultimate destination. Hydrogeology is the term used by geologists and hydrogeologists for this study. Geohydrology is the term most often used by engineers. The two terms are roughly equivalent.

The water contained in materials of Earth's crust is called **groundwater** (sometimes spelled as "ground water" to distinguish ground water from surface water). Groundwater is sometimes defined as water below the earth's surface, but groundwater may occur at the surface especially after heavy rainfall in certain areas.

When groundwater is not at the earth's surface, there is a zone beneath the surface where the majority of pore (open) spaces are filled with air. This is called the vadose zone or zone of aeration. Below a certain depth, all the pore spaces are filled with water. This is known as the phreatic zone or zone of saturation. The zone of saturation extends downward until pressure of the overlying materials is so great that there are no pore spaces available. The **area** separating the vadose and phreatic zones is called the **water table**, which is usually represented on cross sections as a dashed line. The dashed line, as opposed to a solid line, indicates that the water table moves up and down with the **seasons**, being higher and nearer Earth's surface during wet seasons and lower and deeper below Earth's surface during dry seasons.

Modern groundwater studies have their origin in the middle 1850s when a French engineer, Henry Darcy, published a report describing an experiment he conducted with a tube on an incline that he had filled with **sand**. Darcy's experiments led to the first quantitative "law" in hydrogeology, used to determine the rate of flow of groundwater and now known as Darcy's law. It can be expressed mathematically as $v = KIA$, where "v" equals the rate of groundwater flow, "K" equals hydraulic conductivity or **permeability**, "A" equals the area of a cross-section of the water-bearing unit (e.g., cross-section of Darcy's cylinder of sand) and "I" equals the hydraulic gradient.

Hydraulic conductivity or permeability (K) is measured from the material through which the groundwater is flowing and has the dimensions of length per unit time (L/T, e.g., cm per second, feet per year, etc.).

Groundwater occurs underground in bodies of Earth materials called aquifers, which may be of two types: unconfined or confined. Unconfined aquifers are bound at their top by the water table. Confined aquifers are bound both top and bottom by materials through which little or no water flows, (i.e., impermeable materials), or aquicludes (also known as aquitards). The materials holding the water are usually inclined to the horizontal so that pressure builds up in the **aquifer**. When the aquifer is drilled, the water rises to the highest level of water confined within the aquifer or sometimes to the surface. These aquifers represent a special kind of aquifer, called **artesian**, after an area in France where this situation is common.

Hydrogeology is extremely important to mankind because over 100 million people in the United States alone use groundwater for drinking; about a third of the largest U.S. cities have some reliance upon groundwater use in their potable (drinkable) water supply.

It is not always easy to gather data about groundwater in a given area because evidence of underlying water is not usually readily apparent at the surface. Therefore, it is necessary to drill wells into and below the water table into the phreatic zone (zone of saturation). Well drilling is a time-consuming and expensive method for gathering data and is usually not done solely for academic purposes. It is usually done to meet the requirements of state or federal regulatory agencies. Such studies are sometimes financed by the agencies but, in most cases, they are financed by private industry seeking to avoid contamination by the use of proper hydrogeologic profile maps, or to determine the presence and/or extent of existing contamination. In the United States, the federal government, especially the Department of Defense and the Department of Energy, have industrial facilities that are responsible for groundwater cleanup.

The United States Congress passed two major laws that have greatly impacted the study of groundwater: the Resource Conservation and Recovery Act (RCRA) of 1976 and the Comprehensive Environmental Response, Compensation, and Liability Act (CERCLA) of 1980.

RCRA establishes a "cradle to grave" tracking system for hazardous wastes and requires those facilities considered RCRA facilities to define and characterize the uppermost aquifer beneath the facility and to monitor the quality of the groundwater as it flows beneath the facility. Numerous wells are usually required to characterize the uppermost aquifer and, in every case, at least four wells are required (one upgradient and three downgradient from the facility).

CERCLA was originally conceived to allow for cleanup of abandoned hazardous waste sites. It established a fund to be raised and maintained by collecting a tax on the producers of hazardous materials. This fund (and the act) soon became known as "Superfund." Superfund was reauthorized and greatly enhanced in 1986 by the Superfund Amendments and Reauthorization Act (SARA).

Once groundwater is contaminated, it can be very costly and time consuming to remediate (i.e., clean up). Unfortunately, one of the most common groundwater contaminants is also one of the most difficult to remediate—the chemical group known as the chlorinated solvents, especially trichloroethylene (TCE) and perchloroethylene (PCE). These have been widely used as solvents, TCE as a degreaser and PCE in the dry cleaning industry. They do not fully dissolve in groundwater and they tend to form pockets or globules in both the vadose and phreatic zones.

One of the most common methods of cleaning up groundwater is the "pump and treat" method whereby groundwater is pumped to the surface, treated by some method, then reinjected into the aquifer or released as surface water. This does not work well with TCE and PCE because as the groundwater is pumped to the surface, these globules remain in the subsurface and they fail to come to the surface with the groundwater.

See also Drainage basins and drainage patterns; Drainage calculations and engineering; Freshwater; Hydrostatic pressure; Petroleum, detection; Porosity and permeability; Relief; Remote sensing; Runoff; Seismology; Waste disposal; Wastewater treatment; Water pollution and biological purification

HYDROLOGIC CYCLE

The hydrologic, or **water**, cycle is the continuous, interlinked circulation of water among its various compartments in the environment. Hydrologic budgets are analyses of the quantities of water stored, and the rates of transfer into and out of those various compartments. A simplified hydrologic cycle starts with heating caused by **solar energy** and progresses through stages of **evaporation** (or sublimation), **condensation**, **precipitation** (snow, rain, hail, glaze), **groundwater**, and **runoff**.

The most important places in which water occurs are the **oceans, glaciers,** underground aquifers, surface waters, and the atmosphere. The total amount of water among all of these compartments is a fixed, global quantity. However, water moves readily among its various compartments through the processes of evaporation, precipitation, and surface and subsurface flows. Each of these compartments receives inputs of water and has corresponding outputs, representing a flow-through system. If there are imbalances between inputs and outputs, there can be significant changes in the quantities stored locally or even globally. An example of a local change is the **drought** that can occur in **soil** after a long period without replenishment by precipitation. An example of a global change in hydrology is the increasing mass of continental **ice** that occurs during glacial epochs, an event that can remove so much water from the oceanic compartment that sea level can decline by more than 328 ft (100 m), exposing vast areas of **continental shelf** for the development of terrestrial ecosystems.

Estimates have been made of the quantities of water that are stored in various global compartments. By far, the largest quantity of water occurs in the deep **lithosphere**, which contains an estimated 27×10^{18} tons (27-billion-billion tons) of water, or 94.7% of the global total. The next largest compartment is the oceans, which contain 1.5×10^{18} tons, or 5.2% of the total. Ice caps contain 0.019×10^{18} tons, equivalent to most of the remaining 0.1% of Earth's water. Although present in relatively small quantities compared to the above, water in other compartments is very important ecologically because it is present in places where biological processes occur. These include shallow groundwater (2.7×10^{14} tons), inland surface waters such as **lakes** and **rivers** (0.27×10^{14} ton), and the atmosphere (0.14×10^{14} tons).

The smallest compartments of water also tend to have the shortest turnover times, because their inputs and outputs are relatively large in comparison with the mass of water that is contained. This is especially true of atmospheric water, which receives annual inputs equivalent to 4.8×10^{14} tons as evaporation from the oceans (4.1×10^{14} tons/yr) and terrestrial ecosystems (0.65×10^{14} tons/yr), and turns over about 34 times per year. These inputs of water to the atmosphere are balanced by outputs through precipitation of rain and snow, which deposit 3.7×10^{14} tons of water to the surface of the oceans each year, and 1.1×10^{14} tons/yr to the land.

These data suggest that the continents receive inputs of water as precipitation that are 67% larger than what is lost by evaporation from the land. The difference, equivalent to 0.44×10^{14} tons/yr, is made up by 0.22×10^{14} tons/yr of runoff of water to the oceans through rivers, and another 0.22×10^{14} tons/yr of subterranean runoff to the oceans.

The movements of water in the hydrologic cycle are driven by gradients of energy. Evaporation occurs in response to the availability of thermal energy and gradients of concentration of water vapor. The ultimate source of energy for most natural evaporation of water on Earth is solar electromagnetic radiation. Heating from within Earth's mantle and **crust** that results from radioactive decay supplies the other thermal energy requirements. Solar energy is absorbed by surfaces, increasing their heat content, and thereby providing a source of energy to drive evaporation. In contrast, surface and ground waters flow in response to gradients of gravitational potential. In other words, unless the flow is obstructed, water spontaneously courses downhill.

Clouds forming over water. © *Joseph Sohm/Corbis. Reproduced by permission.*

The hydrological cycle of a defined **area** of landscape is a balance between inputs of water with precipitation and upstream drainage, outputs as evaporation and drainage downstream or deep into the ground, and any internal storage that may occur because of imbalances of the inputs and outputs. Hydrological budgets of landscapes are often studied on the spatial scale of watersheds, or the area of terrain from which water flows into a stream, river, or lake.

The simplest watersheds are so-called headwater systems that do not receive any drainage from watersheds at higher altitude, so the only hydrologic input occurs as precipitation, mostly as rain and snow. However, at places where **fog** is a common occurrence, windy conditions can effectively drive tiny atmospheric droplets of water vapor into the forest canopy, and the direct deposition of cloud water can be important.

Vegetation can have an important influence on the rate of evaporation of water from watersheds. This hydrologic effect is especially notable for well-vegetated ecosystems such as **forests**, because an extensive surface area of foliage supports especially large rates of transpiration. Evapotranspiration refers to the combined rates of transpiration from foliage, and evaporation from non-living surfaces such as moist soil or surface waters. Because transpiration is such an efficient means of evaporation, evapotranspiration from any well vegetated landscape occurs at much larger rates than from any equivalent area of non-living surface.

In the absence of evapotranspiration an equivalent quantity of water must drain from the watershed as seepage to deep groundwater or as streamflow.

Forested watersheds in seasonal climates display large variations in their rates of evapotranspiration and streamflow. This effect can be illustrated by the seasonal patterns of hydrology for a forested watershed in eastern Canada. The input of water through precipitation is 58 in (146 cm) per year, but 18% of this arrives as snow, which tends to accumulate on the surface as a persistent snow pack. About 38% of the annual input is evaporated back to the atmosphere through evapotranspiration, and 62% runs off as river flow. Although there is little seasonal variation in the input of water with precipitation, there are large seasonal differences in the rates of evapotranspiration, runoff, and storage of groundwater in the watershed. Evapotranspiration occurs at its largest rates during the growing season and runoff is therefore relatively sparse during this period. In fact, in small watersheds in this region forest streams can literally dry up because so much of the precipitation input and soil water is utilized for evapotranspiration, mostly by trees. During the autumn, much of the precipitation input serves to recharge the depleted groundwater storage, and once this is accomplished stream flows increase again. Runoff then decreases during winter, because most of the precipitation inputs occur as snow, which accumulates on the ground surface because of the prevailing sub-freezing temperatures. Runoff is largest during the early springtime when warming temperatures cause the snow pack to melt during a short period of time, resulting in a pronounced flush of stream and river flow.

Some aspects of the hydrologic cycle can be utilized by humans for a direct economic benefit. For example, the potential energy of water elevated above the surface of the oceans can be utilized for the generation of **electricity**. However, the development of hydroelectric resources generally causes large changes in hydrology. This is especially true of hydroelectric developments in relatively flat terrain, which require the construction of large storage reservoirs to retain seasonal high-water flows, so that electricity can be generated at times that suit the peaks of demand. These extensive storage reservoirs are essentially artificial lakes, sometimes covering enormous areas of tens of thousands of hectares. These types of hydroelectric developments cause great changes in river hydrology, especially by evening out the variations of flow, and sometimes by unpredictable spillage of water at times when the storage capacity of the reservoir is full. Both of these hydrologic influences have significant ecological effects, for example, on the habitat of salmon and other aquatic biota.

Where the terrain is suitable, hydroelectricity can be generated with relatively little modification to the timing and volumes of water flow. This is called run-of-the-river hydroelectricity, and its hydrologic effects are relatively small. The use of geologically warmed ground water to generate energy also has small hydrological effects, because the water is usually re-injecting back into the **aquifer**.

Human activities can influence the hydrologic cycle in many other ways. The volumes and timing of river flows can be greatly affected by channeling to decrease the impediments to flow, and by changing the character of the watershed by paving, compacting soils, and altering the nature of the vegetation. Risks of flooding can be increased by speeding the rate at which water is shed from the land, thereby increasing the magnitude of peak flows. Risks of flooding are also increased if **erosion** of soils from terrestrial parts of the watershed leads to siltation and the development of shallower river channels, which then fill up and spill over during high-flow periods. Massive increases in erosion are often associated with deforestation, especially when natural forests are converted into agriculture.

The quantities of water stored in hydrologic compartments can also be influenced by human activities. An important example of this effect is the mining of groundwater for use in agriculture, industry, or for municipal purposes. The best-known case of groundwater mining in **North America** concerns the enormous Ogallala aquifer of the southwestern United States, which has been drawn down mostly to obtain water for irrigation in agriculture. This aquifer is largely comprised of "fossil water" that was deposited during earlier, wetter climates, although there is some recharge capability through rain-fed groundwater flows from mountain ranges in the watershed of this underground reservoir.

Sometimes industrial activities lead to large emissions of water vapor into the atmosphere, producing a local hydrological influence through the development of low-altitude **clouds**

and fogs. This effect is mostly associated with electric power plants that cool their process water using cooling towers.

A more substantial hydrologic influence on evapotranspiration is associated with large changes in the nature of vegetation over a substantial part of a watershed. This is especially important when mature forests are disturbed, for example, by wildfire, clear-cutting, or conversion into agriculture. Disturbance of forests disrupts the capacity of the landscape to sustain transpiration, because the amount of foliage is reduced. This leads to an increase in stream flow volumes, and sometimes to an increased height of the groundwater table. In general, the increase in stream flow after disturbance of a forest is roughly proportional to the fraction of the total foliage of the watershed that is removed (this is roughly proportional to the fraction of the watershed that is burned, or is clear-cut). The influence on transpiration and stream flow generally lasts until regeneration of the forest restores another canopy with a similar area of foliage, which generally occurs after about 5–10 years of recovery. However, there can be a longer-term change in hydrology if the ecological character of the watershed is changed, as occurs when a forest is converted to agriculture.

See also Alluvial systems; Aquifer; Artesian; Atmospheric composition and structure; Hydrogeology; Hydrologic cycle; Hydrostatic pressure; Hydrothermal processes; Stream capacity and competence; Stream piracy; Troposphere and tropopause; Wastewater treatment; Water pollution and biological purification; Water table; Water

HYDROSTATIC PRESSURE

Hydrostatic pressure is a state of stress characterized by equal principal stresses, $S_1 = S_2 = S_3$. This is the state of stress that exists at any point in a liquid at rest. Units of measurement are pounds per square inch (psi) in the English System and megabars (Mb) in the International System. The concept of hydrostatic pressure or stress is very important to many disciplines of physical science and engineering. Geologists often consider pore fluid pressure in rocks as hydrostatic, as if the fluid is part of a column of **water** open to the surface. Well drilling engineers must know whether the pore fluid pressure at depth is normal hydrostatic, overpressured, or underpressured to design the fluid system used to drill the wells. A correct design minimizes the dangers of blowout or formation damage. The mechanical behavior of rocks depends, in part, on the hydrostatic pressure (also called confining pressure or mean stress) part of the total stress (the sum of the mean stress and shear stresses) acting on the **rock**. In general, brittle rock strength increases with increase in hydrostatic stress.

Hydrostatic pressure is a scalar quantity because it does not vary with direction. The magnitude of hydrostatic pressure P at any point in a liquid is determined by the height of the column of liquid above the point and the density of the liquid. Hydrostatic pressure P varies with depth according to the linear relationship, $P = \rho g h$, where ρ is the fluid density, g is the **gravitational constant**, and h is the depth of the column of fluid to the measured point. If water is the fluid, the hydrostatic pressure gradient is: P = ρgh = 0.4 psi/foot (9.8 kPa/m). The state of stress in liquids is hydrostatic because fluids do not support any shear stress (any differences among the principal stresses). In general, the stress state in rocks of the upper **crust** is not hydrostatic because solids support shear stress. Many **petroleum** engineers assume the state of stress at depth in the earth is hydrostatic to simplify their calculations.

See also Hydrogeology

HYDROTHERMAL PROCESSES

Any subsurface encounter between **water** and heat produces a hydrothermal process. The heat is usually supplied by upwellings of **magma** from the mantle, the water by **precipitation** that percolates downward through surface rocks. Some oceanic water enters the mantle at subduction zones and becomes an important ingredient in upper-mantle magmas.

Most hydrothermal processes are driven by convection. Convection occurs because water, like most substances, expands when heated. The result is that hot water rises and cool water sinks. Convection occurs when any water-permeated part of the earth's **crust** is heated from below: heated water fountains upward over the hot spot and cool water descends around its edges. These movements occur through cracks and channels in the **rock**, forcing the water to move slowly and remain in constant contact with various **minerals**. Water convecting through rock is thus an effective means of dissolving, transporting, and depositing minerals. Most deposits of concentrated minerals, including large, shapely **crystals**, are created by hydrothermal processes.

Some manifestations of hydrothermal processes are dramatic, including the geysers and hot **springs** that sometimes occur where shallow magma is present. However, most hydrothermal circulation occurs inconspicuously in the vicinity of large magmatic intrusions. These can cause water to convect through the rocks for miles around.

Along the **mid-ocean ridges**, for example, the heat of the magma that rises continuously from the mantle to form new oceanic crust causes water to convect through the top mile or two (2–3 km) of oceanic crust over many thousands of square miles. Down-convected ocean water encounters hot rocks at depth, is heated, yields up its dissolved magnesium, and leaches out manganese, copper, calcium, and other **metals**. This hot, chemically altered brine then convects upward to the ocean floor, where it is cooled and its releases most of its dissolved minerals as solid precipitates. This process makes the concentrations of vanadium, cobalt, nickel, and copper in recent sea-floor sediments near mid-ocean ridges 10–100 times greater than those elsewhere, and has formed many commercially important ores.

Two of the metals transported in large quantities by sea-floor circulation (i.e., calcium and magnesium) are important controllers of the **carbon dioxide** (CO_2) balance of the ocean and thus of the atmosphere. A volume of water approximately equal to the world's **oceans** passes through the hydrothermal mid-ocean ridge cycle every 20 million years.

See also Fumarole; Geyser; Mid-ocean ridges and rifts; Sea-floor spreading

HYDROTHERMAL VENTS • *see* HYDROTHERMAL PROCESSES

HYPERSPECTRAL SENSORS • *see* MAPPING TECHNIQUES

I

ICE

Ice is frozen **water**, or in other words, water in solid state. Ice is a transparent, colorless substance with some special properties; it floats in water, ice expands when water freezes, and its **melting** point decreases with increasing pressure. Water is the only substance that exists in all three phases as gas, liquid, and solid under normal circumstances on Earth.

Water, and thus ice molecules, consist of one **oxygen** and two hydrogen atoms. Water is a polar molecule, with a slight negative charge on the oxygen side, and a slight positive charge on the hydrogen side, which makes it possible to interact with other polar molecules or ions. Thus, a loose chemical connection called a hydrogen bond forms between the water molecules, where each water molecule can bind to other water molecules, forming a complex network. These hydrogen bonds are the main reason for the special properties of water and ice.

Water in the solid state forms a highly ordered hexagonal (six-sided) crystal lattice structure, because it is the most stable arrangement of the water molecules. Although the individual molecules can vibrate, they cannot move fast enough to leave the crystal structure, since the opposite electrical polarities hold them together. This lattice crystal can be visualized as layers of hexagonal rings of the oxygen atoms stacked on each other. Ice has eleven known crystal forms, depending on pressure, **temperature**, or how quickly the ice forms. Ice cannot form from liquid water at the **freezing** point, unless there are seeds for the crystal, which dissipate the energy of the colliding water molecules, keeping them locked in the lattice structure. If no seeds are present, spontaneous crystal nucleation begins only if the water is supercooled below the freezing point.

Ice is present in nature in many places and in many forms: **icebergs**, ice sheets, **glaciers**, snow, freezing rain, sleet, ice **crystals**, icicles, hail, rime, graupel, and ice **fog**. Ice plays an important role in **erosion** (water fills the cracks of **rock**, freezes, expands, and breaks the rock), and in atmospheric

Ice forming on water. *Robert J. Huffman. Field Mark Publications. Reproduced by permission.*

energy transport (when water vapor changes into liquid or ice, latent heat is released). The way ice forms in bodies of water (not from the bottom up, but from the top down) protects many organisms in the water from very cold and fast temperature fluctuations.

ICE AGES

The **ice** ages were periods in Earth's history during which significant portions of the earth's surface were covered by **glaciers** and extensive fields of ice. Scientists often use more specific terms for an ice "age" depending on the length of time it lasts. It appears that over the long expanse of Earth history, seven major periods of severe cooling have occurred. These periods are often known as ice eras and, except for the last of these, are not very well understood.

What is known is that the earth's average annual **temperature** varies constantly from year to year, from decade to decade, and from century to century. During some periods, that average annual temperature has dropped to low enough levels for fields of ice to grow and cover large regions of the earth's surface. The seven ice eras have covered an average of about 50 million years each.

The ice era that scientists understand best (because it occurred most recently) began about 65 million years ago. Throughout that long period, the earth experienced periods of alternate cooling and warming. Those periods during which the annual temperature was significantly less than average are known as ice epochs. There is evidence for the occurrence of six ice epochs during this last of the great ice eras.

During the 2.4 million-year lifetime of the last ice epoch, about two dozen ice ages occurred. That means that the earth's average annual temperature fluctuated upwards and downwards to a very significant extent about two dozen times during the 2.4 million-year period. In each case, a period of significant cooling was followed by a period of significant warming—an interglacial period after which cooling once more took place.

Scientists know a great deal about the cycle of cooling and warming that has taken place on the earth over the last 125,000 years, the period of the last ice age cycle. They have been able to specify with some degree of precision the centuries and decades during which ice sheets began to expand and diminish. For example, the most severe temperatures during the last ice age were recorded about 50,000 years ago. Temperatures then warmed before plunging again about 18,000 years ago.

Clear historical records are available for one of the most severe recent cooling periods, a period now known as the Little Ice Age. This period ran from about the fifteenth to the nineteenth century and caused widespread crop failure and loss of human life throughout **Europe**. Since the end of the Little Ice Age, temperatures have continued to fluctuate with about a dozen unusually cool periods in the last century, interspersed between periods of warmer **weather**. Scientists are not certain as to whether the last ice age has ended, or continues to the present.

A great deal of what scientists know about the ice ages they have learned from the study of mountain glaciers. For example, when a glacier moves downward out of its mountain source, it carves out a distinctive shape on the surrounding land. The "footprints" left by continental glaciers formed during the ice ages are comparable to those formed by mountain glaciers.

The transport of materials from one part of the earth's surface to another part is also evidence for the formation of continental glaciers. Rocks and **fossils** normally found only in one region of the earth may be picked up, moved by ice sheets, and deposited elsewhere. The "track" left by the moving glacier provides evidence of the ice sheets movement. In many cases, the moving ice may actually leave scratches on the **rock** over which it moves, providing further evidence for changes that took place during an ice age.

Scientists have been asking what the causes of ice ages are for more than a century. The answer (or answers) to that question appears to have at least two main parts: astronomical factors and terrestrial factors. By astronomical factors scien-

tists mean that the way the earth is oriented in **space**, which can determine the amount of heat it receives and, hence, its annual average temperature.

One of the most obvious astronomical factors about which scientists have long been suspicious is the appearance of sunspots. Sunspots are eruptions that occur on the Sun's surface during which unusually large amounts of **solar energy** are released. The number of sunspots that occur each year changes according to a fairly regular pattern, reaching a maximum about every eleven years or so. The increasing and decreasing amounts of energy sent out during sunspot maxima and minima, some scientists have suggested, may contribute in some way to the increase and decrease of ice fields on the earth's surface.

By the beginning of the twentieth century, however, astronomers had identified three factors that almost certainly are major contributors to the amount of solar radiation that reaches the earth's surface and, hence, the earth's average annual temperature. These three factors are the earth's angular tilt, the shape of its orbit around the **Sun**, and its axial precession.

The first of these factors, the planet's angular tilt, is the angle at which its axis is oriented to the plane of its orbit around the Sun. This angle slowly changes over time, ranging between 21.5 and 24.5 degrees. At some angles, the earth receives more solar radiation and becomes warmer, and at other angles it receives less solar radiation and becomes cooler.

The second factor, the shape of the earth's orbit around the Sun, is important because, over long periods of time, the orbit changes from nearly circular to more elliptical (flatter) in shape. Because of this variation, the earth receives solar radiation in varying amounts depending on the shape of its orbit. The final factor, axial precession, is a "wobble" in the orientation of the earth's axis to its orbit around the Sun. As a result of axial precession, the amount of solar radiation received during various parts of the year changes over very long periods of time.

Between 1912 and 1941, the Yugoslav astronomer Milutin Milankovitch developed a complex mathematical theory that explained how the interaction of these three astronomical factors could contribute to the development of an ice age. His calculations provided rough approximations of the occurrences of ice ages during the earth history.

Astronomical factors provide only a broad general background for changes in the earth's average annual temperature, however. Changes that take place on the earth itself also contribute to the temperature variations that bring about ice ages.

Scientists assert that changes in the composition of the earth's atmosphere can affect the planet's annual average temperature. Some gases, such as **carbon dioxide** and nitrous oxide, have the ability to capture heat radiated from the earth, warming the atmosphere. This phenomenon is known as the **greenhouse effect**. But the composition of the earth's atmosphere is known to have changed significantly over long periods of time. Some of these changes are the result of complex interactions of biotic, geologic and geochemical processes. Humans have dramatically increased the concentration of **carbon** dioxide in the atmosphere over the last century through the burning of fossil **fuels** (**coal**, oil, and **natural gas**). As the concentration of **greenhouse gases**, like carbon dioxide and nitrous oxide, varies over many decades, so does the atmosphere's ability to capture and retain heat.

Other theories accounting for atmospheric cooling have been put forth. It has been suggested that **plate tectonics** are a significant factor affecting ice ages. The uplift of large continental blocks resulting from plate movements (for example, the uplift of the Himalayas and the Tibetan Plateau) may cause changes in global circulation patterns. The presence of large land masses at high altitudes seems to correlate with the growth of ice sheets, while the opening and closing of ocean basins due to tectonic movement may affect the movement of warm **water** from low to high latitudes.

Since **volcanic eruptions** can contribute to significant temperature variations, it has been suggested that such eruptions could contribute to atmospheric cooling, leading to the lowering of the earth's annual temperature. Dust particles thrown into the air during an eruption can reflect sunlight back into space, reducing heat that would otherwise have reached the earth's surface. The eruption of Mount Pinatubo in the Philippine Islands in 1991 is thought to have been responsible for a worldwide cooling that lasted for at least five years. Similarly, the earth's average annual temperature might be affected by the impact of meteorites on the earth's surface. If very large meteorites had struck the earth at times in the past, such collisions would have released huge volumes of dust into the atmosphere. The presence of this dust would have had effects similar to the eruption of Mount Pinatubo, reducing the earth's annual average temperature for an extended period of time and, perhaps, contributing to the development of an ice age.

The ability to absorb heat and the reflectivity of the earth's surface also contribute to changes in the annual average temperature of the earth. Once an ice age begins, sea levels drop as more and more water is tied up in ice sheets and glaciers. More land is exposed, and because land absorbs heat less readily than water, less heat is retained in the earth's atmosphere. Likewise, pale surfaces reflect more heat than dark surfaces, and as the **area** covered by ice increase, so does the amount of heat reflected back to the upper atmosphere.

Whatever the cause of ice ages, it is clear that they can develop as the result of relatively small changes in the earth's average annual temperature. It appears that annual variations of only a few degrees Celsius can result in the formation of extensive ice sheets that cover thousands of square miles of the earth's surface.

See also Earth (planet); Glacial landforms; Glaciation; Historical geology; Polar axis and tilt; Polar ice

ICE HEAVING AND WEDGING

Some 35% of Earth's land **area** undergoes regular **freezing** and thawing. **Ice** heaving and ice wedging are two of the mechanisms by which **water** in **soil** lifts, penetrates, and sorts soils and rocks when repeatedly melted and frozen. Ice heaving is the lifting of soil by horizontal ice layers; ice wedging is the top-down growth into soil of vertical wedges of ice.

Ice heaving is driven by complex molecular interactions between water and soil. The simple result of these complex interactions is that ice forming in soil sucks water to itself by capillary action. The suction exerted by ice upon water in soil is termed cryosuction. Since freezing normally proceeds from the surface down, ice heaving begins with the formation of a layer of ice near the surface. As it grows, this layer draws water to itself from below by cryosuction. This water freezes to the underside of the growing layer. The ice thus formed is termed segregation ice because it grows by segregating previously mixed soil and water. Segregation ice forms from water transported by cryosuction to the upper soil; this imported material, aided slightly by water's 9% expansion upon freezing, raises up the overlying ground surface as segregation ice forms, causing ice heaving.

Segregation ice often forms regularly spaced layers. As each layer forms, it tends to suck dry the soil beneath it. When the force of cryosuction is no longer able to lift water from below, thickening of the current layer ceases and cooling proceeds downward until a new ice layer can begin to form at a greater depth.

Ice wedges form by a simpler process. When soil cools it contracts; this contraction produces cracks. Water trickles into the cracks and freezes, forming an incipient ice wedge. Subsequent cycles of temperature-driven expansion and shrinkage cause the wedge to crack open repeatedly, admitting additional water each time. Wedge ice is termed intrusion ice because its water is not drawn from the surrounding soil, but intrudes into it.

Any flat, smooth coating of particles and liquid (e.g., mud, paint, or soil) tends to crack in a pattern of polygonal shapes when it shrinks, whether by cooling or drying. Large areas of far-northern land are, consequently, covered by ice-wedge polygons, often many meters across. These polygons are an example of patterned ground—that is, terrain marked by natural, repeating, geometric shapes. Most patterned ground is produced by cyclic freezing and thawing, whether by heaving, wedging, or other mechanisms.

Ice wedging is restricted to the far north and high-altitude areas. Ice heaving occurs wherever wet ground freezes even superficially. Pipkrakes—the crunchy, vertical-fibered ice **crystals** that spring up in wet soil on freezing nights—are a small-scale example of ice heaving.

See also Phase state changes

ICEBERGS

An iceberg is a large mass of free-floating **ice** that has broken away from a glacier. Beautiful and dangerous, icebergs wander over the ocean surface until they melt. Most icebergs come from the **glaciers** of Greenland or from the massive ice sheets of **Antarctica**. A few icebergs originate from smaller Alaskan glaciers. Snow produces the glaciers and ice sheets so, ultimately, icebergs originate from snow. In contrast, "sea ice" originates from **freezing** salt **water**. When fragments break off of a glacier, icebergs are formed in a process called calving.

Iceberg. *Photograph by Commander Richard Behn, NOAA Corps. National Oceanic and Atmospheric Administration.*

Icebergs consist of **freshwater** ice, pieces of debris, and trapped bubbles of air. The combination of ice and air bubbles causes sunlight shining on the icebergs to refract, coloring the ice spectacular shades of blue, green, and white. Color may also indicate age; blue icebergs are old, and green ones contain algae and are young. Icebergs come in a variety of shapes and sizes, some long and flat, others towering and massive.

An iceberg floats because it is lighter and less dense than salty seawater, but only a small part of the iceberg is visible above the surface of the sea. Typically, about 80–90% of an iceberg is below sea level, so they drift with ocean currents rather than **wind**. Scientists who study icebergs classify true icebergs as pieces of ice that are greater than 16 ft (5 m) above sea level and wider than 98 ft (30 m) at the water line. Of course, icebergs may be much larger. Smaller pieces of floating ice are called "bergy bits" (3.3–16 ft or 1–5 m tall and 33–98 ft or 10–30 m wide) or "growlers" (less than 3.3 ft or 1 m tall and less than 33 ft or 10 m wide). The largest icebergs can be taller than 230 ft (70 m) and wider than 738 ft (225 m). Chunks of ice more massive than this are called ice islands. Ice islands are much more common in the Southern Hemisphere, where they break off the Antarctic ice sheets.

Because of the unusual forms they may take, icebergs are also classified by their shape. Flat icebergs are called tabular. Icebergs that are tall and flat are called blocky. Domed icebergs are shaped like a turtle shell, rounded, with gentle slopes. Drydock icebergs have been eroded by waves so that they are somewhat U-shaped. Perhaps the most spectacular are the pinnacle icebergs, which resemble mountain tops, with one or more central peaks reaching skyward.

The life span of an iceberg depends on its size but is typically about two years for icebergs in the Northern Hemisphere. Because they are larger, icebergs from Antarctica may last for several more years. Chief among the destructive forces that work against icebergs are wave action and heat. Wave action can break icebergs into smaller pieces and can cause icebergs to knock into each other and fracture. Relatively warm air and water **temperature** gradually melt the ice. Because icebergs float, they drift with water currents towards the equator into warmer water. Icebergs may drift as far as 8.5 mi (14 km) per day. Most icebergs have completely melted by the time they reach about 40 degrees **latitude** (north or south). There have been rare occasions when icebergs have drifted as far south as Bermuda (32 degrees north latitude), which is located about 900 mi (1,400 km) east of Charleston, South Carolina. In the Atlantic Ocean, they have also been found as far east as the Azores, islands in the Atlantic Ocean off the coast of Spain.

One of the best-known icebergs is the one that struck and sank the RMS *Titanic* on April 14, 1912, when the ship was on her maiden voyage. More than 1,500 people lost their lives in that disaster, which occurred near Newfoundland, Canada. As a result of the tragedy, the Coast Guard began monitoring icebergs to protect shipping interests in the North Atlantic sea lanes. Counts of icebergs drifting into the North Atlantic shipping lanes vary from year to year, with little predictability. During some years, no icebergs drift into the lanes; other years are marked by hundreds or more—as many as 1,572 have been counted in a single year. Many ships now carry their own radar equipment to detect icebergs. As recently as 1959, a Danish ship equipped with radar struck an iceberg and sank, resulting in 95 deaths. Some ships even rely on infrared sensors from airplanes and satellites. Sonar is also used to locate icebergs.

Modern iceberg research continues to focus on improving methods of tracking and monitoring icebergs, and on learning more about iceberg deterioration. In 1995, a huge iceberg broke free from the Larsen ice shelf in Antarctica. This iceberg was 48 mi (77 km) long, 23 mi (37 km) wide, and 600 ft (183 m) thick. The iceberg was approximately the size of the country of Luxembourg and isolated James Ross Island (one of Antarctica's islands) for the first time in recorded history. The megaberg was monitored by airplanes and satellites to make sure it didn't put ships at peril. According to some scientists, this highly unusual event could be evidence of **global warming**. Surges in the calving of icebergs known as Heinrich events are also known to be caused by irregular motions of Earth around the **Sun** that cause ocean waters of varying temperatures and salinity to change their circulation patterns. These cycles were common during the last glacial period, and glacial debris was carried by "iceberg armadas" to locations like Florida and the coast of Chile. Scientists have "captured" icebergs for study including crushing to measure their strength. It has been proposed to tow icebergs to drought-stricken regions of the world to solve water shortage problems; however, the cost and potential environmental impact of such an undertaking have so far discouraged any such attempts.

See also Glaciation; Ocean circulation and currents

IGNEOUS ROCKS

The first rocks on Earth were igneous rocks. Igneous rocks are formed by the cooling and hardening of molten material called **magma**. The word igneous comes from the Latin word *ignis*, meaning fire. There are two types of igneous rocks: intrusive and extrusive. Intrusive igneous rocks form within Earth's **crust**; the molten material rises, filling any available crevices, into the crust, and eventually hardens. These rocks are not visible until the earth above them has eroded away. Intrusive rocks are also called plutonic rocks, named after the Greek god Pluto, god of the underworld. A good example of intrusive igneous **rock** is **granite**. Extrusive igneous rocks form when the magma or molten rock pours out onto the earth's surface or erupts at the earth's surface from a **volcano**. Extrusive rocks are also called volcanic rocks. **Basalt**, formed from hardened **lava**, is the most common extrusive rock. **Obsidian**, a black glassy rock, is also an extrusive rock.

Igneous rocks are classified according to their texture and mineral or chemical content. The texture of the rock is determined by the rate of cooling. The slower the cooling, the larger the crystal. Intrusive rock can take one million years or more to cool. Fast cooling results in smaller, often microscopic, grains. Some extrusive rocks solidify in the air, before they hit the ground. Sometimes the rock mass starts to cool slowly, forming larger **crystals**, and then finishes cooling rapidly, resulting in rocks that have crystals surrounded by a fine, grainy rock mass. This is known as a porphyritic texture.

Most of Earth's **minerals** are made up of a combination of up to ten elements. Over 99% of Earth's crust consists of only eight elements (**oxygen**, **silicon**, **aluminum**, **iron**, calcium, sodium, potassium, and magnesium). Most igneous rocks contain two or more minerals, which is why some rocks have more than one color. For example, the most common minerals in granite are **quartz** (white or gray), **feldspar** (white or pinks of varying shades), and mica (black). The amount of a specific element in a mineral can determine a color or intensity of color. Because of the way granite is formed, the different composition of minerals is easy to see. It is difficult to see the distinct composition of some extrusive rocks, like obsidian, due to their extremely fine texture. Igneous rocks contain mostly silicate minerals and are sometimes classified according to their silica content. Silica (SiO_2) is a white or colorless mineral compound. Rocks containing a high amount of silica, usually more than 50%, are considered acidic (sometimes the term **felsic** is used), and those with a low amount of silica are considered basic (or **mafic**). Acidic rocks are light in color and basic rocks are dark in color.

Essentially, Earth's continents are slabs of granite sitting on top of molten rock. The crustal plates of Earth are continually shifting, being torn open by faults, and altered by earthquakes and volcanoes. New igneous material is continually added to the crust, while old crust falls back into the earth, sometimes deep enough to be remelted. Igneous rocks are the source of many important minerals, **metals**, and building materials.

See also Magma chamber; Volcanic eruptions

IMPACT CRATER

An impact crater is a physical scar on a planetary body's surface (topographic depression or geological structure) that is the result of hypervelocity impact by a minor planet, such as an asteroid, comet, or meteorite. Most impact craters are generally circular, although elliptical impact craters are known from very low-angle or obliquely impacting projectiles. In addition, some impact craters have been tectonically deformed and thus are no longer circular. Impact craters may be exposed, buried, or partially buried. Geologists distinguish an impact crater, which is rather easily seen, from an impact structure, which is an impact crater that may be in a state of poor preservation. A meteorite crater is distinguished from other impact craters because there are fragments of the impacting body preserved near the crater. Typically, a meteorite crater is a rather small feature under 0.6 mi (1 km) in diameter.

The impact crater is the most common landform on the surface of most of the rocky and icy planets and satellites in our **solar system**. Impact craters are obliterated or covered over by younger materials where re-surfacing rates are high (e.g., Venus and Io) and where **weathering** and **erosion** are intensive (e.g., Earth and parts of Mars). At present, there are about 150 to 200 impact craters and impact structures on Earth that have been scrutinized sufficiently to prove their origin. There are several hundred other possible impact features that also have been identified. Given Earth's rather rapid weathering and tectonic cycling of **crust**, this is a relatively large preserved crater record. Even though preserved craters are rare on Earth, there is no reason to suspect that Earth has been bombarded any less intensively than the **Moon**, and thus, the vast majority of Earth's impact features must have been erased.

Impact craters are subdivided into three distinctive morphologic classes, which are related to crater size. The simple impact crater is a bowl-shaped feature with relatively high depth to diameter ratio. Most simple impact craters on Earth are less than 1.2 mi (2 km) in diameter. The complex impact crater has a low depth to diameter ratio and possesses a central uplift and a down-faulted and terraced rim structure. Some large complex impact craters possess an uplifted inner ring structure rather than a simple central uplift and they have a down-faulted and terraced rim as well. Complex impact craters on Earth range from the upper limit of simple impact craters to approximately 62 mi (100 km) in diameter. Multi-ring craters (also called multi-ring basins) are impact craters with depth to diameter ratios like complex impact craters, but they possess at least two outer, concentric rings (marked by normal faults with downward motion toward crater center). The five multi-ring impact craters known on Earth range from 62–124 mi (100–200 km) in diameter. On the Moon and other planets and satellites in the solar system, the range of multi-ring crater diameters is from several hundred miles up to 2,485 mi (4,000 km) in diameter. A planet or satellite's **gravity** and the strength of the surface material determine the transition diameter from simple to complex and complex to multi-ring impact crater morphology.

Impact craters go through three stages during formation. Contact and compression is the initial stage. Contact occurs when the projectile first touches the planet's or satellite's surface. Jetting of molten material from the planet's upper crust can occur at this stage and initial penetration of the crust begins. During compression, the projectile is compressed as it enters the target crustal material. Depending upon relative strength of the target and projectile, the projectile usually penetrates only a few times its diameter into the crust. The average velocity of a cosmic projectile is approximately 12.4 mi/sec (20 km/sec) and nearly all the vast kinetic energy of this projectile is imparted to the surrounding crust as shock wave energy. This huge shock wave propagates outward radially into the crust from the point of projectile entry. At the end of compression, which lasts a tiny fraction of a second to two seconds at most (depends upon projectile size), the projectile is vaporized by a shock wave that bounces from the front of the projectile to the back and then forward. At this point, the projectile itself is no longer a factor in what happens subse-

quently. The subsequent excavation stage is driven by the shock wave propagating through the surrounding target crust. The expanding shock wave moves material along curved paths, thus ejecting debris from the opening crater cavity. This is the origin of the transient crater cavity. It may take several seconds to a few minutes to open this transient crater cavity, depending upon the kinetic energy imparted by the projectile. Material cast out of the opening crater during this phase forms an ejecta curtain that extends high above the impact **area**. This ejected material will fall back, thus forming an ejecta blanket in and around the impact crater. During the final modification stage, gravity takes over and causes crater-rim collapse in simple impact craters. In complex and multi-ring impact craters, there is central peak or peak-ring uplift and coincident gravitational collapse in the rim area. Lingering effects of the modification stage may go on for many years after impact.

There is a general relationship between impact-crater diameter, approximate projectile diameter, energy released (in joules (J) and megatons of TNT (MT)). Generally, the ratio 20:1 relates crater diameter to projectile diameter. Kinetic energy imparted to the target may be computed using the formula $KE = \frac{1}{2} mv^2$, where m is projectile mass and v is its velocity.

Further, observational data for **asteroids** and **comets** give us a general idea of impact frequency ($n/10^6$ years), and mean interval between such impacts for projectiles of given sizes. All this can be combined to give scientists an idea of the magnitude of impact energy release and how often it occurs. For example, a .62 mi (1 km) diameter impact crater would be made by a projectile 165 ft (50 m) in diameter, which would release approximately 4.6×10^{16} J (= 11 MT) of energy upon impact. Such an impact would occur approximately 640 times per million (10^6) years, or on average about once per 1,600 years. For a 3.1 mi (5 km) diameter impact crater, an 820 ft (250 m) diameter projectile is required. Approximately 5.7×10^{18} J (= 1,400 MT) energy is released in such an event, which would occur approximately 35 times per million years (or once per 28,500 years on average). For a 6.2 mi (10 km) diameter impact crater, a 1,640 ft (500 m) diameter projectile is needed. Approximately 4.6×10^{19} J (= 11,000 MT) of energy would be released. Scientists expect that such events happen approximately 10 times per million years (or on average once per 100,000 years). For a 31 mi (50 km) diameter crater (made by a 1.6 mi (2.5 km) diameter projectile), we can expect a 5.8×10^{21} J (= 1.3×10^6 MT) energy release. This would happen approximately 0.22 times per million years or on average once per 4.5 million years. For a 62 mi (100 km) diameter crater (made by a 3.1 mi (5 km) diameter projectile), we can expect a 4.6×10^{22} J (= 1.1×10^7 MT) energy release. This would happen approximately 0.04 times per million years, or on average once per 26 million years. To put the energy release in perspective, the largest nuclear weapon ever tested on Earth yielded 58 MT. If all nuclear weapons that existed at the height of the Cold War were exploded at once, the yield would be approximately 10^5 MT. The impact event linked to the dinosaur extinction (Chicxulub impact structure, Mexico) has a diameter of nearly 124 mi (200 km).

It is thought that impact events related to craters greater than 62 mi (100 km) in diameter likely had globally devastating effects. These effects, which may have led to global ecosystem instability or collapse, included: gas and dust discharge into the upper atmosphere (blocking sunlight and causing greenhouse effects); heating of the atmosphere due to re-entry of ballistic ejecta (causing extensive wildfires); seismic sea waves (causing tsunamis); and acid-rain production (causing damage to soils and **oceans**). There is much research currently underway to find the effect of cosmic impact events upon life on Earth during the geological past.

See also Asteroids; Barringer meteor crater; Comets; K-T event; Meteoroids and meteorites; Shock metamorphism

INDIAN OCEAN • *see* OCEANS AND SEAS

INDUSTRIAL MINERALS

Industrial **minerals** is a term used to describe naturally occurring non-metallic minerals that are used extensively in a variety of industrial operations. Some of the minerals commonly included in this category include asbestos, barite, boron compounds, clays, corundum, **feldspar**, fluorspar, phosphates, potassium salts, sodium chloride, and sulfur. Some of the mineral mixtures often considered as industrial minerals include construction materials such as **sand**, gravel, **limestone**, **dolomite**, and crushed **rock**; abrasives and refractories; **gemstones**; and lightweight aggregates.

Asbestos is a generic term used for a large group of minerals with complex chemical composition that includes magnesium, **silicon**, **oxygen**, hydrogen, and other elements. The minerals collectively known as asbestos are often sub-divided into two smaller groups, the serpentines and amphiboles. All forms of asbestos are best known for an important common property—their resistance to heat and flame. That property is responsible, in fact, for the name *asbestos* (Greek), meaning unquenchable. Asbestos has been used for thousands of years in the production of heat resistant materials such as lamp wicks.

Today, asbestos is used as a reinforcing material in cement, in vinyl floor tiles, in fire-fighting garments and fire-proofing materials, in the manufacture of brake linings and clutch facings, for electrical and heat insulation, and in pressure pipes and ducts.

Prolonged exposure to asbestos fibers can block the respiratory system and lead to the development of asbestosis and/or lung cancer. The latency period for these disorders is at least 20 years, so men and women who mined the mineral or used it for various construction purposes during the 1940s and 1950s were not aware of their risk for these diseases until late in their lives. Today, uses of the mineral in which humans are likely to be exposed to its fibers have largely been discontinued.

Barite is the name given to a naturally occurring form of barium sulfate, commonly found in Canada, Mexico, and the states of Arkansas, Georgia, Missouri, and Nevada. One of the most important uses of barite is in the production of heavy muds that are used in drilling oil and gas wells. It is also used in the manufacture of a number of other commercially important industrial products such as paper coatings, battery plates,

paints, linoleum and oilcloth, plastics, lithographic inks, and as filler in some kinds of textiles. Barium compounds are also widely used in medicine to provide the opacity that is needed in taking certain kinds of x rays.

Boron is a non-metallic element obtained most commonly from naturally occurring minerals known as borates. The borates contain oxygen, hydrogen, sodium, and other elements in addition to boron. Probably the most familiar boron-containing mineral is borax, mined extensively in salt **lakes** and alkaline soils.

Borax was known in the ancient world and used to make glazes and hard **glass**. Today, it is still an important ingredient of glassy products that include heat-resistant glass (Pyrex), glass wool and glass fiber, enamels, and other kinds of ceramic materials. Elementary boron also has a number of interesting uses. For example, it is used in nuclear reactors to absorb excess neutrons, in the manufacture of special-purpose alloys, in the production of semiconductors, and as a component or rocket propellants.

Corundum is a naturally occurring form of **aluminum** oxide that is found abundantly in Greece and Turkey and in New York State. It is a very hard mineral with a high **melting** point. It is relatively inert chemically and does not conduct an electrical current very well.

These properties make corundum highly desirable as a refractory (a substance capable of withstanding very high temperatures) and as an abrasive (a material used for cutting, grinding, and polishing other materials). One of the more mundane uses of corundum is in the preparation of toothpaste, where its abrasive properties help in keeping teeth clean and white.

In its granular form, corundum is known as emery. Many consumers are familiar with emery boards used for filing fingernails. Emery, like corundum, is also used in the manufacture of cutting, grinding, and polishing wheels.

The feldspars are a class of minerals known as the aluminum silicates. That is, they all contain aluminum, silicon, and oxygen, as sodium, potassium, and calcium. In many cases, the name feldspar is reserved for the potassium aluminum silicates. The most important commercial use of feldspar is in the manufacture of pottery, enamel, glass, and ceramic materials. The hardness of the mineral also makes it desirable as an abrasive.

Fluorspar is a form of calcium fluoride that occurs naturally in many parts of the world including **North America**, Mexico, and **Europe**. The compound gets its name from one of its oldest uses, as a flux. In Latin, the word *fluor* means flux. A flux is a material that is used in industry to assist in the mixing of other materials or to prevent the formation of oxides during the refining of a metal. For example, fluorspar is often added to an open-hearth steel furnace to react with any oxides that might form during that process. The mineral is also used during the smelting of an ore (the removal of a metal from its naturally occurring ore).

Fluorspar is also the principal source of fluorine gas. The mineral is first converted to hydrogen fluoride which, in turn, is then converted to the element fluorine. Some other uses of fluorspar are in the manufacture of paints and certain types of cement, in the production of emery wheels and **car-**

bon electrodes, and as a raw material for phosphors (a substance that glows when bombarded with energy, such as the materials used in color television screens).

The term phosphate refers to any chemical compound containing a characteristic grouping of atoms, given by the formula PO_4, or comparable groupings. In the field of industrial minerals, the term most commonly refers to a specific naturally occurring phosphate, calcium phosphate, or phosphate rock. By far the most important use of phosphate rock is in agriculture, where it is treated to produce fertilizers and animal feeds. Typically, about 80% of all the phosphate rock used in the United States goes to one of these agricultural applications.

Phosphate rock is also an important source for the production of other phosphate compounds, such as sodium, potassium, and ammonium phosphate. Each of these compounds, in turn, has a very large variety of uses in everyday life. For example, one form of sodium phosphate is a common ingredient in dishwashing detergents. Another, ammonium phosphate, is used to treat cloth to make it fire retardant. Potassium phosphate is used in the preparation of baking powder.

As with other industrial minerals mentioned here, the term potassium salts applies to a large group of compounds, rather than one single compound. Potassium chloride, sulfate, and nitrate are only three of the most common potassium salts used in industry. The first of these, known as sylvite, can be obtained from salt **water** or from fossil salt beds. It makes up roughly 1% of each deposit, the remainder of the deposit being sodium chloride (halite).

Potassium salts are similar to phosphate rocks in that their primary use is in agriculture, where they are made into fertilizers, and in the chemical industry, where they are converted into other compounds of potassium. Some compounds of potassium have particularly interesting uses. Potassium nitrate, for example, is unstable and is used in the manufacture of explosives, fireworks, and matches.

Like potassium chloride, sodium chloride (halite) is found both in sea water and in underground salt mines left as the result of the **evaporation** of ancient **seas**. Sodium chloride has been known to and used by humans for thousands of years and is best known by its common name of salt, or table salt. By far its most important use is in the manufacture of other industrial chemicals, including sodium hydroxide, hydrochloric acid, chlorine, and metallic sodium. In addition, sodium chloride has many industrial and commercial uses. Among these are in the preservation of foods (by salting, pickling, corning, curing, or some other method), highway de-icing, as an additive for human and other animal foods, in the manufacture of glazes for ceramics, in water softening, and in the manufacture of rubber, **metals**, textiles, and other commercial products.

Sulfur occurs in its elementary form in large underground deposits from which it is obtained by traditional mining processes or, more commonly, by the Frasch process. In the Frasch process, superheated water is forced down a pipe that has been sunk into a sulfur deposit. The heated water melts the sulfur, which is then forced up a second pipe to the earth's surface.

The vast majority of sulfur is used to manufacture a single compound, sulfuric acid. Sulfuric acid consistently ranks number one in the United States as the chemical produced in largest quantity. Sulfuric acid has a very large number of uses, including the manufacture of fertilizers, the refining of **petroleum**, the pickling of steel (the removal of oxides from the metal's surface), and the preparation of detergents, explosives, and synthetic fibers.

A significant amount of sulfur is also used to produce sulfur dioxide gas (actually an intermediary in the manufacture of sulfuric acid). Sulfur dioxide, in turn, is extensively used in the pulp and paper industry, as a refrigerant, and in the purification of sugar and the bleaching of paper and other products. Some sulfur is refined after being mined and then used in its elemental form. This sulfur finds application in the vulcanization of rubber, as an insecticide or fungicide, and in the preparation of various chemicals and pharmaceuticals.

See also Geochemistry; Petroleum, economic uses of

INNER CORE · *see* EARTH, INTERIOR STRUCTURE

INOSILICATES

The most abundant rock-forming **minerals** in the **crust** of the earth are the silicates. They are formed primarily of **silicon** and **oxygen**, together with various **metals**. The fundamental unit of these minerals is the silicon-oxygen tetrahedron. These tetrahedra have a pyramidal shape, with a relatively small silicon cation (Si^{+4}) in the center and four larger oxygen anions (O^{-2}) at the corners, producing a net charge of –4. **Aluminum** cations (Al^{+3}) may substitute for silicon, and various anions such as hydroxyl (OH^-) or fluorine (F^-) may substitute for oxygen. In order to form stable minerals, the charges that exist between tetrahedra must be neutralized. This can be accomplished by the sharing of oxygen atoms between tetrahedra, or by the binding together adjacent tetrahedra by various metal cations. This in turn creates characteristic silicate structures that can be used to classify silicate minerals into **cyclosilicates**, inosilicates, **nesosilicates**, **phyllosilicates**, **sorosilicates**, and **tectosilicates**.

Minerals where the silicon-oxygen tetrahedra form chains are called inosilicates. They can take the form of single chains, where tetrahedra line up single-file through the sharing of oxygen atoms, or they can form double chains where the tetrahedra of adjacent single chains also share oxygen atoms. Two important groups of insosilicates are the pyroxenes and the amphiboles. Minerals of the pyroxene group are single-chain ferromagnesian silicates; examples of pyroxene group minerals include enstatite ($MgSiO_3$) and jadeite ($NaAlSi_2O_6$). Minerals of the amphibole group are double-chain ferromagnesian silicates; examples of amphibole group minerals include grunerite ($Fe_7Si_8O_{22}(OH)_2$) and tremolite ($Ca_2Mg_5Si_8O_{22}(OH)_2$). The same cations (such as calcium and sodium) are present in both groups, but the hydroxyl anion is characteristic of amphiboles.

Both pyroxenes and amphiboles are important rock-forming minerals in igneous and metamorphic rocks.

See also Chemical bonds and physical properties

INSOLATION AND TOTAL SOLAR IRRADIANCE

Total solar irradiance is defined as the amount of radiant energy emitted by the **Sun** over all wavelengths that fall each second on 11 ft² (1 m²) outside Earth's atmosphere. Insolation is the amount of **solar energy** that strikes a given **area** over a specific time, and varies with **latitude** or the **seasons**.

By way of further definition, irradiance is defined as the amount of electromagnetic energy incident on a surface per unit time per unit area. Solar refers to electromagnetic radiation in the spectral range of approximately 1–9 ft (0.3–3 m), where the shortest wavelengths are in the ultraviolet region of the spectrum, the intermediate wavelengths in the visible region, and the longer wavelengths are in the near infrared. Total solar irradiance means that the solar flux has been integrated over all wavelengths to include the contributions from ultraviolet, visible, and infrared radiation.

By convention, the surface features of the Sun are classified into three regions: the photosphere, the chromosphere, and the corona. The photosphere corresponds to the bright region normally visible to the naked eye. About 3,100 mi (5,000 km) above the photosphere lies the chromosphere, from which short-lived, needle-like projections may extend upward for several thousands of kilometers. The corona is the outermost layer of the Sun; this region extends into the region of the planets. Most of the surface features of the Sun lie within the photosphere, though a few extend into the chromosphere or even the corona.

The average amount of energy from the Sun per unit area that reaches the upper regions of Earth's atmosphere is known as the solar constant; its value is approximately 1,367 watts per square meter. As Earth-based measurements of this quantity are of doubtful accuracy due to variations in Earth's atmosphere, scientists have come to rely on satellites to make these measurements.

Although referred to as the solar constant, this quantity actually has been found to vary since careful measurements started being made in 1978. In 1980, a satellite-based measurement yielded the value of 1,368.2 watts per square meter. Over the next few years, the value was found to decrease by about 0.04% per year. Such variations have now been linked to several physical processes known to occur in the Sun's interior, as will be described below.

From Earth, it is only possible to observe the radiant energy emitted by the Sun in the direction of our planet; this quantity is referred to as the solar irradiance. This radiant solar energy is known to influence Earth's **weather and climate**, although the exact relationships between solar irradiance and long-term climatological changes, such as **global warming**, are not well understood.

The total radiant energy emitted from the Sun in all directions is a quantity known as solar luminosity. The luminosity of the Sun has been estimated to be 3.8478×10^{26} watts. Some scientists believe that long-term variations in the solar luminosity may be a better correlate to environmental conditions on Earth than solar irradiance, including global warming. Variations in solar luminosity are also of interest to scientists who wish to gain a better understanding of stellar **rotation**, convection, and **magnetism**.

Because short-term variations of certain regions of the solar spectrum may not accurately reflect changes in the true luminosity of the Sun, measurements of total solar irradiance, which by definition take into account the solar flux contributions over all wavelengths, provide a better representation of the total luminosity of the Sun.

Short-term variations in solar irradiation vary significantly with the position of the observer, so such variations may not provide a very accurate picture of changes in the solar luminosity. But the total solar irradiance at any given position gives a better representation because it includes contributions over the spectrum of wavelengths represented in the solar radiation.

Variations in the solar irradiance are at a level that can be detected by ground-based astronomical measurements of light. Such variations have been found to be about 0.1% of the average solar irradiance. Starting in 1978, space-based instruments aboard the *Nimbus 7* Solar Maximum Mission, and other satellites began making the sort of measurements (reproducible to within a few parts per million each year) that allowed scientists to acquire a better understanding of variations in the total solar irradiance.

Variations in solar irradiance have been attributed to the following solar phenomena: Oscillations, granulation, sunspots, faculae, and solar cycle.

Oscillations, which cause variations in the solar irradiance lasting about five minutes, arise from the action of resonant waves trapped in the Sun's interior. At any given time, there are tens of millions of frequencies represented by the resonant waves, but only certain oscillations contribute to variations in the solar constant.

Granulation, which produces solar irradiance variations lasting about 10 minutes, is closely related to the convective energy flow in the outer part of the Sun's interior. To the observer on Earth, the surface of the Sun appears to be made up of finely divided regions known as granules, each from 311—1,864 mi (500–3,000 km) across, separated by dark regions. Each of these granules makes its appearance for about 10 minutes and then disappears. Granulation apparently results from convection effects that appear to cease several hundred kilometers below the visible surface, but in fact extend out into the photosphere, i.e., the region of the Sun visible to the naked eye. These granules are believed to be the centers of rising convection cells.

Sunspots give rise to variations that may last for several days, and sometimes as long as 200 days. They actually correspond to regions of intense magnetic activity where the solar atmosphere is slightly cooler than the surroundings. Sunspots

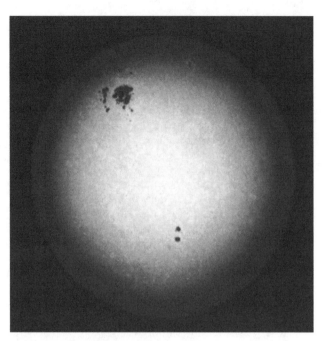

Visible light image of the Sun, showing sunspots. *U.S. National Aeronautics and Space Administration (NASA).*

appear as dark regions on the Sun's surface to observers on Earth. They are formed when the **magnetic field** lines just below the Sun's surface become twisted, and then poke though the solar photosphere. Solar irradiance measurements have also shown that the presence of large groups of sunspots on the Sun's surface produce dips ranging in amplitude from 0.1 to 0.25% of the solar constant. This reduction in the total solar irradiance has been attributed both to the presence of these sunspots and to the temporary storage of solar energy over times longer than the sunspot's lifetime. Another key observation has been that the largest decreases in total solar irradiance frequently coincide with the formation of newly formed active regions associated with large sunspots, or with rapidly evolving, complex sunspots. Sunspots are especially noteworthy for their 11-year activity cycle.

Faculae, producing variations that may last for tens of days, are bright regions in the photosphere where high-temperature interior regions of the Sun radiate energy. They tend to congregate in bright regions near sunspots, forming solar active regions. Faculae, which have sizes on the order of 620 mi (1,000 km) or less, appear to be tube-like regions defined by magnetic field lines. These regions are less dense than surrounding areas. Because radiation from hotter layers below the photosphere can leak through the walls of the faculae, an atmosphere is produced that appears hotter, and brighter, than others.

The solar cycle is responsible for variations in the solar irradiance that have a period of about 11 years. This 11-year activity cycle of sunspot frequency is actually half of a 22-year magnetic cycle, which arises from the reversal of the poles of the Sun's magnetic field. From one activity cycle to the next, the north magnetic pole becomes the south magnetic pole, and

vice versa. Solar luminosity has been found to achieve a maximum value at the very time that sunspot activity is highest during the 11-year sunspot cycle. Scientists have confirmed the length of the solar cycle by examining tree rings for variations in deuterium-to-hydrogen ratios. This ratio is temperature-dependent because deuterium molecules, which are a heavy form of the hydrogen molecule, are less mobile than the lighter hydrogen molecules, and therefore less responsive to thermal motion induced by increases in the solar irradiance.

Surprisingly, the Sun's rotation, with a rotational period of about 27 days, does not give rise to significant variations in the total solar irradiance. This is because its effects are over-ridden by the contributions of sunspots and faculae.

Scientists have speculated that long-term solar irradiance variations might contribute to global warming over decades or hundreds of years. More recently, there has been speculation that changes in total solar irradiation have amplified the **greenhouse effect**, i.e., the retention of solar radiation and gradual warming of Earth's atmosphere. Some of these changes, particularly small shifts in the length of the activity cycle, seem to correlate rather closely with climatic conditions in pre- and post industrial times. Whether variations in solar irradiance can account for a substantial fraction of global warming over the past 150 years, however, remains a highly controversial point of discussion.

See also Electromagnetic spectrum; Greenhouse gases and greenhouse effect; Solar energy; Solar illumination: Seasonal and diurnal patterns

INTERNATIONAL COUNCIL OF SCIENTIFIC UNIONS WORLD DATA CENTER

The International Council of Scientific Unions (ICSU) is a non-governmental organization, founded in 1931, to bring together natural scientists in international scientific endeavor. It comprises multi-disciplinary national scientific members (scientific research councils or science academies) and international, single-discipline scientific unions to provide a wide spectrum of scientific expertise enabling members to address major international, interdisciplinary issues which none could handle alone.

In 1952, the ICSU proposed a comprehensive series of global geophysical activities to span the period July 1957 to December 1958. The International Geophysical Year (IGY), as it was called, was modeled on the International Polar Years of 1882–83 and 1932–33, and was intended to allow scientists from around the world to take part in a series of coordinated observations of various phenomena in Geophysics. A special committee, CSAGI (Comité Spécial de l'Année Géophysique Internationale), was formed to act as the governing body for all IGY activities. Among them, CSAGI established a World Data Center (WDC) system to serve the IGY, and developed data management plans for each IGY's scientific discipline. The data specifications were published in a series of *Guides to*

Data Exchange, originally issued in 1957, and consecutively updated. Data sets were prepared in machine-readable form, which at that time meant punched cards and punched tape. Because of its success, the WDC system was declared permanent at the 22nd General Assembly of ICSU (Beijing, 1988). Since 1999, WDCs are referenced by the type of center rather that by the country operating the center, as for example the World Data Center for Marine Environmental Studies (WDC-MARE at Bremen University, Germany). All centers now have computer facilities and most use electronic networks to meet requests, exchange describing meta-information, and transfer data.

The basic principles and responsibilities include that World Data Centers are operated by national organizations for the benefit of the international scientific community. The resources required to operate WDCs are the responsibility of the host country or institution, which is expected to provide these resources on a long-term basis. If for any reason a WDC is closed, the data holdings shall be transferred to another center. WDCs receive data from individual scientists, projects, institutions, and local and national data centers. Among others, the mechanisms for data acquisition include the WDC Panel's "data rescue" program, which involves all parts of the WDC system and has two main aspects: (1) safeguarding older data sets that may be at risk of loss or deterioration; (2) digitizing old data sets to enable modern techniques to be used for their analysis. WDCs exchange data among themselves, as mutually agreed and whenever possible without charge, to facilitate data availability, to provide back-up copies, and to aid the preparation of higher order data products. They compile specialized data sets for small-scale, regional and global research and combine data from various sources to derive data products, such as indices of solar activity.

WDCs will provide data to scientists in any country free of charge, on an exchange basis or at a cost not to exceed the cost of copying and sending the requested data. Data sets are made available online through the World Wide Web or on media as CD-ROM, enabling users to search large data collections and transfer them to their home laboratory.

Data may be subject to privileged use by their principal investigators, for a period to be agreed beforehand, and not to surpass two years from the date of acquisition by the WDC. Since unpublished data are even more sensitive than published data, the WDC ensures that data be not accessed until they are formally placed in the public domain. In any case, data policy requires the acknowledgement of the original data sources in order to protect the principal investigator.

See also Scientific data management in Earth Sciences

INTERNATIONAL SPACE STATION (ISS)

The International Space Station (ISS) is the most complex international aerospace project in history. Sixteen countries contribute to this massive structure that measures 360 ft (110 m) wide and 289 ft (88 m) long. At Earth's surface **gravity**, the ISS would weigh 503 tons (456,620 kg). Constructed from specialized component modules, the ISS is designed to allow humans to live in space for long durations of time and provide a laboratory for both scientific and engineering experiments. The modular design allows sections to be completed and tested on Earth before being booted into orbit. In addition, the modular design provides a level of security to ISS personnel. Damage from a failure or rupture of a component module can be isolated and the crew evacuated to safe modules. Modular designs are also economical because they allow rapid adaptation to the station to specific uses without having to subject the station to extensive retrofitting. In a engineering sense, the modular design allows maximum safety, design flexibility, and use adaptation at the lowest cost.

Long-range plans include use of the ISS as a spaceport where spacecraft can dock to transfer people, cargo, and fuel without having to re-enter Earth's atmosphere. Use of the ISS as a spaceport would thus, facilitate the construction of a fleet of true space vehicles—craft designed to operate exclusively outside Earth's atmosphere. Such craft would not need to be constructed to withstand the dynamic pressures of reentry, nor would their engine systems need to be designed to provide thrust capable of propelling the craft to high escape velocities.

Although the United States and Russia shoulder the bulk of the technological burden of ISS design and orbital placement, other nations, including Canada, Japan, the 11 nations of the European Space Agency (ESA) and Brazil significantly contribute to ISS development.

The United States is responsible for constructing and operating major ISS elements and systems. The U.S. systems include thermal control, life support, guidance, navigation and control, data handling, power systems, communications and tracking, ground operations facilities, and launch-site processing facilities. Canada is providing a 55-foot-long (16.8 m) robotic arm to be used for the station's assembly and maintenance. The European Space Agency is contributing a pressurized laboratory to be launched on the **Space Shuttle**, and logistics transport vehicles to be launched on an ESA *Ariane 5* launch vehicle. Japan is providing a laboratory to be used for experiments and logistics transport vehicles. Russia is contributing two research modules, the Service Module, which includes early living quarters with life support and habitation systems, a science power platform that can supply 20 kilowatts of electrical power, logistics transport vehicles, and Soyuz spacecraft for crew drop off and pick up. Through agreement with the United States, Italy, and Brazil are also providing ISS components and laboratory research facilities.

Approved by President Ronald Reagan in 1984, ISS (then designated Space Station *Freedom*) development was put on hold by the turmoil and collapse of the Soviet Union in the late 1980s and the subsequent emergence of a revitalized Russian Space Agency in 1993. Broadened in scope to include a true international collaboration, in November of 1998, Russia launched the first part of the developing space station. More than four times as large as the Russian *Mir* space station, ISS assembly will continue until at least 2004.

ISS orbits at an altitude of 250 statute miles with an inclination of 51.6 degrees. This orbit allows maximum accessibility to the station for docking, crew **rotation**, and supply delivery. The orbit also allows for excellent observation of Earth. Orbital dynamics allow observation of up to 85% of Earth's surface and overflight of approximately 95% of Earth's heavily populated areas. Accordingly, the ISS is an ideal platform for the study of dynamic Earth geophysical processes and the long term study of the effects human civilization has upon both the physical and ecological landscape.

In addition to astronomical and **Earth science** research groups, the ISS will support medical and industrial research (e.g., the formation of certain alloys and **crystals** in low gravity environments).

See also History of manned space exploration; Space and planetary geology; Space physiology; Space probe; Spacecraft, manned

INTRUSIVE COOLING

Igneous rocks formed below ground level are termed intrusive, meaning that they originate as **magma** (liquid **rock**) that has intruded itself into preexisting solid rock by squeezing into cracks, eating its way upward from the mantle, or by other means. An intrusive magmatic body begins to cool as soon as it is emplaced, and as it cools, it crystallizes into a mixed mass of mineral grains. Which **minerals** form depends in a complex way on the exact ingredients of the magma and on the speed at which it is cooled. In general, slow cooling permits larger **crystals** to form while fast cooling produces smaller crystals.

Cooling is affected by shape and other factors. Thin or narrow bodies cool faster than globular ones; small bodies cool faster than large ones; convecting bodies cool faster than static (nonconvecting) ones; and bodies surrounded by relatively low-temperature rock cool faster than those emplaced in warm environments. By human standards, cooling time for intruded magma may be quite long. A horizontal, sheet-shaped intrusion of 1,562°F (850°C) magma 2,300 ft (701 m) thick, intruded beneath a cool 77–122°F (25–50°C) cover of rock half as thick, takes 9,000 years to completely crystallize. A vertical sheet of 1,472°F (800°C) magma 6,560 ft (2,000 m) thick emplaced in 212°F (100°C) rock takes 64,000 years to crystallize all through. The largest magmatic intrusions may take a million years to crystallize.

Near-surface intrusive cooling may be speeded by convection of **groundwater** through surrounding rock. In this case, **water** may transport minerals toward and away from the cooling intrusion, further complicating the process of mineral formation.

See also Batholith; Bowen's reaction series; Extrusive cooling; Pluton and plutonic bodies

IONIC BONDS • *see* CHEMICAL BONDS AND PHYSICAL PROPERTIES

IONOSPHERE

The ionosphere is a layer of the earth's atmosphere that is weakly ionized, and thus conducts **electricity**. It is located approximately in the same region as the top half of the **mesosphere** and the entire **thermosphere** in the upper atmosphere, from about 40 mi (60 km), continuing upward to the magnetosphere.

In the ionosphere, the molecules and atoms in the air are ionized mostly by the Sun's ultraviolet, x-ray, and corpuscular radiation, and partially by cosmic rays, resulting in ions and free electrons. The ionization process depends on many factors such as the Sun's activity (e.g., sunspot cycles), time (e.g., seasonal or daily changes), or geographical location (different at polar regions, mid-latitudes or equatorial zones).

The ionosphere can be further divided into sub-regions according to their free electron density profile that indicates the degree of ionization, and these sub-regions are called the D, E, and F layers. The D layer is located lowest among them, and it does not have an exact starting point. It absorbs high-frequency radio waves, and exists mainly during the day. It weakens, then gradually even disappears at night, allowing radio waves to penetrate into a higher level of the ionosphere, where these waves are reflected back to Earth, then bounce again back into the ionosphere. This explains why AM radio signals from distant stations can easily be picked up at night, even from hundreds of miles. Above the D layer, the E layer (or Kennelly-Heaviside layer) can be found, which historically was the first one that was discovered. After sunset, it usually starts to weaken and by night, it also disappears. The E layer absorbs x rays, and it has its peak at about 65 mi (105 km). The F layer (or Appleton layer) can be found above the E layer, above 93 mi (150 km), and it has the highest concentration of charged particles. Although its structure changes during the day, the F layer is a relatively constant layer, where extreme ultra-violet radiation is absorbed. It has two parts: the lower F1 layer, and the higher and more electron-dense F2 layer.

The free electrons in the ionosphere allow good propagation of electromagnetic waves, and excellent radio communication. The ionosphere is also the home for the aurora, a light display mostly in the night sky of the polar areas, caused by excited and light-emitting particles entering the upper atmosphere.

See also Atmospheric composition and structure; Aurora Borealis and Aurora Australialis

IRON

Iron is the fourth-most common element in Earth's **crust**, and the second-most common metal after **aluminum**. Its abundance is estimated to be about 5%. Sampling studies indicate that portions of Earth's core consist largely of iron, and the element is found commonly in the **Sun, asteroids**, and stars.

The chemical symbol for iron, Fe, comes from the Latin name for the element, *ferrum*. The most common ores of iron are hematite and limonite (both primarily ferric oxide; Fe_2O_3)

and siderite iron carbonate ($FeCO_3$). An increasingly important source of iron for commercial uses is taconite, a mixture of hematite and silica. Taconite contains about 25% iron. The largest iron resources in the world are found in China, Russia, Brazil, Canada, **Australia**, and India.

The traditional method for extracting pure iron from its ore is to heat the ore in a blast furnace with **limestone** and coke. The coke reacts with iron oxide to produce pure iron, while the limestone combines with impurities in the ore to form a slag that can then be removed from the furnace: $3C + 2Fe_2O_3 + heat \rightarrow 3CO_2 + 4Fe$.

Iron produced by this method is about 90% pure and is known as pig iron. Pig iron is generally too brittle to be used for most products and is further treated to convert it to wrought iron, cast iron, or steel. Wrought iron is an **alloy** of iron and any one of many different elements, while cast iron is an alloy of iron, **carbon**, and **silicon**. Steel is a generic term that applies to a very wide variety of alloys.

Iron is one of a handful of elements that have been known and used since the earliest periods of human history. In the period beginning about 1200 B.C. iron was so widely used for tools, ornaments, weapons, and other objects that historians and archaeologists have now named the period the Iron Age.

Iron is a silvery white or grayish metal that is ductile and malleable. It is one of only three naturally occurring magnetic elements, the other two being its neighbors in the **periodic table**: cobalt and nickel. Iron has a very high tensile strength and is very workable, capable of being bent, rolled, hammered, cut, shaped, formed, and otherwise worked into some desirable shape or thickness. Iron's **melting** point is 2,797°F (1,536°C) and its boiling point is about 5,400°F (3,000°C). Its density is 7.87 grams per cubic centimeter.

Iron is an active metal that combines readily with **oxygen** in moist air to form iron oxide (Fe_2O_3), commonly known as rust. Iron also reacts with very hot **water** and steam to produce hydrogen gas and with most acids and a number of other elements.

The number of commercial products made of iron and steel is very large indeed. The uses of these two materials can generally be classified into about eight large groups, including (1) automotive; (2) construction; (3) containers, packaging, and shipping; (4) machinery and industrial equipment; (5) rail transportation; (6) oil and gas industries; (7) electrical equipment; and (8) appliances and utensils.

A relatively small amount of iron is used to make compounds that have a large variety of applications, including dyeing of cloth, blueprinting, insecticides, water purification and sewage treatment, photography, additive for animal feed, fertilizer, manufacture of **glass** and ceramics, and wood preservative.

Iron is of critical important to plants, humans, and other animals. It occurs in hemoglobin, the molecule that carries oxygen in the blood. The U.S. Recommended Daily Allowance (USRDA) for iron is 18 mg (with some differences depending on age and sex) and it can be obtained from meats, eggs, raisins, and many other foods. Iron deficiency disorders, known as anemias, are not uncommon and can result in fatigue, reduced resistance to disease, an increase in respiratory and circulatory problems, and even death.

See also Chemical bonds and physical properties; Chemical elements; Earth, interior structure; Minerals

ISLAND ARCS

An island arc is a curving series of volcanic islands that are created through the collision of tectonic plates in an ocean setting. The particular type plate boundary that yields island arcs is called a **subduction zone**. In a subduction zone, one lithospheric (crustal) plate is forced downward under an upper plate. Continual tectonic movement pushes the lower plate deeper until it reaches a depth where temperatures are sufficient to begin to melt the subducted plate and form magmas. These magmas then rise through fractures and melt their way through the overlying **crust** to be extruded in the form of volcanoes. The volcanoes are generally andesitic in composition. If the overriding plate is oceanic, then volcanoes are extruded underwater and may eventually rise high enough to become islands. The volcanoes form in a line because the angle and rate of subduction, and hence the distance to the depth where **melting** occurs is consistent. Because the surface of Earth is curved, the line of volcanoes forms in an arcuate pattern in much the same manner as an arc is produced when a planar surface intersects a sphere.

Island arcs are usually accompanied by rapid **erosion** and **sedimentation** into accompanying basins. A back-arc basin occurs on the side of the overriding plate and a fore-arc basin forms toward the subducted plate side. Typically, a deep oceanic trench, such as the Marianas Trench, bounds an island arc on the oceanic side beyond the fore-arc basin.

The Aleutian Islands, the islands of Japan, and the Lesser Antilles are all examples of island arcs. The term volcanic arc is often interchanged with island arc, although volcanic arc can also refer to land-based volcanoes produced by subduction. The Andes Mountains are the result of a continental volcanic arc.

See also Andesite; Benioff zone; Subduction zone

ISOBARS

Isobars are lines that connect points of equal pressure on **weather** maps. The word originates from Greek, where *isos* means equal and *baros* means weight. Isobars are designed to describe the horizontal pressure distribution of an **area**, and are created from mean sea-level pressure reports. Because most of the weather stations are not located at sea level, but at a certain elevation, the pressure measured at every location has to be converted into sea level pressure before the isobars are drawn. This normalization is necessary because **atmospheric pressure** decreases with increasing altitude, and the pressure difference on the maps has to be due to the weather conditions, not due to the elevation differences of the locations.

Isobars are similar to height lines on a topographic map, and they are defined such that they can never cross each other.

An important consequence of air pressure differences is **wind**, because wind blows from areas of high pressure to areas of low pressure. The greater the pressure contrast and the shorter the distance, the faster the wind will blow, so closer isobars mean faster wind. Although the wind initially is controlled by the pressure differences, it is also modified by the influence of the **Coriolis effect** and friction close to Earth's surface. This is why isobars can only give a general idea about the wind direction and wind strength.

A rule observed first in 1857 by Dutch meteorologist Christoph Buys-Ballott (1817–1890) described the link between isobars and wind: in the Northern Hemisphere, if you stand with your back to the wind, the low pressure area is located on the left. In the Southern Hemisphere, standing with your back to the wind means that the low-pressure area is on the right. This is called Buys-Ballott's law.

Isobars can form certain patterns, making it useful for weather analysis or forecast. A cyclone or depression is an area of curved isobars surrounding a low-pressure region with winds blowing counterclockwise in its center in the Northern Hemisphere. An anticyclone is an area of curved isobars surrounding a high-pressure area, and the wind blows clockwise in the center of an anticyclone in the Northern Hemisphere. Open isobars forming a V-shape define a through of low pressure while high-pressured, N-shaped, open isobars define a ridge of high pressure. These features are usually predictable, and associated with a certain kind of weather, making it easier to forecast weather for a certain area.

See also Atmospheric pressure

ISOMORPH

Crystalline substances with chemical formulas that are similar, and with positively charged cations and negatively charged anions that are similar in size, may form **crystals** with the same structure. These isomorphous groups can be used in mineral classification.

One example is the halite group, which includes halite ($NaCl$), fluorite (CaF_2), and sylvite (KCl), among other **minerals**. The crystals that they form belong to the isometric crystal system, in which the unit cells (the smallest component of a crystal, which can repeat indefinitely in three dimensions) all have three axes of equal length oriented at angles of 90° to each other. Another example of an isomorphous group is the calcite group, which includes the carbonate minerals calcite ($CaCO_3$), magnesite ($MgCO_3$), rhodochrosite ($MnCO_3$), siderite ($FeCO_3$), and smithsonite ($ZnCO_3$). All of these minerals form crystals with the same symmetry, in this case in the hexagonal crystal system, where the unit cells of the crystals have three horizontal axes of equal length and one axis of different length, perpendicular to the other three. Sometimes, different minerals will have the same chemical composition (a situation called polymorphism). Aragonite has the same chemical composition as calcite, but crystallizes in the orthorhombic crystal system (where the unit cells have three axes of unequal length ori-

ented at angles of 90° to each other). The aragonite group includes the minerals cerussite ($PbCO_3$), strontianite ($SrCO_3$), and witherite ($BaCO_3$). Similarity in cation and anion size is more important than chemical composition in isomorphism. Thus, uraninite (UO_2) and chlorargyrite ($AgCl$) both belong to the halite group, although their composition and properties are very different. In a solid solution, an isomorphous group of minerals exhibits of a range of mineral compositions between two end members. The **olivine** group forms such a **solid solution series** in the orthorhombic crystal system. The composition of olivine is usually given as $(Mg,Fe)_2SiO_4$, but it can range from pure forsterite (Mg_2SiO_4) to pure fayalite (Fe_2SiO_4). Isomorphism is also called isostructuralism.

See also Crystals and crystallography

ISOSTASY

Isostasy (also spelled Isotacy) is a geophysical phenomenon describing the force of **gravity** acting on crustal materials of various densities (mass per unit volume) that affects the relative floatation of crustal plates. Isostasy specifically describes the naturally occurring balance of mass in Earth's **crust**.

Continental crust and oceanic crust exist on **lithospheric plates** buoyant upon a molten, highly viscous aethenosphere. Within Earth's crustal layers, balancing processes take place to account for differing densities and mass in crustal plates. For example, under mountain ranges, the crust slumps or bows deeper into the upper mantle than where the land mass is thinner across continental plains. Somewhat akin to how **icebergs** float in seawater, with more of the mass of larger icebergs below the **water** than smaller ones, this bowing results in a balance of buoyant forces termed isostasy.

Isostasy is not a process or a force. It is simply a natural adjustment or balance maintained by blocks of crust of different mass or density.

Within Earth's interior, thermal energy comes from radioactive energy that causes convection currents in the core and mantle. Opposing convection currents pull the crust down into geosynclines (huge structural depressions). The sediments that have collected (by the processes of deposition that are part of the **hydrologic cycle**) are squeezed in the downfolds and fused into **magma**. The magma rises to the surface through volcanic activity or intrusions of masses of magma as batholiths (massive **rock** bodies). When the convection currents die out, the crust uplifts and these thickened deposits rise and become subject to **erosion** again. The crust is moved from one part of the surface to another through a set of very slow processes, including those in Earth's mantle (e.g., convection currents) and those on the surface (e.g. **plate tectonics** and erosion).

With isostasy, there is a line of equality at which the mass of land above sea level is supported below sea level. Therefore, within the crust, there is a depth where the total weight per unit **area** is the same all around the earth. This imaginary, mathematical line is called the "depth of compensation" and lies about 70 mi (112.7 km) below the earth's surface.

Isostasy describes vertical movement of land to maintain a balanced crust. It does not explain or include horizontal movements like the compression or folding of rock into mountain ranges.

Greenland is an example of isostasy in action. The Greenland land mass is mostly below sea level because of the weight of the **ice** cap that covers the island. If the ice cap melted, the water would run off and raise sea level. The land mass would also begin to rise, with its load removed, but it would rise more slowly than the sea level. Long after the ice melted, the land would eventually rise to a level where its surface is well above sea level; the isostatic balance would be reached again, but in a far different environment than the balance that exists with the ice cap weighing down the land.

Scientists and mathematicians began to speculate on the thickness of Earth's crust and distribution of landmasses in the mid-1800s. Sir George Biddell Airy (1801–1892) assumed that the density of the crust is the same throughout. Because the crust is not uniformly thick, however, the Airy hypothesis suggests that the thicker parts of the crust sink down into the mantle while the thinner parts float on it. The Airy hypothesis also describes Earth's crust as a rigid shell that floats on the mantle, which, although it is liquid, is more dense than the crust.

John Henry Pratt (1809–1871) proposed his own hypothesis stating that the mountain ranges (low density masses) extend higher above sea level than other masses of greater density. Pratt's hypothesis rests on his explanation that the low density of mountain ranges resulted from expansion of crust that was heated and kept its volume, but at a loss in density.

Clarence Edward Dutton (1841–1912), an American seismologist and geologist, also studied the tendency of Earth's crustal layers to seek equilibrium. He is credited with naming this phenomenon "isostasy."

A third hypothesis, eventually developed by Finnish scientist Weikko Aleksanteri Heiskanen (1895–1971) was a compromise between the Airy and Pratt models.

The model most accepted by modern geologists is the Hayford-Bowie concept. Advanced by American geodesists John Fillmore Hayford (1868–1925) and John William Bowie (1872–1940), geodesists, or specialists in geodesy, are mathematicians who study the size, shape, and measurement of Earth and of Earth forces (e.g., gravity). Hayford and Bowie were able to prove that the anomalies in gravity relate directly to topographic features. This essentially validated the idea of isostasy, and Hayford and Bowie further established the concept of the depth of isostatic compensation. Both gentlemen published books on isostasy and geodesy. Hayford was the first to estimate the depth of isostatic compensation and to establish that Earth has an oblate spherical shape (a bowed or ellipsoid sphere) rather than a true sphere.

See also Earth, interior structure

ISOTHERM

Isotherms are lines that connect points of equal **temperature** on **weather** maps, so at every point along a given isotherm the temperature values are the same. The word originates from Greek, where *isos* means equal and *therm* means heat. Isotherms are created from regularly scheduled, simultaneous temperature readings at different locations. For a proper comparison between the observation places, the measured temperature values are corrected for each location as if it was located at sea level. Isotherms help to visualize and interpret the horizontal temperature distribution of an **area** by showing patterns of temperature on a weather or **oceanography** map. Constructing a map of isotherms is an elementary step in temperature data analysis, and the process in general is called contouring. It can also be done for other parameters such as barometric pressure (**isobars**), geopotential height (isohypses), **dew point** temperature (isodrosotherms), **wind** speed (isotachs), and salinity (isohalines).

Isotherms are always smooth, labeled with the values, and mostly parallel to each other. Although the interval between the isotherms is arbitrarily chosen, within the same map it is a constant, and usually a round value. The value is selected such that the contour map both contains enough contours to show the patterns, and yet it is not crowded with too many lines. Because data is available only in the temperature observation points, interpolation should be used to create the isotherms between the measurement points. On the other hand, extrapolation to areas where no data is available is not acceptable. An isotherm should never split, cross, or touch another isotherm, because then at the crossing point it would have two different temperature values, which is physically impossible. Sometimes, contour maps are enhanced using color filling, when the area between pairs of isotherms is filled with special colors, so a particular color denotes the range of values between the two temperature values.

The relative spacing of the isotherms indicates the temperature gradient, the amount by which the temperature values vary across each unit of horizontal distance, in a direction perpendicular to the isotherms. The gradient is larger where the isotherms are closer. From the contour maps, areas of large gradient (regions where the temperature is changing quickly), as well as flat fields (regions where the temperature variation is not much) can be easily identified.

See also Isobars; Temperature and temperature scales

ISOTOPE • *see* ATOMIC MASS AND WEIGHT

J

JANSKY, KARL (1905-1950)

American radio engineer

One of the ways modern astronomers study the Universe is by tracing light waves through telescopes; another is by studying radio waves. The man who discovered the existence of these extraterrestrial radio waves, and thus founded radio **astronomy**, was Karl Jansky. Employed as an engineer in Bell Laboratories, New Jersey, Jansky was assigned the job of reducing static noise on transatlantic radio transmissions, and it was while inquiring into the origin of this static that he made his discovery.

The third of six children, Karl Guthe Jansky was born in Norman, Oklahoma, while that region was still a territory. His father, Cyril Jansky, was a college professor who taught electrical engineering and eventually became the head of the School of Applied Science at the University of Wisconsin. Jansky was named after Karl Guthe, a German-born physicist under whom his father had studied at the University of Michigan. Jansky attended the University of Wisconsin, where he played on the **ice** hockey team. He hoped to join the Reserve Officer's Training Program there but was diagnosed with a chronic kidney condition called Bright's disease; Jansky suffered from it all his life. He wrote his senior thesis on vacuum tubes and earned his B.S. in physics in June 1927. He stayed on at the University of Wisconsin for another year and supported himself by teaching while studying to complete the course work for his master's degree. He did not, however, write a thesis, and it would be years before he actually earned the degree.

After leaving the University of Wisconsin, Jansky applied for work at the Bell Communications Laboratories. The company was reluctant to hire him because of possible complications from Bright's disease. But Jansky's older brother, a professor of electrical engineering at the University of Minnesota, knew many Bell personnel. He intervened on behalf of his younger brother and secured the job for Jansky. Fearful of the stress he might suffer if he worked at their head-quarters in New York City, the company assigned Jansky to work at its facilities in New Jersey.

Although transatlantic radio communication was possible in the early 1930s, it was very expensive and poor in quality. It cost 75 dollars to talk for three minutes from New York to London, and the transmissions, which occurred not through cables but through radio waves, were routinely interrupted by static. There were clicking, banging, crackling, and hissing noises that sometimes obliterated the conversation. At Bell, Harald Friis assigned Jansky the job of determining what was causing the static. This was in the summer of 1931, and the first step Jansky took to resolve the problem was to design a new antenna. He built a directional antenna that was capable of receiving a much wider range of wavelengths than conventional antennas of the time. He also developed a receiver that generated as little static as possible, to minimize its interference with his efforts to measure static from outside sources. Last, Jansky developed an averaging device for recording the variations in static. The antenna and the rest of the equipment were installed in Holmdel, New Jersey, a rural **area** where there would be very little interference from man-made radio signals.

The antenna that Jansky assembled at Holmdel was mounted on wheels and moved on a turntable. This allowed it to scan the sky in all directions once every 20 minutes; it could also be pointed at different heights above the horizon. Known as Jansky's "merry-go-round," the antenna is believed to have been the largest of its type at the time. It operated at 20 MHz or 14.6 meters. He categorized the static into three different types: local thunderstorms, distant thunderstorms, and steady static. Jansky was able to establish that thunderstorms were the source of clicks and bangs. But he observed of the last type of static, as quoted in *Mission Communications: The Story of Bell Laboratories*, that it was "a very steady hiss type static, the origin of which is not yet known."

Jansky recorded the intensity of the hiss-type static, and he observed that it peaked when the antenna was pointed at a certain part of the sky. At first, Jansky thought that the point of peak intensity followed the **Sun**, and he initially assumed that the static was solar-generated. However, as he continued to make his observations, he saw that the peaks were moving further and further from the Sun. Indeed, he observed that the peak intensities occurred every 23 hours and 56 minutes. This was perhaps the first time that Jansky truly considered the idea that this static could have an extraterrestrial origin.

Jansky knew little about astronomy, but after consulting some colleagues who did, he learned that while Earth takes 24 hours to rotate once on its axis in relation to the Sun, its **rotation** with respect to the stars is four minutes shorter. Known as a sidereal day, this phenomenon was precisely what Jansky had observed: peak intensities in static readings that occurred at intervals of 23 hours and 56 minutes. Although the existence of radio waves other than those generated by people on Earth had never even been considered as a possibility, Jansky did not doubt his findings. He had made a discovery that was entirely new, and he had done it by accident. He was also fortunate in another respect. His investigations were conducted at a time when the 11-year cycle of solar activity was at a minimum, which rendered the **ionosphere** transparent to 20 MHz wavelengths at night. If this had not been the case, solar flares would have drowned out the weak hisses from **space**, and Jansky would never have been able to measure them.

Jansky had observed that the static was most intense when his antenna was aimed at the center of the Milky Way, the galaxy in which Earth is located. His measurements indicated a direction of 18 hours right ascension and –10 degrees declination. Such a location put the peak static emissions in the constellation of Sagittarius. These observations led Jansky to form two hypotheses concerning the origin of the static; either radio sources are distributed much as the stars are in the galaxy, or the radio emissions come from stars like our own sun. Since Jansky never did pick up such emissions from the Sun (weaker types were found by others), he rejected the second theory; his investigations during a partial solar **eclipse** in 1932 also seemed to support his belief that the Sun was not emitting radio waves. The first hypothesis was supported by the fact that radio emissions were most intense from the center of the Milky Way, which contains the densest clusters of stars. Jansky also reasoned that the emissions from space would be found all along the **electromagnetic spectrum**, a hypothesis confirmed by later researchers.

It was in December 1932 that Jansky realized the extraterrestrial nature of the static he was studying, and he issued his first report on the subject that same month in a paper entitled "Directional Studies of Atmospherics at High Frequencies." He presented it to the Institute of Radio Engineers, but no one made much of his discovery. Indeed, Jansky's boss, Harald Friis, cautioned him against proposing that static came from extraterrestrial sources in case he should be proved wrong. In April 1933, Jansky presented a second paper on these radio signals at a meeting of the International Scientific Radio Union in Washington, D.C. On May 5, 1933, Bell Laboratories issued a press release on the subject, and the next day the *New York Times* headlined his work as "New Radio Waves Traced to the Center of the Milky Way." On May 15, NBC's Blue Network broadcast a sample of Jansky's "star noise" to the nation. It was described by reporters as "sounding like steam escaping from a radiator." Jansky presented his second paper again at the annual convention of the Institute of Radio Engineers (IRE) in June 1933, and it was published the following October.

While researching "star noise," Jansky worked on other projects. He designed a new receiver that could automatically change bandwidths, as well as studied the general effects of bandwidth on an incoming signal. When Bell realized that nothing could be done about the hiss-type static that Jansky was studying, they assigned him to a different project. Jansky wrote to his father in January 1934, as quoted in the *Invisible Universe Revealed*: "I'm not working on the interstellar waves anymore. Friis has seen fit to make me work on the problems of and methods of measuring noise in general. A fundamental and necessary work, but not near as interesting as interstellar waves, nor will it bring near as much publicity. I'm going to do a little theoretical research of my own at home on the interstellar waves, however." Although Jansky presented his findings to astronomers, they largely ignored the implications of his work. One reason was that they did not believe the Milky Way could possibly be such a giant and intensive radio source. Resources were also scarce during the Great Depression of the 1930s, and there was little money for equipment to pursue this discovery. But the primary reason Jansky's work was neglected was that astronomy was then an optical venture. No one had any idea what to do with radio measurements. Jansky was, however, able to use his papers on "star noise" as a thesis for his master's degree. The University of Wisconsin awarded him this degree on June 16, 1936.

Jansky made other contributions to the understanding of radio communications while he worked at Bell. He became adept at detecting the direction of arrival of short-wave transmissions from all over the globe, which led to a better understanding of the effects of radio propagation. The information Jansky gained helped refine the design of both transmitting and receiving antennas. He also conducted research on noise reduction in receivers and other circuits. The outbreak of World War II made it even more difficult for Jansky to pursue his research on "star noise." Still working for Bell Laboratories, he was assigned to a classified project concerning the development of direction finders for German U-boats or submarines. Jansky also worked on identifying particular transmitters by their "signatures," and his contributions led the military to issue him an Army-Navy citation. After the war, Jansky designed and developed frequency amplifiers which met the requirements of wide bandwidth and low noise.

Disappointed by the fact that he never had the time to investigate extraterrestrial radio waves further, Jansky applied for a teaching position at Iowa State University. He hoped that he would be able to use their facilities to further his research, but he was not hired. In 1948, the IRE made Jansky a fellow, but by this time Bright's disease was causing him to suffer from hypertension and heart problems. Although he tried to ward off the effects of his disease with specialized diets and health care, Jansky died at the age of 44 in 1950. He left behind his wife, Alice, to whom he had been married since August 3, 1929, and two children who were still teenagers.

Although never recognized for his contributions to radio astronomy during his lifetime, Jansky's work was honored 23 years later. In 1973, the General Assembly of the International Astronomer's Union adopted the Jansky as a unit of measurement. Defined as 10^{-26} watts per meter squared hertz, the Jansky measures intensity of radio waves.

JEMISON, MAE C. (1956-)
American astronaut

Mae C. Jemison had received two undergraduate degrees and a medical degree, had served two years as a Peace Corps medical officer in West **Africa**, and was selected to join the National Aeronautics and **Space** Administration's astronaut training program, all before her thirtieth birthday. Her eight-day space flight aboard the **space shuttle** *Endeavour* in 1992 established Jemison as the United States' first female African American space traveler.

Mae Carol Jemison was born in Decatur, Alabama, the youngest child of Charlie Jemison, a roofer and carpenter, and Dorothy (Green) Jemison, an elementary school teacher. Her sister, Ada Jemison Bullock, became a child psychiatrist, and her brother, Charles Jemison, is a real estate broker. The family moved to Chicago, Illinois, when Jemison was three to take advantage of better educational opportunities there, and it is that city that she calls her hometown. Throughout her early school years, her parents were supportive and encouraging of her talents and abilities, and Jemison spent considerable time in her school library reading about all aspects of science, especially **astronomy**. During her time at Morgan Park High School, she became convinced she wanted to pursue a career in biomedical engineering. When she graduated in 1973 as a consistent honor student, she entered Stanford University on a National Achievement Scholarship.

At Stanford, Jemison pursued a dual major and in 1977 received a B.S. in chemical engineering and a B.A. in African and Afro-American Studies. As she had been in high school, Jemison was very involved in extracurricular activities, including dance and theater productions, and served as head of the Black Student Union. Upon graduation, she entered Cornell University Medical College to work toward a medical degree. During her years there, she found time to expand her horizons by visiting and studying in Cuba and Kenya and working at a Cambodian refugee camp in Thailand. When she obtained her M.D. in 1981, she interned at Los Angeles County/University of Southern California Medical Center and later worked as a general practitioner. For the next two and a half years, she was the area Peace Corps medical officer for Sierra Leone and Liberia where she also taught and did medical research. Following her return to the United States in 1985, she made a career change and decided to follow a dream she had nurtured for a long time. In October of that year she applied for admission to NASA's astronaut training program. The *Challenger* disaster of January 1986 delayed the selection process, but when she reapplied a year later, Jemison was one of the 15 candidates chosen from a field of about 2000.

When Jemison was chosen in 1987, she became the first African-American woman ever admitted into the astronaut training program. After more than a year of training, she became an astronaut with the title of science-mission specialist, a job that would make her responsible for conducting crew-related scientific experiments on the space shuttle. On September 12, 1992, Jemison finally flew into space with six other astronauts aboard the *Endeavour* on mission STS–47. During her eight days in space, she conducted experiments on weightlessness and motion sickness on the crew and herself. Altogether, she spent slightly over 190 hours in space before returning to Earth on September 20. Following her historic flight, Jemison noted that society should recognize how much both women and members of other minority groups can contribute if given the opportunity.

In recognition of her accomplishments, Jemison received several honorary doctorates, the 1988 *Essence* Science and Technology Award, the *Ebony* Black Achievement Award in 1992, and a Montgomery Fellowship from Dartmouth College in 1993, and was named Gamma Sigma Gamma Woman of the Year in 1990. Also in 1992, an alternative public school in Detroit, Michigan—the Mae C. Jemison Academy—was named after her. Jemison is a member of the American Medical Association, the American Chemical Society, the American Association for the Advancement of Science, and served on the Board of Directors of the World Sickle Cell Foundation from 1990 to 1992. She is also an advisory committee member of the American Express Geography Competition and an honorary board member of the Center for the Prevention of Childhood Malnutrition. After leaving the astronaut corps in March 1993, she accepted a teaching fellowship at Dartmouth and also established the Jemison Group, a company that seeks to research, develop, and market advanced technologies.

See also History of exploration III (Modern era); Spacecraft, manned

JET STREAM

The jet stream is a narrow, fast, upper atmospheric **wind** current, flowing quasi-horizontally at high altitudes around Earth. By definition, the wind speed should be higher than 57 mph (92 kph) for jet streams, although the term is sometimes also erroneously used for all upper-level winds. The jet stream may extend for thousands of miles around the world, but it is only a few hundred miles wide and less than a mile thick. The wind speeds in the core sometimes can reach 200–300 mph (322–483 kph). These wind speeds within the jet stream that are faster than the surrounding regions are called jet streaks. On average, the jet stream flows from east to west, but it often meanders into northern or southern moving loops. Jet streams occur in both hemispheres, but the Southern Hemisphere jet streams show less daily variability. Jet streams can be

The jet stream over the Sahara Desert and the Nile River in northern Africa. NASA/Science Photo Library. Reproduced by permission.

detected by drawing isothachs (the lines connecting points of equal wind speed) on a **weather** map.

Jet streams form in the upper **troposphere**, between 6.2–8.7 mi (10–14 km) high, at breaks in the tropopause, where the tropopause changes height dramatically. Jet streams are located at the boundaries of warm and cold air, above areas with strong **temperature** gradients. For example, the polar front, which separates cold polar air from warmer subtropical air, has a great temperature contrast along the frontal zone, leading to a steep pressure gradient. The resulting wind is the polar jet stream at about 6.2 mi (10 km) high, reaching maximum wind speed in winter. Sometimes the polar jet can split into two jets, or merge with the subtropical jet, which is located at about 8 mi (13 km) high, around 30 degrees **latitude**. A low-level jet stream also exists above the Central Plains of the United States, causing nighttime thunderstorm formation in the summertime. Over the subtropics, there is the tropical easterly jet, at the base of the tropopause in summertime, about 15 degrees latitude over continental regions. Near the top of the **stratosphere** exists the stratospheric polar jet during the polar winter.

Jet streams are well known since World War II. Detailed knowledge about the jet stream's location, altitude, and strength is essential not only for safe and efficient routing of aircrafts, but also for **weather forecasting**.

See also Atmospheric circulation; Coriolis effect; Troposphere and tropopause

JOHNSTON, HAROLD S. (1920-   )

American geochemist

Harold S. Johnston has been recognized as one of the world's leading authorities in **atmospheric chemistry**. He was among

the first to suggest that nitrogen oxides might damage Earth's ozone layer. Johnston's research interests have been in the field of gas-phase chemical kinetics and photochemistry, and his expertise has been employed by many state and federal scientific advisory committees on air pollution, motor vehicle emissions, and stratospheric pollution.

Harold Sledge Johnston was born on October 11, 1920, in Woodstock, Georgia, to Smith L. and Florine Dial Johnston. He graduated with a **chemistry** degree from Emory University in 1941 and, later that year, entered the California Institute of Technology as a graduate student. During the early 1940s, he was a civilian meteorologist attached to a United States Army unit in California and Florida, after which time he returned to graduate studies and earned his Ph.D. in chemistry and physics in 1948. That same year he married Mary Ella Stay, and the couple eventually had four children. Johnston was on the faculty of the chemistry department at Stanford University from 1947 to 1956 and the California Institute of Technology from 1956 to 1957. He then became a professor of chemistry at the University of California, Berkeley, serving as dean of the College of Chemistry from 1966 to 1970.

Johnston's introduction to **meteorology** occurred when he was a civilian scientist working on a defense project in World War II. In 1941, Roscoe Dickinson, Johnston's research director at the California Institute of Technology, was overseeing a National Defense Research Council project, with which Johnston became involved. Dickinson's group tested the effects of poisonous volatile chemicals on charcoals that were to be used in gas masks. Later, they studied how gas **clouds** moved and dispersed under different conditions in order to appraise coastal areas that might be vulnerable to chemical attacks.

In 1943, Johnston moved with the Chemical Warfare Service to Bushnell, Florida, where he worked with, and eventually headed, the Dugway Proving Ground Mobile Field Unit of the U.S. Chemical Warfare Service. This unit carried out test explosions to assess how the dispersion of gas was affected by meterological changes. While he was there, Johnston and John Otvos developed an instrument to measure the concentration of various gases in the air.

Johnston applied his meteorology work to his Ph.D. studies, which he resumed in 1945. He wrote his thesis on the reaction between **ozone**, a naturally occurring form of **oxygen**, and nitrogen dioxide, a pollutant formed during combustion. Later, during his tenure at Stanford, Johnston worked on a series of fast gas-phase chemical reactions. Using photo-electron multiplier tubes left over from the war, he pioneered a method of studying gas phase reactions that was a thousand times faster than existing techniques. Johnston then spent the years 1950 to 1956 researching high and low pressure limits of unimolecular reactions, and for the subsequent ten years, expanded his research to apply activated complex theory to elementary bimolecular reactions.

One of Johnston's most significant research efforts has been on the destruction of the ozone layer. This layer in the earth's upper atmosphere protects people from the Sun's **ultraviolet rays**. Chlorofluorocarbons (CFCs)—gaseous compounds often used in aerosol cans, refrigerants, and air conditioning systems—deplete this ozone layer, resulting in increased amounts of harmful radiation reaching the earth's surface. The Environmental Protection Agency has imposed production cutbacks on these harmful chemicals. Much like CFCs, nitrogen oxides also damage the ozone layer. During the late 1960s, the federal government financed the design and construction of two prototype supersonic transport (SST) aircraft. An intense political debate over whether the program should be expanded to construct five hundred SSTs was waged. Although Congress was split almost evenly, both houses voted to terminate the SST program in March, 1971. Johnston's articles and testimony suggesting the negative effects SSTs could produce on the atmosphere led two senators to introduce the **Stratosphere** Protection Act of 1971, which established a research program concerned with the stratosphere. The resulting 1971 program, with which Johnston was affiliated, was called the Climatic Impact Assessment Program (CIAP). Among other things, CIAP concluded that nitrogen oxides from stratospheric aircraft would further reduce ozone. CIAP recommended that aircraft engines be redesigned to reduced nitrogen oxide emissions.

Throughout his career, Johnston has served on many state and federal scientific advisory committees. In the 1960s, he was a panel member of the President's Science Advisory Board on Atmospheric Sciences and was on the National Academy of Sciences (NAS) Panel to the National Bureau of Standards. Johnston served on the California Statewide Air Pollution Research Center committee and the NAS Committee on Motor Vehicle Emissions during the early 1970s. He also served on the Federal Aviation Administration's High Altitude Pollution Program from 1978 to 1982 and the NAS Committee on Atmospheric Chemistry from 1989 to 1992. He has been an advisor to High Speed Civil Transport Studies for the National Aeronautics and **Space** Administration (NASA) since 1988.

Johnston is the author of the book *Gas Phase Reaction Rate Theory* and the author or coauthor of more than 160 technical articles. He is a member of the NAS, the American Academy of Arts and Sciences, the American Chemical Society, the American Physical Society, the American Geophysical Union, and the American Association for the Advancement of Science. Among Johnston's numerous awards are the 1983 Tyler Prize for Environmental Achievement, the 1993 NAS Award for Chemistry in Service to Society, and an honorary doctor of science degree from Emory University.

See also Global warming; Greenhouse gases and greenhouse effect; Ozone layer and hole dynamics; Ozone layer depletion

JOINT AND JOINTING

Fractures in **rock** are classified according to the type of relative motion that has occurred across the fracture. Extensional fractures, also known as joints, are characterized by movement perpendicular to the fracture. The masses of rock separated by a joint moved away from each other, even if imperceptibly, when the joint was formed. Joints stand in contrast to faults, which are shear fractures across which the opposite sides slide past (rather than away from) each other. Rocks can undergo more than one episode of deformation during their existence, so it is possible for a fracture to begin as a joint and evolve into a fault as the stresses acting on the joint change through **geologic time**. The precise definition of a joint, however, is not universal and some geologists classify fractures as joints if there seems to have been only a small, but measurable, amount of shearing.

Joints formed in **coal** are known as cleats, and joints (or faults) filled with mineral deposits are known as veins. Volcanic rocks such as **basalt** contract as they cool, forming networks of columnar joints that divide the rock into regular polygonal columns such as those seen in Devil's Tower and Devil's Postpile National Monuments, USA.

Joints that are exposed at Earth's surface are often enlarged by chemical dissolution and **weathering**, particularly in soluble rocks such as **limestone** and, to a lesser degree, rocks such as calcite-cemented sandstones. Therefore, the width of joints in outcrops does not necessarily reflect the movement that created the joint.

Groups of joints sharing a similar three-dimensional geometry are known as joint sets. Like all fractures in rock, joints are irregular or wavy features rather than perfect planes. Therefore, joints in a set will have slightly different geometries and the separation of joints into sets can pose a difficult task for geologists. The combination of two or more sets of joints in a rock mass is a joint system.

Joints can also be described as being systematic or non-systematic. Systematic joints are those that are nearly planar (although never perfectly so) and occur in sets with regular spacing and orientation. Non-systematic joints are those with irregular or seemingly random geometry, spacing, and orientation. Although the terms are similar, systematic joints do not necessarily belong to joint systems and joint systems are not necessarily composed of systematic joints. It is possible, for example, for there to be a joint system composed of two sets of non-systematic joints.

The origin of joints and faults is studied by making detailed maps of rock fracture systems in the field and then applying mathematical techniques developed in the discipline of fracture mechanics. In fracture mechanics terminology, joints are known as Mode I fractures. Faults are Mode II or III fractures, depending on the direction of movement parallel to the fault surface. Studies have shown that rocks, like all other materials, contain innumerable microscopic flaws. When a rock is subjected to stress, either within Earth's **crust** or a laboratory-testing device, the fractures that most efficiently dissipate the stress grow in length and combine at the expense of other, less efficiently oriented, fractures. Fractures also perturb the distribution and intensity of stress in the adjacent rock, and the shape of a growing fracture can therefore be strongly influenced by the growth of its neighbors. Evidence for this phenomenon can be found in the field, where neighboring joints curve and then abruptly terminate against each other to form complicated, but understandable, patterns.

Because they are discontinuities in otherwise solid rock masses, joints can influence fluid flow through rocks and rock mass stability. Jointed rock has much higher **porosity and permeability** than intact rock of the same type, and can more easily transmit **petroleum**, **groundwater**, or ore-bearing geothermal fluids. Joints also form mechanical discontinuities that decrease the strength of rock, which is particularly important when rock is excavated during construction projects. Rockslides and rock falls also occur as a result of extensive joint sets or systems. Therefore, the identification and analysis of discontinuities such as joints, faults, and **bedding** planes is a critically important part of many applied geologic studies.

See also Faults and fractures; Field methods in geology

JUPITER • *see* SOLAR SYSTEM

JURASSIC PERIOD

In **geologic time**, the Jurassic Period—the middle of three geologic periods in the Mesozoic Era—spans the time from roughly 206–208 million years ago (mya) to approximately 146 mya.

The Jurassic Period contains three geologic epochs. The earliest epoch, the Lias Epoch, ranges from the start of the Jurassic Period to approximately 180 mya. The Lias Epoch is further subdivided into (from earliest to most recent) Hettangian, Sinemurian, Pliensbachian, and Toarcian stages. The middle epoch, the Dogger Epoch, ranges from 180 mya to 159 mya and is further subdivided into (from earliest to most recent) Aalenian, Bajocian, Bathonian, and Callovian stages. The latest epoch (most recent), the Maim Epoch, ranges from 159 mya to 144 mya and is further subdivided into (from earliest to most recent) Oxfordian, Kimmeridgian, and Tithonian stages.

During the Jurassic Period, the Pangaean supercontinent broke into continents recognizable as the modern continents. At the start of Jurassic Period, Pangaea spanned Earth's equatorial regions and separated the Panthalassic Ocean and the Tethys Ocean. Driven by **plate tectonics** during the Jurassic Period, the North American and European continents diverged, and the earliest form of the Atlantic Ocean flooded the spreading sea floor basin between the emerging continents.

By mid-Jurassic Period, although still united along a broad region, what would become the South American and African Plates and continents became distinguishable in a form similar to the modern continents.

By the end of the Jurassic Period, **North America** and **South America** became separated by a confluence of the Pacific Ocean and Atlantic Ocean. Extensive flooding submerged much of what are now the eastern and middle portions of the United States.

By the end of the Jurassic Period, **water** separated South America from **Africa**, and the Australian and Antarctic continents were clearly articulated. The Antarctic continent began a slow southward migration toward the south polar region.

The Jurassic Period (in popular culture widely recognized as the "Age of the Dinosaurs") was named for the Jura Mountains on the Swiss-French border, an area where the classic formations were first identified and studied.

Large meteor impacts occurred at the start and end of the Jurassic Period (and later intensified during the subsequent **Cretaceous Period**). During the Jurassic Period itself, there is evidence of only one major impact—the Puchezh impact in Russia. The Manicouagan impact in **crust** now near Quebec, Canada, dates to the late **Triassic Period** just before the start of the Jurassic Period. A trio of impacts in areas now located in South Africa, the Barents Sea, and **Australia** occurred near the end of the Jurassic Period and start of the Cretaceous Period.

Although humans and dinosaurs never co-existed—in fact they are separated by approximately 63 million yeas of evolutionary time—the Jurassic Period's wealth of **fossils** have long stirred human imagination about life on Earth during that time. The abundant life of the Jurassic Period also left a legacy of organic remains that today provide an economically important source of fossil **fuels**. Many prominent oilfields date to the Jurassic Period (e.g., the North Sea fields).

See also Archean; Cambrian Period; Cenozoic Era; Dating methods; Devonian Period; Eocene Epoch; Evolution, evidence of; Fossil record; Fossils and fossilization; Geologic time; Historical geology; Holocene Epoch; Marine transgression and marine recession; Miocene Epoch; Mississippian Period; Oligocene Epoch; Ordovician Period; Paleocene Epoch; Paleozoic Era; Pennsylvanian Period; Phanerozoic Eon; Pleistocene Epoch; Pliocene Epoch; Precambrian; Proterozoic Era; Quaternary Period; Silurian Period; Supercontinents; Tertiary Period

K

KARNS · *see* GLACIAL LANDFORMS

KARST TOPOGRAPHY

Karst is a German name for an unusual and distinct **limestone** terrain in Slovenia, called Kras. The karst region in Slovenia, located just north of the Adriatic Sea, is an **area** of barren, white, fretted **rock**. The main feature of a karst region is the absence of surface **water** flow. Rainfall and surface waters (streams, for example) disappear into a drainage system produced in karst areas. Another feature is the lack of topsoil or vegetation. In **geology**, the term karst **topography** is used to describe areas similar to that found in Kras. The most remarkable feature of karst regions is the formation of caves.

Karst landscapes develop where the **bedrock** is comprised of an extremely soluble calcium carbonate rock such as limestone, **gypsum**, or **dolomite**; limestone is the most soluble calcium carbonate rock. Consequently, most karst regions develop in areas where the bedrock is limestone. Karst regions occur mainly in the great sedimentary basins. The United States contains the most extensive karst region of the world. Other extensive karst regions can be found in southern France, southern China, Central America, Turkey, Ireland, and England.

Karst regions are formed when there is a chemical reaction between the **groundwater** and the bedrock. As rain, streams, and **rivers** flow over the earth's surface, the water mixes with the **carbon dioxide** that naturally exists in air. The water and **carbon** dioxide react to form a weak carbonic acid, which causes the **soil** to become acidic and corrode the calcium carbonate rock. The carbonate solution seeps into fissures, fractures, crevices, and other depressions in the rock. **Sinkholes** develop and the fissures and crevices widen and lengthen. As the openings get larger, the amount of water that can enter increases. The surface tension decreases, allowing the water to enter faster and more easily. Eventually, an under-ground drainage system develops. The bedrock is often hundreds of feet thick, extending from near the earth's surface to below the **water table**. Solution caves often develop in karst regions. Caves develop by an extensive enlargement and **erosion** of the underground drainage structure into a system of connecting passageways.

There are many variations of karst landscape, often described in terms of a particular landform. The predominant **landforms** are called fluviokarst, doline karst, cone and tower karst, and pavement karst. Some karst regions were etched during the **Ice** Age and may appear barren and very weathered (pavement karst). Other karst areas appear as dry valleys for part of the year and after seasonal **floods**, as a lake (one example of fluviokarst). In tropical areas, karst regions can be covered with **forests** or other thick vegetation. Sometimes, the underground drainage structure collapses, leaving odd formations such as natural bridges and sinkholes (doline karst). Tall, jagged limestone peaks are another variation (cone or tower karst).

See also Geochemistry; Hydrogeology; Weathering and weathering series

KELVIN SCALE · *see* TEMPERATURE AND TEMPERATURE SCALES

KEPLER, JOHANNES (1571-1630)

German astronomer

Johannes Kepler, the astronomer who determined the laws of planetary motion, was born in Weil in Würtemberg (now southwestern Germany) and seemed destined for a life in the church. He obtained a B.A. degree in theology in 1588 and entered the

•

Johannes Kepler.

University of Tübingen, one of the great centers of Protestant learning, in 1589. He graduated with an M.A. in 1591.

While at Tübingen, Kepler had learned mathematics, and when the high school in Graz, Austria, needed a mathematics teacher, he accepted the job. He turned out to be very poor at teaching; during his first year only a handful of students attended his lectures, and the following year he had no students at all. That gave him considerable free time to further his interest in **astronomy** and produce annual almanacs, which were often heavy on astrological content. He also raised additional income by casting horoscopes.

Kepler published a book in 1596 in which he devised a relationship involving the distances of the planets from the **Sun** with geometric solid objects such as cubes and spheres. His *Mysterium Cosmographicum* (Mystery of the universe) showed considerable knowledge of astronomy and brought him to the attention of **Tycho Brahe** and Galileo.

Beginning in 1597, life for Protestants in the Catholic-dominated **area** in which Kepler lived became unpleasant due to religious persecution, so Kepler moved to Prague (Czech Republic) in 1600, where he began working with Tycho Brahe.

Brahe was the antithesis of Kepler. Brahe had excellent eyesight that he used to compile the most detailed observational data in history; Kepler's eyesight was poor and he suffered from ill health all his life. Brahe was financially secure,

and his extravagance drove Kepler, who always seemed to be in reduced circumstances, to distraction. Their working relationship was not optimal, but Brahe died within 18 months of Kepler's arrival.

Brahe had spent years making accurate observations of Mars with the naked eye and assigned Kepler the task of devising a theory of planetary motion using his observational data. Kepler, the mathematician, was superbly suited for this task, which would end up occupying the majority of his time for the next 20 years.

In 1604, before he had made much progress with Mars, a supernova blazed into view, and Kepler wrote two pamphlets about it. The supernova, like Brahe's Star in 1572, rivaled Venus in brightness and has since come to be known as Kepler's Star. He also wrote about applications of optics in astronomy and proposed a design for a **telescope**. After Galileo discovered the moons of Jupiter, Kepler used a telescope to prove to himself that they did, indeed, exist, and called them satellites.

The task with Mars proved extremely difficult. The circular orbit Kepler calculated did not agree exactly with Brahe's observations. Kepler's creative imagination came up with theory after theory to account for the discrepancy; at one point, after three years of work, he had a geometrical scheme that disagreed with one observation by only eight minutes of arc. (The full **Moon** is 31 minutes of arc in diameter; two objects four minutes of arc apart are barely noticeable to a person with average eyesight.) Kepler did not accept that Brahe's observations were anything less than perfect. He threw his scheme out and started again.

Kepler gave up on circles and epicycles and, out of desperation, tried working with an ellipse (oval). The results matched Brahe's data perfectly. Then he had to devise a law governing the variation of the speed of Mars as it moved along the ellipse. Here he got bogged down and lost his way, but he eventually formulated a simple law that matched observations. Finally, in 1609, Kepler published his first two laws of planetary motion: a planet orbits the Sun in an ellipse, not a circle as Nicholas Copernicus had believed; a planet moves faster when near the Sun, and slower when farther away. Kepler thought, incorrectly, that **magnetism** in the Sun was responsible for the variation.

The third law consumed Kepler's next ten years. In 1619, Kepler determined that the square of the time it takes a planet to orbit the Sun is equal to the cube of its average distance. In other words, once it is known how long it takes a planet to complete an orbit, its relative distance from the Sun can be calculated. However, it is still necessary to have a definite measured distance for one planet to act as a yardstick to determine the distance of the others.

It seems likely that Kepler happened on this formula, not by mathematical calculation, but by accident. He was constantly looking for mystical relations of numerical sequences to explain the "harmony of the heavens." In the same book in which the third law appears, Kepler devotes **space** to the "music of the spheres," assigning individual musical notes that each planet "sings."

Also in 1619, Kepler published a book on **comets** in which he supported Brahe's contention that comets were celestial objects and not manifestations of Earth's atmosphere. Kepler incorrectly believed them to be objects that moved in straight lines, but he had a remarkably accurate explanation of the Sun's part in producing a comet's tail.

Once again, religious persecution of Protestants caused Kepler to move, and he relocated to Ulm (Germany) in 1626. One year later he published his final great work: the *Rudolphine Tables*, a tabular collection of the motions of planets, dedicated to his former patron Emperor Rudolph II. They were used as the standard astronomical tables for the next century. Kepler died after a short illness at the age of 59.

See also Cosmology

KETTLES • *see* GLACIAL LANDFORMS

KIMBERLITIC DIAMOND • *see* DIAMOND

KINETIC AND POTENTIAL ENERGY • *see* ENERGY TRANSFORMATIONS

KRAKATOA • *see* VOLCANIC ERUPTIONS

K-SPAR • *see* FELDSPAR

K-T EVENT

The K-T event (Cretaceous-Tertiary event) refers to the mass extinction of the dinosaurs that took place approximately 65 million years ago (mya). In addition to the dinosaurs, most large land animals perished and an estimated 70% of all species on the planet became extinct.

In the early 1980s, a team of physicists and geologists documented a band of sedimentary **rock** in Italy that contained an unusually high level of the rare metal iridium (usually found on Earth's surface only as a result of meteor impacts). The scientists eventually argued that that the iridium layer was evidence of a large asteroid impact that spewed iridium contaminated dust into the atmosphere. Blown by global **wind** currents, the iridium eventually settled into the present thin sedimentary layer found at multiple sites around the world. Given the generalized dispersion of iridium the researchers argued that the impact was large and violent enough to cause dust and debris particles to reach high enough levels that they seriously occluded light from the **Sun** for a large expanse of Earth.

The subsequent reduction in photosynthesis was sufficient to drastically reduce land plant population levels and eventually drive many plant species to extinction. The reduction in plant population levels also provided evolutionary pressure on species nutritionally dependent upon plant life. Large life forms with high-energy demands (e.g., dinosaurs) were especially sensitive to the depleted dietary base. The adverse consequences of population reductions and extinctions of plant-eating life forms then rippled through the ecological web and food chain—ultimately resulting in mass extinction.

Calculations of the amount of iridium required to produce the observable layer (on average about a centimeter thick) yield estimates indicating that the asteroid measured at least 6 mi (10 km) in diameter. The **impact crater** from such an asteroid could be 100 mi (161 km) or more in diameter. Such an impact would result in widespread firestorms, earthquakes, and tidal waves. Post-impact damage to Earth's ecosystem occurred as dust, soot, and debris from the collision occluded the atmosphere to sunlight.

Based on **petroleum** exploration data, Canadian geologist Alan Hildebrand identified a major impact crater in the oceanic basin near what is now the Yucatan Peninsula of Mexico. The remains of the impact crater, termed the Chicxulub crater, measures more than 105 mi (170 km) in diameter. Argon dating places the Chicxulub impact at the expected Cretaceous-Tertiary **geologic time** boundary, approximately 65 mya.

Other geological markers are also indicative of a major asteroid impact approximately 65 mya (e.g. the existence of shock **quartz**, ash, and soot in sedimentary layers dated to the K-T event). Tidal wave evidence surrounding the **Gulf of Mexico** basin also dates to 65 mya.

Other scientists have argued that it was not a solitary impact that alone caused the mass extinction evidenced by the **fossil record**. At end of the prior **Cretaceous Period** and during the first half of the **Tertiary Period**, Earth suffered a series of intense and large impacts. Geologists have documented more that 20 impact craters greater than 6.2 mi (10 km) in diameter that date to the late Cretaceous Period. Large diameter impact craters were especially frequent during the last 25 million years of the Cretaceous Period (i.e., the Senonian Epoch).

The extinction of the dinosaurs and many other large species allowed the rise of mammals as the dominant land species during the **Cenozoic Era**.

See also Astronomy; Catastrophism; Dating methods; Evolution, evidence of; Fossil record; Fossils and fossilization; Historical geology; Meteoroids and meteorites; Paleocene Epoch

KYOTO TREATY • *see* ATMOSPHERIC POLLUTION

L

LAHAR

Lahars are debris flows associated with volcanoes, and can be further classified as either hot or cold depending on their **temperature**. The word lahar is of Indonesian origin, reflecting the frequency of volcanic debris flows in that region. Lahars can be mobilized by processes similar to those producing non-volcanic debris flows, most notably the transition of saturated **landslide** masses into **debris flow**, but can also be associated with the rapid **melting** of snow and **ice** by hot volcanic (pyroclastic) debris during an eruption. Earthquakes associated with volcanic activity can also trigger landslides that have the potential to mobilize into lahars.

Like non-volcanic debris flows, lahars move as fluid masses with the general consistency of wet concrete. Two characteristics of lahars that can make them particularly hazardous are their potentially very large volume and great velocity relative to many non-volcanic debris flows. For example, a lahar that occurred during an eruption of Cotopaxi **Volcano** (Ecuador) in 1877 traveled more than 186 mi (300 km) at an average velocity of 16.7 mph (27 kph). Lahars triggered by the 1980 eruption of Mount St. Helens traveled at an average velocity of 41.6 mph (67 kph). Eruptions of volcanoes mantled with snow or ice can also produce catastrophic lahars, as in the 1985 eruption of Nevado del Ruiz (Columbia) that killed some 23,000 people. Like non-volcanic debris flows, lahars are able to transport extremely large boulders or other objects because of the density of the flow, which is in most cases nearly the same as the intact debris from which the flow mobilized. Lahars are not necessarily large; some may involve as little as a few cubic centimeters of debris and pose little threat to life and limb.

Mapping and analyzing ancient lahar deposits are an important part of volcanic hazard assessment for volcanoes such as Mount Rainier and Mount Hood in the northwestern United States. Deposits left by past lahars provide insight into the likely size and frequency of future lahars. Trees and other organic material found ancient lahar deposits can be dated using radiometric methods, allowing geologists to assemble a chronology of lahar activity in the **area** around a volcano.

See also Catastrophic mass movements; Debris flow; Erosion; Landslide; Mass movement; Mass wasting; Mud flow

LAKE EFFECT SNOW • *see* BLIZZARDS AND LAKE EFFECT SNOWS

LAKES

Lakes and ponds are bodies of standing fresh **water** impounded in basins and depressions in the earth's continental **crust**. Lakes are temporary catchment basins for flowing surface and **groundwater**. **Freshwater** reservoirs form behind natural and man-made dams, surface water collects in topographic lows, and groundwater discharges into ephemeral lakes, but eventually all continental **runoff** drains to the ocean. Lakes provide humans with fresh drinking water, recreation areas and, in the case of the world's largest lakes, navigable waterways for ship traffic. Regional **climate** strongly affects the chemical and hydrological properties of lakes, and lake sediments often provide high-resolution records of climatic fluctuations. Lake basins typically fill with interlayered coarse and fine sediments, and organic material. Many ancient lacustrine deposits contain **petroleum** reservoirs. Because ponds, lakes, and inland **seas** are smaller and less well-mixed than the **oceans**, they are particularly susceptible to pollution.

Tectonic motion created the crustal basins and sags that contain the world's largest lakes. Elongate, deep lakes fill the axes of incipient divergent plate tectonic boundaries, or rift zones. The lakes of the East African Rift system—Lakes Turkana, Kiva, Tanganyika, and Malawi—fill the central grabens of the rift zone between the African and Somali **Lithospheric Plates**. Lake Baikal, the worlds deepest (5,370 ft,

Lake Powell, Arizona. © Nik Wheeler/Corbis. Reproduced by permission.

or 1,637 m) and most voluminous (Lake Baikal contains about 20% of the earth's fresh surface water) lake, occupies a rift valley in southern Siberia. Lakes also fill broad, shallow intercratonic basins that form during the earliest stages of continental **rifting**. Lake Eyre in central **Australia**, and Lakes Victoria and Chad in **Africa** are examples of lakes in shallow extensional basins.

Many modern lakes, including the **Great Lakes** of **North America**, occupy basins created by Northern Hemisphere **ice** sheets of the **Pleistocene Epoch**. The weight of the Laurentide and Eurasian ice sheets depressed large regions of the continental crust into the mantle, a phenomenon called glacial **isostasy**. Since the ice sheets retreated about 20,000 years ago, meltwater and stream runoff have collected in these broad depressions. Large regions of the northern continents—the Great Lakes region and the Scandinavian Peninsula for example—are presently undergoing rapid uplift, known as isotatic rebound, as these glacially depressed regions continue to readjust. Small ponds and lakes are also common in glacial environments. **Erosion** by moving ice carves **bedrock** depressions where lakes form, and leaves sills that impound glacial streams. Glacial sedimentary **landforms**, including **moraines**, kame terraces, and eskers serve as natural dams for

glacial lakes. Glacial terrains are dotted with small ponds that fill circular depressions called kettles that form when ice blocks buried in glacial **till** deposits melt.

Lake basins also form in a number of other geologic environments. Small lakes and ponds are common in continental fold belts where outcrops of resistant bedrock divert and dam perennial streams. Abandoned meanders along low-gradient streams form circular lakes called oxbow lakes. Groundwater discharge zone lakes form where the top of the **saturated zone**, the **water table**, intersects the land surface. In humid and temperate climates, where the water table is close to the land surface, discharge lakes typically have an outlet stream. In arid regions, ephemeral groundwater discharges into closed, saline playa lakes that fill and dry seasonally.

Man-made lakes are a significant component of the earth's present-day **hydrologic cycle**. Most of the world's **rivers** have been dammed, creating reservoirs for human water supplies, recreation, and generation of electrical power. While reservoirs provide many benefits to human populations, they also force numerous readjustments to natural and artificial systems. Ecosystems must compensate for the loss of drowned habitats, human populations are displaced, and water quality is often compromised. Streams that have been segmented by dams regrade their equilibrium profiles, creating new patterns of erosion and deposition throughout the stream system. In fact, natural stream processes act to remove obstacles like dams by eroding the streambed below them, and depositing sediment in the reservoir above them. Poorly constructed and maintained dams are thus a safety hazard for downstream inhabitants.

Climatic factors control the chemical and hydrological properties of lakes. Regional variations of **temperature**, **precipitation**, and winds determine water levels, circulation patterns, vertical stratification, and the concentration of dissolved materials in lake water. The quantity and seasonality of rainfall in a drainage basin controls the balance between recharge and discharge that maintains lake level. Lake salinity is a function of the relative concentrations of dissolved ions and diluting water. During a **drought**, lake levels fall, salinity increases, and a lake can change from a permanent freshwater reservoir to an ephemeral saline lake. The Great Salt Lake in Utah is all that remains of Lake Bonneville, a much larger freshwater lake that existed during the wet period at the end of the Pleistocene Epoch. A 25% decrease in freshwater flow to the Aral Sea in central **Asia** has led to a 50% decrease in surface **area** and a four-fold increase in salinity since the 1960s.

Seasonal temperature variations, changes in the balance between precipitation and **evaporation**, and **wind** patterns affect lake circulation and stratification. High-latitude lakes that are subject to large diurnal temperature variations and strong winds are typically so well mixed that the water column is unstratified. Warm, stagnant low-latitude lakes are often permanently stratified. Without mixing, the lower water column of these stagnant, oligomictic lakes becomes depleted in **oxygen**, and aquatic plants choke the ecosystem in a process called eutrophication. Human water pollutants that contain phosphates—detergents for example—also encourage eutrophication. Temperate lakes that experience large seasonal temperature fluctuations undergo seasonal overturns in which a layer of cold surface water circulates to the bottom of the lake or the pond. This process of periodic restratification oxygenates the base of the water column and infuses lake-bottom ecosystems with nutrients. Lakebed sediments record these seasonal patterns, and can be used to deduce and date the regional climate history. Lake stratigraphers, or limnologists, use features like preserved pollens and winter-summer couplets of thin sedimentary laminae, called varves, to recreate the geochronology of a lake basin.

See also Hydrogeology

LAND AND SEA BREEZES

Land and sea breezes are **wind** and **weather** phenomena associated with coastal areas. A land breeze is a breeze blowing from land out toward a body of **water**. A sea breeze is a wind blowing from the water onto the land. Land breezes and sea breezes arise because of differential heating between land and water surfaces. Land and sea breezes can extend inland up to 100 mi (161 km) or manifest as local phenomena that quickly

weaken with a few hundred yards of the shoreline. On average, the weather and cloud effects of land and sea breezes dissipate 20–30 mi (32–48 kph) inland from the coast.

Because water has a much higher heat capacity that do sands or other crustal materials, for a given amount of solar irradiation (**insolation**), water **temperature** will increase less than land temperature. Regardless of temperature scale, during daytime, land temperatures might change by tens of degrees, while water temperature change by less than half a degree. Conversely, water's high heat capacity prevents rapid changes in water temperature at night and thus, while land temperatures may plummet tens of degrees, the water temperature remains relatively stable. Moreover, the lower heat capacity of crustal materials often allows them to cool below the nearby water temperature.

Air above the respective land and water surfaces is warmed or cooled by conduction with those surfaces. During the day, the warmer land temperature results in a warmer (therefore less dense and lighter) air mass above the coast as compared with the adjacent air mass over the surface of water. As the warmer air rises by convection, cooler air is drawn from the ocean to fill the void. The warmer air mass returns to sea at higher levels to complete a convective cell. Accordingly, during the day, there is usually a cooling sea breeze blowing from the ocean to the shore. Depending on the temperature differences and amount of uplifted air, sea breezes may gust 15–20 mph (24–32 kph). The greater the temperature differences between land and sea, the stronger the land breezes and sea breezes.

After sunset, the air mass above the coastal land quickly loses heat while the air mass above the water generally remains much closer to it's daytime temperature. When the air mass above the land becomes cooler than the air mass over water, the wind direction and convective cell currents reverse and the land breeze blows from land out to sea.

Because land breezes and sea breezes are localized weather patterns, they are frequently subsumed into or overrun by large-scale weather systems. Regardless, winds will always follow the most dominant pressure gradient.

The updraft of warm, moist air from the ocean often gives rise to daytime cloud development over the shoreline. Glider pilots often take advantage of sea breezes to ride the thermal convective currents (sea breeze soaring). Although most prevalent on the sea coastline, land breezes and sea breezes are also often recorded near large bodies of water (e.g., the **Great Lakes**). In general, land breezes and sea breezes result in elevated **humidity** levels, high **precipitation**, and temperature moderation in coastal areas.

See also Adiabatic heating; Clouds and cloud types; Convection (updrafts and downdrafts); Seasonal winds; Weather forecasting methods

LANDFORMS

Landforms are the mesoscale topographic features that define a regional landscape. **Climate** and **plate tectonics** ultimately determine the system of processes—plate tectonic motion,

This mother and child are taking advantage of the sea breeze. © Tim Kiusalaas/Corbis. Reproduced by permission.

gravity, **erosion**, and deposition by **water**, **wind**, and ice—that interact in complex feedback loops to erect and destroy continental landforms. Climate, the regional pattern of seasonal **precipitation**, **temperature**, and wind-flow, varies with **latitude**, altitude, and physiographic location. Motion of Earth's **lithospheric plates**, called plate tectonics, determines the global geography of **mountain chains**, volcanoes, continental landmasses, and ocean basins, as well as the distribution of **rock** types. The morphology of landforms suggests the processes that created them, and landforms are clues to deciphering the present and past geological setting of a region.

Climate often dictates the mix of sedimentary processes acting to create landforms. For example, eolian, or wind-formed, landforms are common in arid environments, whereas **glacial landforms** like **moraines** and kettle ponds are found in cold regions with adequate snowfall. The abundance and seasonality of annual rainfall determines the intensity of **weathering** and erosion by flowing water. In rainy regions, a dendritic pattern of V-shaped **stream valleys** often dominates the landscape. Low-gradient stream valleys are filled with fluvial landforms like meandering channels, oxbow **lakes**, natural levees, and point bars, while sediment-choked mountain

streams form broad, gravelly plains of intertwined channels and longitudinal bars.

Air currents also shape landscapes by sorting and reworking dry sediments in arid environments. Eolian landforms include numerous types of **sand dunes**: cresent-shaped barchans, merged barchans called transverse dunes, parabolic blowout pits, flow-parallel linear dunes, and large composite dunes called draas. Very large **dune fields** called ergs cover extensive **areas** of the major deserts. Seasonal temperature fluctuations affect the phase of water. Erosion by frozen surface water creates dramatic glacial features like **fjords**, hanging valleys, and cirques in high-latitude regions. **Evaporation** of standing water in hot, arid, **desert** environments forms a unique array of desert landforms, including ephemeral playa lakes and coarse-grained alluvial fans where mountain streams enter arid basins.

The tectonic plate setting and past **geology** of a region also affects its **topography**. Motion along tectonic plate boundaries creates tall mountain belts, volcanic chains, regions of broad uplift, and zones of subsidence. Escarpments, extensional valleys called grabens and pull-apart basins, and linear pressure ridges form along fault planes where rocks bodies move relative to one another. Gravity failure of tectonically over-steepened slopes creates **mass wasting** deposits like mudflows, slumps, and slides. Streams that follow out-crop patterns of folded sedimentary and metamorphic rocks create ridge and valley provinces. Resistant layers of horizontal strata in tectonically uplifted areas form broad tablelands. Caves, **sinkholes**, and other so-called karst features form when acidic rainwater falls on carbonate rock layers. Volcanic activity creates numerous distinctive landforms, including craters, **lava** flows, eroded volcanic necks, and, of course, the conical or shield-shaped volcanoes themselves. **Beach and shoreline dynamics** along continental margins form an array of coastal landforms, from erosional sea cliffs and stacks to depositional deltas, **barrier islands**, and spits.

The continents can be divided into physiographic landform provinces delineated by plate tectonic features and climatic zones. (**Antarctica** is an exception because **ice** covers about 90 percent of its land surface.) The Colorado Plateau, in the southwest United States, is an example of one such province, and a tour of its dramatic landscape illustrates the complex web of processes acting to sculpt its landforms. The Four Corners region is arid because the prevailing Pacific westerly winds have dropped most of their moisture in the high Sierras of California and Mexico before reaching it. Streams flowing onto the Plateau from the Rocky Mountains, however, provide ample erosional and depositional energy. The Plateau is a broad zone of tectonic uplift that has exposed a thick sequence of flat-lying sedimentary strata to deep incision of steep-walled canyons by flood-prone streams. Erosional remnants like flat-topped mesas, conical buttes, arches, and spires remain between the incised fluvial channels. Winds sort and transport grains of siliciclastic sand, creating dunes and sandblasting the landscape.

Humans have also affected **landscape evolution** in the modern southwest desert. Construction of dams along the major **rivers**, especially along the Colorado River, has led to stream regrading and a resulting change in the pattern of fluvial erosional and depositional landforms. Evaporation from man-made reservoirs, like Lake Powell and Lake Mead, has reduced the amount of water in downstream river segments and increased the **humidity** around the lakes. Water use by human cities, industries, and especially irrigated agricultural lands has changed the regional **hydrologic cycle**.

See also Canyon; Desert and desertification; Karst topography; Sedimentation; Stream valleys, channels and floodplains

LANDSCAPE EVOLUTION

A landscape is the cumulative product of interaction among dynamic geological processes over time. A region's **topography** and suite of characteristic **landforms** are, thus, clues to its geologic history. For example, the landscape of rugged, linear **mountain chains**, deep canyons, dry lake beds, and mesas in the United States' **desert** southwest tells a geologic story of fluvial and Eolian **erosion** acting during a period of increasing climatic aridity while plate tectonic forces caused crustal extension and uplift. Earth processes carve a landscape; dynamic interactions between processes control its **evolution** over time.

The earth's internal heat drives plate tectonic motion and influences the related processes of crustal uplift, magmatic intrusion, volcanism, crustal deformation, and seismic activity. External heat from the **Sun** forces circulation of Earth's atmosphere and hydrosphere, which in turn drives sedimentary processes such as weathering, erosion, transportation, and deposition. These forces, interacting under the influence of **gravity**, shape Earth's surface.

Earth processes interact in complex feedback systems. A change in the rate or directional alignment of one process—for example, an increase in rainfall or the abandonment of a river channel—may start a cascade of compensatory changes throughout a region. Plate-tectonic mountain-building and erosion interact in a negative feedback system that regulates the elevation of continental mountain belts. Elevation interacts with **temperature** and rainfall, the components of **climate**, to regulate rates of erosion. Climate interacts with vegetation to create soils. A balance between **precipitation** and temperature maintains a glacier. These are just a few examples of the dynamic processes that shape a regional landscape, and of the interactions that remold an existing array of landforms over time.

See also Eolian processes; Weathering and weathering series

LANDSLIDE

Landslide is a general term used to describe a variety of geologic processes involving the movement of fine-grained earth, coarse-grained debris, or **rock** down a slope under the influence of **gravity**. This broad definition of downslope movements includes falling, toppling, sliding, spreading, and flowing. Landslides can also occur under **water** (submarine landslides) and trigger tsunamis.

Commonly used landslide classification systems rely on separate terms to describe the type of movement (falling, toppling, sliding, spreading, or flowing) and the type of material involved. Unlithified material is classified as debris if it is predominantly coarse-grained (**sand**, gravel, cobbles, or boulders) and earth if it is primarily fine-grained (silt or **clay**). Thus, a **debris flow** would involve the down slope flow of predominantly coarse-grained material, whereas a debris slide would involve the sliding of the same kind of material along a well-defined slip surface. Additional detail can be included by specifying the rate of movement (which can range from several millimeters per year to tens of meters per second) and the water content of the moving mass (which can range from dry to very wet). Landslides moving at velocities faster than a few meters per minute, particularly when they are large, have the potential to cause catastrophic damage and loss of life. The volume of material involved in a landslide, however, is irrelevant to its classification and can range from a few cubic centimeters to several cubic kilometers. Landslides can also change modes as they move. For example, a debris slide may mobilize into a debris flow as the debris begins to move down slope.

The term mudslide, which is often used in news reports, does not exist within the classification systems used

Landslides are commonly considered to be fast-moving rocks rolling down a hillside. However, any sufficiently large section of land moving downhill, no matter how slowly, is considered to be a landslide. *JLM Visuals. Reproduced by permission.*

by most geologists and engineers. It is an imprecise term that is best avoided.

Landslides occur when the forces tending to keep a **soil** or rock mass in place (resisting forces) are exceeded by those promoting movement (driving forces). Resisting forces most commonly arise due to the shear strength of the material acting over an **area**, such as the slip surface beneath a landslide, or as a consequence of engineered works such as retaining walls. The primary driving force—the component of the weight of the earth, debris, or rock mass acting parallel to the slope—and the force occurring when the potential landslide mass is accelerated during an **earthquake** can also trigger landsliding. Changing the geometry of a slope, for example by excavating some areas and placing fill in others during construction, can alter the balance of resisting and driving forces enough to trigger a landslide.

It is a widely held misconception that landslides occur because slopes are lubricated by water. Water does not act as a lubricant in landslides, but instead decreases the shear strength of the earth, debris, or rock by decreasing the normal force acting across a potential slip surface. It is well known from basic **physics** that the sliding of a block down an inclined plane is resisted by the product of the normal force acting on the plane and a coefficient of friction. Similarly, a decrease in the normal force acting across a potential slip surface will decrease the resistance to sliding. Sources of water leading to landsliding can include infiltrating rain and melted snow, leaking water pipes, and irrigation.

It is difficult to estimate the monetary costs of landslides because they can include both direct and indirect costs. Direct costs include damage to structures and roads, whereas indirect costs include items such as decreased property values, lost productivity, and the expense of driving longer distances when roads are blocked. Difficulties aside, in 1985 the National Research Council estimated that landslides cost between $1 billion and $2 billion per year in the United States

alone. Estimates for other countries range from tens of millions to billions of dollars per year.

See also Catastrophic mass movements; Debris flow; Lahar; Mass movement; Mass wasting; Mud flow; Rockfall; Slump; Talus pile or talus slope

LATERAL MORAINE • *see* MORAINES

LATERITE

Laterite is a type of **soil** produced by intense, prolonged weathering, usually in tropical climates. Abundant **oxygen**, **water**, and warmth leach most water-soluble **minerals** from particles of parent **rock** and leave a nonsoluble residue enriched in hydroxides of **aluminum**, **iron**, magnesium, nickel, and titanium. Laterites high in specific **metals** are often stripmined as ores.

Laterite rich in aluminum are termed aluminous laterite or bauxite. Aluminous laterite is formed from **clay** minerals such as kaolinite ($Al_4[Si_4O_{10}][OH]_8$) by the **leaching** of silica (SiO_2). The residue left by the leaching of silica, aluminum hydroxide ($Al[OH]_3$), is termed gibbsite. Gibbsite's dehydrated forms, diaspore and bohemite (both $HAlO_2$), are also common components of aluminous laterite. Aluminous laterite is the world's primary source of aluminum.

Laterite descended from **basalt** is rich in iron and nickel and is termed ferruginous laterite. Laterite formed from rocks particularly rich in nickel may contain a high percentage of the mineral garnierite ($[NiMg]_6Si_4O_{10}[OH]_8$). A continuum of mixed laterites exists between the aluminous and ferruginous extremes. Nickeliferous laterites are an important commercial source of nickel.

See also Soil and soil horizons; Weathering and weathering series

LATITUDE AND LONGITUDE

The concepts of latitude and longitude create a grid system for the unique expression of any location on Earth's surface.

Latitudes—also known as parallels—mark and measure distance north or south from the equator. Earth's equator (the great circle or middle circumference) is designated 0° latitude. The north and south geographic poles respectively measure 90° north (N) and 90° south (S) from the equator. The angle of latitude is determined as the angle between a transverse plane cutting through Earth's equator and the right angle (90°) of the **polar axis**. The distance between lines of latitude remains constant. One degree of latitude equals 60 nautical miles (approximately 69 statute miles, or 111 km).

Longitudes—also known as meridians—are great circles that run north and south, and converge at the north and south geographic poles. As the designation of 0° longitude is arbitrary, international convention, long held since the days of British sea superiority, establishes the 0° line of longitude—also known as the prime meridian—as the great circle that passes through the Royal National Observatory in Greenwich, England (United Kingdom). The linear distance between lines of longitude vary and is a function of latitude. The liner distance between lines of longitude is maximum at the equator and decreases to zero at the poles. There are 360 degrees of longitude, divided into 180° east and 180° west of the prime meridian. The line of longitude measuring 180° west is, of course, the same line of longitude measuring 180° east of the prime meridian and, except for some geopolitical local variations, serves as the international date line. Because Earth completes one **rotation** in slightly less than 24 hours, the angular velocity of rotation is approximately 15° of longitude per hour. This rate of rotation forms the basis for time zone differentiation.

The distance between lines of longitude varies in length at different latitudes, the distance lessening as latitude increases. At the equator, 69.171 statute miles separate lines of longitude, but by 30 degrees latitude, there are only 59.956 statute miles between lines of longitude. At 60 degrees latitude, only 34.697 statute miles separate longitudinal great circles at that latitude. At the poles, all lines of longitude converge.

Every point on Earth can be expressed with a unique set of latitude and longitude coordinates (i.e., lat/lon coordinates). Latitude—specified as degrees north (N) or south (S)—and longitude—specified as degrees east (E) or west (W)—are expressed in degrees, arcminutes, and arcseconds (e.g., a lat/lon of 39:46:05N, 104:52:22W specifies a point in Denver, Colorado).

Lines of latitude and longitude are usually displayed on maps. Although a variety of maps exist, because maps of Earth are two-dimensional representations of a curved three-dimensional oblate spherical surface, all maps distort lines of latitude and longitude. For example, with equatorial cylindrical projections (e.g., a Mercator **projection**), low-latitude regions carry little distortion. Higher latitudes suffer extreme distortion of distance because of erroneously converging lines of latitude (on the surface of the Earth they are parallel). Despite this disadvantage, Mercator projections remain useful in navigation because there is no distortion of direction and vertical lines drawn upon such a map indicate true north or south.

Many maps include inserts showing polar conic projections to minimize the distortion of latitude near the poles.

Although it is relatively easy to ascertain latitude—especially in the Northern Hemisphere where the altitude of the North Star (Polaris) above the horizon gives a fairly accurate estimate of latitude—the accurate determination of longitude proved to be one of great post-Enlightenment scientific challenges. The inability to accurately estimate longitude often proved fatal or costly in sea navigation. It was not until the eighteenth century, when British clockmaker **John Harrison** developed a chronometer that could accurately keep time onboard ship, that the problem of longitude was solved. An accurate clock allows navigators to compare, for example, the local time of observed high noon to the time at Royal National Observatory in Greenwich, England (Greenwich Mean Time

Roald Amundsen, the first man to successfully reach the South Pole, is shown here using a sextant to determine his position. © *Hulton-Deutsch Collection/Corbis. Reproduced by permission.*

or GMT). Knowing that Earth rotates at approximately 15° per hour, the time difference between local noon and GMT local noon is directly related to the degrees of longitude between the prime meridian and the observer's location.

See also Analemma; Astrolabe; Astronomy; Cartography; Celestial sphere: The apparent movements of the Sun, Moon, planets, and stars; Earth (planet); Geographic and magnetic poles; Geography; GPS; Projection; Polar axis and tilt; Time zones

LAVA

Lava is molten **rock** that has been extruded onto Earth's surface. Before it reaches the surface, lava is called **magma**. Magma contains **crystals**, unmelted rock, and dissolved gasses, but it is primarily a liquid. **Oxygen**, silica, **aluminum**, **iron**, magnesium, calcium, sodium, titanium, and manganese are the primary elements found in magma, but other trace elements may be present in small amounts.

Viscosity of lava, or its resistance to flow, is determined by its **temperature** and chemical composition. In general, hotter lava that is low in silica flows much more readily than

Lava from Mt. Kilauea. This kind of lava has little silica in it, and is not explosive. When it dries, it will be referred to as pahoehoe lava. *JLM Visuals. Reproduced by permission.*

cooler, high-silica lava. The composition and viscosity of magma determines both its eruptive style and the rock type that will be formed when it cools. The three main types of lava, named for the rock types that they form, are basaltic (for **basalt**), rhyolitic (for **rhyolite**), and andesitic (for **andesite**).

Basaltic flows are the hottest, erupting at temperatures of 1,832–2,192°F (1,000–1,200°C). Basaltic lavas are high in calcium, magnesium, and iron, and low in silica, sodium, and potassium. Based on its composition, basalt is known as a **mafic** rock. The high temperature and low silica content allow basaltic lavas to flow readily and travel far. The Hawaiian Islands were built up from the seafloor from successive basaltic lava flows, forming large shield volcanoes. When basaltic lava erupts onto relatively flat land, flows known as flood basalts may spread out, with successive flows being piled on top of one another. The Columbia Plateau in Oregon and Washington and the Deccan Traps in northwestern India were formed from such eruptions. Viscosity differences within basaltic lavas result in two distinct flow types, characterized by their surface forms. Pahoehoe (pah-hoy-hoy), formed from less viscous lava, is named for the Hawaiian word for ropy. When solidified, the surface of a pahoehoe flow has a smooth texture that looks like coiled rope, which forms as the outer

layer of the lava cools, then is dragged and folded as the flow continues to move beneath the surface. More viscous basalt flows give rise to aa (ah-ah) lava, characterized by blocky clumps. Aa flows move more slowly than do pahoehoe flows, allowing a thick surface layer to cool as the flow creeps forward. As the flow continues to move, the surface layer is broken into jagged pieces. Pahoehoe is common near the source of a basaltic flow, where the lava is hottest, and aa is normally found farther from the source, where the lava has cooled off significantly. Another unique feature of basaltic lava is the formation of pillow lavas. Mounds of ellipse shaped pillows form when basaltic lava is erupted under **water** or **ice**. As lava is extruded, the water (or ice) quickly chills the outer layer. Molten lava beneath the chilled surface eventually breaks through the skin and the process is repeated, resulting in a pile of lava pillows. Pillow lava deposits found on land indicate that the region was once under water.

Rhyolitic lavas are high in potassium, sodium, and silica, and low in calcium, magnesium, and iron. Rhyolite is a classified as a **felsic** rock. Its felsic composition, in addition to its low eruptive temperature (1,472–1,832°F, or 800–1,000°C), results in highly viscous lava that can just barely flow. Such lavas usually produce a volcanic dome that eventually is destroyed in a massive explosion as the viscous lava tries to escape. Lassen Peak in northeastern California, which last erupted between 1914 and 1917, is an example of a lava dome.

Andesitic lava has a composition that falls in between that of basaltic lava and that of rhyolitic lava. Andesite is hence classified as an intermediate rock. Andesitic lava flows more readily than rhyolitic lava, but not as easily as basaltic lava. Eruptions are characterized by a mixture of explosive activity and lava flows. Such eruptions form composite volcanoes, built up of alternating lava flows and pyroclastic deposits (deposits of debris ejected from the **volcano**). Composite volcanoes such as Mt. Fuji, in Japan, and Mt. Rainier, in Washington, are cone shaped.

See also Volcanic eruptions

LEACHING

Leaching usually refers to the movement of dissolved substances with **water** percolating through **soil**. Sometimes, leaching may also refer to the movement of soluble chemicals out of biological tissues, as when rainfall causes potassium and other ions to be lost by foliage.

Leaching occurs naturally in all soils, as long as the rate of water input through **precipitation** is greater than water losses by evapotranspiration. In such cases, water must leave the site by downward movement, ultimately being deposited to deep **groundwater**, or emerging through **springs** to flow into surface waters such as streams, **rivers**, and **lakes**. As the subterranean water moves in response to gradients of gravitational potential, it carries dissolved substances of many kinds.

Leaching is a highly influential soil-forming process. In places where the **climate** is relatively cool and wet, and the vegetation is dominated by conifers and heaths, the soil-forming process known as podsolization is important. In large part, podsolization occurs through the dissolving of **iron**, **aluminum**, calcium, organic matter, and other chemicals from surface soils and the downward leaching of these substances to lower soil depths, where they are deposited. Some solubilized materials may also be altogether lost from the soil, ending up in deep groundwater or in surface water. A different soil-forming process known as laterization occurs under the warm and humid climatic conditions of many tropical rain **forests**, where aluminum and iron remain in place in the surface soil while silicate is dissolved and leached downward.

The ability of water to solubilize particular substances is influenced to a substantial degree by the chemical nature of the solution. For example, highly acidic solutions have a relatively great ability to dissolve many compounds, especially those of **metals**. Aluminum (Al), for instance, is an abundant metallic constituent of soils, typically present in concentrations of 7–10%, but occurring as aluminum compounds that are highly insoluble, so they cannot leach with percolating water. However, under highly acidic conditions some of the aluminum is solubilized as positively charged ions (or cations). These soluble ions of aluminum are highly toxic to terrestrial plants and animals, and if they are leached to surface waters in large quantities they can also cause biological damage there. Aluminum ions are also solubilized from soils by highly alkaline solutions. A large salt concentration in soil, characterized by an abundance of dissolved ions, causes some ions to become more soluble through an osmotic extraction, also predisposing them more readily to leaching.

Soils can become acidified by various human activities, including emissions of air pollutants that cause acidic precipitation, certain types of agricultural fertilization, harvesting of biomass, and the mining of **coal** and sulfide **minerals**. Acidification by all of these activities causes toxicity of soil and surface waters through the solubilization of aluminum and other metals, while also degrading the fertility and acid-neutralization capacity of soil by causing the leaching of basic cations, especially calcium, magnesium, and potassium.

Another environmental problem associated with leaching concerns terrestrial ecosystems that are losing large quantities of dissolved nitrogen, as highly soluble nitrate. Soils have little capability to bind nitrate, so this anion leaches easily whenever it is present in soil water in a large concentration. This condition often occurs when disturbance, fertilization, or atmospheric depositions of nitrate and/or ammonium result in an availability of nitrate that is greater than the biological demand by plants and microorganisms, so this chemical can leach at relatively high rates. Terrestrial ecosystems of this character are said to be "nitrogen-saturated." Some negative environmental effects are potentially associated with severe nitrogen saturation, including an increased acidification and toxicity of soil and water through leaching of aluminum and basic cations (these positively charged ions move in companion with the negatively charged nitrate), nutrient loading to aquatic systems, potentially contributing to increased productivity there, and possibly predisposing trees to suffer decline and die back. If the nitrogen saturation is not excessive, how-

ever, the growth of trees and other vegetation may be improved by the relatively fertile conditions.

See also Caliche; Soil and soil horizons

LEAD

Lead (Pb) is a relatively common element in Earth crustal materials. Lead is a heavy, soft metal that is a solid at normal atmospheric and crustal pressures. Lead is reactive with **oxygen** and tarnishes and dulls when in contact with oxygen. Lead is not a good conductor of **electricity**, heat, sound, or other pressure vibrations.

Lead is found in Earth's **crust** at an abundance of about 13–20 parts per million. It rarely occurs as a free element, and is found most commonly as a compound in the form of galena (lead sulfide; PbS), anglesite (lead sulfate; $PbSO_4$), cerussite (lead carbonate; $PbCO_3$), and mimetite. Geochemically, lead is a moderately active metal that dissolves very slowly in **water.**

Lead is both ductile and malleable. These properties allow lead to be easily bent, cut, pulled, or otherwise worked to produce specific shapes. The **melting** point of lead is 621.3°F (327.4°C), its boiling point is about 3,180°F (1,749°C), and its density is 11.34 grams per cubic centimeter.

The largest producers of lead in the world are **Australia**, China, the United States, Peru, Canada, Mexico, and Sweden. In the United States, more than 90% of all the lead produced comes from a single state, Missouri. Lead is extracted from its ores by first converting the ore to lead oxide and then heating the oxide with charcoal (pure **carbon**). The lead produced by this process is usually not very pure and can be further refined electrolytically.

Over the past decades, evidence has mounted indicating lead as a significant environmental hazard. Low levels of lead in products (e.g., paint) can accumulate in tissues over time. As a result, many manufactured items (e.g., batteries) now have or seek lead substitutes or provide for contained disposal.

See also Chemical elements; Minerals

LEHMANN, INGE (1888-1993)
Danish geophysicist

Trained as a mathematician and an actuary, Danish geophysicist Inge Lehmann used painstaking analyses, measurements and observations of shock waves generated by earthquakes to propose in 1936 that the earth had a solid inner core. Throughout her long career, which extended far beyond her official retirement in 1953, Lehmann conducted research in **Europe** and **North America** and was active in international scientific organizations including serving as the first president and a founder of the European Seismological Federation.

Lehmann was one of two daughters born to Alfred Georg Ludvig Lehmann, a University of Copenhagen professor of psychology, and Ida Sophie Torsleff. As a child, she attended and graduated from the first coeducational school in

Denmark, an institution founded and run by Hanna Adler, the aunt of future Nobel Prize winning physicist **Niels Bohr**. She began her university education by studying mathematics at the University of Copenhagen from 1907 to 1910. She continued her mathematical studies the following year at Cambridge University in England before returning to Denmark, where she worked as an actuary from 1912 to 1918. She also continued her formal education. In 1920, Lehmann earned her masters degree in mathematics from the University of Copenhagen and later studied mathematics at the University of Hamburg. In 1925, Lehmann began her career in **seismology** as a member of the Royal Danish Geodetic Institute and helped install the first seismographs at her Copenhagen office. "I was thrilled by the idea that these instruments could help us to explore the interior of the earth, and I began to read about it," she was quoted in a 1982 article published in the *Journal of Geological Education*. Lehmann later helped establish **seismograph** stations in Denmark and Greenland.

After further study with seismologists in France, Germany, Belgium, and the Netherlands, and after earning a M.S. degree in geodesy from the University of Copenhagen in 1928, Lehmann was named chief of the Royal Danish Geodetic Institute. In that position, held until her retirement in 1953, Lehmann was Denmark's only seismologist for more than two decades. She was responsible for supervising the Denmark's seismology program, overseeing the operation of the seismograph stations in Denmark and Greenland, and preparing the institute's bulletins.

Despite this heavy workload, Lehmann still found time to explore scientific research. In 1936, she published her most significant finding, the discovery of the earth's inner core, under the simple title of "P." The letter P stood for three types of waves generated by Pacific earthquakes that Lehmann had been carefully observing through the planet for ten years. By studying the shock waves generated by earthquakes, recorded on seismographs as travel-time curves, she theorized that the earth has a smaller solid inner core. Within a few years, work by other scientists, including Harold Jeffreys and Beno Gutenberg, substantiated her findings.

Lehmann continued her research well after her retirement in 1953, exploring the nature of the planet's interior in Denmark, in Canada at the Dominion Observatory in Ottawa and in the United States at the University of California at Berkeley, the California Institute of Technology, and the Lamont Doherty Earth Observatory at Columbia University. She was a named a fellow of both the Royal Society of London and Edinburgh and was named to the Royal Danish Academy of Science and Letters and the Deutsche Geophysikalische Gesellschaft. In 1971, she was awarded the William Bowie Medal of the American Geophysical Union in recognition of her "outstanding contributions to fundamental geophysics and unselfish cooperation in research." She was also awarded honorary doctorates by the University of Copenhagen and Columbia University.

Lehmann remained single throughout her long and productive life. Her interests were not restricted to science. She was concerned with the poor in her native Denmark and the plight of European refugees. Travel in conjunction with her

work also afforded her frequent opportunities to pursue two of her hobbies—visiting art galleries throughout Europe and the United States, and the outdoors. Lehmann enjoyed hiking, mountain climbing, and skiing. She died at the age of 105.

LIBBY, WILLARD F. (1908-1980)
American chemist

Chemist Willard F. Libby developed the radiocarbon dating technique used to determine the age of organic materials. With applications in numerous branches of science, including archaeology, **geology**, and geophysics, radiocarbon dating has been used to ascertain the ages of both ancient artifacts and geological events, such as the end of the **Ice** Age. In 1960, Libby received the Nobel Prize for his radiocarbon dating work. During World War II, Libby worked on the Manhattan Project to develop an atomic bomb and was a member of the Atomic Energy Commission for several years in the 1950s. An outspoken scientist during the Cold War between the United States and the former Soviet Union, Libby advocated that every home have a fallout shelter in case of nuclear war. Libby, however, was a strong proponent of the progress of science, which he believed resulted in more benefits than detriments for the human race.

Willard Frank Libby was born to Ora Edward and Eva May Libby on a farm in Grand Valley, Colorado. In 1913, the family, which included Libby and his two brothers and two sisters, moved to an apple ranch north of San Francisco, California, near Sebastopol, where Libby received his grammar school education. A large boy who would eventually grow to be 6 feet 3 inches tall, Libby developed his legendary stamina while working on the farm. He played tackle for his high school football team and was called "Wild Bill," a nickname used by some throughout Libby's life. After graduating from high school in 1926, Libby enrolled at the University of California, Berkeley. He made money for college by building apple boxes, earning one cent for each box and sometimes $100 in a week. "I was the fastest box maker in Sonoma County," he told Theodore Berland, who interviewed Libby for his book *The Scientific Life*.

Although Libby was interested in English literature and history, he felt obligated to seek a more lucrative career and entered college to become a mining engineer. By his junior year, however, Libby became interested in **chemistry**, spurred on by the discussions of his boarding house roommates, who were graduate students in chemistry. Libby took on a heavy course load, focusing on mathematics, **physics**, and chemistry. After receiving his B.S. in chemistry in 1931, he entered graduate school at Berkeley and studied under the American physical chemist Gilbert Newton Lewis and Wendell Latimer, who were pioneering the physical chemistry field.

Libby received his Ph.D. in 1933 and was appointed an instructor in chemistry at Berkeley. After the Japanese bombed Pearl Harbor in 1941, Libby, who was on a year sabbatical as a Guggenheim Fellow at Princeton University, joined a group of scientists in Chicago, Illinois, to work on the Manhattan

Project, a government-sponsored effort to develop an atomic bomb. During this time, he worked with American chemist and physicist **Harold Urey** at Columbia University on gaseous diffusion techniques for the separation of uranium isotopes (isotopes are different forms of the same element having the same **atomic number** but different atomic weights). After the war, he accepted an appointment as a professor of chemistry at the University of Chicago and began to conduct research at the Institute of Nuclear Studies.

In 1939, scientists at New York University had sent radiation counters attached to balloons into the earth's upper atmosphere and discovered that neutron showers were created by cosmic rays hitting atoms. Further evidence indicated that these neutrons were absorbed by nitrogen, which then decayed into radioactive carbon–14. In addition, two of Libby's former students, Samuel Ruben and Martin Kamen, made radioactive carbon–14 in the laboratory for the first time. They used a cyclotron (a circular device that accelerates charged particles by means of an alternating electric field in a constant **magnetic field**) to bombard normal carbon–12 with neutrons, causing it to decay into carbon–14.

Intrigued by these discoveries, Libby hypothesized that radioactive carbon–14 in the atmosphere was oxidized to **carbon dioxide**. He further theorized that, since plants absorb **carbon** dioxide through photosynthesis, all plants should contain minute, measurable amounts of carbon–14. Finally, since all living organisms digest plant life (either directly or indirectly), all animals should also contain measurable amounts of carbon–14. In effect, all plants, animals, or carbon-containing products of life should be slightly radioactive.

Working with Aristide von Grosse, who had built a complicated device that separated different carbons by weight, and graduate student Ernest C. Anderson, Libby was successful in isolating radiocarbon in nature, specifically in methane produced by the decomposition of organic matter. Working on the assumption that carbon–14 was created at a constant rate and remained in a molecule until an organism's death, Libby thought that he should be able to determine how much time had elapsed since the organism's death by measuring the **half-life** of the remaining radiocarbon isotopes. (Half-life is a measurement of how long it takes a substance to lose half its radioactivity.) In the case of radiocarbon, Libby's former student Kamen had determined that carbon–14's half-life was 5,370 years. So, in approximately 5,000 years, half of the radiocarbon is gone; in another 5,000 years, half of the remaining radiocarbon decays, and so on. Using this mathematical calculation, Libby proposed that he could determine the age of organisms that had died as many as 30,000 years ago.

Because a diffusion column such as von Grosse's was extremely expensive to operate, Libby and Anderson decided to use a relatively inexpensive Geiger counter to build a device that was extremely sensitive to the radiation of a chosen sample. First, they eliminated 99% of the background radiation that occurs naturally in the environment with 8-inch-thick (20 cm) **iron** walls to shield the counter. They then used a unique chemical process to burn the sample they were studying into pure carbon lampblack, which was then placed on the inner walls of a Geiger counter's sensing tube.

Libby first tested his device on tree samples, since their ages could be determined by counting their rings. Next, Libby gathered tree and plant specimens from around the world and discovered no significant differences in normal age-related radiocarbon distribution. When Libby first attempted to date historical artifacts, however, he found his device was several hundred years off. He soon realized that he needed to use at least several ounces of a material for accurate dating. From the Chicago Museum of Natural History, Libby and Anderson obtained a sample of a wooden funerary boat recovered from the tomb of the Egyptian King Sesostris III. The boat's age was 3,750 years; Libby's counter estimated it to be 3,261 years, only a 3.5 % difference. Libby spent the next several years refining his technique and testing it on historically significant, and sometimes unusual objects, such as prehistoric sloth dung from Chile, the parchment wrappings of the Dead Sea Scrolls, and charcoal from a campsite fire at Stonehenge, England. Libby saw his new dating technique as a way of combining the physical and historical sciences. For example, using wood samples from **forests** once buried by **glaciers**, Libby determined that the Ice Age had ended 10,000 to 11,000 years ago, 15,000 years later than geologists had previously believed. Moving on to man-made artifacts from **North America** and **Europe** (such as a primitive sandal from Oregon and charcoal specimens from various campsites), Libby dispelled the notion of an Old and New World, proving that the oldest dated human settlements around the world began in approximately the same era. For many years after Libby's discovery of radiocarbon dating, the journal *Science* published the results of dating studies by Libby and other scientists from around the world. In 1960, Libby was awarded the Nobel Prize in chemistry for his work in developing radiocarbon dating. In his acceptance speech, as quoted in *Nobel Prize Winners*, Libby noted that radiocarbon dating "may indeed help roll back the pages of history and reveal to mankind something more about his ancestors, and in this way, perhaps about his future." Further progress in radiocarbon dating techniques extended its range to approximately 70,000 years.

In related work, Libby had shown in 1946 that cosmic rays produced tritium, or hydrogen–3, which is also weakly radioactive and has a half-life of 12 years. This radioactive form of hydrogen combines with **oxygen** to produce radioactive **water**. As a result, when the United States tested the Castle hydrogen bomb in 1954, Libby used the doubled amount of tritium in the atmosphere to date various sources of water, deduce the water-circulation patterns in the United States, and determine the mixing of oceanic waters. He also used the method to date the ages of wine, since grapes absorb rain water.

In 1954, U.S. President Dwight D. Eisenhower appointed Libby to the Atomic Energy Commission (AEC). Although he continued to teach graduate students at Chicago, Libby drastically reduced his research efforts and plunged vigorously into his new duties. Previously a member of the commission's General Advisory Committee, which developed commission policy, Libby was already acquainted with the inner workings of the commission. He soon found himself embroiled in the nuclear fallout problem. Upon a recommendation by the Rand Corporation in 1953, Libby formed and directed Project Sunshine and became the first person to measure nuclear fallout in everything from dust, **soil**, and rain to human bone.

As a member of the AEC, Libby testified before the U.S. Congress and wrote articles about nuclear fallout. He noted that all humans are exposed to a certain amount of natural radiation in sources such as drinking water. He went on to point out that the combination of the body's natural **radioactivity**, cosmic radiation, and the natural radioactivity of the earth's surface was more hazardous than fallout resulting from nuclear testing. Libby assumed, and most scientists of the day concurred, that the effects of nuclear fallout from careful testing on human genetics were minimal.

See also Chemical bonds and physical properties; Chemical elements; Cosmic microwave background radiation; Dating methods; Nuclear winter

LIFT AND LIFTING BODIES • *see*

AERODYNAMICS

LIGHTNING

Lightning is a large electrical discharge produced by well-developed thunderstorms, a huge spark followed by a rumbling noise of **thunder**. Lightning can happen within the cloud (intra-cloud), between two **clouds** (inter-cloud), or from the cloud to the ground. A lightning bolt can heat the air as much as five times hotter than the surface **temperature** of the **Sun**, or about 54,000°F (30,000°C). This heated air causes expansion in the air as an explosion, starting a shock wave that turns into a sound wave upon reaching the human ear. Thunder travels in all directions (radially) from the lightning at the speed of sound, approximately 738 mph (1,188 kph) at sea level. Because it takes the sound about five seconds to travel each mile (about three seconds for one kilometer), the time between the lightning and the thunder can give a rough estimate of how far an observer is from a thunderstorm.

The quick flash that can be seen as lightning occurs as a complex series of events. In order to have lightning, separate regions of electrical charges must be present in a cumulonimbus cloud. There are several hypotheses as to how this occurs. One mechanism may involve falling **ice** particles within the cloud that transfer ions. This results in a positively charged upper part and a negatively charged middle part in the cloud. The bottom of the cloud is also mostly negatively charged, causing part of the ground underneath to become positively charged. In the insulating dry air an electrical field builds up, and when it reaches a threshold potential, the air is no longer insulating and, as a current flows, lightning occurs.

Cloud-to-ground lightning (arguably the best understood among the different types of lightning) starts inside the cloud when a critical value of the localized electric field is reached along a path, so a surge of electrons will move to the cloud base, then gradually down to the ground. A short (165 ft,

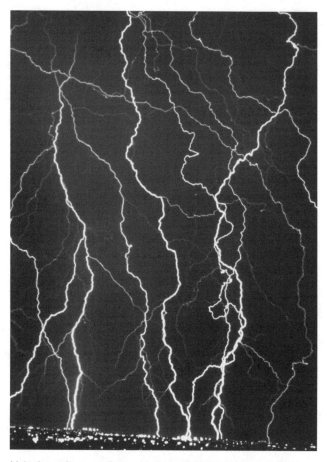

Lightning strikes over Tuscon, Arizona. *Keith Kent/Peter Arnold, Inc. Reproduced by permission.*

or 50 m) and narrow (4 in, or 10 cm) conducting channel is created by ionized air molecules, which are produced by the electron flow out of the cloud. These surges of electrons move downward in a series of steps for about 165–328 ft (50–100 m), then they stop for about 50-millionths of a second, and continue for another 165 ft, creating a stepped leader form of transit. Near the ground, a current of positive charge goes up from the ground to meet the stepped leader, and when they meet, many electrons flow into the ground, and a bright return stroke moves up, following the path of the stepped leader up to the cloud, releasing heat, thunder, and charges. The subsequent leader is called the dart leader and, for subsequent flashes, the same processes reoccur in a similar cycle. Usually, a lightning flash has approximately three or four leaders, each of them accompanied by a return stroke.

To distinguish the several different appearances of lightning, the forms are assigned special names. Heat lightning (also termed clear-air lightning) occurs when lightning can be seen but the following thunder cannot be heard. Forked lightning occurs when a dart leader moving toward the ground diverges from the original path of the stepped leader, so that the lightning seems to be crooked or forked. When the **wind** moves the ionized channel between the return strokes, the lightning looks like a ribbon hanging from a cloud, so it is called a ribbon lightning. Bead lightning looks like a series of beads on a string, and it occurs as the lightning channel disintegrates. Sheet lightning appears as a white sheet, and it occurs either when clouds obscure the lightning, or when the lightning flash happens within a cloud. St. Elmo's Fire, named after the patron of sailors, is a corona discharge, a nonstop supply of sparks in the air, which happens when a positive current moves up on pointed objects. Ball lightning often appears as a luminous, floating sphere in the air. The various mechanisms underlying the varying forms of lightning remain a subject of intensive meteorological research.

During a thunderstorm, usually the tallest object in the **area** is struck because this provides the most rapid form of current transit to lowest energy state. At any moment, there are about 2,000 thunderstorms worldwide, generating about 100 lightning flashes per second. A lightning stroke can deliver a current as great as 100,000 amperes, which can cause electrocution in humans and animals. About 100 people die in a year in the United States alone from lightning, and lightning causes billions of dollars in damage each year.

See also Atmospheric circulation; Atmospheric composition and structure; Atoms; Clouds and cloud types; Meteorology

LIMESTONE

Limestone is a sedimentary **rock** composed almost entirely of the mineral calcite (calcium carbonate, $CaCO_3$). The precursor calcium-carbonate sediment that existed prior to **lithification** of limestone can be of several types. These sediment types include carbonate mud, carbonate fossil fragments, carbonate pellets and rip-up clasts, and ooids. Carbonate mud is made of microcrystalline calcite **crystals** (crystals of a few microns in size) that form directly from seawater and from the disintegration of some calcareous marine algae. Carbonate fossil fragments include all shelly organic debris originally composed of $CaCO_3$ (in the form of calcite or a denser phase, aragonite). Carbonate pellets and rip-up clasts are small lumps of carbonate mud (a few millimeters in size) that have been consolidated either by being eaten and excreted (pellets), or by settling and then being ripped-up by wave energy (rip-up clasts). Ooids are sand-sized, concentrically layered grains that form by inorganic **precipitation** of calcium carbonate during agitation of seawater (usually by rolling on a shallow shoal **area** of the ocean).

The sediment precursor to limestone forms in the shallow marine realm or, less commonly, within carbonate-rich **lakes**. A special kind of marine limestone, composed entirely of the remains of marine micro-plankton, is called chalk. There is also a special kind of limestone, travertine, which forms from fresh-water deposition of dissolved carbonate within **cave** and cavern systems.

The original sediment determines limestone texture (i.e., the size and nature of grains in the rock). Micritic limestone is made of microcrystalline calcite like the original carbonate mud. Fossiliferous limestone is made of a large proportion of fossil fragments and the balance of the rock is either carbonate

These hundred-ton haul trucks look like toys in this limestone quarry near St. Genevieve, Missouri. *AP/Wide World. Reproduced by permission.*

mud, carbonate cement (sparry calcite crystals between grains), or a mixture of both. Pelletal limestone and limestone with rip-up clasts is much like fossiliferous limestone regarding the balance of the rock (mud, spar, or both). Ooid limestone has a high proportion of ooid grains in it. Limestone that is made of sediment formed within or near organic reefs is sometimes referred to as reef rock or boundstone.

Limestones contain extensive fossil records spanning much **geologic time**, including microfossils, megafossils, and trace **fossils** (or ichnofossils). Limestones are particularly common **sedimentary rocks**, representing times in Earth history when globally warm conditions prevailed along with particularly high sea levels. High sea levels during these times promoted development of extensive, shallow **seas** (i.e., epicontinental seas) across much of Earth's continental **space**. Epicontinental seas with extensive limestone deposits characterize the geological time intervals such as Late Cambrian-Ordovician, Mississippian, and Cretaceous.

Limestones are particularly susceptible to dissolution by acidic **groundwater**, and for this reason, extensive subterranean cave and cavern systems are known from within many limestone formations. The calcite in limestones is also susceptible to replacement by **dolomite** where the limestone meets magnesium-rich, briny groundwaters. Where this replacement is extensive, limestone formations are changed to dolostones by this process. Considerable **porosity** (approximately seven percent by volume) is created within the former limestone (now dolostone), and this pore space can host important hydrocarbon and mineral deposits. Limestone is also commonly replaced by **chert** in places where silica has been transported into the limestone by groundwaters.

Limestone is economically important as a source of quicklime (CaO), which can be produced by heating calcite to drive off **carbon dioxide** (CO_2). In addition to agricultural uses, limestone is used in cement making and as a flux in the smelting of **iron**. Limestone can be used also for road construction and as ornamental building stone.

LINEAL DUNES · *see* DUNES

LINEATION

A lineation is any linear feature or element in a **rock**, and can occur as the product of tectonic, mineralogical, sedimentary, or geomorphic processes. Lineations are the one-dimensional counterparts of foliations, and both are part of the fabric (geometric organization of features) of a rock. Lineations and foliations are said to possess preferred orientations, meaning that the spatial orientation of the features comprising the lineation or foliation is similar throughout the rock mass.

The spatial orientation of a lineation is described by two angles known as bearing and plunge. The plunge angle is the inclination of the lineation relative to an imaginary horizontal plane (ranging from 0 to 90 degrees), whereas the bearing angle is the compass direction of the lineation in the direction of the plunge (ranging from 0 to 360 degrees).

Structural lineations are those that are formed by tectonic activity such as folding, faulting, or **metamorphism**. Structural lineations can be either discrete or constructed. Discrete lineations are formed by the deformation and alignment of objects such as **fossils** or initially spherical pebbles. When a rock containing discrete objects (such as fossils or nearly spherical pebbles) is subjected to stress, the objects can be deformed into ellipsoids that share a preferred orientation throughout the rock. Constructed lineations are those that are formed during deformation and therefore do not involve pre-existing objects. Constructed lineations include those formed by the intersections of two planes (e.g., the intersections some combination of foliations, fractures, or **bedding** planes) and slickenlines (also referred to as slickensides) along fault surfaces.

Mineral lineations are formed by the preferred orientation of either individual mineral grains or clusters of mineral grains. Mineral lineations can be formed by the nearly parallel alignment of mineral grains or clusters that have a needle like habit (e.g., amphiboles), by elongation of mineral grains or clusters during deformation, or by the preferentially oriented growth of **minerals** in response to the ambient state of stress during metamorphism. A mineral lineation formed by elongation of grains is similar to a discrete structural lineation.

Sedimentary lineations include pebbles aligned in the direction of stream flow and the crests of ripple marks. They are typically, although not exclusively, found on bedding planes in **sedimentary rocks**.

The slip surfaces of landslides can contain slickenlines similar to those found on fault surfaces, although they are generally considered to be of geomorphic rather than tectonic origin. Likewise, preferentially aligned clasts in sheared glacial **till** are similar to discrete structural lineations in tectonically deformed rocks, but most geologists do not consider them to be tectonic features.

See also Bedforms (ripples and dunes); Metamorphic rock; Shear zones

LINNÉ, CARL VON (1707-1778)
Swedish physician and botanist

Carl von Linné (Linnaeus) decisively broke through centuries of confusion over how to revise the classification system that had been in place since antiquity. With few parallels in the history of science, Linnaeus's contribution to botany will remain intact perhaps as long as the first classification system.

Linnaeus was born in Stenbrohult, Sweden. His father, a clergyman, maintained a small botanical garden on the parsonage grounds, where Linnaeus earned the nickname "little botanist." In 1716, Linnaeus entered a Latin school and began to formalize his interest in botany and the natural sciences. In 1727, he transferred to the University of Lund to study medicine, but he also undertook extensive botanical excursions. One year later, he went on to the University of Uppsala, which was considered a better school for medicine, but Linnaeus was disappointed to find that its facilities were no better than those at the University of Lund. Nevertheless, Uppsala did have something that made up for the shortcoming, a botanical garden containing rare foreign plants.

As his academic ideas started to mature along with his research, Linnaeus constructed a new theory of plant sexuality. In 1735, he published his *System Naturae*, and two years later, *Genera Planetarum*. He also moved briefly to Holland, where he received his M.D. In 1739, Linnaeus began to practice medicine, and two years later, he became the chair of botany, dietetics, and materia medica at the University of Uppsala. For the rest of his life, Linnaeus remained in this position, while his fame as a premier botanist spread throughout the world because of his influence in revising the 2,000-year-old system of classification.

The philosopher Aristotle had devised the first classification system over 2,000 years earlier, when he established the basic principles of dividing and subdividing plants and animals. At that time, only about a thousand species were known. Therefore, he grouped them into simple categories of animals with backbones and animals without backbones. Plants were divided into different categories that dealt more with size and appearance. By the sixteenth century, however, the system was proving to be less and less adequate as the body of knowledge of plants and animals grew. Modification came slowly, often marked with debate and controversy, succeeding only in revealing the complexity of the process. In 1753, Linnaeus published his *Species Planetarium*, in which he replaced the antiquated Aristotelian system with the principles of classification used today.

In creating his system, Linnaeus's primary consideration was the number of observable characteristics of the organism, specifically its anatomy, structures, and details of reproduction. Based on his observations, Linnaeus created a hierarchical system in which living things were grouped according to their similarities, with each succeeding level possessing a larger number of shared traits. He named these levels class, order, genus, and species.

Linnaeus also popularized binomial nomenclature, giving each living thing a Latin name consisting of its genus and species, which distinguished it from all other organisms. For example, the cougar received the scientific name *Felis concolor*, while the lion became *Panthera leo*. This system allowed scientists to communicate worldwide about organisms without having to understand different languages. Also, each type of organism can be fitted into the scheme in a logical and orderly manner, allowing for infinite expansion. The various hierarchical levels in the system provide as well a conceptual framework for understanding the relationships among different organisms or groups of organisms.

Linnaeus's desire to classify all living things often bordered on the compulsive; he believed his work to be divinely inspired and considered those who did not follow his system to be "heretics." However, he was also a skilled and caring instructor who nurtured the interests of his many students, often sending them abroad to the Middle East, China, and the Pacific Islands for new specimens. In 1761, Linnaeus was given the noble title von Linné, and while the king of Spain offered him generous compensation to settle in his country, Linnaeus remained in Sweden at Uppsala until his death after a stroke in 1778.

Today an international commission of scientists maintains the Linnaeus classification system and adheres to the rules for adopting scientific names when newly discovered species or subspecies need to be classified. Although the system depends on the judgments and opinions made by biologists, its concept and general organization are accepted by scientists throughout the world.

LIPPERSHEY, HANS (1570-1619)

German-born Dutch lens maker

Although there is some debate, most historians argue that Hans Lippershey was the first inventor of the **telescope**. Born in Wesel (Germany), Lippershey migrated to Zeeland in the Netherlands. Little is known about Lippershey's personal life except that he married in 1594 and officially became a citizen of his adopted country in 1602.

According to fragmented documentary evidence (including third party diary entries), Lippershey, an eyeglass maker, developed the idea for the telescope after watching two children in his eyeglass shop play with his lenses. The legend holds that children grasping two lenses and, looking through both simultaneously, noticed that a **weather** vane on a nearby church steeple seemed to appear larger and clearer than visible by normal sight. Lippershey realized the potential of this accidental discovery and proceeded to make a telescope by attaching lenses at the two ends of a tube.

Lippershey applied for a patent in 1608, but the device could not be kept a secret and the patent was eventually denied. Others came forth claiming to have invented the new device, including Jacob Metius and Sacharias Janssen, both of the Netherlands. Regardless, Lippershey's application for a patent remains the earliest record of a telescope, which he called a *kijker*, or "looker." Lippershey also benefited financially from his invention through the Dutch government, which paid him to construct several telescopes for military use.

See also Hubble Space Telescope (HST)

LITHIFICATION

When sediments are first deposited, they are unconsolidated and are not considered a **rock**. Lithification is the process of converting unconsolidated sediments into sedimentary rock. Lithification involves primarily the processes of compaction and cementation. Recrystallization is also an important process for some sediments.

Compaction is the rearrangement of sedimentary particles to reduce pore **space** and squeeze out pore **water**. Unlithified sediments generally contain some excess space between grains. This is especially true in the case of very fine-grained sediments such as **clay** and mud. Coarser grained sediments such as **sand** and gravels are heavy enough to settle with a minimum of pore space. Over time, as sediments accumulate in basins, the thickness and weight of the overlying sediments increases. The pressure on the buried sediments causes all the grains to compress together as tightly as possible. Excess interstitial water is also forced out and the sediments are now compacted.

The next step is cementation. Only after cementation has occurred can the sediments be considered a rock. Although compaction has reduced the pore space within sediments, some space remains. In addition, even though the sediments are compacted to a high degree and have been de-watered, the great pressures and higher temperatures in a sedimentary basin can force hot circulating waters, or basinal brines, to permeate through the sediments. These fluids have the ability to carry dissolved **minerals** and deposit them within the available pore space of sediments, binding them into rock.

The cement may be derived either from outside sediments or from within the sediments themselves. Externally sourced cementing material is dissolved from some other rock formation or sediments that the fluids permeated prior to entering the sediments to be cemented. Alternatively, a permeating fluid may dissolve the cementing mineral from within the sediments themselves and then redeposit them before exiting.

The most common cementing materials are calcium carbonate, silica, and **iron** oxides. Calcium carbonate cement is usually in the form of calcite. Silica cement is dominantly **quartz** but also can be **chert** or chalcedony. Iron oxide cements occur in the form of hematite or limonite.

Some sediments may become lithified by the recrystallization of mineral constituents rather than by cementation. This is an important process in the formation of **limestone** and some shales. As the name implies, it involves the *in situ* recrystallization of the sedimentary grains. In this process, minerals will recrystallize as a response to a change in their chemical environment, such as a rise in the **pH**. This is because some minerals are more stable than others are in a given set of conditions. Entire grains may reform, or just the rims of minerals. As mineral grains recrystallize, they grow together, forming new interlocking grain boundaries. These interlocking **crystals** bind the sediments into rock.

See also Sedimentary rocks; Sedimentation

LITHOPHILE · *see* CHALCOPHILES, LITHOPHILES, SIDEROPHILES, AND ATMOPHILES

LITHOSPHERE

The word lithosphere is derived from the word sphere, combined with the Greek word *lithos*, meaning **rock**. The lithosphere is the solid outer section of Earth, which includes Earth's **crust** (the "skin" of rock on the outer layer of planet Earth), as well as the underlying cool, dense, and rigid upper part of the upper mantle. The lithosphere extends from the surface of Earth to a depth of about 44–62 mi (70–100 km). This relatively cool and rigid section of Earth is believed to "float" on top of the warmer, non-rigid, and partially melted material directly below.

Earth is made up of several layers. The outermost layer is called Earth's crust. The thickness of the crust varies. Under the **oceans**, the crust is only about 3–5 mi (5–10 km) thick. Under the continents, however, the crust thickens to about 22 mi (35 km) and reaches depths of up to 37 mi (60 km) under some mountain ranges. Beneath the crust is a layer of rock material that is also solid, rigid, and relatively cool, but is assumed to be made up of denser material. This layer is called the upper part of the upper mantle, and varies in depth from about 31–62 mi (50–100 km) below Earth's surface. The combination of the crust and this upper part of the upper mantle, which are both comprised of relatively cool and rigid rock material, is called the lithosphere.

Below the lithosphere, the **temperature** is believed to reach 1,832°F (1,000°C), which is warm enough to allow rock material to flow if pressurized. Seismic evidence suggests that there is also some molten material at this depth (perhaps about 10%). This zone which lies directly below the lithosphere is called the **asthenosphere**, from the Greek word *asthenes*, meaning weak. The lithosphere, including both the solid portion of the upper mantle and Earth's crust, is carried "piggyback" on top of the weaker, less rigid asthenosphere, which seems to be in continual motion. This motion creates stress in the rigid rock layers above it, forcing the slabs or plates of the lithosphere to jostle against each other, much like **ice** cubes floating in a bowl of swirling **water**. This motion of the **lithospheric plates** is known as **plate tectonics**, and is responsible for many of the movements seen on Earth's surface today including earthquakes, certain types of volcanic activity, and continental drift.

See also Continental drift theory; Earth (planet); Earth, interior structure

LITHOSPHERIC PLATES

Lithospheric plates are regions of Earth's **crust** and upper mantle that are fractured into plates that move across a deeper plasticine mantle.

Earth's crust is fractured into 13 major and approximately 20 total lithospheric plates. Each lithospheric plate is composed of a layer of oceanic crust or continental crust superficial to an outer layer of the mantle. Containing both crust and the upper region of the mantle, lithospheric plates are generally considered to be approximately 60 mi (100 km) thick. Although containing only continental crust or oceanic crust in any one cross-section, lithospheric plates may contain various sections that exclusively contain either oceanic crust or continental crust and therefore lithospheric plates may contain various combinations of oceanic and continental crust. Lithospheric plates move on top of the **asthenosphere** (the outer plastically deforming region of Earth's mantle).

The term "plate" is deceptive. Remembering that Earth is an oblate sphere, lithospheric plates are not flat, but curved and fractured into curved sections akin to the peeled sections of an orange. Accordingly, analysis of lithospheric plate movements and dynamics requires more sophisticated mathematics that account for the curvature of the plates.

In geological terms, there are three types of boundaries between lithospheric plates. At divergent boundaries, lithospheric plates move apart and crust is created. At convergent boundaries, lithospheric plates move together in collision zones where crust is either destroyed by subduction or uplifted to form **mountain chains**. Lateral movements between lithospheric plates create **transform faults** at the sites of plate slippage.

At each of the unique lithospheric plate boundaries there are specific geophysical forces that are characteristic of the plate dynamics. At transform boundaries there are shearing forces between the lithospheric plates. At divergent boundaries, tensional forces dominate the interaction between plates.

At subduction sites, compression of lithospheric plate material dominates.

The dynamics of **plate tectonics**, driven by deeper thermal processes, stress and cause elastic strain on lithospheric materials. Resulting fractures of **rock** in the **lithosphere** cause a release of energy in the form of seismic waves (i.e. an **earthquake**).

Because Earth's diameter remains constant, there is no net creation or destruction of lithospheric plates.

In contrast to the technical definition of lithosphere used by geologists, many geographers use the term lithosphere to denote landmass. This is a distinct concept as the geological definition of lithosphere may include sections containing oceanic crust completely submerged beneath Earth's **oceans**. Using the geographical definition, Earth is approximately 71% hydrosphere (a region covered by **water**) and 21% lithosphere (a region of land).

See also Dating methods; Earth, interior structure; Hawaiian Island formation; Mantle plumes; Mapping techniques; Mid-ocean ridges and rifts; Mohorovicic discontinuity (Moho); Ocean trenches; Rifting and rift valleys; Subduction zone

LOESS · *see* DUNE FIELDS

LONGITUDE · *see* LATITUDE AND LONGITUDE

LONGSHORE DRIFT

Longshore drift is the transport of **sand** along a beach by waves impinging or breaking at an angle to the beach. Longshore drift occurs when a wave breaks, lifts sand into suspension, and then throws a pulse of sand-bearing **water** (swash) up the slope of the beach. If the wave breaks on the beach at an angle, the swash travels simultaneously up the beach and along the beach in the direction of the wave's original motion. Friction and **gravity** slow and halt the upward progress of the swash, and it begins to arc down the beach as backwash (still moving, though more slowly, along the beach). Suspended sand settles out of the backwash when it reaches the waterline and mixes with the relatively stationary water there.

Relative to the beach surface, the path of a typical grain of sand suspended by a breaking wave is therefore a lopsided parabola: up, over, and down. Some grains will be moved similarly by the next wave, or the next, again and again, and so are transported along the beach by steps.

Beaches are continually shaped and shifted by longshore drift. At Cape Cod, for example, long—approximately straight—wave-fronts from the east impinge on a bulging shoreline. Along the northern half of the Cape these westbound waves strike the beach at an angle that moves sand northward toward Provincetown. To the south, the same waves strike at angle that moves sand southward, toward the "elbow" of the Cape. In the last hundred years or so, conse-

quently, the ends of the Cape have gained land while the central beach has receded.

Longshore drift is one of the few processes that can transport sand for long distances along a slope at a fixed altitude.

See also Beach and shoreline dynamics

LYELL, CHARLES (1797-1875)
Scottish geologist

Charles Lyell was a scientist whose ideas were important to the development of theories of geological and evolutionary change. Lyell's most influential textbook was *Principles of Geology*, published in 1833. Other well-known books written by Lyell are *Travels in North America, with Geological Observations* and *The Antiquity of Man*. Because of his great influence on the development of the principles of his discipline, Charles Lyell is sometimes referred to as a father of modern **geology** (along with another Scot, **James Hutton**, 1726–1797).

Lyell's most important theory, which built upon earlier work of Hutton, was the so-called theory of **uniformitarianism**, which largely refuted the previously widely believed doctrine of **catastrophism**.

According to the theory of uniformitarianism, major geological forces that are observed today also occurred in the past, and likely throughout the history of Earth. Examples of such forces include volcanism, earthquakes, and **erosion** by **wind**, **water**, and **gravity**. Moreover, the theory states that these existing causes have been responsible for major changes that have occurred in the structure of the earth during its geological history. A central element of the interpretation of the theory of uniformitarianism is the importance of time (over extremely long periods of time), even relatively slow-acting forces like erosion by wind and water can have an enormous influence on the character of Earth's surface.

Catastrophism is an earlier, very different doctrine that was widely believed by many scientists, but has largely been replaced by Lyell's (this kind of rapid change in scientific understanding is sometimes referred to as a "revolution"). According to catastrophism, major changes in Earth's structure, as implied by rapid changes in geological stratification and in fossil assemblages, were caused by sudden, violent, cataclysmic events, rather than by gradual, evolutionary and environmental changes. Before the influence of Lyell, most scientists and people of western culture believed that Earth and its species only had a history of about 6,000 years, based on a literal interpretation of the Book of Genesis in the *Bible*. However, the observations of Lyell and other geologists of his time were reporting clearly contradictory evidence about the forces influencing the geological character of Earth and the **evolution** of its existing species. These observations of the real, natural world suggested that life was much more ancient than only a few thousand years, and that existing species appeared to have evolved from previous ones, which are now extinct. These ideas of Lyell and his colleagues were extremely influential on other scientists, including Charles

Darwin (1809–1882), who is best known for his theory of the role of natural selection in driving evolutionary change, published in 1859 in his famous book, *On the Origin of Species.*

Another important concept championed by Lyell, also building upon previous work by James Hutton, was that older rocks were generally buried beneath younger ones. As such,

careful excavation and studying of geological layers and the **fossils** they contain could be used to understand the geological and evolutionary history of Earth.

See also Evolution, evidence of; Evolutionary mechanisms; Fossil record; Geologic time; Solar system